World Geomorphological Landscapes

Series Editor

Piotr Migoń, Institute of Geography and Regional Development, University of Wrocław, Wrocław, Poland

Geomorphology—'the Science of Scenery'—is a part of Earth Sciences that focuses on the scientific study of landforms, their assemblages, and surface and subsurface processes that moulded them in the past and that change them today. Shapes of landforms and regularities of their spatial distribution, their origin, evolution, and ages are the subject of geomorphology. Geomorphology is also a science of considerable practical importance since many geomorphic processes occur so suddenly and unexpectedly, and with such a force, that they pose significant hazards to human populations. Landforms and landscapes vary enormously across the Earth, from high mountains to endless plains. At a smaller scale, Nature often surprises us creating shapes which look improbable. Many geomorphological landscapes are so immensely beautiful that they received the highest possible recognition—they hold the status of World Heritage properties. Apart from often being immensely scenic, landscapes tell stories which not uncommonly can be traced back in time for millions of years and include unique events. This international book series will be a scientific library of monographs that present and explain physical landscapes across the globe, focusing on both representative and uniquely spectacular examples. Each book contains details on geomorphology of a particular country (i.e. The Geomorphological Landscapes of France, The Geomorphological Landscapes of Italy, The Geomorphological Landscapes of India) or a geographically coherent region. The content is divided into two parts. Part one contains the necessary background about geology and tectonic framework, past and present climate, geographical regions, and long-term geomorphological history. The core of each book is however succinct presentation of key geomorphological localities (landscapes) and it is envisaged that the number of such studies will generally vary from 20 to 30. There is additional scope for discussing issues of geomorphological heritage and suggesting itineraries to visit the most important sites. The series provides a unique reference source not only for geomorphologists, but all Earth scientists, geographers, and conservationists. It complements the existing reference books in geomorphology which focus on specific themes rather than regions or localities and fills a growing gap between poorly accessible regional studies, often in national languages, and papers in international journals which put major emphasis on understanding processes rather than particular landscapes. The World Geomorphological Landscapes series is a peer-reviewed series which contains single and multi-authored books as well as edited volumes.

World Geomorphological Landscapes—now indexed in Scopus®!

More information about this series at http://www.springer.com/series/10852

Emmanuel Reynard
Editor

Landscapes and Landforms of Switzerland

Springer

Editor
Emmanuel Reynard
Institute of Geography and Sustainability and
Interdisciplinary Centre for Mountain Research
University of Lausanne
Lausanne, Switzerland

ISSN 2213-2090 ISSN 2213-2104 (electronic)
World Geomorphological Landscapes
ISBN 978-3-030-43201-0 ISBN 978-3-030-43203-4 (eBook)
https://doi.org/10.1007/978-3-030-43203-4

This Springer imprint is published by the registered company Springer Nature Switzerland AG
The registered company address is: Gewerbestrasse 11, 6330 Cham, Switzerland

Series Editor Preface

Landforms and landscapes vary enormously across the Earth, from high mountains to endless plains. At a smaller scale, nature often surprises us creating shapes which look improbable. Many physical landscapes are so immensely beautiful that they received the highest possible recognition—they hold the status of World Heritage Sites. Apart from often being immensely scenic, landscapes tell stories which not uncommonly can be traced back in time for tens of million years and include unique geological events such as meteorite impacts. In addition, many landscapes owe their appearance and harmony not solely to the natural forces. For centuries, and even millennia, they have been shaped by humans who have modified hillslopes, river courses and coastlines, and erected structures which often blend with the natural landforms to form inseparable entities.

These landscapes are studied by geomorphology—'the Science of Scenery'—a part of Earth Sciences that focuses on landforms, their assemblages, surface and subsurface processes that moulded them in the past and that change them today. Shapes of landforms and regularities of their spatial distribution, their origin, evolution and ages are the subject of research. Geomorphology is also a science of considerable practical importance since many geomorphic processes occur so suddenly and unexpectedly, and with such a force, that they pose significant hazards to human populations and not uncommonly result in considerable damage or even casualties.

To show the importance of geomorphology in understanding the landscape and to present the beauty and diversity of the geomorphological sceneries across the world, we have launched a book series *World Geomorphological Landscapes*. It aims to be a scientific library of monographs that present and explain physical landscapes, focusing on both representative and uniquely spectacular examples. Each book will contain details on geomorphology of a particular country or a geographically coherent region. This volume presents geomorphological history and landscapes of Switzerland—a country, whose image is created by geomorphology more than anywhere else. A popular view equates Switzerland with the Alps and this book contains many examples of superb Alpine landscapes, shaped by uplift and erosion, glaciers of the past and the present, large-scale mass movements such as huge landslides and rapid debris flows, karstic processes and rivers. However, stories told by geomorphic landscapes outside the Alps are no less fascinating, even if these Fore-Alpine regions lack the grandeur of snow-covered peaks and glacier-filled valleys. The authors of many chapters also remind us about an important role of humans in shaping the landscape, perhaps best exemplified by the vineyards of Lavaux, declared a UNESCO World Heritage. Finally, Switzerland is a country where many important discoveries in the field of geomorphology and Quaternary have been made and these aspects have also been covered.

The World Geomorphological Landscapes series is produced under the scientific patronage of the International Association of Geomorphologists (IAG)—a society that brings together geomorphologists from all around the world. The IAG was established in 1989 and is an independent scientific association affiliated with the International Geographical Union (IGU) and the International Union of Geological Sciences (IUGS). Among its main aims are to promote geomorphology and to foster dissemination of geomorphological knowledge. I believe that this lavishly illustrated series, which keeps to the scientific rigour, is the most appropriate means to fulfill these aims and to serve the geoscientific community. To this end, my great thanks go to Emmanuel Reynard for adding this book to his busy agenda, successfully coordinating the large, multinational team of authors and delivering such an exciting and lavishly illustrated story to read and enjoy. I also acknowledge the excellent work of all individual authors who share their expert knowledge of Switzerland with the global geomorphological community. I am sure that this book, offering the best sample of Swiss geomorphological sceneries, will become both a most useful reference source about the country and a ready list of destinations to visit on a future geomorphological trip.

Piotr Migoń
Series Editor

Foreword

Emmanuel Reynard's email whether I would like to write the foreword to the book on *Landscapes and Landforms of Switzerland* he has been editing reached me while I was doing fieldwork on wind erosion and dust emissions in the Free State in South Africa. Looking through the impressive list of contributions in the Table of Contents for the book made it very clear to me how special the landscapes and landforms of Switzerland are. Here I was working on a flat high plateau during a windy and somewhat cold winter morning in South Africa, trying to measure small amounts of material driven across the land surface by wind, when all the images of Swiss mountains, glaciers and rivers popped up in my mind, a feast for geomorphologists, at least after a few weeks of studying wind erosion on croplands. So I did not hesitate to agree to contribute this short foreword to *Landscapes and Landforms of Switzerland.*

The 29 chapters of *Landscapes and Landforms of Switzerland* comprise a much wider range of landscapes and geomorphic systems than one could expect from a country which is usually associated with its Alpine landscape. Apart from highlighting the great variety of Swiss geomorphology, to me the great number of chapters and topics of *Landscapes and Landforms of Switzerland* also shows two unique elements of the book: on one hand, the classic description of the landforms; on the other, the deep insights into the processes that have formed these landscapes in the past and which shape them today and in the future. This perspective of *Landscapes and Landforms of Switzerland* also conveys a rather unique role of Swiss Geomorphology. In Switzerland, the proximity of humans and the dynamic and changing environment is closer than in many other countries. Geomorphology is, therefore, one of the most important scientific disciplines making the connection between these two spheres by generating an understanding of the material, forms, processes and history of the Earth surface. The efforts of Emmanuel Reynard and the team of authors who contributed to *Landscapes and Landforms of Switzerland*, therefore, illustrate both the relevance of geomorphology in Switzerland and the excellent research on landscapes and landforms conducted in Switzerland.

Bultfontein, South Africa
August 2019

Nikolaus J. Kuhn
President of the Swiss Geomorphological Society
(2017–2019)

Contents

Editor and Contributors

About the Editor

Emmanuel Reynard is Professor of Physical Geography and Director of the Interdisciplinary Centre for Mountain Research at the University of Lausanne, and a member of the International Association of Geomorphologists (IAG) Executive Committee. He has been Chairman of the IAG Working Group on Geomorphosites (2001–2013) and President of the Working Group on Geotopes of the Swiss Academy of Sciences (2006–2017). His research concerns mountain geomorphology, geomorphological heritage, landscape geohistorical analyses and water management in mountains. He has worked in the Alps, the Romanian Carpathians and in the Maghreb (Tunisia, Morocco).

Contributors

Naki Akçar Institute of Geological Sciences, University of Bern, Bern, Switzerland

Christian Ambrosi Institute of Earth Sciences, University of Applied Sciences and Arts of Southern Switzerland (SUPSI), Canobbio, Switzerland

Nico Bätz Institute of Earth Surface Dynamics, University of Lausanne, Lausanne, Switzerland;
Department Surface Waters Research and Management, Swiss Federal Institute of Aquatic Science and Technology (Eawag), Kastanienbaum, Switzerland

Luca Bonzanigo Consultant geologist, geolog.ch SA, Bellinzona, Switzerland

Thomas Buckingham UNESCO-World Heritage Swiss Tectonic Arena Sardona, Sargans, Switzerland

Nancy Calhoun Department of Earth Sciences, Simon Fraser University, Burnaby, BC, Canada;
Oregon Department of Geology and Mineral Industries, Portland, Canada

John J. Clague Department of Earth Sciences, Simon Fraser University, Burnaby, BC, Canada

Emmanuel Estoppey eqiilab, Ollon, Switzerland

Jean-Michel Fallot Institute of Geography and Sustainability, University of Lausanne, Lausanne, Switzerland

Isabelle Gärtner-Roer Department of Geography, University of Zurich, Zurich, Switzerland

Ruedi Haller Swiss National Park, Chastè Planta-Wildenberg, Zernez, Switzerland

Philipp Häuselmann Swiss Institute for Speleology and Karst Studies (SISKA), La Chaux-de-Fonds, Switzerland

Peter Heitzmann Geowissenschaftliche Öffentlichkeitsarbeit und Geologische Beratungen, Bern, Switzerland

Kristina Hippe Laboratory for Ion Beam Physics, ETH Zurich, Zurich, Switzerland; Department of Earth Sciences, Freie Universität Berlin, Berlin, Germany

Martin Hoelzle Department of Geosciences, University of Fribourg, Fribourg, Switzerland

Hanspeter Holzhauser Zürich, Switzerland

Ion Iorgulescu Etat de Genève, Département du Territoire, Genève, Switzerland

Susan Ivy-Ochs Laboratory for Ion Beam Physics, ETH Zurich, Zurich, Switzerland

Pierre-Yves Jeannin Swiss Institute of Speleology and Karst Studies (SISKA), La Chaux-de-Fonds, Switzerland

Beat Keller Keller+Lorenz AG, Luzern, Switzerland

Oskar Keller Lüchingen, Switzerland

Robert Kenner WSL Institute for Snow and Avalanche Research SLF, Davos Dorf, Switzerland

Christophe Lambiel Institute of Earth Surface Dynamics and Interdisciplinary Centre for Mountain Research, University of Lausanne, Lausanne, Switzerland

Stuart N. Lane Institute of Earth Surface Dynamics, University of Lausanne, Lausanne, Switzerland

Hans Lozza Swiss National Park, Chastè Planta-Wildenberg, Zernez, Switzerland

Michel Marthaler Institute of Geography and Sustainability, University of Lausanne, Lausanne, Switzerland

Simon Martin Bureau Relief, Aigle, Switzerland

Brian W. McArdell Swiss Federal Institute for Forest, Snow and Landscape Research WSL, Birmensdorf, Switzerland

Andrea Moscariello Department of Earth Sciences, University of Geneva, Geneva, Switzerland

Samuel U. Nussbaumer Department of Geography, University of Zurich, Zurich, Switzerland

O. Adrian Pfiffner Institute of Geological Sciences, University of Bern, Bern, Switzerland

Marcia Phillips WSL Institute for Snow and Avalanche Research SLF, Davos Dorf, Switzerland

Andreas von Poschinger Bavarian Environment Agency, Geological Survey, Augsburg, Germany

Géraldine Regolini Bureau Relief, Aigle, Switzerland

Emmanuel Reynard Institute of Geography and Sustainability and Interdisciplinary Centre for Mountain Research, University of Lausanne, Lausanne, Switzerland

Henri Rougier University Lyon 3, Lyon, France

Mario Sartori Department of Earth Sciences, University of Geneva, Geneva, Switzerland

Cristian Scapozza Institute of Earth Sciences, University of Applied Sciences and Arts of Southern Switzerland (SUPSI), Canobbio, Switzerland

Christian Schlüchter Institute of Geological Sciences, University of Bern, Bern, Switzerland

Philippe Schoeneich Institute of Urban Planning and Alpine Geography, University Grenoble Alpes, Grenoble, France

Andreas Wipf Swiss Hiking Association, Bern, Switzerland

Heinz J. Zumbühl Institute of Geography, University of Bern, Bern, Switzerland

Introduction

1

Emmanuel Reynard

Abstract

The variety of Swiss landscapes is due to the combination of varied structural and climatic contexts and a rich geological heritage. The book is divided in three main sections. The first section presents the physical context and details the variety of Swiss geomorphological landscapes. The second section proposes 23 examples of Swiss landscapes, in three main subsections: eight examples of structural landscapes; nine cases of glacial, periglacial and high mountain landscapes; and six examples of river landscapes and landscapes related to gravity-driven processes. The last section deals with geoheritage, geoparks and geotourism.

Keyword

Geomorphological heritage • Landscapes • Switzerland

Switzerland is known not only for its typical products (chocolate, cheese, watches) and services (banking), but also for its landscapes. The latter were at the origin of tourism in Europe in the eighteenth century and landscapes are still among the main tourist attractions in Switzerland today. Their diversity is largely due to the influence of the agricultural development, which has given rise to a multitude of rural sceneries, including typical alternating grasslands, alpine pastures, forests and cultivated fields. Today, however, these rural landscapes are increasingly under pressure from urbanisation, especially on the Swiss Plateau.

The variety of Swiss landscapes is also due to the combination of varied structural and climatic contexts and a rich geological heritage. The aim of this book is to present and discuss the influence of the different structural contexts and current and former climates on landforms, and the importance of the landforms in structuring the landscapes of Switzerland.

The book has three main sections. The first section concerns the physical context and details the variety of Swiss geomorphological landscapes. It includes four texts: The first, by O. Adrian Pfiffner, presents the main features of Swiss geology. The second, by Jean-Michel Fallot, does the same for climate. The following text by Christian Schlüchter et al. gives an overview of the evolution of landscapes during the Quaternary. Finally, Emmanuel Reynard et al. provide an overview of the main geomorphological landscapes in Switzerland.

Section two presents 23 characteristic Swiss reliefs and landscapes. It has three subsections. The first one proposes eight examples of structural landscapes, i.e. cases in which structural conditions have strongly influenced morphology. Andrea Moscariello interprets the geomorphological history of the Geneva basin, which has depended to a great extent on structural conditions and has been shaped by repeated passage of glaciers during the Quaternary. Emmanuel Reynard and Philippe Schoeneich portray a typical Jura landscape, the Joux Valley and its surroundings. Emmanuel Reynard and Emmanuel Estoppey offer a geomorphological analysis of the structural landscape (in the Molasse) of the Lavaux World Heritage vineyard. Using the example of the Diablerets Massif, Philippe Schoeneich and Emmanuel Reynard then present the landscapes of the calcareous massifs of the western "Hautes Alpes Calcaires". O. Adrian Pfiffner does the same for the landscapes of central Switzerland, around Lake Lucerne, following Philipp Häuselmann's analysis of the surface and underground landscapes of the Siebenhengste-Hohgant-Schrattenfluh karst system. The following text by Thomas Buckingham and O. Adrian Pfiffner deal with structural landscapes of eastern Switzerland, in the Sardona region, another Swiss World Heritage site. The subsection ends with a geomorphological analysis of one of the most famous peaks in the Alps: the Matterhorn, by Michel Marthaler and Henri Rougier.

E. Reynard (✉)
Institute of Geography and Sustainability and Interdisciplinary Centre for Mountain Research, University of Lausanne, Géopolis, 1015 Lausanne, Switzerland
e-mail: emmanuel.reynard@unil.ch

E. Reynard (ed.), *Landscapes and Landforms of Switzerland*, World Geomorphological Landscapes,
https://doi.org/10.1007/978-3-030-43203-4_1

The second subsection provides nine examples of glacial, periglacial and high mountain landscapes. Switzerland is rich in glaciers. Two examples of regions where glacial processes are still very active were chosen, both of which are part of the Swiss Alps Jungfrau-Aletsch World Heritage site: the Great Aletsch Glacier (text by Hans Holzhauser) and the glaciers on the northern slope of the Bernese Alps in the Jungfrau region (text by Heinz Zumbühl et al.). The next three examples illustrate the rapid evolution of high mountain valleys as a result of global warming, while underlining the importance of the Quaternary heritage: the rockglaciers of the Engadine (text by Isabelle Gärtner-Roer and Martin Hoelzle), the landscapes of the Swiss National Park (text by Christian Schlüchter et al.) and glacial and periglacial landscapes typical of the valleys in the western Alps, such as those of the Hérens Valley (text by Christophe Lambiel). The following two examples deal with Quaternary legacies of two major glaciers on the Swiss Plateau: the Rhone glacier (text by Susan Ivy-Ochs et al.) and the Rhine glacier (text by Oskar Keller). Beat Keller provides a detailed analysis of landscape evolution, highlighting the importance of the Reuss glacier in shaping the lake and the city of Lucerne. Finally, Cristian Scapozza and Christian Ambrosi discuss the importance of glacial, periglacial and fluvial processes in shaping the valleys located in the Southern Alps, emphasising their particularities by comparing them with valleys in the Northern Alps.

The third subsection presents six examples of river landscapes and landscapes related to gravity-driven processes. Peter Heitzmann presents the geomorphological context of the biggest waterfall in Switzerland: the Rhine Falls. Nico Bätz et al. analyse the recent evolution of one of the last braided rivers in Switzerland: the Allondon River in the canton of Geneva. Brian McArdell and Mario Sartori analyse the structural conditions and activity of one of the most active torrential systems in the Alps: the Illgraben. The two following texts concern landslides; the first one, in Campo Vallemaggia (analysed by Luca Bonzanigo) is active, with the most recent major landslide occurring in the 1940s, while the second one in Flims, in the Rhine Valley (studied by Andreas von Poschinger et al.), is an example of a major landslide due to postglacial readjustment of the relief. Here, hillslope landforms caused by gravity combine with fluvial landforms to create an exceptional landscape. The last text (Marcia Phillips, Robert Kenner) concerns the municipality of Pontresina, in Graubünden, and describes the activity of high mountain processes and the new anthropogenic landscapes resulting from measures against natural hazards, very common in the Swiss Alps.

Of course, these examples do not cover all the geomorphological landscapes of Switzerland. Some readers may regret the fact that among the 322 geosites in Switzerland, important sites do not have a chapter to themselves. This is the result of a selection guided not only by scientific criteria (the aim being to present different landscapes influenced by the main structural and climatic contexts throughout the country), but also by practical criteria, including the availability of experts to write a text.

The last section (by Emmanuel Reynard et al.) deals with geoheritage management issues.

This book is a collective work that brings together geomorphologists, geologists and geographers working in different regions of the country. I would like to thank all the colleagues who gave their time to write these summary chapters on their study sites. It may seem easy, but in practice, it is often a time-consuming task searching for research results disseminated in multiple documents, ranging from study reports to scientific articles, student work or geological maps. I would also like to thank Mélanie Clivaz, Research Assistant at the Institute of Geography and Sustainability of the University of Lausanne, who helped produce the figures for some chapters.

The whole book was subject to a review process to ensure the scientific quality of the chapters. I would like to thank the following people who agreed to review one or more chapters: Christian Ambrosi, Philippe Audra, Luca Bonardi, Jean-François Buonchristiani, Dov Corenblit, Philip Deline, Jean-Luc Epard, Monique Fort, Wilfried Haeberli, Peter Heitzmann, Raimund Hipp, Christophe Lambiel, Robin Marchant, Jürg Meyer, O. Adrian Pfiffner, Marcia Phillips, Martine Rebetez, Cristian Scapozza, Philippe Schoeneich, Roberto Seppi and Ronald T. Van Balen.

I also express my warm appreciation to Piotr Migoń, scientific editor of the series "World Geomorphological Landscapes". His careful proofreading of all the chapters and his rich experience in publishing other books in the series made it possible to clarify certain formulations and uses that may be understandable to Swiss researchers, but not necessarily for foreign readers, and to improve the formal and scientific quality of the chapters. I am grateful to him to have invited me to join the admirable editorial project he is coordinating.

I would also like to acknowledge the precious support of Springer Nature, particularly Robert K. Doe, Senior Publisher, and Rema Devi Viswanathan, Manjula Saravanan, Banu Dhayalan and Madanagopal Deenadayalan who coordinated the publication process with endless patience and dedication.

Without claiming to be exhaustive, I hope that reading the general chapters in sections 1 and 3 will help understand the context and reasons for Switzerland's rich geomorphological diversity, while the chapters in section 2 give some concrete examples chosen among the country's many geomorphosites and geomorphological landscapes (Fig. 1.1).

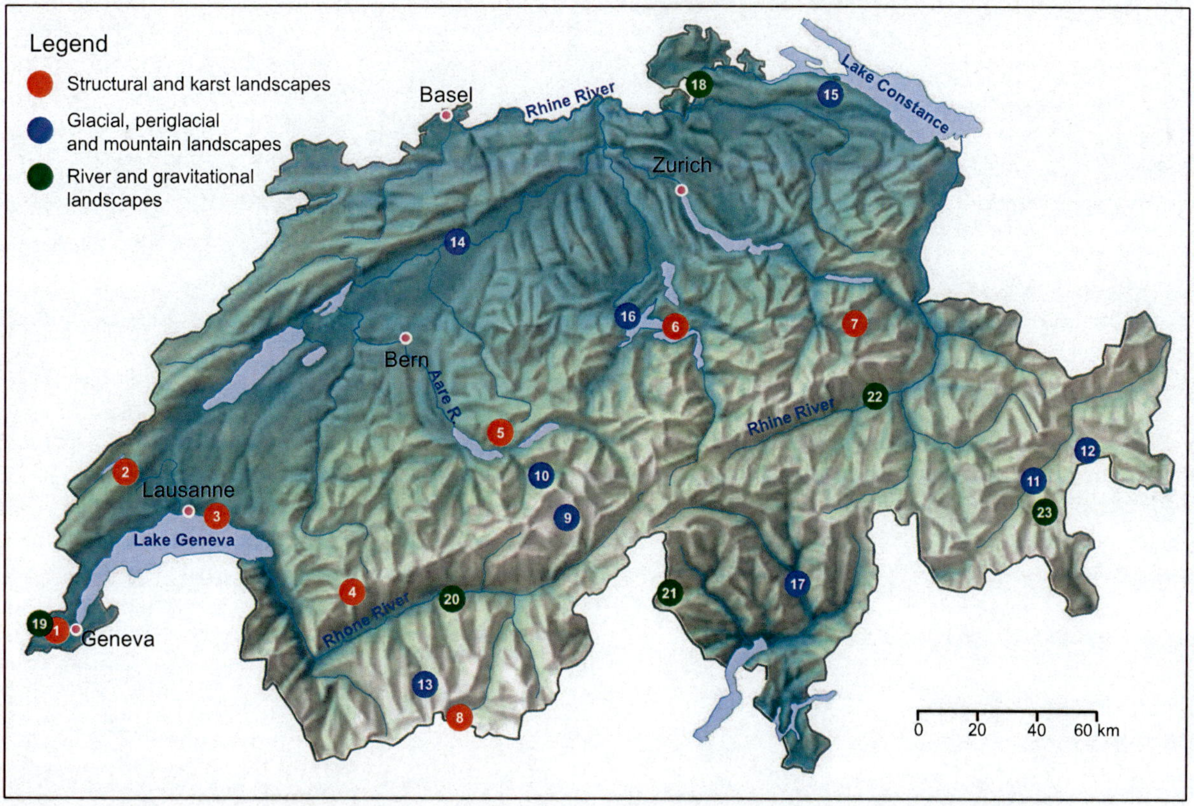

Fig. 1.1 Location of the 23 examples of geomorphological landscapes presented in section 2 of the book. 1. Geneva Basin; 2. Joux Valley; 3. Lavaux terraced vineyard; 4. Diablerets Massif; 5. Siebenhengste-Hohgant-Schrattenfluh karst system; 6. Structural landscapes of Central Switzerland; 7. Tectonic Arena Sardona World Heritage site; 8. Matterhorn; 9. Grosser Aletschgletscher; 10. Finsteraarhorn-Jungfrau glacier landscape; 11. Rockglaciers of the Engadine; 12. Swiss National Park; 13. Hérens Valley; 14. Wangen an der Aare glacier landscape; 15. Lake Constance glacier landscape; 16. Lake Lucerne; 17. Geomorphological landscapes of Ticino; 18. Rhine Falls; 19. Allondon River; 20. Illgraben torrential system; 21. Campo Vallemaggia landslide; 22. Flims and Tamins rockslide landscape; 23. Periglacial landscapes and protection measures above Pontresina. Source of the map: swisstopo

Emmanuel Reynard is Professor of Physical Geography and Director of the Interdisciplinary Centre on Mountain Research at the University of Lausanne, and a member of the International Association of Geomorphologists (IAG) Executive Committee. He has been chairman of the IAG Working Group on Geomorphosites (2001–2013) and president of the Working Group on Geotopes of the Swiss Academy of Sciences (2006–2017). His research concerns mountain geomorphology, geomorphological heritage, landscape geohistorical analyses and water management in mountains. He has worked in the Alps, the Romanian Carpathians and in the Maghreb (Tunisia, Morocco).

Part I

Physical Environment

2 The Geology of Switzerland

O. Adrian Pfiffner

Abstract

The general picture of the physiographic map of Switzerland reflects the tectonic structure rather directly. Local relief in the Jura Mountains is a direct consequence of folding of the detached Mesozoic strata. The Swiss Plateau mimics the Molasse Basin with flat lying sediments while thrusting and tilting of these strata in the Subalpine Molasse amalgamated these units with the Alps. The Alps exhibit nappe stacks of very different origin. Most of them evolved from pre-Triassic crystalline basement rocks and their sedimentary cover. In many cases the cover was detached from its basement and now forms a nappe stack of its own. The Helvetic nappe system derived from the European continental margin contains nappes of cover rocks displaced over 30–50 km; crystalline basement rocks form large-scale domes. The Penninic nappe system is derived from basins that formed in Mesozoic times between the European and Adriatic continents. They contain far travelled nappes of cover rocks, as well as nappes of basement rocks that were transported over considerable distances, too. In addition, nappes of oceanic rocks outcrop as thin slivers at the top. Post-nappe folding within the Penninic nappe stack is reminiscent of their complex formation history. The Austroalpine nappe system was derived from the Adriatic margin and now forms a horizontal layer as the highest unit in eastern and central Switzerland. This nappe system contains crystalline basement as well as Mesozoic cover rocks and was emplaced early in the Alpine history in a ENE direction. The Southalpine nappe system was derived from the Adriatic margin as well. Here thrusting of crystalline basement with its Mesozoic cover was south-directed. The various Alpine nappe piles led to the amalgamation of very different rock types: continental and oceanic basement rocks, shallow marine carbonates, deep marine clastics and radiolarian chert to name the most important. Landforms and landscapes reflect these differences, in addition to the landforms created by fluvial and glacial erosion.

Keywords

Rock types • Tectonic structure • Alpine nappe structure • Jura Mountains • Molasse Basin

2.1 Introduction

The geology of Switzerland covers the Central Alps, a segment of the Alpine chain located between the Western Alps of France and Italy and the Eastern Alps of Italy, Austria and Germany. In this segment the Alpine chain is particularly narrow, yet it contains all the major tectonic units and exhibits landmark landscapes and landforms for which the Alps are known. This chapter aims at giving an overview of the geological structure of Switzerland, the major rock types involved, the geological history related to these rock types and their impact on Alpine landforms and landscapes. An overview of the geology of Switzerland can also be found in Trümpy (1980), Pfiffner (2014, 2019) and geodata are available from the Swiss Geological Survey (https://www.swisstopo.admin.ch).

The digital elevation model in Fig. 2.1 displays the deep valleys carved into the Swiss Alps. The high local relief along the Rhine and Rhone rivers is highlighted by the green color of the low lying valley floors in comparison to the high mountain ranges on either side. Both the Rhine and Rhone rivers follow longitudinal valleys within the Alps but exit the Alps as transverse rivers. The major Alpine rivers drain into various seas: Rhine into the North Sea, Rhone into the Mediterranean Sea, Adda into the Adriatic Sea (via Po River), and Inn into the Black Sea (joining the Danube).

O. A. Pfiffner (✉)
Institute of Geological Sciences, University of Bern,
Baltzerstrasse 1+3, 3012 Bern, Switzerland
e-mail: adrian.pfiffner@geo.unibe.ch

E. Reynard (ed.), *Landscapes and Landforms of Switzerland*, World Geomorphological Landscapes,
https://doi.org/10.1007/978-3-030-43203-4_2

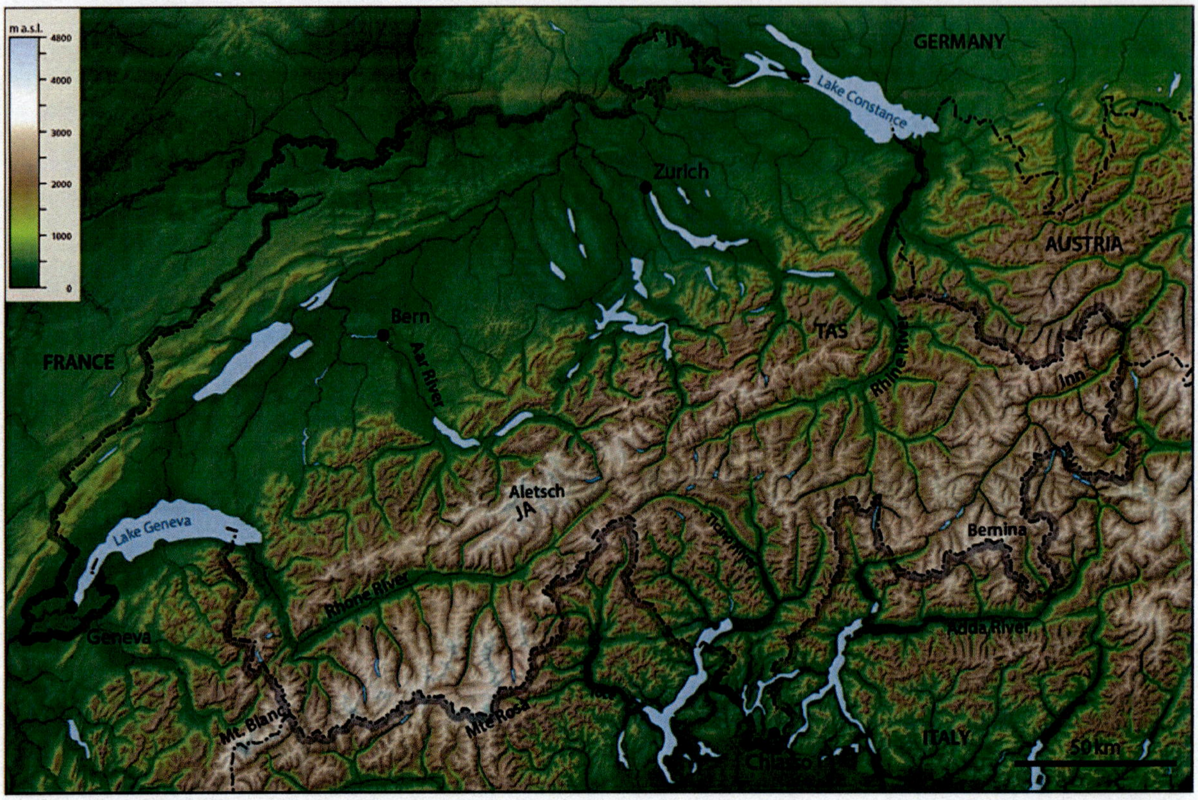

Fig. 2.1 Digital elevation model (DEM) of Switzerland giving the main features pertinent to this chapter. UNESCO World Heritage sites: JA = Swiss Alps Jungfrau-Aletsch, TAS = Swiss Tectonic Arena Sardona

Within the Central Alps three areas of high elevation exceeding 4000 m a.s.l. can be recognized: Mont Blanc—Monte Rosa, Aletsch, and Bernina. These areas are dominated by granitic bedrock. These rocks are resistant to erosion and are therefore worn down more slowly (see Kühni and Pfiffner 2001a) The Po Basin to the south of the Alps (south of Chiasso) lies at around 300 m a.s.l. To the north, the Swiss Plateau extends from Lake Geneva to Bern, Zürich, and Lake Constance tapering off from around 600 to 400 m a.s.l. This plateau corresponds to what is called Molasse Basin discussed later in Sect. 2.2.3. To the NW of this plateau the Jura Mountains straddle along the border between Switzerland and France.

The geological structure of Switzerland is shown as a simplified tectonic map in Fig. 2.2. The nappe systems between the Po Basin and Molasse Basin derive from the continental margins of Europe and Adria (a promontory of Africa) and the ocean basins that once existed between the two margins. These nappe systems consist of two groups of rock types: crystalline basement rocks (granites, gneisses, and schists; stippled in Fig. 2.2) and cover sediments that were deposited on these. The Helvetic nappe system can be attributed to the European margin. Its cover sediments continue in the subsurface of the Molasse Basin and reappear in the Jura Mountains and the Foreland around the basement uplifts of the Vosges and Black Forest. The Penninic nappe system is derived from the marine basins, which formed in Mesozoic times between the two continental margins. The Adriatic margin is found today in three units; the Austroalpine, Southalpine, and Salassic nappe systems. The Insubric Fault is part of the Peri-Adriatic fault system, a major fault zone that marks the northern limit of the Southalpine nappe system. Several magmatic intrusions occurred at 40–30 Ma along this fault system, two of which are shown in red in Fig. 2.2.

The deep geological structure of Switzerland is shown in the simplified cross-section in Fig. 2.3. The MOHO, the crust-mantle boundary, is shown to dip from the ordinary value of 30 km for continental crust to more than 50 km beneath the core of the Alps. As visible in the cross-section the crust is thicker because of thrust faults and folds that affected the entire crust. In case of the Aar massif the top

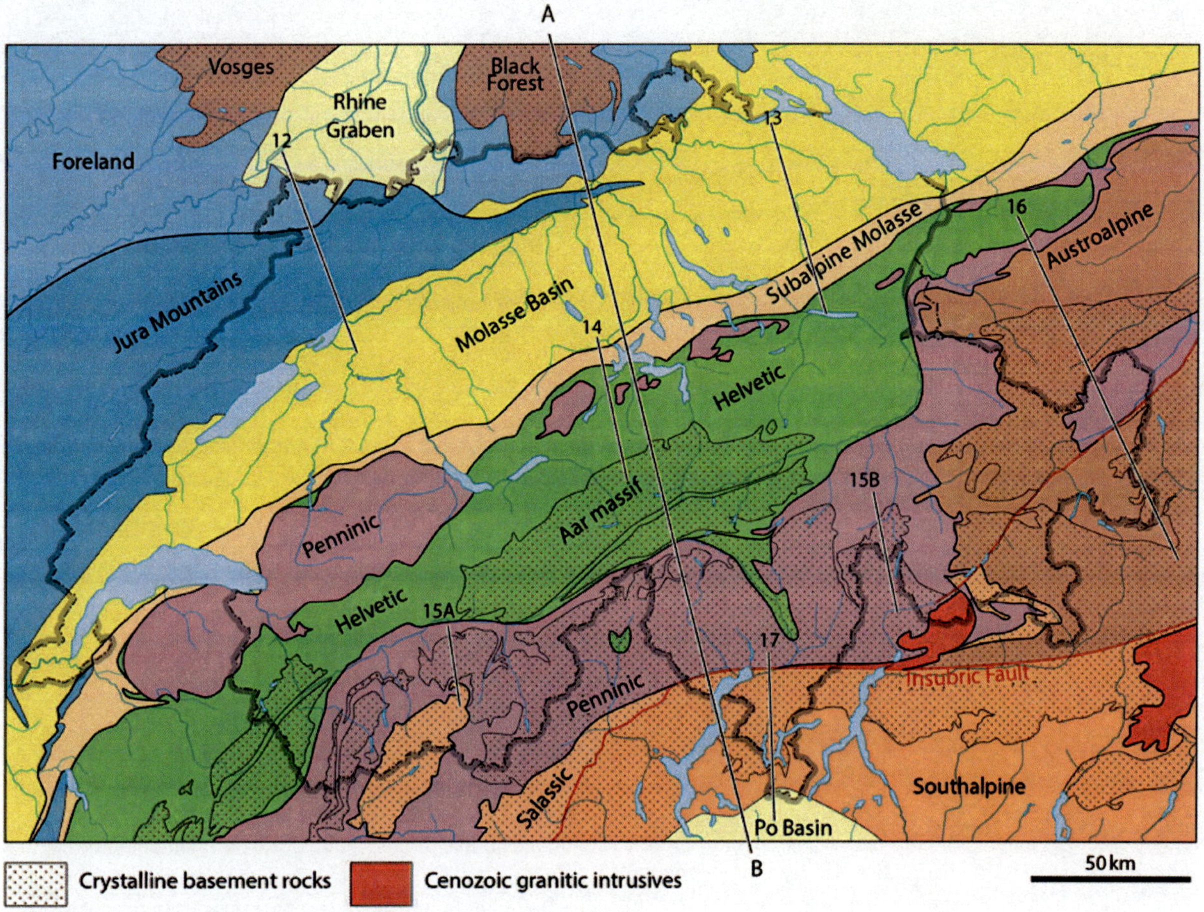

Fig. 2.2 Tectonic map of Switzerland showing the major tectonic units. The line AB is the trace of the cross-section shown in Fig. 2.3, numbers 12–17 denote traces of cross-sections in Figs. 2.12, 2.13, 2.14, 2.15, 2.16 and 2.17

basement was uplifted from 7000 m below sea level to 4000 m above sea level in the process of thickening. The Mesozoic cover sediments are shown to continue from the Aar massif to the NNW beneath the Molasse Basin and link with the Jura Mountains.

The Helvetic and Penninic nappes were transported over distances of 50–200 km as thin sheets of rocks. These sheets, called nappes, are typically 1–3 km thick. It is important to note that most of this nappe pile—a heap of around 30 km—has been eroded in the course of the last 30 million years. The outline of this eroded nappe pile is indicated by the thrust faults shown in the air above the Aar massif (dashed lines in Fig. 2.3). The ancestral rivers, which were responsible for this vast erosion, initiated in the Austroalpine and Penninic nappe systems that once covered the Central Alps. The river network that developed was controlled by the structure and distribution of the different rock types within these nappes. As these rivers incised the bedrock they encountered progressively new rock types and structures which influenced the river network and most likely led to profound changes in the courses of rivers (see also Buckingham and Pfiffner, this volume).

As visible in Fig. 2.3, the Cenozoic sediments of the Molasse Basin extend southward beneath the Helvetic nappe system. The southern part of these sediments was detached, transported northward, and tilted. It now forms the so-called Subalpine Molasse (see Fig. 2.2). The Cenozoic basin fill of the Po Basin to the south of the Alps was also affected by Alpine tectonics. Here thrusting was directed south. Altogether, the Alps form a bivergent mountain belt. The Insubric Fault is a nearly vertical fault that extends deep down into the crust. The units to the north of it were uplifted along this fault and sheared into a large-scale antiformal fold.

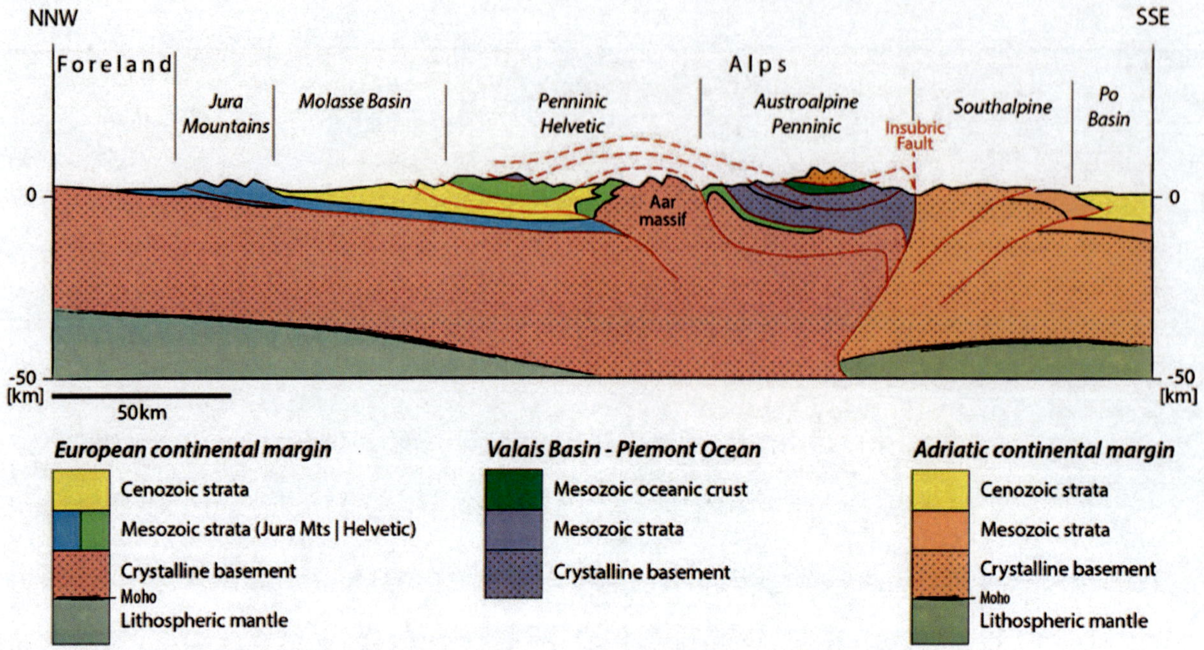

Fig. 2.3 Cross-section from the Northalpine Foreland to the Po Basin. Trace of cross-section is given in Fig. 2.2

2.2 Major Rock Types and Paleogeography

Regarding the geologic structure as well as landforms and landscapes it is useful to distinguish between basement and cover rocks. In the Alpine context basement relates to crystalline basement and cover to its sedimentary cover. In the following the various rock types are discussed in chronological order regarding their age.

2.2.1 Crystalline Basement Rocks

Crystalline basement rocks represent the oldest rock types. They include polymetamorphic gneisses, amphibolites, and granites. Polymetamorphic gneisses are rocks that were heated up to high temperatures (400 °C or more) and deformed. They include granites that intruded the crust more than 400 Ma ago and were transformed (or metamorphosed) to orthogneisses, as well as sediments that were metamorphosed to paragneisses. Amphibolites represent metamorphosed basalts. All these rocks were part of ancient mountain belts that formed 600 and 440 Ma ago and since have been worn down completely. The crystalline basement contains also a rock suite that formed in the course of the Variscan phase of mountain building some 300 Ma ago. They include granites (intruded at 330 and 300 Ma) and late Paleozoic sedimentary and volcanic rocks ranging in age from 300 to 250 Ma. Regarding Switzerland and neighboring areas the crystalline basement was completely worn down by the end of the Paleozoic Era such that shallow seas were able to invade the area subsequently.

In the course of the Alpine cycle of mountain building the crystalline basement was deformed (again), gneisses and granites undergoing another phase of metamorphism. Figure 2.4 shows typical landscapes made of crystalline basement. The photograph in Fig. 2.4a is taken in the UNESCO World Natural Heritage site Swiss Alps Jungfrau-Aletsch (see location in Fig. 2.1). Rugged peaks are at least partly covered with glaciers. The latter were responsible for the development of horns, i.e., peaks with steep flanks created by erosion of cirque glaciers and separated by crests that merge in the summit. Bedrock in this area is predominantly granitic. Figure 2.4b shows the Leventina valley (Ticino River; see location in Fig. 2.1) where the bedrock is made of ortho- and paragneisses. The foliation in these rocks is subhorizontal but does not express itself significantly in the landscape. The strips of agricultural land with the villages of Chironico and Sobrio half way up the mountain flanks are within a homogeneous complex of orthogneisses. The break in slope must be remnant of an ancient valley floor.

2.2.2 Mesozoic Rocks

Mesozoic cover rocks are sediments of variable composition and depositional conditions. The differences arise from the fact that these sediments formed in the course of the breakup

Fig. 2.4 Landscapes in crystalline basement rocks (photos © VBS). **a** Granitic basement in the Swiss Alps region Jungfrau-Aletsch (Bern/Valais) with Bietschhorn, Aletschhorn and Finsteraarhorn. **b** Gneiss basement in the Leventina Valley (Ticino)

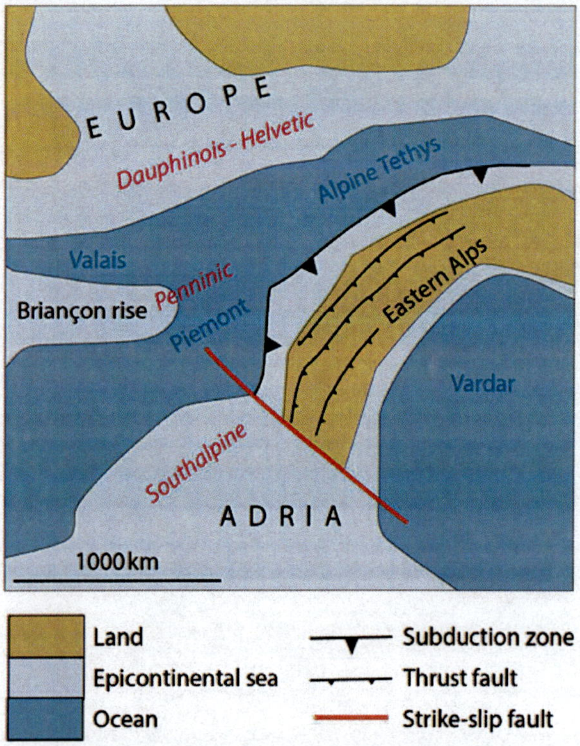

Fig. 2.5 Paleogeographic map in mid-Cretaceous times at around 100 Ma

of Pangea into the continents of Europe, Africa, North, and South America. During this breakup the margins of the continents were first stretched and thinned and as a consequence subsided. Later on complete disintegration of the continents led to the formation of oceanic basins, the Alpine Tethys, between the continental margins. The cover rocks in Switzerland implied the margins of Europe and Adria as well as the Alpine Tethys in between. The paleogeography of the future Alpine region was rather complex as shown in Fig. 2.5, which shows the situation at around 100 Ma. The Alpine Tethys had two branches toward the west: the Piemont Ocean, a true ocean, and the Valais Basin, a deep basin which only locally had true oceanic character. The Briançon rise between the two basins had a continental crust of European affinity. At 100 Ma Adria and Europe converged in a WNW–ESE direction. The Alpine Tethys was subducting beneath Adria and thrusting was going on in the Eastern Alps such that they already rose above sea level. Sedimentation continued in the Southalpine, Penninic, and Helvetic domains.

In the following the most important rock suites are briefly summarized in terms of their importance on landforms and landscapes. Key factors are their thickness and erosional resistance.

Limestone sequences that are 100–500 m thick are widespread, deposited in the shallow epicontinental seas on the European margin and the Briançon rise. Slow subsidence in Late Jurassic and Cretaceous times favored accumulation of a thick sequence of massive limestones, which are resistant to erosion and now form high, laterally continuous cliffs. Figure 2.6 shows the area around Vättis (St. Gallen) in the UNESCO World Heritage site Tectonic Arena Sardona in eastern Switzerland (see Fig. 2.1 for location). Here the gray Jurassic and brownish Cretaceous limestones are stacked on top of each other by thrusting such that cliffs led to a local relief of 2000 m (see also Buckingham and Pfiffner, this volume).

In the Valais basin thick sequences of thin-bedded sandstones and limestone interlayered with shales accumulated. These Late Jurassic–Cretaceous strata (called «Bündnerschiefer» in the local literature) are up to 1000 m thick and are overlain by Cenozoic strata of similar composition that are also a few hundred meters thick. These sequences are also heavily folded and repeated by thrusting such that rivers carved out local relief of more than 1000 m. They appear dark in color in the landscape.

Generally speaking these sandstone-limestone-shale sequences are prone to erosion. Constant dip of these units over larger areas induces asymmetric valleys. The low-angle slopes are parallel to the dip of the beds, including shaly layers. As a consequence, entire slopes slip downhill under the action of gravity, a process that is enhanced by rainfall. An example of such a dip slope is shown in Fig. 2.7a, the western flank of Safiental (Graubünden, eastern Switzerland), which exhibits agricultural activity on the entire flank. Steep slopes on the other hand show gully erosion and large alluvial fans formed at the base of the mountain flanks. Figure 2.7b reveals gully erosion on the northern flank of the Signina Group, including the peaks of Schlüechtli, Tällistock, Piz Riein, and Piz Miezgi.

Oceanic rocks attributed to the Piemont Ocean occur as a sliver of variable thickness that can be traced across the Alps. Their importance lies in the fact that they represent remnants of the subducted ocean floor, which marks the former plate boundary between the Adriatic plate and the Alpine Tethys, which was the leading edge of the European plate. Rock types include serpentinite, gabbro, and basalt. Serpentinite formed by hydration of mantle rocks which appeared at the ocean floor when the ocean opened and the two continents drifted apart. Gabbro and basalts formed from melts within the Earth's mantle, which ascended along fractures that opened in the process of drifting. These oceanic rocks are unfavorable for vegetation owing to their chemical composition (high magnesium content among other) and thus appear as barren rock surfaces even within gentle slopes. They tend to develop reddish-brownish weathering surfaces. Basalts on the other hand can be recognized by their

Fig. 2.6 Limestone cliffs around the village of Vättis (St. Gallen) (photo © IG Tektonik Arena Sardona: R. Homberger)

predominantly greenish color in the landscape. The two photographs in Fig. 2.8 illustrate these observations.

2.2.3 Cenozoic Rocks

Cenozoic rocks in Switzerland encompass mainly sedimentary sequences that are known under the name of Flysch and Molasse. Both are synorogenic sediments, that is to say they were deposited as the Alpine chain was emerging. And both Flysch and Molasse accumulated in foreland basins that were adjacent to the front of the growing chain. The basins migrated outward as the chain became wider.

Flysch sediments were deposited in narrow but deep marine troughs by turbidity flows, i.e., submarine mass flows that were possibly triggered by earthquakes on the steep sea bottom. Turbidite deposits are fining upward clastic sediments, with conglomerates or coarse-grained sandstone at the base and shale at the top. Flysch deposits consist of hundreds of turbidites.

Molasse sediments were deposited under conditions changing from marine to continental (Kempf et al. 1999). The oldest unit, the Lower Marine Molasse (*Untere Meeresmolasse* in German, UMM) evolved in time and space from Flysch deposits. As the basin was filled, conditions became continental and gave way to deposition of the Lower Freshwater Molasse (*Untere Süsswassermolasse*, USM). The latter encompasses conglomerates and sandstones deposited by rivers emanating from the early Alpine chain as alluvial fans, as well as mudstones and marls which were deposited on floodplains outside these fans. The incursion of a shallow sea from the east led to the accumulation of sand on tidal flats (Upper Marine Molasse; *Obere Meeresmolasse*, OMM). Once more the basin was filled and a new set of alluvial fans with conglomerates and sandstones were built and accompanied by mudstone and marl deposits in flood plains (Upper Freshwater Molasse; *Obere Süsswassermolasse*, OSM).

The palaeogeographic map in Fig. 2.9a depicts the situation during the deposition of the Lower Freshwater Molasse (UMM) in Oligocene times (25 Ma). The ancestral rivers that drained the Central Alps to the north flowed eastward into the Vienna Basin once they left the mountain chain, while the rivers draining to the south flowed into the persisting Po Basin. The situation changed in the following 10 million years. In Miocene times (15 Ma) the rivers of the

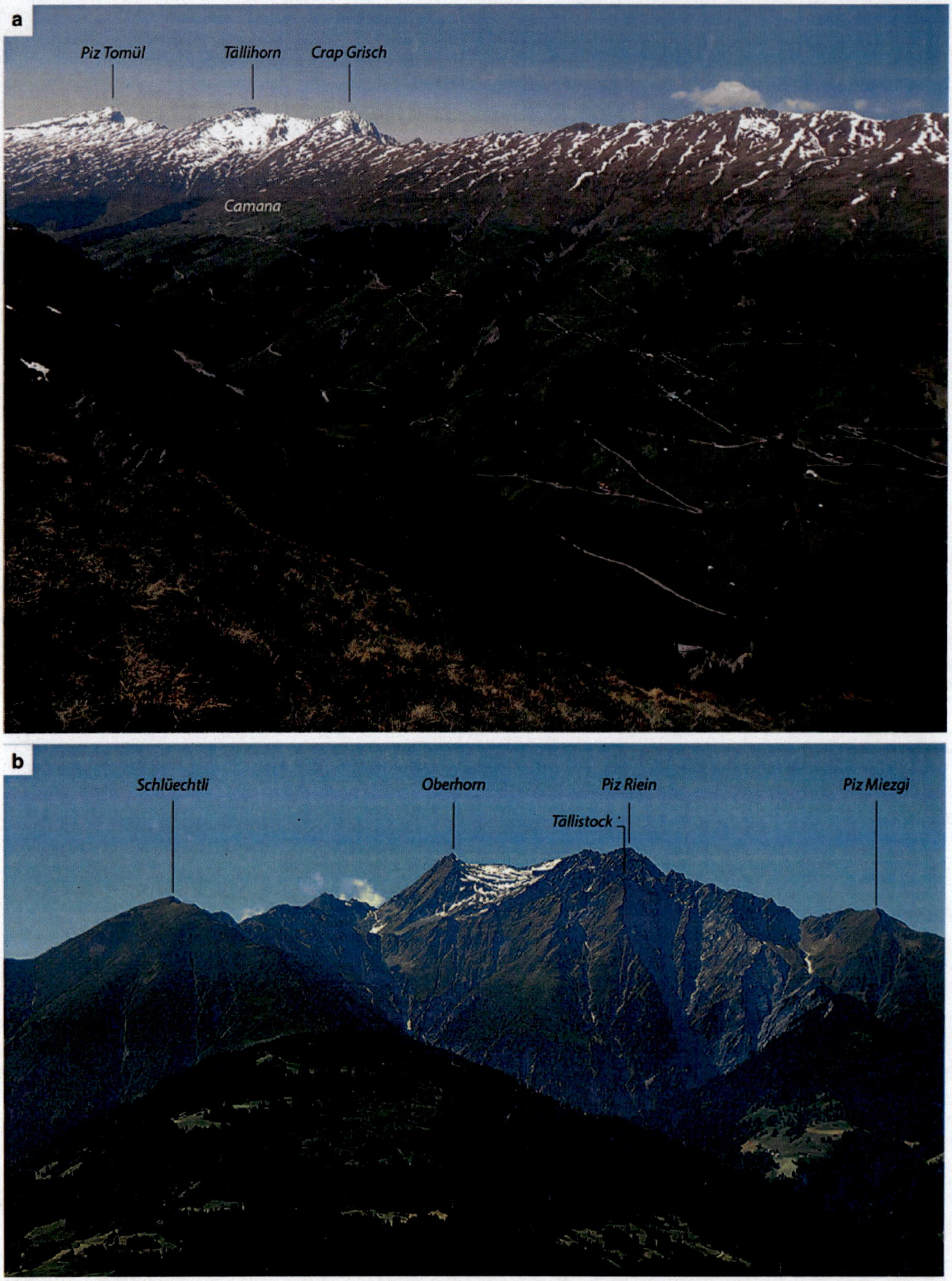

Fig. 2.7 "Bündnerschiefer" landscapes in Graubünden (photos A. Pfiffner). **a** Safiental (Graubünden). The smooth flank is a dip slope gently moving downhill. **b** Signina Group (Graubünden). The steep slopes incised by numerous channels bear witness to gully erosion

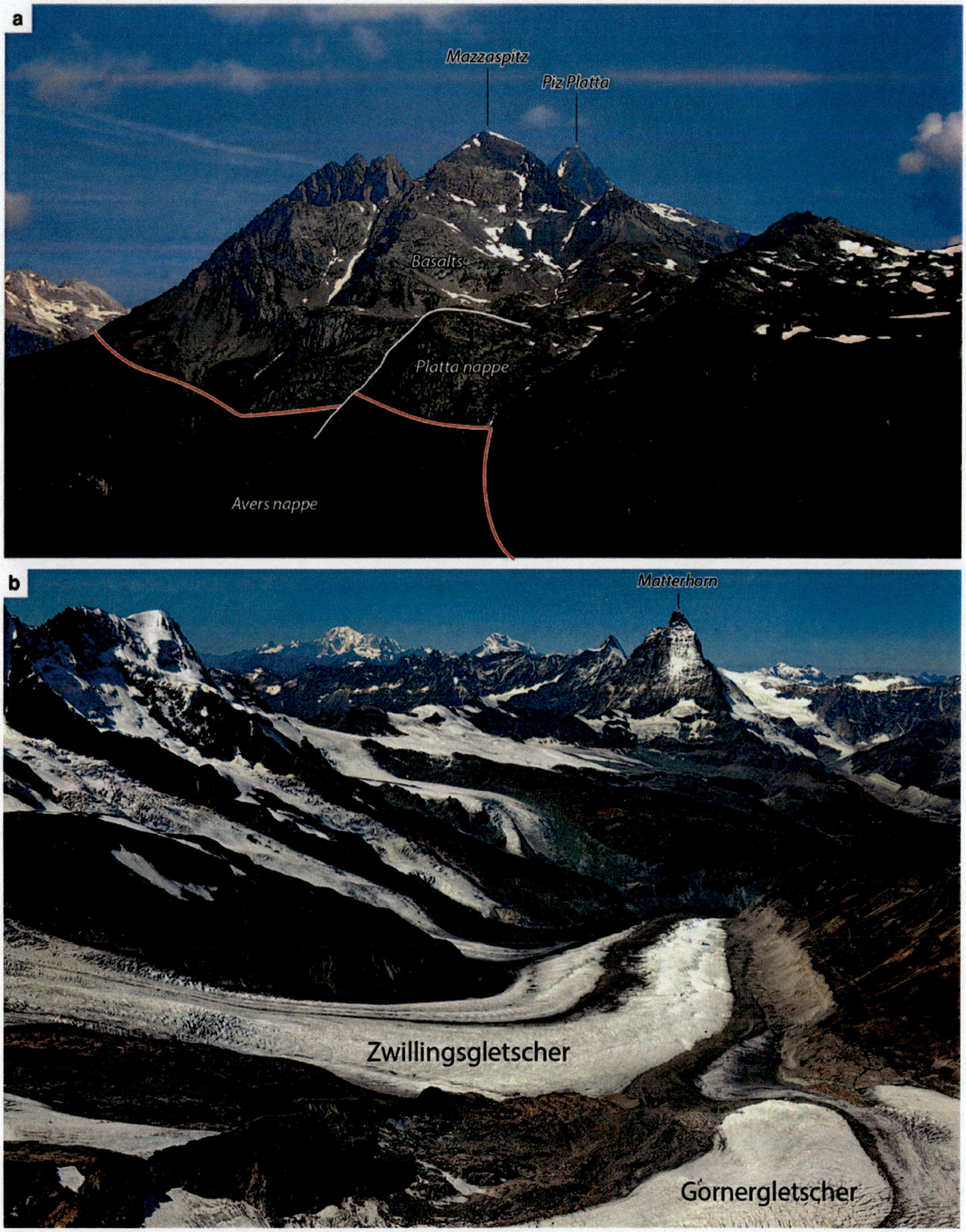

Fig. 2.8 Oceanic rocks in the Alps. **a** Avers Valley (Graubünden). The Avers nappe is made of "Bündnerschiefer" beneath the grass covered slope. The Platta nappe contains basalts and, visible in the summit of Mazzaspitz, serpentinite (photo A. Pfiffner). **b** Zermatt area (Valais) with Matterhorn. View is towards the west. The reddish-brownish ridges next to the glaciers are made of serpentinite (photo © VBS)

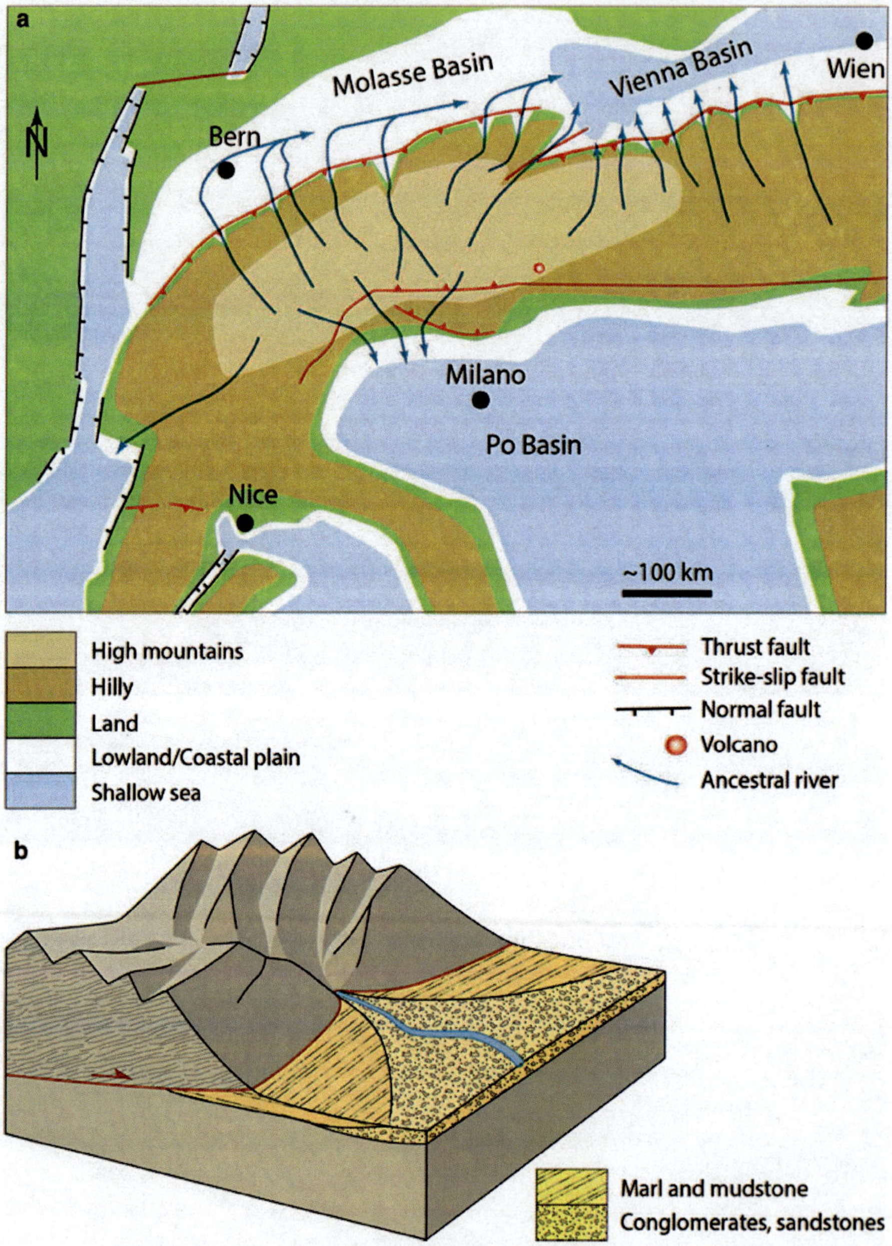

Fig. 2.9 Oligocene paleogeography of the Alps. **a** Paleogeographic map in Oligocene times at 25 Ma. **b** Architecture of an alluvial fan in the Molasse Basin

Upper Freshwater Molasse (OSM) draining to the north took a westward course after leaving the mountain chain. Figure 2.9b illustrates the landscape and depositional environment of an alluvial fan of the Lower or Upper Freshwater Molasse. The mountain chain was incised by the rivers and was contemporaneously uplifted by active thrusting. The observation that some of the alluvial fans kept their position over a long time suggests that the rivers had a constant course at least in their lower reaches. Here it has to be noted that locally the bedrock of the lower reaches was uplifted and eroded simultaneously with incision (Pfiffner 2014). It thus seems that incision was efficient, possibly due to steep channel gradients and/or high discharges. Numerical modeling of tectonic forcing on river network evolution confirms the persistence of rivers in such a dynamic environment (Kühni and Pfiffner 2001b). As reported by Schlunegger and Castelltort (2016), changes in the sedimentary architecture and pebble composition did occur and suggest that in the upper reaches new source rocks were tapped at times suggesting a dynamic landscape evolution.

2.2.4 Quaternary Deposits and Landforms

In the course of the Quaternary the Swiss Alps and their foreland were covered by ice more than a dozen times. Fluvio-glacial gravels deposited in the foreland allowed us to date the sequence of these drastic climatic changes (see Schlüchter et al., this volume). Several times the glaciers advanced way out into the foreland in the north, but barely reached the southern foreland as for example evident on the map showing ice distribution during the Last Glacial Maximum, LGM (see Schlüchter et al., this volume).

Shaping of the landscapes by glaciers was particularly important in the inner Alpine valleys. The glaciers over-deepened and widened the valleys. But after the melt down of the ice running water incised the slopes and existing valleys rather efficiently. The bed load of these rivers accumulated in the over-deepened valleys. As a consequence we find valleys of different type throughout the Alps (see Fig. 2.10a). V-shaped river valleys are the youngest and occur as tributaries of trunk rivers and on the higher areas in the absence of glaciers. In the Illgraben (Valais; see McArdell and Sartori, this volume) shown in Fig. 2.10b incision was aided by a crumbling rock type and high river gradient. A huge fan (Pfyn fan) formed where the river joins the Rhone River. Such fans exist in many other places along the Rhine and Rhone valleys. On the south side of the Alps, a number of V-shaped canyons was carved out by rivers when their base level was lowered in Messinian times (at around 6 Ma) because the Mediterranean had dried out completely (see Scapozza and Ambrosi, this volume). Bedrock incision reached to 500 and even 800 m below sea level. The canyons were later backfilled by Pliocene and Quaternary sediments and present themselves now as flat-floored valleys.

Two types of U-shaped valleys need to be distinguished (see Fig. 2.10a): flat-floored valleys and trough valleys. Flat-floored valleys emanate from glacially over-deepened valleys that are backfilled by glacio-fluvial and fluvial sediments. Figure 2.10c shows the example of an inner Alpine valley, the Gastern valley. It became famous for a misinterpretation by Albert Heim who at the time did not believe that glaciers were able to erode downward. As a consequence the railroad tunnel, which was intended to cross the valley at roughly 1200 m a.s.l., was flooded when it arrived in the water-saturated Quaternary fill. If the bedrock surface is considered only, the Gastern valley resembles more a V- than a U-shaped valley. In case of the Rhone and Rhine valleys (Pfiffner et al. 1997), as well as many other valleys (Wildi 1984), the bedrock surface beneath the valley floors is at and below sea level. These over-deepened valleys were backfilled during the meltdown of the glacier by fine-grained clastics that accumulated at the bottom of lakes underlying the floating dead ice. In trough valleys the bedrock is (still) exposed on the valley floor. Figure 2.10d illustrates a classic trough valley, the Morteratsch valley (Graubünden). The river flows on bedrock and the lateral moraines (in gray) of the Little Ice Age (LIA) testify to the level reached by the glacier in the 1850s. The photograph also highlights the scoured and rounded ridges, which date from the LGM.

Quaternary deposits outside the Alps include glacio-fluvial gravels which accumulated at the front of the glaciers. These gravels occur at several levels, the oldest (*Deckenschotter*) being the topmost (see Schlüchter et al., this volume, for a detailed discussion). The various gravels were incised after their deposition and are now visible as terraces in the landscape.

Several large rock avalanches went down from mountain flanks and covered valley floors in Holocene times. The largest one, the Flims rock avalanche (Graubünden) occurred ca. 9500 years ago (see von Poschinger et al., this volume). It covers an area of 40 km^2 and dammed a lake upstream. Another rock avalanche is located immediately downstream at Reichenau/Tamins (Graubünden). It went down earlier and also dammed a lake. Outbursts of these lakes significantly modified the valley floor of the Rhine River. Notable landforms that were created by the outbursts are "tumas", i.e. hills composed of rock avalanche material that was remobilized and transported over large distances upstream and downstream before being deposited (V. Poschinger et al., this volume, Pfiffner 2019). Other examples of a pre-historic rock avalanche that significantly modified the valley floor can be found in the Kander valley (Bern), near Davos (Graubünden), near Lenzerheide (Graubünden), and around Sierre (Valais). Besides, a number of younger rock avalanches destroyed villages and/or caused heavy damages: Biasca (Ticino), Corbeyrier (Valais), Piuro (Graubünden), Derborence (Valais), Goldau (Schwyz), Elm (Glarus), Randa (Valais), and Bondo (Graubünden).

2.3 Tectonic Structure of the Swiss Alps

The Alps are a classic example for the study of mountain building. Generations of geologists mapped the many valleys and peaks in such great detail that today we can study the interplay between plate tectonics and the creation of the Alpine relief. Plate tectonics involves the breakup of the supercontinent Pangea with the opening of the Alpine Tethys, followed by the subduction of this ocean beneath the Adriatic continent and finally the collision of the European with the Adriatic continent. The convergence between the two continents led to the formation of tectonic nappes, i.e., slices of the Earth's crust piled up on top of each other. One of the advocates of these nappe piles was Emil Argand. He illustrated his conception in a cross-section across the entire

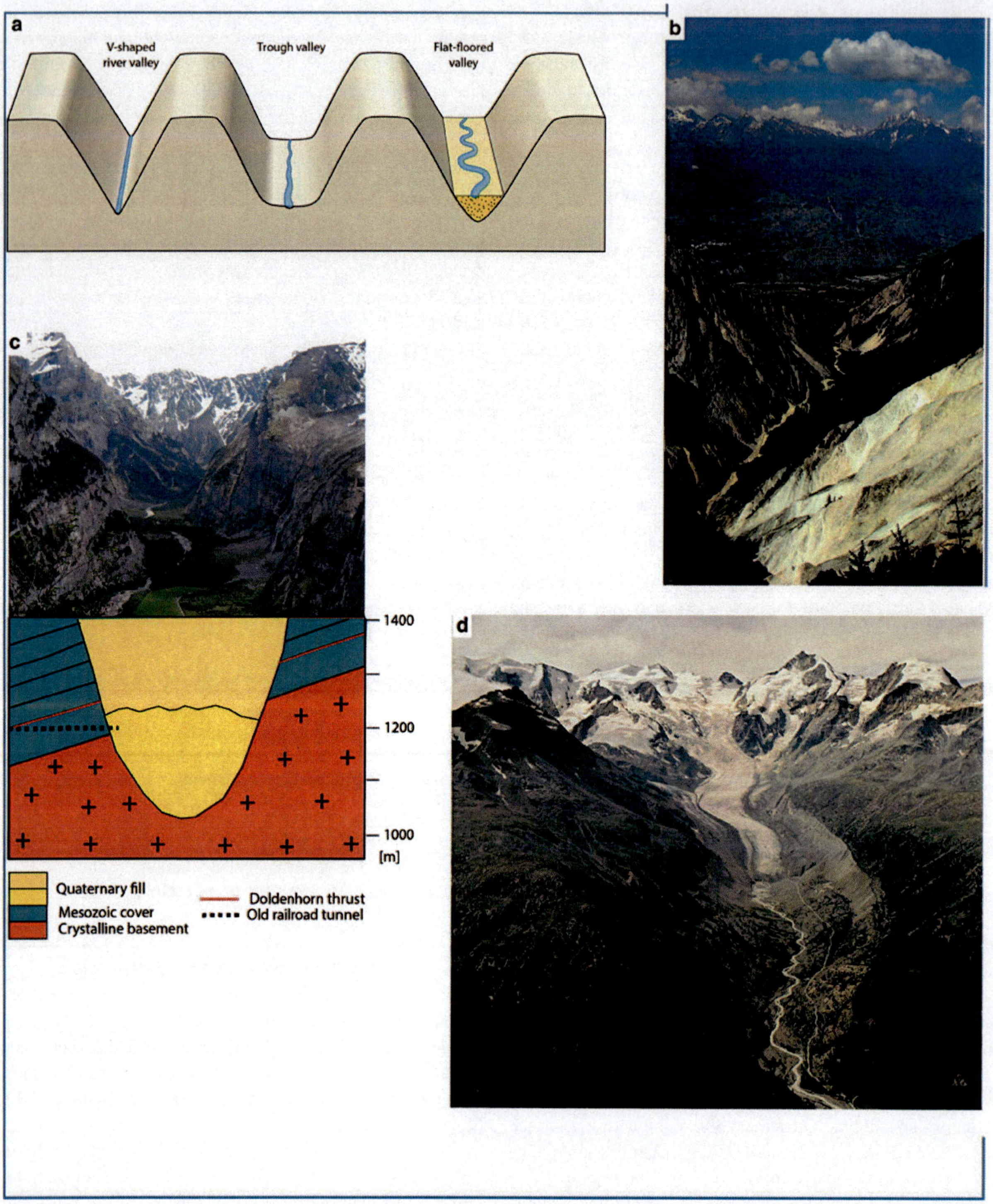

Fig. 2.10 Quaternary landforms. **a** Valley types. **b** V-shaped valley Illgraben (photo A. Pfiffner). **c** Overdeepened flat-floored Gastern valley (photo A. Pfiffner). **d** Trough valley Morteratsch (photo © VBS)

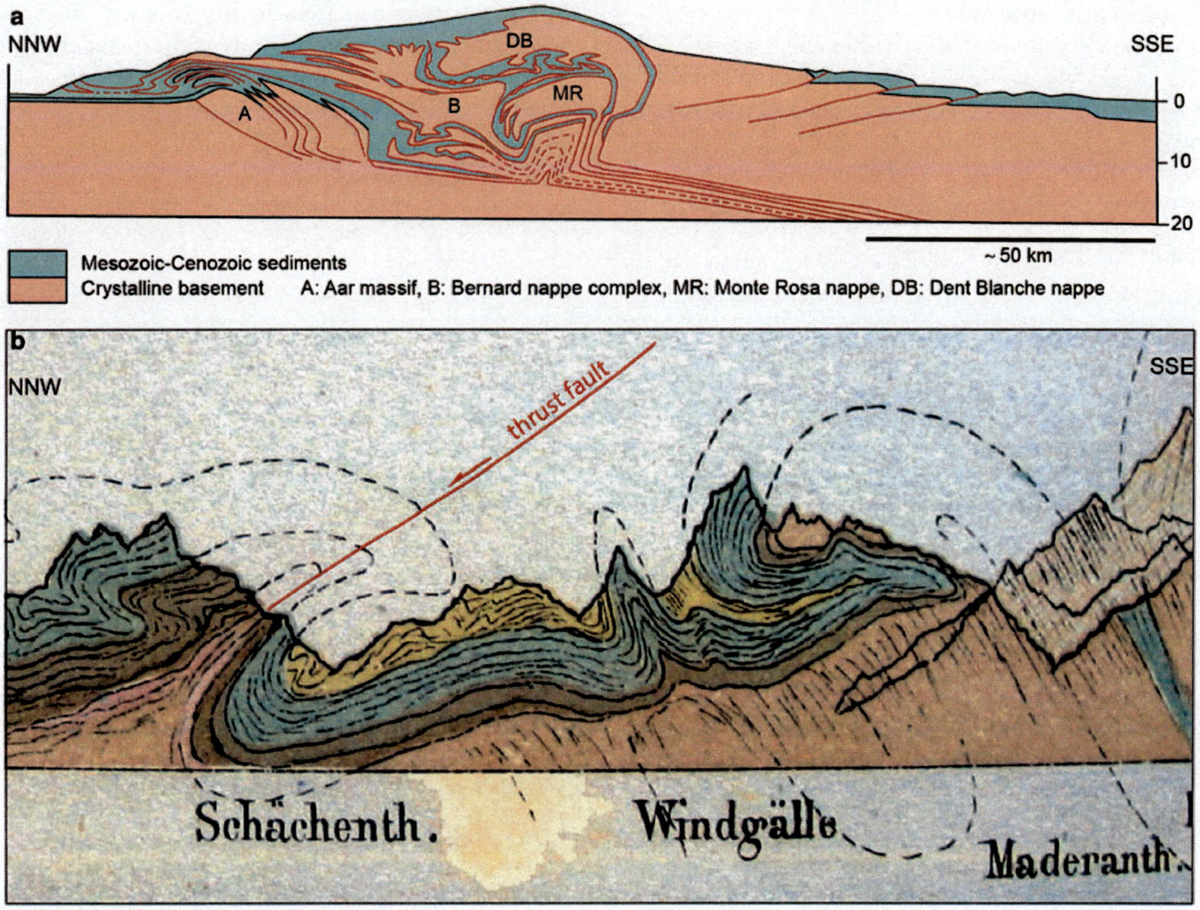

Fig. 2.11 Tectonic style of the Alps as conceived by early workers. **a** Cross-section redrawn after Argand (1928). **b** Detail of cross-section taken from Heim (1919–1922). The thrust fault marked in red was added by the author to indicate the modern understanding of nappe tectonics

Alps (Fig. 2.11a). Nappes of crystalline basement rocks separated by thin slivers of Mesozoic-Cenozoic sediments are piled on top of each other; the contacts between basement and cover is shown to be intricately folded. Albert Heim worked in the Windgälle area (Fig. 2.11b) and was impressed by the pervasive folding he saw there, to a point where he rejected the idea of nappes for a long time. The anticline north of Schächenthal in Fig. 2.11b, with a thin tongue of crystalline basement sticking upward to the south (and not reaching the surface) was drawn to avoid a thrust fault explaining that older rocks are lying on top of younger rocks in the northern flank of Schächenthal. Today we know that this fault (the Axen thrust) is one of the important thrust faults, which moved the Axen nappe over a distance of some 50 km.

Thrusting and folding resulted in a complex nappe stack and the juxtaposition of very different rock types deep in the crust and near the surface. The removal of the top part of this nappe stack by erosion rendered the situation even more complex because deeply incised valleys make the original structure discontinuous. Thus a geological map seems like a puzzle of differently shaped and colored pieces difficult to interpret. In order to make this chaos intelligible, the major geological units are discussed using representative cross-sections and photographs. Generally speaking we distinguish three major nappe systems within the Alpine chain. The Helvetic nappe system consists of units that are derived from the margin of the European continent. The Penninic nappe system encompasses nappes derived from the Alpine Tethys that once existed between the European and Adriatic margin. The Adriatic margin in the Alps is represented by the Austroalpine nappe system in eastern and central Switzerland, by the Southalpine nappe system in southern Switzerland and northern Italy, and the Salassic nappes that outcrop mainly in western Switzerland and neighboring Italy. Apart from the Helvetic nappe system, the former European margin includes also the Jura Mountains and the Molasse Basin of the Swiss Plateau. The discussion of these tectonic provinces will proceed from north to south.

2.3.1 Jura Mountains

The Jura Mountains are located in northwestern Switzerland and extend into neighboring France (see Fig. 2.2). They are made of Mesozoic and Cenozoic sediments, which were detached from their crystalline basement and pushed to the northwest for up to 10–15 km (Laubscher 1965). The detachment occurred mostly along a Triassic layer of evaporites. The sediments above the evaporites, mainly Jurassic limestones, were folded as the entire package was pushed to the northwest. The cross-section in Fig. 2.12a shows the general structure with folds dominating in the SSE between Biel and Glovelier and thrust faults in the NNW from Mont-Terri to Miécourt. This change in style is likely to be caused by thickness changes of the evaporite layer. If this layer is thick the evaporites can flow and fill the cores of the anticlines. Alternatively if they are thin they can act as a detachment for a thrust fault. If absent altogether the cover sediments remain attached to the basement as is the case at Seppois-le-Bas. The Mesozoic sediments continue to the SSE beneath the sediments of the Molasse Basin to Aarberg and beyond. Several patches of Permo-Carboniferous sediments are drawn in the subsurface. The one beneath Hermrigen is based on seismic data and the other ones are extrapolated from an area farther west and are more speculative. In any case the detachment horizon is not a planar surface. It has been deformed in a later stage of deformation after the main detachment (Pfiffner 2014, 2015). The Jura Mountains formed mainly from Miocene to Pliocene times between 10 and 3 Ma (Bolliger et al. 1993; Kälin 1997).

The photograph in Fig. 2.12b shows a sequence of ranges each representing an anticline. The one in the center of the image is dissected by a gorge cut by the Suze River. In the sidewall of the gorge one can see how the massive limestones of the Late Jurassic bend around the topography outlining the fold nearly perfectly.

2.3.2 Molasse Basin

The Molasse Basin extends from Geneva toward the northeast where it widens considerably and continues beyond Lake Constance all along the Alps (see Fig. 2.1). The cross-section in Fig. 2.13a illustrates the structure of the Molasse Basin. The Molasse sediments discussed in Sect. 2.2.3 overly the Mesozoic cover sediments, which in turn overly the crystalline basement. The Mesozoic sediments dip gently to the SSE from a depth of 1000 m below sea level in the foreland to 7000 m below sea level beneath the Helvetic nappes. In the Plateau Molasse the sediments are more or less horizontal and increase in thickness toward the Subalpine Molasse. The Subalpine Molasse consists of three thrust sheets that dip to and extend to the SSE beneath the Alps (the Helvetic nappes in this cross-section), from where the term "Subalpine" is derived. The transition from the Plateau to the Subalpine Molasse is characterized by an antiformal structure (also called triangle zone): near Wattwil the Plateau Molasse is tilted to a northerly dip, near Krummenau the Subalpine Molasse is tilted with a southerly dip. Within the Subalpine Molasse only the older Molasse sediments occur (Lower Marine and Lower Freshwater Molasse, UMM, and USM), whereas in the Plateau Molasse also the younger sequences (Upper Marine and Upper Freshwater Molasse, OMM and OSM) are present. The UMM pinches out toward the NNW near the transition between Plateau and Subalpine Molasse, a clear sign that the foreland basin migrated to the NNW with time (see also Kempf and Pfiffner 2004).

The photograph in Fig. 2.13b illustrates how the USM strata of the Subalpine Molasse dip to the SSE to disappear beneath the Helvetic nappes (Säntis nappe). A thin sheet of sandstones and shales referred to as Subalpine Flysch makes a cushion between the Säntis nappe and the USM strata.

2.3.3 Helvetic Nappe System

The Helvetic nappe system extends from Austria across Switzerland to France (see Pfiffner 2011 for a detailed description). Here a cross-section through central Switzerland is presented. The Helvetic nappe system is subdivided into a lower and upper unit. The Lower Helvetics include the Aar massif and its sedimentary cover. This massif is a basement uplift where the crystalline basement was thrust upward and northward along a thrust fault as visible in the cross-section of Fig. 2.14a. The Mesozoic cover sediments (Triassic–Cretaceous) are folded with narrow synclines pinched into the basement rocks. The Cenozoic cover consists of Flysch (North Helvetic Flysch) and Molasse (Subalpine Molasse, mainly UMM). The transition from Flysch to Molasse is in the subsurface and nowhere exposed. The zigzag line underlines this uncertainty. Nevertheless it has to be kept in mind that age wise the Flysch deposits grade smoothly into the Molasse deposits.

The Upper Helvetics are made up of two major nappes in this cross-section. The lower one, the Axen nappe, consists mainly of Jurassic strata, which are intensively folded and faulted. Folds contain plunging anticlines, that is isoclinal folds that became rotated such that the oldest, Triassic strata close downward to the NNW (ca. 6 km to the NNW of the southern end of the cross-section). Several steeply dipping normal faults affect the NNW part of this nappe. The upper of the two nappes, the Drusberg nappe, is composed of essentially Cretaceous limestones. These limestones represent the stratigraphic roof of the Jurassic limestones of the Axen nappe. They have been detached along a lowermost

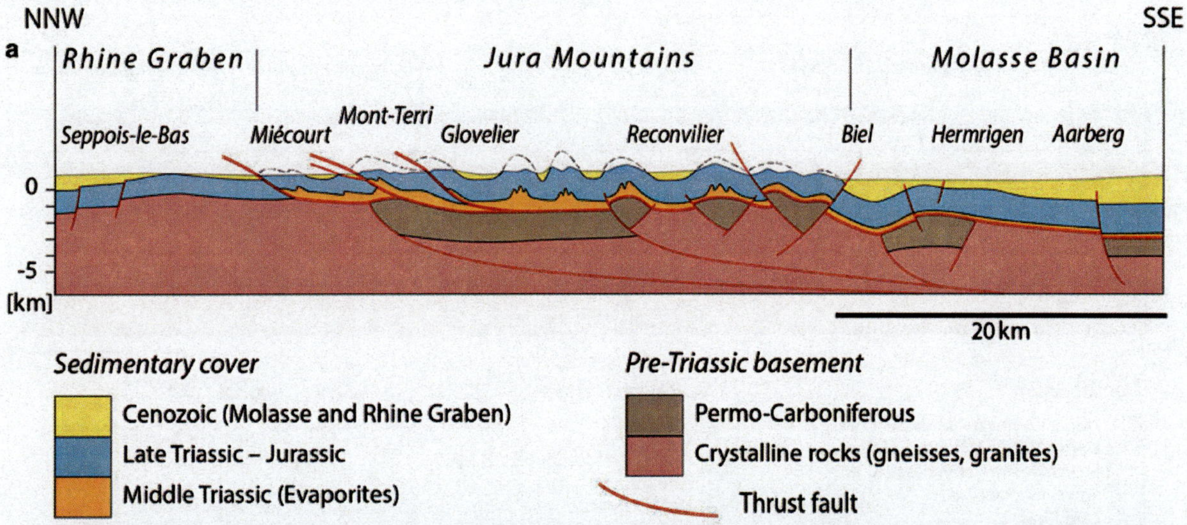

Fig. 2.12 Structure of the Jura Mountains. **a** Cross-section from the Molasse Basin to the Rhine Graben (from Pfiffner 2014, 2015). **b** Areal view of the Reuchenette cluse and adjacent anticlines (photo © VBS)

Cretaceous marl (the Palfris Formation) and traveled nearly 20 km farther than the Jurassic limestones. In this process the Cretaceous limestones were intensively folded as can be seen between Melchtal and Pilatus. In this segment there are also Cenozoic strata overlying the Cretacous, and in addition a thin sliver of Ultrahelvetic sediments, which represent an Eocene mélange with various blocks of older sediments. This mélange underlies the Penninic nappes which is undifferentiated in the cross-section but consist of Mesozoic sediments (see Pfiffner, this volume, for a detailed discussion

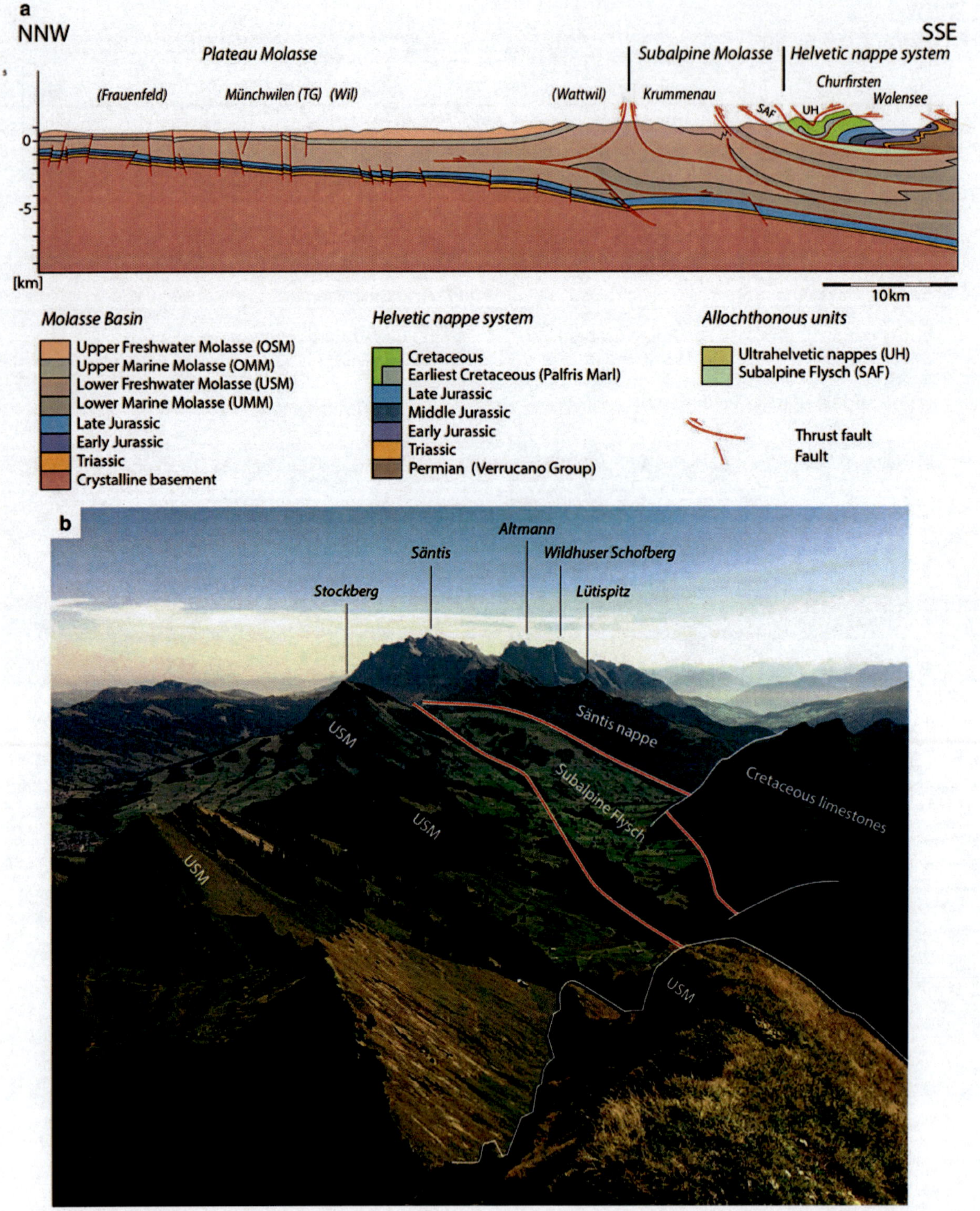

Fig. 2.13 Structure of the Molasse Basin. **a** Cross-section Frauenfeld-Walensee in eastern Switzerland (after Pfiffner 2014, 2015). **b** Photograph looking ENE from Speer towards Säntis (photo A. Pfiffner)

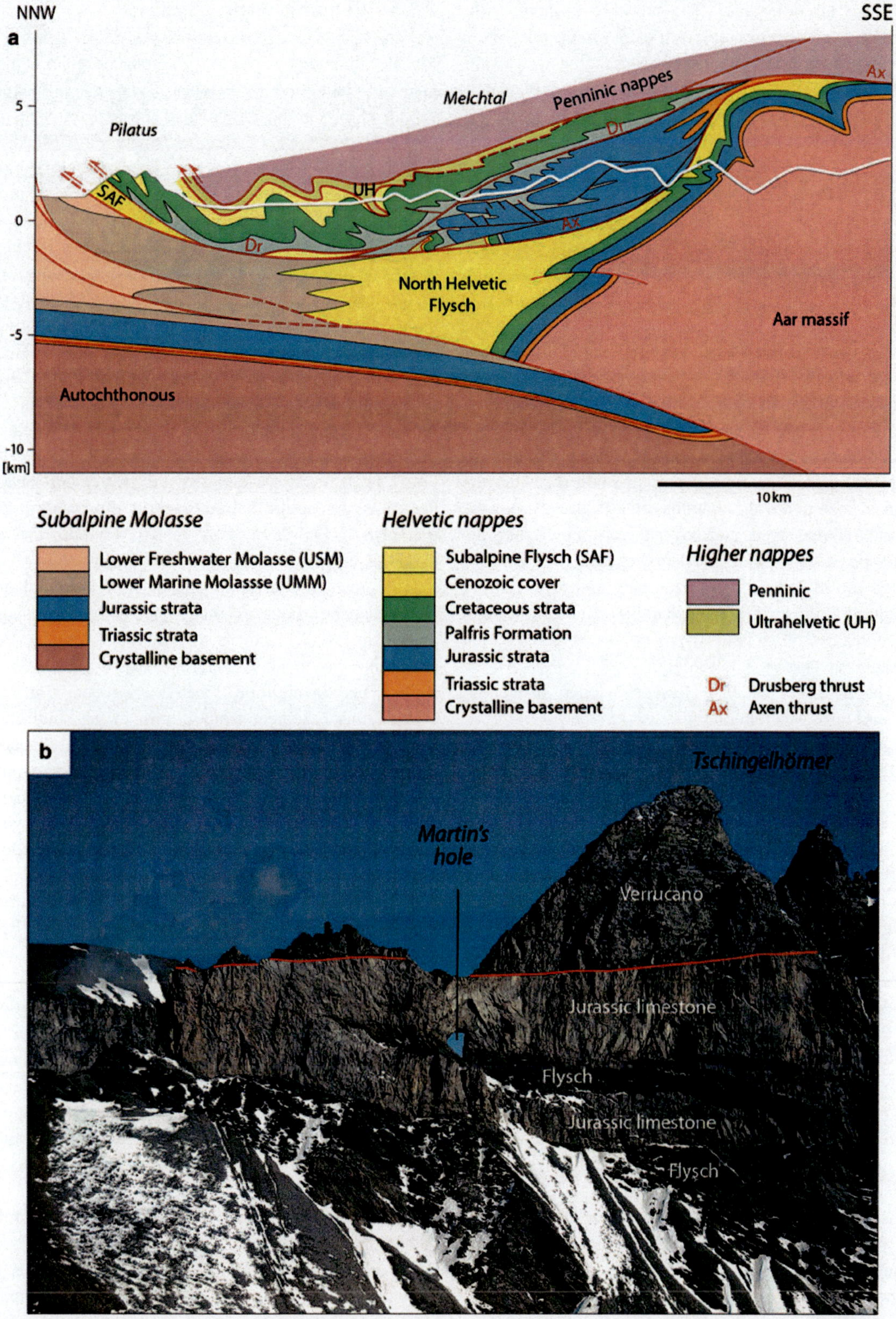

Fig. 2.14 Structure of the Helvetic nappe system. **a** Cross-section through Central Switzerland (Luzern, Obwalden, Nidwalden) based on Pfiffner (2014, 2015). **b** Photograph showing the Glarus thrust in the Tschingelhörner (Glarus/Graubünden) (photo © IG Tektonik Arena Sardona: Ruedi Homberger)

on the Lake Lucerne area). The mélange formed as the Penninic nappes were thrust on top of the future Helvetics. Pieces scraped off the footwall and the hanging wall of the active Penninic thrust were intensively mixed and sheared. The wild nature of this mélange led the original workers to speak of "wildflysch" (e.g., Kaufmann 1886). Note that the Penninic thrust, together with the Ultrahelvetic unit, is folded just as the Cretaceous limestones of the Drusberg nappe. The leading edge of the Drusberg nappe in Pilatus is underlain by a lens of Subalpine Flysch.

It is obvious that a thrust displacement along the Axen or Drusberg thrust, as well as the Penninic thrust in its present curved geometry is not possible. The solution to this dilemma is that in a first step the Penninic and Ultrahelvetic units were thrust onto the Cretaceous strata of the Helvetic nappes. As deformation attained these limestones, the thrusts were no longer active but passively folded. Emplacement of the Penninic as well as the Helvetic nappes occurred upward along thrust surfaces that were dipping southward into the crust. In a later phase the Aar massif was thrust onto the Autochthonous basement and cover thereby upwarping the entire nappe stack of Helvetic and Penninic nappes. The associated uplift created a large-scale antiform with a north-dipping limb south of Melchtal. So once again folding deformed inactive thrust faults, a process that is referred to in the literature as "post-nappe folding."

The basal thrust of the Helvetic nappes in central Switzerland (cross-section of Fig. 2.14a) is the Axen thrust. Toward the east this major thrust fault is replaced by the Glarus thrust. The Glarus thrust is famous for the historical role it played as birthplace of nappe tectonics (see Buckingham and Pfiffner, this volume). A spectacular view of this thrust fault can be gained in the Tschingelhörner (see Fig. 2.14b). The importance of this locality is underlined by the fact that it is the core area of the UNESCO World Heritage site Tectonic Arena Sardona. The red line drawn in Fig. 2.14b follows the Glarus thrust and is also referred to as the "Magic line." The rugged peaks, above the Tschingelhörner, are made of Permian sediments (so-called Verrucano). Beneath is a layer of Jurassic limestones, which contains a natural opening, the Martin's hole. Twice a year, in spring and autumn, the sun shines through this hole and casts its light onto the church tower in the village of Elm (Glarus). The layer of Jurassic limestones is actually repeated by a thrust fault. This fault is responsible for the dark Flysch layer that runs more or less through the Martin's hole. The layer(s) of Jurassic limestones represent a sliver which was plucked off the footwall of the Glarus thrust farther south, where Verrucano was thrust over Mesozoic limestones, and transported northward until it came to lie on Flysch.

2.3.4 Penninic Nappe System

The Penninic nappe system is the most complex in the Alps. One can distinguish three types of nappes: (1) nappes consisting of crystalline basement rocks and remnants of their sedimentary cover, (2) nappes made of Mesozoic cover rocks which were detached from their crystalline basement, and, (3) nappes containing oceanic basement and cover rocks. The tectonic evolution of the Penninic nappe stack encompasses several phases of thrusting and folding. Generally speaking the cover sediments were detached in a first phase from their crystalline basement and piled up to a nappe stack of their own. The crystalline basement was then deformed and piled up to a nappe stack; this phase was in many places associated with post-nappe folding, i.e., earlier formed thrust faults were passively folded. This section focuses on type (1) nappes and discusses the structural style that had intrigued E. Argand in 1928 already (see Fig. 2.11a). Type (2) and (3) are only mentioned briefly here; their impact in the landscape is illustrated in the photographs of Fig. 2.8a, b. A type (2) nappe, the Klippen nappe, is discussed in Pfiffner, this volume.

The cross-section running from the Rhone Valley to the Dent Blanche traverses the Siviez-Mischabel nappe. As evident in Fig. 2.15a there are Triassic sediments above the crystalline basement from Bella Tola to Le Toûno and Le Boudri, and in addition Triassic sediments can be found beneath this basement from Illhorn to the south. This led researches to postulate that the Siviez-Mischabel nappe forms a huge recumbent fold where these sediments are wrapped around the basement block (e.g., Escher et al. 1993). However, detailed structural analyses revealed that the sediments below the crystalline basement are made up of a number of upright thin nappes (see Markley et al. 1999, Scheiber et al. 2013) as indicated by the arrows in Fig. 2.15a. The sediments on top of the crystalline basement are thoroughly folded as can be seen beneath the peak Le Boudri.

The Siviez-Mischabel nappe is underlain by two type (2) nappes, the Houillère nappe consisting of a normal sequence in this cross-section, and the Sion-Courmayeur nappe. Further along strike, the Houillère nappe also contains Carboniferous and Permian sediments. On top of the Siviez-Mischabel nappe, the Tsaté nappe is of type (3), which is in turn overlain by crystalline basement rocks of the Salassic Dent Blanche nappe.

In the case of the Suretta nappe (Fig. 2.15b) the kilometer-scale structure is even more complex. The crystalline basement made of polymetamorphic gneisses and Permian orthogneisses (Rofna Porphyry complex) is

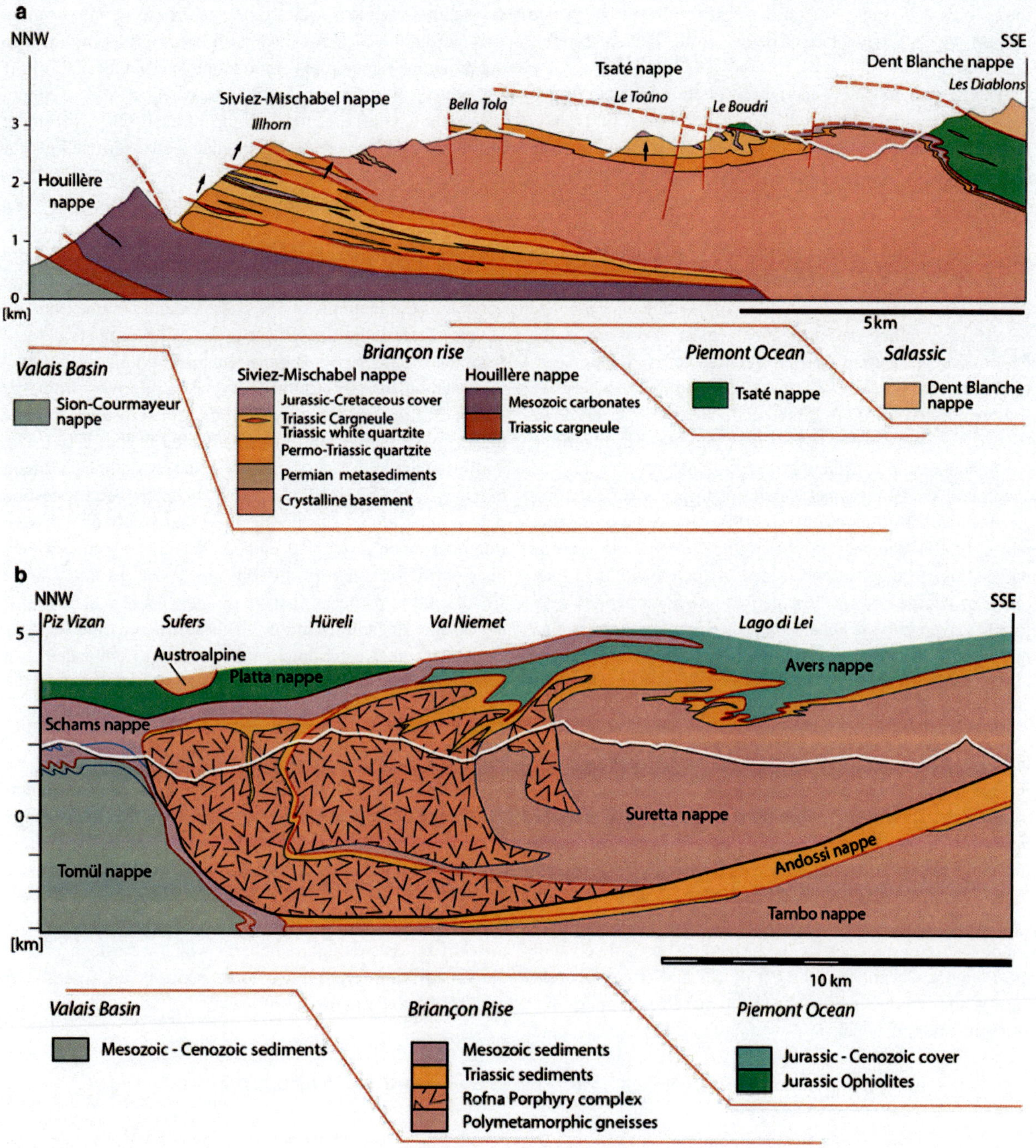

Fig. 2.15 Structure of the Penninic nappes. **a** Cross-section in western Switzerland (Valais) simplified from Scheiber et al. (2013). **b** Cross-section in eastern Switzerland based on Scheiber et al. (2012) and amended

repeated by a thrust fault, which is almost vertical beneath Hüreli and then becomes nearly horizontal deeper down. The thrust fault is outlined by a thin sliver of Triassic sediments, which represent the oldest (Triassic) part of the sedimentary cover of the lower basement unit, and is obviously folded. The upper basement unit has also retained the oldest part of the sedimentary cover, and these Triassic sediments have been tightly folded together with the basement rocks. Interestingly the folds look the "wrong way": anticlines close toward the SSE instead of the NNW closing that is normally observed in the nappe stacks formed by NNW directed thrusting. Structural analysis clearly indicates that these folds had been refolded or backfolded in a later stage of post-nappe folding (see Scheiber et al. 2012 and references therein).

The Suretta nappe is underlain by another type (1) nappe, the Tambo nappe. Between these two a thin layer of Triassic sediments can be followed over tens of kilometers. The lowermost part of these sediments represent remnants of the sedimentary cover of the Tambo basement, but the main part is an allochthonous sequence, the Andossi nappe in Fig. 2.15b, which has been scraped off the Tambo basement in its rear some 40–50 km south of the area shown in the cross-section. Such thin layers of sediments pinched between crystalline basement blocks exist elsewhere in the Penninic nappe system and are referred to as "nappe separators" in the literature. The question then arises where the sedimentary cover of the Tambo and Suretta nappes are to be looked for. The answer is linked to the question regarding the Schams nappes, which are type (2) nappes that outcrop all around the Tambo and Suretta nappes (see Fig. 2.15b). As discussed by Schmid et al. (1990) the Schams nappes represent the detached cover of the Tambo and Suretta nappes and formed a nappe stack of their own. In a later phase of Alpine deformation the two basement nappes were jointly driven into the Schams nappes wrapping them around the basement. This later phase was also responsible for the backfolding of the top of the Suretta nappe described above.

Another nappe that merits attention is the Avers nappe. As can be seen in Fig. 2.15b, this nappe is separated from the underlying Suretta nappe by a thrust fault which is affected by the backfolds. The Avers nappe consists of Jurassic–Cretaceous sediments, which were detached from their oceanic basement (the Piemont Ocean) in an early phase of nappe formation and came to lie on the Suretta and Schams nappes. In the ensuing nappe building processes, it was passively deformed. Figure 2.15b also depicts the position of the Platta nappe that represents a detached piece of the oceanic crust of the Piemont Ocean. Detachment of oceanic crust occurred during the subduction of the Piemont Ocean under the Adriatic margin, which is now found in the Austroalpine nappe system. An erosional remnant of Austroalpine nappes is present in the north of the cross-section of Fig. 2.15b.

2.3.5 Austroalpine Nappe System

The structure of the Austroalpine nappe system is discussed based on a cross-section located in easternmost Switzerland and neighboring Austria. In this transect the Austroalpine nappes form a sheet on top of the Penninic nappe system (see Fig. 2.16a), which has been partly removed by erosion. Denudation was more complete around the Engadin Window where the underlying Penninic nappes appear at the surface. North of the Engadin Window the Silvretta nappe consists of crystalline basement rocks, its detached cover now forming the Northern Calcareous Alps situated farther north. The Austroalpine nappe pile is more complete to the south of the Engadin Window where several basement nappes (Campo, Sesvenna, Ötztal) are piled on top of each other, including also Mesozoic sediments. The detailed structure and evolution is rather complex and involves faults active at depth as well as near-surface faulting (see Schmid and Haas 1989). The Penninic nappe system in Fig. 2.16a is not differentiated because it is mostly buried at depth. Within the Engadin Window oceanic rocks from the Piemont Ocean make the top surface of the Penninic nappe complex and are underlain by slivers of crystalline basement rocks and sediments derived from the Briançon rise. But the remainder within the window is made of Jurassic–Cretaceous clastics derived from the Valais Basin, similar to the outcrops at the erosional front of the Penninic nappes north of Misthaufen in Fig. 2.16a. The antiformal updoming and position of the crystalline basement deep beneath the Engadin Window is constrained by seismic data (Hitz 1996). Considering the basal thrust of the Austroalpine nappe system it is evident that this fault underwent a phase of deformation post-dating its activity as thrust fault: small-scale folds near Misthaufen and a large-scale updoming in the Engadin Window. This testifies to the tectonic evolution: the Austroalpine nappe pile was emplaced in a ENE direction in Cretaceous times and redeformed in the course of north–south convergence in the Cenozoic (see Pfiffner 2014, 2015 and references therein).

Figure 2.16b illustrates the landscape north of Julierpass with differently colored rocks of the Err nappe. The mountain range with Corn Suvretta consists of yellow-orange Triassic cornieules to the right and greenish crystalline basement to the left. Il Nes is built of grayish Jurassic breccias that were accumulated at the base of a steep submarine slope. The red patch near the center of the image is made of Jurassic radiolarian cherts, which were deposited on the ocean floor. The light colored rocks in the foreground correspond to Triassic dolostones.

2.3.6 Southalpine Nappe System

The Southalpine nappe stack in southern Switzerland is shown in Fig. 2.17. The Insubric Fault, which is part of the Peri-Adriatic Fault System, is dipping steeply to the north. It has two components: a reverse (or thrust) component that raised the northern block and a strike-slip component that moved the southern block to the west. The units to the north of the Insubric Fault, located in the southernmost part of the Penninic nappe stack, all display penetrative foliation dipping parallel to the fault and is therefore referred to as "Southern Steep Belt." Strictly speaking Sesia and Tonale Zone are considered as Salassic units.

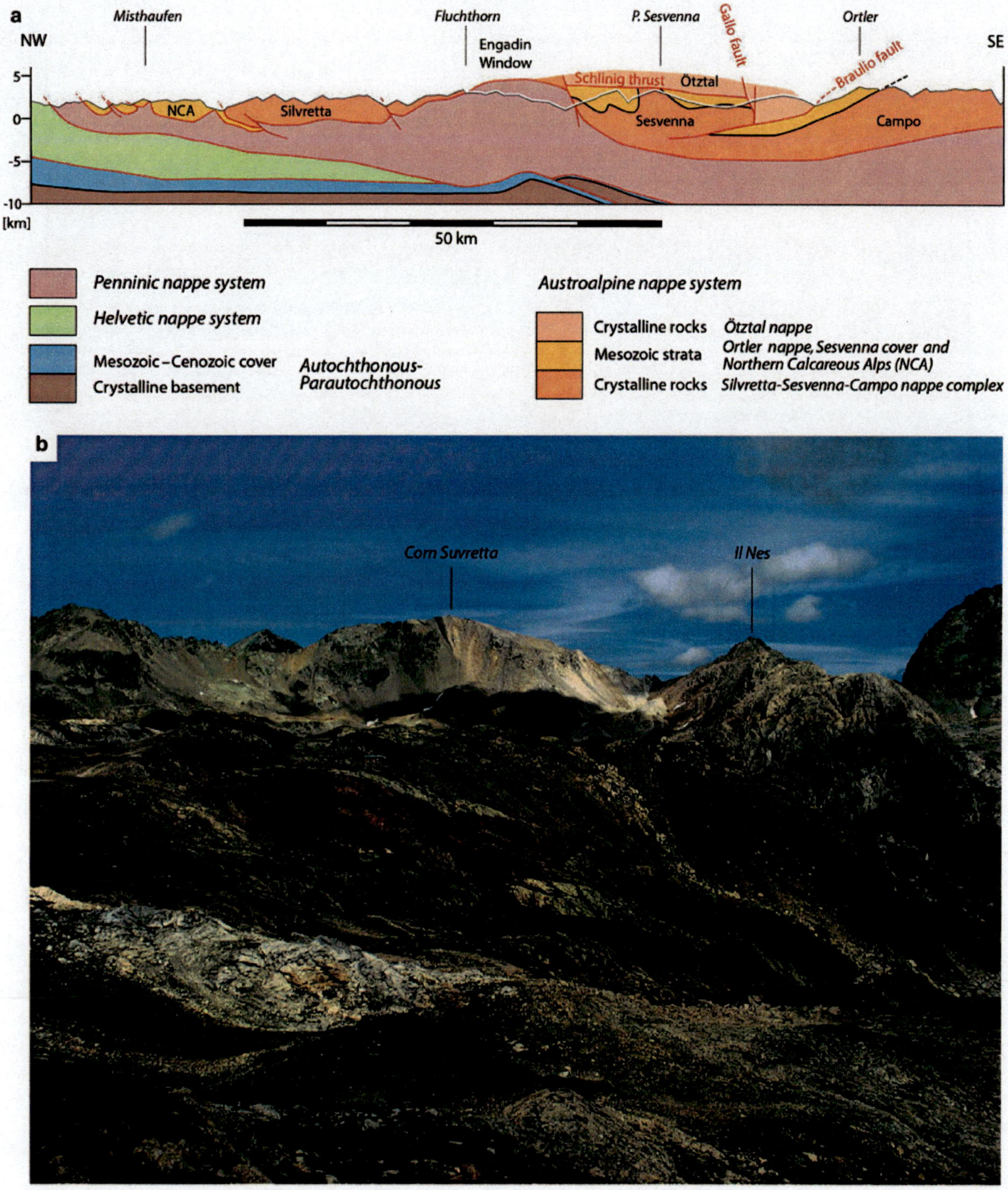

Fig. 2.16 Structure of the Austroalpine nappe system. **a** Cross-section in easternmost Switzerland (Graubünden) based on Pfiffner (2016). **b** Photograph showing the colorful landscape north of Julierpass (Graubünden) (photo A. Pfiffner)

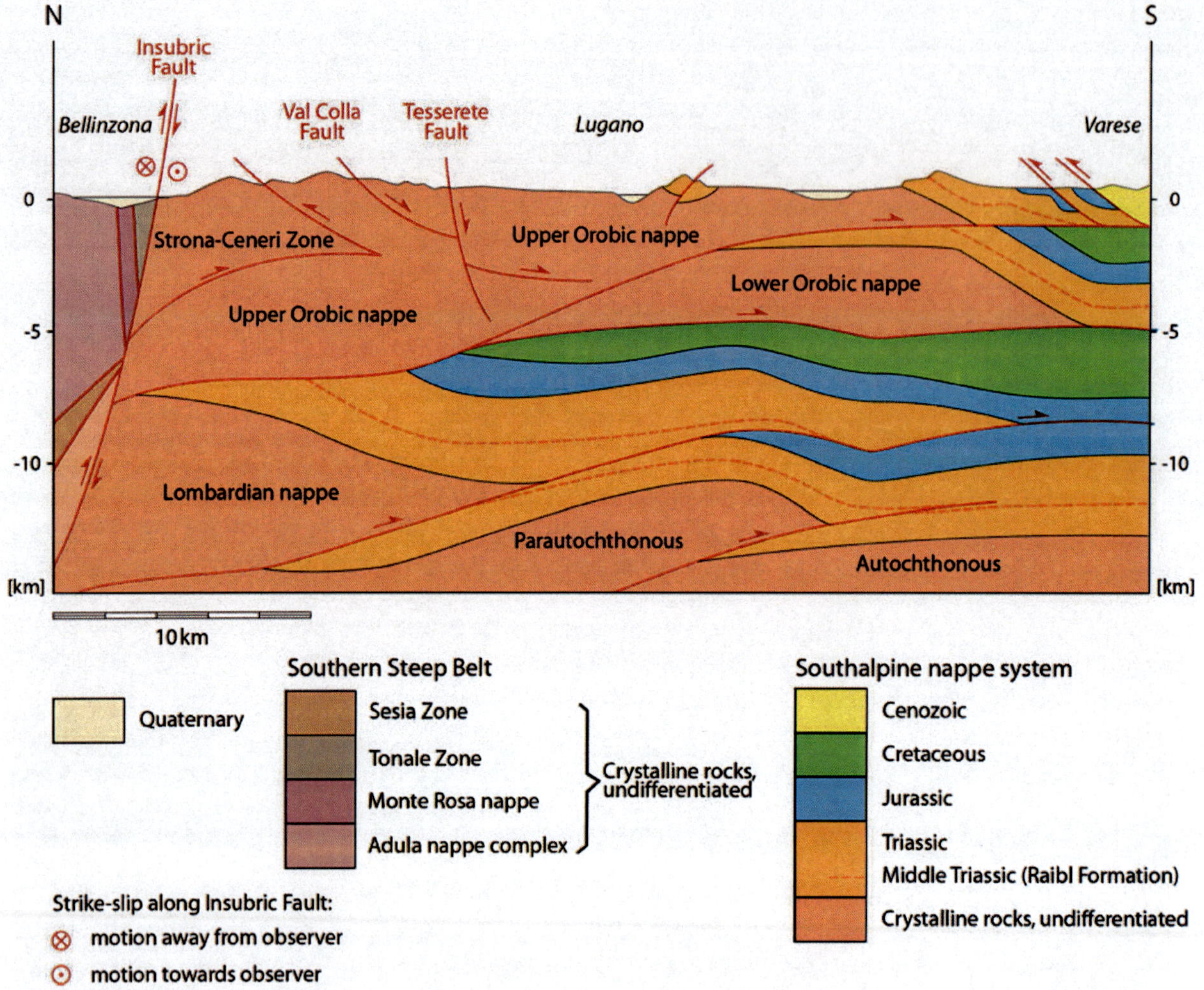

Fig. 2.17 Structure of the Southalpine nappes in a cross-section from the Insubric Fault to the Po Basin at Varese (modified from Schumacher 1997)

The Southalpine nappe stack itself consists of a number of thrust sheets made of crystalline basement rocks and sediments. They were thrust toward the south. The structure shown for the nappes from the Autochthonous up to the Upper Orobic nappe is somewhat constrained by seismic data (see Schumacher 1997). The detailed structural style reflects also the findings gained farther east from boreholes and reflection seismics (see Schönborn 1992). South-directed thrusting affected the Parautochthonous only slightly and the Lombardian nappe moderately. In contrast, the Lower Orobic nappe was moved over at least 30 km toward the south such that its crystalline basement rests over a long distance on the Mesozoic sedimentary cover of the Lombardian nappe. Higher up in the nappe pile the Upper Orobic nappe was moved over a moderate distance to the south. Its sedimentary cover was forced as a wedge into the Cenozoic sediments of the Po Basin as indicated by the north directed thrusting around Varese. Similarly, the Strona-Ceneri Zone was forced into the crystalline basement of the Upper Orobic nappe, splitting it apart. The younger Adamello Intrusive complex truncates the thrust faults of the Orobic and Lombardian nappes. Thus the thrusts were active prior to 30 Ma (see Schönborn 1992, Pfiffner 2014, 2015) and therefore shown to be also truncated by the Insubric Fault.

2.4 The Making of the Alps

Mountain chains grow along the boundary of convergent tectonic plates due to deformation at the margin of the plates where material is either subducted or escapes upward. In the case of the Alps the collision of two continental margins thickened the continental crust. Thereby shallow parts of the crust were buried deeply in the crust, heated up and metamorphosed. The descending crustal pieces were specifically light and replaced denser mantle rocks. The effect of buoyancy tended to return these light rocks toward the surface again. The nappe structures discussed in the preceding sections may be considered an expression of such buoyant rise. Analysis of pressure-temperature conditions in the metamorphic rocks now found at the surface reveals that some of these rocks had been heated up to more than 500 °C before returning to the surface. But also motion along thrust faults active within the shallow part of the growing orogen resulted in rock uplift.

But the Alpine chain did not grow vertically only. It became wider in time, with younger thrust faults developing at the outer margins of the orogen (see orogenic time tables in Pfiffner 2014, 2015). As a consequence also the foreland basins migrated outward as shown by Kempf and Pfiffner (2004), Schlunegger et al. (1997, 1998).

Buoyant rise and thrusting raised rock units together with the paleo-landsurface. Once a positive relief was created, the embryonic mountain chain immediately experienced enhanced precipitation owing to the cooling of the rising air. The ensuing erosion lowered the load the mountain chain exerted on the rocks in the subsurface. This in turn triggered buoyant rise of the buried crust and facilitated the upward motion of rock units along thrust faults. Thus rock uplift by tectonic activity and erosion are both acting simultaneously and the question is which of the two—if at all—dominates. If we consider the present day situation it seems likely that the two processes are in some sort of balance. Precise leveling suggests that the core of the Alps is rising at 1 mm/a, with two maxima of 1.5 mm/a in western and eastern Switzerland (Brig, Valais, and Chur, Graubünden, respectively). Toward the foreland these values taper off to zero in the north and the south. Fission-track ages, which give us information on the cooling (unroofing by erosion), indicate that this pattern was already active 5 million years ago. On the other hand the growth of the deltas of large rivers (Rhine, Rhone, Ticino) in peri-Alpine lakes suggests that denudation rates that are slightly lower but comparable in magnitude (Pfiffner 2019). These deltas accumulated in the past 12 000 years only and they do not record material removed by dissolution. Interestingly, erosion rates determined from smaller catchments within the mountain range give similar magnitudes but are tentatively higher in the areas of maximum uplift (Wittman et al. 2007). Thus it seems that today uplift and erosion are in balance. Whether this was true 10–30 million years ago when the Alpine orogeny was fully active is questionable. As Schlunegger and Castelltort (2016) argue, the slab breakoff at 30 Ma seemingly had a pronounced influence on sediment supply to the foreland; a secondary pulse may be attributed to a change in the bedrock exposed in the orogen. Once the crystalline rocks of the Austroalpine nappes were worn down, the underlying weaker sediments of the Penninic nappes led to rapid incision and thus higher erosion rates.

2.5 Conclusions

The large-scale Alpine landscape resembles a puzzle of rock types juxtaposed seemingly in arbitrary order. Detailed analyses reveal though that this puzzle reflects the rock types created during the breakup of the former large continent Pangea, the stacking of nappes during the convergence between the European and Adriatic continents and the erosional removal of a thick rock layer. Erosion occurred by river incision, which created local relief and rendered the tectonic structure visible to the observer. The evolution of the landscape was dynamic because of the varying rock types submitted to river incision. The final landscape left by the rivers was subsequently modified during the many Quaternary glaciations. The impact is visible today in the form of glacial landforms which in turn were modified again by fluvial incision and mass movements.

Acknowledgments The author thanks Jean-Luc Epard and Emmanuel Reynard for their constructive comments.

References

Argand E (1928) Sur l'arc des Alpes Occidentales. Eclogae Geol Helv 14:145–190

Bolliger T, Engesser B, Weidmann M (1993) Première découverte de mammifères pliocènes dans le Jura neuchâtelois. Eclogae Geol Helv 86(3):1031–1068

Buckingham T, Pfiffner OA (this volume) Mountain building and valley formation in the UNESCO World Heritage Tectonic Arena Sardona Region

Escher A, Masson H, Steck A (1993) Nappe geometry in the Western Swiss Alps. J Struct Geol 15(3–5):501–509. https://doi.org/10.1016/0191-8141(93)90144-Y

Heim A (1919–1922) Geologie der Schweiz, I, II/1, II/2. Tauchnitz, Leipzig

Hitz L (1996) The deep structure of the Engadine window: evidence from deep seismic data. Eclogae Geol Helv 89(2):657–675

Kälin D (1997) Litho- und Biostratigraphie der mittel- bis obermiozänen Bois de Raube-Formation (Nordwestschweiz). Eclogae Geol Helv 90(1):97–114

Kaufmann FJ (1886) Zentralgebiet der Schweiz: Emmen- und Schlierengegenden nebst Umgebungen Brünigstrasse und Linie Lungern-Grafenort. Beiträge zur geologischen Karte der Schweiz 24(1). Francke, Bern, 608 pp

Kempf O, Pfiffner OA (2004) Early Tertiary evolution of the North Alpine Foreland Basin of the Swiss Alps and adjoining areas. Basin Res 16:549–567. https://doi.org/10.1111/j.1365-2117.2004.00246.x

Kempf O, Matter A, Burbank DW, Mange M (1999) Depositional and structural evolution of a foreland basin margin in a magnetostratigraphic framework: the eastern Swiss Molasse Basin. Int J Earth Sci 88:253–275

Kühni A, Pfiffner OA (2001a) The relief of the Swiss Alps and adjacent areas and its relation to lithology and structure: topographic analysis from a 250-m DEM. Geomorphology 41:285–307
Kühni A, Pfiffner OA (2001b) Drainage pattern and tectonic forcing: a model study for the Swiss Alps. Basin Res 13:169–197
Laubscher HP (1965) Ein kinematisches Modell der Jurafaltung. Eclogae Geol Helv 58(1):231–318
Markley MJ, Teyssier C, Caby R (1999) Re-examining Argand's view of the Siviez-Mischabel nappe. J Struct Geol 21(8–9):1119–1124. https://doi.org/10.1016/s0191-8141(99)00063-2
McArdell B, Sartori M (this volume) The Illgraben torrent system
Pfiffner OA (2011) Explanatory notes to the structural map of the Helvetic zone of the Swiss Alps, including Vorarlberg (Austria) and Haute Savoie (France). Geological Special Map 128, Text and 10 plates. Federal Office of Topography swisstopo
Pfiffner OA (2014) Geology of the Alps. Wiley Blackwell, Chichester, 376 pp
Pfiffner OA (2015) Geologie der Alpen, 3rd edn. utb/Haupt Verlag, Bern, 400 pp
Pfiffner OA (2016) Basement-involved thin-skinned and thick-skinned tectonics in the Alps. Geol Mag 152:1085–1109. https://doi.org/10.1017/S0016756815001090
Pfiffner OA (2019) Landschaften und Geologie der Schweiz. Haupt Verlag, Bern, 360 pp
Pfiffner OA (this volume) The structural landscapes of Central Switzerland
Pfiffner OA, Heitzmann P, Lehner P, Frei W, Pugin A, Felber M (1997) Incision and backfilling of Alpine valleys: Pliocene, Pleistocene and Holocene processes. In: Pfiffner OA, Lehner P, Heitzmann P, Mueller S, Steck A (eds) Deep structure of the Swiss Alps: results of NRP 20. Birkhäuser, Basel, pp 265–288
Scapozza C, Ambrosi C (this volume) Between glaciers, rivers and lakes: the geomorphological landscapes of Ticino
Scheiber T, Pfiffner OA, Schreurs G (2012) Strain accumulation during basal accretion in continent-continent collision—a case study from the Suretta nappe (Eastern Switzerland). Tectonophysics 529:56–73. https://doi.org/10.1016/j.tecto.2012.03.009
Scheiber T, Pfiffner OA, Schreurs G (2013) Upper crustal deformation in continent-continent collision: a case study from the Bernard nappe complex (Valais, Switzerland). Tectonics 32:1320–1342. https://doi.org/10.1002/tect.20080
Schlüchter C, Akçar N, Ivy-Ochs S (this volume) The Quaternary Period in Switzerland
Schlunegger F, Castelltort S (2016) Immediate and delayed signal of slab breakoff in Oligo/Miocene Molasse deposits from the European Alps. Sci Rep 6:31010. https://doi.org/10.1038/srep31010
Schlunegger F, Jordan TE, Klaper EM (1997) Controls of erosional denudation in the orogen on foreland basin evolution: the Oligocene central Swiss Molasse Basin as an example. Tectonics 16(5):823–840
Schlunegger F, Slingerland R, Matter A (1998) Crustal thickening and crustal extension as controls on the evolution of the drainage network of the central Swiss Alps between 30 Ma and the present: constraints from the stratigraphy of the North Alpine Foreland Basin and the structural evolution of the Alps. Basin Res 10:197–212
Schmid SM, Haas R (1989) Transition from near-surface thrusting to intrabasement decollement, Schlinig thrust, Eastern Alps. Tectonics 8:697–718
Schmid SM, Rück P, Schreurs G (1990) The significance of the Schams nappes for the reconstruction of the paleotectonic and orogenic evolution of the Penninic zone along the NFP-20 East traverse (Grisons, eastern Switzerland). Mém Soc géol France 156:263–287
Schönborn G (1992) Alpine tectonics and kinematic models of he central Southern Alps. Memorie di Scienze Geologiche (Memorie degli Istituti di Geologia e Mineralogia dell'Università di Padova) XLIV:229–393
Schumacher M (1997) Geological interpretation of the seismic profiles through the Southern Alps (lines S1-S7 and C3-south). In: Pfiffner OA, Lehner P, Heitzmann P, Mueller S, Steck A (eds) Deep structure of the Swiss Alps: results of NRP 20. Birkhäuser, Basel, pp 101–114
Trümpy R (ed) (1980) Geology of Switzerland. Birkhäuser, Basel, 104 + 334 pp
von Poschinger A, Clague JJ, Calhoun N (this volume) The Flims and Tamins rockslide landscape
Wildi W (1984) Isohypsenkarte der quartären Felstäler in der Nord- und Ostschweiz mit kurzen Erläuterungen. Eclogae Geol Helv 77:541–551
Wittman H, von Blanckenburg F, Kruesmann T, Norton K, Kubik PW (2007) Relation between rock uplift and denudation from cosmogenic nuclides in river sediment in the Central Alps of Switzerland. J Geophys Res 112:F04040. https://doi.org/10.1029/2006jf000729

O. Adrian Pfiffner is Professor Emeritus of the University of Bern, president of the Scientific Council of the UNESCO World Heritage site "Tectonic Arena Sardona" and member of the Federal Commission of Geology. His main scientific interests are tectonics, structural geology, geophysics, geomorphology, and mountain building processes. He has worked in the Alps and the Andes of Peru.

Climate Setting in Switzerland

3

Jean-Michel Fallot

Abstract

This chapter presents the main climate features in Switzerland for several parameters that might influence geomorphological elements and processes. In addition to altitude, the complex topography of Switzerland strongly influences air temperature and frost frequency by favouring cold air accumulation in the bottom of valleys and depressions. The Alps and Jura Mountains also generate cloud accumulations by orographic forcing and foehn effects that strongly influence mean precipitation and snowfall at a local scale. This chapter also presents variations of these climatic parameters in connection with the global climate warming observed in Switzerland for about a hundred years and particularly since the late 1970s as well as some projections for the future.

Keywords

Climate setting • Topography • Climate change • Switzerland

3.1 Introduction

Due to its location in central Europe, climate in Switzerland is influenced by maritime air masses flowing from the Atlantic Ocean and the Mediterranean Sea, as well as by continental air masses originating from northern and eastern Europe. This country is characterised by a temperate semi-continental climate with mean annual temperature ranging from 15–20 °C at low altitudes and with maximum rainfall during the warm season, without any dry month. The complex topography of Switzerland, in particular in the Alps, also strongly influences the climate at the local and regional scale. This chapter describes features of present climate and some effects of topography on several parameters measured in different regions in Switzerland such as air temperature, frost, precipitation, snowfall and sunshine duration. It also presents variations of these parameters in connection with the global climate warming observed in Switzerland for about a hundred years, which might influence several geomorphological elements or processes such as glaciers, permafrost, gelifraction, floods, debris flows, mudslides, rock falls and landslides. Figure 3.1 shows the main regions and all locations in Switzerland mentioned in this chapter.

3.2 Air Temperature

At low altitudes (200–500 m a.s.l.), average monthly air temperatures range from –0.5 to 3.5 °C for the coldest month (January) and from 18 to 22 °C for the warmest month (July). Absolute maximum and minimum air temperatures have been recorded at Grono (382 m a.s.l.) near Bellinzona (+41.5 °C on 11 August 2003) and La Brévine (1050 m a.s.l.) in a wide depression of the Jura Mountains (–41.8 °C on 12 January 1987) (MeteoSwiss 2019).

Mean air temperatures are measured in about a hundred climatic stations. They strongly vary according to regional and local topography (Fig. 3.2) and they primarily depend on altitude. Bouët (1985) determined mean monthly and annual vertical temperature gradients to reproduce this effect. He distinguished two air layers, one below and one above 1500 m a.s.l. (Table 3.1).

Mean vertical temperature gradients are lower in the boundary layer below 1500 m a.s.l., especially during the cold season, consecutively to cold air accumulation in the bottom of valleys and depressions during anticyclonic situations with clear sky and weak winds. In such situations, infrared long wave radiation strongly cools the earth surface

J.-M. Fallot (✉)
Institute of Geography and Sustainability, University of Lausanne, Géopolis, 1015 Lausanne, Switzerland
e-mail: jean-michel.fallot@unil.ch

E. Reynard (ed.), *Landscapes and Landforms of Switzerland*, World Geomorphological Landscapes,
https://doi.org/10.1007/978-3-030-43203-4_3

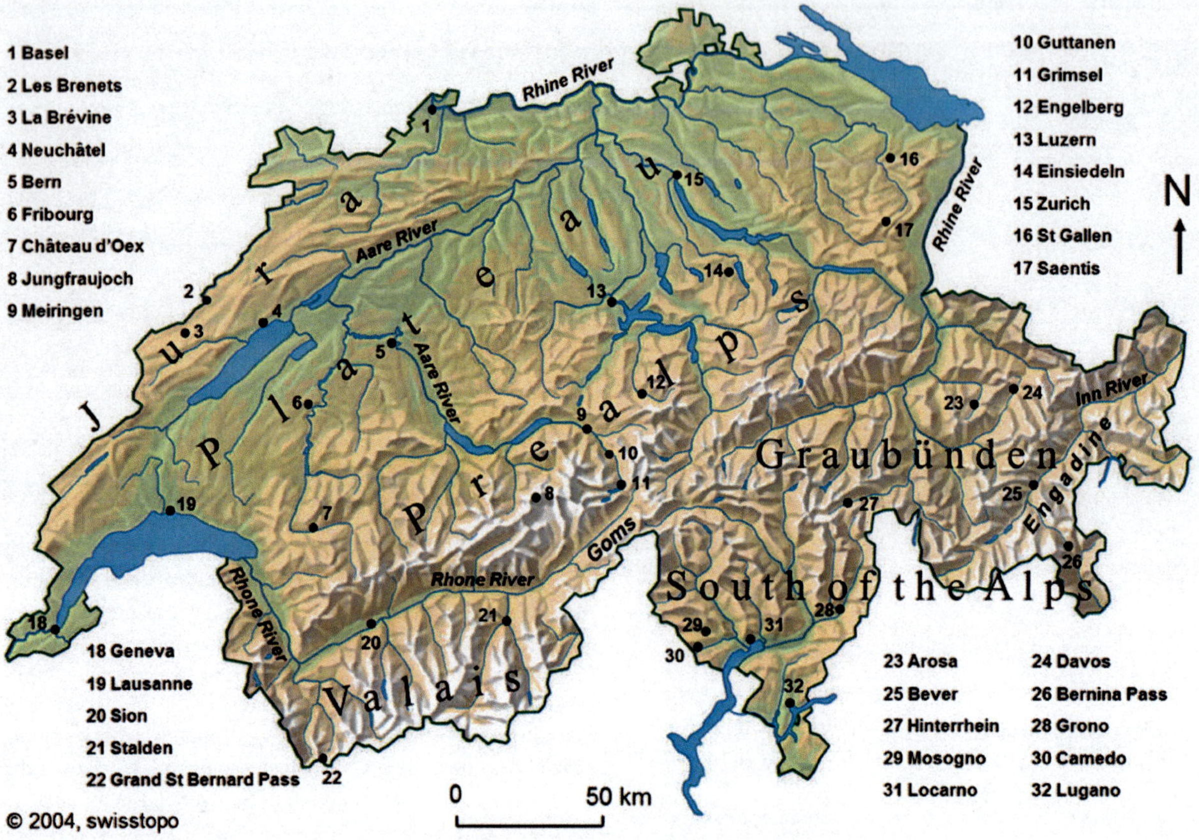

Fig. 3.1 Location of places and regions in Switzerland mentioned in this chapter

and air near the surface through conduction during the night (Thompson 1998). In mountainous terrain, heavier cold air near the surface flows down along the slopes and then fills the bottoms of valleys and basins. Cold air accumulations are more important in winter due to longer nights and they can reach a thickness of several hundred metres in Alpine Valleys and on the Swiss Plateau (Fallot 1992).

Mean monthly and annual temperatures in Switzerland are affected by these cold air accumulations, especially in the first hundred metres above the ground (Fallot 2012a). Mean annual temperatures are then about 1 °C lower in the valley bottoms than on slopes and summits at the same altitude. Mean annual temperature deficit can even reach 2–3 °C in some wide Alpine Valleys very exposed to cold air accumulation such as the Inn Valley in Upper Engadine (Bever) or Goms Valley in Valais (near the source of the Rhone River). Similar temperature deficits may also be measured in some large and shallow depressions in the Jura Mountains such as La Brévine, well known for its cold records in Switzerland (Fig. 3.3a, b). Cold air accumulations are less pronounced in narrow valleys because more sensitive night-time thermal winds carry cold air away.

Mean air temperatures strongly vary from one location to another at the same altitude (Table 3.2). They mainly depend on latitude in summer (July) and increase from north to south. Mean annual and winter (January) temperatures are more affected by regional and local topography, especially by cold air accumulation at the bottom of valleys and depressions.

A global warming trend is recorded in Switzerland since the late nineteenth century as elsewhere in the world (Fig. 3.4). Average annual temperatures have increased by +2.07 °C from 1901 to 2018 (Table 3.3), about twice the measured planetary warming (+0.85 °C from 1880 to 2012; IPCC 2013). A mean decadal trend of +0.57 °C was measured during the three decades from 1975 to 2004 (Rebetez and Reinhard 2008), more than twice the corresponding averaged temperature trend in the Northern Hemisphere. The warming trend shows the highest value in summer since 1901 whereas the lowest one typifies winter (Table 3.3). But the recent warming trend shows the highest value in spring and summer since the late 1970s (Rebetez and Reinhard 2008).

Global warming trend should increase in Switzerland during the twenty-first century, according to MeteoSwiss

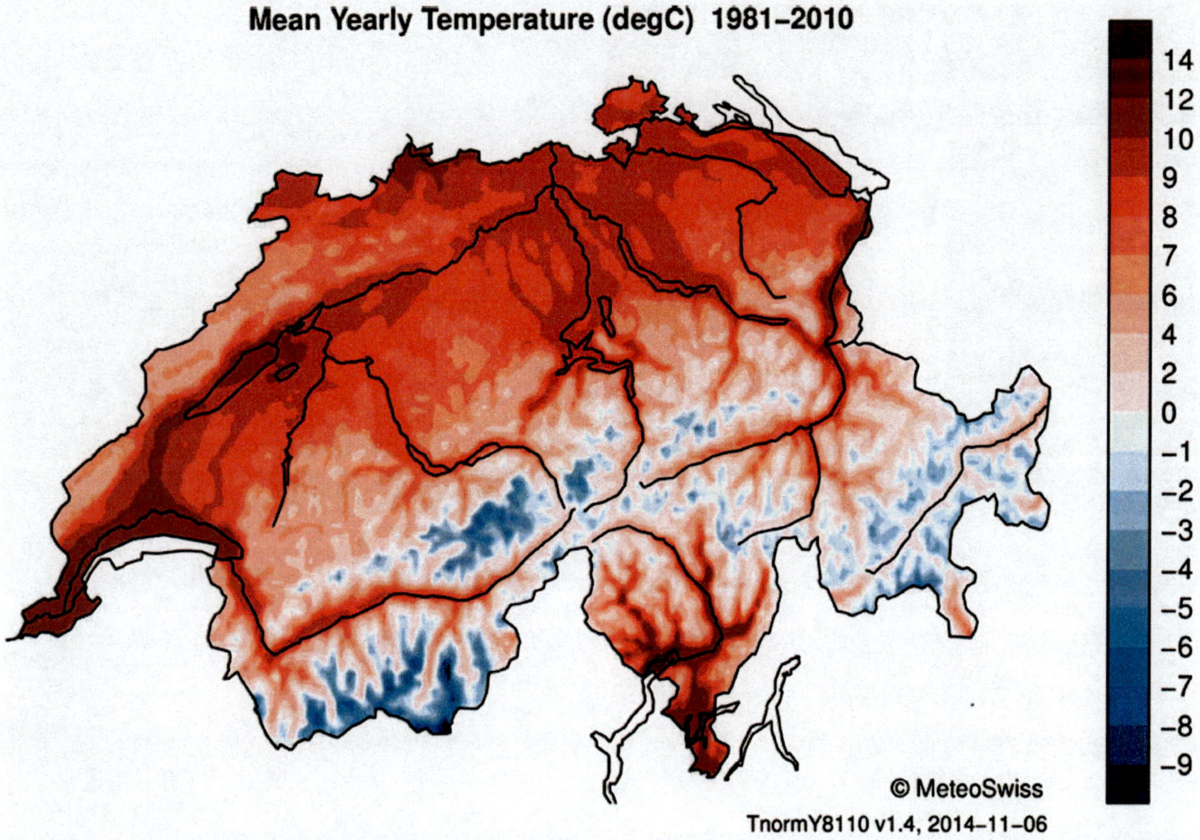

Fig. 3.2 Mean annual temperature measured in Switzerland for the period 1981–2010 (*Source* MeteoSwiss 2019)

Table 3.1 Mean monthly and annual vertical temperature gradients (°C/100 m) calculated by Bouët (1985) for two different air layers

Air layer (m a.s.l.)	Jan	Feb	March	April	May	June	July	Aug	Sept	Oct	Nov	Dec	Year
500–1500	0.24	0.33	0.50	0.61	0.63	0.63	0.63	0.61	0.49	0.39	0.33	0.26	0.46
1500–4000	0.54	0.56	0.58	0.61	0.63	0.63	0.63	0.61	0.59	0.57	0.55	0.54	0.59

projections (CH2018 2018). It should be more marked in summer and in the Alps (Table 3.4). Heat waves as in summer 2003, 2015, 2017, 2018 and 2019 will be more frequent in the future (Beniston et al. 2007; Fischer and Schär 2009). On the other hand, cold waves such as those in February 1956, January 1963 or more recently in February 2012 will become scarcer during the twenty-first century (Orlowsky and Seneviratne 2012).

3.3 Frost/Thaw

Frost frequency increases with altitude and also depends on topography (valley, slope) and type of surface (lake). Mean numbers of days with frost (when minimum temperature is below 0 °C) and days without thaw (when maximum temperature remains below 0 °C) strongly vary from one location to another at the same altitude: frost frequency is higher in the bottom of valleys and depressions than on slopes and summits because of cold air accumulation during clear nights. South of the Alps and Valais are less affected by icy air masses brought by northerly to easterly winds (called 'Bise' in Switzerland and eastern France) in winter: days without thaw are much less frequent in these regions than elsewhere in Switzerland at the same altitude, while numbers of days with frost are rather similar (Table 3.5).

Climate warming has caused a significant decrease in frost frequency in Switzerland since the early twentieth century up to at least 2500 m a.s.l. (Fig. 3.5). A similar decreasing trend is also observed for days without thaw at mid-altitude (Saentis 2502 m a.s.l.) and it appears at least up to 3600 m a.s.l. (Jungfraujoch). Furthermore, the number of

Fig. 3.3 **a** La Brévine village (1050 m a.s.l.) where the lowest temperature record in Switzerland was measured on 12 January 1987 (photo J.-M. Fallot, 9 May 2016); **b** La Brévine valley, also called 'Siberia of Switzerland' because it is a wide depression in the Jura Mountains favourable to cold air accumulation by anticyclonic situation (photo J.-M. Fallot, 9 May 2016); **c** Stratus cloud over the Swiss Plateau due to anticyclonic situation during winter: view from Grenchenberg in the Jura Mountains towards the Alps (photo J.-M. Fallot, 26 December 2006); **d** Stratus cloud in the Alpine Valleys due to anticyclonic situation during the cold season when Bise wind pushes cold air towards the Alps: view from Les Attelas above Verbier in Valais (photo J.-M. Fallot, 14 October 2010)

Table 3.2 Mean air temperatures (period 1981–2010) at several altitudes in different regions of Switzerland from field measurements and vertical temperature gradients calculated by Bouët (1985). Values into brackets correspond to those measured at the bottom of wide Alpine Valleys (Goms, Inn Valley in Engadine, Hinterrhein, Landwasser at Davos, upper Saane Valley at Gstaad near Château d'Oex) and in some Jura depressions (La Brévine) where cold air accumulations are the largest

Altitude (m a.s.l.)	Annual temperature (°C)	January temperature (°C)	July temperature (°C)
500	8.5 to 11.9	–1.3 to +3.1	18.0 to 21.1
1000	6.2 to 9.6 (5.1 to 6.1)	–2.5 to +1.9 (–6.8 to –2.6)	14.9 to 18.0
1500	3.9 to 7.3 (2.8 to 3.8)	–3.7 to –0.7 (–8.0 to –3.8)	11.7 to 14.8
2000	1.0 to 4.3 (–0.1 to +0.9)	–6.4 to –2.0 (–10.7 to –6.3)	8.6 to 11.7
2500	–2.0 to +1.4	–9.1 to –4.7	5.4 to 8.5
3000	–4.7 to –2.2	–11.2 to –7.9	2.5 to 5.4
3500	–7.5 to –5.5	–13.3 to –11.2	–0.4 to +2.1
4000	–10.5 to –8.5	–16.0 to –14.0	–3.5 to –1.0

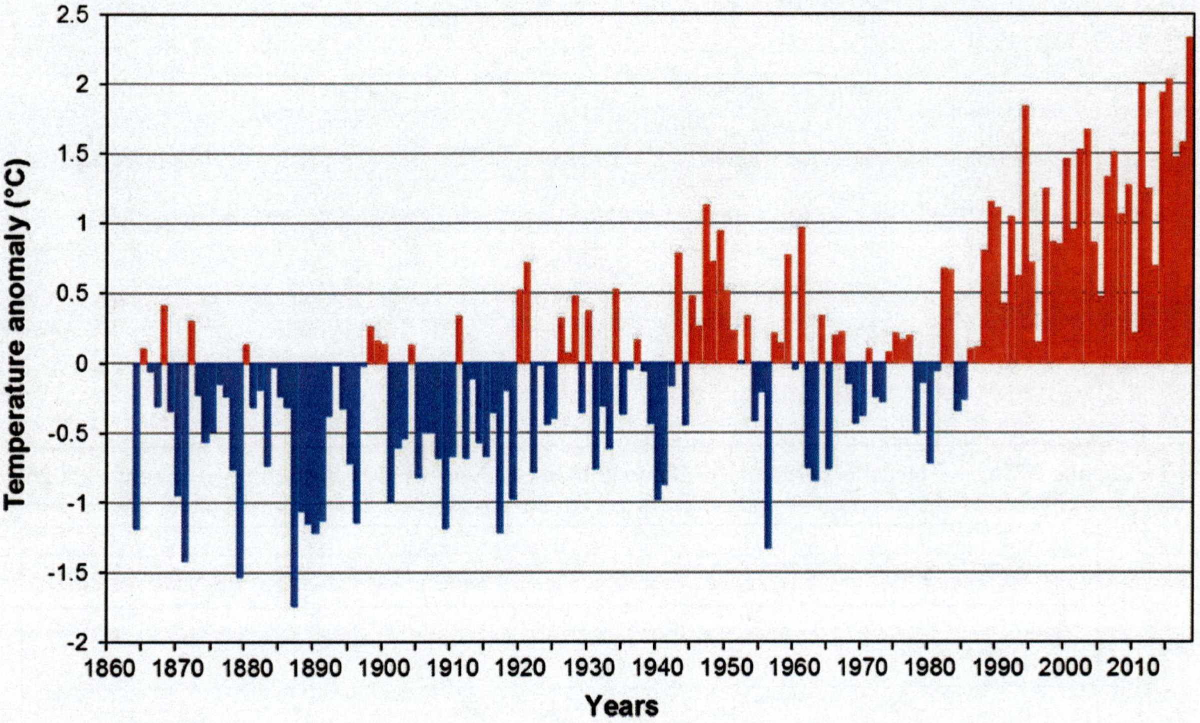

Fig. 3.4 Mean annual temperature measured in Switzerland since 1864. Annual temperature deviation (anomaly) relative to the period 1901–2000 in Switzerland: average of 13 stations with homogeneous values from MeteoSwiss (*Data source* MeteoSwiss 2019)

Table 3.3 Mean temperature change measured in Switzerland from 1901 to 2018: average for 26 locations with homogeneous time series from MeteoSwiss (Begert et al. 2005)

	Winter (°C)	Spring (°C)	Summer (°C)	Autumn (°C)	Year (°C)
Global warming from 1901 to 2018	+1.73	+1.99	+2.30	+2.12	+2.07
Mean decadal trend from 1901 to 2018	+0.146	+0.169	+0.195	+0.179	+0.175

Table 3.4 Projected future change of temperature (°C) in Switzerland for the late twenty-first century (2085) relative to the period 1981–2010 for two greenhouse gas emission scenarios of IPCC (CH2018 2018). RCP8.5 = high greenhouse gas emission scenario of IPCC (2013); RCP4.5 = intermediate greenhouse gas emission scenario of IPCC (2013)

Region in Switzerland	Winter		Spring		Summer		Autumn	
	RCP4.5 (°C)	RCP8.5 (°C)	RCP4.5 (°C)	RCP8.5 (°C)	RCP4.5 (°C)	RCP8.5 (°C)	RCP4.5 (°C)	RCP8.5 (°C)
Northeastern Switzerland	+2.0	+3.8	+1.6	+3.3	+2.3	+4.5	+2.0	+3.9
Western Switzerland	+2.0	+3.7	+1.5	+3.1	+2.4	+4.6	+2.0	+3.9
South of the Alps	+2.1	+3.8	+1.7	+3.4	+2.6	+5.0	+2.2	+4.1
Eastern Alps	+2.4	+3.9	+1.8	+3.8	+2.8	+5.3	+2.4	+4.5
Western Alps	+2.2	+3.9	+1.8	+3.9	+2.8	+5.5	+2.3	+4.6

Table 3.5 Mean number of days with frost and days without thaw per year at several altitudes in some regions of Switzerland (period 1981–2010). Numbers into brackets correspond to the frequency of days without thaw in Valais and south of the Alps when they are different to those in other regions in Switzerland

Region	Altitude (m a.s.l.)	Days with frost	Days without thaw
South of the Alps: lakes	200–300	25–30	0.5–1.5
South of the Alps: inland	200–300	35–105	1–2
Lakes in western Switzerland	400	40–60	8–15
Plateau, Jura and Alps	500	70–105	12–28 (4–8)
Jura and Alps	1000	110–185	20–45 (15–25)
Jura and Alps	1500	130–205	45–70 (30–45)
Alps	2000	170–250	65–100
Alps	2500	230–260	145–160
Alps	3500	340	260

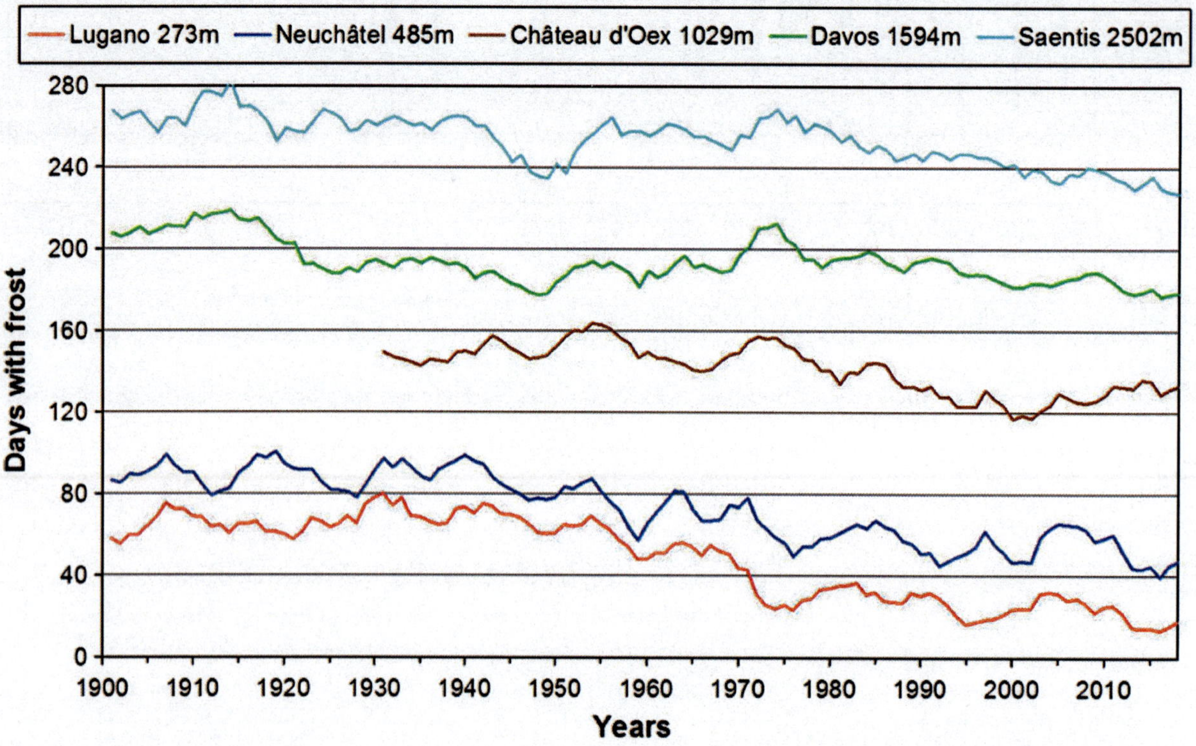

Fig. 3.5 Frost frequency measured at some locations in Switzerland since 1901: 5-year moving average (*Data source* MeteoSwiss 2019)

days without frost has increased in the Alps at mid- and high altitudes (2500 to at least 3600 m a.s.l.). Such trends have effects on glaciers and permafrost in particular (CH2014-Impacts 2014; Bauder et al. 2016; PERMOS 2019).

3.4 Precipitation

Precipitation amounts also strongly vary according to the regional and local topography. Mean annual rainfall increases with altitude, by 30–80 mm per 100 m elevation on average. They also depend on exposition of mountains to moist winds and foehn effects they can produce. Barrier and foehn effects on windward and lee sides are observed when relative stable and moist air is forced to ascend a mountain range like the Alps. Because ascending air on windward slopes cools down and becomes saturated, water vapour condensates. Thus clouds will form and rain is falling on windward slopes (= barrier effect). Condensation of water vapour releases heat and ascending air cools slower at a rate between 0.5 and 0.65 °C/100 m according to the saturated adiabatic lapse rate instead of the dry adiabatic lapse of 1 °C/100 m. After reaching the top of the mountain range, the air flows down on the opposite side (lee side), warms up by compression and becomes dry. Thus for every 100 m drop, temperature rises by 1 °C, according to the dry adiabatic lapse of 1 °C/100 m. For this reason, the wind blowing on the lee side (called 'Foehn' in the Alps) gets warm and dry when the air arrives at the foot of the mountain range. This air is significantly warmer and drier on the lee side than on the windward side of the mountains at the same altitude; this is called the 'foehn effect'.

Such effects can be observed in Switzerland in the north of the Alps, with moist air blowing from south-west to south-east (South Foehn wind; Fig. 3.6) or in the south of the Alps with moist air blowing from north-west to north-east (North Foehn wind; Fig. 3.6). Jura Mountains can also produce a small foehn effect (and a katabatic wind called 'Joran') on the southeastern side with moist air blowing from west to north (Fig. 3.6).

With these barriers and foehn effects, mean annual precipitation ranges with a ratio of 1–6 between the driest and wettest locations in Switzerland (Fig. 3.7). The driest places are located in the valleys within the Alps, especially in Valais: they are sheltered from moist winds by surrounding high mountain ranges. Mean annual precipitation amounts range from 600 to 800 mm in the Rhone Valley and tributary valleys in Central Valais. The driest spot is located at Stalden near Visp (Valais). Valleys in Central Graubünden and Engadine (Inn Valley) are also rather dry with precipitation amounts ranging from 700 to 1000 mm per year. Mean rainfall in the Swiss Plateau varies from 900 to 1300 mm per year and it gradually increases from the southern piedmont of the Jura Mountains towards the northern side of the Alps (Prealps).

Mean rainfall is much higher in the south and north of the Alps, as well as in the Jura Mountains. Mean annual precipitation amounts range from 1500 to 2300 mm in the southern Alpine Valleys and from 1200 to 2200 mm in the northern Alpine Valleys and in the Jura Mountains. These amounts exceed 2000 mm per year in the Prealps and reach 3000 mm per year or even more on the wettest summits in the Bernese Alps (HADES 2014; Fig. 3.7).

The southern side of the Alps is mainly exposed to fluxes of Mediterranean moist air from south-east to south-west, whereas the northern side of the Alps is mostly exposed to fluxes of moist air originating from the Atlantic Ocean or North Sea. Southern and northern slopes of the Alps, as well

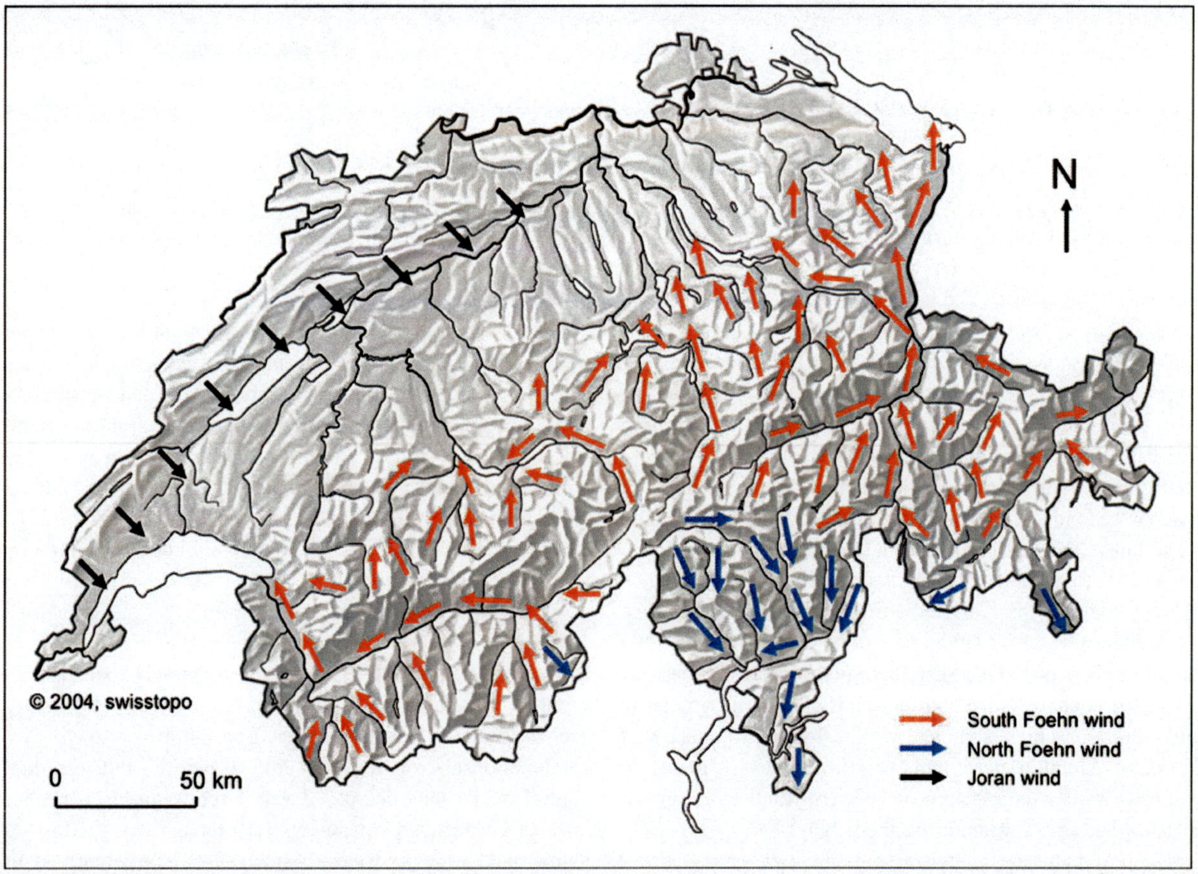

Fig. 3.6 Schematic air flows in Switzerland, observed with South and North Foehn wind as well as with Joran wind

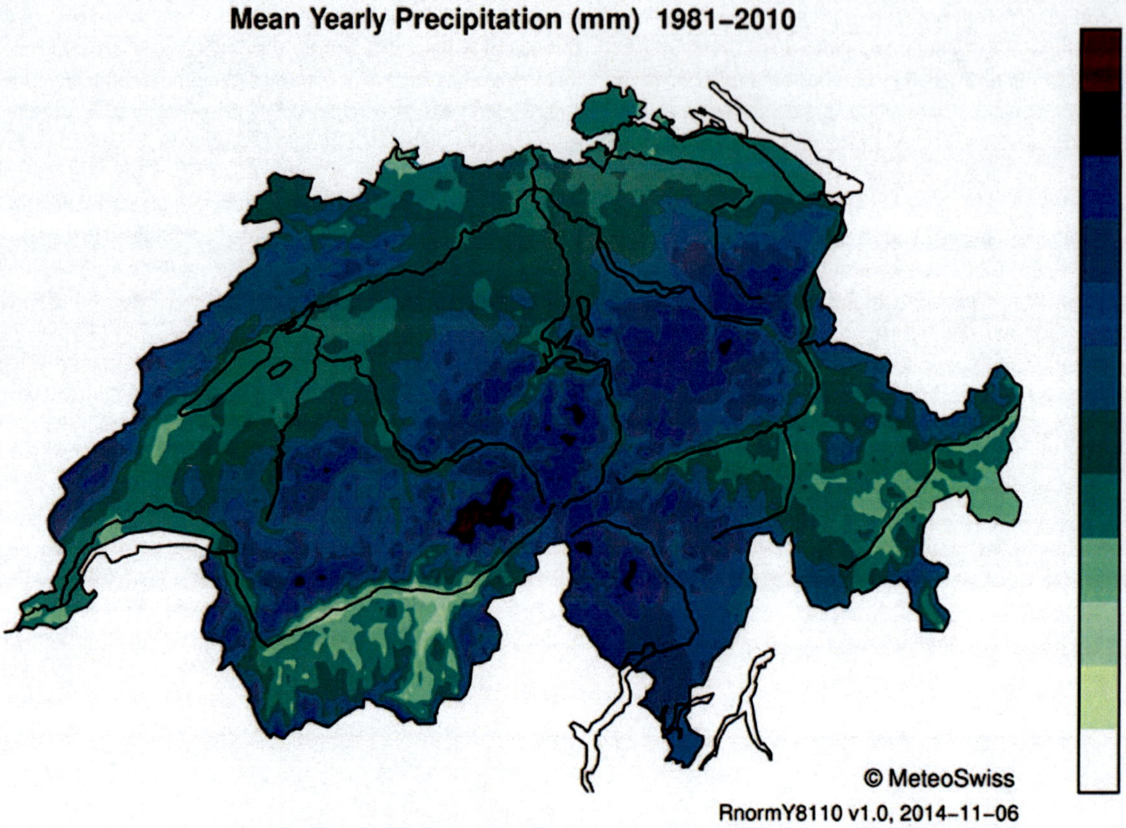

Fig. 3.7 Mean annual precipitation amounts measured in Switzerland for the period 1981–2010 (*Source* MeteoSwiss 2019)

as the Jura Mountains, are also more affected by convective rainfall (thunderstorms) during the warm season.

Due to its location in central Europe, Switzerland is partly influenced by oceanic, continental and Mediterranean climates. The area to the north of the Alps has a semi-continental climate with rainfall maximum during the warm season as for continental climates. This maximum is less pronounced in the Jura Mountains (Les Brenets; Fig. 3.8) than on the northern side of the Alps and a large area of the Swiss Plateau (Fribourg). Furthermore, a more marked secondary maximum appears in winter in the Jura Mountains. This corresponds to a feature of oceanic climate under an attenuated form, which gradually disappears when moving towards the Alps and eastern Switzerland.

Sheltered valleys within the Alps (Sion) receive little rainfall during the whole year. The southern side of the Alps (Lugano) is characterised by two rainfall maxima, in spring and autumn, which coincide with a higher frequency of influx of Mediterranean moist air from the south-east to south-west. This is a feature of northern Mediterranean climate which also appears in northern Italy.

Rainfall patterns in Switzerland not only change from north to south, but also gradually from west to east: rainfall maximum in summer is more marked when moving eastwards just like rainfall minimum in winter. Such a change reflects a weakening influence of oceanic climate in favour of continental climate.

Annual rainfall shows a high interannual and spatial variability (Fig. 3.9). Five-year moving average values (Fig. 3.9) smooth this high variability, but allow showing the main trends. There is no significant precipitation trend since the end of the nineteenth century, except in the north-east of Switzerland (St Gallen; Fig. 3.9), where annual precipitation has significantly increased from 1901 to 2018 at several locations according to annual trends calculated by MeteoSwiss (2019).

Table 3.6 shows that rainfall averaged for the whole country from 110 stations increased more in winter from 1900 to 2010, while it remained relatively constant in summer; it increased a little in the Alps, but slightly decreased elsewhere in summer. The number of days with precipitation also increased a little in winter, but less than rainfall, while they decreased very slightly during the other seasons. Moreover, mean rainfall intensity is rising in Switzerland for each season since 1900, consistently with the global climate warming (Fallot 2010).

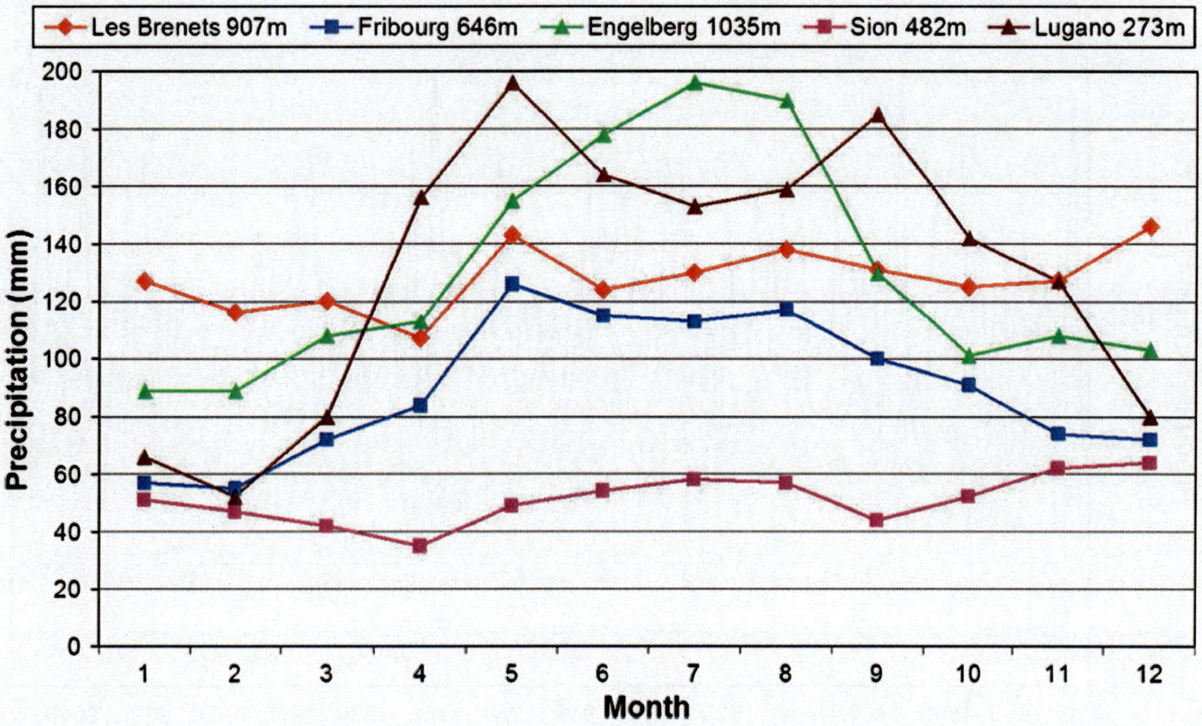

Fig. 3.8 Mean monthly precipitation measured at some locations in Switzerland over a period of 30 years (1981–2010). Locations were chosen to illustrate as best as possible the variations of rainfall patterns from north to south and from west to east (*Data source* MeteoSwiss 2019)

According to MeteoSwiss projections (CH2018 2018), mean rainfall should strongly decrease in summer during the twenty-first century over the whole Switzerland, as a result of the strengthening and extent of the Azores High (IPCC 2013). It is expected to markedly increase in winter. Future changes of precipitation should not exceed 10% for the other seasons (Table 3.7).

Heavy rainfall and flood events produce the greatest damages due to natural hazards in Switzerland (Hilker et al. 2009; PLANAT 2017). They also affect several geomorphological processes like rock falls, landslides, mudslides and debris flows. Rainfall records measured in Switzerland since 1864 are the following (MeteoSwiss 2019):

- 455 mm in 24 h at Camedo (Ticino) on 26 April 1935;
- 612 mm in 48 h at Mosogno (Ticino) on 23–24 September 1924;
- 768 mm in 72 h at Camedo from 3 to 5 September 1948;
- 1239 mm in one month at Camedo in April 1986;
- 4173 mm in one year at Saentis in 1922.

Heaviest rainfall during 24, 48 and 72 h occurs in the south of the Alps, in effect of influx of moist and warm Mediterranean air from south-west to south-east and with a stationary front over the Alps. The most intense rainfall with strong thunderstorms was measured in the south of the Alps, with 91.2 mm in one hour at Locarno-Monti on 28 August 1997, and in the west of Switzerland, with 41.0 mm in 10 min at Lausanne on 11 June 2018 (MeteoSwiss 2019).

The frequency of heavy rainfall events has increased in Switzerland and in the Alps (Schmidli and Frei 2005), like in numerous regions elsewhere, since about 1950 in connection with global warming (IPCC 2013). The highest daily rainfall measured each year is also increasing during the last 100 years for most stations (90%) in Switzerland (Fallot 2010). According to global and regional modelling, such trends should continue during the twenty-first century in Europe, and therefore in Switzerland (Frei et al. 2006; Beniston et al. 2007; Orlowsky and Seneviratne 2012; IPCC 2013).

Consequently, extreme precipitation amounts for enough long return periods need to be assessed for calculating the dimensions of protection structures against floods or other natural hazards. Daily (24-hour) rainfall could be estimated for a return period of 500 years from Gumbel analyses carried out on rainfall time series of 100 years at 151 locations in Switzerland (Fig. 3.10). Previous studies revealed that Gumbel analyses are globally efficient in Switzerland and Central Europe for such estimations (Trömel and Schönwiese 2007; Fallot 2012b; Fallot and Hertig 2013).

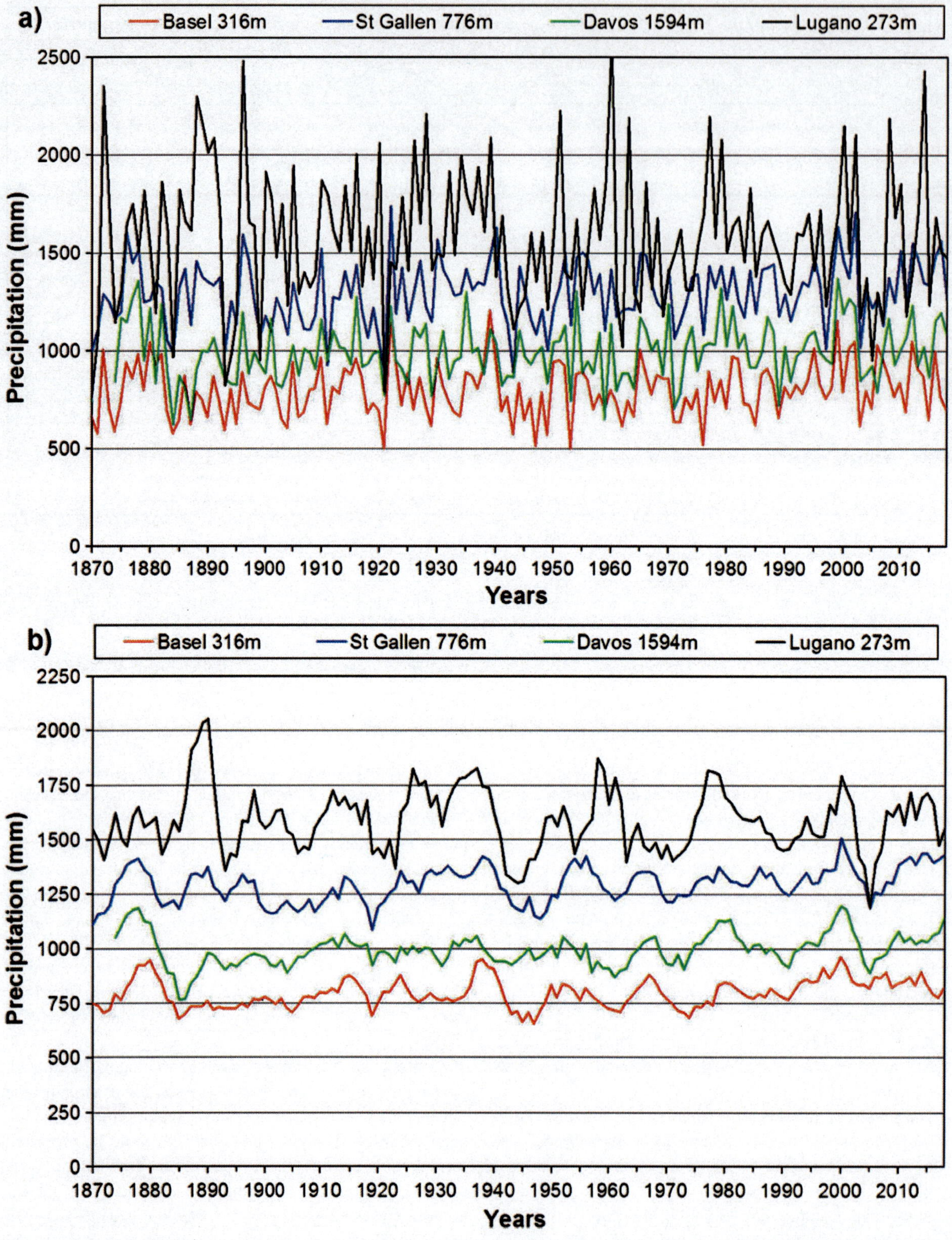

Fig. 3.9 Annual precipitation amounts measured at some locations in Switzerland since 1870: **a** annual values; **b** 5-year moving average values (*Data source* MeteoSwiss 2019)

Table 3.6 Mean trend (% per decade) averaged for the whole Switzerland from 1900 to 2010 for rainfall and number of days with precipitation (≥ 1 mm)

	Winter (%)	Spring (%)	Summer (%)	Autumn (%)	Year (%)
Rainfall trend (% per decade)	+1.59	+1.03	0.0	+1.02	+0.80
Number of days with precipitation	+0.48	–0.14	–0.32	–0.05	–0.05

Table 3.7 Projected future changes of precipitation (%) in Switzerland for the late twenty-first century (2085) relative to the period 1981–2010 for two greenhouse gas emission scenarios of IPCC (CH2018 2018). RCP8.5 = high greenhouse gas emission scenario of IPCC (2013); RCP4.5 = intermediate greenhouse gas emission scenario of IPCC (2013)

Region in Switzerland	Winter		Spring		Summer		Autumn	
	RCP4.5 (%)	RCP8.5 (%)	RCP4.5 (%)	RCP8.5 (%)	RCP4.5 (%)	RCP8.5 (%)	RCP4.5 (%)	RCP8.5 (%)
Northeastern Switzerland	+10.8	+14.5	+5.9	+8.2	-7.3	–16.9	+3.3	-2.2
Western Switzerland	+8.2	+15.4	+4.3	+4.0	-8.5	-24.2	+1.1	-3.4
South of the Alps	+8.6	+21.8	+1.8	+1.0	-8.2	-22.8	+0.3	0.0
Eastern Alps	+8.3	+16.6	+3.3	+4.8	–0.8	-9.8	+0.9	-2.9
Western Alps	+7.4	+12.3	–0.6	+0.1	-8.7	-20.9	-4.2	-6.0

Daily rainfall with a return period of 500 years is the highest in the south of the Alps (up to 560 mm in 24 h), which is very exposed to the influx of warm and moist air coming from the Mediterranean Sea (Fig. 3.10).

3.5 Snowfall and Snowpack

Mean snowfall amounts increase with altitude and also depend on exposition to moist winds, similarly to rainfall. Mean annual snowfall amounts measured during the period 1972–2002 reach about

- 1.3–2 m at 700 m a.s.l. on the Swiss Plateau in front of central and eastern Prealps;
- 2.5–3.5 m at 1000 m a.s.l. in the Alps and the Jura Mountains, but 3.5–4.5 m in central and eastern Prealps;
- 4.5–5.5 m at 1500 m a.s.l. in the Alps, but 5.5–6.5 m in central and eastern Prealps;
- 12–14 m on exposed summits and passes between 2000 and 2500 m a.s.l. such as Saentis in eastern Prealps, Grand St-Bernard pass on the crest of Valais Alps or Grimsel pass in the Gothard area. Snowfall amounts up to 25 m for one year can be recorded at these locations.

On the other hand, mean annual snowfall amounts are much lower in sheltered valleys within the Alps, especially in Engadine (Inn Valley) and central Valais south of the Rhone River, where they do not exceed 2.5–3 m at 1500 m a.s.l.

As a result of these variations in snowfall, mean snowpack depths recorded in winter are also lower at the same altitude within the Alps than on their northern or southern sides. At the end of winter (February, March or April depending on the location), they reach an average thickness of (data for the period 1982–2002; HADES 2014):

- 30–80 cm at 1000 m a.s.l. on the northern side of the Alps and 20–50 cm on the southern side of the Alps and the Jura Mountains;
- 70–130 cm at 1500 m a.s.l. on the northern and southern sides of the Alps, but only 30–90 cm in central Valais, central Graubünden and Engadine;
- 150–280 cm at 2000 m a.s.l. on the northern and southern sides of the Alps, but only 80–130 cm in central Valais, central Graubünden and Engadine at 2000 m a.s.l.;
- up to 485 cm on the northern side of the Alps at 2500 m a.s.l. (Saentis), but only 150–300 cm in central Valais and central Graubünden between 2500 and 3000 m a.s.l.

Mean snowpack depths are on average the highest on the northern side (Prealps) of central and eastern Alps. The record depth of snowpack (8.16 m) was measured at Saentis in April 1999, after exceptional snowfall. In February 1999 snowfall amounted up to 4–5 m on the northern side of the Alps at about 2000 m a.s.l.

Record snowfall amounts measured in Switzerland since 1931 are the following (MeteoSwiss 2019):

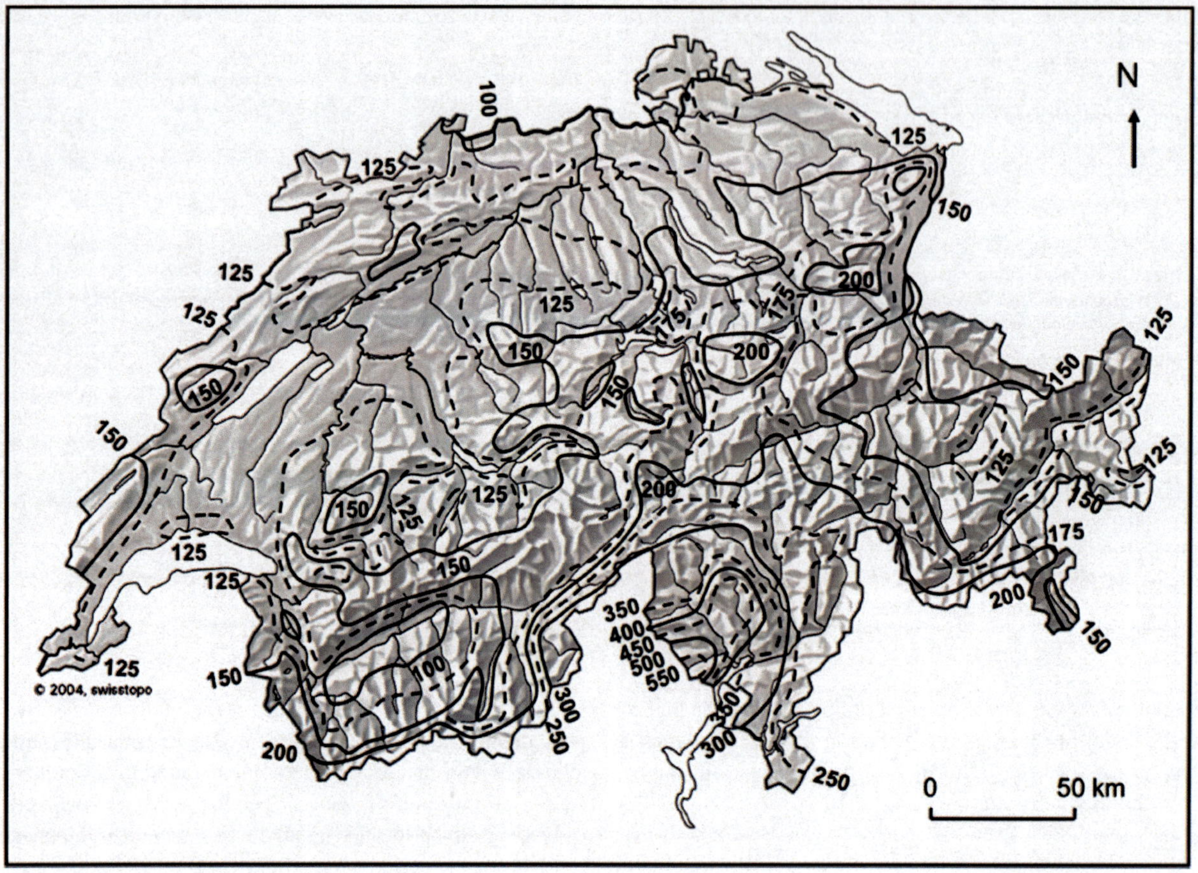

Fig. 3.10 Daily (24-hour) extreme precipitation amounts for a return period of 500 years estimated in Switzerland from Gumbel analyses carried out on rainfall time series at 429 locations over the period 1961–2010 (Fallot and Hertig 2013)

- 130 cm in 24 h at Bernina pass (2307 m a.s.l.) on 15 April 1999 and at Grimsel Hospiz (1980 m a.s.l. on 30 March 2018;
- 215 cm in 48 h at Bernina pass (2307 m a.s.l.) on 15 and 16 April 1999;
- 229 cm at Weissfluhjoch (2691 m a.s.l.) above Davos in 72 h from 13 to 15 February 1990.

Daily snowfall can reach 40–60 cm in 24 h at low altitude on the Swiss Plateau or south of the Alps during some particular meteorological situations.

Annual snowfall amounts have fluctuated since 1901 but they are lower from 1990 in the north of the Alps between 600 and 1800 m a.s.l. (Fig. 3.11). These variations are more marked on Alpine summits like Saentis at 2500 m a.s.l. and a decrease in annual snowfall appears at the beginning of the twenty-first century as at Weissfluhjoch (2691 m a.s.l.) above Davos (not shown here). Snowfalls have diminished at mid-altitude in the south of the Alps and in Engadine (Bever) from 1950 to 1990. The number of days with snow cover shows rather similar variations since 1960 at low and mid-altitudes in the Alps (Marty 2008; CH2014-Impacts 2014). Continuous snow cover duration has decreased at low and mid-altitudes, by 8.9 days per decade on average since 1970 (Klein et al. 2016). As mean rainfall slightly increased during the twentieth century, especially in winter, the part of snowfall in annual precipitation has decreased since the 1980s at low and mid-altitudes (Serquet et al. 2011).

As a result of global warming, the mean height of zero degree line in winter rose about 300 metres, from 600 to 900 m a.s.l. between 1958 and 2003. This line roughly corresponds to the height of snow line. It should rise by about 360 metres in winter by 2050, if consistent with a mean global warming of 1.8 °C (OcCC 2007; CH2011 2011). Annual maximum snow depth has also decreased since 1970 at low and mid-altitudes in the Alps (Klein et al. 2016) and the mean snowpack depth could reduce by 60% toward the end of the twenty-first century for an intermediate greenhouse gas emission scenario (A1B) of IPCC (2007) (Marty et al. 2017).

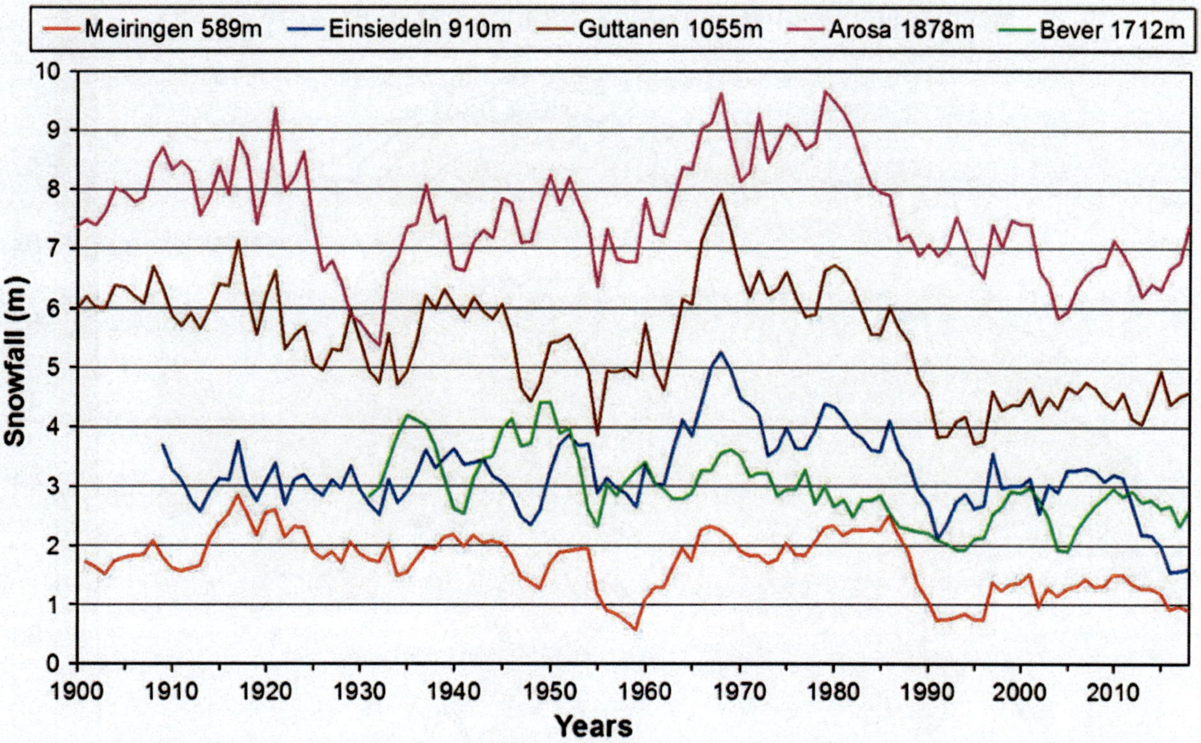

Fig. 3.11 Annual snowfall amounts measured at some locations in front and in the Swiss Alps since 1900: 5-year moving average (*Data source* MeteoSwiss 2019)

3.6 Sunshine

Sunshine duration also strongly varies according to the topography and exposition to moist winds. By clear sky sunshine duration is shorter in the valley bottoms than on flat terrain or summits. Calculation of relative sunshine duration (%) includes this effect of topography: it is equivalent to a ratio between measured sunshine duration and maximum possible sunshine duration at one location if sky was clear every day. Mean annual relative sunshine duration increases from northern to southern Switzerland: it is the highest in Valais, south of the Alps and Engadine (50–60%), while it is the lowest on the Swiss Plateau and the northern areas of the country (35–40%; Fig. 3.12). Fog and stratus rather often cover the Plateau by anticyclonic situation in autumn and winter (Fig. 3.3c, d) when sun is shining in the Alps and the Jura Mountains. These low clouds reduce sunshine duration on the Plateau, in particular in the northern and eastern areas, during the cold season of the year. Mean sunshine duration decreased in Switzerland from 1960 to 1980, before increasing again from 1980 to 2010 (CH2014-Impacts 2014; CH2018 2018).

3.7 Conclusion

In Switzerland, topography has a great influence on several climate parameters at a local and regional scale, in addition to altitude. It contributes to cold air accumulation, colder mean temperatures and higher frost frequency in the valley bottoms than on slopes at the same altitude. Mountains also produce barrier and foehn effects that strongly influence mean annual rainfall and snowfall, which much vary at a local scale. Global warming measured during the twentieth century has also influenced some climate parameters: in particular, frost frequency and snowfall have decreased, while mean rainfall intensity has increased, with also higher extreme values. These local and regional climate variations produced by topography may have effects on several geomorphological elements or processes, similar to the global warming which is expected to strengthen during the twenty-first century.

Glacier melting and permafrost thawing in the Alps during the last decades are likely to continue and even accelerate in the future (CH2014-Impacts 2014). This may lead to greater ground instabilities and destabilisation of

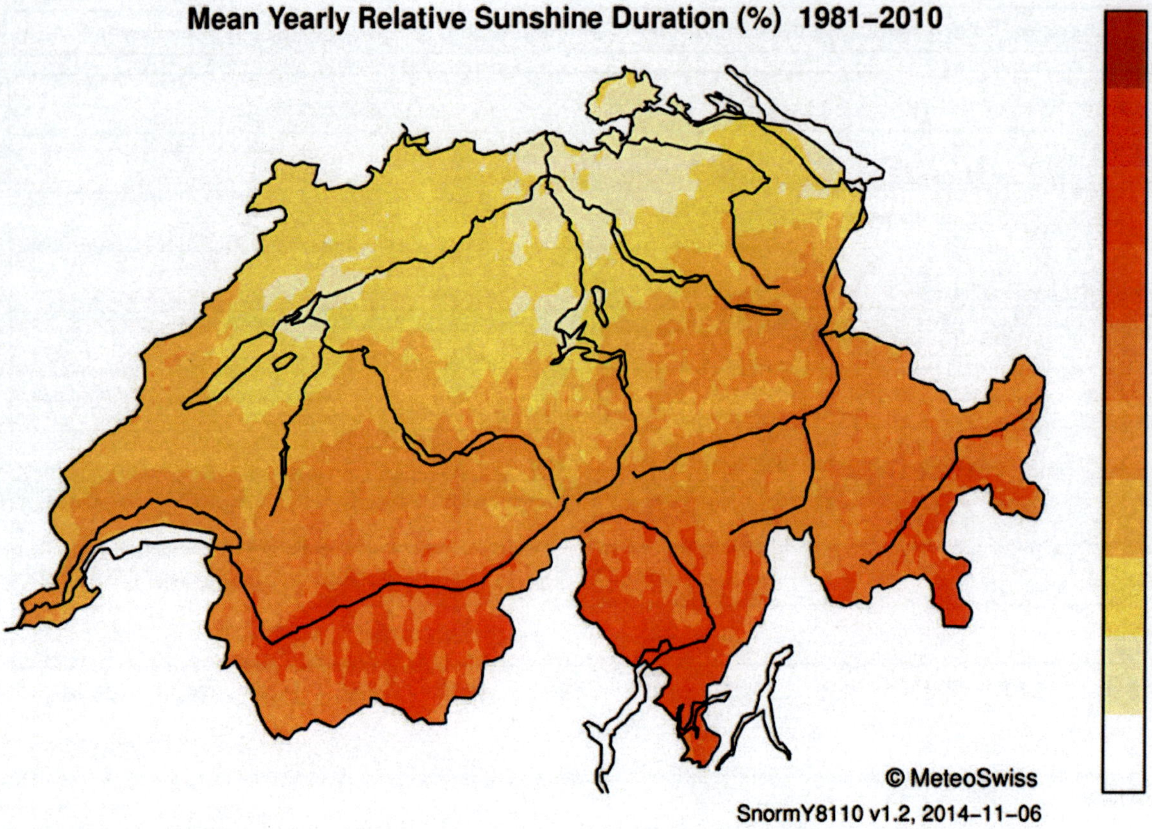

Fig. 3.12 Mean annual sunshine duration (%) measured in Switzerland for the period 1981–2010 (*Source* MeteoSwiss 2019)

slopes (Stoffel and Huggel 2012) and rock falls, landslides, mudslides and debris flows are expected to become more frequent in the future, especially as the frequency of heavy rainfall is expected to increase with global warming.

References

Bauder A, Fischer M, Funk M, Huss M, Kappenberger G (2016) The Swiss glaciers 2011/12 and 2012/13. Glaciological report (Glacier) No 133/134. Publication of the Cryospheric Commission of the Swiss Academy of Sciences, Bern. http://doi.org/10.18752/glrep_133-134

Begert M, Schlegel T, Kirchhofer W (2005) Homogeneous temperature and precipitation series of Switzerland from 1864 to 2000. Int J Climatol 25:65–80

Beniston M, Goyette S, Stephenson DB, Christensen OB, Frei C, Schöll R, Halsnaes K, Holt T, Palutikof J, Jylhä K, Koffi B, Semmler T, Woth K (2007) Future extreme events in Europe climate: an exploration of regional climate model projection. Clim Chang 81:71–95

Bouët M (1985) Climat et météorologie de la Suisse romande. Payot, Lausanne, 170 pp

CH2011 (2011) Swiss climate change scenarios CH2011. C2SM, MeteoSwiss, ETH, NCCR Climate and OcCC, Zurich, 88 pp

CH2014-Impacts (2014) Toward quantitative scenarios of climate change Impacts in Switzerland. OCCR, FOEN, MeteoSwiss, C2SM, Agroscope and ProClim, Bern, 136 pp

CH2018 (2018) CH2018 – climate scenarios for Switzerland. Technical report, National Centre for Climate Services, Zurich, 271 pp

Fallot JM (1992) Etude de la ventilation d'une grande vallée préalpine: la vallée de la Sarine en Gruyère. PhD Thesis, University of Fribourg. Editions Copyta, Fribourg, 475 pp

Fallot JM (2010) Evolution de l'intensité moyenne et de la fréquence des précipitations en Suisse de 1901 à 2009. In: Actes du 23ème colloque de l'Association Internationale de Climatologie (AIC), Conference proceedings online, Rennes, France, 1–5 September 2010, pp 249–254. http://www.climato.be/aic/colloques.html

Fallot JM (2012a) Influence de la topographie et des accumulations d'air froid sur les températures moyennes mensuelles et annuelles en Suisse. In: 25ème colloque de l'Association Internationale de Climatologie (AIC), Conference proceedings online, Grenoble, France, 5–8 September 2012, pp 273–278. http://www.climato.be/aic/colloques.html

Fallot JM (2012b) Efficiency of Gumbel analyses for determining extreme daily precipitation in Switzerland. In: 4th conference on air and water components of the environment, Cluj-Napoca, Romania, 23–24 March 2012. DOAJ 2067743X, Issue 2012, pp 1–8

Fallot JM, Hertig JA (2013) Détermination des précipitations extrêmes en Suisse à l'aide d'analyses statistiques et augmentation des valeurs extrêmes durant le 20ème siècle. Mém Soc Vaud Sc Nat 25:13–24

Fischer SI, Schär C (2009) Future changes in daily summer temperature variability: driving processes and role of temperature extremes. Clim Dyn 33:917–935. https://doi.org/10.1007/s00382-008-0473-8

Frei C, Schöll R, Futukome S, Schmidli J, Vidale PL (2006) Future change of precipitation extremes in Europe. J Geophys Res 111: D06105. https://doi.org/10.1029/2005JD005965

HADES (2014) Hydrological Atlas of Switzerland. http://www.hades.unibe.ch

Hilker N, Badoux A, Hegg C (2009) The Swiss flood and landslide damage database 1972–2007. Nat Haz Earth Syst Sci 9:913–925. http://www.nat-hazards-earth-syst-sci.net/9/913/2009/

IPCC (2007) Climate Change 2007. The physical science basis. Contribution of working group I to the fourth assessment report of the Intergovernmental Panel on Climate Change (IPCC). Cambridge University Press, Cambridge, 996 pp

IPCC (2013) Climate Change 2013. The physical science basis. Contribution of working group I to the fifth assessment report of the Intergovernmental Panel on Climate Change (IPCC). Cambridge University Press, Cambridge, 1535 pp

Klein G, Vitasse Y, Rixen C, Marty C, Rebetez M (2016) Shorter snow cover duration since 1970 in the Swiss Alps due to earlier snowmelt more than to later snow onset. Clim Chang 139:637–649. https://doi.org/10.1007/s10584-016-1806-y

Marty C (2008) Regime shift of snow days in Switzerland. Geophys Res Lett 35:L12501. https://doi.org/10.1029/2008GL033998

Marty C, Schlögl S, Bavay M, Lehning M (2017) How much can we save? Impact of different emission scenarios on future snow cover in the Alps. The Cryosphere 11:517–529. https://doi.org/10.5194/tc-11-517-2017

MeteoSwiss (2019). http://www.meteosuisse.admin.ch

OcCC (2007) Climate change and Switzerland 2050. Expected impacts on environment, society and economy. OcCC and ProClim, Bern, 168 pp

Orlowski B, Seneviratne SI (2012) Global changes in extreme events: regional and seasonal dimension. Clim Chang 110:669–696. https://doi.org/10.1007/s10584-011-0122-9

PERMOS (2019) Permafrost in Switzerland 2014/2015 to 2017/2018. Nötzli J, Pellet C, Staub B (eds). Glaciological glaciological report permafrost No. 16–19 of the cryospheric commission of the swiss academy of sciences, 104 pp. https://doi.org/10.13093/permos-rep-2019-16-19

PLANAT (2017) National platform for natural hazards. http://www.planat-ch/en/knowledge-base/flood-high-water/flood-inundation

Rebetez M, Reinhardt M (2008) Monthly air temperature trends in Switzerland 1901–2000 and 1975–2004. Theor Appl Climatol 91:27–34

Schmidli J, Frei C (2005) Trends of heavy precipitation and wet and dry spells in Switzerland during the 20th century. Int J Climatol 25:753–771. https://doi.org/10.1002/joc.1179

Serquet G, Marty C, Dulex JP, Rebetez M (2011) Seasonal trends and temperature dependence of the snowfall/precipitation-day ratio in Switzerland. Geophys Res Lett 38:L07703. https://doi.org/10.1029/2011GL046976

Stoffel M, Huggel C (2012) Effects of climate change on mass movements in mountain environments. Prog Phys Geogr 36:421–439

Thompson RD (1998) Atmospheric processes and systems. Routledge, London, 224 pp

Trömel S, Schönwiese CD (2007) Probability change of extreme precipitation observed from 1901 to 2000 in Germany. Theor Appl Climatol 87:29–39

Jean-Michel Fallot is Senior Lecturer at the Institute of Geography and Sustainability, University of Lausanne. He is a member of the International Association of Climatology (AIC). His main scientific interest is the influence of topography on climate parameters, in particular, air flows in relation to air quality, as well as the assessment of some climate trends and related hazards in Switzerland.

4 The Quaternary Period in Switzerland

Christian Schlüchter, Naki Akçar, and Susan Ivy-Ochs

Abstract

The evolution of the Swiss landscape during the Quaternary Period over the past 2.6 million years is controlled by multiple glaciations and deglaciations with at least 15 drastic environmental changes between glacier advances (with yearly average temperatures of –16° compared with today) and warm phases (with yearly average temperatures of +2° compared with today). During the Most Extensive Glaciation (MEG) several hundred thousand years ago, Switzerland was almost completely ice-covered with the exception of the area around Basel and the most external parts of the Jura Mountains. During the warmest interglacial periods the glaciers were, most likely, completely gone. The feeding mechanism for the inner-alpine ice accumulation was a southerly (foehn) circulation. Vegetation cover during interglacials was comparable to today except for the Last Interglacial when *Fagus* (beech) was missing and during at least one older interglacial when a *Fagus/Petrocary*-forest was growing in the Central Plateau. The age for the deep valley erosion in the northern Alpine Foreland is several millions of years younger than in southern Alpine Insubria.

Keywords

Quaternary • Glaciations • Interglacials • Chronology • Paleoclimate • Switzerland

C. Schlüchter (✉) · N. Akçar
Institute of Geological Sciences, University of Bern, Baltzerstrasse 1 + 3, 3012 Bern, Switzerland
e-mail: christian.schluechter@geo.unibe.ch

N. Akçar
e-mail: naki.akcar@geo.unibe.ch

S. Ivy-Ochs
Laboratory for Ion Beam Physics, ETH Zurich, 8093 Zurich, Switzerland
e-mail: ivy@phys.ethz.ch

4.1 Introduction

The evolution of the Swiss landscape during the Quaternary Period over the past 2.6 million years is controlled by multiple glaciations (e.g., Fig. 4.1) and deglaciations with at least 15 drastic environmental changes between glacier advances, with yearly average temperatures as low as –16° compared with today; and warm phases, with yearly average temperatures of +2° compared with today. After discussing dating issues and the question of long sedimentary records, this chapter presents the main glaciations and interglacials and their legacy.

This chapter is neither considered to be a review focusing on the history of research on the paleoglaciations or on the evolution of Quaternary sciences in Switzerland, nor a fundamental discussion on the validity of the classical Quaternary stratigraphy of the Alpine Forelands by Penck and Brückner (1909). There is nothing wrong with using the term "Würmian" for the last glaciation. However, going further back in time, correlations and chronostratigraphic orders are more complex and mostly unsolved. An initiative to solve some of these issues has been taken by the Federal Office of Topography swisstopo for the unification of the legend of the Geological Atlas of Switzerland. This initiative, called HARMOS (Strasky et al. 2016), is summarized in Fig. 4.2.

Throughout the chapter, radiocarbon dates are given as uncalibrated calculations as given by the laboratories.

4.2 The Dating Issues

Advances in knowledge on the Quaternary Period in Switzerland is directly related to the progress in physical and physico-chemical dating of different geological products and to the availability of large outcrops and, therefore, long sedimentary records.

E. Reynard (ed.), *Landscapes and Landforms of Switzerland*, World Geomorphological Landscapes,
https://doi.org/10.1007/978-3-030-43203-4_4

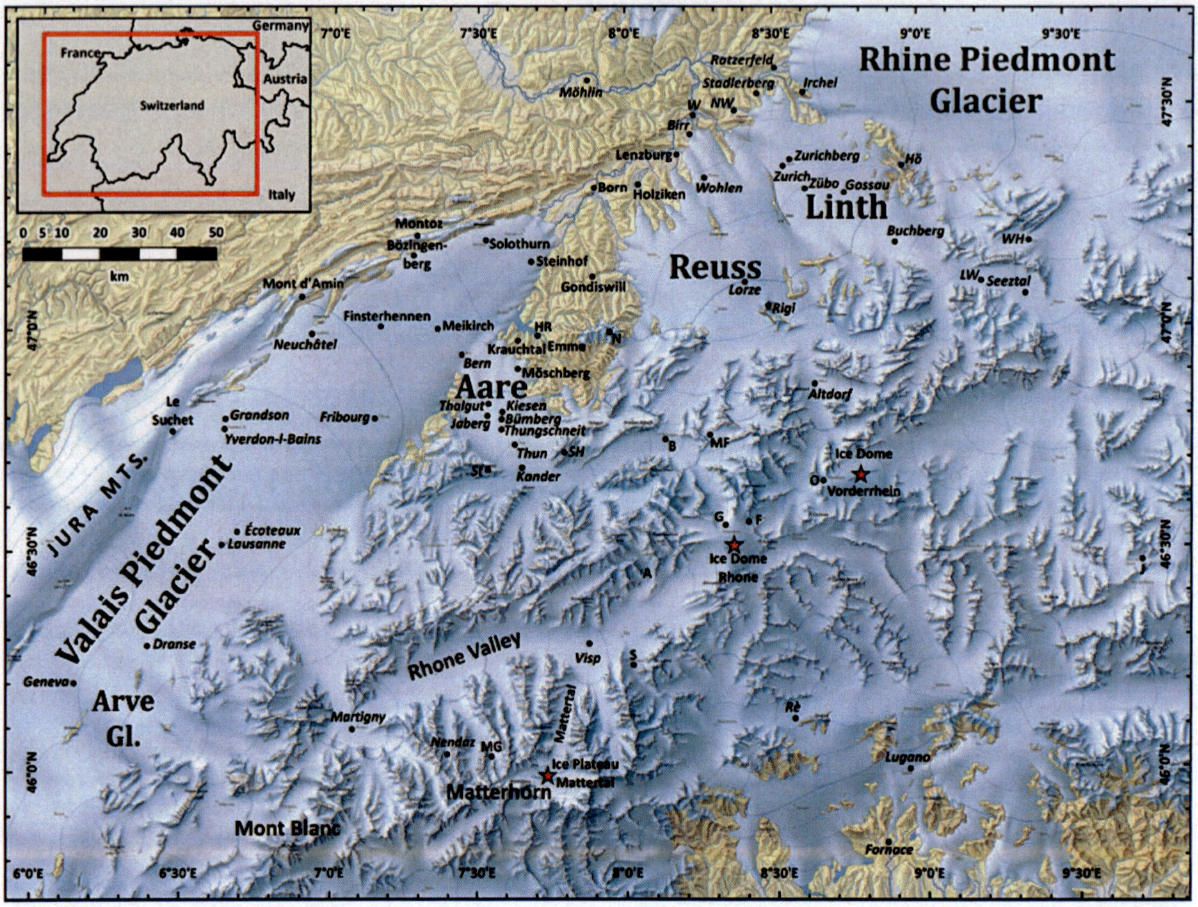

Fig. 4.1 The extent of the Last Glacial Maximum (LGM) in Switzerland with locations of key sites referred to in the text: MG = Mont Miné Glacier, S = Simplonpass, G = Grimselpass, F = Furkapass, O = Oberalppass, J = Julierpass, A = Grosser Aletschgletscher, B = Brünigpass, SH = Sieben Hengste, NW = Niederweningen, W = Nordalpines Wasserschloss (confluence of Aare, Reuss and Limmat rivers), LW = Lake of Walenstadt, WH = Wildhaus, N = Napf, Hö = Hörnli, MF = Melchsee-Frutt, HR = Hasle-Rüegsau, Emme = Emme River

The classical dating approach has been by comparative morphostratigraphy, as it has been ingeniously elaborated by Penck and Brückner (1909) who referred to defined levels of aggradational terraces as time and event markers, and by mapping soil depth (Conradin 1991). These methods are still part of the Quaternary stratigraphy repertoire (Mailänder and Veit 2001) and applicable, if appropriately evaluated.

Absolute, radiometric dating started to be an important contribution after the mid-1960s with the establishment of a radiocarbon laboratory at the University in Bern by Hans Oeschger. Today, this method is fully developed and at the core of radiometric dating of Quaternary age organic material (Hajdas 2008). The methodological refinement of the method to work with small samples is the key to many chrono- and ecostratigraphic uncertainties of the past ca. 30 ka (Hajdas 2008).

It is only in the early 1990s that a method, developed in meteorite research, emerged and became the counterpart of radiocarbon dating in terrestrial Quaternary chronology: cosmogenic nuclide dating of non-organic geomorphological surfaces (Ivy-Ochs 1996). This method allowed an old dream in geosciences to become true: to determine not only the age of the rock as a material but of a rock surface and thus of a landscape element as a morphological unit. The signal used is the measured concentration of a given "cosmogenic nuclide" in a sample. The following terrestrial cosmogenic nuclides are used: ^{3}He, ^{10}Be, ^{14}C, ^{21}Ne, ^{26}Al, ^{36}Cl, and, rarely, ^{54}Mn. The standard application is the dating of a rock surface, i.e., to measure the length of time elapsed since a specific rock surface at a specific place has been formed (e.g., the resting time of an erratic boulder at a stable place). This application is called *surface exposure*

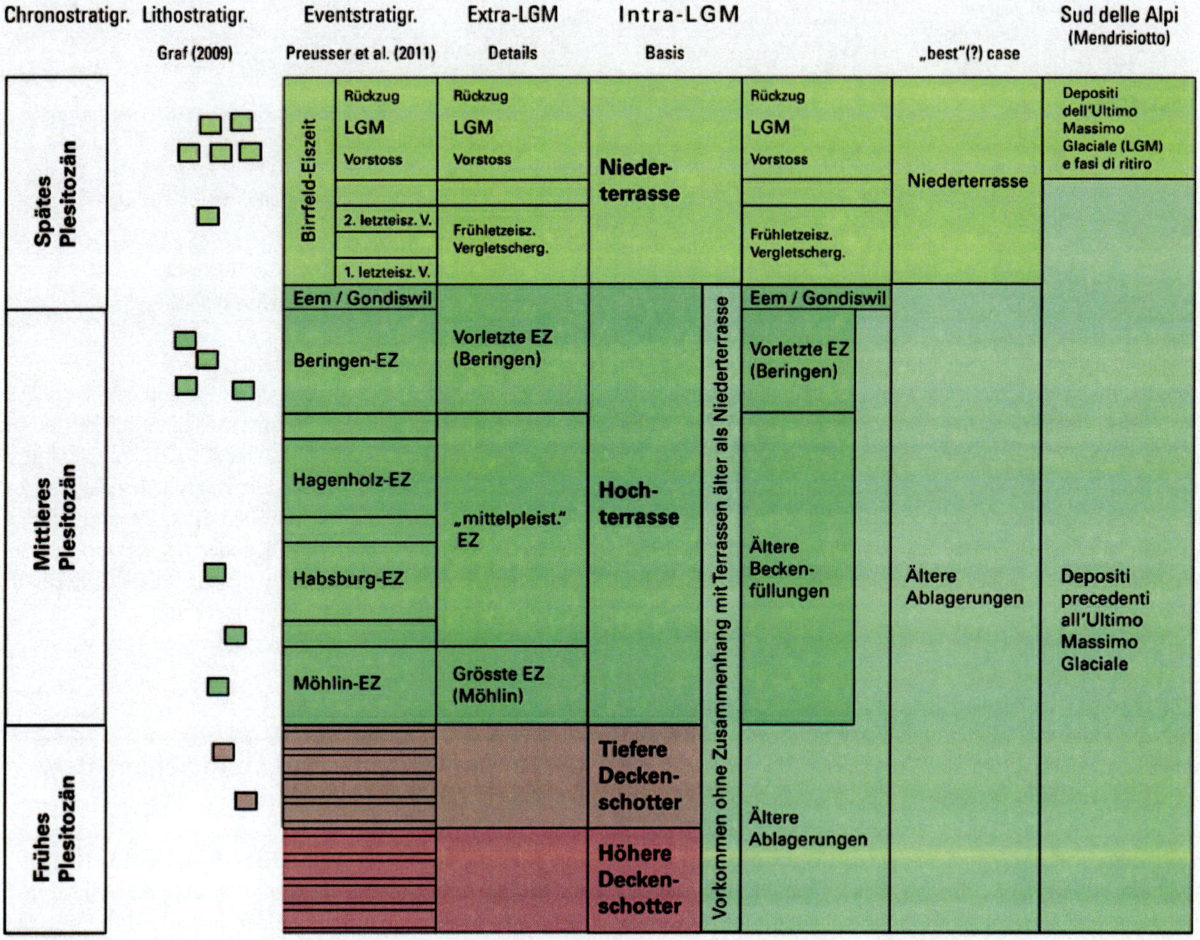

Fig. 4.2 HARMOS: the pragmatic approach to the Quaternary stratigraphic system of Switzerland by swisstopo. *Note* the table is in national languages (German, Italian) to show the comparison between different terms used

dating. A promising new trend in the application of cosmogenic nuclides is *depth profile dating* of a vertical sediment column and *isochrone burial dating* of well-defined beds in an aggradational sequence (Dehnert and Schlüchter 2008). Of challenging interest is still the combination of both, strict burial and isochrone burial dating, ideally applied to the dating of terraces. Terrestrial cosmogenic nuclide methodology has moved Quaternary geochronology from painting the landscape in a still more or less open timeframe to a quantitative geochronological system.

The method of luminescence dating with all more advanced stimulating techniques, e.g., thermo-, optical-, infrared-, laser-stimulation, is operating at a high, but still experimental stage. Additional isotope-based standard dating techniques, e.g., K/Ar on selected minerals, are not commonly used in Switzerland because appropriate material is lacking (e.g., volcanic ash). For young sediments (<100 years), ^{137}Cs has been successfully used on several occasions (Blass et al. 2003; Gilli et al. 2003). Finally, paleomagnetic dating in Switzerland is restricted to very small time windows due to a lack of long, continuous, suitable sedimentary records. However, interesting secondary magnetic overprint has been detected in *Deckenschotter* accumulations (Graf 1993), which allowed the interpretation of minimum ages, and at Ecoteaux/VD,[1] where the Matuyama/Brunhes paleomagnetic boundary was detected (Pugin et al. 1993).

Biostratigraphy, namely palynostratigraphy, is an important contribution to put sections into evolutionary frameworks of vegetation and in interpreted versions of ecology, climate, and time (Welten 1982, 1988; Ammann et al. 1994). Palynological records in combination with radiocarbon or paleomagnetic dating are indispensable for an understanding of succeeding colder/warmer or more

[1]For the key study sites, we indicate the local name and the abbreviation of the canton in which it is located. Locations are indicated in Fig. 4.1.

Table 4.1 Internationally accepted reference time levels in the Quaternary Period, valid for Switzerland

Reference time level	Boundary
11.7 ka (end of the Younger Dryas cold spike)	Pleistocene-Holocene boundary
22.1 ± 4.3 ka	Northern hemisphere Last Glacial Maximum (LGM); Shakun and Carlson (2010); maximum cold and maximum ice volume on the continents during the last glacial/interglacial cycle
115 ka	Last interglacial/glacial transition (=end of last interglacial)
774 ka	Matuyama/Brunhes magnetic reversal
890 ka	End of main phase of Middle Pleistocene Revolution (MPR)
2.6 ma (2.588 ma)	Pliocene (Neogene)/Pleistocene (Quaternary) boundary

specifically stadial/interstadial and glacial/interglacial windows in the history of the Quaternary Ice Age (Wegmüller 1992; Bezat 2000). It must be kept in mind that palynostratigraphy has been the unique approach for dating non-glacial sediment sequences in the absence of more varied dating methods at that time (Welten 1982, 1988; Wegmüller 1992; Bezat 2000; Drescher-Schneider et al. 2007).

A short list of internationally accepted reference time levels in the Quaternary Period is given in Table 4.1 (see also Figs. 4.3 and 4.4). These chronostratigraphic boundaries are valid for Switzerland. Some are even locally constrained (e.g., 11.7, 115 ka).

Figure 4.4 gives a summary of the Quaternary stratigraphy of Switzerland, based on records and available dates from the northern Alpine Foreland. Comparison with the southern Alpine area is schematically shown and Fig. 4.3 gives an indication of likely links with the global oxygen isotope record. These comparisons are not ascertained correlations. They should stimulate discussion and further research. However, with more radiometric dates available now for the *Deckenschotter* (Swiss Cover Gravels, Fig. 4.3; Claude et al. 2013, 2017), the Early/Middle Pleistocene boundary is likely to be correct and being associated with the main morphotectonic event of the Middle Pleistocene Revolution (MPR). The term "Middle Pleistocene Revolution" is somehow misleading as it suggests one severe morphodynamic event only in the Early Middle Pleistocene. In reality, it represents a stepwise lowering of the erosive base level for the drainage in the northern Alpine Foreland or a stepwise uplift of the headwaters. The first step defines the end of deposition of the *Höhere Deckenschotter* (Akçar et al. 2014; Claude et al. 2017). The second step is marked by deep downcutting into the aggradation surfaces of the *Tiefere Deckenschotter* and makes it a prominent well-defined landscape element. The complex morphological situations make it likely that this second step consisted of a series of smaller steps as well. The Most Extensive Glaciation (MEG, which may not have been the Most Voluminous Glaciation, MVG) is placed in the present chronological framework just following the deepest erosion of the valleys in the Swiss Midlands and, in this way time-like related to the most severe erosion (Figs. 4.3 and 4.4). By far the largest volume of sediments in these deeply eroded valleys is of glaciolacustrine facies (e.g., waterlain tills). Despite the fact that the age of the MEG is still under debate and that its relationship to the *Deckenschotter* glaciations still needs to be worked out, there is no doubt that it did exist. The red bars in Fig. 4.4 are not fake news; however, they will need to be adjusted in the future.

4.3 Large Outcrops and Long Sedimentary Records

The construction boom of the late 1960s and early 1970s opened up large and deep windows into the Quaternary sediments and, suddenly, the hidden interior of the Ice Age landscapes became visible, measurable, and accessible to analysis. Before that time, the largest outcrops were along deeply incised rivers as was the case of the rivers Dranse (south of Lake Geneva, Haute-Savoie, France), Kander (west of Thun, artificially initiated), Lorze (area of Zug) (see location in Fig. 4.1) and a great number of relatively small extraction sites of gravel for local use. The new large outcrops were and are mainly gravel pits, with sections hundreds of meters long and tens of meters of vertical extension. Unfortunately for Quaternary research, most of these outcrops are subject to environmental protection policy and have to be re-naturalized, which means refilling and re-greening after the end of extraction. Therefore, archives of former photograph coverage and of selected samples become important. In many cantons, some outcrops are supplemented by extensive drilling projects in the context of groundwater studies and management policies. Over the past 40 years, Quaternary research in Switzerland became, therefore, 3D research.

One of the most striking and unexpected results was the discovery of hidden paleolandscapes covered by more recent deposits. The Quaternary paleorelief predating the last

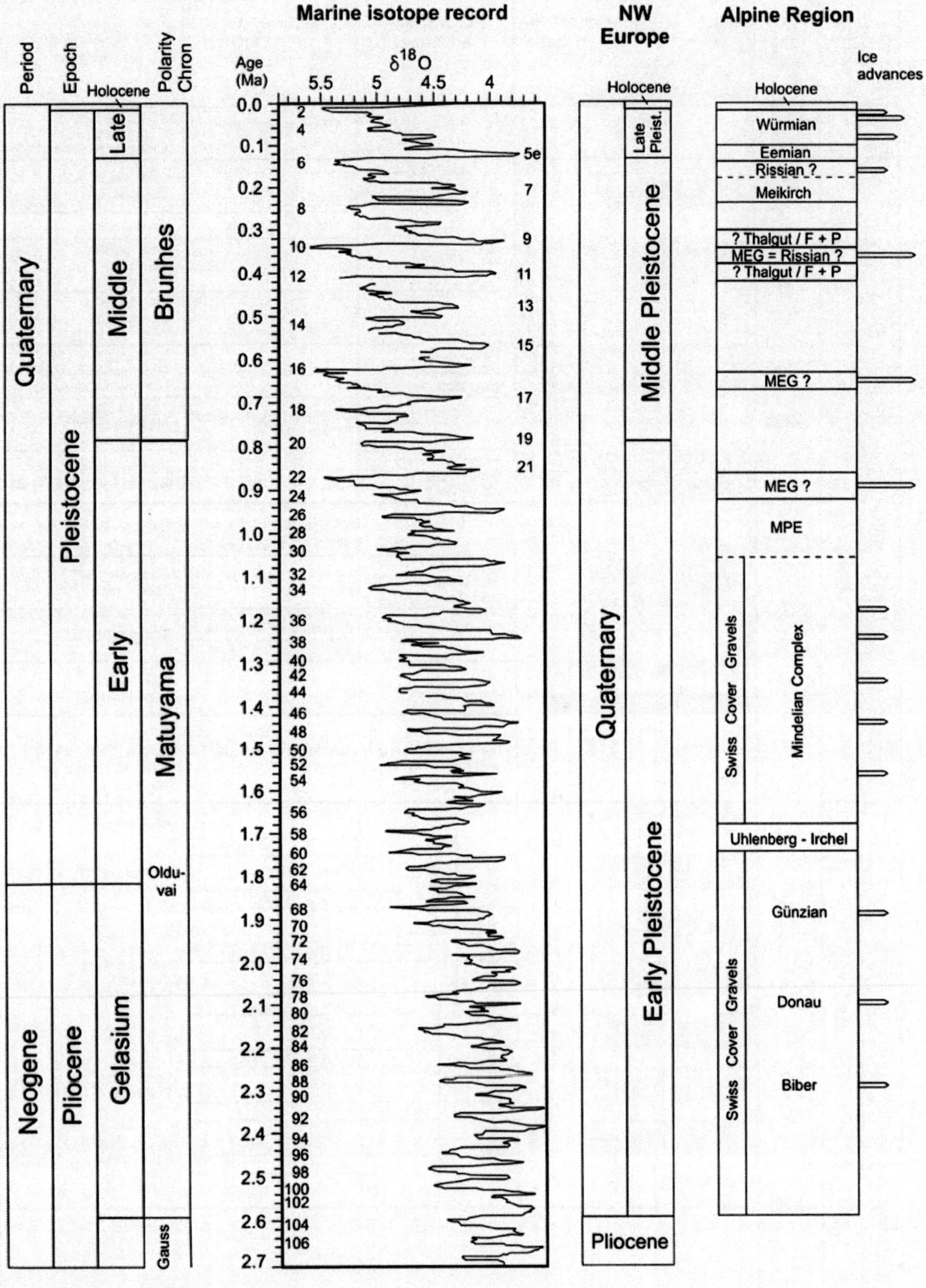

Fig. 4.3 The Alpine Quaternary history in a global context (from Litt et al. 2008)

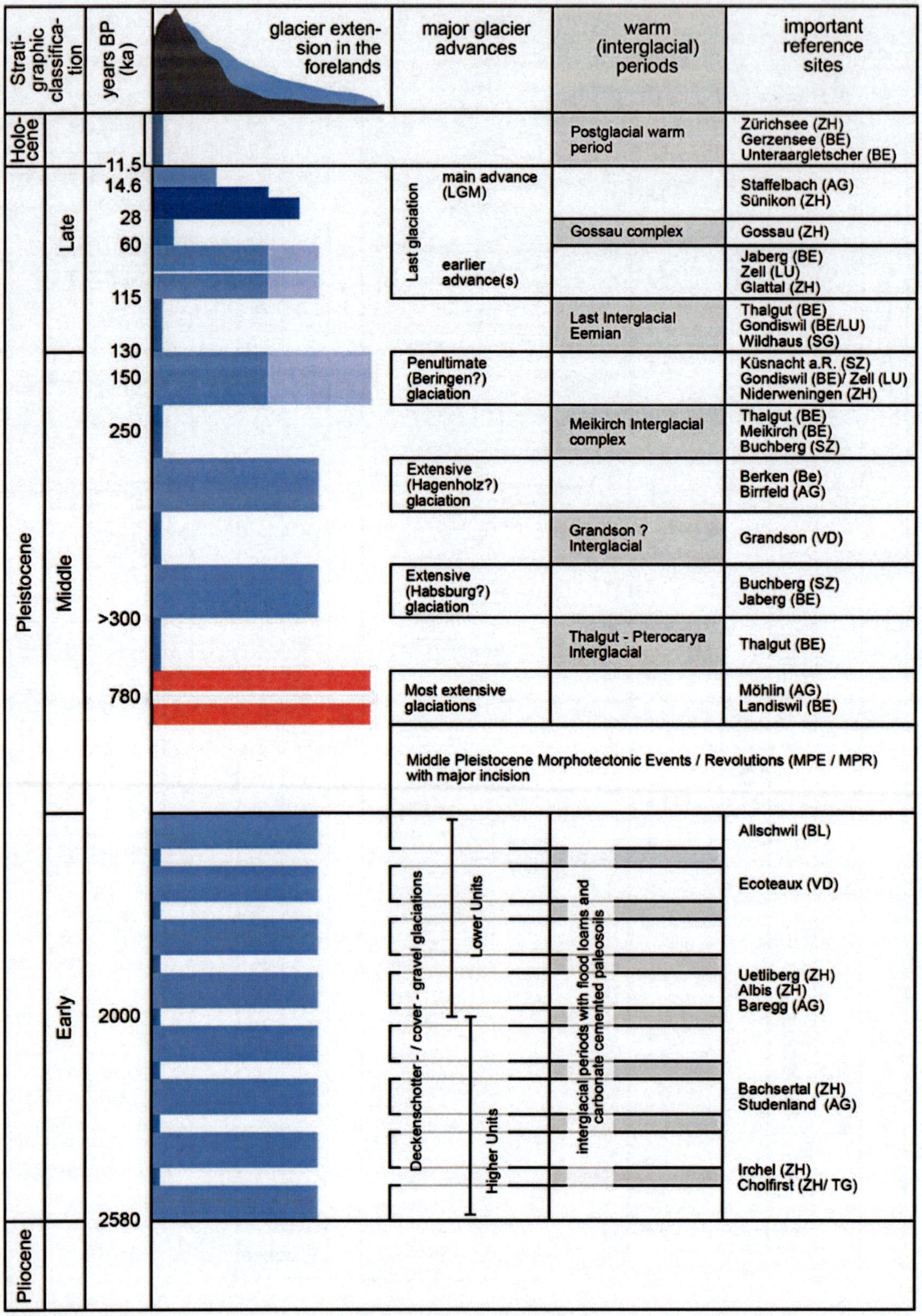

Fig. 4.4 Summary table of the Quaternary in the northern Alpine Foreland (revised from Schlüchter 2017, Preusser et al. 2011). The stratigraphic position of the Most Extensive Glaciation(s) is given in red

Fig. 4.5 Gravel pit at Thalgut/BE. **a** General view of the "core of the section", with major unconformity in the central part of the picture and with the drilling rig; **b** detail view of the major unconformity (photos C. Schlüchter 1991)

glaciation was as accentuated as the present-day landscape (e.g., Thalgut/BE, Fig. 4.5, and Buchberg/SZ). A number of specifically planned or extended drillings carried out within research projects produced important results, in particular, in Kiesen/Rotachewald/BE (Schlüchter 1984), Thalgut/BE (Welten 1982, 1988; Schlüchter 1989), Meikirch/BE (Welten 1982, 1988; Preusser et al. 2005), Grandson/VD (Welten 1988), Zübo/Lake Zürich/ZH (Hsü and Kelts 1984), Buchberg/SZ (Welten 1988), Gondiswil-Zell/BE-LU (Wegmüller 1992), Ecoteaux/VD (Pugin et al. 1993; Bezat 2000) and the Seeztal/SG-GR, between Lake of Walenstadt and Sargans (Müller 1995) (see location in Fig. 4.1). The two most important results of these drillings were (i) the discovery of superimposed interglacial deposits with different paleovegetation signals and (ii) the lithostratigraphic discrepancy between records in the main thalwegs of the valleys and those in higher lateral (terrace) positions.

4.4 Large Quaternary Landscapes in Switzerland

A striking contrast in the Quaternary landscape in Switzerland is between the soft and smoothened areas covered by the glaciers of the Last Glacial Maximum (LGM) and the external areas where vertical erosion processes dominated, like the prominent examples of landscapes sculptured by fluvial processes of the Napf and the Hörnli (Fig. 4.1) or the high-mountain landscapes above the trimline, where mechanical weathering dominated (e.g., Grimsel, Fig. 4.6). The reconstruction of the ice cover in the Central Alps, e.g., for the LGM (Fig. 4.1), relies on the spectacular trimlines in the crystalline rocks of the central massifs which delineate the maximum elevation of actively moving ice (Fig. 4.6). A magnificent example of a landscape in the Molasse bedrock, partially drowned by more recent accumulations (here of the LGM) is the "crest and earth pillar" morphology at Krauchthal/BE (Lauber 2003, Fig. 4.7).

Important landscape elements, however, neither visible nor directly accessible, are deeply eroded valleys of the Alpine Foreland (e.g., Dürst Stucki et al. 2010; Dehnert et al. 2012; Ivy-Ochs et al., this volume) and of the inner-Alpine areas, with the most important examples of the Insubric (=the morphological transition between the southern border of the Alps and the Po Plain in Italy) valleys and lakes, south of the Alps, where bedrock lies as deep as more than 600 m below the present level of the Adriatic sea (Finckh 1978; Finckh et al. 1984). The age of the deepest erosion is not, necessarily, the same in the Insubric area, where it most likely goes back to the Messinian salinity crisis in the Mediterranean, 5–7 million years ago (see Scapozza and Ambrosi, this volume), and in the northern Alpine Foreland, where it is related to the Middle Pleistocene Events (MPE) or Revolution (MPR). A revolution is usually a multi-step event, also in geology. The deep incision of the main valleys into bedrock is, therefore, about 6 million years younger to the north of the Alps than to the south.

Some areas are first-order Quaternary landscapes as they control the contemporary (and did control in the past) landscape evolution:

(1) The mesa-type landscape of the *Deckenschotter* in northeastern/northern Switzerland (see Keller, this volume) connects to the classical "*Eiszeitlandschaft*" (Ice age landscape) in Southern Germany. The Swiss *Deckenschotter* sector is part of the same (paleo) tectonic area, of the so-called "*Süddeutsche Grossscholle.*"

Fig. 4.6 The trimline in the Central Alps: Vorder Zinggenstock near Grimselpass, with spectacular *roches moutonnées* landscape (photo C. Schlüchter 2017)

This landscape has no equivalent neither in the central nor in the western sector of the Swiss Midlands.

(2) The position of the sub-Jurassic lakes (Neuchâtel, Murten, and Biel) is delicately arranged in the tectonic structures parallel to the structural elements of the Jura Mountains (Gorin et al. 2003) and most likely related to subrecent subsidence of the western part of the Swiss Midlands and of the Lake Geneva Basin.

(3) The drainage of the Swiss Midlands is directly connected to the southern front of the Jura Mountains, where the River Aare cuts through the Born anticline (Fig. 4.1). River downcutting in the Midlands is obviously balanced by uplift of the Jura. This situation is repeated further to the east, where the merged waters of the Aare, Reuss, and (Linth-) Limmat rivers cut the entire folded and tabular Jura Mountains to reach the southern border of the Black Forest Massif ("*nordalpines Wasserschloss,*" northern Alpine water tower) to join the Rhine drainage system.

4.5 Glaciations

Two questions are of concern: (i) When did the first glaciers develop in the Alpine region at the onset of the ongoing Ice Age (Fig. 4.3) and (ii) When did the last glaciation end? The answer to the first question is complex and difficult; the answer to the second question is rather straightforward (Figs. 4.2 and 4.3).

4.5.1 The First Glaciation(s)

The first glaciation(s) in the Insubric area are recorded in sections in the Valle della Fornace (Fig. 4.1) in the Swiss/Italian border area. The lower part of the section contains brackish to lacustrine sediments and voluminous delta deposits (*Formazione della Valle della Fornace* (Valle della Fornace Formation); Uggeri et al. 1997) as an infill of

Fig. 4.7 An example of landforms beyond the limit of LGM glaciers: rock spires at Krauchthal/BE (photo C. Schlüchter 2003)

one of the deep Messinian Insubric valleys. The Valle della Fornace Formation is grading upward into glaciolacustrine facies (Vivirolo Formation). Below the accessible outcrops in the valley are marine clays (*Argille di Castel di Sotto*) recorded in drill cores. They are attributed, through pollen analyses, to the Middle Pliocene, which puts the overlying *Formazione della Valle della Fornace* to the Upper Pliocene (Uggeri et al. 1997; Martinetto and Ravazzi 1997). However, this chronostratigraphic positioning was inferred under the philosophy of the previously fixed Pliocene/Pleistocene boundary at 1.65 ma, without radiometric dating at this specific site. By adjusting the Pliocene/Pleistocene boundary to 2.588 ma, following recent research, the Valle della Fornace Formation and, mainly, the Vivirolo Formation are considered to be part of the Pleistocene series. The glaciolacustrine sediments of the Vivirolo Formation, therefore, document of the onset of glaciation in the Insubric area of the Alps in the early Pleistocene (>2 ma). Martinetto and Ravazzi (1997) classified some of the sediments of the Vivirolo Formation as tills; a glacier was obviously there at that time.

The discussion about the earliest glaciation in the northern Alpine Foreland concentrates on facies interpretation of the *Deckenschotter* (Cover gravel) sequences of northern Switzerland. These coarse gravel aggradations rest unconformably on the Cenozoic (Molasse) or Mesozoic bedrock and contain in their oldest units at Irchel/ZH or Stadlerberg/ZH boulders of Alpine provenance and proximal fluvioglacial sandy-silty gravel, which asks for a near-glacier facies (Graf 1993, 2009). The minimum chronostratigraphic position of the oldest units of *Deckenschotter* is given by the mammal findings at Irchel-Steig, which belong to the Neogene Mammalian Stage NM 17 (Bolliger et al. 1996). The precise chronostratigraphic position of the Pliocene/Pleistocene boundary is of direct relevance here, as it makes the oldest unit of *Deckenschotter*, older than 2 ma. Ongoing dating efforts by using cosmogenic nuclides confirm that the oldest Ice Age deposits here are older than 2 ma (Akçar et al. 2014; Claude et al. 2017).

The current level of knowledge allows one for the safe assumption that the first glaciation of the Quaternary Ice Age did occur more than 2 ma ago in the Alps and that it did produce glacial sediment facies associations in both the northern and the southern Alpine Forelands. However, the pre-glacial landscape in the north was different from that in the south. In the north, broad valleys existed beyond the border of the Alps and far-reaching piedmont aggradational fans were fed by the advancing glaciers. In the Insubric area to the south of the Alps, the glaciers were confined to the much narrower, steeper, and deeper valleys of the Messinian fjords. This N-S asymmetry has consequences in the visibility of landscape elements today. In the north, the *Deckenschotter* landscape is characterized by mesa-type hills more than 150 m above the present valleys. It is a typical erosional relict of the Ice Age landscape, survived after a major phase of relief inversion and, most likely, of stepwise uplift and, as a consequence, stepwise incision. In the south, the records of the early glaciations are buried below younger sediments in deep valleys, with the exception of the brackish clays at Balerna/TI, a sedimentary unit of the post-Messinian high stand of the Adriatic sea. The question about the pre-glacial landscape and, more specifically, about perialpine lakes, as they exist today, remains largely open for the time being. Large slabs of highly compacted lacustrine deposits, including compressed peat, occur occasionally like erratic boulders in delta deposits of the Most Extensive Glaciation (e.g., Thalgut/BE). The original position of these old lake sediments as well as their age is not known. However, they exist somewhere along the border of the Alps where the sediment slabs were picked up by an advancing glacier later. The only areas where speculations on the pre-glacial landscape are possible are the base levels of the *Deckenschotter*. There, it looks like a weakly structured, slightly inclined erosional landscape on Molasse sandstone or Mesozoic limestone.

Old landscape elements are reported by Gnägi and Schlüchter (2012) in the Berner Oberland (at Sieben Hengste/BE and at Stockhorn/BE) in the form of relict coarse gravel lenses within weathered slope deposits, containing indicator lithologies from the Penninic part of the Alps. They may be of fluvioglacial origin. However, the glacial component is not ascertained and they are, most likely, relict sediments in an unroofed cave. Silty gravel with erratic boulders were mapped at the surface and within the deposit on Piz Starlex at 3100 m in the Val Müstair/GR (not represented in Fig. 4.1) by Kocher (2012). The petrographic composition of the boulders and of the coarse gravel fraction is unique and points to east–west transport which required substantial changes in the drainage pattern of the Eastern Alps in the Early Pleistocene.

4.5.2 The Last Glaciation

The last glaciation has not been a uniform and permanently cold period for 100,000 years. However, it was initiated by a severe short-term cooling event (see "Interglacials and Interstadials" section). The comparison with the global oxygen isotope stratigraphy (Fig. 4.8a) is ascertained to an acceptable level for the time window of the last glacial cycle (Figs. 4.2, 4.3, and 4.8b). The climate pattern behind the evolution of the oxygen isotope composition of the atmosphere is supported by stratigraphy, despite the fact that first-order correlations are missing. After the first cold spell of the Marine Oxygen Isotope Stage (MIS) 5d, a long and complex, oscillating climate pattern occurred, which is difficult to characterize using lithostratigraphy (Fig. 4.8b). This is possible only through pollen records, e.g., at Gossau/ZH (Welten 1982; Burga and Wynistorf 1987) or at Gondiswil-Zell/BE-LU (Wegmüller 1992). In the voluminous gravel aggradations predating the final LGM advance there is evidence for two main fluvioglacial phases in Thalgut/BE and Jaberg/BE, Gossau/ZH, Buchberg/SZ, as well as in the Lake Geneva Basin (Arn 1984). The exact timing of the earlier advance phase is still open. There are chronological points, e.g., at Finsterhennen/BE (Preusser et al. 2011), at Gossau/ZH (Schlüchter et al. 1987), as well

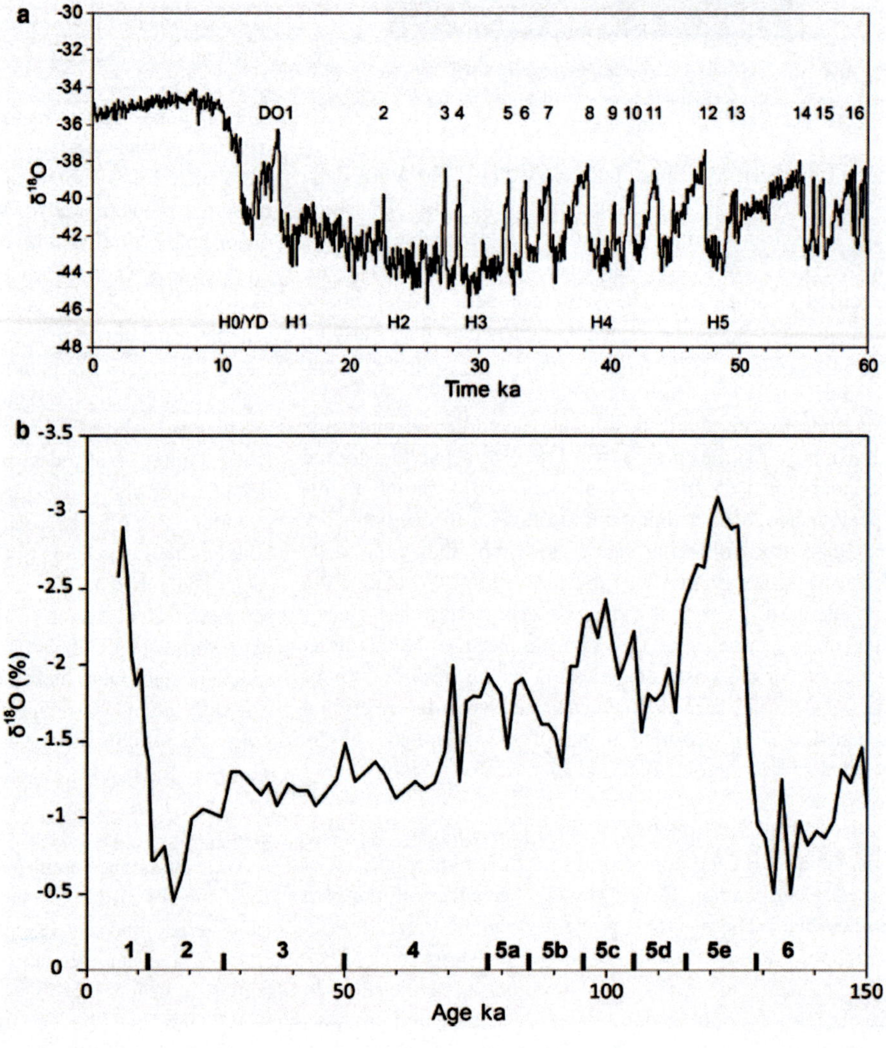

Fig. 4.8 a The Greenland Isotope Record over the past 60 ka with the denomination of DO- and H-events (from Anderson et al. 2004); **b** The Marine Oxygene Isotope Stratigraphy of the Last Glacial Cycle (from Bassinot et al. 1994)

as along Lake of Walenstadt/SG (Schindler et al. 1985). The onset of the "final" advance, which produced the LGM-extent (Fig. 4.1), is sufficiently constrained. The Valais Paleoglacier reached the Midlands after 30 ka in the west as did the Rhine Paleoglacier in the east. For the Lyonnais Lobe of the "Valais Glacier" which is, in reality, the Arve Glacier, the final advance passed the border of the Alps not later than 25,470 years BP (Moscariello 1996, Schoeneich 1998); the same timing holds for the Swiss Lobe of the "Valais Glacier", with 25,370 years BP for the Mammoth Tusk at Finsterhennen from the fluvioglacial aggradation below the basal till of the LGM. For the Linth Glacier the arrival time to the greater Zürich area is less than 28,550 years BP at Gossau/ZH and less than 28,060 years BP at Zürichberg. This last and main LGM advance to the Alpine Forelands is not an oscillation only out of an ice reservoir "just inside the border of the Alps." Considerable areas in the high Alps must have remained ice-free before. This is documented by subfossil bear bones from caves in the Melchsee Frutt/OW area, with a radiocarbon age of about 33,000 years BP. It can be concluded that the LGM advance was a well-defined event, which started to develop in the higher inner-alpine valleys. This interpretation is supported by many sites containing organic sediments, with radiocarbon ages ranging between 30 ka and about 40 ka (e.g., van der Meer 1982, Wegmüller 1992, Hajdas 2008).

The reconstruction of the LGM glacier system in the Alps and its maximum position in the Alpine Forelands has been subject of active research in the recent years, with some far-reaching results (Schlüchter 2009; Luetscher et al. 2015; Becker et al. 2016, 2017): (i) moisture forming the inner-alpine ice masses was transported by southerly air flow (Foehn); (ii) this southerly moisture transport has been so important that accumulations were forming ice domes to the south of the main weather divide (Ice Dome Engiadina, Ice Dome Vorderrhein, Ice Dome Rhone, and the Ice Plateau Mattertal (Florineth and Schlüchter 1998, 2000; Kelly et al. 2004). This inner-alpine configuration has caused massive outflows to form the Valais Piedmont Glacier in the west and the Rhine Piedmont Glacier in the east (Fig. 4.1), with corresponding transport and distribution of erratic material in the forelands (Becker et al. 2017).

With maximum glacier extension to the western foreland a lake was dammed in the Emme Valley (Fig. 4.1). It was bound upvalley by a large fan complex. With the glacier recession the lake drained and the base level for the Emme River was lowered. As a consequence, the upvalley fan was dissected and extensive terraces formed. The final aggradation of the fan, and therefore the very maximum strength of the LGM ice dam, was radiocarbon dated at Hasle-Rüegsau/BE (Fig. 4.1). The most spectacular dated specimen is a humerus of the Woolly Rhino (Fig. 4.9), whose uncalibrated radiocarbon age is 17,448 ± 46 years BP (Laboratory reference ETH-55'673).

4.5.3 The End of the Last Glaciation

The end of the last glaciation is the most recent major environmental change, which fundamentally transformed the landscape: valleys and lakes formed and the topography became again very varied. The down-melt of the last large glaciers was an environmental catastrophe because the glaciers vanished fast. There are two striking pieces of evidence for this scenario: (i) radiocarbon dates on basal organic matter (macroremains, peat) in lakes and ponds which are synchronous within 1000 years in the northern and southern forelands and in the inner-alpine pass areas (Schlüchter 1988), and (ii) the "missing moraines" between the inner LGM ridges and the high valleys which is explained only by rapid, non-oscillating down-melt. Deglaciation from the LGM positions started as early as 24.0 ± 1.1 ka based on surface exposure dates on the erratic boulders at Steinhof/SO, which are part of the LGM system (Ivy-Ochs et al. 2004). Confirming dates on erratic boulders exist for the Reuss Glacier with 22.2 ± 1.0 ka (Reber et al. 2014) at Lenzburg/AG and for the Aare Glacier with 21.5 ± 1.0 ka in lateral positions at Möschberg/BE (Akçar et al. 2011). As early as 18 ka more than 80% of the LGM ice volume had disappeared (Ivy-Ochs et al. 2008), and the northern Alpine Foreland was already ice-free by that time (van Husen 1997; Hadorn et al. 2002; Keller and Krayss 2005). The same time range exists for the moraines of the "stade lémanique" in the Lake Geneva Basin (Arn 1984; Schoeneich 1998).

The glaciers disappeared by that time not only from the forelands but from long reaches of the alpine valleys as well (Ammann et al. 1994). This is inferred from the first reorganization of the back- and downwasting valley glaciers which occurred around 16–17 ka, producing the terminal moraine of the Gschnitz Stadial after a period of considerable retreat further upvalley before (Kerschner et al. 2002). The Gschnitz Stadial could be as old as 17.5 ka (Vescovi et al. 2007), relying on a number of surface exposure dates on boulders from the type locality morainic ridge in the Gschnitz valley in Austria, with a mean age of 16.7 ± 1.5 ka (Ivy-Ochs et al. 2006). The oscillations, formerly named Clavadel and Daun (Maisch 1987) and extensively correlated, are most likely not reflecting true and paleoclimatically important re-advances. The Alpine glaciers retreated way up into the cirque areas with reduced extensions comparable to the Little Ice Age already during the Bølling–Allerød warm interstadial period. The last major cold phase followed, with glaciers extending beyond the cirques to form the Egesen Moraine Complex. Surface

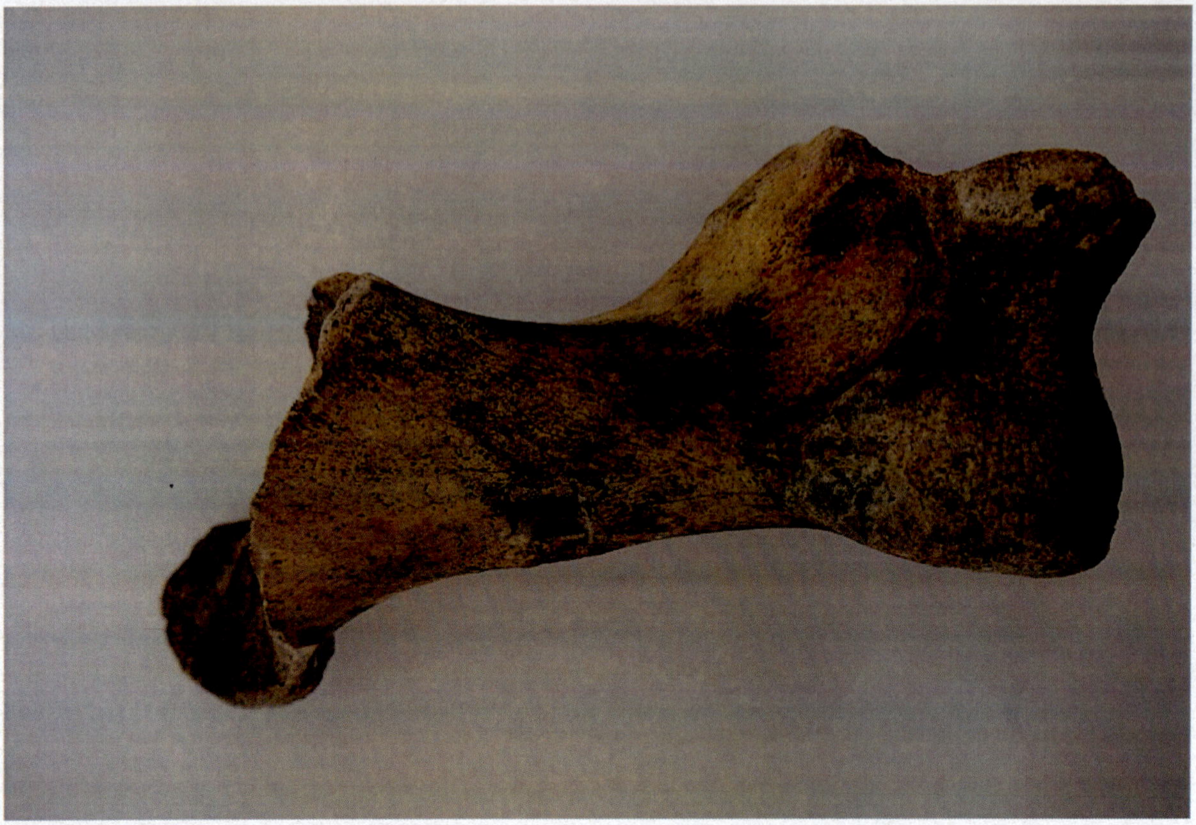

Fig. 4.9 Humerus of the Woolly Rhino from the gravel pit at Dicki/Emmental/BE (photo Dr. A. Rehazek, Natural History Museum Bern)

exposure ages on boulders from the Egesen moraines at Julierpass/GR give ages of 12.0–13.5 ka for moraine stabilization (Ivy-Ochs et al. 2008). The spectacular lateral moraine systems of the Grosser Aletschgletscher/VS and Tortin in the high valley of Nendaz/VS (Scapozza 2015 and Fig. 4.10) confirm the Egesen advance in the western part of the Swiss Alps by ^{10}Be-dates as well (Kelly et al. 2006; Schindelwig et al. 2011). The response of Grosser Aletschgletscher to the Egesen cooling resulted in a simple, however, voluminous lateral moraine (Kelly et al. 2006). Next to Grosser Aletschgletscher, a small cirque glacier at Belalp produced a delicate oscillative lateral moraine pattern during the Egesen Stadial (Schindelwig et al. 2011). Retreat from Egesen moraines started between 11.7 and 10.8 ka in the Grimsel area (Kelly et al. 2006; Wirsig et al. 2016). The Egesen moraines are most likely, as given by the surface exposure dates on boulder surfaces, the physical product of the Younger Dryas cold phase, which ended in the Greenland Ice Core Records at 11.5 ka. With the glaciers melting back from the Egesen moraines, the lateglacial phase of the last glaciation terminated in the high Alps.

4.5.4 The Holocene

The Holocene glacier record is based on two lines of evidence: (i) direct and simple moraine record as the physical–morphological product of glacier advances and (ii) more complete climate record of the past 11.5 ka based on subfossil trees and peat as evidence for glacier retreat in between the moraine ridges. Figure 4.11 summarizes the Holocene record, demonstrating the highly dynamic glacier response to climate forcing (Hormes et al. 2001; Holzhauser et al. 2005; Joerin et al. 2006; Goehring et al. 2011; Nicolussi and Schlüchter 2012). The most spectacular results are:

(1) Confirmation of the 8.2 ka event at the Mont Miné Glacier/VS, linking directly and precisely the Greenland ice core to an alpine glacier record (Nicolussi and Schlüchter 2012). The 8.2 ka advance of the Mont Miné Glacier occurred suddenly after a several centuries long retreat phase.
(2) The data from Rhone Glacier bedrock ridge (next to the Belvedere Hotel at Furkapass/VS) show pre-Little Ice

Fig. 4.10 The lateral moraine of the Egesen advance (i.e., morphological testimony of the Younger Dryas cold phase) in the upper valley of Nendaz/VS (photo C. Schlüchter 2009)

Age exposure time of the local bedrock and thus demonstrate, in combination with data as shown in Fig. 4.11, reduced Holocene glacier expansion (compared to the 2005 position) for >50% of time since glacier retreat from the Egesen positions. A unique piece of evidence for reduced glaciers is found at Unteraargletscher, in the samples of compressed peat washed from underneath the actual glaciers by meltwater outbursts: a completely preserved wasp (*Attractodes mesoleptus*) in a peat fragment has been discovered (Schürch et al. 2015; Fig. 4.12).

4.5.5 Glaciations *in-between* and the Most Extensive Glaciation

The earliest and the last glaciations are the reference time-and-event-frames for the glaciations during the Quaternary (Figs. 4.2 and 4.3). The assemblage of available lithosequences allows for the reconstruction of a total of 15 different glacier advances over the past 2.6 ma for the northern Alpine Foreland, with a comparable complexity for the south (Bini 1997; Graf 2009). The two levels of *Deckenschotter* (Upper *Deckenschotter* and Lower *Deckenschotter*) contain lithostratigraphic units of at least eight glacier advances (Graf 1993) far beyond the border of the Alps. The upper (i.e., older) units are separated from the lower (i.e., younger) units by an important phase of incision (Fig. 4.13, Akçar et al. 2014). It is likely that this major incision occurred in the main phase of the MPR/MPE. The second revolutionary incision phase of the Middle Pleistocene Event is cutting down into the Lower *Deckenschotter*. Based on a few long records (e.g., Thalgut, Meikirch, Buechberg), the deepest erosion can roughly be related to the MEG (Fig. 4.3), which follows the incision phase of the Lower *Deckenschotter*, and which makes the *Deckenschotter* sequence older than the MEG (Figs. 4.2 and 4.3), i.e.,

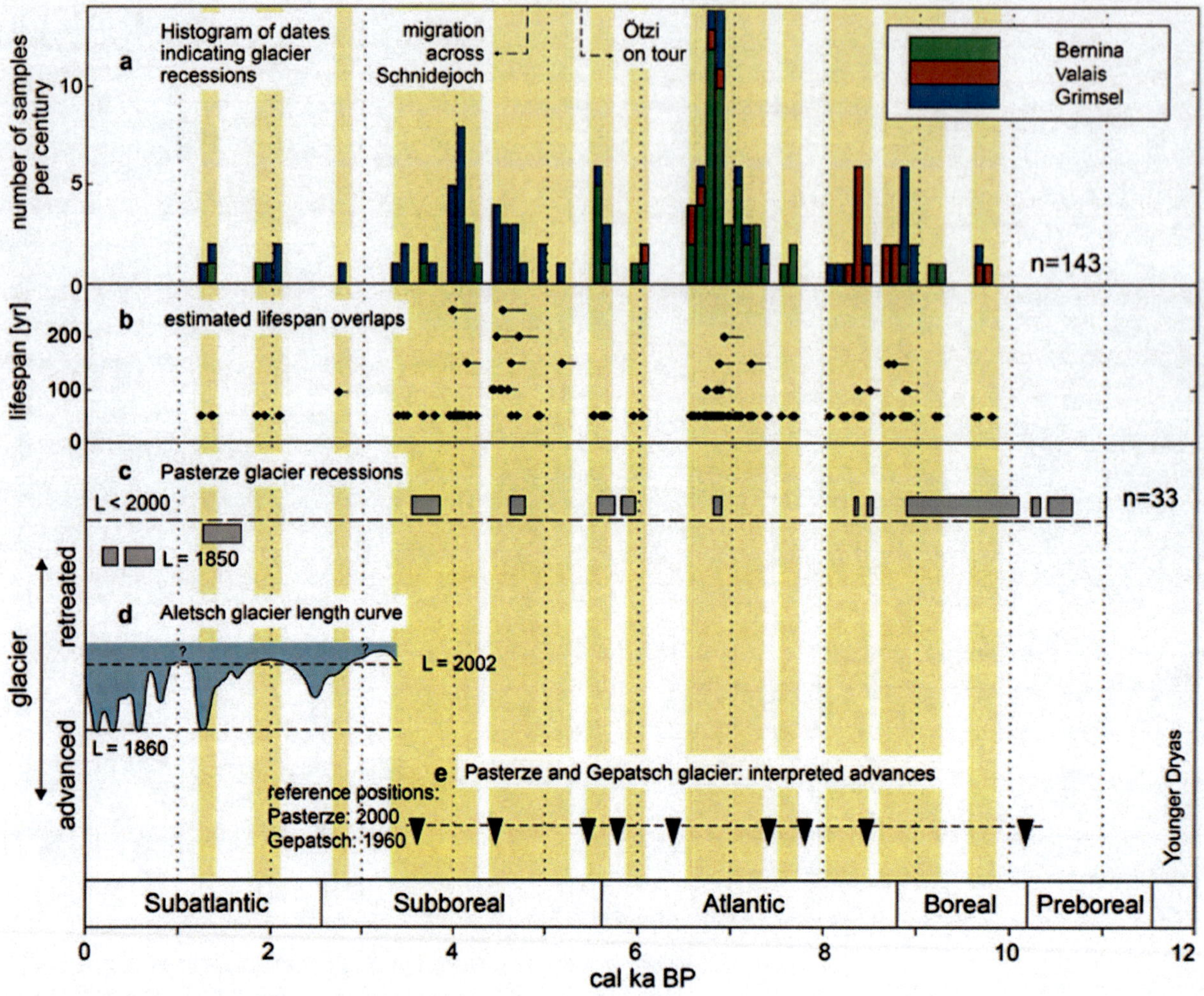

Fig. 4.11 Summary table of Holocene Alpine glacier variations (after Joerin et al. 2006)

Fig. 4.12 Subfossil wasp (*Attractodes mesuleptus*) from compressed subglacial peat, Unteraargletscher (from Schürch et al. 2015)

older than 780 ka. Field evidence is excellent and shows glacial deposits often in complex lithological relationship between LGM and MEG ice extents, corresponding to at least three major glacier advances beyond the limits of LGM but not reaching the extent of MEG (=Complex of the *Hochterrasse*, Figs. 4.3 and 4.4). Erratic boulder evidence for the MEG is found in the Jura Mountains and comprises solitary boulders (Fig. 4.14, Graf 1993) and a hilly transverse-ridge dominated area in the valley of the Rhine River between Rheinfelden and Säckingen (=Möhlin Terminal Moraines, Dick et al. 1996; Gaar 2014). This area is traditionally considered the terminal area of the Most Extensive Alpine Glacier extent, which has been correlated with the Riss Glaciation of the classical glacier chronology of Southern Germany (Penck and Brückner 1909). As this Most Extensive Glaciation is, most likely, several hundred thousand years old, the term "Riss" needs re-definition: if the

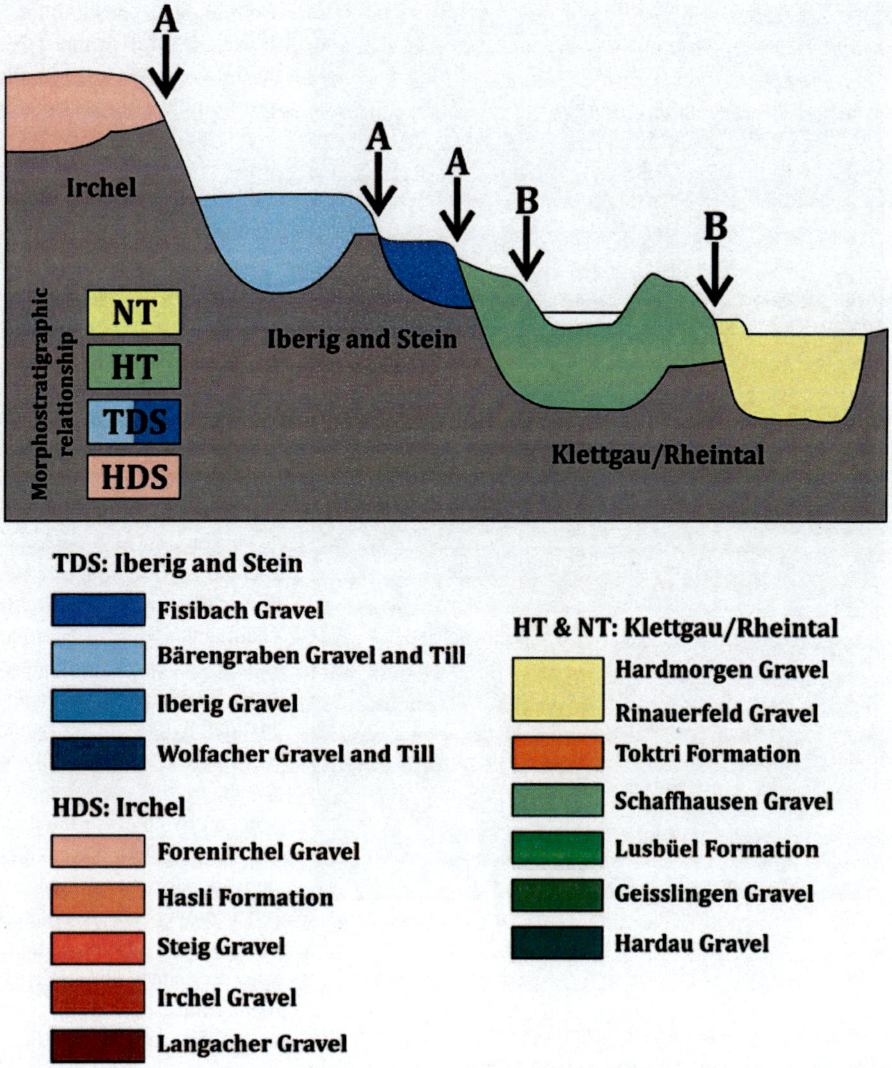

Fig. 4.13 Schematic NW-SE cross section from Irchel to Klettgau and Rhine Valley. HDS = *Höherer Deckenschotter*, TDS = *Tieferer Deckenschotter*, HT = *Hochterrasse* (high terrace) and NT (*Niederterrasse*) multiphase aggradation complexes. A = incisions related to Middle Pleistocene Revolution (MPR) events; B = incisions related to the Most Extensive Glaciation (MEG) event(s). Stratigraphic relationships: oldest aggradation complex (HDS) is in highest morphostratigraphic position

term is kept for the Most Extensive Glacier advance, then it names an event in the order of >780 ka. If the term is kept for the penultimate glaciation, then it stands for an event of less extent than the MEG. In such a classification dilemma it is recommended not to use the term “Riss” any longer until an unbiased definition is accepted. The mammoth finding site of Niederweningen/ZH is a key in this discussion (Welten 1988; Dehnert et al. 2012).

Geometry and ice flow configurations of the Most Extensive Glaciation (MEG) need to be addressed as the relationship between the Möhlin Terminal Moraines and the *Deckenschotter* is not sorted out in detail. There are observations in the Möhlin area that ice from the Black Forest and not solely from the Alps has played a constructive role in the formation of the landscape (Dick et al. 1996). Optically stimulated luminescence (OSL) tests on loess-like sediments at Möhlin question the old age (Gaar 2014). In addition, lithostratigraphy of the Quaternary deposits contains a complex aggradation/paleosoil sequence representing at least two glaciations (Dick et al. 1996). This observation is of fundamental paleoclimatic importance, as it makes the MEG at least a two-phase event (Fig. 4.3).

Are there any more arguments to pin down the age of the MEG? If it is reasonable to assume that the till directly overlying bedrock in the deep valleys belongs genetically to the ice extent of the MEG far out to the foreland, then it belongs to a pre-MIS-11 event (e.g., Thalgut, Meikirch, Buechberg, see below “Interglacials and Interstadials” section). This age position goes in synchrony with advanced weathering and evolution of paleosoils, e.g., on *Höhenschotter* (Upper gravels) in the western Napf area or on tills in the external part of the Jura Mountains (Hildbrand 1990; Christen 1999; Schlüchter 2015). Direct dating of the vestiges of the MEG is still in progress. First sample sets for

Fig. 4.14 Pre-LGM solitary crystalline erratic boulder at Vorberg/SO (photo C. Schlüchter 2011)

surface exposure dating of MEG erratic boulders did produce minimum ages of >100 ka and >150 ka, respectively (Graf et al. 2015). The uncertainties are as elsewhere: how stable were these boulders morphologically and how severe was the human impact, at least during and since Roman times.

4.6 Interglacials and Interstadials

Interglacials are periods when vegetation is characterized by the presence of broadleaf trees, jointly with *Abies sp.*; interstadials are periods when vegetation is characterized by grassland with limited tree growth (e.g., *Betula, Salix*). Fine-grained lacustrine deposits within complex lithostratigraphic successions with tills and coarse gravelly aggradations, typical for the Alpine area, are sediments with a realistic potential for detailed palynostratigraphic information. In the Swiss Alpine Forelands, a number of interglacial records of paleoclimatic and stratigraphic importance exist. Stratigraphic positioning by palynostratigraphy needs, however, some further considerations. There are two plants in the northern Alpine milieu which allow for stratigraphic positioning of the sections investigated: if *Fagus* is present it is not the Last Interglacial and if *Pterocarya* is present, then it is an "older" interglacial corresponding to MIS 9 or older. The debate centered around the characterization of the Last Interglacial, which finally led to the correlation with the Nordic Eemian. The palynological diversity of older interglacials and their comparison with the Nordic Holsteinian is the matter of an ongoing debate. Four interglacial sites are referenced in more detail: Thalgut/BE, Meikirch/BE, Ecoteaux/VD and Gondiswil-Zell/BE, LU.

The section at Thalgut, between Bern and Thun (Figs. 4.1, 4.5 and 4.15), is an open pit section extended by drilling, where bedrock has been reached at 147 m below the floor of the pit. The lithostratigraphy is summarized as follows (Fig. 4.15): at least four glacial advances are documented by sediments. Bedrock is directly overlain by waterlain till which grades upward into non-glacial lake sediments (=Jaberg-Seetone) containing a full interglacial pollen sequence: from lateglacial to interglacial and again to glacial vegetation character. This lake was finally directly invaded by the glacier, which deposited an ice-contact delta. This unit is followed by waterlain till which, in turn, is overlain by an early lateglacial varve-sequence. This unit is then cut by an erosional unconformity (Fig. 4.5), forming an important paleorelief, which allowed subaerial weathering and soil formation which is, unfortunately, preserved as relict only. This paleolandscape was drowned by a lake again and a voluminous delta unrelated to direct glacial input was deposited. However, the drainage system in the Aare Valley must have been changed by the presence of a glacier at that time, as the delta is a document of the final phase of a glaciation. The sandy-gravelly foreset beds stop suddenly and are conformably overlain by clayey bottomsets of interglacial character (=Thalgut Seetone, Fig. 4.15). The interglacial beds are truncated by erosion and followed by two fluvioglacial aggradational cycles, separated by a paleosoil. The upper gravel grades into the most recent till. This lithostratigraphy at Thalgut is more complete than at other sites and it is situated in a lateral "terrace position". It demonstrates that the valley has been much broader before the sequence was deposited. This observation is valid for all sections in the Alpine Foreland with substantial lithostratigraphic complexity and completeness. They are in lateral valley positions and they are in striking contrast to the simple stratigraphy of the drilling, e.g., in Lake Zurich (Hsü and Kelts 1984).

The lithostratigraphic complexity and morphostratigraphic position at Thalgut are just one aspect of relevance. The other aspect is the palynostratigraphic positioning of the two lacustrine units, the Jaberg- and Thalgut-Seetone. The

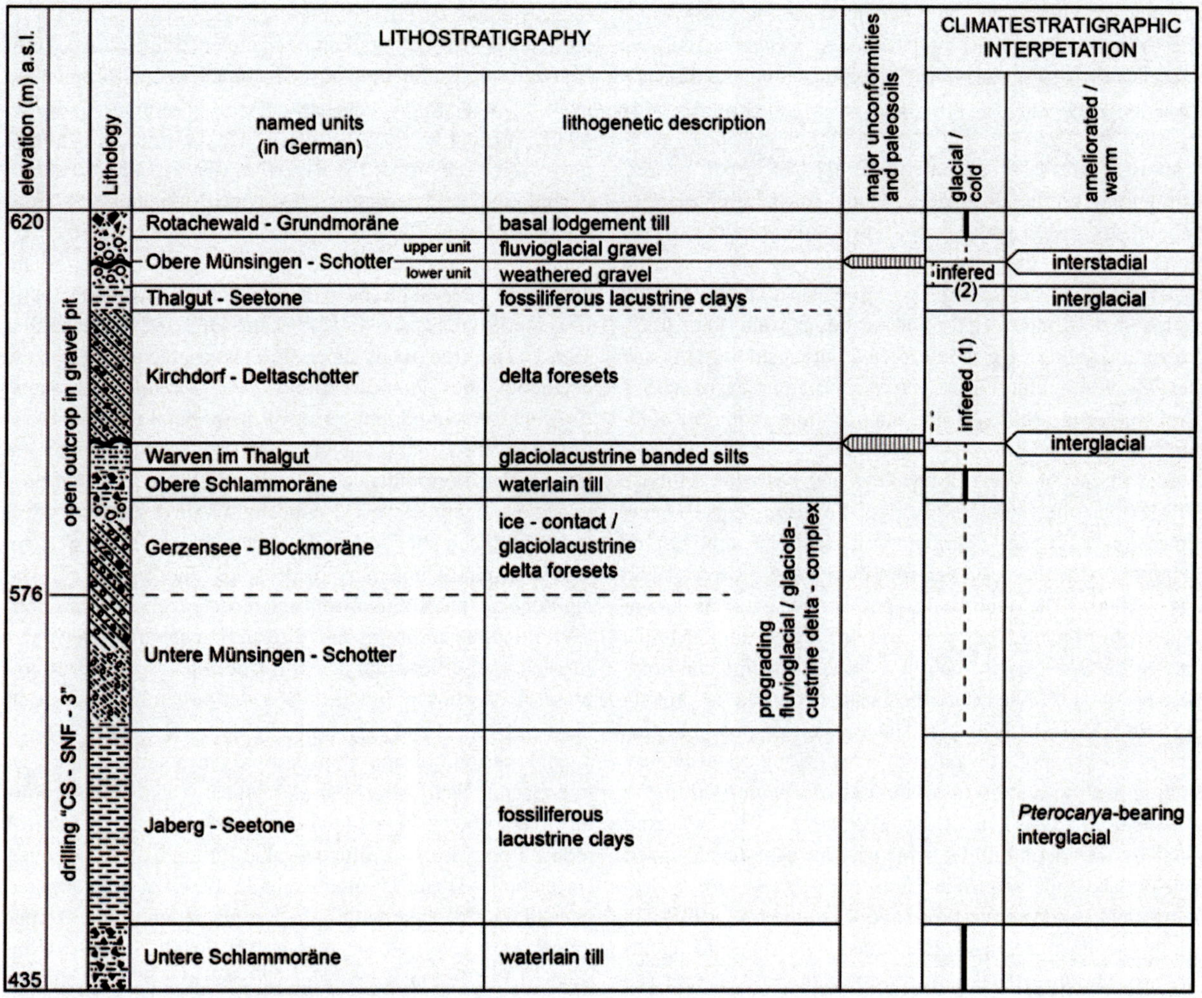

Fig. 4.15 Gravel pit at Thalgut/BE: stratigraphic summary table (adapted from Schlüchter 1989)

lower Jagerg-Seetone, above the basal waterlain till, contains an interglacial pollen sequence culminating in a *Fagus-Pterocarya* forest (Welten 1988; Drescher-Schneider et al. 2007). Comparative paleoclimatic reconstructions and palynostratigraphy allow us to conclude that this interglacial represents MIS 11 or is—more likely—even older. The more recent lacustrine bed of the sequence, the Thalgut Seetone, is characterized again by evolution from a lateglacial to an interglacial vegetation with broadleaf trees and with a strong presence of *Abies sp.* Through comparative evaluations with other interglacial sequences the conclusion was reached that the Thalgut-Seetone belongs to the early phase of the Last Interglacial and that it can be considered as equivalent of the northern European Eemian (Welten 1982). No units in between the two interglacials could be positioned palynostratigraphically; however, the important erosional unconformity in the middle of the succession (Fig. 4.5) with a preserved relict soil is interpreted as product and result of at least one interglacial time period. This interpretation needs to be seen in comparison with the reference succession at Meikirch, about 15 km further to the north of Thalgut (Fig. 4.1).

At Meikirch, several drillings were performed for groundwater prospecting and were complemented by a final scientific drilling operation to 112 m below ground surface. The lithostratigraphy is, in summary, simple: a complex of coarse glacial/fluvioglacial sediments is underlain by a uniform sequence of lacustrine, thinly bedded clayey silt grading at 110 m depth into more diamictic facies (waterlain till?). Original pollen analysis by Welten (1982, 1988) has brought evidence for three distinct vegetation zones of interglacial character, separated by two colder intervals recorded by the clayey silt. The youngest warm phase at Meikirch cannot be compared to the Eemian succession at Thalgut because of the presence of *Fagus* here. And two older warm phases at Meikirch are different from the *Fagus-*

Pterocarya Interglacial at Thalgut. The Meikirch warm phases are palynologically positioned between the two interglacials at Thalgut, namely, representing the time, at least partially, of the important erosional unconformity recorded in the latter (Fig. 4.5). Meikirch has been palynostratigraphically re-evaluated recently and tested by OSL with an interesting conclusion, ending, for the time being, a difficult discussion about direct correlation with the Nordic Holsteinian: the lithostratigraphy of the lacustrine beds at Meikirch is uniform and a sedimentary subdivision is excluded. It is interpreted as a pre-Eemian interglacial. So the combination of the re-interpreted palynostratigraphy and the OSL tests points to the interglacial complex of MIS 7 (Preusser et al. 2005), which supports the proposed comparison with Thalgut.

The timing of erosion of valleys and basins in Molasse bedrock in the northern Alpine Foreland gets a new dimension with the sedimentary fill and its stratigraphic position in the Ecoteaux Basin (Pugin and Rosetti 1992; Bezat 2000). The lithostratigraphy at Ecoteaux as reconstructed from natural outcrops and drillings is similar to the succession at Meikirch: below a fluvioglacial/glacial complex of the LGM a lacustrine basin evolution is directly overlying Molasse bedrock. The lowermost post-molassic unit is till grading upward into a waterlain till sequence, grading itself into uniform rhythmites (*Formation inférieure d'Ecoteaux* or Lower Ecoteaux Formation). They are truncated by an unconformity, which is the base of an upper, coarser lacustrine sequence (*Formation supérieure d'Ecoteaux* or Upper Ecoteaux Formation, Pugin et al. 1993). So far, this sediment sequence is unique for the Swiss Midlands for two reasons: (i) the lower formation is of reversed magnetic polarity, which makes it older than 774 ka; (ii) the palynostratigraphy of the lower formation indicates important oscillations between cooler and warmer phases. This sequence is defined as the Ecoteaux Interglacial with the presence of *Carya* and *Pterocarya* (Bezat 2000). Both arguments, paleomagnetic and palynostratigraphic, point to a Cromerian sequence. The Cromerian Complex is positioned "somewhere between 850 and 456 ka" (Litt et al. 2005 among others). However, with the available paleomagnetic data, the sequence at Ecoteaux is older than 774 ka, or pre-MIS 19 (Fig. 4.3). The stratigraphic content of the Lower and Upper Ecoteaux Formations not only broadens the view on interglacial diversity but is asking for careful considerations about vertical movements of the Swiss Midlands throughout Quaternary time.

Additional records with proven interglacial levels are not abundant in the Swiss forelands. They are reported from Grandson/VD, Thungschneit/BE, Wildhaus/SG, and Rè, close to the Swiss/Italian border (Sidler and Hantke 1993; see location in Fig. 4.1) with palynostratigraphic, however, incomplete records. In addition, a few important paleosoils (e.g., Bümberg/BE or Holziken/AG; see location in Fig. 4.1) are crucial regional stratigraphic reference levels. A significant interglacial succession is recorded in the compressed peats (*Schieferkohle*) at Gondiswil-Zell (Fig. 4.1, Küttel 1989; Wegmüller 1992), interbedded between two coarse gravel aggradations of the River Luthern and deposited in a meandering and swampy river environment. The Last Interglacial (i.e., Gondiswil Interglacial) is documented by detailed palynosequences. It is the only site in Switzerland where this interval has been directly dated radiometrically by U/Th-series at 115.7 +4.4/–4.8 ka (Küttel 1989; Wegmüller 1992). The true interglacial flora is characterized by the deciduous trees *Quercus, Ulmus, Tilia, Fraxinus,* and some *Carpinus*. An important result of the palynological analysis is the severe collapse of the vegetation at the end of the interglacial. Wegmüller (1992) observed that the Napf area must have been deforested radically within a short period; it is a well-documented example of rapid paleoclimate change.

The Alpine Quaternary records are relict by nature. The high-energy environment during repeated glacier advances to varying positions in the forelands, changing meltwater streams, and differential soil formation, leave only partially preserved sections. Units available for pollen analysis are in most cases short and may contain only the very late phase of a cold period, or the transition to an interstadial or an interglacial. Many palynostratigraphically analyzed sections turn out to be of interstadial character and therefore are of reduced or at least of difficult value for comparison or even correlation. There is one example of a more complete interstadial sequence within the last glaciation, in the former gravel and sand pit at Gossau/ZH in the Glatt Valley (Figs. 4.1, 4.16 and 4.17). This classical site is important for the multi-method dating efforts (radiocarbon, U/Th-series, and advanced luminescence; Preusser 1999) carried out there and for the paleoclimatic significance of the palynostratigraphic record. Between an important delta with fining up-section and downvalley gravelly-sandy foresets and proglacial coarse gravel overlain by basal lodgement till, a complex of compressed peat and weathered gravel was available for analysis. The lower (=main) peat is a set of two seams and is genetically developed as the final organic phase of the topset accumulation in the delta. This lower peat is dated radiometrically to 54,000 ± 3000 years BP at its base and to 33,400 ± 500 years BP at the top. The upper peat has produced radiocarbon dates of 28,500 ± 200 years BP. The beds in between the dated peat show evidence of periglacial conditions comparable to the Niederweningen mammoth finding site (Schlüchter et al. 1987). The sequence with peat and sandy-clayey silt was analyzed for its pollen content and the results produced the picture of dramatically changing vegetation pattern between 5 and 80% tree cover. Between about 30,000–60,000 years ago the altitude of Gossau corresponded to a dramatically fluctuating upper forest limit.

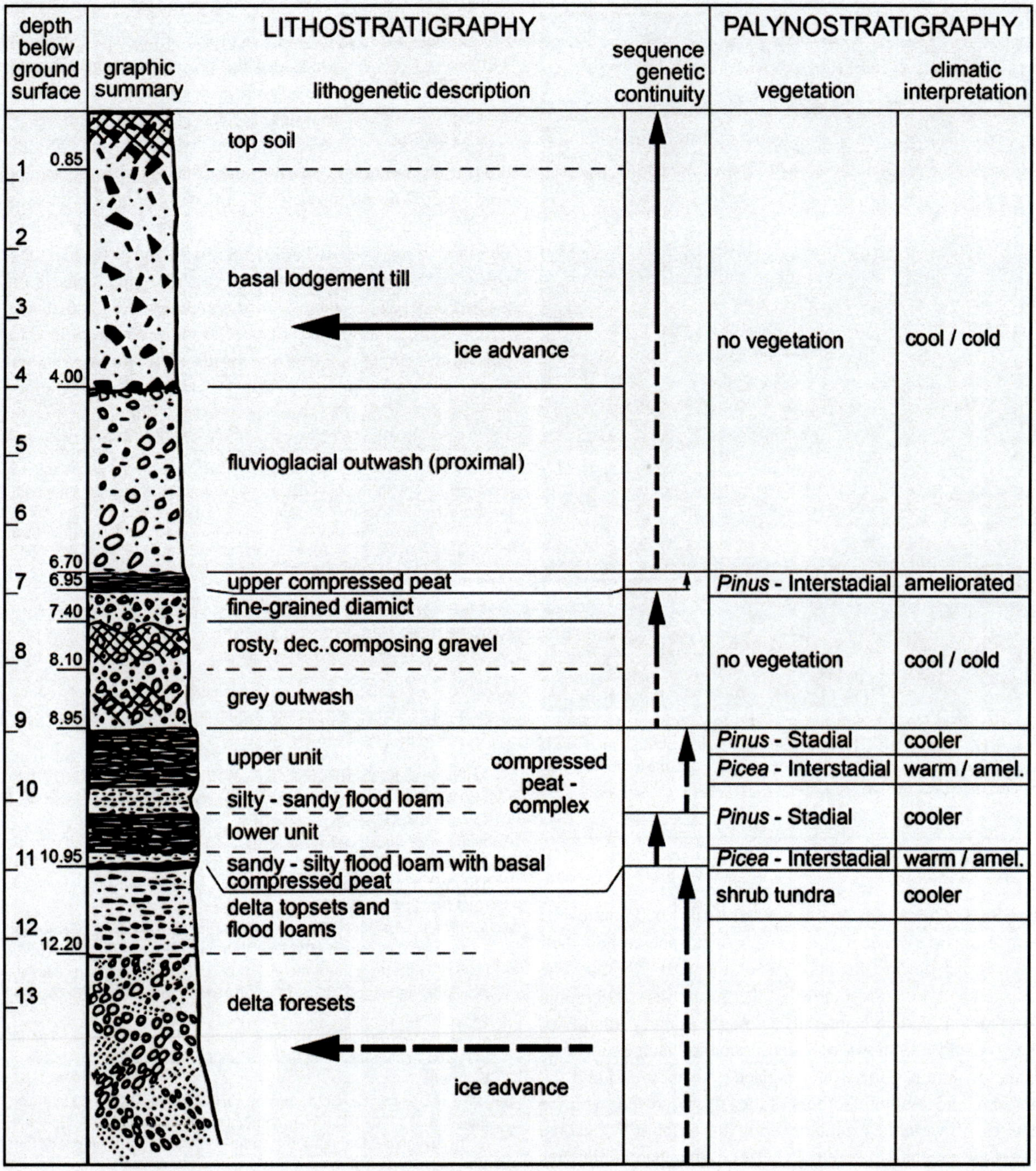

Fig. 4.16 Former gravel pit at Gossau/ZH: stratigraphic summary table (adapted from Schlüchter et al. 1987)

Compared to the oxygen isotope record of the Greenland ice cores, this is the period of the pronounced Dansgaard–Oeschger Events (Fig. 4.8b). The heart of the Gossau Interstadial Complex with the peat layers can be considered, therefore, as the continental paleoclimatic mirror-picture of the paleoclimate signal in the Greenland ice core. The severe temperature changes recorded in the Gossau sediments are also reflected in the beetle assemblages in these beds (Jost-Stauffer et al. 2005), which produce quantitative temperature references for this interval.

Fig. 4.17 Former gravel pit at Gossau: historical document of a section logging expedition with the upper part of the section in the picture (photo C. Schlüchter 1986)

4.7 Conclusions

The Swiss Quaternary record is a terrestrial document of the paleoclimatic impact on the geo- and biospheres and their interactions during the Ice Age of the past 2.6 ma. The record as it is known now is the result of stratigraphic comparisons and combinations across the country, mainly of the forelands. It shows repeated dramatic changes in landscape evolution. Fifteen major glacier advances beyond the northern and southern borders of the Alps occurred, with ice masses of variable extent covering the Alpine Forelands. The Last Glacial Maximum (LGM) is mapped more precisely than earlier events and is usually set in comparison to the Most Extensive Glaciation (MEG), which in itself, most likely, has been a two-phase event. There is a distinct difference in landscape style between the western and northeastern, more external part of the northern Alpine Foreland. The northeastern part to the north and northeast of the Lägern is characterized by the "Table Mountains of the *Deckenschotter*" (in transition to Southern Germany), whereas the more western and interior part represents a more subdued landscape. The Quaternary products rest on dissected and highly variable bedrock topography (mostly Molasse), which controls also the present-day surface morphology in some regions. Valleys deeply eroded into bedrock may contain complex Quaternary accumulations of several hundred meters in thickness and up to several hundred thousand years in age. An important issue is the north-south disharmony in the timing of bedrock incision which produced the deep valleys. In the northern Alpine Foreland it was, most likely, related to the erosional processes of the MEG in combination with the final phase(s) of the MPR/MPE(s), whereas in the southern foreland it was the Messinian lowstand of the Mediterranean Sea. The present-day topography is the result of multiple erosional and accumulative processes and represents a combination of pre-last glaciation and younger morphodynamics.

Acknowledgments All best thanks go to Olivia Kronig for her professional help with the final design of the figures. Special thanks go to the Bern Natural History Museum and to Dr. A. Rehazek for taking care of the bone of the Woolly Rhino and for the permission to publish Fig. 4.9.

References

Akçar N, Ivy-Ochs S, Kubik PW, Schlüchter C (2011) Postdepositional impacts on "Findlinge" (erratic boulders) and their implications for surface exposure dating. Swiss J Geosci 104:445–453

Akçar N, Ivy-Ochs S, Alfimov V, Claude A, Graf HR, Dehnert A, Kubik PW, Rahn M, Kuhlemann J, Schlüchter C (2014) The first major incision of the Swiss Deckenschotter landscape. Swiss J Geosci 107:337–347

Ammann B, Lotter AF, Eicher U, Gaillard MJ, Wohlfarth B, Haeberli W, Lister G, Maisch M, Niessen F, Schlüchter C (1994) The Würmian late-glacial in lowland Switzerland. J Quat Sci 9 (2):119–125

Andersen KK, Azuma N, Barnola JM, Bigler M, Biscaye P, Caillon N, Chappellaz J, Clausen HB, Dahl-Jensen D, Fischer H, Flückiger J (2004) High-resolution record of Northern Hemisphere climate extending into the last interglacial period. Nature 431(7005):147–151

Arn R (1984) Contribution à l'étude stratigraphique du Pleistocène de la région lémanique. PhD thesis, Faculty of Sciences, University of Lausanne

Bassinot FC, Labeyrie LD, Vincent E, Quidelleur X, Shackleton N, Lancelot Y (1994) The astronomical theory of climate and the age of the Brunhes-Matuyama magnetic reversal. Earth Planet Sci Lett 126(1–3):91–108

Becker P, Seguinot J, Jouvet G, Funk M (2016) Last Glacial Maximum precipitation pattern in the Alps inferred from glacier modeling. Geogr Helv 71:173–187

Becker P, Funk M, Schlüchter C, Hutter K (2017) A study of the Würm glaciation on the Valais region (Alps). Geogr Helv 72:421–442

Bezat E (2000) Exkursion A2 "Warmzeiten" 5. September 2000. In: Kelly M, Linden U, Schlüchter C (eds) Exkursionsführer DEUQUA 2000. Bern, pp 137–147

Bini A (1997) Stratigraphy, chronology and paleogeography of Quaternary deposits of the area between the Ticino and Olona rivers (Italy-Switzerland). Geol Insubrica 2(2):21–46

Blass A, Anselmetti FS, Ariztegui D (2003) 60 years of glaciolacustrine sedimentation in Steinsee (Sustenpass, Switzerland) compared with historic events and instrumental meteorological data. Eclogae Geol Helv 96(Suppl 1):59–71

Bolliger T, Feifar O, Graf HR, Kälin D (1996) Vorläufige Mitteilung über Funde von pliozänen Kleinsäugern aus den höheren Deckenschottern. Eclogae Geol Helv 89:1043–1048

Burga CA, Wynistorf E (1987) Pollenanalytische Untersuchungen. In: Schlüchter C, Maisch M, Suter J, Fitze P, Keller WA, Burga CA, Wynistorf E (eds) Das Schieferkohlen-Profil von Gossau (Kanton Zürich) und seine stratigraphische Stellung innerhalb der letzten Eiszeit. Vierteljahrsschrift der Naturforschenden Gesellschaft in Zürich 132(3):135–174

Christen G (1999) Pedogenese im solothurnischen Gäu. Diploma thesis, University of Bern

Claude A, Akçar N, Ivy-Ochs S, Christl M, Vockenhuber C, Schlüchter C (2013) Cosmogenic nuclide dating of Swiss Deckenschotter. ENSI Erfahrungs- und Forschungsbericht 2013:267–272

Claude A, Akçar N, Ivy-Ochs S, Schlunegger F, Kubik PW, Dehnert A, Kuhlemann J, Rahn R, Schlüchter C (2017) Timing of early quaternary gravel accumulation in the Swiss Alpine Foreland. Geomorphology 276:71–85

Conradin H (1991) Pedologische und sedimentpetrographische Untersuchungen im Quartär der Nordschweiz. PhD thesis, ETH Zurich

Dehnert A, Schlüchter C (2008) Sediment burial dating using terrestrial cosmogenic nuclides. Quat Sci J (Eiszeitalter und Gegenwart) 57(1–2):210–225

Dehnert A, Lowick SE, Preusser F, Anselmetti FS, Drescher-Schneider R, Graf HR, Heller F, Horstmeyer H, Kemna HA, Nowaczyk NR, Züger A (2012) Evolution of an overdeepened trough in the Northern Alpine Foreland at Niederweningen, Switzerland. Quat Sci Rev 34:127–145

Dick KA, Graf HR, Müller BU, Hartmann P, Schlüchter C (1996) Das nordalpine Wasserschloss und seine eiszeitgeologische Umgebung. Eclogae Geol Helv 89(1):635–645

Drescher-Schneider R, Jacquat C, Schoch W (2007) Palaeobotanical investigations at the mammoth site of Niederweningen (Kanton Zürich), Switzerland. Quat Int 164(165):113–129

Dürst Stucki M, Reber R, Schlunegger F (2010) Subglacial tunnel valleys in the Alpine foreland: an example from Bern, Switzerland. Swiss J Geosci 103:363–374

Finckh P (1978) Are southern Alpine lakes former Messinian canyons? Geophysical evidence for preglacial erosion in the southern Alpine lakes. Mar Geol 27:289–302

Finckh P, Kelts K, Lambert A (1984) Seismic stratigraphy and bedrock forms in perialpine lakes. Bull Geol Soc Am 95:1118–1128

Florineth D, Schlüchter C (1998) Reconstructing the Last Glacial Maximum (LGM) ice surface geometry and flow lines in the Central Swiss Alps. Eclogae Geol Helv 91:391–407

Florineth D, Schlüchter C (2000) Alpine evidence for atmospheric circulation patterns in Europe during the LGM. Quat Res 54:295–308

Gaar D (2014) Establishing chronologies of Late Pleistocene glacier advances in Northern Switzerland. PhD thesis, University of Bern, 168 pp

Gilli A, Anselmetti FS, Ariztegui D, McKenzie JA (2003) A 600-year sedimentary record of flood events from two sub-alpine lakes (Schwendiseen, Northeastern Switzerland). Eclogae Geol Helv 96 (Supplement 1):49–58

Gnägi C, Schlüchter C (2012) High-altitude erratics in the Bernese Alps (Switzerland). Swiss J Geosci 105:401–415

Goehring BM, Schaefer JM, Schlüchter C, Lifton NA, Finkel RC, Jull T, Akçar N, Alley RB (2011) The Rhone Glacier was smaller than today for most of the Holocene. Geology 39(7):679–682

Gorin G, Morend D, Pugin A (2003) Bedrock, Quaternary sediments and recent fault activity in central Lake Neuchâtel, as derived from high-resolution reflection seismics. Eclogae Geol Helv 96(Suppl 1):3–10

Graf HR (1993) Die Deckenschotter der zentralen Nordschweiz. PhD thesis, ETH Zürich

Graf HR (2009) Stratigraphie von Mittel- und Spätpleistozän in der Nordschweiz. Beiträge zur Geologischen Karte der Schweiz, N.F. 168, Bundesamt für Landestopografie swisstopo, Wabern, 198 pp

Graf A, Akçar N, Ivy-Ochs S, Strasky S, Kubik PW, Christl M, Burkhard M, Wieler R, Schlüchter C (2015) Multiple advances of Alpine glaciers into the Jura Mountains in the Northwestern Switzerland. Swiss J Geosci 108:225–238

Hadorn P, Thew N, Coope GR, Lehmdahl G, Hajdas I, Bonani G (2002) A Late-Glacial and early Holocene environment and climate history for the Neuchâtel Region (CH). In: Richard H, Vignot A (eds) Equilibres et ruptures dans les écosystèmes depuis 20 000 ans en Europe de l'Ouest, Collections Annales Littéraires, Série Environnement, Sociétés et Archéologie 3, pp 75–90

Hajdas I (2008) Radiocarbon dating and its applications in Quaternary studies. Eiszeit Gegenw 57(1–2):2–24

Hildbrand K (1990) Das Endmoränengebiet des Rhonegletschers östlich von Wangen a.A. Diploma thesis, University of Zurich, 164 pp

Hormes A, Müller BU, Schlüchter C (2001) The Alps with little ice: evidence for eight Holocene phases of reduced glacier extent in the Central Swiss Alps. The Holocene 11:255–265

Holzhauser H, Magny M, Zumbühl HJ (2005) Glacier and lake level variations in west-Central Europe over the last 3500 years. The Holocene 15:789–801

Hsü KJ, Kelts KR (1984) (eds) Quaternary geology of Lake Zurich: an interdisciplinary investigation by deep-lake-drilling. Schweizerbart'sche Verlagsbuchhandlung, Stuttgart (Contributions to Sedimentology 13), 210 pp

Ivy-Ochs S (1996) The dating of rock surfaces using in situ produced 10Be, 26Al and 36Cl, with examples from Antarctica and the Swiss Alps. PhD thesis, ETH Zurich, 196 pp

Ivy-Ochs S, Schaefer J, Kubik PW, Synal HA, Schlüchter C (2004) The timing of deglaciation on the northern Alpine foreland (Switzerland). Eclogae Geol Helv 97:47–55

Ivy-Ochs S, Kerschner H, Reuther A, Maisch M, Sailer R, Schaefer J, Kubik PW, Synal HA, Schlüchter C (2006) The timing of glacier advances in the northern European Alps based on surface exposure dating with cosmogenic ^{10}Be, ^{26}Al, ^{36}Cl, and ^{21}Ne. Geol Soc Am Spec Pap 415:43–60

Ivy-Ochs S, Kerschner H, Reuther A, Preusser F, Heine K, Maisch M, Kubik PW, Schlüchter C (2008) Chronology of the last glacial cycle in the European Alps. J Quat Sci 23(6–7):559–573

Ivy-Ochs S, Hippe K, Schlüchter C (this volume) The glacial landscape at Wangen an der Aare

Joerin U, Stocker TF, Schlüchter C (2006) Multicentury glacier fluctuations in the Swiss Alps during the Holocene. The Holocene 16:697–704

Jost-Stauffer M, Coope GR, Schlüchter C (2005) Environmental and climatic reconstructions during Marine Oxygen Isotope Stage 3 from Gossau, Swiss Midlands, based on coleopteran assemblages. Boreas 34:53–60

Keller O (this volume) The landscape of the Rhine Glacier in the Lake Constance Area

Keller O, Krayss E (2005) Der Rhein-Linth-Gletscher im letzten Hochglazial. Vierteljahresschrift der Naturforschenden Gesellschaft in Zürich 150:19–32, 69–85

Kelly MA, Buoncristiani C, Schlüchter C (2004) A reconstruction of the last glacial maximum (LGM) ice surface geometry in the western Swiss Alps and contiguous Alpine regions in Italy and France. Eclogae Geol Helv 97:57–75

Kelly MA, Ivy-Ochs S, Kubik PW, von Blanckenburg F, Schlüchter C (2006) Exposure ages of glacial erosional features in the Grimsel Pass region, central Swiss Alps. Boreas 35:634–643

Kerschner H, Ivy-Ochs S, Schlüchter C (2002) Die Moräne von Trins im Gschnitztal. Innsbrucker Geographische Studien 33:185–194

Kocher M (2012) Die Lockergesteine am Piz Starlex. Master thesis, University of Bern

Küttel M (1989) Jungpleistozän-Stratigraphie der Zentralschweiz. In: Rose J, Schlüchter C (eds) Quaternary type sections: imagination or reality?. Balkema, Rotterdam, pp 179–191

Lauber O (2003) Geologie, insbesondere Quartärgeologie, von Krauchthal und Umgebung. Diploma thesis, University of Bern

Litt T, Ellwanger D, Villinger E, Wansa S (2005) Das Quartär in der stratigraphischen Tabelle von Deutschland 2002. Newsl Stratigr 41 (1–3):385–399

Litt T, Schmincke HU, Frechen M, Schlüchter (2008) Quaternary. In: McCann T (ed) The geology of Central Europe, vol 2. Mesozoic and Cenozoic. The Geological Society, London, pp 1287–1347

Luetscher M, Boch R, Sodermann H, Spötl C, Cheng H, Edwards RL, Frisia F, Hof F, Müller W (2015) North Atlantic storm track changes during the Last Glacial Maximum recorded by Alpine speleothems. Nat Commun 6. https://doi.org/10.1038/ncomms7344

Nicolussi K, Schlüchter C (2012) The 8.2 ka event – calendar-dated glacier response in the Alps. Geology 40(9):819–822

Mailänder R, Veit H (2001) Periglacial cover beds on the Swiss Plateau: indicators of soils, climate and landscape evolution during the Late Quaternary. CATENA 45:251–272

Maisch M (1987) Zur Gletschergeschichte des alpinen Spätglazials: analyse und interpretation von Schneegrenzdaten. Geogr Helv 42:63–71

Martinetto E, Ravazzi C (1997) Plant biochronology of the Valle della Fornace succession (Varese) based on the Plio-Pleistocene record in Northern Italy. Geol Insubrica 2(2):81–98

Moscariello A (1996) Quaternary geology of the Geneva Bay (Lake Geneva, Switzerland): Sedimentary record, paleoenvironmental and paleoclimatic reconstructions since the Last Glacial Cycle. PhD thesis, University of Geneva (Terre et Environnement 4), 230 pp

Müller BU (1995) Das Walensee-/Seeztal – eine Typusregion alpiner Talgenese. PhD thesis, ETH Zurich, 219 pp

Penck A, Brückner E (1901/1909) Die Alpen im Eiszeitalter. Tauchnitz, Leipzig, 1200 pp

Preusser F (1999) Luminescence dating of fluvial sediments and overbank deposits from Gossau, Switzerland: fine grain dating. Quat Sci Rev 18:217–222

Preusser F, Drescher-Schneider R, Fiebig M, Schlüchter C (2005) Reinterpretation of the Meikirch pollen record, Swiss Alpine Foreland and implications for middle Pleistocene chronostratigraphy. J Quat Sci 20:607–620

Preusser F, Graf HR, Keller O, Krayss E, Schlüchter C (2011) Quaternary glaciation history of northern Switzerland. Eiszeitalter und Gegenwart/ Quat Sci J 60:282–305

Pugin A, Rossetti S (1992) Acquisition of land based high resolution seismic profiles in glacial basins, two case studies in the Alpine foreland of Switzerland. Eclogae Geol Helv 85:491–502

Pugin A, Bezat E, Weidmann M, Wildi W (1993) Le bassin d'Ecoteaux (Vaud, Suisse): Témoin de trois cycles glaciaires quaternaires. Eclogae Geol Helv 86(2):343–354

Reber R, Akçar N, Ivy-Ochs S, Tikhomirov D, Burkhalter R, Zahno C, Lüthold A, Kubik PW, Vockenhuber Ch, Schlüchter C (2014) Timing of retreat of the Reuss Glacier (Switzerland) at the end of the Last Glacial Maximum. Swiss J Geosci 107:293–307

Scapozza C (2015) Glaciers and permafrost evolution since the Last Glacial Maximum in the Mont Gelé-Mont Fort region (Valais Alps, Switzerland): chronology, modalities of the last deglaciation, and Schmidt hammer exposure-age dating. Quaternaire 26(2):141–173

Scapozza C, Ambrosi C (this volume) Between glaciers, rivers and lakes: the geomorphological landscapes of Ticino

Schindelwig I, Akçar N, Kubik PW, Schlüchter C (2011) Lateglacial and early Holocene dynamics of adjacent valley glaciers in the Western Swiss Alps. J Quat Sci 27:114–124

Schindler C, Fisch W, Streiff P, Ammann B, Tobolski K (1985) Vorbelastete Seeablagerungen und Schieferkohlen südlich des Walensees – Untersuchungen während des Baus der Nationalstrasse N3. Eclogae Geol Helv 86(3):643–665

Schlüchter C (1984) Die wissenschaftliche Kernbohrung im Aaretal bei Kiesen, Kt. Bern. In: Lang G (ed) Festschrift Max Welten. Cramer, Vaduz (Diss Bot 72), pp 255–273

Schlüchter C (1988) The deglaciation of the Swiss Alps: a paleoclimatic event with chronological problems. Bull de l'Association française pour l'étude du Quat 1988(2/3):141–145

Schlüchter C (1989) Thalgut. Ein umfassendes eiszeitstratigraphisches Referenzprofil im nördlichen Alpenvorland. Eclogae Geol Helv 82 (1):277–284

Schlüchter C (compil) (2009) Die Schweiz während des letzteiszeitlichen Maximums (LGM), 1:500 000. GeoKarten 500. Bundesamt für Landestopografie swisstopo, Wabern

Schlüchter C (2015) Boden als geologisches Archiv. Umwelt Aargau, Sondernummer 45:42–43

Schlüchter C (2017) La Période glaciaire en Suisse, un résumé schématique. Stiftung Landschaft und Kies, Rubigen, 4 pp

Schlüchter C, Maisch M, Suter J, Fitze P, Keller WA, Burga CA, Wynistorf E (1987) Das Schieferkohlen-Profil von Gossau (Kanton Zürich) und seine eiszeitstratigraphische Stellung innerhalb der letzten Eiszeit. Vierteljahresschrift der Naturforschenden Gesellschaft in Zürich 132:135–174

Schoeneich P (1998) Corrélation du dernier maximum glaciaire et de la déglaciation alpine avec l'enregistrement isotopique du Groenland. Quaternaire 9:203–215

Schürch N, Hajdas I, Schlüchter B, Schlüchter C, Drescher-Schneider R (2015) Fruits, beetles and the Unteraargletscher. Annual report, Laboratory of Ion Beam Physics, ETH Zurich, 96 pp

Shakun JD, Carlson AE (2010) A global perspective on Last Glacial Maximum to Holocene climate change. Quat Sci Rev 29:1801–1816

Sidler C, Hantke R (1993) La flora del Lago interglaciale Riss/Würm di Rè (Val Vigezzo). Boll Soc Tic Sci Nat 81(1):11–20

Strasky S, Morard A, Möri A (2016) Harmonising the lithostratigraphic nomenclature: Towards a uniform geological dataset of Switzerland. Swiss J Geosci 109:123–136

Uggeri A, Felber M, Bini A, Bignasca C, Heller F (1997) The Valle della Fornace succession. Geol Insubrica 2(2):69–80

Van der Meer JJM (1982) The Fribourg area, Switzerland – A study in Quaternary geology and soil development. Akad. Proefschrift, Universiteit van Amsterdam, Fysisch Geografisch em Bodemkundlich Laboratorium, 32, 203 pp

Van Husen D (1997) LGM and Late-glacial fluctuations in the eastern Alps. Quat Int 38–39:109–118

Vescovi E, Ravazzi C, Arpenti E, Finsinger W, Pini R, Valsecchi V, Wick L, Ammann B, Tinner W (2007) Interactions between climate and vegetation during the Lateglacial period as recorded by lake and mire sediment archives in Northern Italy and Southern Switzerland. Quat Sci Rev 26:1650–1669

Wegmüller S (1992) Vegetationsgeschichtliche und stratigraphische Untersuchungen an Schieferkohlen des nördlichen Alpenvorlandes.

Denkschriften der Schweizerischen Akademie der Naturwissenschaften 102:1–82
Welten M (1982) Pollenanalytische Untersuchungen im Jüngeren Quartär des nördlichen Alpenvorlandes der Schweiz. Beiträge zur Geologischen Karte der Schweiz, NF 156:1–174
Welten M (1988) Neue pollenanalytische Ergebnisse über das jüngere Quartär des nördlichen Alpenvorlandes der Schweiz (Mittel- und Jungpleistozän). Beiträge zur Geologischen Karte der Schweiz, NF 162:1–40
Wirsig C, Zasadni J, Ivy-Ochs S, Christl M, Kober F, Schlüchter C (2016) A deglaciation model of the Oberhasli, Switzerland. J Quat Sci 31(1):46–59

Christian Schlüchter is Professor Emeritus at the University of Bern. He is an active member of several professional organizations and editorial boards of international peer-review journals. He has been a member of the Executive Committee of the International Union for Quaternary Research (INQUA) and actively engaged in its commission work. The XVIII International Congress of INQUA was held in Bern in 2011 under his presidency. He is a former president of the German Quaternary Association (DEUQUA) and recipient of the Albrecht Penck Medal. He has served for many years as President of the Commission for Quaternary Research of the Swiss Academy of Sciences. His research activities in glacial geology, environmental and engineering geology have brought him to mountains on all continents.

Naki Akçar (PD Dr.) is a sedimentologist and stratigrapher by training with a strong research focus on the study of Quaternary sequences and facies analysis in the field. In 2012, he obtained the Venia Docendi title in Geology when his habilitation thesis was accepted by the University of Bern. His main research focus lies on exploring the interaction between geosphere, climate, and human societies during the Quaternary based on the study of geological archives. He is interested in contributing to understanding this interaction by producing quantitative sedimentological and chronological data based on detailed fieldwork, quantitative methods in the lab, time calibration with cosmogenic nuclide chronologies, and modeling. He has participated in research projects in different parts of the world (e.g., Alps, Scandinavia, Eastern Mediterranean, Africa, and Antarctica).

Susan Ivy-Ochs is a Professor at the Earth Sciences Department of ETH Zurich and in the Laboratory of Ion Beam Physics of ETH Zurich. She is a Quaternary geologist who studies landscape change, especially in the European Alps. She and her students use field mapping, remote imagery interpretation, and dating with cosmogenic nuclides to understand the evolution of the shape of the Alps. Her BSc, MSc, and Ph.D. degrees are all in Geology. In 2010 her habilitation was accepted at the Earth Science Department of ETH Zurich, with emphasis in Quaternary Geology and Geomorphology.

5 Geomorphological Landscapes in Switzerland

Emmanuel Reynard, Philipp Häuselmann, Pierre-Yves Jeannin, and Cristian Scapozza

Abstract

Switzerland's geomorphological diversity is the result of the combination of (i) a complex geological structure, marked by the prevalence of Alpine orogeny, (ii) Quaternary glaciations, during which Alpine glaciers invaded a large part of the Swiss Plateau and (iii) the country's climatic diversity, driven by its position in the Alps and the confluence of climates from southern, northwestern, and eastern Europe. Four main types of geomorphological landscapes are present: glacial, fluvial, karstic and gravitational.

Keywords

Geomorphology • Landscapes • Switzerland

5.1 Introduction

Despite its small size (41,285 km^2), Switzerland has a very varied geomorphological landscape. This is primarily due to its geographical location at the heart of the Alps (Fig. 5.1), meaning that the entire country has been influenced by Alpine orogeny: the Alps themselves, which account for 60% of the country's surface area, but also the Swiss Plateau (or Central Plateau), a Molasse sedimentation basin resulting from the erosion of the Alpine chain forming during the Cenozoic, and the Jura Mountains, which, from a geological point of view, are nothing less than the external, slightly deformed, part of Alpine orogeny. The presence of the Alps is also responsible for the wide variety of climates (see Fallot this volume) which depend equally on altitude and on their geographical position in relation to the Alpine barrier. In addition to these geological and climatic factors, the role of climatic variations over time, particularly the major influence of Quaternary glaciations, also has to be taken into account (see Schlüchter et al. this volume). It is thus the combination of the diversity of structural conditions (see Pfiffner this volume), climate variability and palaeoclimatic evolution that explains the great variety of geomorphological landscapes in Switzerland. In this chapter, we present the main types of geomorphological landscapes (in the sense of Reynard 2005).

5.2 Physiography and Hydrology

Switzerland has traditionally been divided into three main physiographic regions based on two criteria, altitude and relief: The Alps (60% of the country's surface area), the Swiss Plateau (30%) and the Jura Mountains (10%) (Fig. 5.1). The highest point of the country is the Dufourspitze (4,634 m a.s.l.) in the Monte Rosa Massif, at the border with Italy (Fig. 5.1), whose name was given in memory of General Guillaume Henri Dufour (1787–1875), who, between 1845 and 1865, published the first topographic map of Switzerland, called the Dufour Map, at 1:100,000 scale. The lowest altitudes are the shores of Lake Maggiore (193 m a.s.l.), also located at the border with Italy (Fig. 5.1).

E. Reynard (✉)
Institute of Geography and Sustainability and Interdisciplinary Centre for Mountain Research, University of Lausanne, Géopolis, 1015 Lausanne, Switzerland
e-mail: emmanuel.reynard@unil.ch

P. Häuselmann · P.-Y. Jeannin
Swiss Institute of Speleology and Karst Studies (SISKA), Rue de la Serre 68, 2301 La Chaux-de-Fonds, Switzerland
e-mail: praezis@speleo.ch

P.-Y. Jeannin
e-mail: pierre-yves.jeannin@isska.ch

C. Scapozza
Institute of Earth Sciences, University of Applied Sciences and Arts of Southern Switzerland (SUPSI), Campus Trevano, 6952 Canobbio, Switzerland
e-mail: cristian.scapozza@supsi.ch

E. Reynard (ed.), *Landscapes and Landforms of Switzerland*, World Geomorphological Landscapes,
https://doi.org/10.1007/978-3-030-43203-4_5

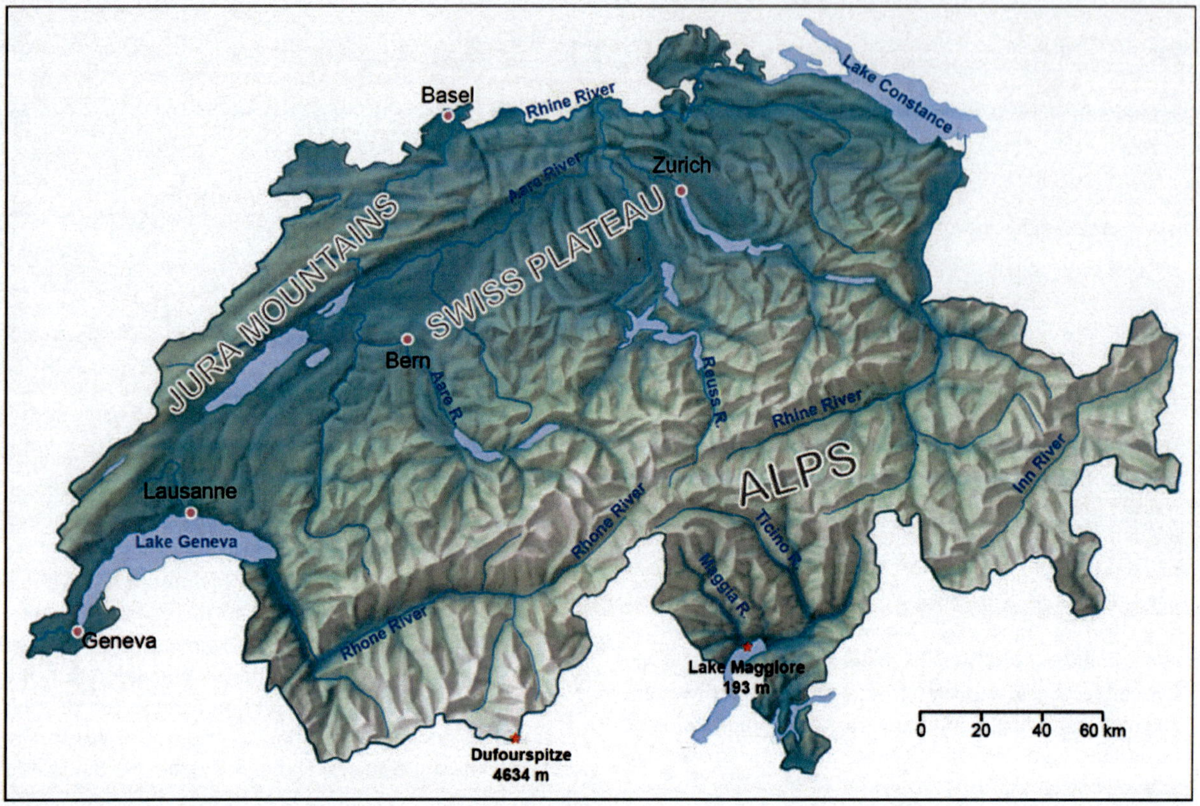

Fig. 5.1 Main physiographic regions of Switzerland (map: swisstopo)

The three physiographic regions are quite different. The average altitude of the Alps is 1,700 m and 48 mountains reach 4,000 m a.s.l. or higher. As a result of Alpine orogeny, which was more intense in the south, the chain is dissymmetric, the highest part being in the south, where the main glaciers are also concentrated. The Swiss Alps are considered as the water tower of Europe (Spreafico and Weingartner 2005; Blanc and Schädler 2014). Not only is the Gotthard area the source of some of the main rivers of Europe (Rhine, Rhone, and Ticino), but these rivers also contribute significantly to the water balance in the lowlands. For example, although the Alps account for only 15% of the total Rhine River catchment, they contribute 34% of the annual discharge at Rees, close to the confluence with the North Sea (Viviroli and Weingartner 2004). The Central Plateau extends from Lake Geneva to Lake Constance. It is sharply delimited by the Jura Mountains to the north and northwest; in the south, its border with the Alps is less clear. The mean altitude of the Central Plateau is around 580 m a.s.l., and it concentrates two-thirds of the Swiss population (around 8.5 million inhabitants). Except around Lake Geneva, most of the rivers in the Central Plateau flow northward into the Aare River or the Rhine River. The Jura is a middle mountain region with an average altitude of 700 metres and summits at less than 1,700 metres (Mont Tendre, 1,679 m a.s.l.). The topography and hydrography are highly influenced by the succession of parallel anticlines and synclines, oriented SW-NE. Two small areas do not belong to the three main physiographic regions. In the north of Switzerland, the region of Basel is part of the Rhine Graben, delimited by the Vosges and Black Forest horsts, and the southern part of Ticino with the hills belongs to the Gonfolite Lombarda group, which are part of the Po River plain.

However, this simple subdivision of the country's morphology does not enable us to understand the geomorphological diversity merely by accounting for the combined influence of geological structures, and present and past climates. This is why, after a brief presentation of the hypsometry, physiography and hydrography, we prefer to use a morphogenetic approach to describe the country's major geomorphological features. The chapter is organised in the four main morphogenetic categories, glacial, fluvial, karst, and gravitational landscapes, used in Plate 8 of the Geographical Atlas of Switzerland (Annaheim 1975; Fig. 5.2), which can be consulted on the Swiss geoportal (https://map.geo.admin.ch) or the Atlas of Switzerland (www.atlasderschweiz.ch). For a presentation of the main structural landscapes, please refer directly to the chapter written by Pfiffner (this volume).

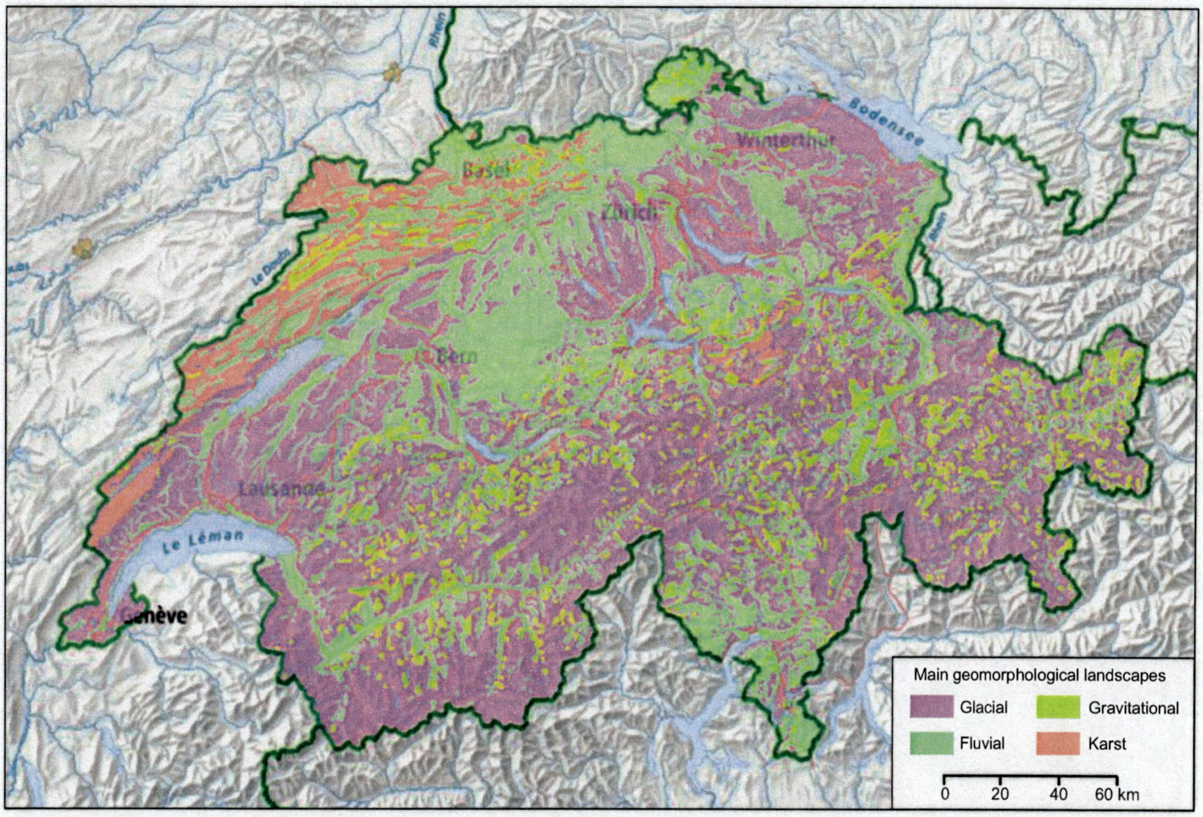

Fig. 5.2 Main geomorphological landscapes of Switzerland (modified after Annaheim 1975; simplified version available on the Swiss geoportal https://map.geo.admin.ch, Plate "Geomorphology"). Since this map was drawn in 1975, no new geomorphological mapping project has covered the whole country. This map should therefore be considered as indicative. In most cases, several processes have occurred in a given area, and the choice between one or the other category can be discussed. In particular, the karst landscapes in the Swiss Prealps are under-represented (see Fig. 5.4). Permafrost-related gravitational landscapes are also under-represented as in the 1970s, permafrost research was only beginning. We use the term "Gravitational" in preference to "Denudative", which was used in the original map

5.3 Glacial Landscapes

Given the importance of the Quaternary glaciations (Schlüchter et al. this volume), when glaciers covered almost the whole of Switzerland, glacial landforms are found almost everywhere. However, glacial landscapes dominate in two regions of Switzerland: the high Alpine valleys and the north-eastern and south-western parts of the Swiss Plateau (Fig. 5.2). Although their morphogenetic origin—the presence of cold climatic conditions—is the same, their landscapes are very different (Fig. 5.3).

Both erosion landforms of Quaternary glaciations, very well preserved in crystalline regions (such as the *roches moutonnées* in the Grimsel or Gotthard areas) and limestone regions (such as the glacial trough of Lauterbrunnen), and accumulation landforms of the Late Glacial Period, such as the multiple type-localities described by Maisch (1982) in Grisons valleys, are visible in the high Alpine valleys. In addition to the forms created by the glaciers themselves, periglacial landforms (rock glaciers, gelifluxion forms, high altitude screes) linked to the occurrence of freeze-thaw cycles and/or permafrost, are also visible in these high mountain environments (see e.g. Noetzli et al. 2019). Due to recent global warming (CH2018 2018), the upper parts of Alpine valleys are currently subject to accelerated cryospheric processes (melting glaciers, accelerating rock glaciers, increasing frequency of rockfalls and debris flows; e.g. Gruber and Haeberli 2007; Delaloye et al. 2010).

On the Swiss Plateau and in the southern part of Ticino, landforms associated with the last two glaciations dominate, in the form of either extended areas of glacial and fluvio-glacial accumulations, with alternating mainly moraine or fluvial deposits, often organised in terraces due to postglacial river incision, or landforms that were deposited (moraine crests) or shaped (drumlins) by glaciers. Readers should refer to the chapter by Schlüchter et al. (this volume) for a detailed description of the Quaternary history of these

Fig. 5.3 Examples of glacial landscapes in the Alps. **a** Current glacial processes in the Gorner Glacier area, Zermatt, 2015; **b** Erosional glacial landforms on crystalline rocks in the Gotthard Pass, 2017; **c** Quaternary glacial landscape north of Lausanne, 2018 (photos E. Reynard)

deposits, which are an important source of building materials. Their intense exploitation led to a parallel increase in the number of research projects, which in turn, substantially improved our knowledge of the conditions that determined the development of these complex landscapes, often difficult to interpret by simple observation.

5.4 Fluvial Landscapes

Rivers are widespread throughout Switzerland, except in karstic regions. River discharge is strongly influenced by water stored as snow or ice, and as a result, 16 different types of hydrological regimes have been defined (Weingartner and Aschwanden 1992; Hydrological Atlas of Switzerland, plate 5.2, https://hydrologicalatlas.ch). Alpine rivers are mainly fed by glacier or snowmelt, with high flows in summer and low flows in winter; in the Jura and on the Central Plateau, the rivers are fed by rainfall or snowmelt, depending on the altitude; finally, the valleys in the southern Alps have particular regimes caused by the presence of the Alpine barrier and by Mediterranean influences. The result is a wide variety of natural flows in Swiss rivers.

It is worth mentioning that the influence of hydropower plants has significantly modified these natural regimes. In many Alpine valleys, flows are reduced, which led the Confederation to introduce the principle of quantitative water protection (minimum residual flows) in legislation in 1991. In the plains, downstream of hydroelectric plants, flows are strongly influenced by hydropeaking phenomena resulting from hydroelectric production. Finally, most of the rivers were systematically channelled from the middle of the nineteenth century onward to protect the population against floods and to enable the development of agriculture (Vischer 2003). Since the new law on the management of watercourses came into force in 1991, public policy has attempted to reconcile the objectives of protecting the population while maintaining the hydrogeomorphological and ecological functions of watercourses by promoting their enlargement. This is the case, for example, of the major training works on the Rhone River that have been carried out since the early 2000s.

Fluvial landscapes are widespread in Switzerland (Fig. 5.2). They are particularly entrenched in the centre and the north of the Swiss Plateau, where major postglacial fluvial dissection occurred, as well as along the main Alpine valleys modelled by glaciers and then infilled by fluvial sedimentation (e.g. the Rhine, Rhone, Ticino rivers). The presence of highly carved valleys in the southern Swiss Alps is also the result of intense river erosion preceding the Quaternary (see Scapozza and Ambrosi this volume). River style depends on the slope, discharge and solid load. The rivers of the Central Plateau often form meanders, sometimes entrenched in the Molasse and which, in the Middle Ages, were used to defend cities (e.g. Bern (Fig. 5.4a) and Fribourg). Eroded fluvial landscapes are common in the Alps (Fig. 5.4b) and braided rivers, which were common in the nineteenth century, have almost all been channelled, the few exceptions being, for example, the Maggia River and the Rhone River in the Finges area; (Fig. 5.4c). Torrential

Fig. 5.4 Examples of fluvial landscapes. **a** The historical centre of Bern was in a meander of the Aare River (aerial photograph: swisstopo; https://map.geo.admin.ch); **b** Erosional river landscape in the Verzasca Valley, Ticino, 2017; **c** Braided section of the Rhone River in the Finges area, near Sierre, Valais, 2015: it is one of the few remaining braided sections of a lowland river in Switzerland (photos E. Reynard)

systems are a particular case of mountain watercourses that have often created large alluvial cones at the outlet into the main valleys. These fans have been systematically engineered, but often remain very active, particularly due to debris flow processes (e.g. Illgraben).

5.5 Karst Landscapes

Karst occupies roughly 20% of the surface area of Switzerland (Wildberger and Preiswerk 1997). Most karst areas are found in the Jura Mountains, where roughly 90% of the surface area is karstified, in the Helvetic Alps, in the Penninic Prealps and in the South Alpine Alps (Fig. 5.5). Other karst areas exist but they are much smaller than those in these four regions. Karst landscapes are linked to the presence of water-soluble rocks (limestone, gypsum, anhydrite). Water dissolves the rock at the surface and along fissures, and infiltrates into the ground; the absence of surface water is therefore typical of karst landscapes. Infiltration may be concentrated and clearly visible like in ponors (river sinkholes), or more diffuse, like in most karren fields. Blind valleys, closed depressions, large springs and caves are other typical landforms of karst regions. Karst landscapes include both surface and underground forms.

Several typical Swiss surface karst landscapes are presented in the different chapters of this book that deal with particular sites in the Jura Mountains (Fig. 5.6a) and in the Alps. One of the largest and most spectacular karst areas in Switzerland is mentioned here as an example: the Charetalp (canton Schwyz, Fig. 5.6b, c). The harsh climate and inaccessibility of this plateau mean it is less known than other karst areas. Its elevation ranges between 1,800 and 2,600 m a.s.l., annual precipitation is around 2,000 mm and the mean annual air temperature is 1.4 °C; this humid climate facilitated the development of large karren fields. The lunar landscape, stretching over several dozen square kilometres, is unique in Switzerland.

Underground landscape is worth to be mentioned as caves can be quite large. For example, a 3D-view of the *Grotte de la Crête de Vaas* (canton Valais, Fig. 5.7) shows the size of the underground voids, which may contain sediments, lakes and other features such as stalagmites and stalactites. This underground environment can thus be

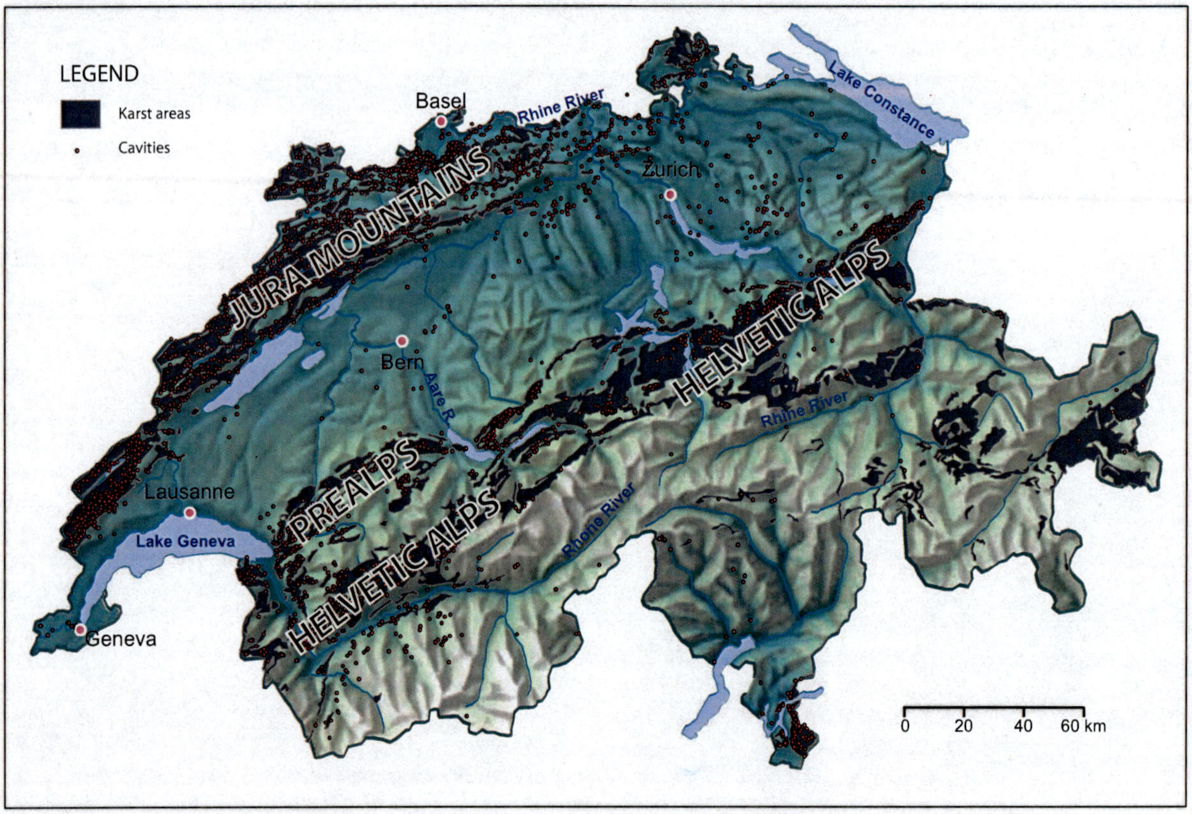

Fig. 5.5 Karst areas and cave entrances in Switzerland. Entrances can be seen in non-karst areas because artificial cavities are also shown (map: swisstopo)

Fig. 5.6 Examples of karst landscapes. **a** Typical karst landscape in the Jura Mountains: Combe des Begnines, Vaud (photo E. Reynard); **b** The karst plateau of Charetalp, Schwyz (photo R. Wenger). In the foreground, non-karstified rocks produce superficial brooks which enter the karst upon contact with limestone; **c** Another lunar view of the Charetalp karst area (photo R. Wenger)

Fig. 5.7 View of the Grotte de la Crête de Vaas, Valais, produced by a 3D-scanner. The underground landscape is clearly variegated

described by geologists and geomorphologist in exactly the same way as the world aboveground. Over geological time, the position of the karst springs may move, most often downward due to valley incisions, and form a succession of phreatic/epiphreatic conditions at different elevations in the massif. The evolution of the surface is reflected in the morphology of cave passages (Häuselmann et al. 2002). Since caves contain sediments that can be dated, the rate of valley deepening can be quantified (Häuselmann et al. 2007). Such information is often no longer available on the surface, where erosional processes have erased it. In this respect, caves are one of the only ways to quantify and assess landscape-forming processes over the last few million years. A nice example is the Siebenhengste-Hohgant cave system (Häuselmann this volume).

Despite rather humid climatic conditions (1,300–1,400 mm/year in the Jura Mountains, 1,600–1,800 mm/year in the Prealps), water is very scarce in karst regions because most water infiltrates into the ground. This fact made (and still makes) it difficult for people to settle there. In former times, rainwater had to be stored in cisterns (tanks) to be available for human (and animal) consumption. Larger settlements could only flourish where water was available all year round, i.e. mainly close to springs and perennial rivers. The town of La Chaux-de-Fonds is a good example. Originally situated at the outlet of a small perennial spring, it grew quickly with the development of the watch industry. However, both industry and households used the natural watercourse as a sewer. In the nineteenth century, most of the water had become unsuitable for consumption. Finally, water from a karst spring was captured 15 km away and 400 m lower, pumped up by hydropower stations and led to La Chaux-de-Fonds (ISSKA 2008). Two lessons can be learned from this example. First, water in karst areas may have to be sought far away. Second, karst springs may be a welcome source for drinking (and hydroelectric) purposes—as long as pollution within the catchment area is restricted.

5.6 Gravitational Landscapes

Except in the Central Plateau where the Molasse is covered with thick Quaternary deposits, gravitational landscapes (i.e. landforms resulting from alteration and erosion under the influence of gravity; e.g. permafrost creeping, sagging, landslides, rockslides, rockfalls) are widespread in all regions of Switzerland (Fig. 5.2). In Alpine valleys, landscapes created by sudden events are common, like the Randa rockslides, which happened on the 18 April and 9 May 1991, causing the closure of busy road and railway line to Zermatt and the creation of a steep debris cone (Fig. 5.8a; Sartori et al. 2003). Another example is the Goldau rockslide

Fig. 5.8 Examples of gravitational landscapes. **a** Deposits of the two rockslides in Randa (Valais) that occurred on the 18 April and 9 May 1991 representing a total of 30 million cubic metres; **b** Accumulation of debris on cones along the Flüela Pass road; **c** Tsarmine slope in the Arolla Valley (Valais) combining different gravitational processes such as rockfalls, permafrost creep, debris flows; **d** Muragl rock glacier (Graubünden) (photos E. Reynard)

(35 million cubic metres) which happened in 1806, and had made 457 victims. Even more frequent are gravitational landscapes made of recurrent accumulation of blocks on debris cones at the foot of slopes (Fig. 5.8b). On high slopes, permafrost creep forms rock glaciers (Delaloye et al. 2010) whose morphology (Fig. 5.8c, d) depends to a great extent on slope, the availability of debris and micro-climatic conditions. Slopes are very frequently prone to multiple gravitational landforms (debris flows, rock glaciers, talus screes, etc.), as shown in Fig. 5.8c. At lower elevations, because high cliffs are less frequent, gravity produces landslides and mudflows. Both may form superficially during heavy rainfall or intense snowmelt events. Large-scale landslides also permanently affect some slopes (e.g. La Frasse on the road Aigle—Les Diablerets, canton Vaud; Campo Vallemaggia, Ticino), and can have catastrophic effects, as was the case at Fälli Hölli (Fribourg), where a holiday village was completely destroyed in 1994 by the unexpected acceleration of the landslide. Flysch lithology, common in the Prealps, is prone to landslides.

Today most of the gravitational landscapes are prone to accelerated changes related to climate warming, a new challenge for geomorphologists.

References

Annaheim H (1975) (ed) Geomorphologie I. Übersicht. Plate 8, Atlas der Schweiz Eidgenossische Landestopographie, Wabern

Blanc P, Schädler B (2014) Water in Switzerland—an overview. Swiss Hydrological Commission, Bern, 27 pp

CH2018 (2018) Climate Scenarios for Switzerland, Technical Report. National Centre for Climate Services, Zurich, 271 pp

Delaloye R, Lambiel C, Gärtner-Roer I (2010) Overview of rock glacier kinematics research in the Swiss Alps. Seasonal rhythm, interannual variations and trends over several decades. Geogr Helv 65(2):135–145

Fallot JM (this volume) Climate setting in Switzerland

Gruber S, Haeberli W (2007) Permafrost in steep bedrock slopes and its temperature-related destabilization following climate change. J Geophys Res, 112:F02S18, https://doi.org/10.1029/2006jf000547

Häuselmann P (this volume) The karst system Siebenhengste-Hohgant-Schrattenfluh

Häuselmann P, Jeannin PY, Monbaron M, Lauritzen SE (2002) Reconstruction of Alpine Cenozoic paleorelief through the analysis of caves at Siebenhengste (BE, Switzerland). Geodin Acta 15:261–276

Häuselmann P, Granger DE, Lauritzen SE, Jeannin PY (2007) Abrupt glacial valley incision at 0.8 Ma dated from cave deposits in Switzerland. Geology 35(2):143–146

ISSKA (2008) De l'Areuse au Doubs – Sur les traces de l'eau de La Chaux-de-Fonds. Guides Karst ISSKA, La Chaux-de-Fonds, 46 pp

Maisch M (1982) Zur Gletscher-und Klimageschichte des alpinen Spätglazials. Geogr Helv 37:93–104

Noetzli J, Pellet C, Staub B (2019) (eds) PERMOS 2019. Permafrost in Switzerland 2014/2015 to 2017/2018. Glaciological Report Permafrost No. 16–19, Cryospheric Commission of the Swiss Academy of Sciences, Fribourg, 104 pp

Pfiffner (this volume) The geology of Switzerland

Reynard E (2005) Géomorphosites et paysage. Géomorphologie. Relief, processus, environnement 3:181–188

Sartori M, Baillifard E, Jaboyedoff M (2003) Rouiller JD (2003) Kinematics of the 1991 Randa rockslides (Valais, Switzerland). Nat Haz Earth Syst Sci 3:423–433

Scapozza C, Ambrosi C (this volume) Between glaciers, rivers and lakes: the geomorphological landscapes of Ticino

Schlüchter C, Akçar N, Ivy-Ochs S (this volume) The Quaternary Period in Switzerland

Spreafico M, Weingartner R (2005) The hydrology of Switzerland – Selected aspects and results. FOWG Reports, Water Series 7, Berne, 139 pp

Vischer DL (2003) Histoire de la protection contre les crues en Suisse. Rapports de l'OFEG, Série Eaux 5, Bienne, 208 pp

Viviroli D, Weingartner R (2004) The hydrological significance of the European Alps. Hydrological Atlas of Switzerland, Bern, Plate 6:4

Weingartner R, Aschwanden H (1992) Discharge regime – the basis for the estimation of average flows. Hydrological Atlas of Switzerland, Plate 5.2, Bern, https://hydrologicalatlas.ch

Wildberger A, Preiswerk C (1997) Karst and Caves of Switzerland. SpeleoProjects, Allschwil, 208 pp

Emmanuel Reynard is Professor of Physical Geography and Director of the Interdisciplinary Centre for Mountain Research at the University of Lausanne, and a member of the International Association of Geomorphologists (IAG) Executive Committee. He has been chairman of the IAG Working Group on Geomorphosites (2001–2013) and president of the Working Group on Geotopes of the Swiss Academy of Sciences (2006–2017). His research concerns mountain geomorphology, geomorphological heritage, landscape geohistorical analyses and water management in mountains. He has worked in the Alps, the Romanian Carpathians and in the Maghreb (Tunisia, Morocco).

Philipp Häuselmann is Scientific Collaborator at the Swiss Institute for Speleology and Karst studies SISKA and Lecturer at University of Life Sciences, Vienna. He is chairman of the International Union of Speleology Working Group Topography and Mapping, and an active caver. His main scientific interest is in speleogenesis and dating cave deposits, both by U/Th and by cosmogenic nuclides. His main focus is research on caves in mountain belts.

Pierre-Yves Jeannin is leading the Swiss Institute for Speleology and Karst Studies (SISKA) he created in 2000 in La Chaux-de-Fonds. He is an External Professor of the University of Neuchâtel, and member of several national and international commissions. His main domain of research is karst hydrogeology, and related topics such as karst deposits, karst genesis and cave climate. Landscape is a major boundary condition for groundwater flow, and is thus an important aspect of the research at SISKA, especially concerning karst genesis.

Cristian Scapozza is Senior Researcher and Lecturer in Physical Geography and Geomorphology at the University of Applied Sciences and Arts of Southern Switzerland (SUPSI), where he leads the Geology Research Group, and Director of the Valle di Blenio Museum. He is president of the Swiss Geomorphological Society (SGmS) and member of the Scientific Committee of the Swiss Permafrost Monitoring Network PERMOS. His main scientific interests are in the field of geomorphological evolution of Alpine valleys, with a particular focus on Quaternary geological mapping, glacial and periglacial processes, hillslope processes and dynamics, natural and historical evolution of Alpine floodplains, and geoheritage promotion.

Part II

Landscapes and Landforms

The Geomorphological Landscapes in the Geneva Basin

6

Andrea Moscariello

Abstract

The landscape of the Geneva Basin, located in the southwesternmost part of the Swiss Plateau between the northern end of the Prealps and the Jura, results from the long-term, continuous tectonic deformation of the Alpine Foreland and surface processes reflecting alternation of glacial and interglacial periods. The main geomorphological elements of the Geneva Basin can be thus divided into (i) tectonically controlled features, associated with the north-westernmost compressive Alpine tectonics represented by SW-NE oriented ridges interrupted by NW-SE lineaments; (ii) glacially-related erosional and depositional features; and (iii) postglacial establishment of the present-day hydrographic network. Responsible land-use planning over the last centuries resulted in a region internationally known for a balance between urban and industry expansion, agricultural activities and nature preservation areas, which is ensuring both economic and landscape attractiveness of the Geneva Basin.

Keywords

Morphotectonics • Glacial erosion • Postglacial • Frontal moraines • Fluvial incision • Geneva

6.1 Introduction

The city of Geneva (46°12′N, 6°09′E) is the second largest city in Switzerland and the capital of Geneva Canton. It is situated at the southwestern termination of Lake Geneva (or Lake Leman), here known as the *Petit Lac* (small lake) (372 m a.s.l.), where the Rhone River exits Lake Geneva and meets the Arve River generated from the Mont Blanc massif in France (Figs. 6.1 and 6.2).

Situated at the southwestern termination of the Swiss Plateau set between the Alps and the Jura, the city of Geneva and its surroundings have been for centuries appreciated for their scenery and location at the heart of Europe, which made this region a crossroad of culture and trading since the first human occupation of the area around 11,000 years ago.

This chapter describes the key geomorphological features of the city of Geneva and its basin by highlighting the importance of the interplay between a long, continuous active tectonic history associated with the Alpine orogenesis that occurred especially over the last 30 million years, and the profound landscape changes associated with shorter and higher frequency glacial and interglacial cycles over approximately the last million years.

6.2 Geographical and Geological Setting

The Geneva Basin is located in the southwesternmost termination of the Swiss Plateau and constitutes a prominent geomorphological element of western Alpine Foreland region. The Geneva Basin itself, as defined in this work, comprises a central part with an area of ca. 600 km^2 characterized by a generally undulating morphology with low relief, consisting of hills, terraces and slopes with gently dipping surfaces and the surrounding higher ground formed by Mt Salève (a 15 km long ridge culminating at 1379 m a.s.l. at the Grand Piton oriented NE-SW, plunging below the Arve Valley bottom to the NE (Petit Salève, Fig. 6.2) and gradually merging with Mont de Sion hills to the southwest), the Folded Jura chain (culminating in the area with the Crêt de la Neige at 1720 m a.s.l.) and Mt Vuache (a NW-SE oriented, 14 km long and 1.5 km wide ridge culminating at 1105 m a.s.l.; Fig. 6.1).

The climate of Geneva is temperate with cool winters, usually with light frosts at night and thawing conditions

A. Moscariello (✉)
Department of Earth Sciences, University of Geneva, Rue des Maraîchers 13, 1205 Geneva, Switzerland
e-mail: andrea.moscariello@unige.ch

E. Reynard (ed.), *Landscapes and Landforms of Switzerland*, World Geomorphological Landscapes,
https://doi.org/10.1007/978-3-030-43203-4_6

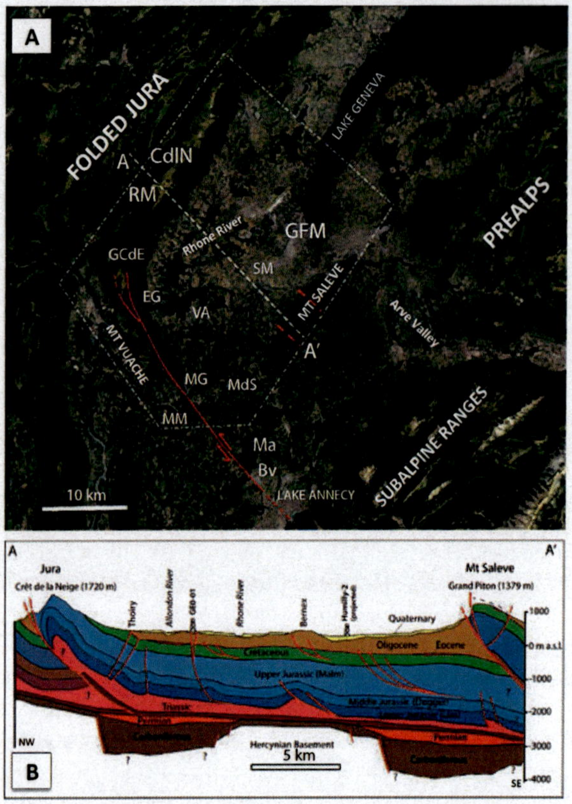

Fig. 6.1 **a** Satellite photo of the western part of the Swiss Plateau and the Geneva Basin. Location of places described in the text is indicated as follows: SM: Soral moraine; GFM: Geneva frontal moraine; EG: Ecluse gateway; MG: Malpas gateway; MM: Mt Musiège; MdS: Mont de Sion; GCdE: Grand Crêt d'Eau Mt; RM: Reculet Mt; CdlN: Crêt de la Neige; Bv: Bromines village; Ma: Mt Mandellaz; VA: Valleiry; A-A': location of geological cross section shown on Fig. 6.1b. B. Simplified geological cross section across the Geneva Basin based on outcrop, borehole, and seismic data (Moscariello 2019). Elevation are in meters a.s.l.

during the day, and warm summers. Precipitation is relatively well distributed throughout the year, with the highest rainfall in autumn and annual rainfall reaching 1005 mm (1981–2010 period; MeteoSwiss data). The mean yearly temperature is 10.5 °C (1981–2010 period; MeteoSwiss data).

Geologically, in the study area the Swiss Plateau consists of an Oligocene sedimentary succession of mixed fluvial and lacustrine deposits of variable thickness, unconformably overlying the eroded Lower Cretaceous limestone layers ("Urgonian" facies, Barremian age). Here, this unit, known as the Molasse Formation (Oligocene age), onlaps to NW and SW against the reliefs of the Jura and Vuache, while to the SE, it is overthrusted by the Salève fold (Fig. 6.1b). This indicates that the morphological highs were already existing at the time of deposition of the Molasse as inferred from the youngest rocks underlying the Molasse continental sediments, of presumed Eocene age (*"Sidérolitique"*) resting on Lower Cretaceous rocks (Charollais et al. 1998; Mastrangelo et al. 2013). The same elevations evolved considerably after the deposition of the Molasse and the underlying Eocene sediments until reaching approximately the present-day configuration.

The deformation in the vicinity of Geneva results from compressional tectonics, which generated NW-verging thrust planes (Mt Salève) and NW verging anticlinal folds (Folded Jura and Mt Salève—Musiège ridge; Fig. 6.3). In addition to these features, important SE-NW striking, sinistral strike-slip faults are well documented in outcrops (Mt Vuache and Mt Musiège) and in the subsurface through the study of several two-dimensional reflection seismic lines calibrated with boreholes (Clerc et al. 2015). These strike-slip features at the surface develop typical flower structures as the Grand Crêt d'Eau (Charollais et al. 2013; Fig. 6.4).

It is in this active tectonic setting and morphostructural context that the large Pleistocene glaciers occupied the region on several occasions throughout the last million years. In the Lake Geneva Basin, the evidence of an oldest Pleistocene glaciation likely associated with an old Rhone Glacier has been found in the Canton Vaud near Palézieux at an altitude of 800 m a.s.l. (Pugin et al. 1993). Here, glaciolacustrine delta deposits record a time older than the last inversion of the terrestrial magnetic field some 860,000 years ago, making these sediments likely related to the so-called "Günz Glaciation" (Pugin et al. 1993). Up to five younger glacial advances during the Middle and Late Pleistocene time are documented in the Aubonne valley, 40 km east of Geneva (Arn 1984) and in time-equivalent deposits on the southern shore of Lake Geneva descending the Dranse Valley in France (Brun 2000). In the Geneva area, the large glacier mass occupying the region was formed by the junction of the main Swiss Plateau Rhone Glacier, the Arve Glacier, generated from the Mont Blanc massif and tributary valleys, and the Jura ice cap (Coutterand 2010). Vestiges of ancient glaciations of undefined age are represented by deeply buried long, narrow and deep incisions (Fig. 6.5) cut into the Molasse bedrock (Amberger 1978; Wegmüller et al. 1995; Moscariello et al. 1998; Fiore 2007; Fiore et al. 2010). The oldest known glacial deposits have been attributed to the "Riss Glaciation" on the basis of stratigraphic evidence. They underlie in fact a thick package of well-sorted and often-cemented coarse sandy gravels known as the "*Alluvion ancienne*", thought to represent a large braided fluvial system developed during the last interglacial (Eemian). This in turn is overlain by the youngest glacial deposits, which refer to the last glacial occupation of the Geneva Basin, associated with the Last Glacial Maximum (LGM), known as "Würm Glaciation" in the Alps (Fig. 6.6). Most of the large erratic

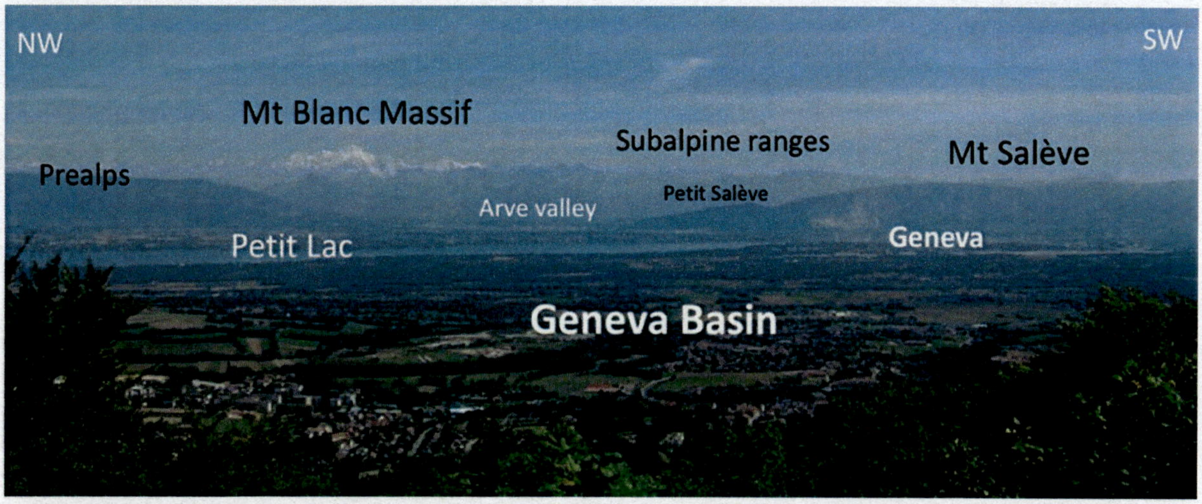

Fig. 6.2 Overview of the flat, glacial valley of the Geneva Basin with the termination of the Lake Geneva *(Petit Lac)* and Mt Salève at its SE boundary. In the far background the Mt Blanc massif where the Arve River is generated (photo B. Le Clerc)

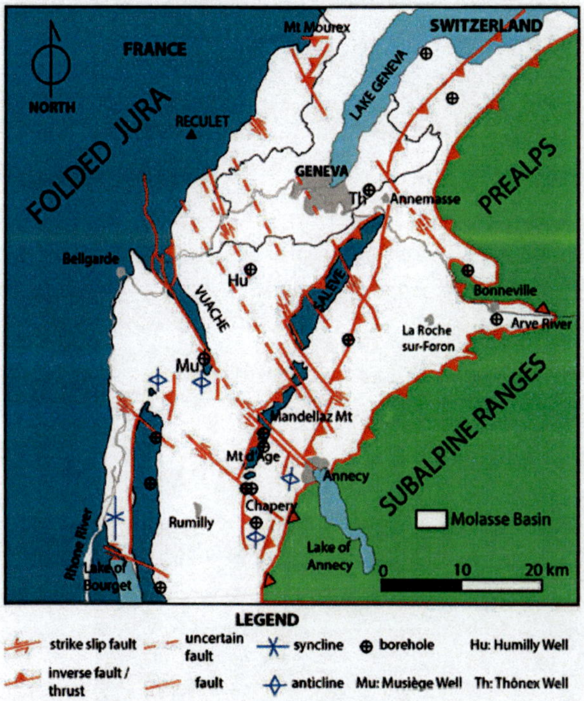

Fig. 6.3 Tectonic framework of the Greater Geneva area inferred from surface geology and subsurface mapping using 2D seismic calibrated to boreholes (from Paolacci 2012 and Clerc et al. 2015)

blocks found in the Geneva Basin both near the lake, such as the famous "Pierres du Niton" (Sesiano et al. 2011), and at high elevations in the surroundings (e.g., top of Salève at 1200 m a.s.l.) and glacial landforms such as moraines ridges and drumlins are genetically associated with this last glacial advance (Fig. 6.7).

According to various reconstructions (see Moscariello et al. 1998 and Coutterand 2010 for references), during the LGM in the Geneva area (Fig. 6.6) the junction between the surfaces of Alpine ice and the Jura would have been at an altitude of about 1200 m a.s.l.; the prealpine glaciers joined the Rhone Glacier from the south between 1250 m and 1400 m a.s.l. (Blavoux 1988), while the Arve Glacier joined the Rhone Glacier at about 1200 m a.s.l. on Mt Salève (Monjuvent and Nicoud 1988).

6.3 Landforms and Landscapes

6.3.1 Tectonics and Landscape

The compressive Alpine orogenesis strongly determined the present-day prominent geomorphological lineaments of the Geneva Basin which evolved during approximately the last 50 million years. The Jura Mountain chain, elongated in a SW-NE direction, forms a uniform monocline toward the Geneva Basin (Fig. 6.1), and represents the innermost folded anticline of the Folded Jura, an extensive complex of north-west verging thrusted anticlines, extending for several tens of kilometers to the NW (Fig. 6.1).

To the SW, the Geneva Basin is bounded by Mt Salève, which is made of SE dipping strata of mainly carbonates of Upper Jurassic–Lower Cretaceous age onlapped by Tertiary Molasse beds on its southeastern flank. The sub-horizontal strata mostly control the relief profile. They plunge to the NE in correspondence of the relief termination, (Fig. 6.8) whereas they are vertical in correspondence of the NW-facing sharp and vertical cliff forming ca. 1000 m of relief from the Lake Geneva level. Mt Salève is a textbook

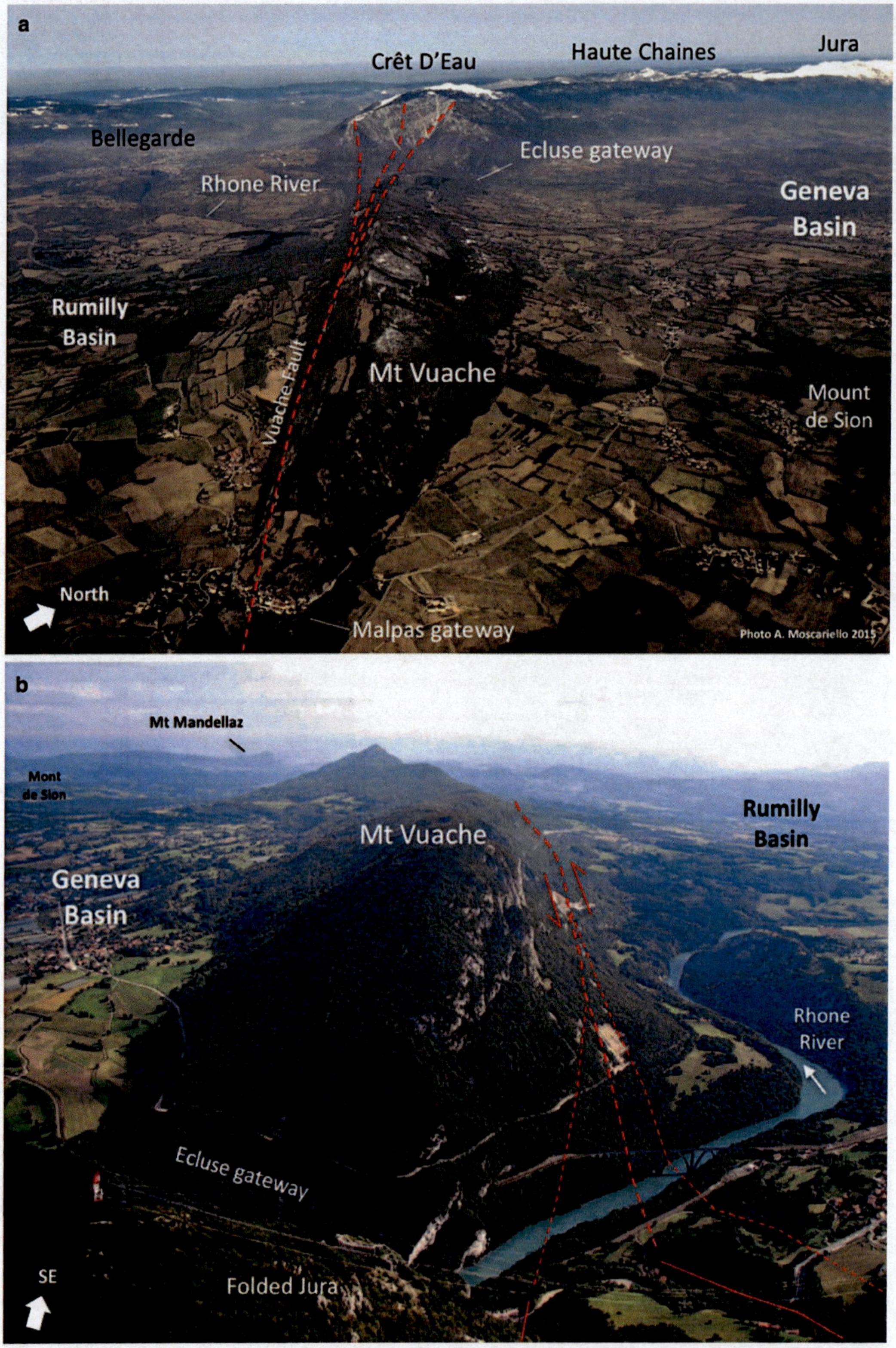

Fig. 6.4 Aerial views of Mt Vuache defining the southwestern end of the Geneva Basin. **a** Aerial photo taken from southeast showing the entire NW-SE extension of Mt Vuache ridge, a morphostructural expression of a sinistral transpressive fault system (photo A. Moscariello 2015); **b** View toward the SE, taken from Mt Grand Crêt d'Eau (see Fig. 6.1a for location). The entire Upper Jurassic and Lower Cretaceous section steeply dipping to the NE can be observed on the northern flank of Mt Vuache (modified after Charollais et al. 2013; photo courtesy of D. Ernst)

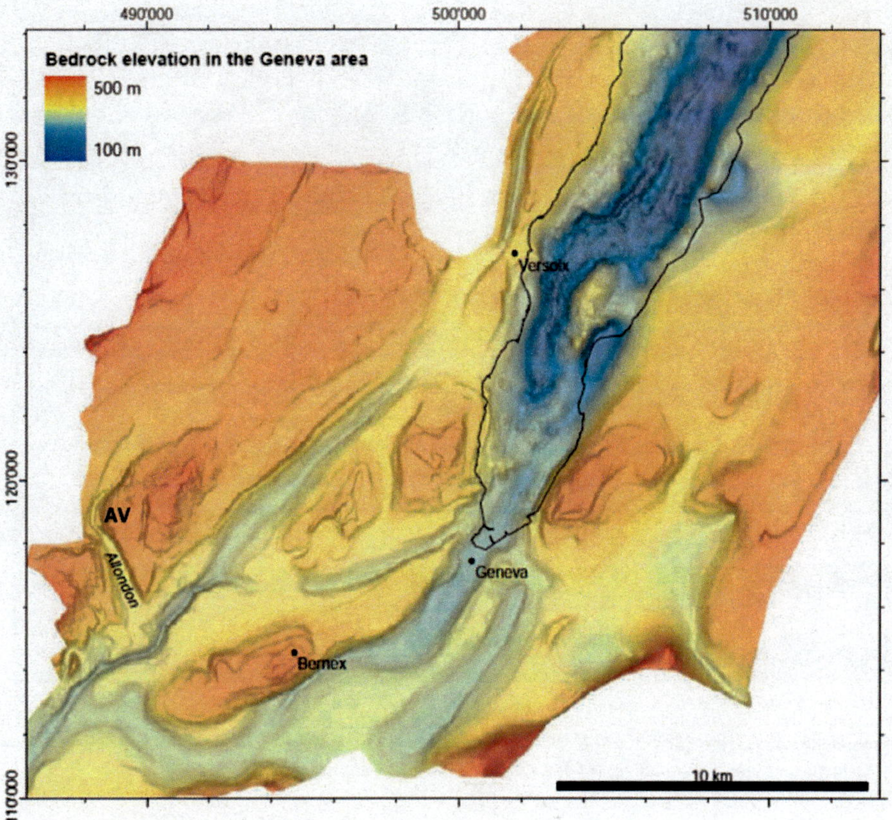

Fig. 6.5 Map of bedrock elevation in the Geneva area (elevation-dependent color scale blended with slope). Black line indicates the present-day Lake Geneva shoreline. Note the valley anastomosing pattern and undulating longitudinal profile. The only exception is the Allondon Valley (AV): its longitudinal profile presents a regularly dipping slope (0.7%) typical of a subaerial stream (modified after Fiore 2007)

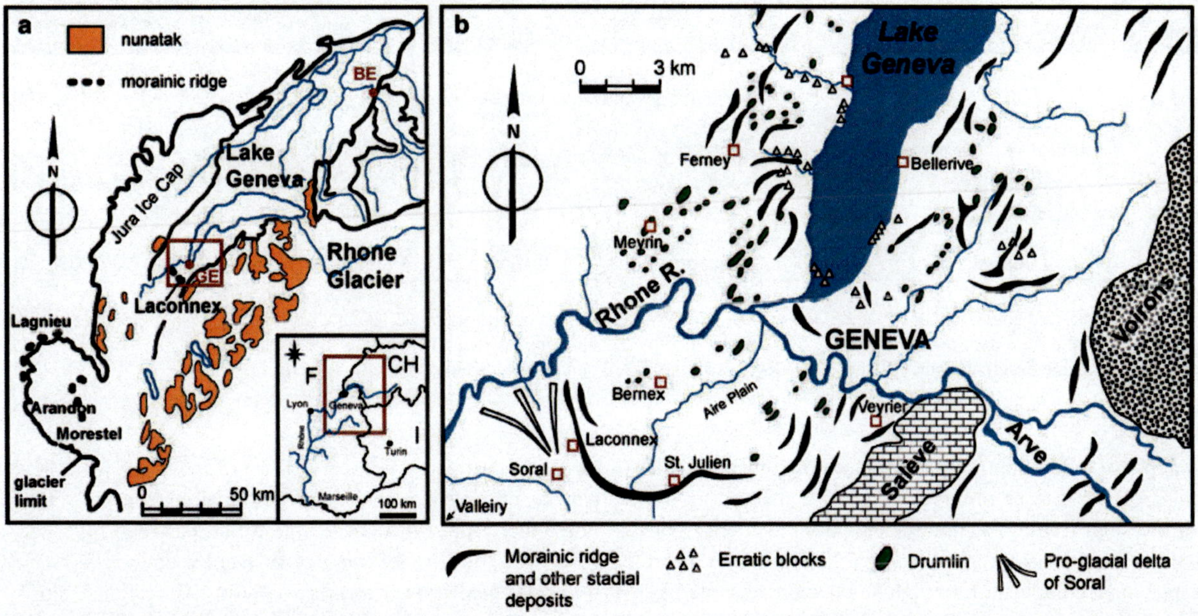

Fig. 6.6 **a** Extension of the Rhone Glacier and Jura ice cap during the Last Glacial Maximum. **b** Summary of main glacial features within the Geneva Basin (modified from Moscariello et al. 1998). GE: Geneva; BE: Bern

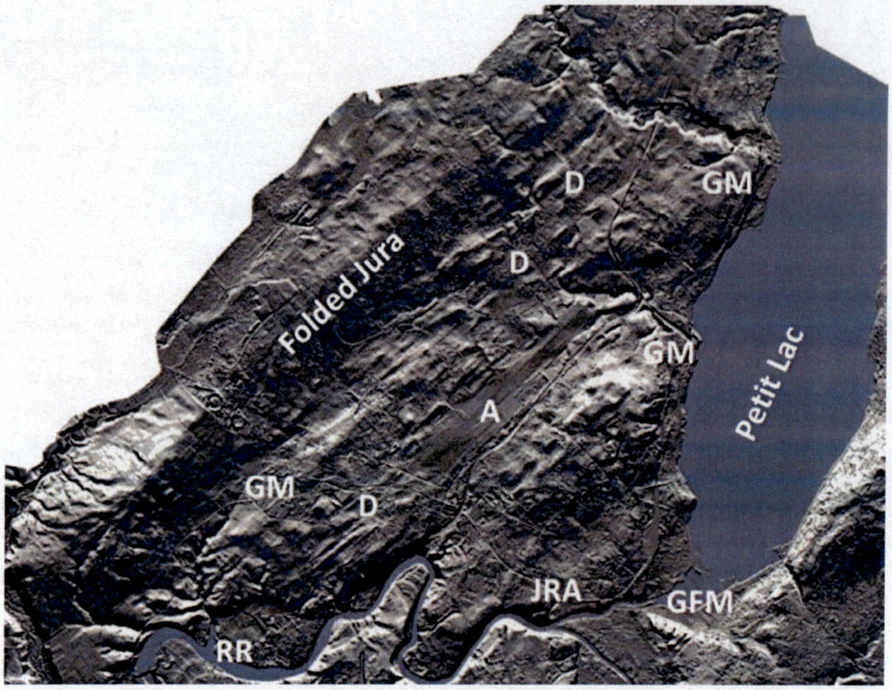

Fig. 6.7 Digital elevation model of the northern side of Lake Geneva and Rhone River in the Geneva area (modified from Fiore 2007 and Wildi et al., 2017). A: large man-made morphology of the Geneva International Airport; GM: glacial morphology such as polished outcrops, grooves, *roches moutonnées*, etc.; D: drumlins; RR: Rhone River; JRA: Junction between the Rhone and Arve rivers; GFM: Geneva front moraine

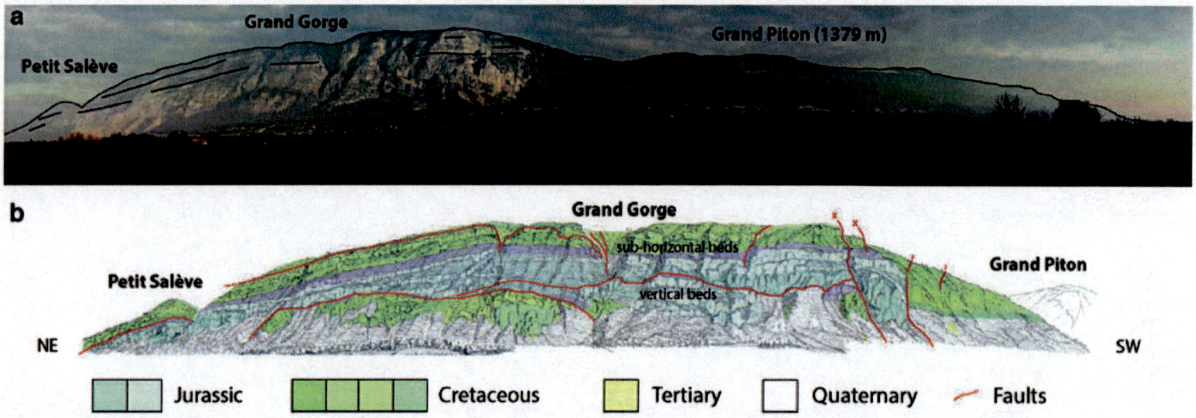

Fig. 6.8 Mt Salève defining the SE border of Geneva Basin. **a** NW-SE 18 km long profile from the Petit Salève to the Grand Piton (photo F. Mondino); **b** geological panorama of the Mt Salève vertical cliff, modified from Lombard (1965), showing the upper part formed by sub-horizontal stratified Cretaceous and Jurassic limestones gently plunging to the NW and the lower part formed by the same rocks plunging vertically, parallel to the cliff orientation

example of outcropping thrust fault-scarp along which several segments can be observed (Figs. 6.1, 6.3 and 6.8): the northwestern fault tip represented by the Petit Salève plunging to the NW; a faulted drag fold exposed in the cliff displaying a combination of high and low angle strata likely generated by a high-angle inverse fault where the maximum vertical offset and relief (ca. 550 m) of the entire Mt Salève can be observed; a strike-slip deformation zone offsetting horizontally the Mt Salève ridge by at least 600 m; a segment with lower relief (ranging in average between 200 and 400 m), characterized at present by a NW irregular slope. The latter segment has been investigated by deep reflection

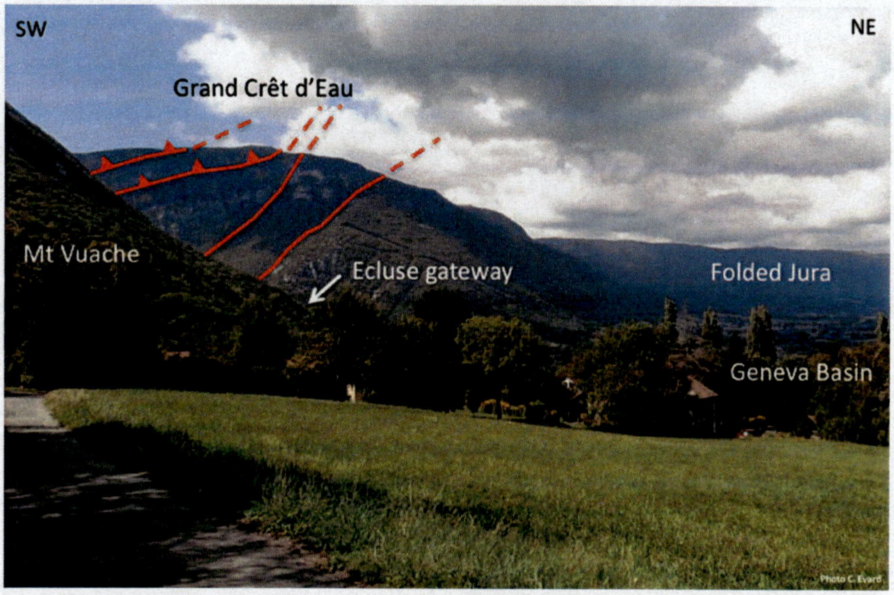

Fig. 6.9 View of the northeastern flank of Mt Vuache and its connection to the Folded Jura chain through a complex set of strike-slip and inverse faults overall forming a flower structure which culminates with the high relief of the Grand Crêt d'Eau Mount (see Charollais et al. 2013 for details) (photo C. Evard)

seismic (Gorin et al. 1993; Moscariello 2016), which indicates the presence of a low angle thrust fault at the base of Mt Salève (Fig. 6.1b). The south-eastern flank of the Mt Salève is affected by faults, kinks, and folds oriented parallel to the axis of the chain (Mastrangelo et al. 2013).

In the Geneva Basin, vertical strike-slip faults, such as the one mentioned above, have been also recognized in other areas, cutting through the Jura relief and in the subsurface. One of these structural elements, with compressional strike-slip sinistral movement (i.e., transpressive) is the Vuache fault associated with the homonymous relief whose activity caused the formation of a NW-SE striking, narrow faulted anticlinal ridge plunging to the SE and connecting to the Jura (Fig. 6.4).

The Vuache relief is a morphostructural element representing the abrupt termination of the Western Swiss Plateau against the French Subalpine range and the Folded Jura structural complex. This is a crucial location for the Geneva area as this natural barrier has influenced the geological history of the Geneva Basin since Tertiary times (e.g., Molasse deposition and connection between the Mediterranean and Pannonian basin), during the Pleistocene, forming a natural dam to the major ice streams descending the Rhone valley. Since human occupation of this region, the Vuache has represented a geographical divide between the "Genevois" region and France to the west. The northern and southern termination of Mt Vuache (Ecluse and Malpas gateways, respectively, Figs. 6.1 and 6.9) consist in fact of narrow and deep incisions formed in postglacial times by fluvial network downcutting (Fig. 6.5), which represented important strategic gateways for merchandise trading and people passage, whose control often caused wars and deaths.

These deep incisions and clear morphotectonic surficial expression of the Vuache fault are most likely associated with the strong repeated seismic activity (Sambeth and Pavoni 1988; Baize et al. 2011). The most recent earthquake associated with the Vuache fault reactivation was recorded in 1996 with local magnitude (M_L) of 5.3, and was followed by an important increase in warm water springs discharge (Bromines village, Fig. 6.1; Thouvenot et al. 1998), suggesting together with geophysical evidence a possible connection with deeply rooted basement fault (Baize et al. 2011).

Deep-seated faults, reaching the basement and probably related to older lineaments reactivated during the Alpine orogeny (Gorin et al. 1993; Signer and Gorin 1995) with transpressive movement, led to the formation of the above-described flower structures and gentle anticlines in the subsurface. One of these structures, the Humilly anticline, located in the SW of the Geneva Basin on the northern foothills of Mt Salève, was drilled by oil exploration wells (Humilly-2; Fig. 6.3) (Paolacci 2012; Moscariello 2016). The northern continuation of the Humilly structure is topographically expressed in the Folded Jura by the existence of the highest tops in the Geneva region, called the *"Haute Chaîne"* (e.g., the Reculet Mount; Fig. 6.1). These sinistral transpressive faults offset Mt Salève range and the subalpine Molasse frontal thrust.

The expression of deep-seated fault structures with normal component has also been suggested to explain some of the peculiar characteristics of the Rhone River valley immediately to the eastern side of Mt Vuache (Gorin et al. 1993). Here, the large triangular-shaped, flat surface where the Rhone River expands its extensive floodplain inset in

Fig. 6.10 The Rhone River at the termination of the Swiss Plateau upstream of the Ecluse gateway, developing a wide floodplain and wetland known as the Etournel marshes (photo Foxalpha Forums)

older terraced Quaternary deposits (Fig. 6.10) is bounded in the subsurface by a couple of short, deeply rooted vertical faults, which are associated with a basement low in this area (Gorin et al. 1993). The large floodplain and low gradient river flow, just upstream of the narrow gorge of the Ecluse gateway are interpreted as the surface expression of an area with higher subsidence rate, associated with an extensional basin.

The Allondon Valley (see Bätz et al., this volume) has a longitudinal profile with a regularly dipping slope (0.7%) typical of a subaerial stream. However, the longest continuous segments of the flanks of the valley are straight and parallel. They follow the same NW-SE direction as the main tectonic lineaments of the Vuache fault and those observed in the subsurface (Fig. 6.3). No direct link between these morphological features and the tectonics has been proved yet but this striking similarity points to the importance of morphostructural elements in the Geneva geomorphological landscape (Fig. 6.11).

6.3.2 Glacial Landscape

Glacial erosion and deposition predominately associated with glacial occupation of the region during the LGM have been the prominent landscape shaping processes in the central part of the Geneva Basin.

The largest geomorphological feature of the Geneva Basin is the southwestern termination of the *Petit Lac*. This narrow and elongated feature represents the remains of a subglacial tunnel valley (Moscariello 1996; Moscariello et al. 1998; Fiore 2007; van der Vegt et al. 2012) formed underneath the Rhone Glacier by subglacial meltwater flowing downstream toward the SW. Borehole data indicate that the Rhone Glacier during the LGM was responsible for incising this large groove and removing older Quaternary sediments reaching the Tertiary Molasse bedrock. This incision continues beyond the present-day lake margin to the SW as indicated by borehole and geophysical data (Figs. 6.5 and 12). In the subsurface, this incision terminates a few kilometers before Mt Vuache where the glacier was dammed and deviated toward the SE to pass over the Mont de Sion hill area (Fig. 6.13) and continue its course toward Lyon. The surface expression of this buried incision can be inferred by the flat and narrow Aire valley, a small tributary of the Arve River (Fig. 6.12). This is located to the SE of a bold elongated Tertiary Molasse bedrock ridge (Bernex Ridge) and follows a NE-SW direction possibly continuing through a large gap in the northern slope of the Mont de Sion hills (Figs. 6.1 and 6.13). The latter consist of an irregular thickness of glacial deposits accumulated in subglacial and subaqueous environment overlying the Tertiary Molasse bedrock.

The Rhone Glacier tunnel valley was thus left unfilled in the present-day lake portion and subsequently filled by a thick succession of proglacial lake deposits (Moscariello 1996), accumulated in the large lake formed at the front of the melting LGM glacier. On the other hand, the downstream segment, today deeply buried, was filled with a mixed succession of subglacial, lacustrine, and proglacial deposits generated also by the Arve Glacier. Wider and low relief ridges, elongated perpendicular to the main valley axis

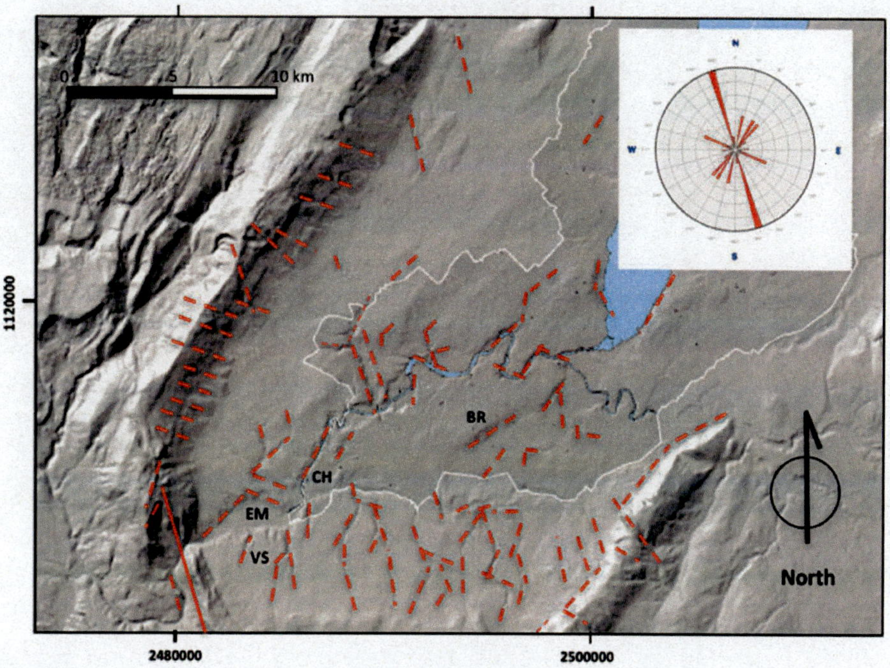

Fig. 6.11 Digital Elevation Model (DEM) of the Geneva Basin with indication of geomorphological lineaments as observed on the map (i.e., river incision and terrace escarpments, rock cliff escarpments). Note that, as the rose diagram demonstrates, a considerable number of segments corresponds to the NW-SE direction of major strike-slip faults occurring in the area (e.g., Vuache fault). CH: Chancy terrace; VS: Vulbens stable terrace; EM: Etournel marshes; BR: Bernex Ridge

such as the Soral–Laconnex hills (Fig. 6.14), represent the subaqueous frontal moraines formed during temporary re-advances of the glacier during the overall withdrawal phase. It is likely that these moraines were associated with the Arve Glacier, which most probably outlived the Rhone Glacier after the latter started its fast withdrawal on a deep lacustrine environment (Moscariello 1996). Thus, during temporary advances, the Arve Glacier would have reached the Geneva area and occupied portion of the Rhone River valley. The large, flat-bottomed incision forming a 90-degree angle bordering the Soral–Laconnex subaqueous moraine results from the large meltwater discharge generated by the Arve Glacier, which carved a long and narrow spillway channel (Arande spillway channel; Figs. 6.12 and 6.14) between the glacier and the northern slope of Mt Salève and then widened in the Soral area to merge with the proglacial/lacustrine deposits. The latter form the flat and wide morphology of the SW part of the Geneva Canton, which is currently one of the main areas for agriculture and gravel and sand exploitation.

The natural separation between the buried tunnel valley and its partly filled portion *(Petit Lac)* is a small transversal and narrow ridge, which is today occupied by the old town of Geneva (Fig. 6.15). To the SE this ridge widens up forming a large apron gently dipping to the SW. This is interpreted as one of the few evidences of frontal moraine genetically related to the Rhone Glacier in the Geneva area, most likely formed in subaqueous and possibly subglacial conditions. The sharp upstream margin of the ridge, which continues on the SW margin of the lake (Fig. 6.7), suggests a resting ice margin of a relatively small glacier, during the overall withdrawal phase, again pointing to a very fast melting and retreating phase of the Rhone Glacier. Wood fragments discovered in the gravels and sands overlying directly the moraine (Moscariello 1996) yielded an age of around 22,500 years cal BP.

In addition, a number of subglacial erosional and depositional features such as *roches moutonnées* and drumlin grooves are well preserved in the topography and easily recognizable by an accurate examination of the digital elevation model of the area (Fig. 6.7).

6.3.3 Postglacial Landscape

The large and deep proglacial and paraglacial lake generated by the melting of the Rhone, Arve, and Jura glaciers progressively reduced its elevation and volume as indicated by the terraced surfaces occurring on both sides of Lake Geneva (Burri 1981). Kame terraces are located between 730 and 420 m a.s.l. and are thought to be genetically related to the LGM glacier. Kame terraces formed in ephemeral, local, ice-dammed lakes generally developed at the front of the ice ('Valleiry Lake' in Donzeau et al. 1997; Figs. 6.1 and 6.6) and in correspondence with lateral tributaries or supraglacial stream mouths (Moscariello et al. 1998. Lacustrine terraces

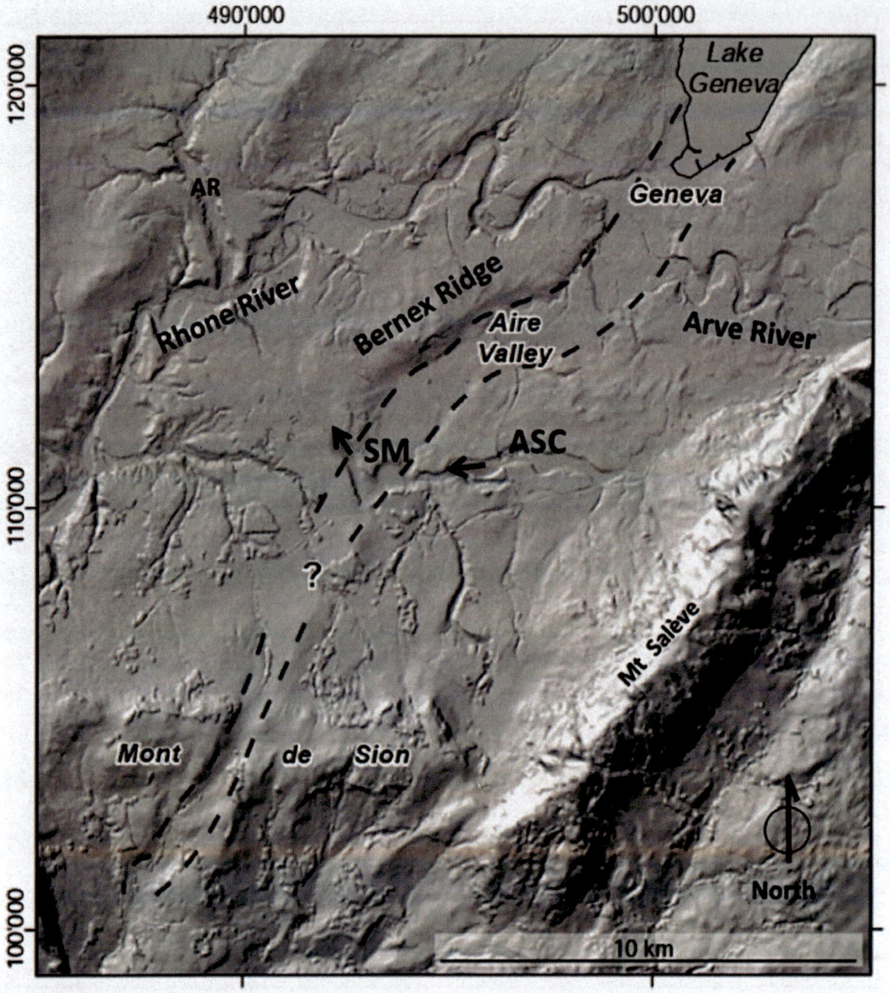

Fig. 6.12 Hillshaded view of the Geneva and Haute-Savoie (data from LiDAR) showing the continuation of the *Petit Lac* tunnel valley through the Geneva Basin (modified from Fiore 2007) passing to the SE of the bedrock (Tertiary) bold headed Bernex Ridge. SM: Soral moraine; AR: Allondon River; ASP: Arande spillway channel

Fig. 6.13 View of the Mont de Sion hill area situated between the termination of Mt Salève and Mt Vuache and thus representing the SW border of the Geneva Basin. This area represented the spill point of the Rhone Glacier flowing out of the Geneva Basin toward the SW. This irregular hilly morphology is related to glacial deposition, most likely in subglacial conditions (photo C. Evard)

Fig. 6.14 The frontal moraine of Soral consisting of mixed till and outwash deposits accumulated in a subaqueous/subglacial conditions. The Salève Mountain in the background. See Fig. 6.3 for location (photo A. Moscariello)

are preserved at approximately +30 m, +10 and +3 m (Chambers 1848; Morlot 1858; Favre 1867; Burri 1981; Arn 1984; Gabus et al. 1987) above the present-day lake level (372 m a.s.l.) and below the level located at 402 m a.s.l., which represents the last glaciolacustrine deposit (Moscariello, 1996). They are related to the evolution of syn- and postglacial lake levels developed after complete vanishing of the Rhone Glacier from the Lake Geneva region and the progressive downcutting of the natural barrier represented by Mt Vuache. It is at this time that the Ecluse gateway formed (Figs. 6.1, 6.5 and 6.6) allowing the Rhone River to develop its valley as we know it today. The Rhone River forms a highly sinuous talweg with most of its course entrenched within older deposits, mostly within the *"Alluvion Ancienne"* cemented gravels, forming incisions up to 70 m deep. At the end of its course within the Geneva Basin, in the proximity of Mt Vuache ridge, the Rhone River has a tendency to develop low and extended terraces on its southeastern side (e.g., Chancy and Vulbens stable terraces; Fig. 6.11) elevated around 18-20 m above the present-day average river water level. In this location, the Rhone River also develops an unusual extended floodplain (2.7 km long and 700 m wide) characterized by interfluves islands, marshes, and wetland development known as the Etournel site (Fig. 6.10), an important bird natural reserve site of the region.

The Arve River also displays a highly sinuous entrenched talweg, indicating a relatively constant and slow incision operating in response to change in base level of the Rhone River (e.g., incision of the Ecluse gateway).

The progressive incision caused the entrenchment of the tributary fluvial network generated from the surrounding higher ground (Jura and Mt Salève). As a result, the central Geneva Basin is characterized by a combination of gentle slopes of glacial origin dipping toward the main axial drainage system represented by the *Petit Lac* and the Rhone River and steep slopes and escarpments associated with the incision of the tributaries rivers and streams (Fig. 6.2).

Spring generated rivers such as the Allondon River (see Bätz et al., this volume) occur on the SE slope of the Jura and are most likely associated with karst and fractured network. This again may support a genetic link between the Allondon Valley flanks geometry and the major lineaments detected in the subsurface and in outcrops (Fig. 6.3).

At present, large dams regulate the natural course of the Rhone River, Lake Geneva level, and the downstream sediment transport. The latter is almost entirely supplied by the Arve River, which, on the other side, has no artificial damming and flood control system in place.

Postglacial processes also include large landslides such as those occurring on the northwestern slope of Mt Salève (Fig. 6.3). Some of these gravity phenomena may have also affected collapses of cavities and caves (e.g., Grande Gorge on Mt Salève; Fig. 6.8), which was formed in the past by karstic dissolution processes of Mesozoic carbonate rocks.

Smaller landslides have also affected the Quaternary deposits cut by the Rhone and Arve rivers. These phenomena may have also occurred in the past as suggested by archeological remains indicating a temporary rise of the Lake Geneva level during the Late Bronze time (ca. 2800 years BP), which may have been caused by damming of the Rhone River outlet (Moscariello 1997; Corboud 2012).

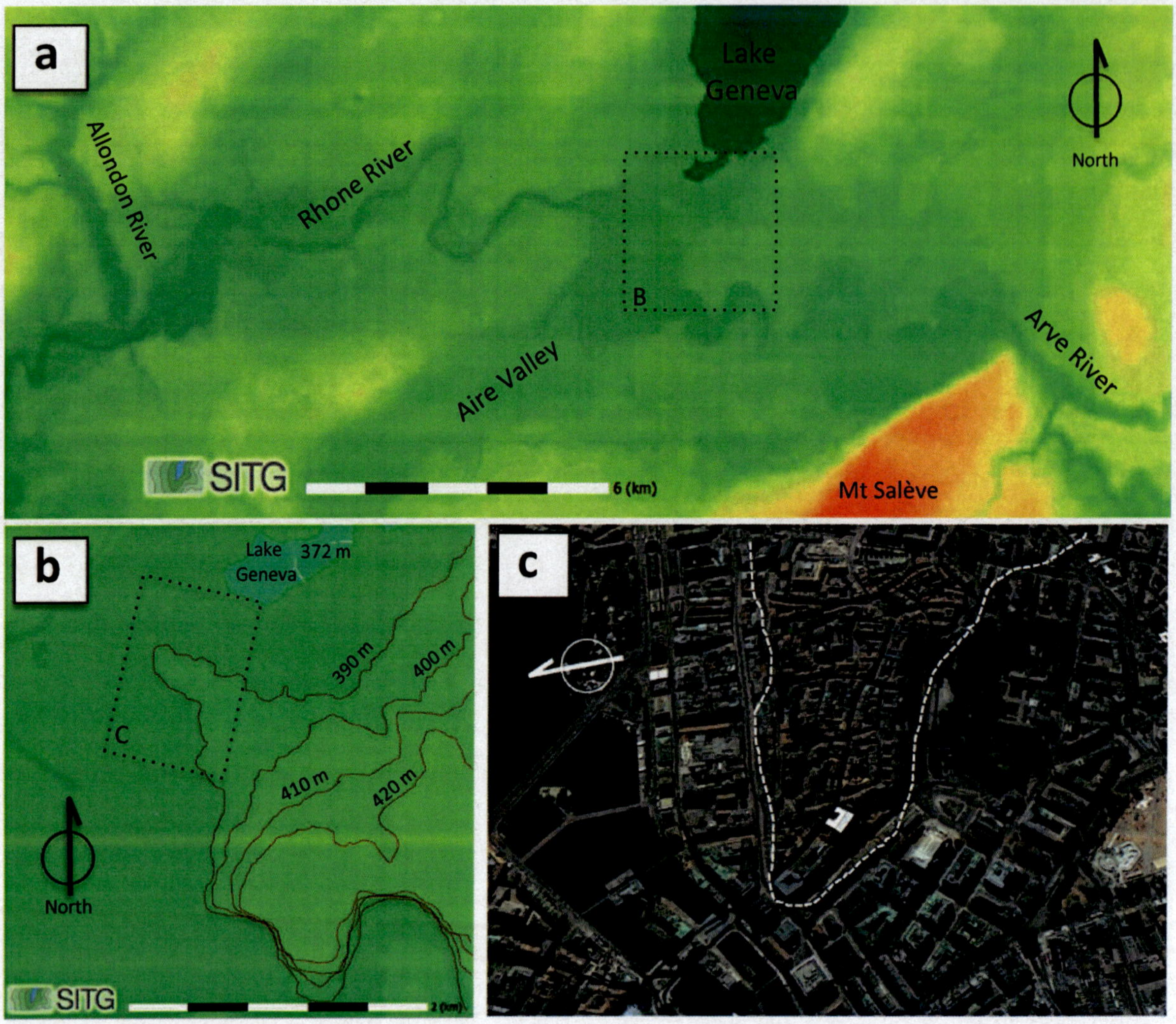

Fig. 6.15 Geomorphology of the city center of Geneva. **a** Detailed digital elevation model of the area surrounding the City of Geneva (source: SITG; from ge.ch/sitg, accessed 27.06.2017) showing the location of the city at the confluence between the Arve River and the Rhone River at the southwestern termination of Lake Geneva; **b** Distribution of contour lines corresponding to the city center of Geneva showing a NW-SE oriented ridge connected to a larger plateau to the SE formed mostly by deposits, accumulated in subaqueous environment during the last phase of occupation of the Pleistocene glacier in the region (Moscariello et al. 1998; **c** detailed satellite photo of the frontal moraine ridge of the Rhone Glacier (see Fig. 6.6 for location) where the old town of Geneva is located (source Google Earth)

6.4 Conclusions

The Geneva Basin geomorphology results from a long tectonic history, which has involved this part of the Alpine Foreland at least from the Eocene, thus shaping the main outlines of the area. On this tectonic imprint, subsequent glacial, interglacial, and ultimately Holocene postglacial processes formed the landscape as we know it today. The tectonic lineaments, in places still active today, have developed the key structural boundaries of the basin within which glacial processes likely operated on several occasions during the last 1 million years. The glaciated landscape and the establishment of the postglacial hydrographic network form the main character of the central part of the Geneva Basin where low rise, gentle relief left by subaqueous and subglacial deposits is today incised by the tributary streams, which connect with either the lake or the Rhone River, forming the main axial hydrographic system. The latter has been strongly controlled by the opening of the Ecluse gateway, which presumably occurred in early postglacial time during the progressive emptying of the proglacial lake formed by the melting of glaciers once occupying the Geneva area.

The result of the long- and short-term sculpturing processes described in this chapter made the Geneva Basin a very attractive geographical location, which in addition to the reputation as a worldwide center for diplomacy, and an international center for business and culture makes Geneva the place of quite unique location. This was also possible thanks to responsible land-use planning over the last 50 years, which established a fine balance between urban and industry development, agricultural activities and nature preservation, which ensure both the economic and landscape attractiveness of the Geneva Basin.

Acknowledgments The author would like to thank several colleagues and students (J. Charollais, C. Chelle-Michou, N. Clerc, M. Brentini, E. Davaud, D. Do Couto, S. Girardclos, G. Gorin, B. Mastrangelo, M. Mayer, E. Rusillon, M. Sartori, R. Wernli, W. Wildi) with whom discussions and ideas sharing on the genesis and evolution of the Geneva Basin never had a dull moment. Special thanks are for Dominique Ernst and Camille Evard for making their photographs available for this publication. The editors and J.-F. Buoncristiani are especially thanked for their meticulous review which improved the earlier version of this manuscript.

References

Amberger G (1978) Contribution à l'étude du Quaternaire de la région lémanique: résultats de quelques sondages profonds exécutés à Genève. Eclogae Geol Helv 71(1):193–206

Arn R (1984) Contribution à l'étude stratigraphique du Pléistocène de la région lémanique. PhD thesis, University of Lausanne, 307 pp

Bätz N, Iorgulescu I, Lane SN (this volume) The Allondon River: Decadal planform changes under changing boundary conditions

Baize S, Cushing M, Lemeille F, Gelis C, Texier D, Nicoud G, Schwenninger JL (2011) Contribution to the seismic hazard assessment of a slow active fault, the Vuache fault in the southern Molasse basin (France). Bull Soc géol France 182(4):347–365

Blavoux B (1988) L'occupation de la cuvette lémanique par le glacier du Rhône au cours du Würm. Bull AFEQ 2(3):69–79

Brun A (2000) Révision des données polliniques dans le Pléistocène du Bas-Chablais (Haute-Savoie): implications chronostratigraphiques et pluralité des glaciations. Quaternaire 11(1):41–51

Burri M (1981) Les terrasses lémaniques: géologie. Arch suisses Anthropol gén 45(2):107–115

Chambers R (1848) Notes géologiques sur les vallées du Rhin et du Rhône. Arch Soc Phys et Hist nat Genève 11:143–144

Charollais J, Plancherel R, Monjuvent G, Debelmas J (1998) Annemasse. Carte géologique de la France au 1:50 000, feuille 654 et notice explicative. BRGM, Orléans

Charollais J, Wernli R, Mastrangelo B, Metzger J, Busnardo R, Clavel B, Conrad M, Davaud E, Granier B, Saint Martin M, Weidmann M (2013) Présentation d'une nouvelle carte géologique du Vuache et du Mont de Musièges (Haute-Savoie, France): stratigraphie et tectonique. Archives des Sciences 66:1–64

Clerc N, Rusillon E, Moscariello A, Renard P, Paolacci S, Meyer M (2015) Detailed structural and reservoir rock typing characterisation of the Greater Geneva Basin, Switzerland, for geothermal resource assessment. Proceedings World Geothermal Congress 2015, Melbourne, 10 pp (https://pangea.stanford.edu/ERE/db/WGC/papers/WGC/2015/16075.pdf)

Corboud P (2012) L'archéologie lémanique un siècle après F.A. Forel: quelques questions encore à résoudre. Archives des Sciences 66:237–248

Coutterand S (2010) Etude géomorphologique des flux glaciaires dans les Alpes nord-occidentales au Pléistocène récent. Du maximum de la dernière glaciation aux premières étapes de la déglaciation. PhD thesis, University of Savoie, 468 pp (https://tel.archives-ouvertes.fr/tel-00517790v3)

Donzeau M, Wernli R, Charollais J, Monjuvent G (1997) Saint-Julien-en-Genevois Carte géologique de la France au 1:50 000, feuille 653 et notice explicative. BRGM, Orléans

Favre A (1867). Recherches géologiques dans les parties de la Savoie, du Piémont et de la Suisse voisine du Mt. Blanc. Masson, Paris, 3 vol

Fiore J (2007) Quaternary subglacial processes in Switzerland: Geomorphology of the Plateau and seismic stratigraphy of Western Lake Geneva. PhD thesis, University of Geneva, 169 pp (https://archive-ouverte.unige.ch/unige:714)

Fiore J, Girardclos S, Pugin A, Gorin G, Wildi W (2010) Würm deglaciation of western Lake Geneva (Switzerland) based on seismic stratigraphy. Quat Sci Reviews 30:377–393

Gabus JH, Lemdal G, Weidmann M (1987) Sur l'âge des terrasses lémaniques au Sud-Ouest de Lausanne. Bull Soc vaud Sci nat 78:419–429

Gorin GE, Signer C, Amberger G (1993) Structural configuration of the western Swiss Molasse Basin as defined by reflection seismic data. Eclogae Geol Helv 86:693–716

Lombard A (1965) Notice explicative de la feuille 1301 Genève. Atlas géologique de la Suisse au 1:25 000. Commission géologique suisse, Berne

Mastrangelo B, Charollais J, Wernli R, Guyonnet D, Widemer F, Blondel T (2009) Carte géologique du Vuache, 1:25 000. In: Bordon J, Charollais J (eds) Le Vuache, montagne insolite. SIPCV, Vulbens

Mastrangelo B, Charollais J, Wernli R, Metzger J (2013) Accidents longitudinaux dans la Molasse rouge auct. et dans son substratum sur le versant oriental du Salève (Haute-Savoie, France). Swiss J Geosci 106(2):253–263

Monjuvent G, Nicoud G (1988) Interprétation de la déglaciation rhodanienne au Würm, des moraines « internes » à la cuvette lémanique. Bull AFEQ 2(3):129–140

Morlot A (1858) Sur le terrain quaternaire du bassin du Léman. Bull Soc vaud Sci nat 6:101–108

Moscariello A (1996) Quaternary geology of the Geneva Bay: sedimentary record, palaeoclimatic and palaeoenvironmental reconstruction since the Last Glacial Cycle. PhD thesis, University of Geneva, 230 pp

Moscariello A (1997) Lacustrine ooidal sands in Lake Geneva (Switzerland): sedimentological evidence for high energy conditions and lake-level rise in the Late Bronze Age. Climatic implications and constraints on the location of lake-dwellings. Eclogae Geol Helv 90(1):143–150

Moscariello A (2016) Geothermal exploration in SW Switzerland. Proceeding European Geothermal Congress 2016, Strasbourg, 9 pp (https://archive-ouverte.unige.ch/unige:90788)

Moscariello A. (2019) Exploring for geo-energy resources in the Geneva Basin (Western Switzerland): opportunities and challenges. Swiss Bulletin für angewandte Geologie 24(2):105–124

Moscariello A, Pugin A, Wildi W, Beck C, Chapron E, De Batist M, Girardclos S, Ivy Ochs S, Rachoud-Schneider AM, Signer C, van Clauwenberghe T (1998) Déglaciation würmienne dans des conditions lacustres à la terminaison occidentale du bassin lémanique (Suisse occidentale et France). Eclogae Geol Helv 91:185–201

Paolacci S (2012) Seismic facies and structural configuration of the Western Alpine Molasse basin and its substratum (France and Switzerland). PhD thesis, University of Geneva (unpublished)

Pugin A, Bezat E, Weidmann M, Wildi W (1993) Le bassin d'Ecoteaux (Vaud, Suisse): Témoin de trois cycles glaciaires quaternaires. Eclogae Geol Helv 86(2):343–354

Sambeth U, Pavoni N (1988) A seismotectonic investigation in the Geneva Basin, southern Jura Mountains. Eclogae Geol Helv 81:433–440

Sesiano J, Schnyder C, Proz PA, Gnos E, Schaltegger U (2011) Les Pierres du Niton revisitées: soubassement, minéralogie, datation et origine. Archives des Sciences 64:81–90

Signer C, Gorin GE (1995) New geological observations between the Jura and the Alps in the Geneva area, as derived from reflection seismic data. Eclogae Geol Helv 88(2):235–265

Thouvenot F, Frechet J, Taponnier P, Thomas JC, Le Brun B, Menard G, Lacassin R, Jenatton L, Grasso JR, Coutant O, Paul A, Hatzfeld D (1998) The ML 5.3 Épagny (French Alps) earthquake of 1996 July 15: a long-awaited event on the Vuache fault. Geophys J Internat 135:876–892

van der Vegt P, Janszen A, Moscariello A (2012) Tunnel valleys: current knowledge and future perspectives. In: Huuse M, Redfern J, Le Heron DP, Dixon RJ, Moscariello A, Craig J (eds) Glaciogenic reservoirs and hydrocarbon systems. Geol Soc Spec Publ 368(1):75–97

Wegmüller S, Amberger G, Vernet JP (1995) La formation de Montfleury près de Genève: Etude palynologique et sédimentologique d'une séquence du Pléistocène moyen. Eclogae Geol Helv 88(3):595–614

Wildi W, Corboud P, Girardclos S, Gorin G (2017). Guidebook: geology and archaeology of Geneva. Société de physique et d'histoire naturelle de Genève, Geneva, 93 pp (http://www.unige.ch/forel/en/services/visitegeologique/)

Andrea Moscariello is Full Professor and head of the Geo-Energy/Reservoir Geology and Basin Analysis Group at the Department of Earth Sciences of the University of Geneva. He spent 15 years in the energy industry as geoscientist in several locations worldwide after obtaining a PhD at the University of Geneva on the Quaternary geology of the Lake Geneva bay and surrounding area. His research activities encompass a large variety of subsurface and outcrop studies internationally and in Switzerland aimed at assisting geo-resources (e.g. geothermal and hydrocarbons) exploration and development. He is currently member of the executive committee of energy-related professional associations such as the American Association of Petroleum Geologists (AAPG) and the Swiss Association of Energy Geoscientists (SASEG).

Structural and Karstic Landscapes of the Joux Valley (Southwestern Jura)

7

Emmanuel Reynard and Philippe Schoeneich

Abstract

The Joux Valley is located in the Jura Mountains at the border between France and Switzerland. It has a cold damp climate and during the last glaciation, it was covered by a large ice cap, whose decay resulted in a particular esker morphology. The geological structure is characterised by simple concentric folds and conformal structural relief. The lithology is composed of Jurassic to Cretaceous limestones and marls, which strongly influence the hydrogeological circulation and karst erosion. The Vallorbe–Pontarlier strike-slip fault is responsible for the closed tectonic Joux Valley and associated Vallorbe caves. The anticlines are rich with karst landforms (karren, sinkholes, caves) and closed depressions of karstic origin functioning as a cold air trap responsible for the presence of dwarf spruce trees. Some caves are frozen all year round, but are now suffering from climate change.

Keywords

Folded Jura • Closed depressions • Karst • Jura ice cap

7.1 Introduction—Geography and Climatic Context

The Jura is a mountain range located at the border between France and Switzerland (Fig. 7.1). It is a young limestone mountain range, whose formation is tectonically related to Alpine orogenesis. The chain forms an arc parallel to the Alpine arc. South of Geneva, the direction of the fold axes is S–N; in the north, south of the Rhine graben, the direction of the axes is E–W, and in between, it is more or less SW-NE. The Jura Range is traditionally divided into four parts moving from the SE to the NW (Bichet and Campy 2009): the high chain *(Haute-Chaîne* in French*)*, the plateaux, the beams *(les faisceaux)* and the foothills (*les Avant-Monts)*.

The entire Swiss part of the Jura Mountains is located in the high chain—also called 'folded Jura'—characterised by simple concentric folds and mostly conformal structural relief, resulting in long parallel chains and valleys. In the southwestern part, with which we are concerned here (Fig. 7.1), the direction of fold axes is SW-NE. Anticline swells (1500–1700 m a.s.l.) alternate with longitudinal syncline valleys (900–1000 m a.s.l.). The highest summit in the Swiss Jura is Mont Tendre (1679 m a.s.l.). The parallel SW-NE folds are cut by important tear faults which have been exploited by erosion and are thus the preferred axes of communication through the Jura (e.g. Givrine Pass along the Morez–St-Cergue fault; the Vallorbe–Pontarlier fault used by the Paris–Lausanne railway line; Fig. 7.1).

The climate in the Jura is cold and damp. Even though the Jura range is not very high, it is nevertheless the first obstacle in the way of depressions originating from the Atlantic Ocean, which explains the relatively high precipitation. Annual precipitation for the period 1981–2010 was 1273 mm in Couvet (728 m a.s.l.), 1597 mm in La Brévine (1050 m a.s.l.), 1410 mm in Vallorbe (748 m a.s.l.), 1781 mm in Les Bioux (1025 m a.s.l.) and 1888 mm in La Dôle (1669 m a.s.l.) (source: MeteoSwiss). In the Risoux and on the Mont Tendre anticline, annual rainfall is around

E. Reynard (✉)
Institute of Geography and Sustainability and Interdisciplinary Centre for Mountain Research, University of Lausanne, Géopolis, 1015 Lausanne, Switzerland
e-mail: emmanuel.reynard@unil.ch

P. Schoeneich
Institute of Urban Planning and Alpine Geography, University Grenoble Alpes, 14 bis av. Marie-Reynoard, 38100 Grenoble, France
e-mail: philippe.schoeneich@univ-grenoble-alpes.fr

E. Reynard (ed.), *Landscapes and Landforms of Switzerland*, World Geomorphological Landscapes,
https://doi.org/10.1007/978-3-030-43203-4_7

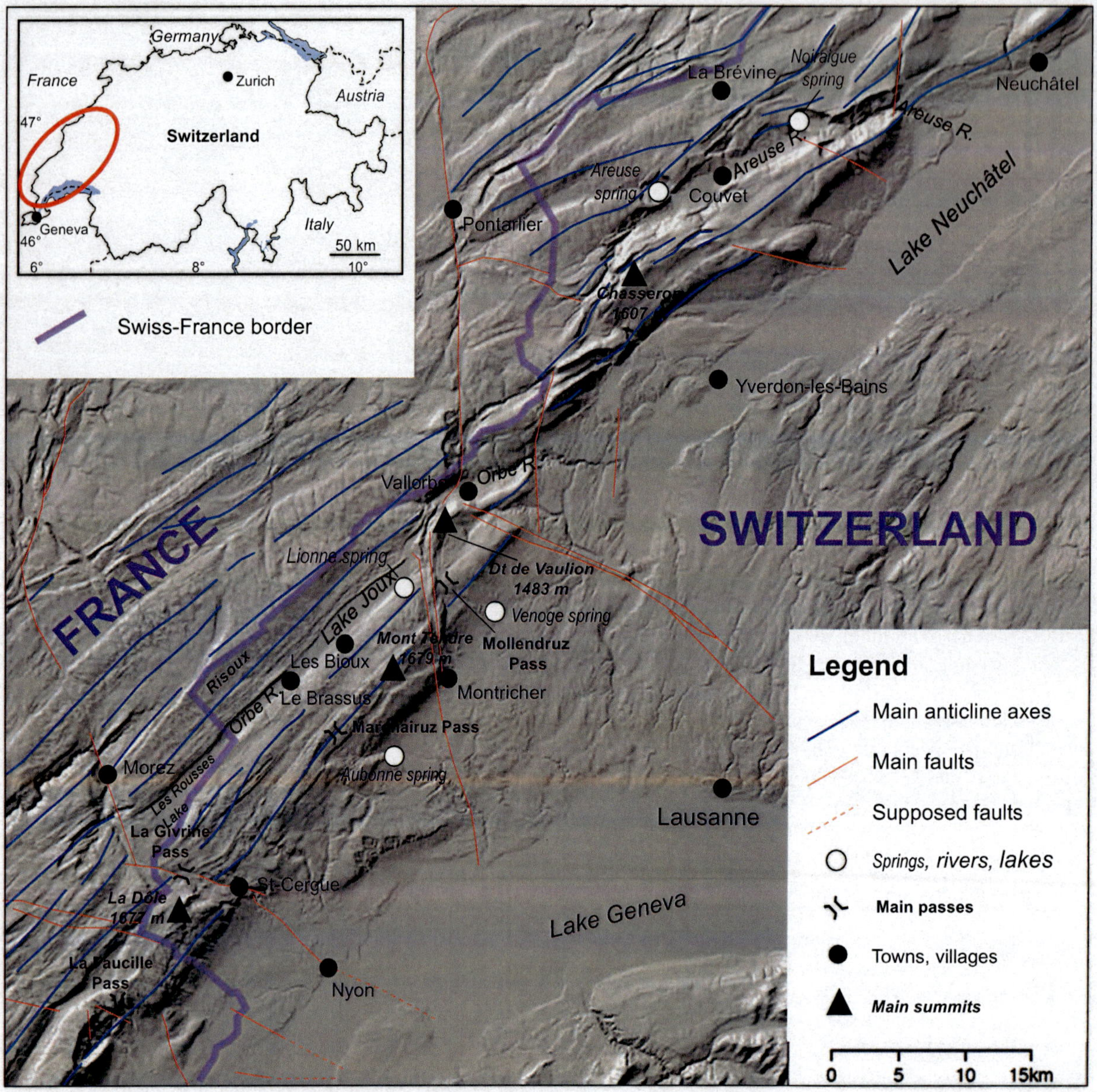

Fig. 7.1 Map of the study area showing the location of main sites mentioned in the text (shaded relief map: swisstopo)

1800 mm and 2000 mm, respectively (Perrin 2002). On the anticlines, snow cover duration is 140–150 days per year (Perrin 2002). The inhabited valleys are relatively high (900–1000 m a.s.l.) and the presence of numerous topographic depressions favours temperature inversions, which explain both the relatively low annual temperature and the record low temperatures in some places. The mean annual air temperature for the period 1981–2010 was 4.9 °C in La Brévine (1050 m a.s.l.), 6.5 °C in La Chaux-de-Fonds (1017 m a.s.l.), 3.9 °C in La Dôle (1669 m a.s.l.) (source: MeteoSwiss). The 0 °C isotherm is estimated to be at approximately 2200 m a.s.l. (Luetscher et al. 2005). The lowest temperature officially recorded in Switzerland was in La Brévine (–41.8 °C on 12 January 1987). These climatic conditions are favourable for both karstic dissolution and mechanical weathering.

Hydrography and hydrogeology are influenced by lithology and tectonics (see below and Perrin 2002). Because of the frequency of limestone outcrops, rivers are rare on the anticlines and, depending on the elevation, 50–75% of the total precipitation infiltrates the karst (Luetscher and Perrin 2005) and drains towards large karstic springs inside the massif (the Brassus and the Lionne springs in the Joux Valley, the Areuse and the Noiraigue springs in the Areuse Valley) or at the foot of the massif, at the border with the Central Plateau (e.g. Aubonne River spring, Venoge River spring) (see location in Fig. 7.1). The main rivers flow along the syncline valleys and some of them (Orbe River, Areuse River) cross anticlines along transverse faults.

7.2 Geological Setting

The Jura mountain belt is characterised by thrusting and folding that occurred between 10 and 3 Ma, from the Miocene to the Pliocene, above a weak Triassic detachment zone composed of evaporites and clays (Sommaruga 1997, 1999). The tectonics of the Jura belt and the Molasse Basin are intimately linked, the latter being much less deformed than the Jura massif because of the thickness of the Tertiary sediments. During the Triassic, the Jura and the Molasse Basin became part of the Alpine Tethys passive margin, and up to 1 km of evaporites and shales accumulated in a basin in the region of the future Jura arc (Sommaruga 1997). It is this Triassic formation that acted as a decollement zone for the displacement of Jurassic and Cretaceous series towards the NW during the orogenesis.

Horizontal displacement was up to 15 km. Folding can be divided into high amplitude folds (e.g. Risoux anticline, Dent de Vaulion—Mont Tendre anticline, Joux Valley syncline) and low amplitude folds, at the local scale (Audétat and Heiss 2002). High amplitude folds are related to large NW-vergent thrusts with at least one kilometre of dip-slip displacement, which resulted in the doubling of the Jurassic and Cretaceous series, as evidenced by a core drilled in the Risoux in 1960, which reached a depth of 1958 m and revealed the presence of three thrusts in the Jurassic series (Bichet and Campy 2009). The thrusts include both flats and ramps (Sommaruga 1999).

The stratigraphic series of the high chain are mainly composed of Jurassic to Cretaceous limestones and marls (Sommaruga 1997, 1999; Perrin and Audétat 2002a; Figs. 7.2 and 7.3). The crystalline basement is not exposed anywhere in the Jura Mountains. The Triassic and lower Jurassic levels are also poorly visible in this part of the Jura Mountains, whereas they are visible in the French part of the massif (e.g. Salins-les-Bains, Lons-le-Saunier).

Middle Oxfordian marls (Argovian facies) can only be seen in some anticline depressions. This layer is hydrogeologically important as it is a thick impermeable layer below the thick series of massive Upper Oxfordian (Rauracian facies), Kimmeridgian (including Sequanian facies) and Tithonian (Portlandian facies) limestone, which form most of the outcrops and are up to 500 m thick (Figs. 7.2 and 7.3). The Cretaceous series are much thinner and are mainly represented by Berriasian/Valanginian limestones, yellowish Hauterivian limestones and Barremian white limestones, separated by shallow marly levels (Fig. 7.2). In the Areuse

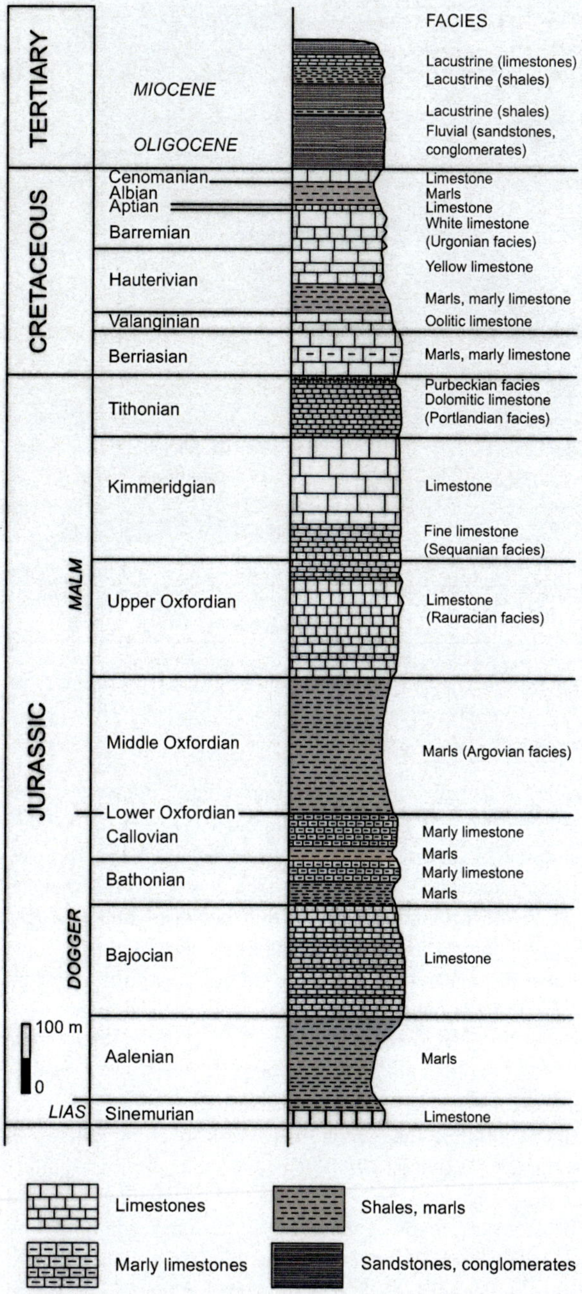

Fig. 7.2 Stratigraphy of the western Jura Mountains (after Sommaruga 1997; Bichet and Campy 2009; Guglielmetti et al. 2013). There may be regional variations in the thickness of various layers and facies

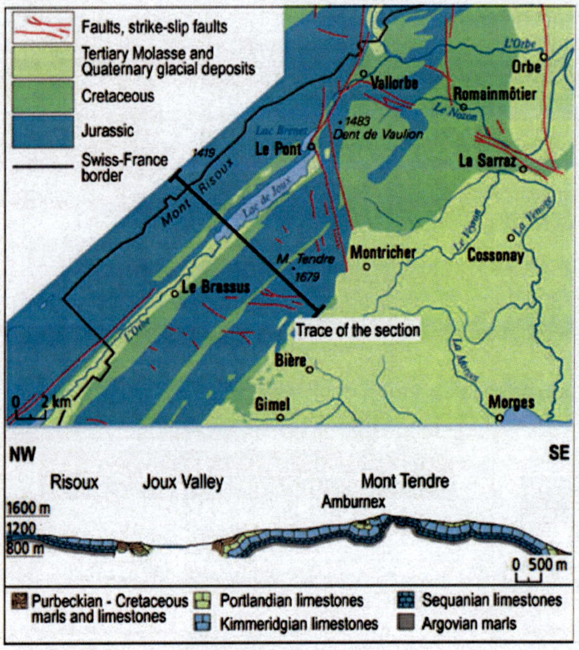

Fig. 7.3 Simplified geological map and geological cross-section of the Mont Tendre area (modified after Bissig and Reynard 2006)

Valley, asphalt trapped in Barremian limestones was exploited from 1711 to 1986 (Jelmini 1987) and the locality is now classified as a Swiss geosite (n° 31; Reynard et al. 2012). During the Tertiary, the area was forming the distal part of the Molasse basin and thin layers of detrital rocks were deposited (Weidmann 2008). These deposits have almost disappeared, since due to folding and uplift, the Jura emerged from the sea early, and most of the sediments were eroded by Quaternary glaciers.

Hydrogeological circulation depends to a great extent on stratigraphy and tectonics. The karstified series are relatively thin, resulting in the predominance of horizontal galleries in the endokarst (Nicod 1990). The two main Jurassic aquifers are limestones and marly limestones of the Dogger, dammed by Aalenian marls, and Malm limestones, dammed by Oxfordian marls (Argovian; 150 m thick) (Guglielmetti et al. 2013), which form the karstic base level in the region (Perrin and Audétat 2002a); smaller Cretaceous aquifers are Valanginian limestones, dammed by Tithonian and Berriasian marls and marly limestones, and Hauterivian/Barremian limestones, dammed by Lower Hauterivian marls. Small local aquifers are also present in Tertiary and Quaternary sediments. Several tracing surveys allowed mapping of karst aquifers (see Perrin 2002 for an overview) revealing the complexity of the hydrogeological circulation and the presence of several diffluences. Most municipalities draw their drinking water from karst sources, which makes them relatively vulnerable to accidental pollution, as was the case in Le Brassus in 1989 (Parriaux and Mayoraz 1990; Perrin 2002).

7.3 The Jura Ice Cap

During the last glaciation, the southern part of the Jura was covered by a large ice cap (Arn and Campy 1990; Campy and Arn 1991; Fig. 7.4). It was centred approximately on the Joux Valley and extended far west to France. Its thickness is unknown, but its maximum elevation was estimated at 1800–2000 m a.s.l.

The existence and the shape of the Jura glaciation remained highly controversial for decades since no evidence of local glaciers could be found in the higher parts of the massif, although exceptionally well-preserved frontal morainic systems are known along its western edge, mainly in the Combe d'Ain in France (Campy and Arn 1991). The existence of an ice cap covering the Joux Valley area was first demonstrated by Aubert (1965). Evidence includes smoothed erosional landforms, glacial striae, upslope transport of erratics up to some of the highest crests and overflow of ice tongues over all low sections of the main chain eastwards into the Swiss lowlands. During the Last Glacial Maximum (LGM), the boundary between the Rhone glacier and the Jura ice cap was approximately 1200 m a.s.l. along the eastern flank of the main chain and is now represented on the LGM map of Switzerland (Schlüchter 2009). After the LGM, the lowering of the Rhone glacier surface allowed the outlet tongues of the Jura ice cap to extend towards the Swiss lowlands (Arn and Campy 1990; Campy and Arn 1991; Campy 1992).

The ice cap is thought to have decayed very fast after its top dropped below the equilibrium line altitude. As a matter of fact, there is no evidence of subsequent morainic ridges inside the massif, and, except for some thin discontinuous till covers, there is almost no geomorphological evidence of past glaciers at all, except in the centre of the Joux Valley and its upper reach in France (Aubert 1938). However, an exceptional and very rare feature, an esker, can be observed in Praz-Rodet, near Le Brassus. A sinuous and discontinuous ridge can be traced along the valley bottom extending from the French border down to Lake Joux, and probably continues into the lake. A gravel pit opened in the 1990s and ground-penetrating radar investigations showed that it is an esker, deposited during a decay phase with a glacier snout ending in the lake (Fiore et al. 2002; Fig. 7.5). In 1972, the skeleton of a young mammoth was found on top of the esker deposit, in an older gravel pit (Weidmann 1969, Aubert 1970). It is one of the best-preserved mammoth skeletons

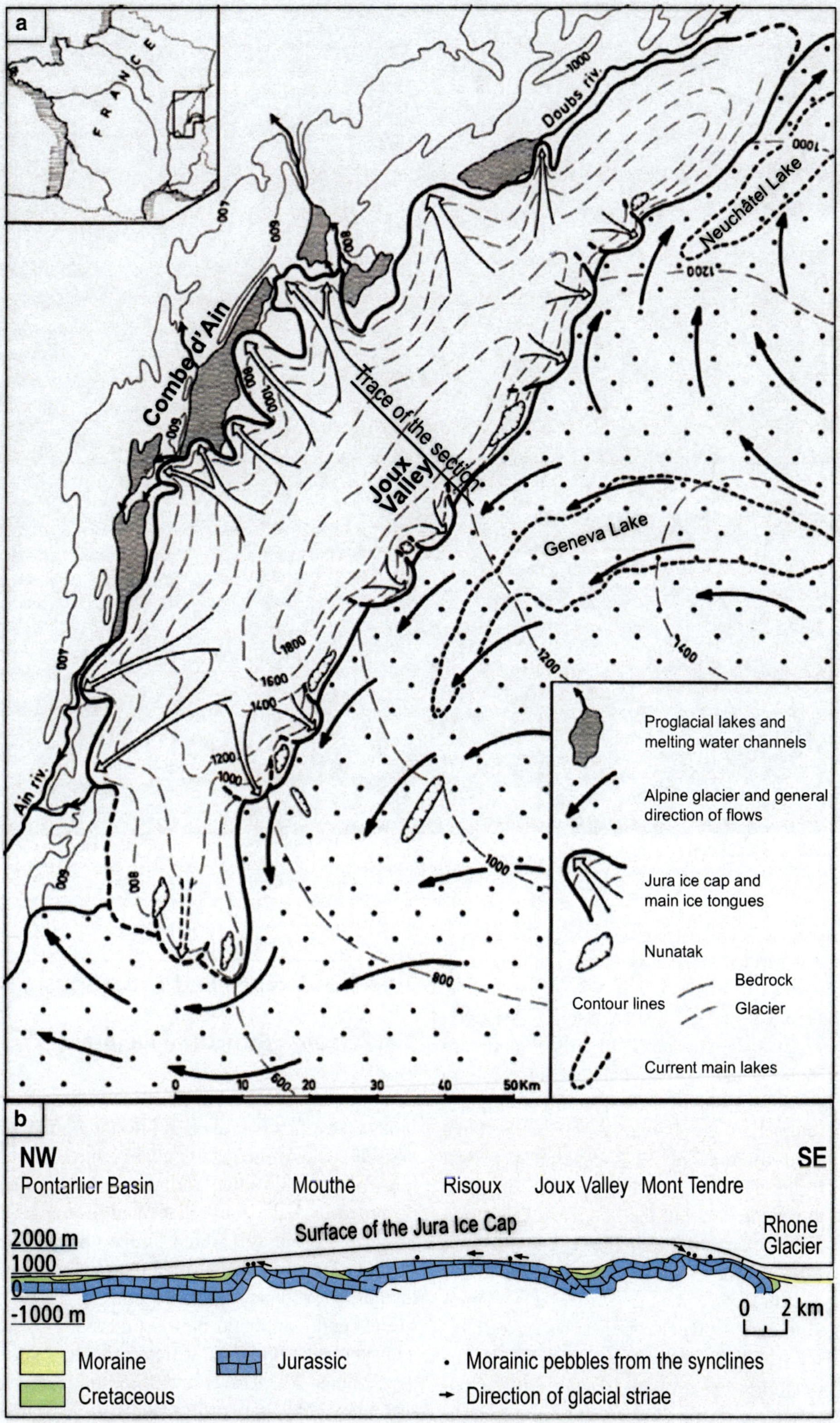

Fig. 7.4 **a** Extent of the Jura ice cap during the last glaciation (modified after Arn and Campy 1990); **b** Cross-section of the Jura ice cap in the Mont Tendre—Joux Valley area (modified after Bissig and Reynard 2006)

Fig. 7.5 Esker infilling in the Praz-Rodet area near Le Brassus (photo E. Reynard)

known so far in Switzerland and can be seen in the Geology Museum in Lausanne. It was dated to 13,705 ± 55 ^{14}C years BP (16,348 ± 189 cal BP) (Lab. No OxA-12982; Barnes et al. 2007), which is assumed to be the minimum age of the esker deposit and most probably corresponds to the final decay of dead ice bodies or to a solifluction phase that covered the animal. A few kilometres upstream, Lake Rousses is dammed by the only well-developed frontal moraine known in the massif, suggesting that the centre of the Jura glaciation was probably in the SW part of the Joux Valley. Several long pollen profiles indicate that the area was totally ice-free before Bølling–Allerød (Wegmüller 1966). One of the first paleomagnetic secular variation profiles performed in continental Europe was in Lake Joux; it reached ca. 15 ka and for many years was the paleomagnetic reference profile for the Late Glacial (Creer et al. 1980).

7.4 Landscapes and Landforms

7.4.1 Large Structural Landforms

The structure of the area can be described at two scales. At a large scale, it can be divided into three main parts (Figs. 7.1 and 7.3): the Mont Tendre chain, which limits the valley to the SW, is a wide anticlinorium; the Joux Valley itself, partly occupied by Lake Joux, is located in a large deep syncline; the Risoux massif, which limits the valley to the NW towards France, is again a wide anticlinorium, partly thrust towards NW over a similar series of rocks. In detail, the Mont Tendre anticlinorium is folded into distinct anticlines and synclines (Fig. 7.3), which form long parallel ridges and depressions. The Risoux anticlinorium is also folded, but the secondary folds are hardly visible in the relief.

The structural relief is conformal, meaning the anticlines roughly correspond to the ridges and the synclines to the valleys, even if the relief has been subject to erosion since the uplift of the entire range. The Cretaceous rocks are therefore only preserved in synclines (Fig. 7.3), and most of the structural surface is developed on or within the Jurassic limestones (mainly Kimmeridgian and Tithonian). Neither deep anticlinal valleys nor any transversal gorges—called *cluses* in French—are visible; in this sense, the area can be regarded as one of the most conform parts of the Jura chain.

7.4.2 The Joux Valley: A Closed Tectonic Valley

One major structural feature is the Vallorbe–Pontarlier strike-slip fault (Aubert 1959; Figs. 7.1 and 7.3). The movement is to the left, i.e. the northeastern part moved north relative to the southwestern part. The fault crosses the Mont Tendre anticlinorium from Montricher in the south-west through the Mollendruz Pass, then along the foot of the Dent de Vaulion escarpment to Vallorbe. The pass leading from Vallorbe to Pontarlier follows the continuation of the fault. The Dent de Vaulion was moved by the fault to the north and today represents the continuation of the Mont Tendre anticlinorium, with lateral displacement of several kilometres (Trümpy 1980; Figs. 7.3 and 7.6). This major tectonic accident blocked the outlet of the Joux Valley and dammed Lake Joux.

The displacement not only modified the geological structure and anticlinal arrangements but also significantly modified the water flows. The Orbe River, which flows into Lake Joux, has no surface outlet. Several swallow holes (sinkholes, called locally *'entonnoirs'*, from Lake Brenet and Lake Joux, e.g. *entonnoir du Moulin du Rocheray, entonnoir du Bon-Port*) feed the underground network that follows the fault system and joins the resurgence of the river Orbe in Vallorbe (Vallorbe caves) (Audétat and Heiss 2002). The hydrogeological links between the closed basin of the Joux Valley and the caves of Vallorbe were long known implicitly, but were recognised in 1776 when the accidental emptying of an artificial basin led to a rapid flow of water through the Bon Port swallow hole and caused a spectacular increase in the turbidity of the Vallorbe spring. Several tracing operations using various techniques then took place in the second half of the nineteenth century and made it possible to map the underground watershed of the resurgence of the Orbe River.

The resurgence of the Orbe River is the biggest karstic spring in the Swiss Jura (Audétat and Heiss 2002). Flows vary from 2 to 3 m^3/s at low water to 70–80 m^3/s in peak

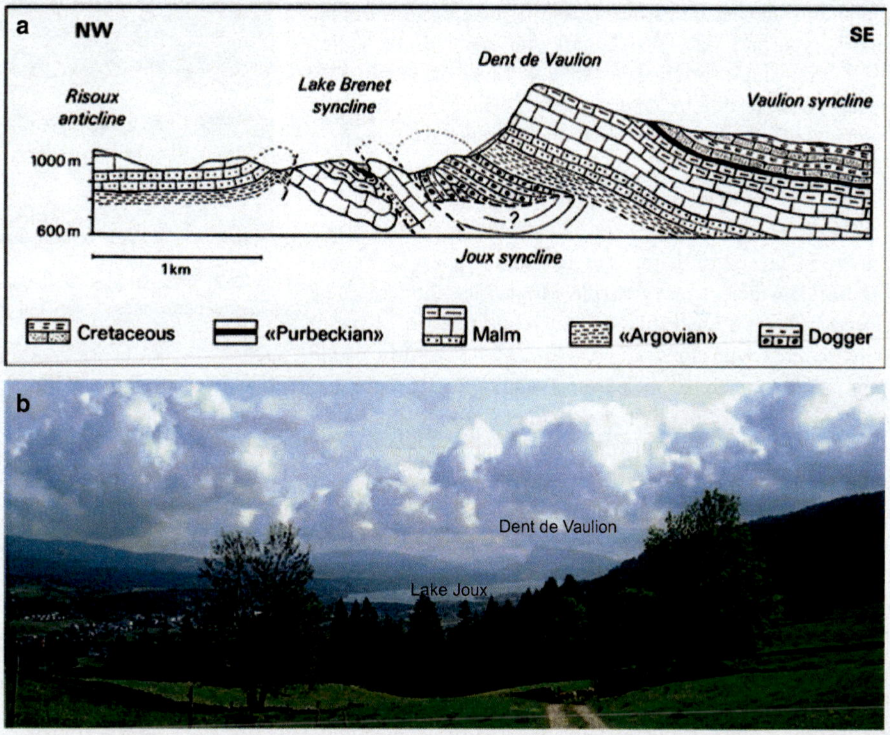

Fig. 7.6 **a** Geological section of the Dent de Vaulion (modified after Trümpy 1980); **b** A view of the damming of Joux Lake by the Dent de Vaulion anticline (photo E. Reynard)

flood conditions. At low water, half the flow comes from the losses of the lakes in the Joux Valley. Part of the underground network—the Vallorbe caves, rich in speleothems—has been open to the public since 1974. Several other independent underground networks collect the region's waters, including the *Petite Grotte aux Fées* and *Grande Grotte aux Fées* (Audétat and Heiss 2002); the resulting springs have no or very little flow at low water but can present spectacular cascade flows during exceptional floods, as was the case on the 15 February 1990.

Because of temperature inversion, Lake Joux freezes almost every winter. It then becomes one of the largest natural skating rinks in Switzerland. The railway between Paris and Lausanne via Vallorbe opened in 1875, and from winter 1879/1880 onwards, ice from Brenet Lake was exploited and sent to Paris and to the South of France by train to supply hotels with ice (Duvoisin 1887). The ice was cut with floating saws on the lake and sent to the Vallorbe railway station first by horse and cart and then, after the construction of the Vallorbe—Le Pont line, from 1886 on by train. The industrial exploitation of ice ended in 1942.

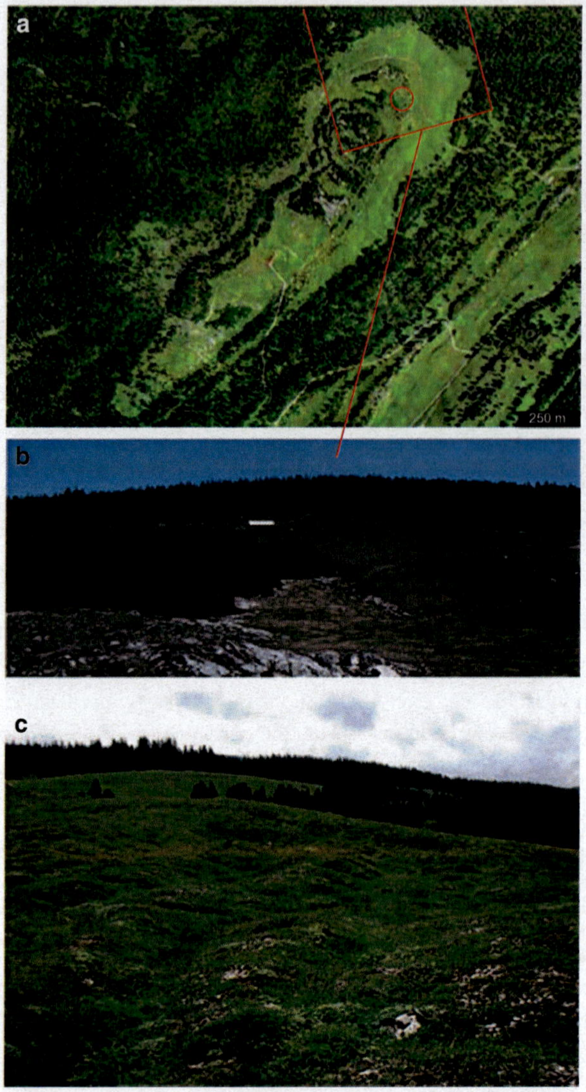

Fig. 7.7 **a** Aerial view of the *Sèche des Amburnex* closed depression (Google Earth image). The red square and the red circle indicate the location of Figs. 7.7b and 7.7c, respectively; **b**. A view of the northeastern part of the depression (photo E. Reynard); C. Covered lapiés (photo E. Reynard)

7.4.3 Closed Depressions of Karstic Origin

Among the main geomorphological features of the Jura mountains are the numerous closed depressions of various shape, type and size. The largest one is the Joux Valley itself, of tectonic origin (see above). Most of the smaller depressions are due to karst evolution. Three of them have been classified as a geosite of national significance (geosite n ° 296; Reynard et al. 2012): the *Sèche des Amburnex*, the *Combe des Begnines* and the *Creux du Croue*.

The *Sèche des Amburnex* cuts through the massive limestone layers of Kimmeridgian and Upper Oxfordian (Purbeckian facies) rocks of the Mont Sâla anticline. The closed basin is remarkably delimited by a series of rocky steps covered with forest (Figs. 7.7a and 7.7b). Most of the depression is occupied by pastures (Fig. 7.7b). The central part, which is very flat, is rich in kamenitsas and sinkholes. Around it are karren, with orthogonal shapes due to differentiated pedogenesis (poor and stony soils on the lapiés tables and richer soils in fractures) (Fig. 7.7c).

The *Combe des Begnines* is an anticline depression dug in the Kimmeridgian and Upper Oxfordian (Purbeckian facies) series of the Mont Sâla anticline, which reaches the marl levels of the Middle Oxfordian (Argovian facies). The depression, which is more than 5 km long, is relatively narrow (on average 500 m) and asymmetric (Fig. 7.8a). The northwestern ridge is marked by a series of indentations, formed by SE-NW faults, parallel to the Jura compression. The southeastern flank is more regular, and the edge of the valley is formed by two ridges, separated by a depression, thanks to alternating limestone and marl levels. The bottom of the closed depression is enamelled with a series of funnel-shaped sinkholes (local name *'emposieux'*), aligned either in the direction of the fold axis (due to the presence of more permeable rocks; Fig. 7.8b) or along faults perpendicular to the axis. The sinkholes are shaped like funnels, evidence for the efficient removal of residual clays by karst extraction. The shape of the funnels is often asymmetrical: snow spots persist longer on the north-facing flank, thus promoting dissolution. This flank is therefore steeper than the south-facing slope (Fig. 7.8b).

Fig. 7.8 **a** General view of the *Combe des Begnines* towards the east; **b** Sinkholes in the Middle Oxfordian (Argovian facies) marls. Persistence of snow patches in the north-facing side of sinkholes increases karstic erosion (photos E. Reynard)

The *Creux du Croue* is an anticline depression, dug through the geological layers of the Noirmont anticline, to the heart of the anticline (Middle Oxfordian (Argovian facies) layers). The depression reaches the soft Argovian layers and widens into a large hollow, several tens of metres deep (Fig. 7.9). The limestone layers form high walls facing the centre of the depression. Below the walls, the marl layers underlie gently on inclined slopes and an impervious bottom, enabling the formation of a large marsh which occupies the entire centre of the depression. The sector is classified as a marsh of national importance. The southwestern part of the valley is also covered with scree. The sinkholes surrounding the marsh drain both the waters flowing from the slopes and those of the marsh. The stream that forms in the marsh is lost in one of them and joins the underground network.

From a structural point of view, several types of depressions can be distinguished according to their setting within the folded structure: they may be located in a syncline (like Lake Joux), in the axis of an anticline (so-called *'combes anticlinales'*, like the *Sèche des Amburnex*, the *Combe des Begnines* and the *Creux du Croue*), or on the flank of an anticline (*'combe de flanc'*, like the *Combe des Amburnex*, Fig. 7.10a or the *Pré de Bière*; Fig. 7.10b).

From a microclimatic point of view, these closed depressions are known as cold air traps: negative temperatures can occur at their bottom all year round and can reach –25 to –30°C several times each winter (Bloesch and Calame 1994; Schoeneich 2012). A study conducted since 2008 in the anticline depression of La Plateforme—La Perrausaz (46°34'30"N/3°54'30"E), at 1330 m a.s.l., shows that temperature inversion occurs every cold night (Schoeneich 2012). Temperature inversion has an effect on vegetation: in the lower parts of many closed depressions dwarf spruce trees can be seen, whose growth is hindered by the cold microclimate (Vittoz 1998; Fig. 7.11).

7.4.4 Karren

Surface karstic landforms were studied in detail by Aubert (1969, 1974, 1975) who highlighted the importance of structural factors in explaining exokarst and endokarst features. The presence of the Jura ice cap during the last glaciation favoured soil stripping and the exposure of limestone rocks, which accentuated the effects of dissolution (Aubert 1965). The spatial distribution of karren fields is closely dependent on lithology (mainly on Malm limestone outcrops), and their forms depend on fracturing (which gives the karren a chequered morphology), and on the fact that dissolution occurs mainly under vegetation, which gives the lapiés their rounded shape (Fig. 7.12).

The Jura sedimentary basin has experienced several emersions from the Tithonian on and signs of karstification can be detected from the Eocene, when continental Siderolithic deposits filled cavities in Cretaceous limestones (Perrin and Audétat 2002b). The stripping of the impermeable cover of the chain (Cretaceous deposits) probably happened during folding, which favoured karstification that continued throughout the Quaternary and is still active today. It was most efficient during the melting periods of the glaciers (Perrin and Audétat 2002b).

Surface dissolution was particularly active in anticlinal vaults, weakened by fractures, which facilitated the creation of the anticlinal depressions described above and flat-bottomed dolines and uvalas (Aubert 1969). The depressions were widened by a particular karst dissolution process, which affected both the heads of the limestone banks and their surface, and which Aubert (1974) called 'regressive karst erosion'. Ablation by dissolution has a

Fig. 7.9 *Creux du Croue* closed depression. The centre of the depression is occupied by a marsh of national importance (photo E. Reynard)

weak vertical component but a strong horizontal component, favoured by the frequency of marl banks and by the impurity of some limestones; the residues of dissolution thus moderate the process of vertical erosion. The Perrausaz karren field (near the Marchairuz Pass) is classified in the inventory of Swiss geosites (n° 256; Reynard et al. 2012) as a representative example of a covered karren field, rich in lapiés, dolines and uvalas of various shapes.

7.4.5 Sinkholes and Caves

Anticlinal surfaces and the bottoms of major structural depressions are perforated with dolines of various shapes (e.g. Fig. 7.8), some of which function as swallow holes and are connected to underground networks (Perrin and Audétat 2002b). Most of them certainly date from before the last glaciation as proved by the presence of morainic material in a doline in the *Grande Chaumille* pasture (Aubert 1966).

The density of caves is high (1.8 per km^2) and is explained by the presence of large karren fields and the high altitude (Perrin and Audétat 2002b). They are not distributed uniformly; the areas with the highest concentrations are located on the Mont Tendre anticline north-east of the Marchairuz Pass and along the Mont-Pelé anticline, south-west of the Marchairuz Pass (*Combe des Begnines - Creux du Croue* region). The Risoux has few caves, which is explained by the presence of large areas of Kimmeridgian marly limestone and by intense fracturing mainly resulting in diffuse infiltration. Three-quarters of the caves open in the Malm limestones (Perrin and Audétat 2002b).

Longirod cave (Swiss coordinates 507.642/153.896) is the deepest in the Jura (–509 m) and as such, is classified on the list of Swiss geosites (n° 146; Reynard et al. 2012). It is connected to a collector (underground river) carved in the limestones and marly limestones of the lower Sequanian. Tracing tests located the catchment area in the Amburnex valley thereby confirming a link with the Aubonne and Toleure springs (Perrin and Luetscher 2001).

Fig. 7.10 Two closed depressions on anticline flanks: the *Combe des Amburnex* (**a**) and the *Pré de Bière* depression (**b**). Arrows indicate the direction of dip of strata (photos E. Reynard)

Fig. 7.11 **a** Digital Terrain Model of La Perrausaz area showing the links between closed depressions (source: swisstopo); **b** *La Plateforme* closed depression functioning as a cold air trap responsible for the presence of dwarf spruce trees (photo: P. Schoeneich)

7.4.6 Ice Caves

The presence of frozen caves at low elevations in the Jura has long puzzled scientists. Indeed, as the 0°C isotherm is located above the highest crests (2200 m), freezing conditions are not expected in caves (Luetscher et al. 2005). Twenty-five caves presenting perennial ice are listed in the whole Jura Range (Luetscher et al. 2005); nine of which in the Marchairuz area, south-west of Lake Joux. Most of the Jura ice caves act as cold air traps where no air is exchanged with the outside atmosphere in summer, which prevents the ice from melting and favours refreezing of infiltration waters. The persistence of ice is favoured by air circulation in winter and the snow precipitation regime (Luetscher et al. 2005; Luetscher and Jeannin 2018), which influence the positive ice mass balance. Because of relatively high ice turnover, the age of cave ice certainly does not exceed several hundred years (Luetscher et al. 2005). Volumes vary from a few cubic metres to more than 6 000 m^3 in the Monlési ice cave in canton Neuchâtel. In the past, ice caves were used as cool places to store food and ice was also extracted. Systematic analysis of historical data (photographs, topographic maps, in situ inscriptions) showed a general trend to a decrease in ice volumes throughout the twentieth century in most caves in the Jura Mountains and acceleration of ice melting since the beginning of the 1990s (Luetscher et al. 2005).

Two ice caves in the Marchairuz area have been studied in great detail. The St-George ice cave (Swiss coordinates 508.075/153.430, alt. 1290 m a.s.l.; Figure 7.13a) is protected by a forest. During the two last centuries (Brulhart 2001), it was marked by a period of intense exploitation of ice in the first part of the nineteenth century, followed by a period of regeneration due to decline in ice exploitation. The ice has been regressing since the mid-twentieth century (in 1942, the volume of ice was estimated to be 2300 m^3; Audétat and Heiss 2002), for climatic reasons, with a period of stagnation in the 1980s and high rates of debris production by gelifraction of the cave walls. St-George ice cave is a geosite of regional importance (site n° 31 in the inventory of Vaud geosites; Pieracci 2008).

Fig. 7.12 Two examples of karren in the Marchairuz area: **a** Round-shaped lapiés; **b** Chequered round-shaped lapiés (photos E. Reynard)

The St-Livres ice cave (Swiss coordinates 512.408/157.750, alt. 1370 m a.s.l.; Figure 7.13b and c) is a geosite of national importance (site n° 257; Reynard et al. 2012). Ice was dated to 1200 years BP by Luetscher (2005), making it the oldest known ice in the Jura Mountains. Like in St-George, ice was highly exploited in the nineteenth century. In 1960, the total volume was estimated to be 3500 m^3 since when the mass balance has been negative and ice volume is decreasing; at the beginning of the twenty-first century it was around 1500 m^3 (Luetscher et al. 2005).

7.5 Conclusions

The structural and karstic landforms of the Joux Valley and its surroundings, together with pastures and forests, form the backbone of the Jura landscapes. It was to preserve the quality of these landscapes that the *Parc jurassien vaudois* was founded in 1973, bringing together local authorities, landowners, individuals and the Swiss League for the Protection of Nature (now named Pro Natura), who were committed to protecting the landscape and environmental quality, while allowing coexistence with agricultural, forestry and tourist activities (Capt et al. 1994). The park, which covers 532 km^2 and comprises territory belonging to 30 municipalities, has been recognised as a regional park of national importance since 2012 (*Parc Jura vaudois*; https://parcjuravaudois.ch). Let us hope that this national recognition will make it possible to better promote the value of its rich geomorphological heritage (Durussel et al. 1994; Perret 2008; Perret and Reynard 2011).

Fig. 7.13 **a** Ice cascade in the St-George ice cave (photo D. Brulhart, 28 April 1998); **b** St-Livres ice cave entry protected by trees in the middle of a pasture (photo D. Brulhart, 20 July 1997); **c** Layering of ice in the St-Livres ice cave (photo D. Brulhart, 20 July 1997)

Acknowledgments We are grateful to David Brulhart for his photographs of ice caves and Felix Hauser for Figs. 7.3 and 7.4.

References

Arn R, Campy M (1990) Un problème de paléogéographie glaciaire au maximum würmien dans la zone circumalpine: le glacier jurassien. Bull Soc neuch Sci Nat 113:115–131

Aubert D (1938) Les glaciers quaternaires d'un bassin fermé: la vallée de Joux (Canton de Vaud). Bull Soc vaud Sci Nat 60:117–130

Aubert D (1959) Le décrochement de Pontarlier et l'orogenèse du Jura. Mém Soc vaud Sci Nat 12(4):93–152

Aubert D (1965) Calotte glaciaire et morphologie jurassiennes. Eclogae Geol Helv 58(1):558–578

Aubert D (1966) Structure, activité et évolution d'une doline. Bull Soc neuch Sci Nat 89:115–120

Aubert D (1969) Phénomènes et formes du Karst jurassien. Eclogae Geol Helv 62(2):325–399

Aubert D (1970) Les graviers du Mammouth de Praz Rodet (Vallée de Joux, Jura vaudois). Bull Soc vaud Sci Nat 70(334):421–430

Aubert D (1974) L'érosion karstique régressive dans le Jura. In: Phénomènes karstiques II. Mémoires et Documents CNRS 15, pp 71–80

Aubert D (1975) L'évolution du relief jurassien. Eclogae Geol Helv 68 (1):1–64

Audétat M, Heiss G (2002) (eds) Inventaire spéléologique de la Suisse. Tome IV Jura vaudois, partie ouest. Commission de spéléologie de l'Académie suisse des sciences naturelles, La Chaux-de-Fonds

Barnes I, Shapiro B, Lister A, Kuznetsova T, Sher A, Guthrie D, Thomas MG (2007) Genetic structure and extinction of the Wooly Mammoth, *Mammuthus primigenius*. Current Biol 17:1072–1075 (supplemental data)

Bichet V, Campy M (2009) Montagnes du Jura. Géologie et paysages. Néo Editions, Besançon, 2nd edn, 303 pp

Bissig G, Reynard E (2006) L'eau en zone calcaire: Le Brassus. Atlas hydrologique de la Suisse, Berne (Excursion 3.3), 42 pp

Bloesch B, Calame F (1994) L'air du temps. In: Capt G, Jean-Petit-Matile O, Reymond J (eds) Le Parc jurassien vaudois. Editions 24 Heures, Lausanne, pp 35–45

Brulhart D (2001) La glacière naturelle de Saint-Georges (VD, Suisse): histoire et fluctuations d'un glacier souterrain. Bull Soc vaud Sci Nat 76(3):261–276

Campy M, Arn R (1991) The Jura glacier: paleogeography in the Würmian circum-alpine zone. Boreas 20:17–27

Campy M (1992) Palaeogeographical relationships between Alpine and Jura glaciers during the two last Pleistocene glaciations. Palaeo 93 (1–2):1–12

Capt G, Jean-Petit-Matile O, Reymond J (1994) Le Parc jurassien vaudois. Editions 24 Heures, Lausanne, 182 pp

Creer KM, Hogg TE, Readman PW, Reynaud C (1980) Palaeomagnetic secular variation curves extending back to 13,400 years B.P. recorded by sediments deposited in Lac de Joux, Switzerland; comparison with U.K. records. J Geophys 48:139–147

Durussel N, Schoeneich P, Reynard E (1994) Relief et paysage. In: Capt G, Jean-Petit-Matile O, Reymond J (eds) Le Parc jurassien vaudois. Editions 24 Heures, Lausanne, pp 23–33

Duvoisin G (1887) Exploitation de la glace des lacs de la Vallée de Joux. Bull Soc vaud Ing Arch 13:54–55

Fiore J, Pugin A, Beres M (2002) Sedimentological and GPR studies of subglacial deposits in the Joux Valley (Vaud, Switzerland): Backset accretion in an esker followed by an erosive jökulhlaup. Géog Phys Quatern 56(1):19–32

Guglielmetti L, Negro F, Mauri G, Vuataz FD, Abdelfettah Y, Clerc N, Giroud N, Marthet L, Schill E (2013) GeoNE: an integrated project for the exploration of low enthalpy deep aquifers in the canton of Neuchatel, western Switzerland. In: European geothermal congress 2013, Pisa, Italy, 3–7 June 2013, 13 pp

Jelmini JP (1987) Les mines d'asphalte du Val-de-Travers (1711-1986). Brève histoire d'une industrie neuchâteloise. Nouvelle Revue Neuchâteloise 14:1–45

Luetscher M (2005). Processes in ice caves and their significance for paleoenvironmental reconstructions. PhD thesis, Swiss Institute for Speleology and Karst Studies, La Chaux-de-Fonds, 154 pp

Luetscher M, Perrin J (2005) The Aubonne karst aquifer (Swiss Jura). Eclogae Geol Helv 98:237–248

Luetscher M, Jeannin PY (2018) Ice caves in Switzerland. In: Persoiu A, Lauritzen SE (eds) Ice caves. Elsevier, Amsterdam, pp 221–235

Luetscher M, Jeannin PY, Haeberli W (2005) Ice caves as an indicator of winter climate evolution: a case study from the Jura Mountains. The Holocene 15(7):982–993

Nicod J (1990) Le karst jurassien. Modèle géomorphologique spécifique. Bull Soc neuch Sci Nat 113:13–25

Parriaux A, Mayoraz R (1990) Fragilité des ressources en eau du Jura ou le jeu des probabilités à la vallée de Joux. Bull Soc neuch Sci Nat 113:107–114

Perret A (2008) Inventaire de géomorphosites du Parc jurassien vaudois. Essai d'intégration des géotopes spéléologiques et valorisation géomorphologique des réserves naturelles. MSc thesis, University of Lausanne, 285 pp

Perret A, Reynard E (2011) Inventaire des géomorphosites du Parc jurassien vaudois (Col du Marchairuz, Vaud). In: Lambiel C, Reynard E, Scapozza C (eds) La géomorphologie alpine: entre patrimoine et contrainte. Institut de Géographie, Lausanne, Géovisions 36, pp 19–34

Perrin J (2002) Hydrogéologie. In: Audétat M, Heiss G (eds) Inventaire spéléologique de la Suisse. Tome IV Jura vaudois, partie ouest. Commission de spéléologie de l'Académie suisse des sciences naturelles, La Chaux-de-Fonds, pp 18–22

Perrin J, Audétat M (2002a) Aperçu géologique. In: Audétat M, Heiss G (eds) Inventaire spéléologique de la Suisse. Tome IV Jura vaudois, partie ouest. Commission de spéléologie de l'Académie suisse des sciences naturelles, La Chaux-de-Fonds, pp 12–17

Perrin J, Audétat M (2002b) Géomorphologie et phénomènes karstiques. In: Audétat M, Heiss G (eds) Inventaire spéléologique de la Suisse. Tome IV Jura vaudois, partie ouest. Commission de spéléologie de l'Académie suisse des sciences naturelles, La Chaux-de-Fonds, pp 23–29

Perrin J, Luetscher M (2001) Observations géologiques et hydrogéologiques dans les gouffres de Longirod et du Narcoleptique (Jura vaudois). In: Actes du 11e Congrès National de Spéléologie 2001, pp 99–106

Pieracci K (2008) Inventaire des géotopes du canton de Vaud. University of Lausanne, Lausanne

Reynard E, Berger JP, Constandache M, Felber M, Grangier L, Häuselmann P, Jeannin PY, Martin S (2012) Révision de l'inventaire des géotopes suisses: rapport final. Groupe de travail pour les géotopes en Suisse, Lausanne, 45 pp

Schlüchter C (compil) (2009) Die Schweiz während des letzteiszeitlichen Maximums (LGM), 1:500 000. GeoKarten 500. Bundesamt für Landestopografie swisstopo, Wabern

Schoeneich P (2012) Microclimat d'une combe froide du Jura (région du Marchairuz, Vaud, Suisse). In: Bigot S, Rome S (eds) Les climats régionaux: observation et modélisation, 25ème Colloque de l'Association Internationale de Climatologie, 5–7 septembre 2012, Grenoble, pp 703–708

Sommaruga A (1997) Geology of the central Jura and the Molasse Basin: new insights into an evaporite-based foreland fold and thrust belt. Mem Soc neuch Sci Nat 12:1–176

Sommaruga A (1999) Décollement tectonics in the Jura foreland fold-and-thrust belt. Mar Pet Geol 16(2):111–134

Trümpy R (ed) (1980) Geology of Switzerland. A guide-book, 2 vol. Swiss Geological Commission and Wepf, Basel, 104 + 334 pp

Vittoz P (1998) Flore et végétation du Parc jurassien vaudois: typologie, écologie et dynamique des milieux. PhD thesis, University of Lausanne, 446 pp

Wegmüller S (1966) Über die spät- und postglaziale Vegetationsgeschichte des südwestlichen Jura. Beitr Geobot Aufnahme Schweiz 48, Bern, 148 pp

Weidmann M (1969) Le mammouth de Praz-Rodet (Le Brassus, Vaud). Note préliminaire. Bull Soc vaud Sci Nat 70(6):229–240

Weidmann M (2008) La Molasse de la Vallée de Joux (Jura, Suisse et France). Bull Soc vaud Sci Nat 91(1):69–101

Emmanuel Reynard is a Professor of Physical Geography and Director of the Interdisciplinary Centre for Mountain Research at the University of Lausanne, and a member of the International Association of Geomorphologists (IAG) Executive Committee. He has been chairman of the IAG Working Group on Geomorphosites (2001–2013) and president of the Working Group on Geotopes of the Swiss Academy of Sciences (2006–2017). His research concerns mountain geomorphology, geomorphological heritage, landscape geohistorical analyses and water management in mountains. He has worked in the Alps, the Romanian Carpathians and in the Maghreb (Tunisia, Morocco).

Philippe Schoeneich is a Professor of Physical Geography at the University Grenoble Alpes. He has been member of the Swiss Quaternary Commission, and is a member of the Global Terrestrial Network for Permafrost (GTN-P) steering committee. His research concerns mountain geomorphology, Quaternary, mass movements and mountain permafrost.

8 The Lavaux World Heritage Terraced Vineyard

Emmanuel Reynard and Emmanuel Estoppey

Abstract

Lavaux is a vineyard region located on the northern shore of Lake Geneva, between Lausanne and Vevey. Named a UNESCO World Heritage site as a cultural landscape in 2007, Lavaux testifies to centuries of wine growing and adaptation by man to natural processes. Conglomeratic hard banks and softer sandstones dating back to Oligocene sedimentation of large alluvial fans in the foreland molassic basin were differentially eroded by the Rhone glacier that occupied the Geneva Lake basin several times during the Quaternary. The result is a steep stair-like slope (in some places > 30°) that forms the skeleton of the vineyard terraces. Because of the steep slopes, winegrowers developed complex systems to manage hydrological processes and sediment fluxes. From a geomorphological point of view, the Lavaux cultural landscape is a structural landscape modelled by glacier erosion. Because the natural processes have been significantly modified by man, Lavaux can be considered as a cultural geomorphosite, i.e. a place where geomorphological processes and human activities form a symbiosis.

Keywords

Structural landforms • Glacier erosion • World Heritage site • Lavaux

E. Reynard (✉)
Institute of Geography and Sustainability and Interdisciplinary Centre for Mountain Research, University of Lausanne, Géopolis, 1015 Lausanne, Switzerland
e-mail: emmanuel.reynard@unil.ch

E. Estoppey
eqiilab, Ruelle du Four à Pains 6, 1867 Ollon, Switzerland
e-mail: e.estoppey@eqiilab.com

8.1 Introduction

Lavaux vineyard (46° 29′ N, 6° 45′ E; minimum altitude: 373 m a.s.l., maximum altitude: 650 m a.s.l.) is located in south-western Switzerland, between the urban zones of Lausanne and Vevey-Montreux (Fig. 8.1). This south-facing vineyard was developed on the Cenozoic conglomerates and sandstones that form the steep slopes of Mont-Pèlerin (alt. 1080 m a.s.l.) overlooking Lake Geneva. Cistercian monks were the first to start cultivating grapevines in the middle of the twelfth century. Lavaux vineyard was classified as a UNESCO cultural World Heritage site in 2007.

We first describe the geographical characteristics—including inscription on the World Heritage list—focussing on the impact of glaciation on landscape shaping, on structural landforms (alternating conglomerate and sandstone), on erosional processes and on the mitigation measures taken by winegrowers, and on the landslides, which affect parts of the area. This is followed by a presentation of the geological and climatic contexts.

8.2 Geographical Setting and UNESCO Recognition

Lavaux vineyard (Fig. 8.2) is located in the Vaud canton and stretches about 15 km along the south-facing northern shores of Lake Geneva from Vevey (in the east) to Lutry in the eastern suburbs of Lausanne (in the west). Fourteen historical villages with strict building regulations are scattered in and around the vineyard area. “It is an outstanding example of a centuries-long interaction between people and their environment, developed to optimise local resources so as to produce a highly valued wine that has always been important to the economy” (http://whc.unesco.org/fr/list/1243, accessed 10.12.2019).

The vineyard is divided into more than 10,000 terraces—some of which do not exceed a few dozen square metres in

E. Reynard (ed.), *Landscapes and Landforms of Switzerland*, World Geomorphological Landscapes,
https://doi.org/10.1007/978-3-030-43203-4_8

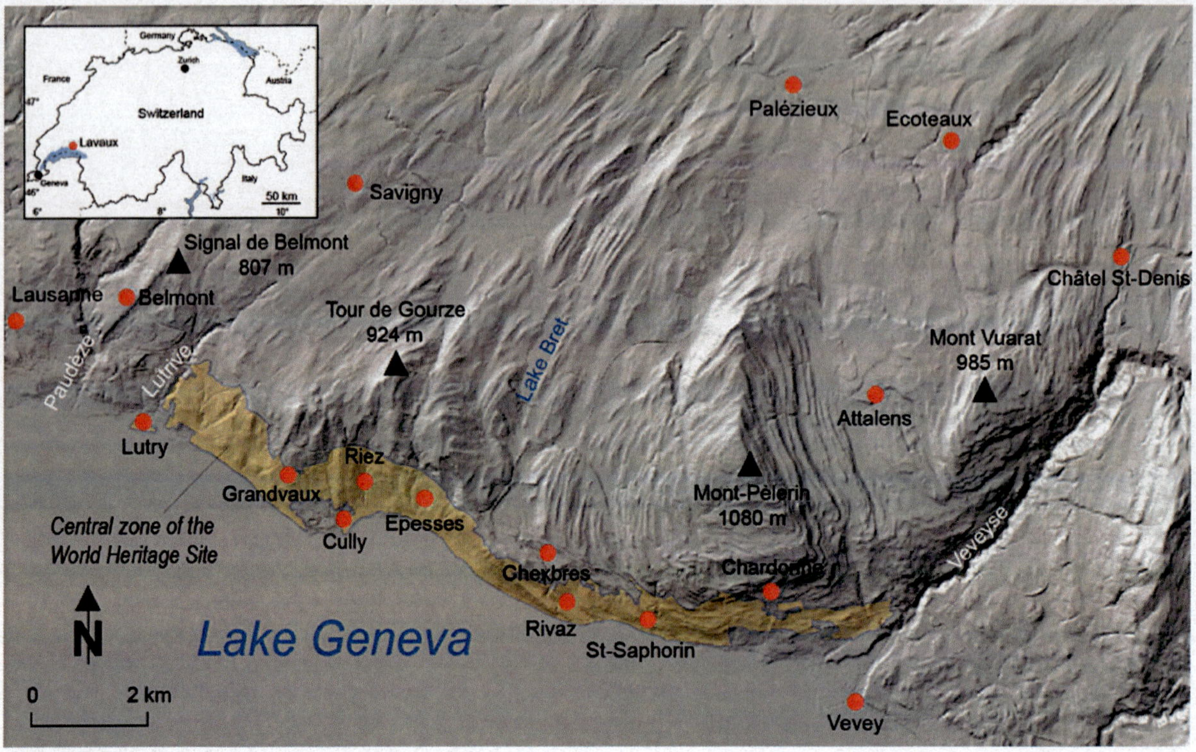

Fig. 8.1 Map of Lavaux and surroundings (hillshade relief: swisstopo)

Fig. 8.2 A view of Lavaux vineyard (photo Montreux-Vevey Tourisme)

size—separated by more than 400 km of high walls. Today's landscape is the result of several centuries of human works and changes in the landscape (AILU 2006; Dresco et al. 2007). Although there is some evidence for wine growing in the area in Roman times, the roots of the present vineyard can be traced back to the twelfth century, when Cistercian monasteries were given land by the Bishop of Lausanne and began to build the terraces. By the fourteenth century, most of the land was being worked by tenants, who, in exchange, owed part of the yield to the monasteries. At that time, and until the 1950s, mixed farming was the rule; in addition to vines, farmers planted orchards and grazed livestock on the higher land, i.e. above 600 m a.s.l. By the beginning of the nineteenth century, like in other parts of Switzerland, the landscape had become a patchwork of very small plots that were difficult to manage efficiently. After 1803, when Vaud joined the Swiss Federation as a new canton, a period of agricultural improvement commenced. In particular, terraces were widened, taller walls were built and new drainage systems were installed to limit erosion (see below). Like in other European vineyards, by 1866 phylloxera disease threatened the survival of Lavaux vineyard

Table 8.1 Changes in the extent of Lavaux vineyard between 1891 and 1998 (Demaurex 2006)

	1891	1934	1958	1980	1998
Surface area (ha)	1166	885	784	679	708
Surface area (%)	100	75.9	67.2	58.2	60.7

and new agricultural techniques were introduced to save it. Two railway lines were built in the upper and the lower part of the vineyard in the 1860s. After World War II, urban sprawl became an acute issue for this rural zone between two expanding urban zones, Lausanne and Vevey-Montreux (Table 8.1). As early as the 1950s, some municipalities introduced measures to protect the vineyard against land speculation but the whole area was not protected until the end of the 1970s. During that period, mixed farming disappeared, and farmers in Lavaux specialised in wine growing and cattle grazing in the uplands. The last major change to the landscape occurred in 1970 with the construction of the A9 motorway which runs along the upper boundary of the western part of the vineyard and crosses the eastern part (Chardonne).

The outstanding universal value of Lavaux was recognised by UNESCO based on three out of the ten selection criteria, *iii*: in Lavaux, evidence for the nearly a thousand years of wine growing is clearly visible in the landscape, *iv*: the vineyard landscape that has evolved since the Middle Ages illustrates the long-term development of wine-growing activities and the importance of Lavaux for the historical development of the Lausanne area, and *v*: Lavaux is both an example of the symbiosis between humans and nature based on farming, and of effective regulations to protect the highly vulnerable vineyard from the expansion of the Lausanne and Vevey-Montreux urban zones. The UNESCO World Heritage site is divided into two zones: a central zone covering 898 ha and a 1408-ha buffer zone. An association (*Association Lavaux Patrimoine mondial*) gathering local municipalities, various stakeholders and ordinary inhabitants manages the site with the help of a management plan, which provides "an analysis of socio-economic data, and a series of management strategies for research and culture, economy, land-use planning and tourism" (http://whc.unesco.org/fr/list/1243, accessed 10.12.2019). UNESCO also classified Lavaux as a cultural landscape, i.e. a place where the symbiosis between nature and humans expresses "a long and intimate relationship between peoples and their natural environment" (http://whc.unesco.org/en/culturallandscape/, accessed 10.12.2019).

In addition, Lavaux is protected by several cantonal and federal regulations, in particular the Federal Inventory of Landscapes and Natural Monuments (IFP), the Federal Inventory of Swiss Heritage Sites (ISOS), which protects several villages, and a specific cantonal law (*Loi sur le plan de protection de Lavaux*), adopted in 1979 after the Vaud citizens accepted a cantonal initiative launched by the famous ecologist Franz Weber to protect Lavaux against urban sprawl.

8.3 Geology and Climate

The Lavaux bedrock consists of Tertiary sediments, called the Lower Freshwater Molasse (*Untere Süsswasser-Molasse* or *USM*), deposited in a foreland basin when the Alps were forming. The Lower Freshwater Molasse sediments were deposited in three phases (Weidmann 1988; Platt and Keller 1992; Weidmann et al. 1993; Gorin et al. 1993; Borel and Marchant 2007; Baud et al. 2014).

- During the lower Chattian (28.5–26 Ma) marls and sandstones—called red molasse—were deposited in river plains under warm climate conditions;
- During the upper Chattian (26–23 Ma) conglomerates—called *poudingues* due to their resemblance to the dried fruit in English Christmas pudding—and sandstones were laid down in a large alluvial fan. Due to the decrease in slope (from about 1% in the Vevey area to a flat plain in the Lausanne region), the grain size of the deposits decreased from the proximal (*Mont-Pèlerin* conglomerates) to the distal (*La Cornalle* and *Lausanne* sandstones) part of the fan (Trümpy and Bersier 1954). The series is 1000 m thick and the conglomerates represent less than half the volume (Burri and Bersier 1972). Coal and gypsum also formed in lacustrine and palustrine depressions (Weidmann 1988);
- During the Aquitanian (23–20.5 Ma) sandstone and marl sedimentation were alternated in a vast alluvial plain to form the grey molasse of Lausanne (*molasse grise de Lausanne*).

Marls and coal strata in the Lower Freshwater Molasse are particularly rich in fossils—plants, vertebrates (crocodiles, turtles) and invertebrates (snails, mussels). Thousands of specimens belonging to 145 species, of which 44 were previously unknown, were found in the *Moulin-Monod* deposits (Heer 1855). The site is classified as a geosite of cantonal importance.

From a tectonic point of view, Lavaux is part of the folded and thrusted Subalpine Molasse, limited to the west by the Paudèze fault (local name of the frontal thrust of the Subalpine Molasse). The sediments were folded and thrusted approximately 10 Ma years ago (Baud et al. 2014) and are now divided into several southwest dipping rock segments, separated by thrust faults (Fig. 8.3). The Mont-Pèlerin forms a large perched syncline (Fig. 8.4).

The Rhone glacier occupied the Lake Geneva region several times during the Quaternary. Up to five glacier advances are documented in the Geneva Lake basin during the middle and late Pleistocene (Wildi et al. 2014). In the Lavaux area, deposits of three glaciations were recognised by Pugin et al. (1993) in Ecoteaux (Fig. 8.1), the earliest one being more than 780,000 years old (last paleomagnetic inversion).

Climatic conditions were summarised for the World Heritage candidature as follows (AILU 2006; Rebetez 2007). Mean annual air temperature is 10.5 °C and the minimum monthly mean temperatures in January is 2 °C, the maximum in July is 20 °C (data from Montreux weather station, 1978–2004). Temperature gradients are 0.6 °C/100 m in summer and 0.5 °C/100 m in winter. Mean annual precipitation is around 1200 mm (data from Montreux weather station, 1931–2004) and summer storms can be particularly intense, especially in the eastern part of the area close to the Prealps. Current climate warming is beneficial for wine growing in Lavaux and the upper limit of the vineyard could even shift 100 m upwards per decade if the current temperature increase continues (Rebetez 2007); on the other hand, heavy rainfall events, which favour soil erosion, could become more frequent in the future.

Lavaux is renowned for the famous "three suns" believed to facilitate wine growing: first, direct radiation; second, indirect radiation due to reflection of light by the lake surface, which is particularly efficient between September and March and mitigates the risk of frost (Bouët 1946); and third, at night, restitution of heat accumulated by high walls and conglomerate escarpments during the day.

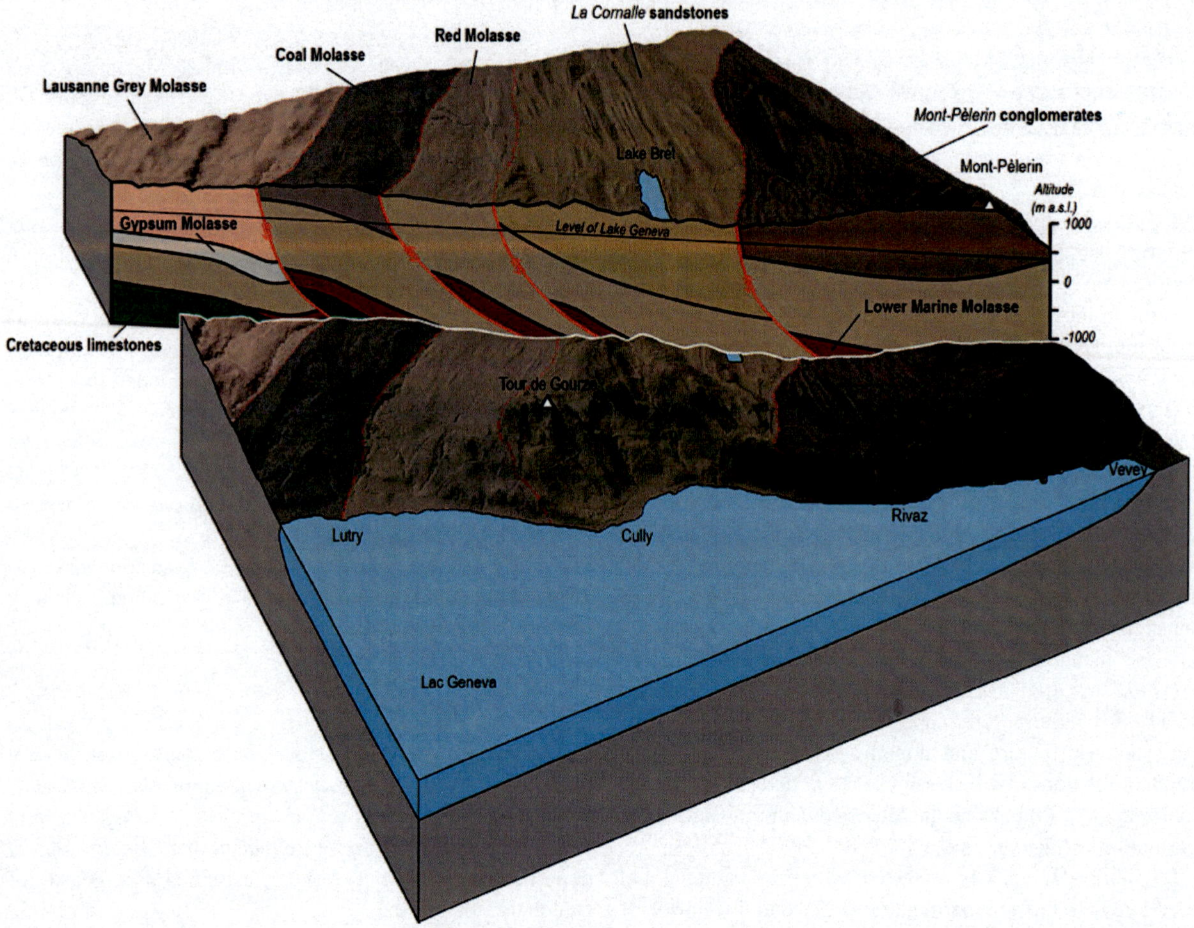

Fig. 8.3 Block diagram of the Lavaux area showing the stratigraphy and deformation of the Molasse (modified after Borel and Marchant 2007)

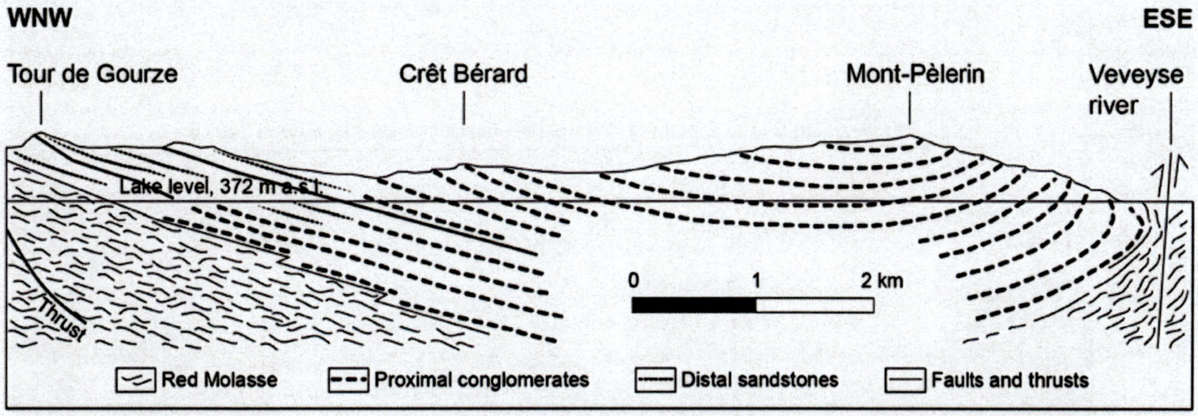

Fig. 8.4 Simplified cross section through the Molasse Basin in the Lavaux area (modified after Trümpy 1980)

8.4 Landscapes and Landforms

8.4.1 Glacial Landforms

The first geomorphological impact of Quaternary glaciations was the formation of the Lake Geneva basin (Fiore 2007; Wildi et al. 2014). Glacial incision was more marked in the upper part of the depression (in the Chablais area, where the top of the bedrock is located 400 m below sea level), than in the Molasse Basin. This reflects the transition of the Rhone glacier from an alpine to a piedmont glacier (Fiore 2007). In the upper Rhone valley, meltwater drainage was concentrated at the bottom of the main valley (one glacier flow), whereas on the Central Plateau the glacial flows were divided between several wider, shallower valleys, as is the case in the Geneva area and in the northern part of the Vaud canton.

In Lavaux, both systems are present. Because of the hard Mont-Pèlerin conglomerates and the relatively high altitude of the hill (1080 m a.s.l.), the main glacier flow was concentrated south of Mont-Pèlerin hill. This was particularly the case at the end of the glaciation periods, when the volumes of ice were decreasing and the elevation of the top of the glacier was lower, and the glacier incised a deep depression subsequently filled by Lake Geneva sediments and water (Figs. 8.5 and 8.6). Here, the top of the Molasse bedrock is about 200 m below sea level. Ice also flowed northwards (the Swiss lobe of the Rhone glacier) and eroded shallow south-north valleys filled with moraines, fluvio-glacial and fluvial sediments, and sometimes occupied by lakes (e.g. Lake Bret, Fig. 8.1). The hard Mont-Pèlerin conglomerates formed an obstacle to ice flows that were consequently channelled eastwards and westwards (Bret and Puidoux depressions) (Fig. 8.1). It was in one of these shallow valleys (Palézieux depression) that Pugin et al. (1993) found evidence for the three glacial cycles.

Unlike in Lausanne, moraine ridges were not deposited because of the steep slopes, only lodgement till. The till was deposited on sandstone terraces eroded by the glacier and is now partly incorporated in the vineyard soils. All the morainic deposits visible in the vineyard area date back to the last glaciation (Würm, 115,000–11,500 BP).

Crystalline erratic boulders transported from the Alps (Mont Blanc massif, Penninic domain) were certainly quite numerous in the past, but were nearly all used to build wine presses, for example. An erratic block located in Grandvaux, outside the vineyard area, was classified by the Vaud canton in 1985. Some blocks are also preserved along river courses (e.g. Forestey) because they were too difficult to extract.

8.4.2 Structural Landforms

The impressive Lavaux landscape, characterised by thousands of terraces separated by high walls, and the steepness of the slopes, is linked with the structural context. The hard conglomerates resisted lateral glacial erosion. The concentration of conglomerates in the sector between Epesses and Rivaz (Dézaley) explains why the slopes are so steep in this sector (Fig. 8.7), they are the remnants of the former U-shaped glacial valley. Note that altitude of the hills decreases from the east (Mont-Pèlerin, 1080 m a.s.l., hard conglomerates, proximal facies) to the west (Signal de Belmont, 807 m a.s.l., sandstones, distal facies of the coal molasse). The cuesta morphology (Pralong and Reynard 2004; Reynard 2007) fades towards the west, reflecting the decreasing concentration of hard conglomerates.

Differential glacial erosion shaped the slope into alternating vertical cliffs in zones filled with conglomerates and into more gentle slopes in sandstone areas (Fig. 8.8). Differential erosion was more marked where the ice flow ran sub-parallel to the structural trend (e.g. east and west of

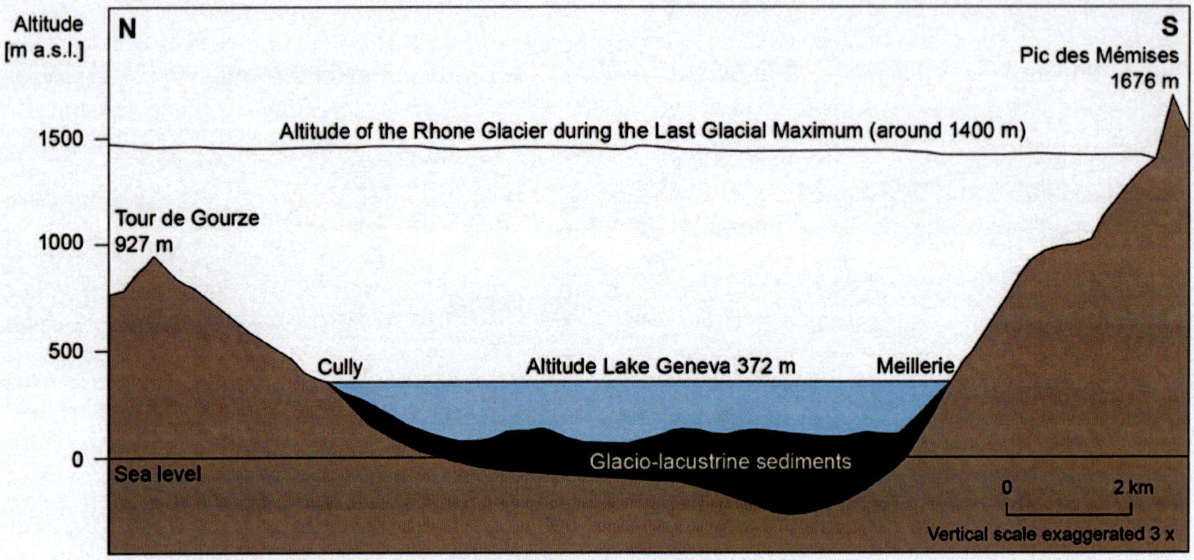

Fig. 8.5 Simplified cross section through Lake Geneva (modified after Borel and Marchant 2007)

Fig. 8.6 View of Geneva Lake towards the east. The arrow indicates the remnants of the U-shaped glacial valley in Dézaley area (photo M. Rosset)

Mont-Pèlerin, where a typical cuesta morphology is visible (Fiore 2007; Fig. 8.1). People used this natural configuration of sandstone terraces covered with moraines to grow grapevines while the conglomerate cliffs were used as natural walls to support the terraces (Fig. 8.8). Poorly cemented and fractured conglomerates are particularly sensitive to weathering and small rockslides are frequent. After World War II, the winegrowers, organised in syndicates and subsidised by the cantonal administration, undertook intensive, large-scale stabilisation works (Fig. 8.8). The techniques they used evolved with time: concrete was used in the first works, with no attention paid to aesthetic and ecological considerations; today, such works are limited because of their negative impacts on the landscape value of Lavaux.

Fig. 8.7 View of terraces and conglomerate banks in Dézaley area (photo E. Reynard)

8.4.3 Hydrology and Soil Erosion Management Systems

The many watercourses which flow from the Tour de Gourze (924 m a.s.l.) and Mont-Pèlerin (1080 m a.s.l.) summits to the lake have incised relatively deep valleys because of the steepness of the slopes—typically more than 15°, reaching maximum in the Dézaley area. Their longitudinal profile is a steep stair-like slope with some spectacular waterfalls.

The steep slopes also cause soil erosion and have resulted in the development of sophisticated surface water collection systems (Fig. 8.9). First, high walls (Fig. 8.7) were built with the aim of reducing the slope. In the steepest parts of the vineyard, the total volume of the walls represents more than 1100 m^3/ha, and their total length is more than 400 km (AILU 2006). The upper part of the walls is slightly elevated relative to the terrain (Fig. 8.9, letter a), meaning surface waters can be transported by gravity through artificial channels (Fig. 8.9, letter b; Fig. 8.10 letter c) towards the lake. These small elevations also trap eroded soil, which can then be transported back to the upper part of the terrace.

To avoid a splash effect during heavy storms the soil has a grass cover (Fig. 8.10, letter a) and after pruning, the vine shoots are left on the ground (b) thereby creating a shallow step parallel to the rows of vines that reduces the angle of the slope and consequently erosion. The aim of orienting the rows of vines parallel to the contour lines was also to reduce erosion.

All these techniques are an example of man's adaptation to natural processes and justify considering the Lavaux terraces as a "built cultural landscape".

8.4.4 Landslides

Postglacial decompression favoured rockslides in the Dézaley area, where there are numerous conglomerate cliffs, and large-scale landslides in the western part of Lavaux, where the rocks are Chattian sandstones and marls gently dipping to the south-east (Weidmann 1988). Landslides result from the alteration of the sandstone cement (due to dissolution) and the sensitivity of fractured sandstones to frost weathering. The slides are favoured by the presence of impervious marls dipping toward the lake and that deform plastically, as well as by water infiltration (Parriaux 1998; Fig. 8.11).

Fig. 8.8 Conglomerates banks structuring the vineyard slopes and maintenance works (photos E. Reynard)

Most of these landslides are now stabilised. However, some remain active, particularly west of Lavaux, around Belmont (Lausanne suburbs; Noverraz and Bonnard 1992) and near Epesses, where a large landslide (*La Cornalle – Les Luges* landslide, Bersier et al. 1975; Parriaux 1998) occurred in the sandstone area and slid into the lake. The landslide is divided into three parts (Fig. 8.11): (a) the erosional escarpment (35 m high, 115 m long) situated above the motorway and characterised by intense rock sagging (Fig. 8.12); (b) a zone of accumulation (*La Cornalle*) corresponding to a slow debris slide moving south-eastwards on a dip slope; (c) a landslide channel (*Les Luges*, French for *sledges*) leading to Lake Geneva. The location of the channel is controlled by a vertical fault and the direction of the slope. Terrestrial laser scanning measurements estimated cliff retreat at 12 mm/year (Carrea et al. 2015). This active landslide was a serious obstacle to geotechnical works for the construction of two railway lines in the 1850s and for the construction of the motorway one century later (Parriaux 1998).

8.5 Conclusions: Lavaux, a Cultural Geomorphosite

The character of the Lavaux landscape (vineyard terraces) is closely linked to local geomorphology. Both the geological structure and glacial processes explain the shape of the slope. The bedrock is Lower Freshwater Molasse, characterised by conglomerates alternating with sandstones in the east (proximal part of the Mont-Pèlerin Chattian alluvial fan) and sandstones alternating with marls, and in some places with coal, in the west (distal position). This explains the steep slopes in the Dézaley area and the frequent landslides in the western part of the vineyard. Lavaux was classified as a World Heritage site by UNESCO in 2007 and was recognised as a cultural landscape, i.e. an area in which nature and man interact harmoniously.

In this paper, we demonstrate how geomorphology forms the skeleton of the landscape. In this sense, Lavaux can be

Fig. 8.9 Hydrological and sediment flux management. **a** Upper elevated part of the wall; **b** artificial channels collecting water (photo E. Reynard)

Fig. 8.10 Techniques for mitigating soil erosion. **a** grass soil cover; **b** dead branches left on the soil; **c** system for water collection (photo E. Reynard)

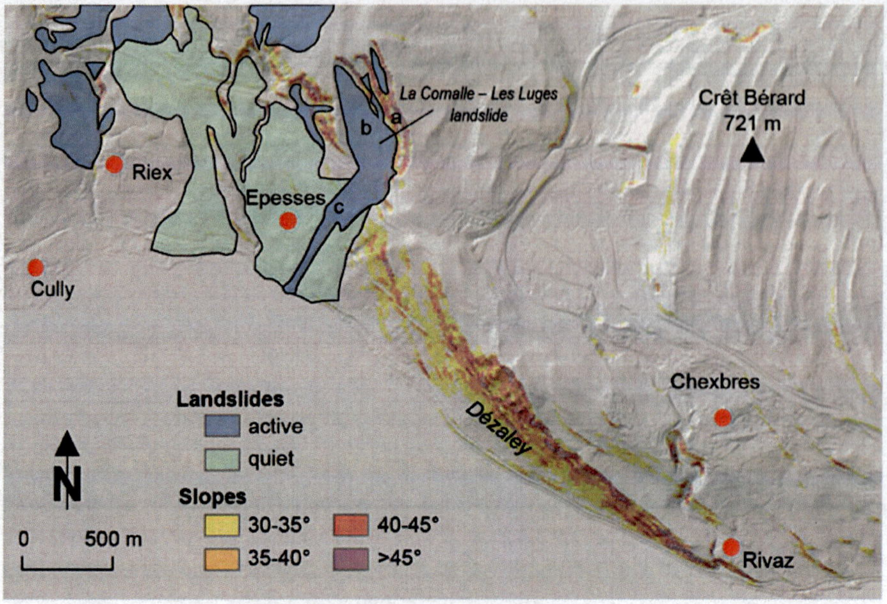

Fig. 8.11 Active and relict landslides in the western part of Lavaux area and steepest part of the vineyard (Dézaley) due to the presence of hard conglomerates. La Cornalle – Les Luges landslide: **a** = source area (escarpments), **b** = accumulation zone (slow debris slide); **c** = landslide channel (*source* of DTM and data: swisstopo)

Fig. 8.12 La Cornalle escarpment: sandstone weathering and sagging (photo J.P. Pralong)

considered as a cultural geomorphosite, i.e. a geomorphological system where human activities (in this cultivating vineyards) are in symbiosis with the geomorphological context and processes (Reynard and Giusti 2018).

Acknowledgments We thank Robin Marchant for making Figs. 8.3 and 8.5 available. We are grateful to Lavaux Patrimoine mondial, Montreux-Vevey Tourisme, Manon Rosset and Jean-Pierre Pralong for their photographs.

References

AILU (2006) Lavaux, vignoble en terrasses face au lac et au Alpes. World Heritage candidature report, Cully, 264 pp https://whc.unesco.org/en/list/1243/documents/

Baud A, Marchant R, Meisser N, Testaz G (2014) Sedimentology and wine, a cross road from early evaporite to carbonate platform and foreland basin terroirs of the best Swiss wines – a Rhone Valley to Lavaux (UNESCO Site) taste transect. In: 19th International

Sedimentological Congress 2014, 18–22 Aug 2014, Geneva, Switzerland, Field trip FTB 2 guidebook, 33 pp
Bersier A, Blanc P, Weidmann M (1975) Le glissement de terrain de La Cornalle – Les Luges (Epesses, Vaud, Suisse). Bull Soc vaud Sc Nat 72(4):165–191
Borel G, Marchant R (2007) Géologie. In: Dresco JP, Chuard C, Nicod C, Villiger D (eds) Lavaux, vignoble en terrasses. Favre, Lausanne, pp 93–99
Bouët M (1946) Note sur le climat de Lavaux. Bull Soc vaud Sc Nat 63:305–314
Burri M, Bersier A (1972) Les complications tectoniques du Mont-Pèlerin. Eclogae Geol Helv 65(1):21–26
Carrea D, Abellan A, Derron MH, Jaboyedoff M (2015) Automatic rockfalls volumes estimation based on terrestrial laser scanning data. In: Lollino G et al (eds) Geology for society and territory, vol 2. Springer, Berlin, pp 425–428
Demaurex G (2006) Evolution du paysage de Lavaux au cours du 20e siècle. Master's thesis, University of Lausanne
Dresco JP, Chuard C, Nicod C, Villiger D (eds) (2007) Lavaux, vignoble en terrasses. Favre, Lausanne
Fiore J (2007) Quaternary subglacial processes in Switzerland: geomorphology of the plateau and seismic stratigraphy of Western Lake Geneva. PhD thesis, University of Geneva, 169 pp
Gorin GE, Signer C, Amberger G (1993) Structural configuration of the western Swiss Molasse Basin as defined by reflection seismic data. Eclogae Geol Helv 86(3):693–716
Heer O (1855–1859) Flora tertiaria Helvetiae: Die tertiäre Flora der Schweiz. Wurster Verlag, Winterthur, 3 vol
Noverraz F, Bonnard C (1992) Le glissement rapide de la Chenaula. In: Interpraevent 1992 – Bern, Tagungspublikation, Band 2, pp 65–76
Parriaux A (1998) Glissement de la Cornalle. Bull Géol appl 3(1):49–56
Platt NH, Keller B (1992) Distal alluvial deposits in a foreland basin setting. The Lower Freshwater (Miocene), Switzerland: sedimentology, architecture and palaeosols. Sedimentology 39:545–565
Pralong JP, Reynard E (2004) Lecture et analyse d'un paysage: Lavaux (Vaud, Suisse). In: Reynard E, Pralong JP (eds) Paysages géomorphologiques, Compte-rendu du séminaire de 3ème cycle CUSO 2003. Institut de Géographie, Lausanne, pp 35–51
Pugin A, Bezat E, Weidmann M, Wildi W (1993) Le bassin d'Ecoteaux (Vaud, Suisse): témoin de trois cycles glaciaires quaternaires. Eclogae Geol Helv 86(2):343–354
Rebetez M (2007) Climatologie. In: Dresco JP, Chuard C, Nicod C, Villiger D (eds) Lavaux, vignoble en terrasses. Favre, Lausanne, pp 101–105
Reynard E (2007) Géographie. In: Dresco JP, Chuard C, Nicod C, Villiger D (eds) Lavaux, vignoble en terrasses. Favre, Lausanne, pp 89–91
Reynard E, Giusti C (2018) The landscape and the cultural value of geoheritage. In: Reynard E, Brilha J (eds) Geoheritage. Assessment, protection, and management. Elsevier, Amsterdam, pp 147–166
Trümpy R (1980) Geology of Switzerland, a guide-book, 2 vol. Wepf, Basel/New York, 334 pp
Trümpy R, Bersier A (1954) Les éléments des conglomérats oligocènes du Mont-Pèlerin. Pétrographie, statistique, origine. Eclogae Geol Helv 47:119–166
Weidmann M (1988) Atlas géologique de la Suisse, feuille 1243, Lausanne, Notice explicative. Service hydrologique et géologique national, Berne
Weidmann M, Homewood P, Morel R, Berchten JD, Bucher H, Burri M, Cornioley JR, Escher P, Rück P, Tabotta A, Zahner P (1993) Atlas géologique de la Suisse, feuille 1244, Châtel-St-Denis, Notice explicative. Service hydrologique et géologique national, Berne
Wildi W, Corboud P, Girardclos S, Gorin G (2014) Geological and archaeological visit of Geneva. University of Geneva, Geneva, 63 pp https://archive-ouverte.unige.ch/unige:37233

Internet Resources

UNESCO World Heritage List, Lavaux. https://whc.unesco.org/en/list/1243
Lavaux vineyard terraces UNESCO World Heritage Site. https://www.lavaux-unesco.ch

Emmanuel Reynard is Professor of Physical Geography and Director of the Interdisciplinary Centre for Mountain Research at the University of Lausanne, and a member of the International Association of Geomorphologists (IAG) Executive Committee. He has been chairman of the IAG Working Group on Geomorphosites (2001–2013) and president of the Working Group on Geotopes of the Swiss Academy of Sciences (2006–2017). His research concerns mountain geomorphology, geomorphological heritage, landscape geohistorical analyses and water management in mountains. He has worked in the Alps, the Romanian Carpathians and in the Maghreb (Tunisia, Morocco).

Emmanuel Estoppey is Co-founder and Director of eqiilab, a private office specialised in territorial management and sustainable development. He was formerly Site Manager of Lavaux Patrimoine mondial (2009–2018) and a member of ViTour network that includes the vineyards of the World Heritage List. He is the president of World Heritage Experience Switzerland and is active in several projects aiming at managing and promoting World Heritage sites. He has also participated in the development of educational paths on geology and develops projects on the promotion and understanding of landscape issues.

9 Structural Landscapes and Relative Landforms of the Diablerets Massif

Philippe Schoeneich and Emmanuel Reynard

Abstract

The Diablerets Massif is a striking structural landscape, rising over 2000 m above the valley bottom of Les Diablerets. The massif makes it possible to observe the complex folded structure of a tectonic nappe, as well as the sharp topographical contrast between the resistant limestone rock walls and the surrounding softer rocks. The area is still glacierised, and a very well preserved Lateglacial moraine sequence can be observed in the Pillon Pass area. An extended mountain karst—the Tsanfleuron glacio-karst—developed on the south-eastern side of the massif, and spectacular gypsum karst features can be observed all around the massif. The historical Derborence rock avalanches (1714 and 1749 AD) started from the southern rock walls of the summit area and form a spectacular rock avalanche deposit landscape.

Keywords

Structural landscape • Glaciers • Lateglacial • Karst • Rock avalanche • Diablerets

9.1 Introduction

The Diablerets Massif (46°20′ N, 7°10′ W; maximum altitude: 3 210 m a.s.l.) is located in south-western Switzerland in the "Hautes Alpes Calcaires", a limestone massif, first described by the Swiss geologist Maurice Lugeon in the early twentieth century (Lugeon 1914/1918). The Massif is famous for its impressive structural landscapes, for its cultural renown ("Diablerets" means "the realm of devils") and for its geohistorical importance, specifically the development of the cover nappe theory.

In this chapter, we describe five main parts of the massif (Fig. 9.1): the Diablerets-Pillon area (Ormonts valley), the Tsanfleuron glacio-karstic area, the Derborence cirque and rock avalanche deposits, the Anzeindaz Valley, and the Col de la Croix gypsum pyramids.

9.2 Geographical and Geological Setting

The Diablerets Massif includes three cantons: Vaud, Valais and Bern. The Oldenhorn peak (written *Oldehore* on the map in Fig. 9.1) (3123 m a.s.l.) is located at the junction of the three cantons. The massif is drained by six rivers: the Grande Eau, the Gryonne, the Avançon, the Lizerne and the Morge, tributaries of the river Rhone, and the Saane River, a tributary of the river Rhine.

The Diablerets Massif is sparsely inhabited. The biggest commune is Ormont-Dessus (1468 inhabitants in 2018), including Les Diablerets tourist resort. On the Bern side of the massif, the village of Gsteig has a population of 987 inhabitants (2018). Only these two areas, situated on the two sides of Pillon Pass, and La Barboleusaz on the south-western side, are inhabited all year round. The rest of the massif (Anzeindaz, Derborence, Tsanfleuron) can only be reached in summer. Agriculture (cattle breeding, milk production and cheese making) and tourism, focused on Les Diablerets tourist resort and Gsteig, part of the Gstaad tourist destination and involving both summer (hiking) and winter (skiing) activities, are the main branches of the local economy.

Tectonically, the Diablerets Massif is part of the Helvetic Alps. These are formed by the so-called Helvetic nappes—

P. Schoeneich (✉)
Institute of Urban Planning and Alpine Geography, University Grenoble Alpes, 14 bis av. Marie-Reynoard, 38100 Grenoble, France
e-mail: philippe.schoeneich@univ-grenoble-alpes.fr

E. Reynard
Institute of Geography and Sustainability and Interdisciplinary Centre for Mountain Research, University of Lausanne, Géopolis, 1015 Lausanne, Switzerland
e-mail: emmanuel.reynard@unil.ch

E. Reynard (ed.), *Landscapes and Landforms of Switzerland*, World Geomorphological Landscapes,
https://doi.org/10.1007/978-3-030-43203-4_9

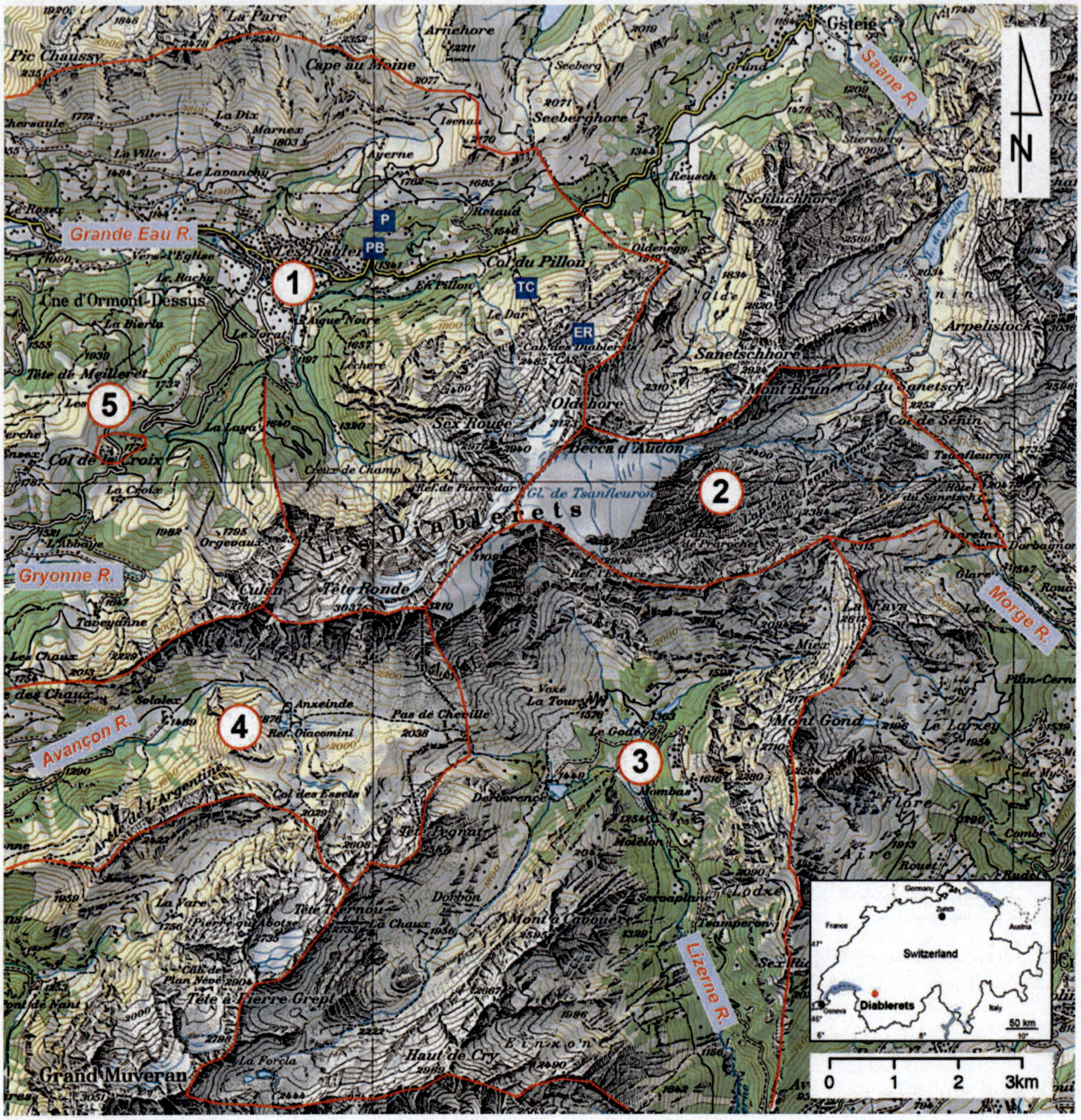

Fig. 9.1 Topographic map of the Diablerets Massif (source: swisstopo). Numbers refer to the five areas described (outlined in red): (1) Les Diablerets—Pillon Pass; (2) Tsanfleuron; (3) Derborence; (4) Anzeindaz; (5) Col de la Croix. *P* Les Parchets slide; *PB* Pont Bourquin earth flow; *TC* Tête aux Chamois deep rockslide; *ER* Entre la Reille

Morcles nappe, Wildhorn nappe, itself subdivided into the Diablerets, Mt Gond and Sublage nappes, and the Ultrahelvetic nappes which were folded with the Helvetic nappes and are, therefore, partly inserted between the Morcles and Wildhorn nappes, for example in Derborence (Fig. 9.2; Escher et al. 1993, 1997; Epard 2001; Steck et al. 2001). The Diablerets Massif itself mainly comprises two tectonic nappes: the Mont-Gond nappe, which covers the Diablerets nappe. They are separated from the underlying Morcles nappe by Ultrahelvetic terrains.

At the front of the Helvetic nappes, the Ultrahelvetic nappes occupy large areas. As the rocks that form these nappes are quite soft and easy to erode (marls, flysch, evaporites), they formed the so-called “Zone des Cols”, a relatively low altitude area with several passes (Col de la Croix, 1778 m a.s.l., Col du Pillon, 1546 m a.s.l.) (Fig. 9.1).

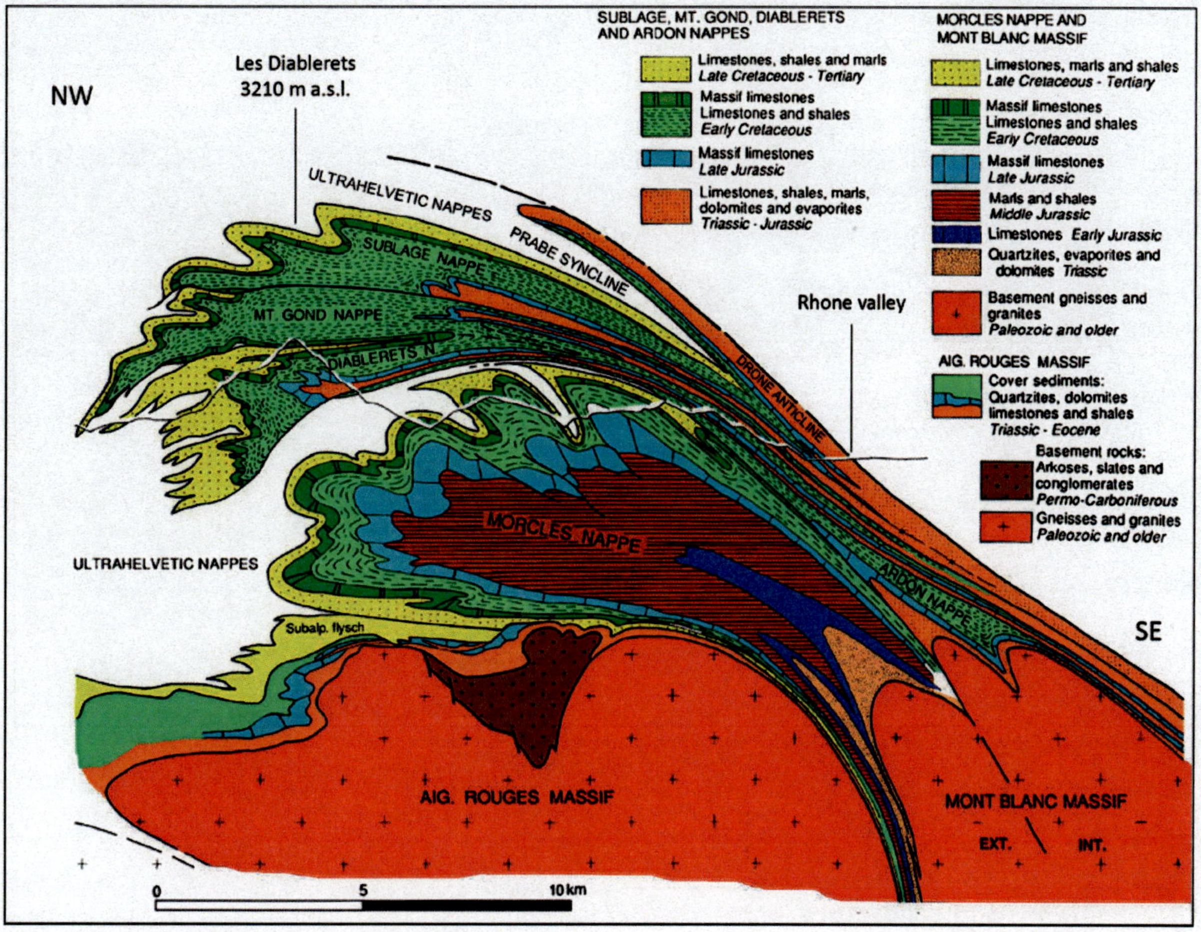

Fig. 9.2 Cross section of the Helvetic nappe system in Western Switzerland (after Escher et al. 1997, coloured by J.-L. Epard, University of Lausanne)

The stratigraphy of the Helvetic nappes is characterised by alternating schists and limestones of varying degrees of purity. Due to selective erosion, the result is a stair-like morphology, particularly in Les Ormonts valley and in Anzeindaz and Derborence. Four thick limestone strata form high rock walls: the late Jurassic, the Hauterivian, the Valanginian, and the Barremian (Urgonian/Schrattenkalk facies) (Fig. 9.3). In the Tsanfleuron area, which is at the back of the Diablerets nappe, these extensive limestone outcrops form a 10 km^2 karstified area. Marls, flysch and evaporitic rocks, e.g. salt, gypsum, cornieules (or rauwackes), predominate in the Ultrahelvetic nappes.

The Diablerets Massif is an obstacle in the path of western meteorological depressions, meaning annual precipitation is particularly high in the northern part of the massif: Les Diablerets (1162 m a.s.l.): 1577 mm; Gsteig (1195 m a.s.l.): 1689 mm; Diablerets Hut (2485 m a.s.l.): 2379 mm (Fallot 2014, based on the measurements for the period 1994–2007). Even if characterised by a rain-shadow effect, the upper part of the southern valleys is also quite humid due to precipitation overflows through the Pas de Cheville and Sanetsch passes. Annual rainfall is estimated at 1600 mm at Derborence (1580 m a.s.l.) (Fallot 2014) and 2400 mm at Tsanfleuron (2500 m a.s.l.) (Maire 1976). The mean annual temperature is 6.7 °C at 1500 m a.s.l. and 0.8 °C at 2500 m a.s.l. (Fallot 2014). The annual 0 °C isotherm is at about 2600 m a.s.l. The relatively wet climate explains why catchments are glacierised at relatively low altitudes and why permafrost is not extended, although not absent (Lambiel et al. 2009).

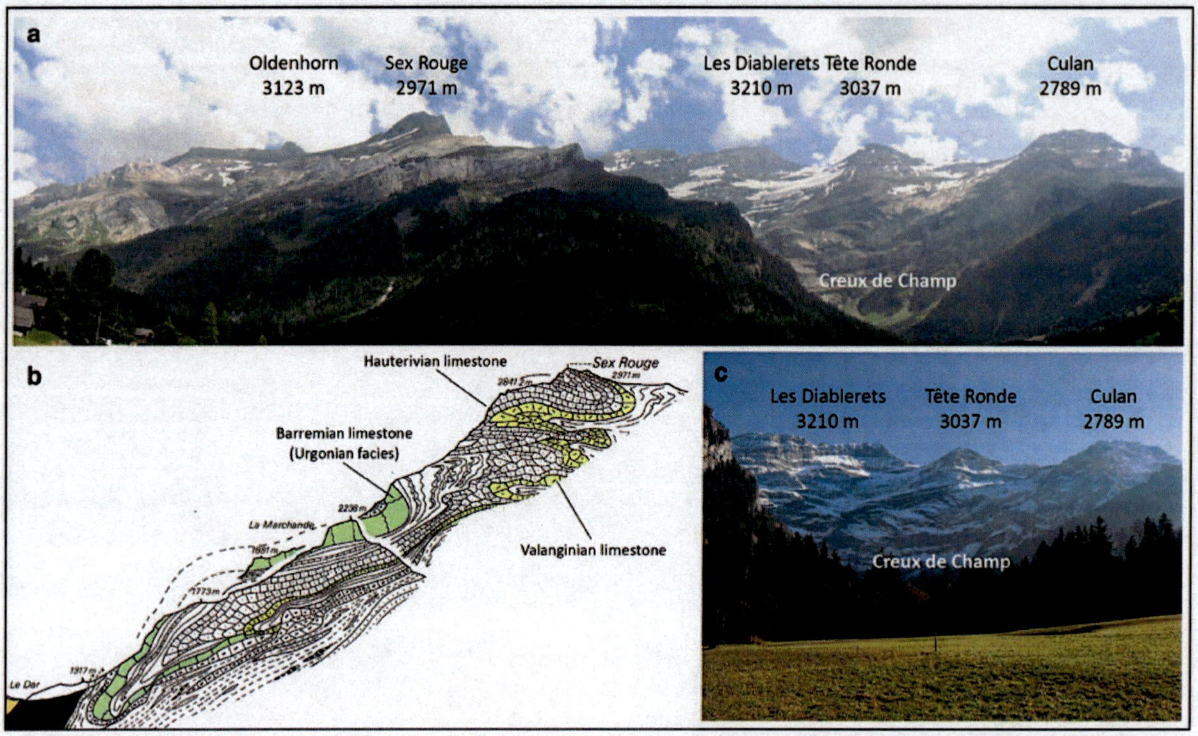

Fig. 9.3 **a** The Diablerets Massif, viewed from Les Diablerets (photo P. Schoeneich); **b** Geological cross-section along the Sex-Rouge—Le Dar crest (from Badoux and Gabus 1991, modified); **c** Structural landscape of the Creux de Champ cirque, with effects of selective erosion highlighted by snow (photo P. Schoeneich). See location of place names and summits in Fig. 9.1

9.3 Landforms and Landscapes

9.3.1 Diablerets and Pillon Pass

The tourist resort Les Diablerets offers the most famous and spectacular view of the Diablerets Massif (Fig. 9.1, site 1). The impressive barrier (Fig. 9.3a) formed by the highest peaks of the Sex Rouge (2971 m a.s.l.), Diablerets (3210 m a.s.l.), Tête Ronde (3037 m a.s.l.) and Culan (2789 m a.s.l.), dominates the famous Creux de Champ cirque, with its 600- to 800-m high cliff, and the wide flat valley bottom of Les Diablerets at 1200 m a.s.l. (see Fig. 9.1 for location). The Creux de Champ cirque (Fig. 9.3c) is often compared to similar cirques of Gavarnie in the Pyrenees or Sixt/Fer à Cheval in the French Alps.

Tectonically, this side of the massif is formed by the frontal folds of the Helvetic nappes, in sharp contrast to the less resistant Ultrahelvetic nappes in the foreground. In fact, there are two superimposed nappes, formed by similar stratigraphic series. The cliffs and peaks dominating Creux de Champ itself belong to the Diablerets nappe, whereas the Sex Rouge and the rock slopes dominating the valley towards Pillon Pass are formed by the overlying Mont-Gond nappe. In both units, the structure is underlined by the vertical cliffs of the Urgonian limestones. The axial dipping towards the east is distinctly visible in the upper cliffs, whereas the folds, seen here in a frontal view, are more difficult to observe.

Schoeneich et al. (2013) produced a detailed geomorphological map. In addition to the dominant structural landforms, extended glacial erosional features are present, such as Holocene and Lateglacial moraine systems, two karren fields developed on Urgonian and Paleogene limestones (at Pierredar and La Marchande, with numerous caves), several large slides, gypsum karst and fluvial terraces. In particular, the Pillon Pass area and the valley between the pass and the village show some striking landforms (see below).

The complex debris-covered glacial system of Entre la Reille is visible close to the Diablerets mountain hut (Swiss Alpine Club) and the *Glacier 3000* cable car midway station (Fig. 9.4; see Fig. 9.1 for location). It has been extensively studied by Bosson (2016). This complex landform combines

Fig. 9.4 Complex glacial/periglacial system of Entre la Reille (photo C. Lambiel)

glacial and periglacial features (Lambiel et al. 2009; Bosson 2016; Bosson and Lambiel 2016): geophysical, thermal and kinematic investigations revealed that the lower part is a talus-derived rock glacier whose upper part is covered by a buried glacier, present at the surface during the Little Ice Age (LIA) and probably during other cold phases of the Holocene.

A spectacular deep-seated slope deformation affects the lower slope of the Diablerets cliff below Tête aux Chamois, along the axis of the lower section of the cable car. The structural landforms were completely destroyed by the deformation, and a 400-m downslope shift of the stratigraphic boundaries between the limestones and the underlying marls is clearly visible from the cable car (Fig. 9.5; see Fig. 9.1 for location).

Between the pass and the village, the valley follows the foot of the massif. Whereas the left orographic side is dominated by the very steep limestone slopes of the Helvetic nappes, it contrasts sharply with the orographic right side developed on the much softer terrains of the Ultrahelvetic nappes. The talweg almost exactly follows the tectonic boundary. A narrow zone of evaporitic rocks (gypsum and cornieules) is present along the valley axis, from Les Diablerets to Gsteig on the other side of the pass. It bears an almost hidden, forest-covered, but sometimes spectacular gypsum karst (see Sect. 9.3.5 for more details). Three gypsum caves on both sides of the pass have been explored (Dutruit 1986; Schoeneich and Imfeld 1997). The upper right valley sides are dominated by marl and flysch, resulting in smooth slopes, often marshy and very prone to sliding. The Parchets slide (see Fig. 9.1 for location), situated in the uppermost part of the building zone of Les Diablerets and which crosses the Pillon Pass road, is one of the best-dated slides in Switzerland: dendrochronological dating of over 50 fossil tree trunks enabled five catastrophic flow slide events to be dated between the first and the thirteenth century AD (Schoeneich et al. 1997). On the 2nd of July 2007, due to heavy rainfall, a rapid earth flow occurred on a slope adjacent to Les Parchets (Pont Bourquin landslide) and cut the road linking Les Diablerets to Gstaad (Jaboyedoff et al. 2009). The volume was estimated to be 11 000 m^3.

A well preserved Lateglacial moraine system is visible in the Pillon Pass area. It was formed by the Dar glacier, which originated in a cirque between Oldenhorn and Sex Rouge summits (see Fig. 9.1 for location). Several stadials can be inferred from the morphological record (Schoeneich 1998a, b; Fig. 9.6). Historical and Holocene moraines are preserved in the upper cirque and can be seen only from the Cabane des Diablerets area or from the cable car. The lower moraines are visible almost from the road. The youngest

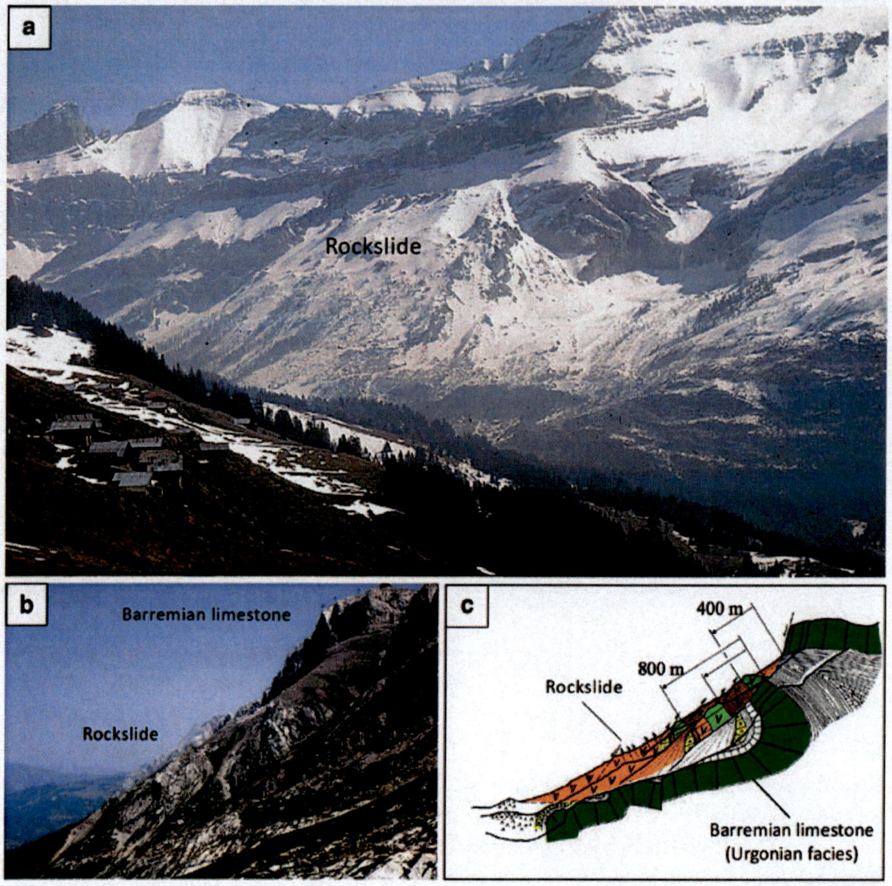

Fig. 9.5 The Tête aux Chamois rockslide. **a** General view from the opposite slope showing the destructuring of the parallel structural limestone bars by the slide; **b** Lateral view showing the displacement of the Urgonian limestone bar and stratigraphic contacts; **c** Stacked geological cross-sections of the slide. The slide mass (in red) displaces the Urgonian (green) and the Nummulitic (yellow) limestones and the stratigraphic contacts with the marls. Two embedded sliding slices shift the features respectively 400 m and 800 m downslope. The slide is almost dormant (40 cm/50 years) (photos and figure P. Schoeneich)

Lateglacial stadial is marked mainly by a sharp right lateral moraine above the lower Dar cascade, allowing us to reconstruct an overhanging front above the cascade. It is attributed to the Egesen stadial (Younger Dryas) based on the calculated ELA (equilibrium line altitude) (–350 m relative to Little Ice Age). During older extensions, the glacier tongue took a sharp left turn and followed the valley axis. Several right lateral moraines are preserved. The first moraine is visible below the road and connects with a very distinct frontal moraine. The second system can be seen a little further down, close to the valley bottom. The outermost lateral moraine was deposited on top of the gypsum zone, above the road, and can be followed down to Les Diablerets. These moraines are considered to be older than the Bølling-Allerød interstadial. This valley section west of the Pillon Pass is the only area around the Diablerets massif where it was possible to reconstruct a true Lateglacial moraine sequence. Isolated moraines can be also found in the Creux de Champ and Orgevaux cirques and are hypothetically correlated with the Egesen stadial.

Older Lateglacial moraine systems of local glaciers are exceptionally well preserved further west, on the northern side of the Pic Chaussy chain, in the Col des Mosses and Hongrin areas, as well as at the confluence area of the Grande-Eau River with the Rhone River. They allowed Schoeneich (1998a, b) to define a detailed morphostratigraphic framework of the Prealpine Lateglacial.

Finally, the Diablerets massif is included in the inventory of landscapes of national importance, while the existing glaciers and the Lateglacial moraine systems in the Dar valley (Pillon Pass) are included in the geosite inventory of Vaud Canton.

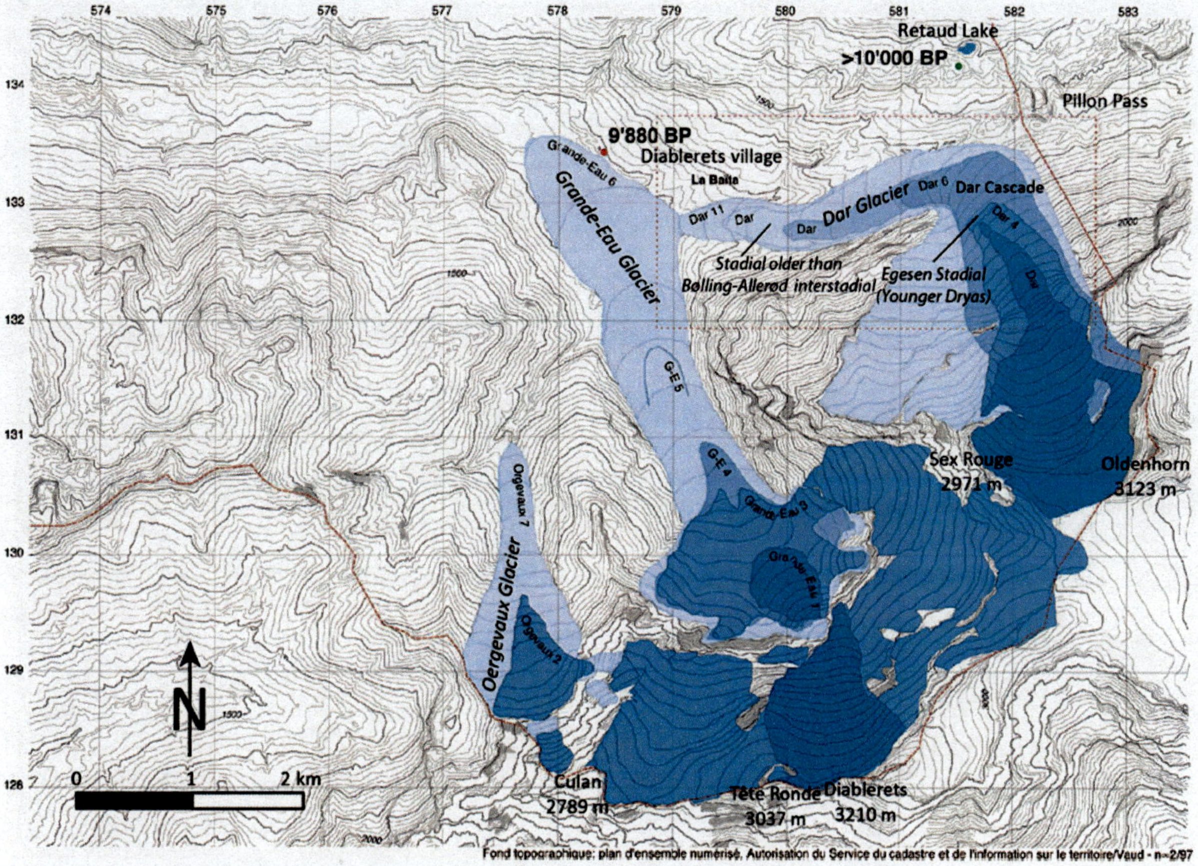

Fig. 9.6 Reconstruction of the Lateglacial and Holocene moraine sequence of the Dar Valley and adjacent glaciers (from Schoeneich 1998a, modified). Each colour represents a stadial reconstructed based on mapping of the moraine ridges and on ELA (equilibrium line altitude) calculations

9.3.2 The Tsanfleuron Area

The Tsanfleuron glacio-karstic area (Fig. 9.1, site 2) is a large karren field developed mainly on limestones of the Diablerets nappe, and partly on limestones of the overlaying Mont-Gond nappe (Gremaud et al. 2009; Fig. 9.7). A small isoclinal syncline connects the two nappes and plays an important role in underground water circulation: tracer experiments demonstrated that most circulating water drains to this syncline (Fig. 9.8) and forms a large karstic spring (Glarey spring) which is currently used for drinking water production (Gremaud et al. 2009; Gremaud 2011). The nappe is gently folded, forming a huge anticlinorium, whose axis plunges 5° to 10° to the ENE (Steck et al. 2001). The nappe is estimated to be about 1200 m thick in the karren field (Gremaud et al. 2009).

The main karstified formations are the Cretaceous limestones (Urgonian facies) in the upper part of the karren field and Palaeogene limestones in the lower part. A stratigraphic gap due to uplift and associated marine regression separates the two formations (Gremaud et al. 2009) and large palaeokarst sinkholes (Fig. 9.9a), partly filled with "Siderolithic" deposits (detrital rocks, with conglomeratic or sandstone facies, rich with iron minerals), testify to an important karstification during the regression (Upper Cretaceous—Lower Palaeogene).

Fractures of different ages exist in the area (Franck et al. 1984). The oldest was active before and during the deposition of "Siderolithic" sediments during the Eocene (Gremaud et al. 2009), but the main fractures and faults occurred after the formation of the nappes, evidence for 'a stress field with a NW–SE (130–160°) direction' (Gremaud et al. 2009). Fractures influence underground water circulation (Gremaud and Goldscheider 2009) and explain the extensive development of fracture-related karren fields (Fig. 9.10a).

The rapidly retreating Tsanfleuron glacier (VAW/ETH Zürich 2017) overlies the upper part of the karren field (Fig. 9.11). Its thickness and volume were investigated by

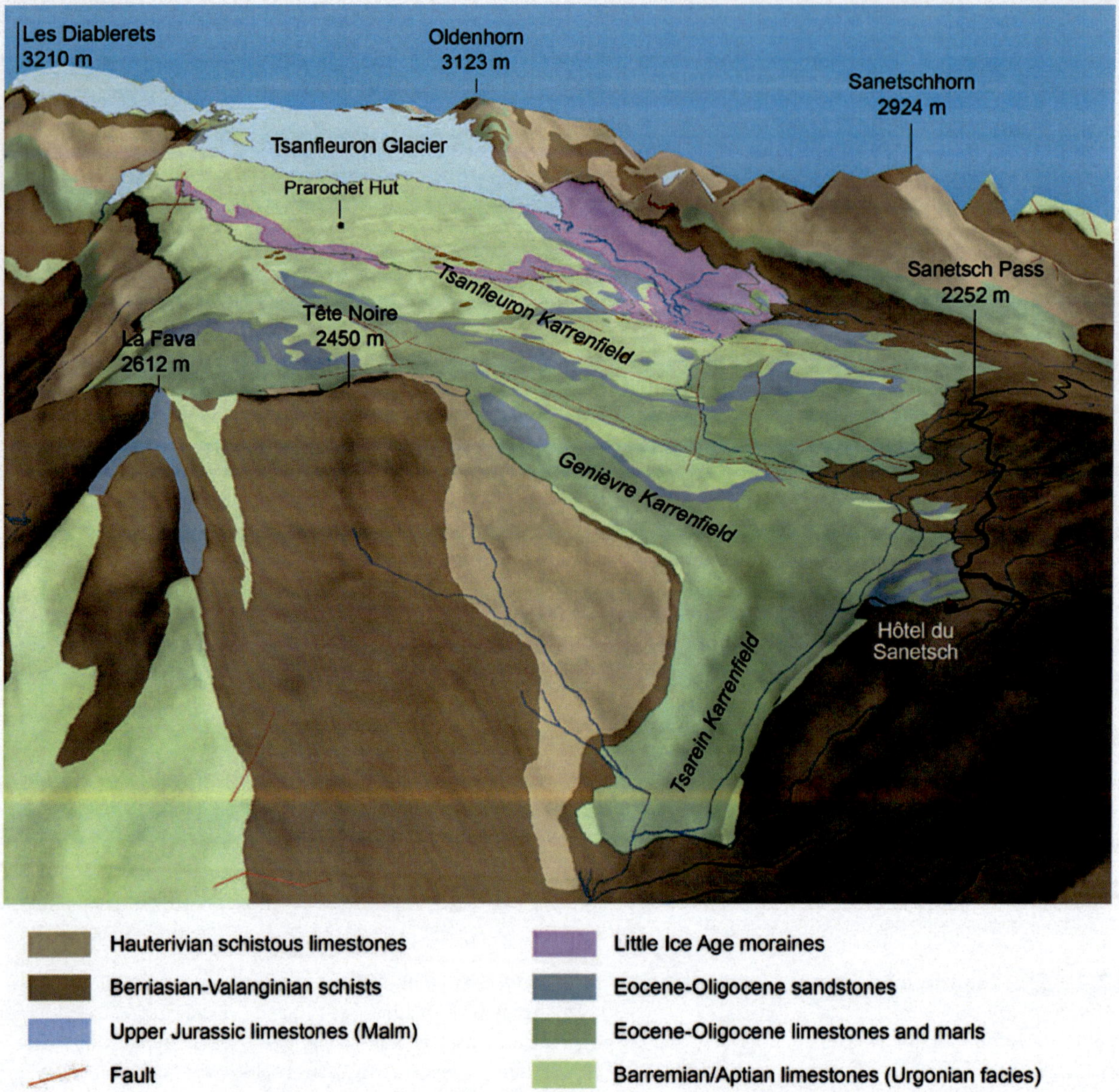

Fig. 9.7 Simplified geological sketch of the Tsanfleuron karren field (after Martin and Reynard 2008, modified)

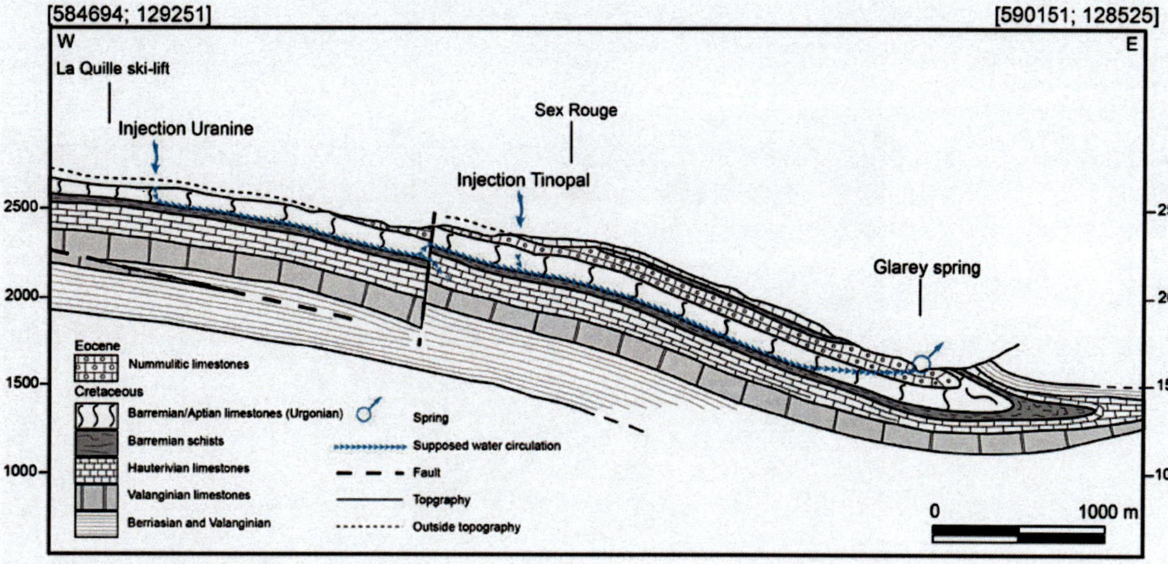

Fig. 9.8 Geological section (Swiss coordinates) across the Tsanfleuron karren field, showing the underground water flowing toward the Glarey karstic spring (after Gremaud 2008, modified)

Gremaud and Goldscheider (2009) using the radio-magnetotelluric (RMT) method. The maximum depth measured was 138 m and the ice volume was estimated at 1.0×10^8 m^3 (±10%), but the glacier is currently losing 1.5 m in thickness per year (Gremaud and Goldscheider 2009).

After the pioneering work of Maire (1976, 1977), several authors studied the hydrological and geomorphological processes at the interface between the ice and the bedrock (e.g. Hallet et al. 1978; Hubbard and Hubbard 1998; Hubbard 2002, Hubbard et al. 2003), and showed that basal ice is partly formed by refreezing processes that force out part of the $CaCO_3$ trapped in the basal water, leading to the formation of thin carbonate deposits on the surface of the bedrock.

The karren field can be divided into two parts (Maire 1976) (Fig. 9.11b): above and below the Little Ice Age moraine belt (1860). In the lower part, the karst surface has been deglaciated since the beginning of the Holocene and karstic landforms including different types of karren (Fig. 9.10a,b), kamenitsas, some dolines or karstic valleys predominate. In the upper part, the glacier has been retreating since the 1860s and the area has consequently been free of ice for a maximum of 150 years. Here, karstic landforms, e.g. runoff karren along the walls of large sinkholes, are combined with glacial landforms, e.g. deposits of erratics and moraines, *roches moutonnées* (Fig. 9.10c). Large depressions, partly filled with "Siderolithic" or morainic deposits and sometimes containing a lake (Fig. 9.9b), are frequent and, in the recently deglaciated area, large glacio-karstic amphitheatres (*Schichttreppenkarst)* (Fig. 9.10c) have formed due to the combination of glacial (plucking), fluvioglacial (evorsion) and karstic (dissolution) processes. Spectacular micro-landforms are also visible including glacial striae or ephemeral carbonate deposits (Hallet et al. 1978, Hubbard and Hubbard 1998), which are very frequent in the vicinity of the glacier front.

Maire (1976) proposed a bioclimatic classification of karren fields based on the Tsanfleuron karst. The lower part (Tsarein karren field; see Fig. 9.1 for location) is a mountain karren field with smoothed landforms, partially covered with vegetation. The central part is a typical alpine or nival karren field, characterised by the absence of vegetation and sharp forms; the upper part is a glacial karren field, characterised by the combination of glacial and karstic processes. Toth and Reynard (2011) proposed a hierarchical classification of alpine karren in four levels: individual forms, complex forms, karren cells and karren fields. The shape of the karren cells depends on factors such as slope inclination or fracture pattern.

The Tsanfleuron karren field has been proposed as a geosite of national importance, as it is a unique site to study the interactions between glacial and karstic processes in Switzerland. Long-term research in karstology, glaciology, geomorphology by several Swiss and foreign universities

Fig. 9.9 Typical sinkholes of the Tsanfleuron karst. **a** Palaeokarstic sinkholes partly filled with "Siderolithic" sediments (Eocene); **b** Large glacio-karstic depression occupied by a lake (photos E. Reynard)

Fig. 9.10 Various types of karren in the Tsanfleuron glacio-karst. Fracture karren (**a**) and nival karren (**b**) typical of the area deglaciated almost since the beginning of the Holocene; *roches moutonnées* karren (**c**) and *Shichttreppenkarst* formed by the combination of glacial, fluvioglacial and karstic processes (**d**) typical of the upper part of the karst deglaciated only after the glacier advance in the Little Ice Age (photos E. Reynard)

also motivated the inscription of this complex geomorphosite (Reynard 2008, 2009). Unfortunately, several threats (building of tourist infrastructures, especially on the glacier, use of four-wheel drive vehicles, water pollution) are of concern in the area and since the inventory of Swiss geosites has no legal value (see Reynard et al. this volume), the area is still not protected (Reynard 2008).

9.3.3 The Derborence Cirque and Rock Avalanche

Derborence (Fig. 9.1, site 3) is located south of the Tsanfleuron karst and the high cliffs clearly reveal the stratigraphy of the Diablerets nappe (Sartori 2014; Fig. 9.12). Selective erosion has resulted in a stair-like structural landscape, particularly visible below the Diablerets (Figs. 9.12 and 9.13a) and the Mont-Gond peaks (Fig. 9.13b). Derborence and the Lizerne Valley are located at the interface of the Diablerets nappe and the underlying Morcles nappe (Fig. 9.2), separated by cornieules and evaporites belonging to the Ultrahelvetic nappes and forming spectacular gypsum karst features in La Tour area (Fig. 9.14a).

The morphology, therefore, depends on the structural context. The morphology was also modelled by glacial, fluvial and gravitational phenomena (Reynard and Martin 2014; Maret and Reynard 2015). Fluvial erosion has followed the principal contact between the Diablerets and Morcles nappes (the flow direction of the river Lizerne) and it is currently particularly active (torrential erosion) in schist formations. Although glaciers have nearly completely disappeared from the Derborence cirque, they formed the U-shaped Lizerne Valley and eroded the Derborence depression during the Pleistocene.

This depression is partially filled with deposits of two historical rock avalanches, one in 1714 and the other in 1749. The scar is situated south of the Diablerets glacier and

Fig. 9.11 **a** View of the Tsanfleuron glacio-karst, with the fast retreating Tsanfleuron glacier (numbers indicate the year AD of the position of the glacier snout) (after Martin and Reynard 2008, modified); **b** View of the two parts of the Tsanfleuron glacio-karst; downstream from the Little Ice Age moraines, the limestone bedrock has been deglaciated since the beginning of the Holocene and the karstic features are well developed, whereas upstream from the moraine ridges, the glacial landforms, in particular *roches moutonnées*, predominate (photos E. Reynard)

Fig. 9.12 View of the 1000-m high Diablerets wall dominating the Derborence cirque. Selective erosion highlights variations in lithology. The red arrows indicate the detachment niches and the trajectories of the two eighteenth century rock avalanches (after Martin and Reynard 2008, modified)

concerns nearly all the strata of the Diablerets nappe (Fig. 9.13a). The deposits form a tongue between 2250 and 1150 m a.s.l. The estimated volume ranges from 50 million m^3 (Becker 1883) to 57 million $m^3 \pm 20\%$ (Garazzi and Moret 1999). The first event (23rd of September 1714) is estimated to have moved 35 million m^3 as a rock avalanche and covered 4.6 km^2 (Sartori 2014). The Derborence lake probably appeared after this event, due to the dam formed by the rock avalanche deposits (Fig. 9.13b, c). The rock avalanche was certainly the consequence of an earthquake that occurred two years previously (1712) in Valais which could have destabilised the rock wall. The destabilisation could also have combined with ice melting at the end of a particularly cold phase of the Little Ice Age which could have increased water pressure in the rock fractures (Sartori 2014). The second rock avalanche (23rd of June 1749) can be explained by the destabilisation of the rock wall during the first event. It covered the central part of the first deposit (Fig. 9.13c).

As the area was used for alpine pasture, the first event was catastrophic: 14 people and 120 animals died, and 55 buildings were destroyed. The second event killed five people and several infrastructures were destroyed (Rey Carron 2014). The 1714 catastrophe inspired the Swiss writer Charles-Ferdinand Ramuz, who published a novel entitled *Derborence* in 1934, which relates the story of a shepherd who remained a prisoner of the rock avalanche deposit for several weeks before re-appearing in the village where everybody thought he was dead. In 1985, the Swiss filmmaker Francis Reusser directed the eponym film drama inspired by Ramuz's novel. The fame of the rock avalanches, as well as the legend of Diablerets devils bowling with stones in the Diablerets massif (explaining the multiple small rockfalls that occur in Derborence) make Derborence a geocultural site (Reynard and Giusti 2018), i.e. place where cultural elements (here a legend, a novel and a film) are deeply connected with the geological and geomorphological features (here the gravitational processes).

One of the most spectacular impacts of the rock avalanches was the damming of the Derbonne River that led to the formation of the Derborence lake. The intense torrential activity (Fig. 9.13a) in the southern part of the Diablerets

Fig. 9.13 **a** View of the detachment areas of the two eighteenth century rock avalanches. The dashed line shows the path of the two rock avalanches. The white circle indicates the position of gypsum pyramids (see Fig. 9.14); **b** The Derborence lake looking east. This shallow lake is threatened by silting due to torrential inflows from schistous layers in the Diablerets wall (see Fig. 9.13a); **c** General view of the paths of the two historical rock avalanches and their contribution to the Derbonne River damming and the Derborence lake formation. The dashed green lines indicate the alluvial fan which is increasingly reducing the surface area of the lake (after Garazzi and Moret 1999; Martin and Reynard 2008; Sartori 2014, modified)

cliff currently threatens the lake, whose surface area and depth have been decreasing rapidly in recent decades. As a result, dams and dikes have been built along the channel to stop the infilling of the lake (Fig. 9.14b).

This area is now a tourist attraction and the lake and its surroundings—particularly a forest that was no longer exploited after the rock avalanches—have been protected by the Canton of Valais since 1961 (Pro Natura natural reserve). In 1977, the Swiss Confederation recognised the site as a landscape of national importance and the area has also been a geosite of national significance since 1998.

Fig. 9.14 a Karstic landforms in gypsum outcrops; in the background, the sedimentary succession of the Diablerets wall (photo S. Martin); b Artificial dikes divert debris flows and reduce sedimentation in Lake Derborence (photo E. Reynard)

9.3.4 Anzeindaz

Anzeindaz (Fig. 9.1, site 4) shows the 'backside' of the folded Diablerets massif (Fig. 9.2). The whole folded package has been cut almost vertically by erosion, exposing the lower and older stratigraphic units that form the heart of the nappe. These are mainly Jurassic and Lower Cretaceous limestones and marls. Huge coalescent compound debris, avalanche and torrential fans cover the lower slopes. The very wide depression of Anzeindaz and Pas de Cheville developed on the Ultrahelvetic terrains separating the Morcles nappe to the south from the Diablerets nappe to the north. Anzeindaz offers a very contrasted and impressive landscape, with subvertical, over 1000-m high cliffs dominating a wide smooth pasture (Fig. 9.15). The valley rises gently to the Pas de Cheville, which dominates the deep depression of Derborence. During glacial ages, this very large area situated between 1800 and 2100 m a.s.l.

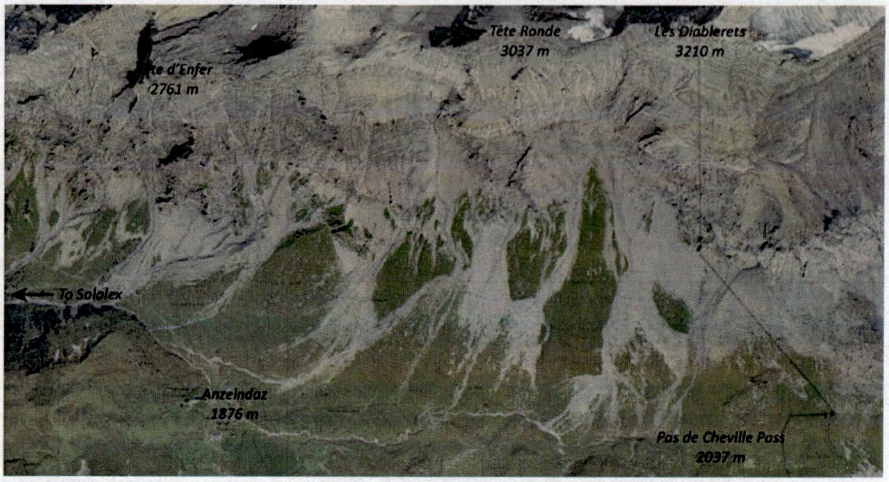

Fig. 9.15 Aerial photograph of the Anzeindaz area showing the large composite fans at the foot of the subvertical cliffs (orthophoto Swisstopo, geoplanet.vd)

represented the main accumulation zone of the Avançon glacier. To the south, the Paneirosse glacier occupies a cirque on the northern slopes of the Morcles nappe and is the only remnant of this former glacier. Dominating Solalex, the 'Miroir de l'Argentine' (written 'Arête de l'Argentine' in Fig. 9.1) is a famous spot for climbers: a long, steep dip-slope has developed on the surface of the Urgonian limestone on the frontal slope of the Morcles nappe.

9.3.5 The 'Col de la Croix' Gypsum Karst

As mentioned above, the Diablerets massif is almost surrounded by Triassic evaporitic rocks, and gypsum karst has developed in several places. The most spectacular and most easily accessible gypsum karst can be observed on the Col de la Croix (Fig. 9.1, site 5). It has developed on both sides of the car park, on top of the pass, and is one of the best examples of a very characteristic feature of Alpine and Prealpine gypsum karst, which some authors called a 'dôme écumoire' ('skimmer dome') (Chardon 1992).

Tectonically, the Col de la Croix belongs to what is called 'Zone des Cols' and forms the continuation of the Pillon Pass, towards the area of Bex and its salt mines. In the Alps, gypsum is always a secondary rock resulting from the rehydration of anhydrite. Gypsum forms a shallow around 30-m thick 'hydration crust' over anhydrite. The gypsum outcrops form elongated patches, surrounded and sometimes separated by zones of cornieules (or rauwackes). Gypsum outcrops often form domes on passes, like on the Col de la Croix, or crests along glacially shaped valleys, like near the Pillon Pass. This suggests that anhydrite is quite resistant to glacial erosion in cold conditions.

Large asymmetric dolines often develop in the contact zone between cornieules and gypsum, but the most striking features are found on the gypsum itself (Schoeneich and Imfeld 1997; Fig. 9.16a). Coalescent funnel-shaped sinkholes are separated by very sharp conical peaks and narrow crests (Fig. 9.16b,c). The popular name 'Pyramides du Col de la Croix' evokes these residual peaks. At the Col de la Croix, the paths resulting from high frequentation makes it relatively easy to walk through the features. Similar gypsum karsts that are not degraded are almost impossible to cross. An example of an intact karst can be observed 'from the air' (i.e. from the chairlift) between Les Diablerets village and Les Mazots halfway down to Les Diablerets. Similar, forest-covered karsts are present all along the Pillon Pass axis (see Sect. 9.3.1). Sinkholes on slopes are elongated in shape (Fig. 9.16d). Observations made in road cuts and in a quarry on the Pillon Pass showed that the sinkholes are connected to vertical chimneys down through the gypsum layer to the top of the impermeable anhydrite, where the water flow follows the interface between the gypsum and anhydrite towards small springs at the bottom of gypsum outcrops (Schoeneich and Imfeld 1997). A dendrogeomorphological survey enabled the calculation of a mean average denudation rate of 5.6 mm/y (Bollati et al. 2017). Gypsum karst supports a very contrasted vegetation mosaic: megaphorbiae or even trees can grow in the more humid sinkholes, whereas only drought-resistant shrubs can survive on the very dry crests and peaks, or the soil may even be bare. A detailed pedologic study of two gypsum soils on Col de la Croix was published in 2014 (Biedermann et al. 2014).

Between the sharp Helvetic limestone relief and the gypsum karst, smoother intermediate landforms developed in flysch of the Ultrahelvetic nappes. A single very characteristic

Fig. 9.16 Karstic landforms at Col de la Croix. **a** Sketch of the karstic system on a "dôme écumoire". Adjacent sinkholes determine residual "pyramids". Sinkholes elongate with increasing slope and form monoliths on the edge of the gypsum dome (after Schoeneich and Imfeld 1997, modified). **b** Sinkhole (photo I. Bollati); **c** Pyramid feature (photo I. Bollati); **d** Monoliths (photo M. Pellegrini)

sandstone layer (Taveyannaz sandstone) is embedded in the flysch and forms the small peaks and cliffs of this zone: Taveyannaz sandstone's type locality is just southwest of the Col de la Croix. It is a coarse greenish volcano-sedimentary sandstone, with remarkable white sand grains.

The gypsum karst of Col de la Croix is classified as a landscape of cantonal importance, and like the Taveyannaz sandstone type locality, it is included, in the geosite inventory of the Canton of Vaud. It is also of geosite of national significance.

9.4 Conclusions

A very large panel of structural landforms shaped by selective erosion of folded sedimentary rocks, as well as glacial, karstic and mass wasting landforms, can be seen in the Diablerets Massif. Most sites have an important landscape and geoheritage value, making them part of the nature conservation hotspots in Switzerland. In 1977, almost the entire area described in this chapter was classified as a landscape of national importance (objects 1503 and 1713: Les Diablerets—Vallon de Nant—Derborence; see Beutler and Gerth 2015). Moreover, three sites: the Tsanfleuron glacio-karst (geosite 219), the Derborence cirque and rock avalanche deposits (geosite 57), the gypsum pyramids of Col de la Croix (geosite 49) are part of the Swiss inventory of geosites (Reynard et al. 2012). The area covered by the Morcles nappe is also a geosite of national importance (geosite 314). This value has been promoted for geotourism and several products have been developed, including leaflets, panels, geotourist maps and guided tours (see Martin 2013), as well as a guided itinerary along the Via GeoAlpina (Schlup et al. 2009a, b), an initiative of the International Year of Planet Earth (Cayla and Hobléa 2011).

Acknowledgments This chapter benefitted from useful comments by Christophe Lambiel and Jean-Luc Epard, University of Lausanne. We thank Simon Martin, who created some figures when he was a member of the Institute of Geography and Sustainability of Lausanne University, and Irene Bollati for providing some photographs and figures of the Col de la Croix karstic landforms.

References

Badoux H, Gabus J (1991) Les Diablerets. Sheet 88 of the Geological Atlas of Switzerland, 1:25 000. Swisstopo, Bern

Becker F (1883) Der Bergsturz der Diablerets. Jahrbuch SAC 18:310–316

Beutler R, Gerth A (2015) Patrimoine naturel de suisse. Les paysages, sites et monuments naturels d'importance nationale. Haupt, Berne (in French, also a version in German)

Biedermann Y, Gobat JM, Vittoz P (2014) Typologie des sols sur gypse et végétation associée en Suisse. Bull Soc vaud Sc Nat 94 (1):107–129

Bollati I, Pellegrini M, Reynard E, Pelfini M (2017) Water driven processes and landforms evolution rates in mountain geomorphosites: examples from Swiss Alps. Catena 158:321–339

Bosson JB (2016) Internal structure, dynamics and genesis of small debris-covered glacier systems located in alpine permafrost environments. PhD thesis, University of Lausanne

Bosson JB, Lambiel C (2016) Internal structure and current evolution of very small debris-covered glacier systems located in alpine permafrost environments. Frontiers in Earth Science: Cryospheric sciences 4:39

Cayla N, Hobléa F (2011) La Via GeoAlpina. Itinérance géologique à travers les Alpes. In: Reynard E, Laigre L, Kramar N (eds) Les géosciences au service de la société. Institut de géographie, Université de Lausanne, Géovisions 37, pp 119–130

Chardon M (1992) Evolution récente des karsts de la Vanoise orientale. In: Salomon JN, Maire R (eds) Karst et évolutions climatiques. Presses Universitaires de Bordeaux, Bordeaux, Hommage à Jean Nicod, pp 293–307

Dutruit J (1986) Grotte du Creux du Pillon. Le Trou 41:26–27

Epard JL (2001) The Helvetic cover units. In: Stampfli G (ed) Geology of the western Swiss Alps, a guide-book. Mémoires de Géologie 36, Lausanne, pp 87–94

Escher A, Masson H, Steck A (1993) Nappe geometry in the western Swiss Alps. J Struct Geol 15(3):501–509

Escher A, Hunziker J, Marthaler M, Masson H, Sartori M, Steck A (1997) Geologic framework and structural evolution of the western Swiss-Italian Alps. In: Pfiffner OA, Lehner P, Heitzmann P, Müller S, Steck A (eds) Deep structure of the Swiss Alps: Results of the NRP 20. Birkhäuser, Basel, pp 205–221

Fallot JM (2014) La climatologie. In: Rey Carron C and S (eds) Derborence. La nature et les hommes. Monographic, Sierre, pp 23–31

Franck P, Wagner JJ, Escher A, Pavoni N (1984) Evolution des contraintes tectoniques et sismicité dans la région du col du Sanetsch. Alpes valaisannes helvétiques. Eclogae Geol Helv 77 (2):383–393

Garazzi G, Moret A (1999) Eboulement de Derborence: étude géologique et géomécanique (Valais, Suisse). Master's thesis, University of Geneva

Gremaud V (2008) Géologie du karst de Tsanfleuron. In: Hobléa F, Reynard E, Delannoy JJ (eds) Karsts de montagne. Géomorphologie, patrimoine et ressources. Collection Edytem, Chambéry, Cahiers de Géographie, 7, pp 127–134

Gremaud V (2011) Relations between retreating alpine glaciers and karst aquifer dynamics. Tsanfleuron-Sanetsch experimental test site, Swiss Alps. PhD thesis, University of Neuchâtel

Gremaud V, Goldscheider N (2009) Geometry and drainage of a retreating glacier overlying and recharging a karst aquifer, Tsanfleuron-Sanetsch. Swiss Alps. Acta Carsologica 39(2):289–300

Gremaud V, Goldscheider N, Savoy L, Favre G, Masson H (2009) Geological structure, recharge processes and underground drainage of a glacierised karst aquifer system, Tsanfleuron-Sanetsch. Swiss Alps. Hydrogeol J 17(8):1833–1848

Hallet B, Lorrain R, Souchez R (1978) The composition of basal ice from a glacier sliding over limestones. Geol Soc Am Bull 89 (2):314–320

Hubbard B (2002) Direct measurement of basal motion at a hard-bedded, temperate glacier: Glacier de Tsanfleuron. Switzerland. J Glaciol 48(160):1–8

Hubbard B, Hubbard A (1998) Bedrock surface roughness and the distribution of subglacially precipitated carbonate deposits: implications for formation at Glacier de Tsanfleuron. Switzerland. Earth Surf Proc Land 23(3):261–270

Hubbard B, Hubbard A, Tison J-L, Mader HM, Nienow P, Grust K (2003) Spatial variability in the water content and rheology of temperate glaciers: Glacier de Tsanfleuron, Switzerland. Ann Glaciol 37:1–6

Jaboyedoff M, Pedrazzini A, Loye A, Oppikofer T, Pons Güell I, Locat M (2009) Earth flow in a complex geological environment: the example of Pont Bourquin, Les Diablerets (Western Switzerland). In: Malet JP, Remaître A, Bogaard TA (eds) Landslide processes, from geomorphologic mapping to dynamic modelling. CERG Edition, Strasbourg, pp 131–137

Lambiel C, Bardou E, Delaloye R, Schuetz P, Schoeneich P (2009) Extension spatiale du pergélisol dans les Alpes vaudoises; implication pour la dynamique sédimentaire locale. Bull Soc vaud Sc Nat 91(4):407–424

Lugeon M (1914–1918). Les Hautes Alpes calcaires entre la Lizerne et la Kander (Wildhorn, Wildstrubel, Balmhorn et Torrenthorn). Francke, Berne (3 vol)

Maire R (1976) Recherches géomorphologiques sur les karsts haut-alpins des massifs de Platé, du Haut-Giffre, des Diablerets et de l'Oberland Occidental. PhD thesis, University of Nice
Maire R (1977) Les karsts haut-alpins de Platé, du Haut-Giffre et de Suisse Occidentale. Rev Géogr Alpine 65(4):403–425
Maret H, Reynard E (2015) Géomorphologie de la région de Derborence. Bull Murithienne 132:45–55
Martin S, Reynard E (2008) Tsanfleuron: on-site interpretive panels. University of Lausanne and Municipality of Savièse
Martin S (2013) Valoriser le géopatrimoine par la médiation indirecte et la visualisation des objets géomorphologiques. PhD thesis, University of Lausanne, Gévisions 41
Rey Carron C and S (2014) Derborence. La nature et les hommes. Monographic, Sierre
Reynard E (2008) Le lapiaz de Tsanfleuron. Un paysage glacio-karstique à protéger et à valoriser. In: Hobléa F, Reynard E, Delannoy JJ (eds). Karsts de montagne. Géomorphologie, patrimoine et ressources. Collection Edytem, Chambéry, Cahiers de Géographie, 7, pp 157–168
Reynard E (2009) Geomorphosites: definitions and characteristics. In: Reynard E, Coratza P, Regolini-Bissig G (eds) Geomorphosites. Pfeil, München, pp 9–20
Reynard E, Giusti C (2018) The landscape and the cultural value of geoheritage. In: Reynard E, Brilha J (eds) Geoheritage. Assessment, Protection, and Management. Elsevier, Amsterdam, pp 147–166
Reynard E, Martin S (2014) La géomorphologie. In: Rey Carron C and S (ed) Derborence. La nature et les hommes. Monographic, Sierre, pp 49–54
Reynard E, Berger JP, Constandache M, Felber M, Grangier L, Häuselmann P, Jeannin PY, Martin S (2012) Révision de l'inventaire des géotopes suisses : rapport final. Groupe de travail pour les géotopes en Suisse, Lausanne
Reynard E, Buckingham T, Martin S, Regolini G (this volume) Geoheritage, geoconservation and geotourism in Switzerland
Sartori M (2014) La géologie/Description des deux grands éboulements. In: Rey Carron C and S (ed) Derborence. La nature et les hommes. Monographic, Sierre, pp 33–40 and 90–94
Schlup M, Kozlik L, Marthaler M (2009a) Le vallon de la Vare. Du Pont de Nant à Anzeindaz: au coeur des plis de l'Argentine. Swisstopo, Berne. https://www.swisstopo.admin.ch/fr/connaissances-faits/geologie/geologie-quotidien/geologie-pour-tous/via-geoalpina.html
Schlup M, Kozlik L, Marthaler M (2009b) Les Diablerets et le cirque de Derborence. D'Anzeindaz à Derborence: une histoire longue de 250 millions d'années se terminant par une catastrophe toute récente. Swisstopo, Berne (https://www.swisstopo.admin.ch/fr/connaissances-faits/geologie/geologie-quotidien/geologie-pour-tous/via-geoalpina.html)
Schoeneich P (1998a) Le retrait glaciaire dans les vallées des Ormonts, de l'Hongrin et de l'Etivaz. PhD thesis, University of Lausanne, Travaux et Recherches 14
Schoeneich P (1998b) Les stades tardiglaciaires des Préalpes vaudoises et leur corrélation avec le modèle des Alpes orientales. ETHZ, VAW-Mitteilungen 158:192–206
Schoeneich P, Imfeld P (1997) Les karsts du gypse de la zone des Cols (Préalpes vaudoises). Bull Soc Neuch Géogr 41:39–54
Schoeneich P, Tercier J, Hurni JP, Orcel C (1997) Datation par dendrochronologie du glissement des Parchets (Les Diablerets, Alpes vaudoises). Eclogae Geol Helv 90(3):481–496
Schoeneich P, Lambiel C, Bosson JB (2013). Geomorphological map of the Diablerets massif – Swiss Alps. Poster, 8th International Conference on Geomorphology, 27-31 August 2013, Paris
Steck A, Epard JL, Escher A, Gouffon Y, Masson H (2001) Carte tectonique des Alpes de Suisse occidentale et des régions avoisinantes 1:100 000. Notice explicative, Service géologique national, Berne
Toth G, Reynard E (2011) Développement et spécificités des cellules lapiazées sur le Lapiaz de Tsanfleuron (Alpes Bernoises, Suisse). Zeitschrift für Geomorphologie 55(2):231–245
VAW/ETH Zürich (2017) The Swiss Glaciers. Glaciological reports of the Cryospheric Commission of the Swiss Academy of Sciences (SCNAT). http://www.glamos.ch

Philippe Schoeneich is Professor of Physical Geography at the University Grenoble Alpes. He has been member of the Swiss Quaternary Commission, and is a member of the Global Terrestrial Network for Permafrost (GTN-P) steering committee. His research concerns mountain geomorphology, Quaternary, mass movements and mountain permafrost.

Emmanuel Reynard is Professor of Physical Geography and Director of the Interdisciplinary Centre for Mountain Research at the University of Lausanne, and a member of the International Association of Geomorphologists (IAG) Executive Committee. He has been chairman of the IAG Working Group on Geomorphosites (2001–2013) and president of the Working Group on Geotopes of the Swiss Academy of Sciences (2006–2017). His research concerns mountain geomorphology, geomorphological heritage, landscape geohistorical analyses and water management in mountains. He has worked in the Alps, the Romanian Carpathians and in the Maghreb (Tunisia, Morocco).

10 The Karst System Siebenhengste-Hohgant-Schrattenfluh

Philipp Häuselmann

Abstract

The karst region Siebenhengste-Hohgant-Schrattenfluh is located in west-central Switzerland, north of Lake Thun. Although it is not well known outside the karst and caving community, it is one of the important karstic sites of the world. Surface landscape is impressive, with vast karren fields with distinct micromorphologies, stark contrast of sandstone and limestone pavements and the impressive view on the high Bernese Alps. The underground landscape is rich as well, with more than 340 km of cave passages developed below three adjacent massifs. Caves evolved in response to valley deepening processes, which could be dated by cave sediments. Soft tourism (hiking, walking) developed in the last two decades, and protected zones have multiplied in recent years.

Keywords

Karst landscape • Cave system • Speleogenesis • Dating methods

10.1 Introduction

The karstic area of Siebenhengste-Hohgant-Schrattenfluh lies in the Prealps of western central Switzerland, north of Lake Thun and Interlaken (47°N, 8°E; Fig. 10.1). It extends from the spring area at Lake Thun, at 558 m a.s.l., up to the summits of Hohgant, at 2197 m a.s.l. When looking at the area from a plane or with Google Earth, it is not typically karstic at first sight, since only the karren fields of Siebenhengste and Schrattenfluh are bare. However, below the green landscape, covered by non-carbonate rocks, the limestone is also karstified. The area is of prime interest because of the vast and spectacular karren fields of Siebenhengste and Schrattenfluh, and of the many different microkarren forms of Innerbergli (Hohgant). The underground cave network is one of the largest in the world. The Réseau Siebenhengste-Hohgant has a length of 160 km, ranging over a vertical extent of 1340 m, and altogether, the area up to Schrattenfluh comprises more than 340 km of cave passages.

The paper presents a short overview of the karren fields and karren forms that are found in the area, and gives hints for further research in that domain. The cave network morphology and sediments contain archives of useful information about the past landforms and climate conditions. Cave morphology makes possible to infer speleogenetical phases, and sediments provide dates to establish the timing of landscape-forming processes of this region. Such information is generally lost outside of caves because of glacial and/or river erosion as well as vegetation overgrowth.

10.2 Geographical, Geological and Hydrogeological Setting

10.2.1 Geography and Climate

The mountain range belongs to the frontal alpine range (Helvetic Prealps), forming the first alpine rocks facing the Swiss Plateau. Most of the area belongs to the southeast-dipping limb of an anticline. Its northwestern limit is formed by high cliffs (Fig. 10.2). Elevations of the summits located along the cliffs range between 1950 and 2197 m a.s.l. Denuded karren fields occur above 1700 m a.s.l., in the area where limestone is exposed. Below, sandstone mainly crops out; therefore, the area is mainly covered by forest, meadows and swamps. The climate is humid and temperate, dominated by western winds. The average annual temperature is about 2 °C at 1700 m a.s.l., and annual precipitation ranges between 1500 and 2000 mm.

P. Häuselmann (✉)
Swiss Institute for Speleology and Karst Studies, Rue de La Serre 68, 2301 La Chaux-de-Fonds, Switzerland
e-mail: praezis@speleo.ch

E. Reynard (ed.), *Landscapes and Landforms of Switzerland*, World Geomorphological Landscapes,
https://doi.org/10.1007/978-3-030-43203-4_10

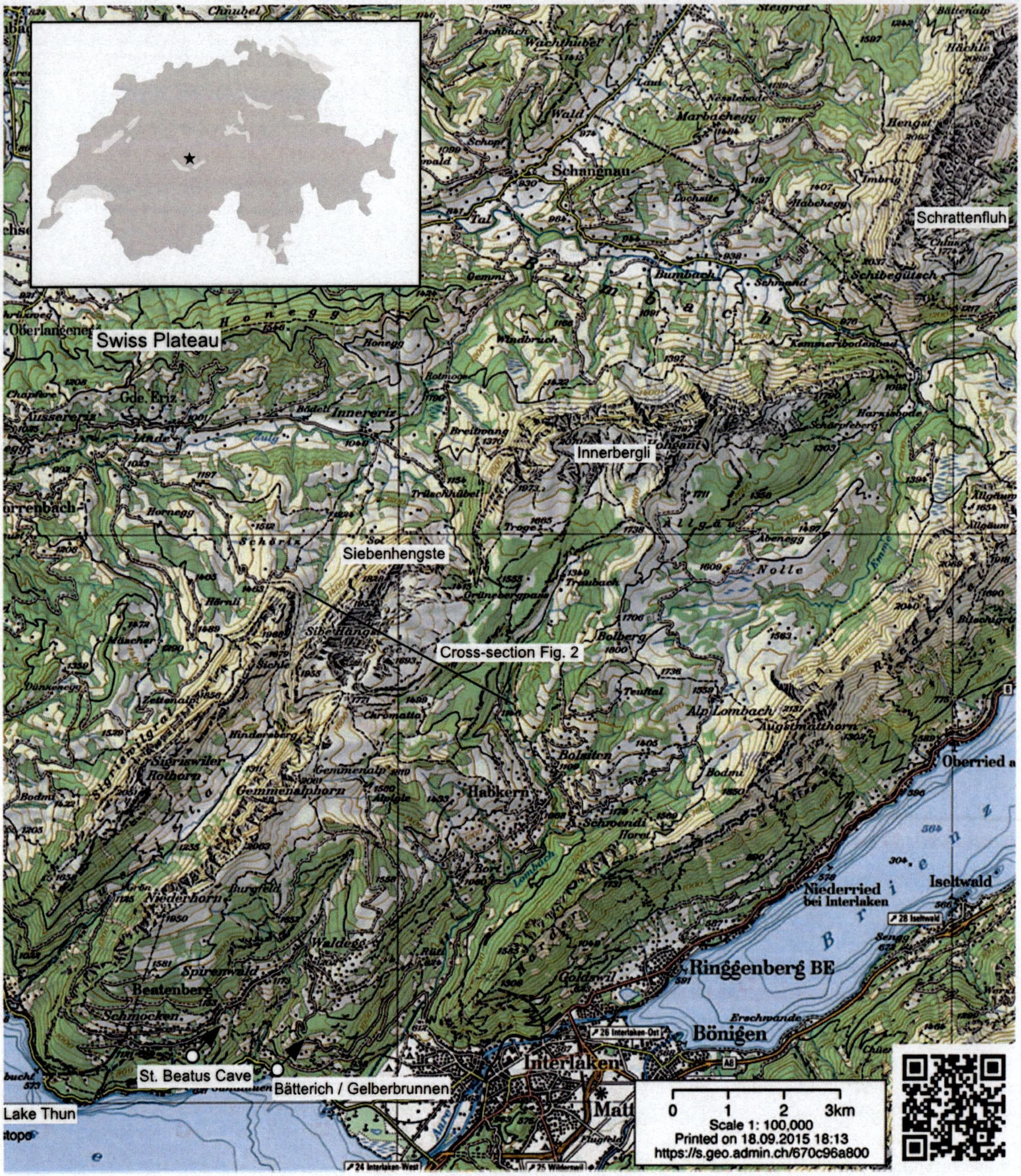

Fig. 10.1 Map of the Helvetic Prealps between Lake Thun and Schrattenfluh (swisstopo)

10.2.2 Geology

Karst features are developed predominantly within the Schrattenkalk Formation (Barremian to Aptian, Cretaceous, Urgonian facies), which is 150–340 m thick (Fig. 10.2). It is underlain by the Drusberg marls (Lower Barremian). These marls are 30 to 50 m thick. Most of the underground rivers are located at the bottom of the limestone, and follow the dip on top of the impervious Drusberg marls, often along main faults (Jeannin 1989). The Hohgant Formation (Eocene),

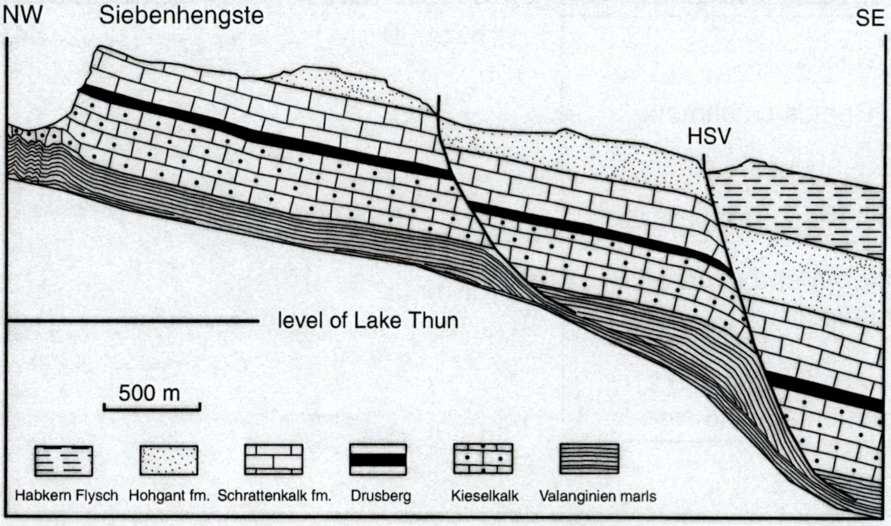

Fig. 10.2 Geological cross-section across the Siebenhengste. The monocline and the normal faults are easily visible. HSV is the Hohgant-Sundlauenen fault. The geology is as follows: Habkern Flysch, of mixed lithology (limestone, sandstone, marl) and age (Cretaceous to Oligocene); Hohgant formation, of mixed lithology (sandstones to subordinate pure limestone), of Eocene age; Schrattenkalk formation, pure limestone of Middle Cretaceous age; Drusberg marls, Lower Cretaceous, Kieselkalk formation, siliceous limestone, Lower Cretaceous, Valanginien marls, Lower Cretaceous

which is locally more than 200 m thick, overlies the Schrattenkalk Formation (Fig. 10.2). The Hohgant Formation is not a singular uniform sandstone body, but comprises rock sequences deposited in several cycles starting with limy sandstone and ending with purely quartzitic sandstone (Breitschmid 1976). Locally, bioherms of pure lithothamnia limestone are found. Therefore, parts of the Hohgant formation may be well karstified and contain caves. In most of the region, faults enable surface waters to eventually flow through the sandstones down into the Urgonian limestones. The Hohgant Formation is covered by Globigerina marls, upon which thick flysch deposits of the Pennine Nappes are overthrusted. Both marls and flysch are impervious.

An important normal fault, stretching from Lake Thun to Schrattenfluh (the Hohgant-Sundlauenen Fault HSV), disrupts the continuity of the southeastward dipping monocline of the Siebenhengste range (Fig. 10.2). The offset of the fault is 200–1000 m, depending on the location. Several subparallel normal faults are also present. The normal faults in the Siebenhengste region mainly developed during the Lower Cretaceous to the Eocene.

10.2.3 Hydrogeology

The hydrogeology of the area is somehow complicated, because surface rivers (flowing on flysch and sandstone) and underground karst rivers are interwoven. Much of the region is drained by underground karst systems emerging at two main locations: the St. Beatus spring and the Bätterich/Gelberbrunnen springs.

The St. Beatus spring with a discharge ranging between 10 l/s and 2–3 m^3/s (average 72 l/s) has a catchment area of about 10.5 km^2 that lies in the southeast part of the region (Häuselmann et al. 2004). The Bätterich/Gelberbrunnen spring catchment extends at least 21 km to the northeast, reaching the Schrattenfluh massif, as proven by a tracing experiment (Fig. 10.3). As the main spring (Bätterich) lies below Lake Thun, the discharge of the system is very difficult to measure. It probably exceeds 20 m^3/s during floods. The catchment area of the Bätterich/Gelberbrunnen system is around 32 km^2 and is largely covered by sandstone where surface flow may occur locally. Recharge rates are therefore difficult to assess.

10.3 Landforms and Landscapes

10.3.1 General Overview of the Landscape

The karst area of Siebenhengste-Hohgant-Schrattenfluh is not the largest in size of Switzerland. However, the moderate altitude (less frost weathering compared to high-alpine karren fields), the very pure limestone and the gently sloping monocline are the reasons why beautiful karren fields formed. The photographs of Schrattenfluh (Fig. 10.4) show the vastness and the desert-like appearance of the landscape. The Siebenhengste, albeit smaller, occurs in the same

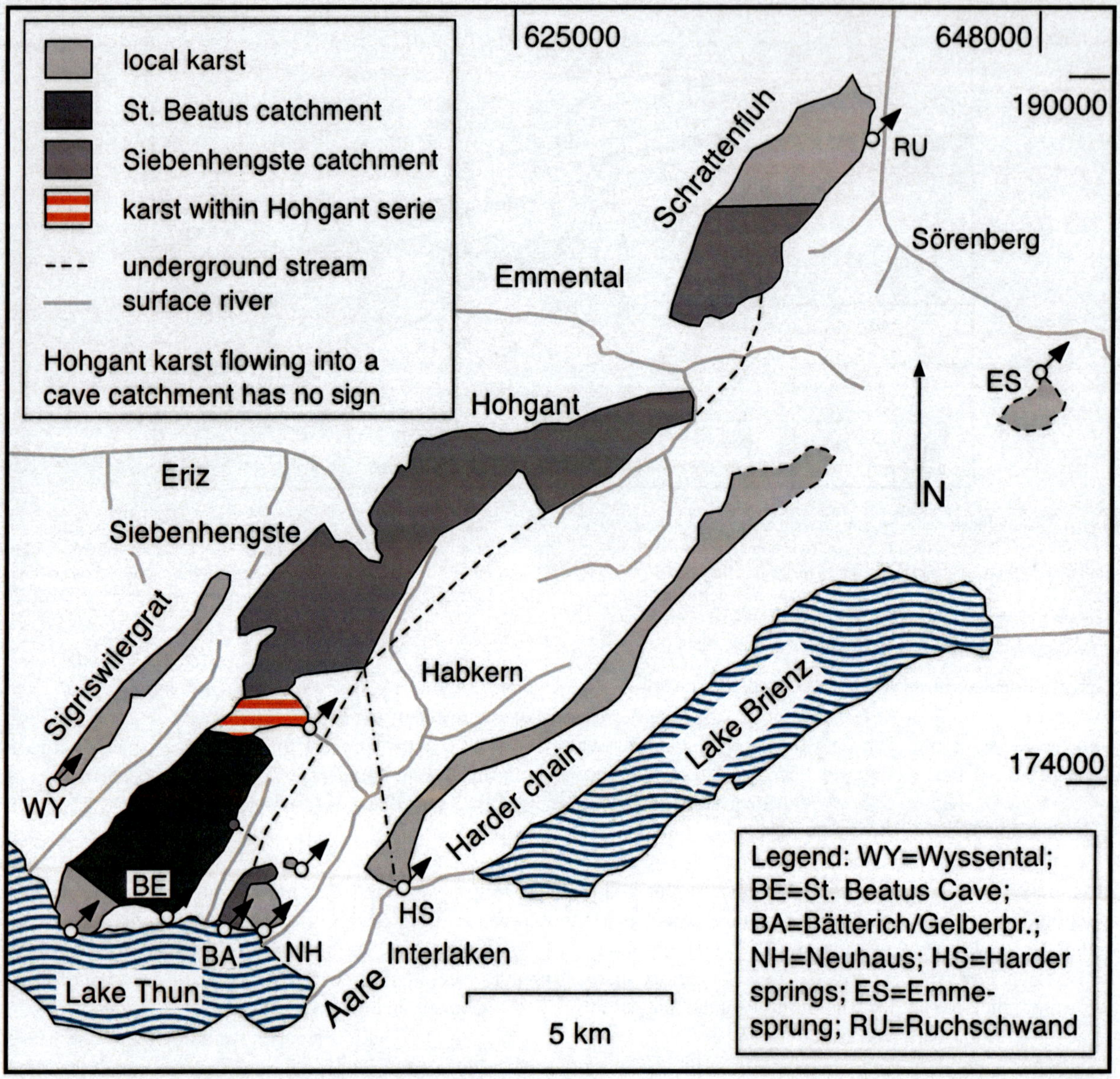

Fig. 10.3 Catchment areas of the wider region. The most important catchment is the one of Siebenhengste-Hohgant-Schrattenfluh, followed by the one of Beatus Cave. All other catchment areas are much less known. The link of Harder Springs to the Siebenhengste catchment was proved in a 1959 dye tracing experiment that never could be repeated

setting. In Siebenhengste, *Schichttreppenkarst* and *Schichtrippenkarst* (following Bögli 1964) prove the passage of ancient glaciers (Fig. 10.5a). Both forms result from the glacial scraping off of single limestone beds along bedding planes. If the remaining beds are towards the mountain top, the morphology looks like steps, so it is called Schichttreppenkarst; if the remaining beds are towards the mountain base, they are called Schichtrippenkarst. *Spitzkarren* (Fig. 10.5b), on the other hand, could only develop in areas, which were not covered by the main glaciers of the last glaciation; otherwise, they would have been destroyed by the glacier. Therefore, the summit area of Siebenhengste with Spitzkarren was a nunatak during the Last Glacial Maximum (LGM; Schlüchter 2009).

The karren fields are bordered by Hohgant sandstones. The more or less impervious character of these rocks, together with a significant part of low-solubility material, enhancing soil generation, explains why most of the sandstone areas are grassy, swampy, and in lower areas, forested (Fig. 10.6). Much of the waters collected in the sandstone area eventually disappears underground (Fig. 10.7) and joins the main karst body. Especially in the border between

Fig. 10.4 Schrattenfluh karren field. **a** The northern part of the Schrattenfluh is more rugged and partially overgrown by vegetation; **b** The middle part of the karren field is homogenous. With respect to the Siebenhengste, it is more steeply inclined (photos R. Wenger)

limestone and sandstone, many brooks emerge from the sandstone and disappear underground when reaching the limestone. In Innerbergli karren field, ancient ponors (visible as shafts) are prominently seen along a line from the present stream course towards the karren field (Fig. 10.8). This shows that the karren field in limestone was gradually exposed, and not uncovered at once by a glacier scraping off the sandstone cover. This does not exclude (partial) erosion from a glacier per se; the general topography of the karren field and adjacent areas reveal the presence of typical glacial forms such as *roches moutonnées* (Fig. 10.9) and glacio-karstic forms such as *Schichttreppenkarst*. Most of these forms are probably related to the geomorphic action of local glaciers.

Fig. 10.5 **a** *Schichtrippenkarst* in the sense of Bögli (1964) is proof of an ancient glaciation; **b** *Spitzkarren* on the summit of Siebenhengste. Such acicular structures could not have survived the last glaciation. The largest Spitzkarren reach almost one metre in height (photos Association française de karstologie)

10.3.2 Influence of Glaciations

Franz Knuchel tried, already back in the 1960s, to quantify surface corrosion at the Siebenhengste karren field (Fig. 10.10). Taking into account corroded limestone, present-day precipitation and corrosion rates, he concluded that the last glaciation terminated 14,500 years ago—a value that was confirmed by Quaternary scientists only much later (Schlüchter and Kelly 2000). Unfortunately, Knuchel never published these results.

Fig. 10.6 Area covered with sandstone (wet, partially swampy, vegetated and forested) No traces of karstification are seen on the first sight, but the brook in the picture disappears underground 30 m further (photo P. Häuselmann)

Fig. 10.7 A ponor within the sandstone area where a river disappears underground. It will only reappear at an elevation 1000 m lower, at Lake Thun (photo D. Sanz)

Knuchel, as well as a Belgian caver, André Minet (1970), found granite pebbles in the SW of Siebenhengste, near Wagenmoos, at an altitude of 1700 m a.s.l. Since it is assumed that the Aare glacier during LGM did not rise above 1400 m a.s.l. in this region (Schlüchter 2009), these pebbles are supposed to come from previous glaciations, as also proven by findings in caves by Jeannin (1991). A detailed analysis of these (and other) pebbles by Gnägi and Schlüchter (2012) showed not only that they really are of glacial origin, but that they have their source area in the southern side of the Upper Valais area! Thus, these pebbles travelled across one of the present-day deep troughs, crossed the second most important alpine ridge, crossed again the present-day Lake Thun area, and finally were deposited in the Siebenhengste! Dating of such pebbles with cosmogenic nuclides, found in A201 cave (Siebenhengste), revealed an age of 1.87 ± 0.21 Ma (Häuselmann et al. 2007). This gives the minimum age of deposition in the area. The first glaciations seem therefore to present S-N ice-flow directions, which are completely disconnected from the present-day valleys direction with a W-WNW trend. These old pebbles were conserved on Siebenhengste because of the presence of efficient traps in the karst system, and to a lower degree of mechanical erosion at the surface than in non-karstic massifs.

10.3.3 Micromorphology of Karren Fields

Although all karren fields in the area present beautiful examples of karren forms, the Innerbergli karren field is distinctive, because almost all morphologies (with the exception of *Spitzkarren*) are found within a short distance (Bitterli and Häuselmann 2010). As in most parts of the other karren fields, all the micromorphologies present in Innerbergli had been chiselled out since the last glaciation. Only larger karst features, such as shaft entrances, could be conserved after the erosion related to local glaciers. Research on dissolution around letters/numbers painted by cavers 30 to 40 years ago evidenced a corrosion rate of 0.014 mm/a, which is in good agreement with other sites worldwide (Häuselmann 2008).

As in many alpine karren fields, *rillen-* and *rinnenkarren* are well developed in the region (Fig. 10.11), especially in Innerbergli. *Mäanderkarren*, *hohlkarren*, grikes, *flachkarren*, and *trittkarren* are also present. So far, the Innerbergli karren field was only researched by Franz Knuchel, trying to decipher the influence of karstification since the last glaciation, but due to the well-developed forms, the clear geomorphologic context, the easy access and the good knowledge on climate conditions, future research projects on karren genesis are in discussion.

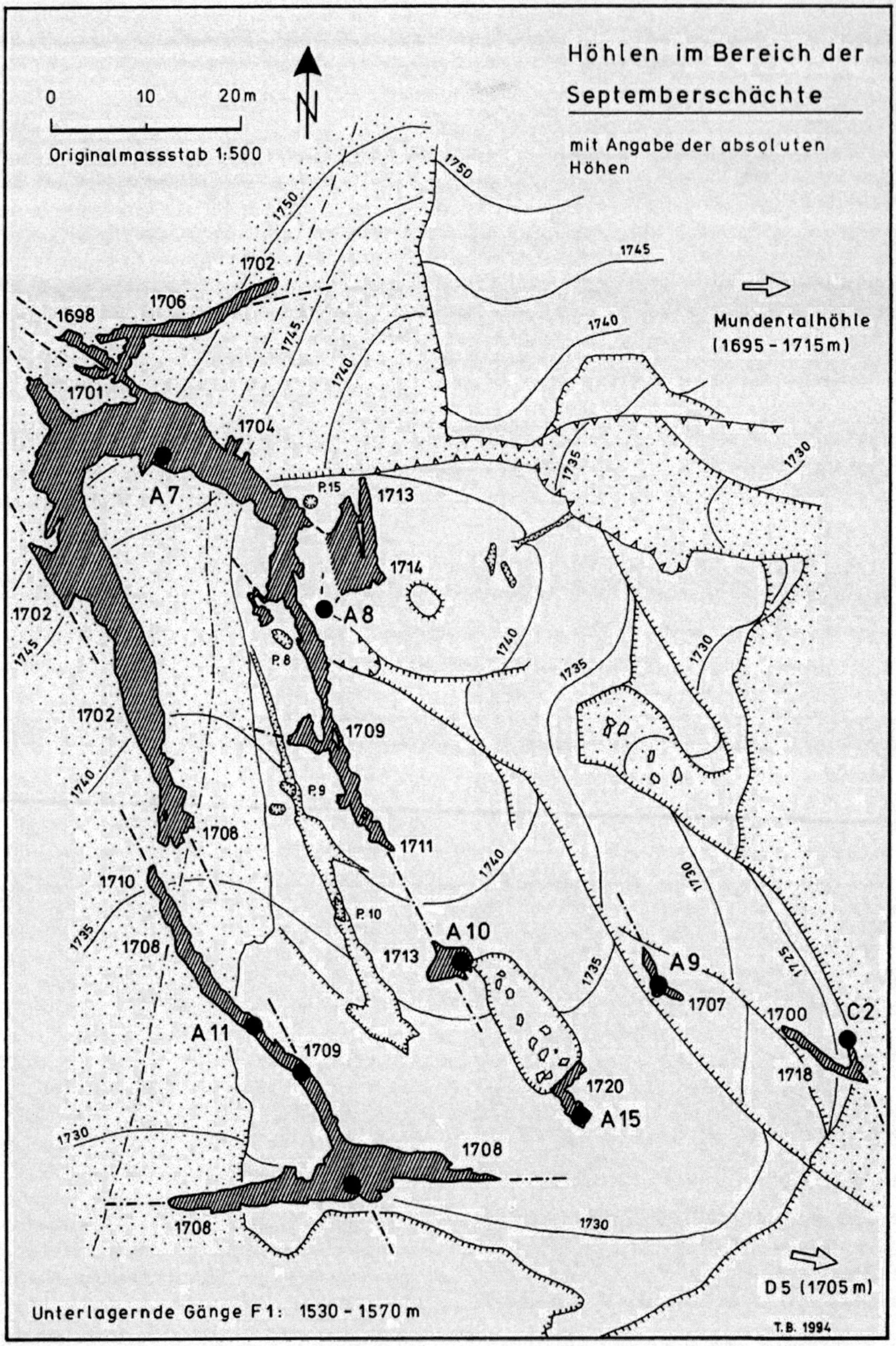

Fig. 10.8 Plot of the entrances of Septemberschächte in Innerbergli. The shafts A7, A8, P9, P10, A10 and A15 are thought to be progressively older ponors proving a backweathering of the sandstone covering the karren field. Dotted areas: limestone overgrown with vegetation; white: limestone at the surface; grey: underground passages

Fig. 10.9 View to a sort of a "*roche moutonnée*" in Innerbergli karren field. Although not very typical, the position and shape of the limestone hump tells that it was remodelled by a glacier (photo C. Prenez)

10.3.4 Tourism Impacts

The karren fields, although spectacular, are not necessarily of major interest for an average tourist. Furthermore, the area was spared from the big ski development, experienced by the destinations south of Interlaken (Grindelwald, Lauterbrunnen)—the village of Habkern has only one single ski lift. In recent years, tourism promoters discovered that the region is suitable for cross-country skiing and hiking, and they deliberately decided to develop soft tourism. However, the area comprises many wetlands in the flysch- and sandstone-covered areas. Therefore, in parallel to growing tourism, vast areas of wetlands and many forested areas were set aside for either undisturbed development or wildlife

Fig. 10.10 Franz Knuchel measuring corrosion activity on the Siebenhengste karren field. He was a precursor of the present-day scientists working in the area (photo taken by unknown person)

refuge. This protection of the landscape has already led to restrictions for accessing some caves.

10.3.5 The Underground Landscape

The word "landscape" refers to the visible features of an area. Nevertheless, caves are part of the territory, which can be visited, described and interpreted by humans in a similar way to landscapes at the surface. The underground landscape of Siebenhengste is mainly characterised by two types of cave passages: meanders (high and narrow canyons) and tube passages (cave passage with an elliptical cross-section). The size of canyons typically ranges between 0.4 and 4 m in width and between 2 and 40 m in height. The size of tubes is rarely larger than 5–7 m in diameter. The minimal size of cave passages is mainly given by the size a person can penetrate (about 0.5 m). Sometimes both profiles are combined, forming a key-hole cross-section (tube entrenched by a canyon). Along the contact with Drusberg marls, the entrenchment of the marls, which are mechanically less stable than limestones, often formed larger "square" passages ("*Kastenprofil*") and sometimes even rooms. Their size typically ranges between 5 and 50 m in width and/or height. Another typical feature are shafts (vertical passages) with diameters ranging between 1 and 50 m and depths between 5 and 180 m in the Siebenhengste region. All these types of cave passages can be more or less filled with sediments, decorated with speleothems, and can be traversed by flowing water or being fossil since millions of years, leading to a large variety of underground landscapes.

The cave region between Lake Thun and Schrattenfluh encloses four main massifs: Niederhorn, Siebenhengste, Hohgant and Schrattenfluh. With 160 km, the longest cave system is the Réseau Siebenhengste-Hohgant, which combines the two central massifs and reaches down to the level of Lake Thun, with 1340 m total height difference. Figure 10.12 displays all the caves of the area between Niederhorn and Hohgant, summing up a total length of 340 km. Caves of the Schrattenfluh massif are less extended, summing up about 30 km, but further discoveries may change the picture.

Such a vast cave system was not carved out of the rock in a single event. The morphological analysis of cave passages gives us hints to discover the genetic phases. If the water can circulate freely in the rock (i.e. without geological constraints, such as folds or faults), it flows vertically downwards until reaching the water table or the bottom of the aquifer (the top of the Drusberg marls in the Siebenhengste region). In this latter situation, water then follows the top of the marls down-dip until it reaches the water table. Then, it flows more or less horizontally towards the spring along the most karstifiable way, which is often a bedding plane. The elevation of the spring is generally close to the valley bottom (Häuselmann et al. 2003), and the spring itself defines the height of the water table inside the karstified massif. Therefore, elevations of ancient karst water tables (thus of ancient valley bottoms) are seen in the morphology of cave passages. A careful interpretation of the "underground landscape" thus directly tells us about the evolution of the landscape above ground (Häuselmann et al. 2002). In Siebenhengste cave system, 14 speleogenetic phases were identified so far. During the first, oldest and uppermost five phases the system had their spring in Eriz Valley (Fig. 10.13). At that time the Aare Valley did obviously not exist, as also proved by the exotic pebbles found by Gnägi and Schlüchter (2012) (see above). In the following nine phases underground flows were directed towards the area of present Lake Thun in the Aare Valley.

Such a complete information is no more available at the surface due to glacial erosion. However, caves enclose even more information: they contain speleothems and other sediments which can be dated. The most common dating methods are U/Th on speleothems (range up to 500 ka), and concentrations of cosmogenic nuclides on quartz (range up to 5 Ma). Both techniques were applied in the Siebenhengste area, providing ages of the respective phases, i.e. of the valley incision rate, as well as on the presence or absence of glaciers at a given time (Häuselmann et al. 2007, 2008). Such a complete reconstruction makes the Siebenhengste massif to be a renowned example among karst scientists and geomorphologists.

Discrepancies between the generally accepted model of cave genesis, the Four State Model of Ford and Ewers (1978), and observations in Siebenhengste caves and in

Fig. 10.11 a *Rinnenkarren* on a steep fault line in Innerbergli; b *Rillenkarren* "mountains" are very typical of many alpine karren fields, but in Innerbergli they are very spectacular (photos C. Prenez)

other alpine cave systems, led to the formulation of another speleogenetic model. This new model postulates that horizontality (or undulation) of the cave is related to uniform (or contrasting) recharge (Gabrovsek et al. 2014). This shows the possibility that the geometry of the cave system gives hints to paleoclimatic conditions: constant rain would give rather horizontal passages, contrasting climate would yield undulating passages. In any case, paleoclimatic research is

Fig. 10.11 (continued)

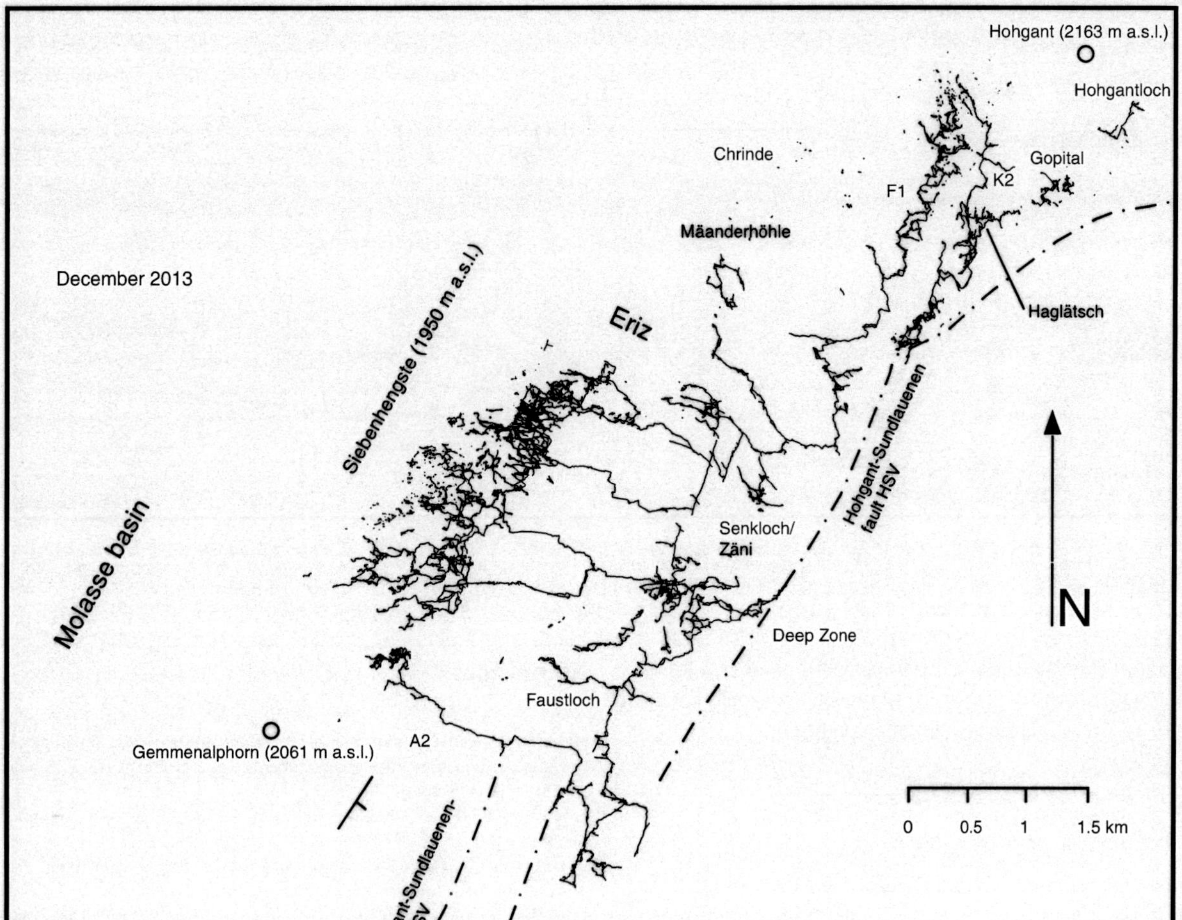

Fig. 10.12 Overview of the caves between Lake Thun and Hohgant. The density of caves is impressive

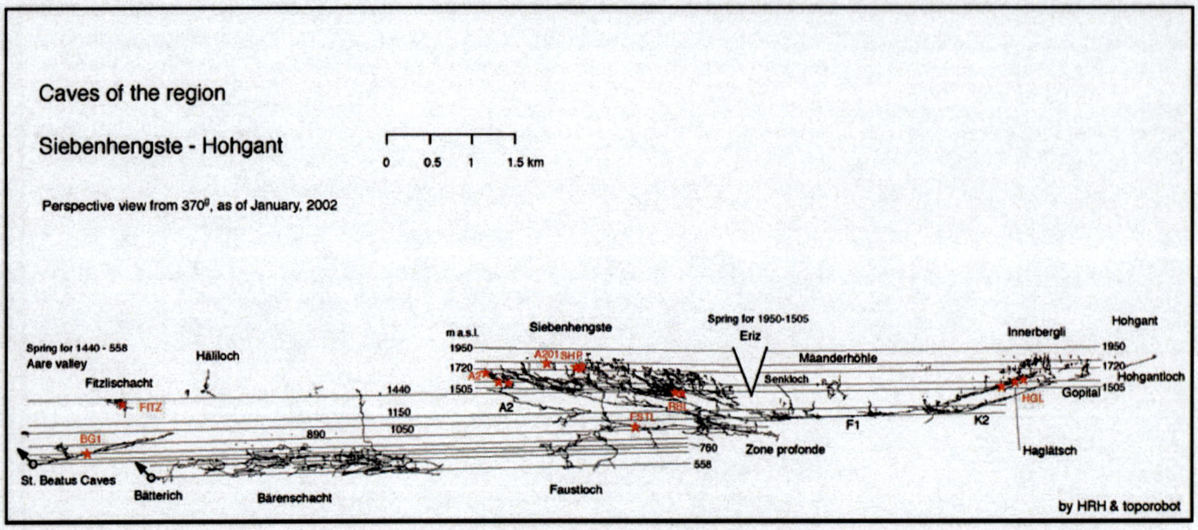

Fig. 10.13 Projection (370^{gr}) of the Siebenhengste caves. The lines represent speleogenetic phases related to ancient springs (and thus to paleovalley bottoms). Note that the altitude indicated is present-day altitude, not taking into account recent uplift of the Alps

going on (Luetscher et al. 2015), and still, there is a lot to be done with classical analysis of sediments and speleothems in order to investigate paleoclimatic conditions.

10.4 Conclusions

Karst areas are interesting regions for the study of landforms, landscape, landscape evolution and palaeoenvironments. The Siebenhengste region shows interesting karren forms, indicators of ancient glaciations, indicators of landscape evolution as well as palaeoclimate. Such a rich information about the last 2 to 5 millions of years of alpine history can almost only be found in karst regions because erosion removed most comparable indicators elsewhere. Caves are natural archives, which are not always easy to study and to decipher, but which often produce rewarding results.

Acknowledgments First, I would like to thank the cavers. Their relentless efforts in mapping the caves were the base for all ulterior scientific work. Alex Hof, Thomas Bitterli, Pierre-Yves Jeannin, Stein-Erik Lauritzen and Darryl Granger believed in the underground richness and were instrumental in reading the archive. Franz Knuchel, Christian Gnägi, Nicolas Fauquex and Emmanuel Reynard, on their turn, did great work on the surface.

References

Bitterli T, Häuselmann P (2010) Die Höhlen des Innerberglis. Höhlenforschung im Gebiet Siebenhengste-Hohgant Nr. 8, Speleo Projects, Allschwil, 440 pp

Bögli A (1964) Le Schichttreppenkarst. Revue Belge de Géographie 88:63–81

Breitschmid A (1976) Geologie im Gebiet des Gemmenalphorns. Unpublished diploma thesis, University of Bern

Ford DC, Ewers RO (1978) The development of limestone cave systems in the dimensions of length and depth. Can J Earth Sci 15:1783–1798

Gabrovsek F, Häuselmann P, Audra P (2014) "Looping caves" versus "watertable caves": The role of base-level changes and recharge variations in cave development. Geomorphology 204:683–691

Gnägi C, Schlüchter C (2012) High-altitude erratics in the Bernese Alps (Switzerland). Swiss J Geosci 105(3):401–415

Häuselmann P, Jeannin PY, Monbaron M, Lauritzen SE (2002) Reconstruction of Alpine Cenozoic paleorelief through the analysis of caves at Siebenhengste (BE, Switzerland). Geodin Acta 15:261–276

Häuselmann P, Jeannin PY, Monbaron M (2003) Role of epiphreatic flow and soutirages in conduit morphogenesis: the Bärenschacht example (BE, Switzerland). Zeitschrift für Geomorphologie N.F. 47 (2):171–190

Häuselmann P, Bitterli T, Höchli B (2004) Die St. Beatushöhlen: Entstehung, Geschichte, Erforschung. Höhlenforschung im Gebiet Siebenhengste-Hohgant Nr. 7, Speleo Projects, Allschwil, 256 pp

Häuselmann P, Granger DE, Lauritzen SE, Jeannin PY (2007) Abrupt glacial valley incision at 0.8 Ma dated from cave deposits in Switzerland. Geology 35(2):143–146

Häuselmann P, Lauritzen SE, Jeannin PY, Monbaron M (2008) Glacier advances during the last 400 ka as evidenced in St. Beatus Cave (BE, Switzerland). Quatern Int 189:173–189

Häuselmann P (2008) Surface corrosion of an Alpine karren field: recent measures at Innerbergli (Siebenhengste, Switzerland). Int. Journal of Speleology 37(2):107–111

Jeannin PY (1989) Etude géologique de la région Burst-Sieben Hengste. Unpublished diploma thesis, University of Neuchâtel

Jeannin PY (1991) Mise en évidence d'importantes glaciations anciennes par l'étude des remplissages karstiques du Réseau des Siebenhengste (chaîne bordière helvétique). Eclogae Geol Helv 84 (1):207–221

Luetscher M, Boch R, Sodemann H, Spötl C, Cheng H, Edwards RL, Frisia S, Hof F, Müller W (2015) North Atlantic storm track changes during the Last Glacial Maximum recorded by Alpine speleothems. Nat Commun 6:6344. https://doi.org/10.1038/ncomms7344

Minet A (1970) Etude préliminaire de la région des Siebenhengste (Eriz, BE). In: Actes du 4e Congrès national de spéléologie, pp 35-47

Schlüchter C (compil) (2009) Die Schweiz während des letzteiszeitlichen Maximums (LGM), 1:500 000. GeoKarten 500. Bundesamt für Landestopografie swisstopo, Wabern

Schlüchter C, Kelly M (2000) Das Eiszeitalter in der Schweiz. Publikation des Geologisches Instituts der Universität Bern, Bern, 4 pp

Philipp Häuselmann is Scientific Collaborator at the Swiss Institute for Speleology and Karst studies SISKA and Lecturer at University of Life Sciences, Vienna. He is chairman of the International Union of Speleology Working Group Topography and Mapping, and an active caver. His main scientific interest is in speleogenesis and dating cave deposits, both by U/Th and by cosmogenic nuclides. His main focus is research on caves in mountain belts.

11 The Structural Landscapes of Central Switzerland

O. Adrian Pfiffner

Abstract

Both, landforms and landscape of Central Switzerland mimic the structure of the underlying bedrock rather closely. As an example, Stanserhorn, Buochserhorn, and Mythen can be mentioned where Jurassic carbonates of the Klippen nappe form steep peaks above the slopes underlain by the marly sequence of the Subpenninic mélange. At the contact between the Helvetic nappes and the Subalpine Molasse, the massive Cretaceous carbonates of the Drusberg nappe form a long through going cliff, which marks this tectonic contact. The steep SE face of Rigi corresponds to the SE boundary of the Rigi alluvial fan. A major tear fault east of the Pilatus massif explains the juxtaposition of this mountain range to the basin of Lake Lucerne, and a tear fault running along the Uri branch of Lake Lucerne is held responsible for the position and orientation of this valley.

Keywords

Klippen • Helvetic nappes • Penninic nappes • Molasse

11.1 Introduction

Several rugged mountain peaks dominate the landscape of Central Switzerland. These peaks tower above Lake Lucerne (8° 30' E / 47° N), a long but narrow lake winding between these mountains. The mountains are all made of sedimentary rocks, but the sedimentary sequences are of very different nature owing to their distinct geological history. The spectacular views across the lake onto the mountains on the opposite side have caught the eye of early visitors of the area and laid the foundation for tourism. Already in the nineteenth century tourist paths were opened. In the second half of the nineteenth century, several of the peaks were made accessible to tourists by cogwheel trains (Rigi in 1871 from Vitznau and in 1875 from Arth-Goldau, Pilatus in 1889) and funicular trains (Stanserhorn in 1893) (see location of the peaks in Fig. 11.1). The transport system was later replaced or complemented by cable cars. Pilatus, Stanserhorn, and Rigi can now be reached by spectacular aerial cable cars (aerial panorama gondola and aerial cableway toward Pilatus, Cabrio to Stanserhorn and panorama cable car to Rigi). Several smaller cable cars reach Rigi, also called "Queen of the Mountains," from the south (Brunnen) and from the north (Küssnacht). As for the Mythen you have to climb to the summit. The first climb occurred in 1790 and in 1864 a path to the summit was constructed under the auspices of the Mythen Society.

11.2 Geographical and Geological Setting

The landscape of central Switzerland discussed in this chapter is located within the first range of mountains on the northern slope of the Alps (see Fig. 11.1). The rivers draining this part of the Alps carved out N-S oriented valleys. From west to east these are the Sarner Aa valley and its extension toward Lake Zug, the Engelberger Aa, and the Reuss valleys, as well as its extension to Brunnen and Zug. All these valleys were deepened and widened by the Pleistocene glaciers as discussed by B. Keller (this volume). Two valley segments occupied by Lake Lucerne run E-W, one from the town of Lucerne toward the east, the other from Brunnen toward the west (Fig. 11.1).

Two major geological domains can be distinguished in the study area, the Molasse Basin in the north and an Alpine nappe stack to the south. With regard to landforms, two aspects are particularly important: the rock types present in each of the tectonic units and the structure of the rocks' body

O. A. Pfiffner (✉)
Institute of Geological Sciences, University of Bern, Baltzerstrasse 1 + 3, 3012 Bern, Switzerland
e-mail: adrian.pfiffner@geo.unibe.ch

E. Reynard (ed.), *Landscapes and Landforms of Switzerland*, World Geomorphological Landscapes,
https://doi.org/10.1007/978-3-030-43203-4_11

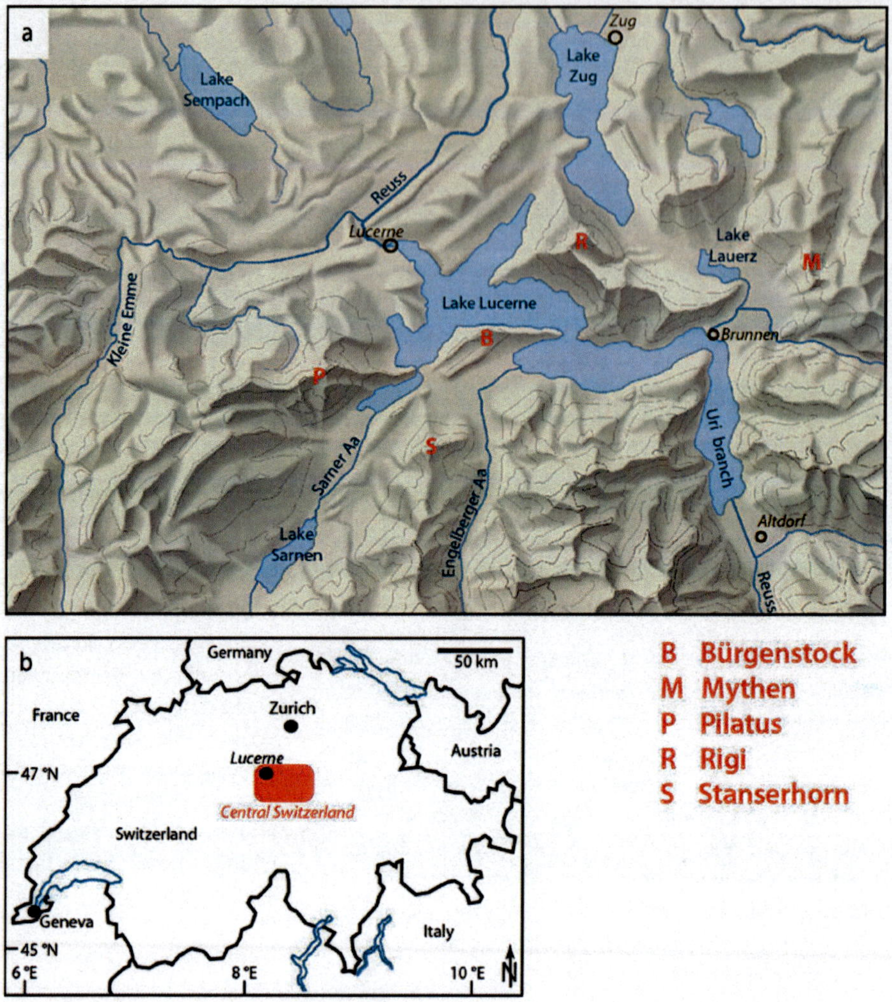

Fig. 11.1 **a** Topographic map of Central Switzerland (based on AdS) giving the main features pertinent to this chapter. **b** General map of Switzerland showing location of Central Switzerland

within each unit. In the following paragraphs the tectonic units are discussed individually, covering both lithologic and structural aspects.

The Molasse Basin represents a foredeep, which formed along the northern rim of the growing Alps during the formation of the nappe stack. The basin fill consists of predominantly clastic rocks, conglomerates, sandstones, and mudstones of Oligocene–Miocene age. The entire sequence includes a succession of marine alternating with fluvial deposits. The latter are sediments transported into the foredeep by rivers draining the ancestral Alps. For a more detailed discussion of the stratigraphic succession the reader is referred to Kempf et al. (1999) and Pfiffner (2014). Structurally, two superunits must be distinguished. The Plateau Molasse located in the north represents a flat-lying sequence affected by local very open folds. However, toward the contact to the second superunit, the Subalpine Molasse, the strata are tilted to a nearly vertical dip. The tilted strata form a bundle of ridges clearly visible in the tectonic map to the NE and WSW of Lucerne (Fig. 11.2). Within the Subalpine Molasse the strata dip at various angles to the SSE. Near the contact with the Plateau Molasse these strata also strongly tilted to a nearly vertical dip. The steep dips along this contact are opposite in direction and form an antiformal structure referred to as "triangle zone" (visible in the cross-section of Fig. 11.3 discussed later). The term "Subalpine Molasse" refers to the structural position of this superunit relative to the overriding Alpine nappe stack. The most prominent landform within the Subalpine Molasse is Rigi (labeled R in Figs. 11.1 and 11.2), located between Lake Lucerne and Lake Zug.

Within the Alpine nappe stack two nappe systems are recognized in the study area: the Helvetic nappes and the overlying Penninic nappes. In the tectonic map (Fig. 11.2) the Helvetic nappes form a broad green band from the SW to the NE corner. The Helvetic nappes are thrust upon the

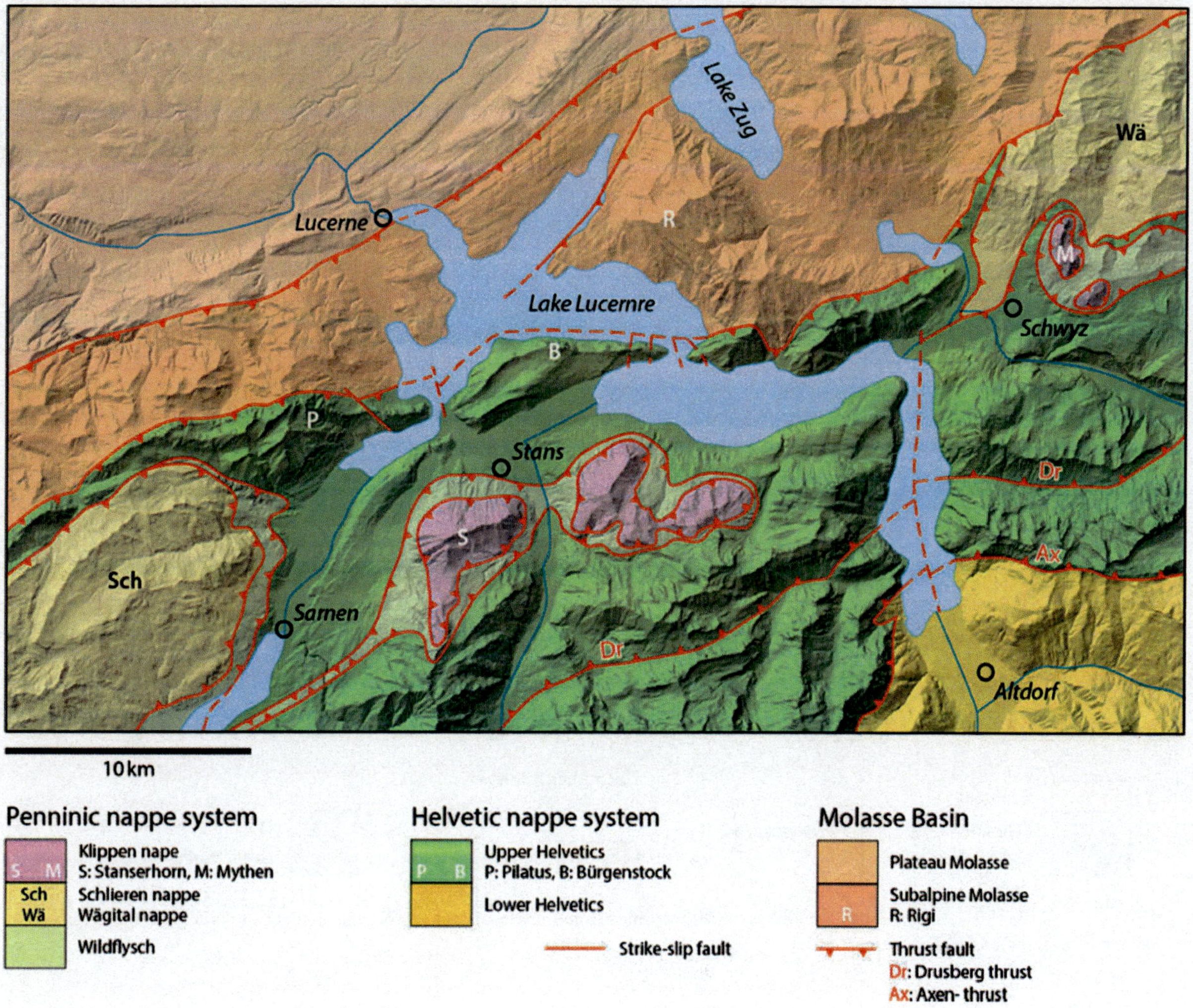

Fig. 11.2 Tectonic map of Central Switzerland showing the major tectonic units on a shaded relief map (shaded relief map swisstopo)

Subalpine Molasse along a major thrust fault that dips to the SSE. The displacement that the Helvetic nappes experienced along this thrust fault was of the order of 50 km (Pfiffner 1980; Kempf and Pfiffner 2004). The Helvetic nappes are made of Mesozoic sediments. Jurassic and Cretaceous limestones form the thickest strata and are responsible for much of the impressive local relief. As is evident in Fig. 11.2, a chain of mountains that displays a steep topographic slope marks the northern front of the Helvetic nappes. This chain straddles the thrust fault making the contact with the underlying Subalpine Molasse. Structurally, the Helvetic nappes can be subdivided into two major nappes, the Axen nappe below and the Drusberg nappe above. The thrust fault separating the two, the Drusberg thrust, follows a relatively thick layer of marl interlayered between Late Jurassic and Early Cretaceous limestones. The trace of the Drusberg thrust is highlighted by valleys (Fig. 11.2), which owe their existence to these mechanically weak and erodible marls. Two mountains within the Helvetic nappes, or the Drusberg nappe more precisely, are dominant landforms bordering Lake Lucerne: Pilatus and Bürgenstock (labeled P and B in Figs. 11.1 and 11.2).

In the southeastern corner of the study area (Fig. 11.2), the Helvetic nappes overly the Infrahelvetic complex. This complex is made up of crystalline basement rocks and their Mesozoic and Cenozoic sedimentary cover, of which only the youngest, Eocene–Oligocene strata, outcrop in the study area. These youngest sediments are a Flysch sequence consisting mainly of sandstone and shale, deeply incised by tributaries of the Reuss River. The overlying Jurassic limestones of the Helvetic nappes (the Axen nappe) form a string of steep cliffs following the tectonic contact. A more detailed

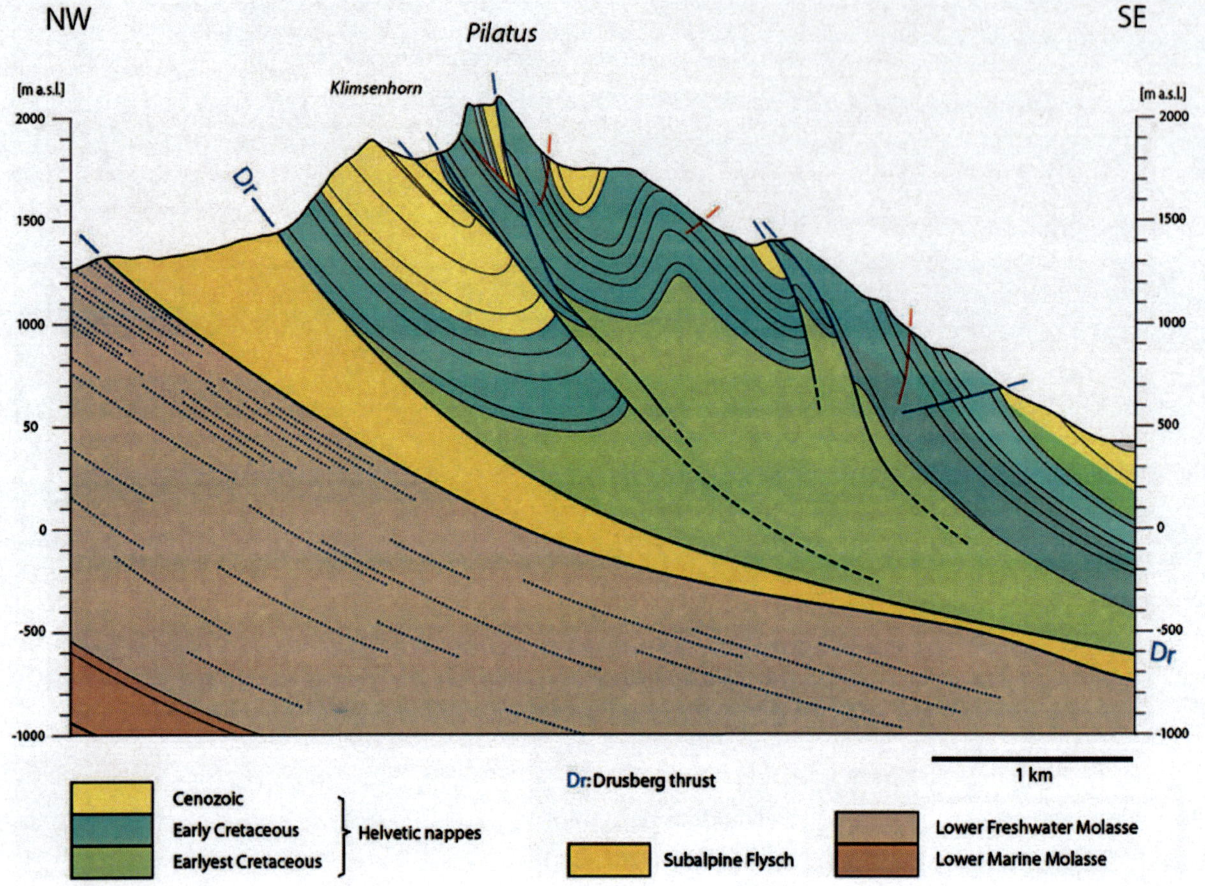

Fig. 11.3 Geological cross-section through the Pilatus massif (modified from Funk et al. 2013). The blue dotted pattern in the Lower Freshwater Molasse indicates the dip of conglomerate beds

discussion of the Helvetic nappes in general can be found in Pfiffner (2011, 2014) and references therein.

The Penninic nappes that overly the Helvetic nappes outcrop as erosional remnants or klippen only in this area of the Alps. Two nappes with very different lithologies may be distinguished here. The Schlieren nappe, in the SW corner of Fig. 11.2, consists of a thick Flysch sequence with sandstones and shales of Paleocene–Eocene age. The relatively homogeneous sequence manifests itself in a smooth topography incised by a deep valley. The other nappe, the Klippen nappe, got its name from its occurrence as erosional remnants (klippen). The Klippen nappe is made of Mesozoic sediments, with a thick sequence of Late Jurassic limestones being responsible for many steep cliffs and high local relief. Two outstanding landforms bordering Lake Lucerne are the Stanserhorn and the Mythen (labeled S and M in Figs. 11.1 and 11.2). Two much larger klippen can be found in western Switzerland (the Préalpes romandes) and neighboring France (Chablais).

The contact of these Penninic nappes to the underlying Helvetic nappes is marked by a highly complex mélange where a variety of blocks of limestones and sandstones swim in a tightly folded matrix of shales cut by many veins. This unit is referred to as Subpenninic mélange. This mélange formed in response to shearing during the emplacement of the Penninic nappes onto the (future) Helvetic nappes whereby pieces of mechanically strong units were plucked off the footwall and the hanging wall of the thrust contact. Mixing owing to shearing explains the variety of blocks included as well as the intense folding and vein formation by fracturing. These mélanges are referred to as "Wildflysch." Of interest in the geomorphological context are the large-scale homogeneous composition and the laterally highly variable thickness of the Subpenninic mélange. For a more general discussion of the Penninic nappes the reader is referred to Pfiffner (2014) and references therein.

11.3 Landforms and Landscapes

The landscape of Central Switzerland is dominated by Lake Lucerne, a relatively narrow lake curving around and surrounded by impressive mountains. The geomorphological characteristics of these mountains share a high local relief but are otherwise quite distinct. In the following five of these mountains are discussed in more detail, proceeding from west to east. Each case is illustrated with a geological cross-section and photographs.

11.3.1 Pilatus

Pilatus is the name of the entire mountain. Its highest peak is Tomlishorn, 2128.5 m high. The name Pilatus possibly derives from the latin *mons pileatus*, the "mountain with pillars." The Pilatus massif forms the erosional front of Helvetic nappes. In the course of the collision between the European and Adriatic plates limestone strata once deposited in a shallow shelf sea along the southern shore of the European continent were horizontally shortened and pushed toward the north over a distance of ca. 50 km until they came to lie on top of the Subalpine Molasse. Contemporaneously to their movement to the north, the limestone package was folded, and in some instances dismembered into smaller packages, which in turn were stacked on top of each other by thrusting. The cross-section displayed in Fig. 11.3 shows such folds just south of Tomlishorn and a thrust fault at the summit and one north of it. The Cretaceous limestone formations are made up of massive carbonates, which resist weathering and erosion. Hence they form steep cliffs where the limestones dip into the topographic slope, whereas where the strata dip parallel to the topographic slope, the slope angle is lower. The Cenozoic formations, which contain marl, are more easily eroded as can be seen by the depressions between Tomlishorn and Klimsenhorn and in the two synclines south of Tomlishorn. The major thrust fault at the base of the Drusberg nappe followed a thick marl sequence (earliest Cretaceous in age) and accumulated to great thickness in the core of the Pilatus massif. It has to be remembered, however, that the steep northern slope of the Pilatus massif is an erosional feature. This is proved by the fact that the rocks of the Subalpine Molasse just north of Pilatus show a metamorphic grade that can only be explained by a pile of nappes several kilometers thick (von Hagke et al. 2012), which were later removed by erosion.

A lense of Subalpine Flysch separates the Drusberg nappe from the Subalpine Molasse. Such Flysch cushions are found all along this contact in the Central Alps. They contain slices of sediments scraped off the footwall of the thrust fault as well as pieces from the hanging wall that were overridden by the nappe. In some instances the equivalent units contain blocks of very different origin and the entire unit is actually a mélange similar to the mélange at the base of the Penninic nappes. In any case, marls, shales, and sandstones are the prevailing rock types within the Subalpine Flysch. Carbonates are hardly present and this explains why this unit was easily eroded and led to low surface gradients as evident in the cross-section of Fig. 11.3.

In the east, the Pilatus massif is delimited by a tear fault along which Pilatus was pushed north relative to Bürgenstock located east of this fault (Fig. 11.2). As a consequence, the high topography of the Pilatus massif ends abruptly toward the east and produces the spectacular shape of this mountain when viewed from the east (Fig. 11.4).

11.3.2 Bürgenstock

Bürgenstock (1128 m) is the frontal range of the Drusberg nappe bordering Lake Lucerne. The range consists of carbonates that overlie easily erodible siltstones and sandstones of Subalpine Flysch and the Subalpine Molasse (Fig. 11.5). The internal structure of this range is governed by a syncline–anticline pair. Similar to Pilatus, the south-dipping Cretaceous carbonates in the northern flank of Bürgenstock represent an erosional front. Between Bürgenstock and Obbürgenberg, the shallow syncline with Cenozoic (Eocene) strata in the core and covered by Quaternary sediments is responsible for the topographic depression. The anticline of Obbürgenberg, on the other hand, expresses itself in the surface topography owing to the massive Early Cretaceous carbonates, which withstood erosion. The Drusberg nappe containing all these folds has been overthrust onto a sliver of Cretaceous carbonates, which is to be sought beneath Lake Lucerne; this imbricated slice outcrops a few kilometers east of the cross-section on the shore of Lake Lucerne and is taken to represent an initially frontal part of the Drusberg nappe that later was overridden by the main nappe body.

A thin cushion of Subalpine Flysch is interpreted to be present beneath the Helvetic nappes of Bürgenstock and overlying the SSW-dipping Subalpine Molasse sediments. Such a cushion can be expected judging from the regional outcrop pattern of this contact. The discordance drawn between the Subalpine Flysch and the Subalpine Molasse in the cross-section corresponds to surface patterns observed at outcrop. Within the Subalpine Molasse a thrust fault puts Lower Marine Molasse onto Lower Freshwater Molasse in the NNW of the cross-section. This thrust fault breaks

Fig. 11.4 Pilatus viewed toward the west (photo Pilatus-Bahnen AG)

surface farther to the NNW where it shows the same repetition of strata. The thrust fault indicated in the SSE of the cross-section is speculative.

The fold axes within the Drusberg nappe are more or less parallel to the strike of the Drusberg thrust and thus follow the trace of this thrust fault at the surface. As a consequence, the Bürgenstock range is a narrow-elongated ridge parallel to the nappe contact. Its width is controlled by the wavelength of the folds, while its elevation reflects the amplitude of the folds. Its western end is located at the tear fault bordering the Pilatus range discussed above. Toward the east the fold axes plunge eastward into Lake Lucerne and the Bürgenstock range consequently disappears. The range on the northern shore of Lake Lucerne farther to the east lacks folds and is made up of straight SSE-dipping strata (Ochsenalp in Fig. 11.8). The range thus is a simple cuesta as shown in the cross-section of Fig. 11.5. In Fig. 11.6, the Bürgenstock range is seen to narrow toward the lower left as the Obbürgenberg anticline plunges into the subsurface.

11.3.3 Stanserhorn

The Stanserhorn (1898 m) and the neighboring Buochserhorn are erosional remnants of the Klippen nappe, a part of the Penninic nappes. These klippen bear witness to the fact that the Helvetic nappes were once covered by a pile of Penninic nappes. This pile was 5–7 km thick just south of Lake Lucerne and explains the anchizonal metamorphic overprint of the Helvetic nappes (Breitschmid 1982). It was eroded in the course of the past 30 million years. The Klippen nappe consists mainly of Jurassic–Cretaceous limestones, which were deposited on a rise (the Briançon rise) within the Alpine Tethys (Baud and Septfontaine 1980; Wissing and Pfiffner 2002; Pfiffner 2014). At the base, a thin layer of Triassic evaporites and marl served as detachment horizon for the Klippen nappe.

The cross-section (Fig. 11.7) is based on Christ (1920), Geiger (1956), Funk et al. (2013), and author's own interpretation. The internal structure of the Klippen nappe

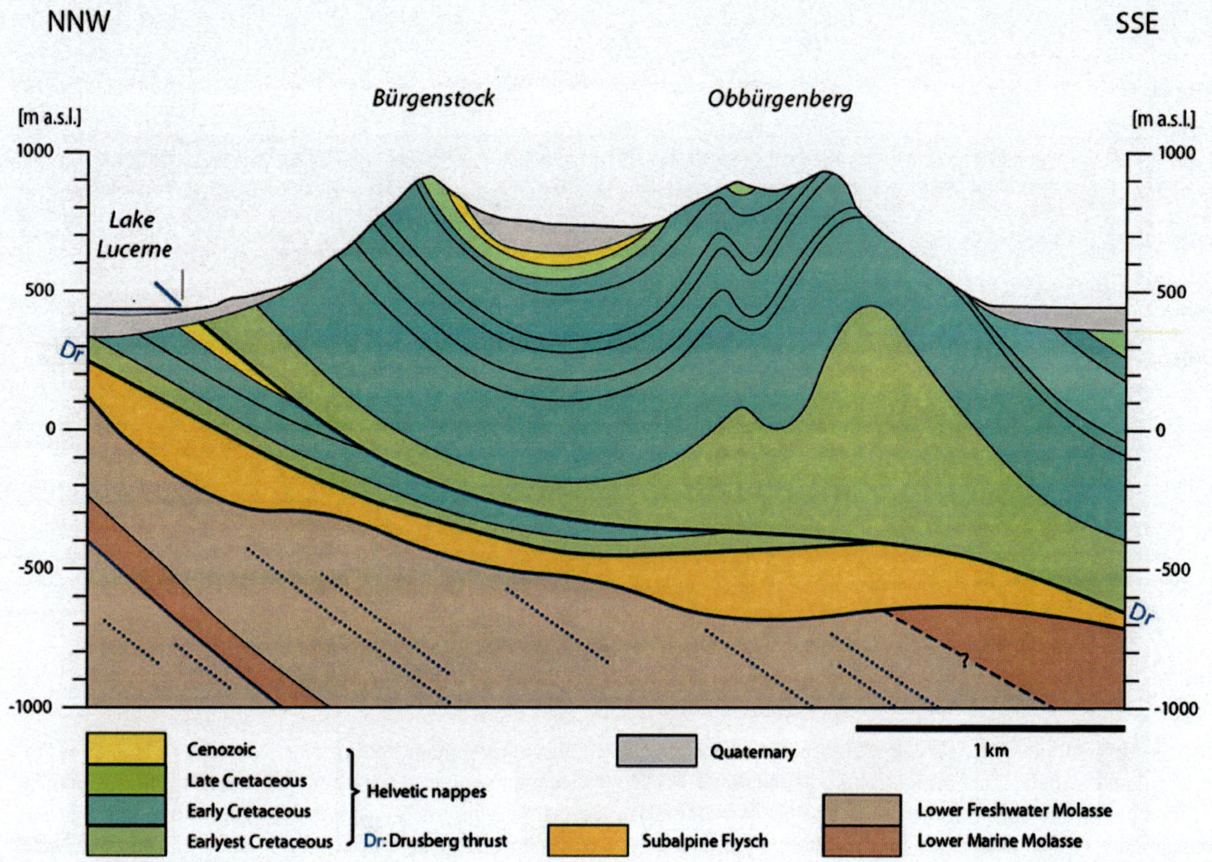

Fig. 11.5 Geological cross-section through the Bürgenstock range (modified from Funk et al. 2013). The blue dotted pattern in the Lower Freshwater Molasse indicates the dip of conglomerate beds

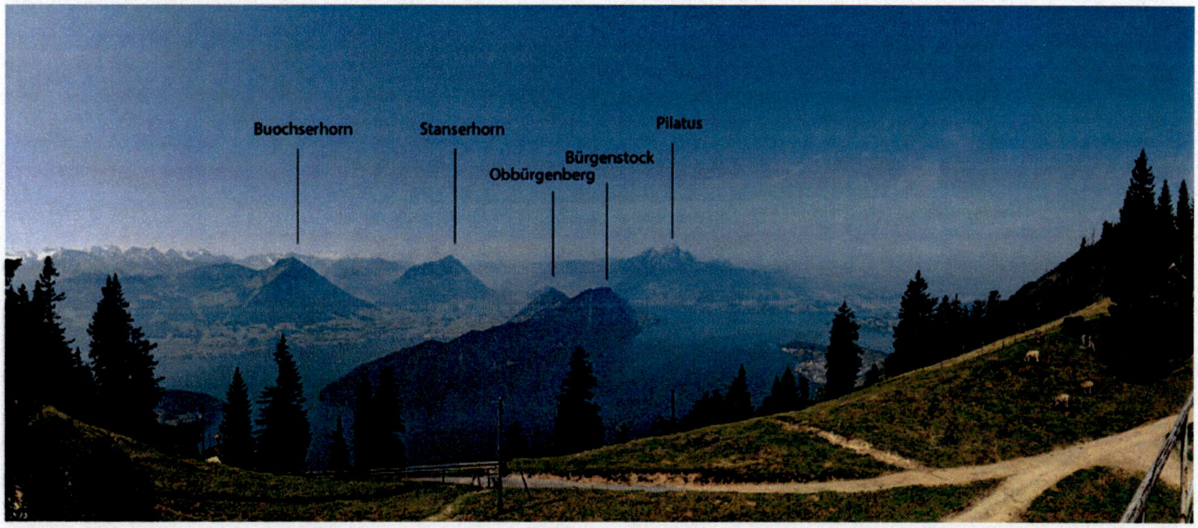

Fig. 11.6 Lake Lucerne with Buochserhorn, Stanserhorn, Bürgenstock range, and Pilatus massif viewed from Rigi (photo Rigi-Bahnen AG)

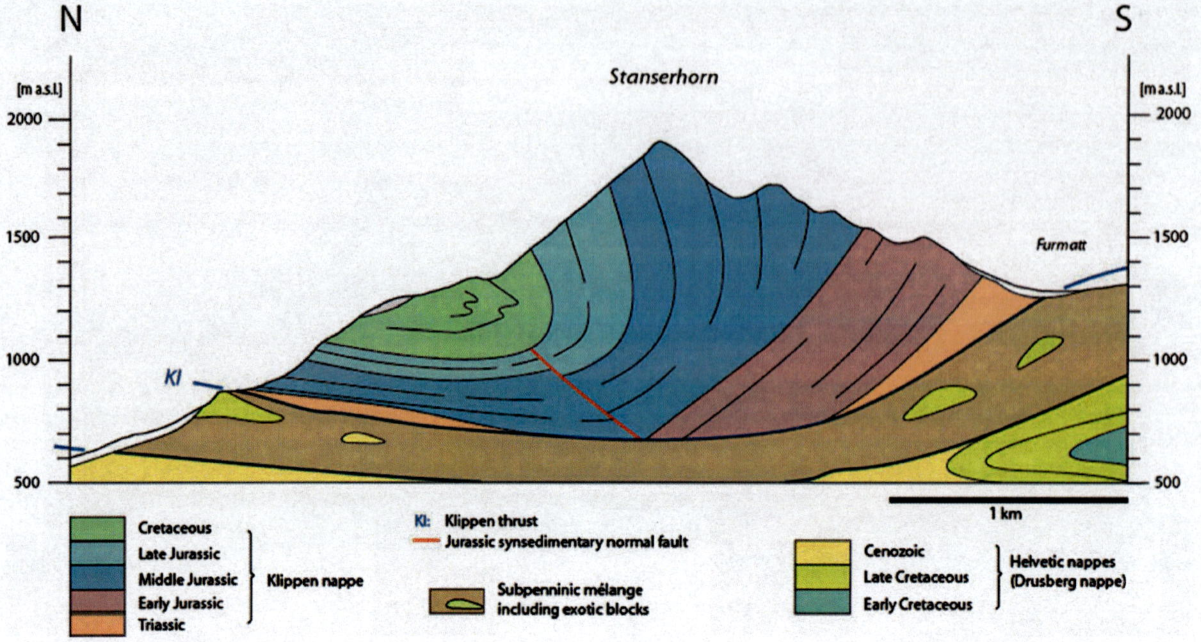

Fig. 11.7 Geological cross-section through the Stanserhorn (modified from Funk et al. 2013)

consists of a single syncline with a thin normal limb and a much thicker upper limb, which is in part overturned. The difference in thickness of the strata on the two limbs is explained by a synsedimentary normal fault which then also acted as nucleus for the fold hinge (see Wissing and Pfiffner 2002, for a discussion of this topic). The summit of Stanserhorn is formed by steeply dipping overturned Jurassic limestones (Fig. 11.7). A more marly and thin-bedded sequence 300 m south of the summit is the reason for the notch in the topography. The depression at Furmatt, on the other hand, owes its existence to the more easily erodible strata of the Triassic and the Subpenninic mélange. The latter is a nearly 500 meters thick layer underlying the Klippen nappe. Its internal structure is difficult to decipher owing to the poor outcrop conditions (Geiger 1956). Nevertheless the slope angles associated with this mélange are relatively low.

11.3.4 Rigi

Rigi (1798 m) consists of an entirely different lithology as the ones discussed in the previous examples. Thick conglomerate beds interlayered with marly-silty layers of Oligocene age are the main constituents. They were deposited as an alluvial fan (Rigi fan) by a paleo-river draining the early Alps. The conglomerates are cliff forming whereas the marly-silty layers weather back and are responsible for the many stripes covered with vegetation that can be traced along the mountain flanks. This feature is visible from distance and is referred to by the local population as "Riginen" (steps) giving Rigi its name. The conglomerate–marl sequence makes up a succession called Lower Freshwater Molasse (USM, for *Untere Süsswassermolasse*) in Alpine literature. It overlies a sequence of marine sandstones and shales, which make up the Lower Marine Molasse (UMM, for *Untere Meeresmolasse*).

The cross-section (Fig. 11.8) is redrawn after Hantke (2006) and spans from Rigi Kulm (the highest peak of Rigi) to Rigi Scheidegg and on across the Helvetic Drusberg nappe to Lake Lucerne. The Lower Freshwater Molasse beneath Rigi Kulm and Rigi Scheidegg dips gently to the SE. As witnessed by the underlying UMM, the USM is repeated by a bedding-plane parallel thrust that steepens to where it breaks the topographic surface. An additional thrust fault still deeper down makes the contact with the tilted Plateau Molasse. Along the cross-section discussed here, the Plateau Molasse is also made of USM, but younger strata can be found farther to the north. To the south, at the contact with the Helvetic nappes, a small imbricate of USM, and a cushion of Subalpine Flysch straddle the Drusberg thrust. The marly and shaly nature of the Subalpine Flysch explains the notch observed along this contact in cross-sectional as well as in map view. The monocline of Cretaceous limestones pertaining to the Drusberg nappe on the other hand form a cuesta (Ochsenalp) that can also be traced along the strike (Rigi Hochfluh and Vitznauerstock). The valley

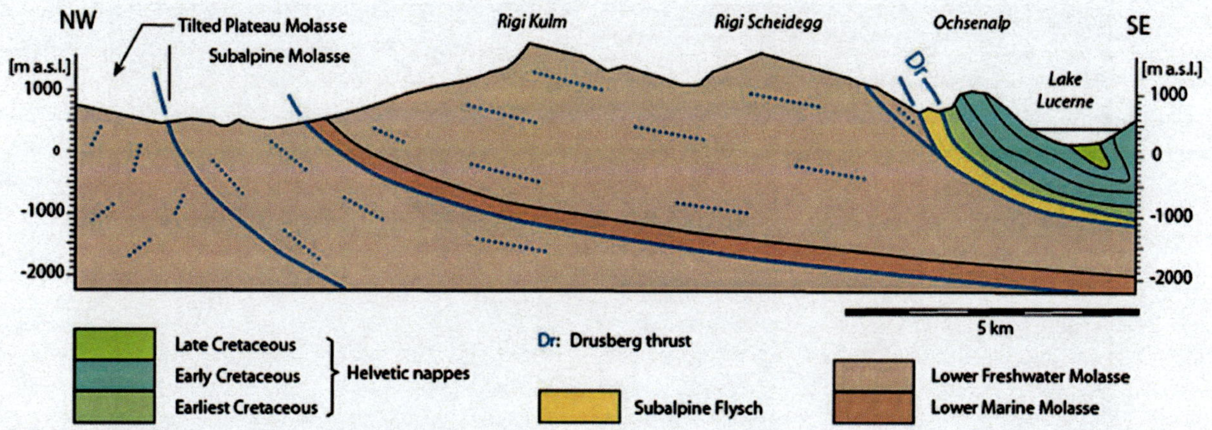

Fig. 11.8 Geological cross-section through Rigi Kulm-Rigi Scheidegg-Lake Lucerne (modified from Hantke 2006). The blue dotted pattern in the Lower Freshwater Molasse indicates the dip of conglomerate beds

occupied by Lake Lucerne in this transect was obviously carved along the axis of a large-scale syncline in the Drusberg nappe.

The USM sediments of Rigi were deposited in the proximal part of the Rigi fan as indicated by the presence of coarse conglomerates lacking mudstones. These rocks resisted erosion, which explains the topographic high. Toward the west, these conglomerates end abruptly and pinch out north of Bürgenstock. This feature may be explained by a morphologic inversion of the alluvial fan terminating laterally toward the west. Figure 11.9 demonstrates lateral discontinuity of the USM toward the viewer as well as the gentle southeasterly dip of these beds.

11.3.5 Mythen

The Mythen is formed of two peaks, Kleiner Mythen (1811 m) and Grosser Mythen (1898 m), and has a similar geological framework as Stanserhorn and Buochserhorn, being an erosional remnant of the once mightier Klippen nappe. Several other klippen outcrop slightly farther to the east. In those klippen, higher units of the nappe pile, including oceanic units and units derived from the Austroalpine domain of the Adriatic margin are preserved. Trümpy (2006) presented a thorough discussion of these klippen. Massive Late Jurassic limestones are the thickest formation and form major cliffs. The cross-section (Fig. 11.10) is based on the work by Jeannet and Leupold (1935) and Felber (1984). The internal structure of the Klippen nappe is characterized by isoclinal anticlines cored by Triassic strata. The anticline beneath Grosser Mythen developed a thrust fault on its lower, inverted limb, and the anticline in Kleiner Mythen has a core of Middle Jurassic strata that thin out markedly in both limbs. It is difficult to make out effects of Jurassic synsedimentary tectonics in this cross-section other than that one notices the complete absence of Early Jurassic strata. The basal thrust of the Klippen nappe is rather irregular; the antiform indicated at the southern end of the section correlates to a fold in the Helvetic nappes located deeper down. The folded Klippen thrust thus indicates that the Penninic nappes were emplaced onto the Helvetic domain prior to the formation and folding of the Helvetic nappes. The Subpenninic mélange is about 500 m thick and contains dispersed blocks of Late Cretaceous and Eocene strata derived from the southernmost Helvetic (Ultrahelvetic) domain.

Regarding the surface expression of the structure of the Klippen nappe, differential erosion of the easily erodible Middle Jurassic and Triassic strata and the massive Late Jurassic limestones is the cause for the notch in Kleiner Mythen and the saddle between Kleiner and Grosser Mythen (Fig. 11.10). Once more the Infrapenninic mélange is associated with low surface slopes in comparison to the Late Jurassic limestones. These morphologic features are also evident on Fig. 11.11, on which one also recognizes the Cretaceous Couches Rouges (red beds) in the summit of the Grosser Mythen.

11.3.6 Lakes and Valleys

Considering Lake Lucerne in the maps of Figs. 11.1 and 11.2, the question concerning processes responsible for its complex shape arises. The easternmost branch of the lake, the Uri branch (see Fig. 11.1), lines up with the Reuss Valley to the south. Within the Uri branch a major tear fault must be assumed because of the very different internal structure of the

Fig. 11.9 Rigi viewed from Lucerne (photo Rigi-Bahnen AG)

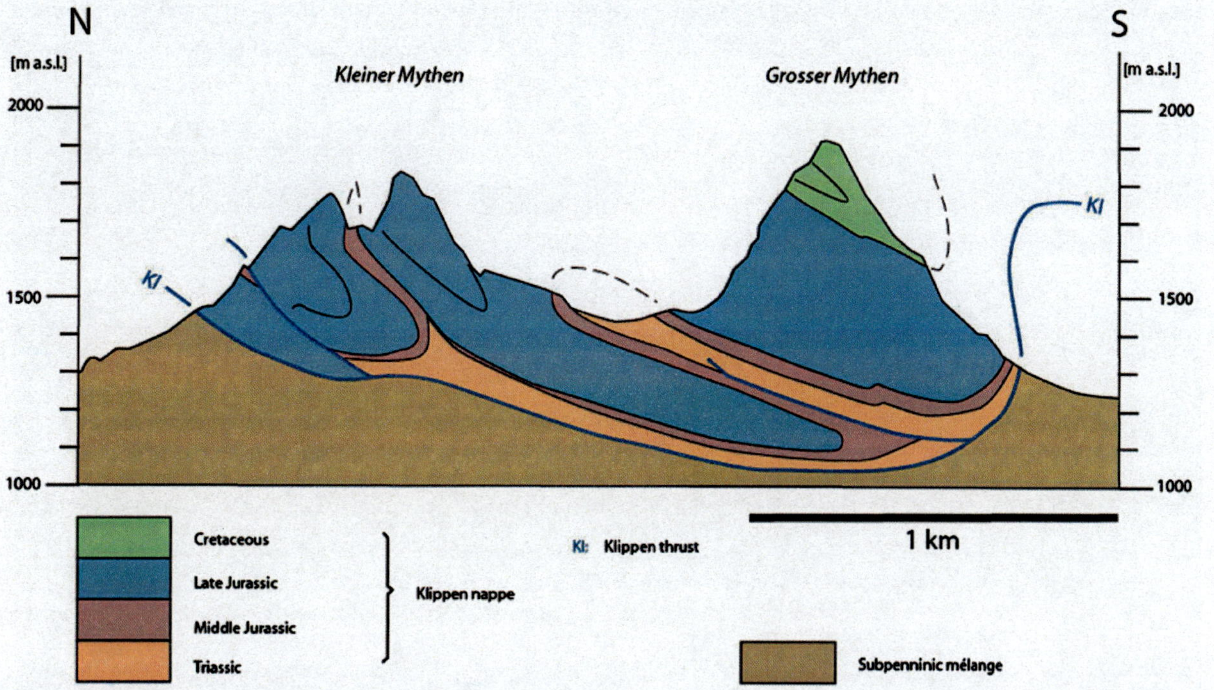

Fig. 11.10 Geological cross-section through Kleiner and Grosser Mythen (modified from Jeannet and Leupold 1935 and Felber 1984)

Helvetic nappes on either side. Although this fault does not extend southward along the Reuss Valley, the tear fault is likely to have controlled the orientation of the valley during incision.

The next segment of the lake to the west (west of Brunnen) continues westward along the valley between the Bürgenstock range and Stanserhorn. The entire elongate depression follows a syncline within the Drusberg nappe. This syncline has a core of Eocene marl and, higher up in the eroded part, the Subpenninic mélange, which is also dominated by marl. Hence it seems that erosion of easily erodible units was controlling the formation of the elongate depression.

The next basin of Lake Lucerne north of the Bürgenstock range is oriented E-W and follows the tectonic contact between the Drusberg nappe and the Subalpine Molasse (Fig. 11.2). As mentioned in conjunction with the cross-sections, a sliver of Subalpine Flysch straddles this contact. The shales within this unit, together with the western end of the Rigi fan made of massive conglomerates explain the location of this lake basin.

Three major rivers drain valleys in a northerly direction within the study area: Reuss River in the east, Engelberger Aa River in the center, and Sarner Aa River in the west. At present all three run into Lake Lucerne. Considering that a pile of Alpine nappes several kilometers thick has been eroded in the past 25 myr the present-day river courses must be considered as the end stage of long-term evolution. It may well be possible that paleo-rivers, located 500 m and more above the present-day valley floors had very different courses. For example, a Paleo–Reuss River might have connected the Reuss Valley south of Altdorf with the valley of the Uri branch of Lake Lucerne and the valley from Brunnen to Lake Lauerz and Lake Zug.

Going further back in time, a fluvial valley draining the Alps and building the Rigi fan must have existed in Oligocene times (see also the chapter on Lavaux and Mont-Pèlerin by Reynard and Estoppey, this volume, which shows a similar case in Western Switzerland). The bedrock of this river was in the Austroalpine nappes, and the Helvetic nappes were located in a position some 10 km farther south and buried beneath the Austroalpine and Penninic nappes.

The ensuing glacial overprint of the valleys and mountains that occurred in the past 2.6 million years is discussed in more detail by B. Keller (this volume).

Fig. 11.11 Grosser and Kleiner Mythen viewed from Rigi Scheidegg (photo A. Pfiffner)

11.4 Conclusions

Landforms at the scales of 10 km–100 m strongly reflect structure and lithology of the bedrock. Thick massive limestones in the Helvetic and Pennnic nappes form high cliffs, while thick marly and shaly units like the Subpenninic mélange and parts of the Molasse manifest themselves by relatively modest slope angles. The stacking of nappes in the course of the formation of the Alps led to a succession of lithologies with alternating contrasting erodibilities. These contrasts, and the subsequent differential erosion, are clearly recognizable in the landforms. In addition, tilting and folding of lithological units, as well as faults within the tectonic units express themselves in the landscape. If stratigraphic and structural data help to understand the morphology, careful analysis of geomorphologic features deepens the understanding of bedrock structure of this impressive landscape.

Acknowledgments Thanks go to Rigi-Bergbahnen AG and Pilatus-Bergbahnen AG for helping to find and providing photographs. The paper benefitted from comments by Peter Heitzmann and Emmanuel Reynard.

References

Baud A, Septfontaine M (1980) Présentation d'un profil palinspastique de la nappe des Préalpes médianes en Suisse occidentale. Eclogae Geol Helv 73:651–660

Breitschmid A (1982) Diagenese und schwache Metamorphose in den sedimentären Abfolgen der Zentralschweizer Alpen (Vierwaldstättersee, Urirotstock). Eclogae Geol Helv 75:331–380

Christ P (1920) Geologische Beschreibung des Klippengebietes Stanserhorn-Arvigrat am Vierwaldstättersee. Beitr Geol Karte Schweiz [N.F.] 12

Felber PJ (1984) Der Dogger der Zentralschweizer Klippen. PhD thesis, ETH Zürich (Diss. ETH Nr. 7506)

Funk HP, Pfiffner OA, Menkveld-Gfeller U, Kempf O (2013) Blatt 1170 Alpnach. Geol. Atlas Schweiz 1:25'000, Erläuterungen 137 [Explanatory notes to the geological atlas sheet Alpnach]

Geiger M (1956) Die Unterlage der zentralschweizerischen Klippengruppe Stanserhorn-Arvigrat, Buochserhorn-Musenalp und Klewenalp. Eclogae Geol Helv 47:407–452

Hantke R (2006) Blatt 1151 Rigi. Geol. Atlas Schweiz 1:25’000, Erläuterungen 116 [Explanatory notes to the geological atlas sheet Rigi]

Jeannet MA, Leupold W (1935) Excursion de la Société géologique suisse dans les Alpes de Schwytz (Mythen-Iberg-Sihltal) du mardi 20 au samedi 24 août 1935. Eclogae Geol Helv 28:627–640

Keller B (this volume) Lake Lucerne and its spectacular landscape

Kempf O, Pfiffner OA (2004) Early Tertiary evolution of the North Alpine Foreland Basin of the Swiss Alps and adjoining areas. Basin Res 16:549–567

Kempf O, Matter A, Burbank DW, Mange M (1999) Depositional and structural evolution of a foreland basin margin in a magnetostratigraphic framework: The eastern Swiss Molasse Basin. Int J Earth Sci 88:253–275

Pfiffner OA (1980) Displacements along thrust faults. Eclogae Geol Helv 78(2):13–333

Pfiffner OA (2011) Explanatory notes to the structural map of the Helvetic zone of the Swiss Alps, including Vorarlberg (Austria) and Haute Savoie (France). Geological Special Map 128, Text and 10 plates. Swiss Geological Survey, swisstopo, Bern

Pfiffner OA (2014) Geology of the Alps. Wiley-Blackwell, Chichester

Reynard E, Estoppey E (this volume) The Lavaux World Heritage terraced vineyard

Trümpy R (2006) Geologie der Iberger Klippen und ihrer Flysch-Unterlage. Eclogae Geol Helv 99:79–121

von Hagke C, Cederbom CE, Oncken O, Stöckli DF, Rahn MK, Schlunegger F (2012) Linking the northern Alps with their foreland: The latest exhumation history resolved by low-temperature thermochronology. Tectonics 31:TC5010

Wissing SB, Pfiffner OA (2002) Structure of the eastern Klippen nappe (BE, FR): Implications for its Alpine tectonic evolution. Eclogae Geol Helv 95:381–398

O. Adrian Pfiffner is Professor Emeritus of the University of Bern, President of the Scientific Council of the UNESCO World Heritage site “Tectonic Arena Sardona” and Member of the Federal Commission of Geology. His main scientific interests are tectonics, structural geology, geophysics, geomorphology, and mountain building processes. He has worked in the Alps and the Andes of Peru.

Mountain Building and Valley Formation in the UNESCO World Heritage Tectonic Arena Sardona Region

12

Thomas Buckingham and O. Adrian Pfiffner

Abstract

Situated in eastern Switzerland the UNESCO World Heritage Tectonic Arena Sardona and its surrounding region isvery well known due to its most spectacular structural feature, the Glarus thrust, visible as a clear line, cutting along mountain faces. The Glarus thrust represents one key for the inscription into the UNESCO World Heritage List; other key features include clearly exposed examples of mountain building phenomena and spectacular geomorphological features highlighting the formation of the local relief. The prerequisite for their appearance on the Earth's surface, besides differential uplift of the region from great depths, is its erosive history. Deep valleys cut through the entire region, providing spectacular outcrops of an entire range of stratigraphic units, tectonic structures, and Quaternary landforms.

Keywords

Tectonic Arena Sardona • UNESCO World Heritage • Glarus thrust • Mountain building • Valley formation • River deflection

12.1 Introduction

The Swiss UNESCO World Heritage Tectonic Arena Sardona offers a unique view into the formation of Alpine mountains and valleys. The World Heritage site Sardona covers almost 330 km^2 in an Alpine landscape. Within its boundary no settlements that are inhabited year-round are present and the site was mainly defined on the basis of land use planning. The area is accessible to hikers and alpinists through public transport and/or mountain railways. The World Heritage is surrounded by the Swiss Geoparc Sardona to all sides, covering a total of 1800 km^2, established already in 1998 (Imper 2004). Together they are referred to as the "Sardona region" in this article. The site was included into the World Heritage List in July 2008. UNESCO hereby recognizes that the area is invaluable to human understanding of the formation of mountains through its unique visible features, its geohistoric relevance and ongoing scientific research (UNESCO 2008). The center of the region is marked by Piz Sardona (N 46° 55′ 00″/E 09° 15′ 00″, Swiss National Coordinates 2738,060/1198,470; 3055.8 m a.s.l.), which lends its name to the entire region. It lies near the point where the three cantons of Graubünden, St. Gallen, and Glarus meet and stands for long-term collaboration, while also providing great visibility of the site's most striking feature—the Glarus thrust. Due to the exceptional tectonic features, such as thrusts, imbricates, faults, and folds, a vast amount of stratigraphic units come to lie within a relatively small region. This leads to a large diversity of outcrops. Due to a pronounced mountain relief as a consequence of deep incision by rivers and glaciers, the visibility and quality of geologic and geomorphologic features are striking.

12.2 Geographical and Geological Setting

Located in eastern Switzerland, the larger outline of the Sardona region can be described by the Churfirsten group to the north, the Linth and Sernf valleys to the west, the Surselva to the southwest, and the Rhine Valley in the south and east (Fig. 12.1). In mountaineering terms the region is referred to as the Glarus Alps and the southern Sankt Gallen Alps. The highest peak, 3247 m high, is Ringelspitz. The

T. Buckingham (✉)
UNESCO-World Heritage Swiss Tectonic Arena Sardona, Städtchenstrasse 45, 7320 Sargans, Switzerland
e-mail: thomas.buckingham@unesco-sardona.ch

O. A. Pfiffner
Institute of Geological Sciences, University of Bern, Baltzerstrasse 1+3, 3012 Bern, Switzerland
e-mail: adrian.pfiffner@geo.unibe.ch

E. Reynard (ed.), *Landscapes and Landforms of Switzerland*, World Geomorphological Landscapes,
https://doi.org/10.1007/978-3-030-43203-4_12

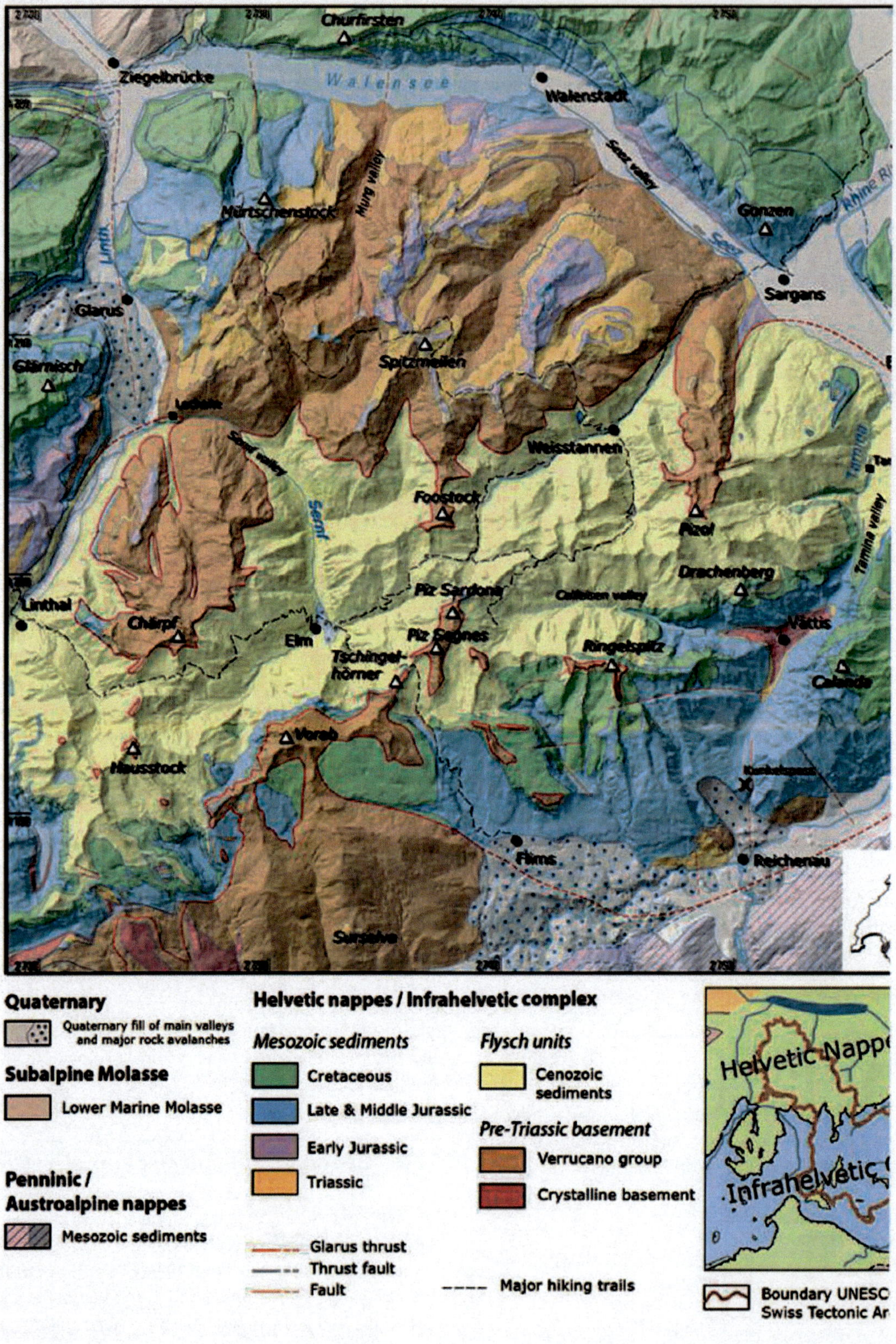

2

Fig. 12.1 A simplified geologic–tectonic map of the Sardona region (Geoparc Sardona) with the World Heritage in its center (modified after Pfiffner 2010; shaded relief swisstopo) geological map shown in Fig. 12.1 illustrates the geographic setting and the distribution of the different rock types.

In the northern Sardona region, the mountains consist of Permian, Mesozoic, and Cenozoic sediments belonging to the Helvetic nappes (Fig. 12.1). This allochthonous nappe pile is locally referred to as the Glarus nappe complex. An important thrust, the Säntis thrust, outcrops in the northernmost region at the base of the Churfirsten group. Most peaks here consist of limestones, marking the dominantly visible features in the landscape.

The famous Glarus thrust separates the Helvetic nappes in the hanging wall from the Infrahelvetic complex in the footwall (Pfiffner 1977). The thrust fault can be seen cutting across steep mountain faces as a sharp line throughout the region (the "magic line"). The scenic exposure of this great structure allows for a clear three-dimensional visibility of the thrust surface as well as its outcrops above and below (Fig. 12.2). Hereby the phenomenon of thrusting can be studied and experienced in an arena-like view. This is how the World Heritage received the name: Tectonic Arena Sardona.

The footwall of the Glarus thrust, the Infrahelvetic complex, comprises a crystalline basement (Aar massif), which itself has been affected by Alpine folding and thrusting, as well as a sedimentary cover, which is shaped into folds and several small-scale nappes. In the central part, Flysch units of late Cretaceous to Eocene age are exposed, dominantly consisting of alternating marl and sandstone sequences. To the south, Mesozoic cover sediments, built up mostly of limestones of Jurassic to Cretaceous age, substitute the Flysch units. Both Flysch units and the Mesozoic cover, are cut discordantly by the Glarus thrust (Schmid 1975; Herwegh et al. 2008; Pfiffner 2011) as is clearly visible in the geological cross-section shown in Fig. 12.3.

The Glarus thrust formed during Oligocene to Miocene times and put older rocks onto younger rocks in the entire region. Shearing along the thrust fault took place at a maximal depth interval of 16 km in the south of the Sardona region to 9 km in the north. The minimal displacement along this large-scale shear zone is estimated to be more than 50 km (Pfiffner 2014). Peak metamorphic conditions vary along the thrust from anchizonal facies with maximum

Fig. 12.2 Photograph from southern Sernf Valley toward Piz Segnes and Tschingelhörner showing the trace of the "magic line," the thrust contact between the Flysch units below and the Verrucano Group above (photo R. Homberger)

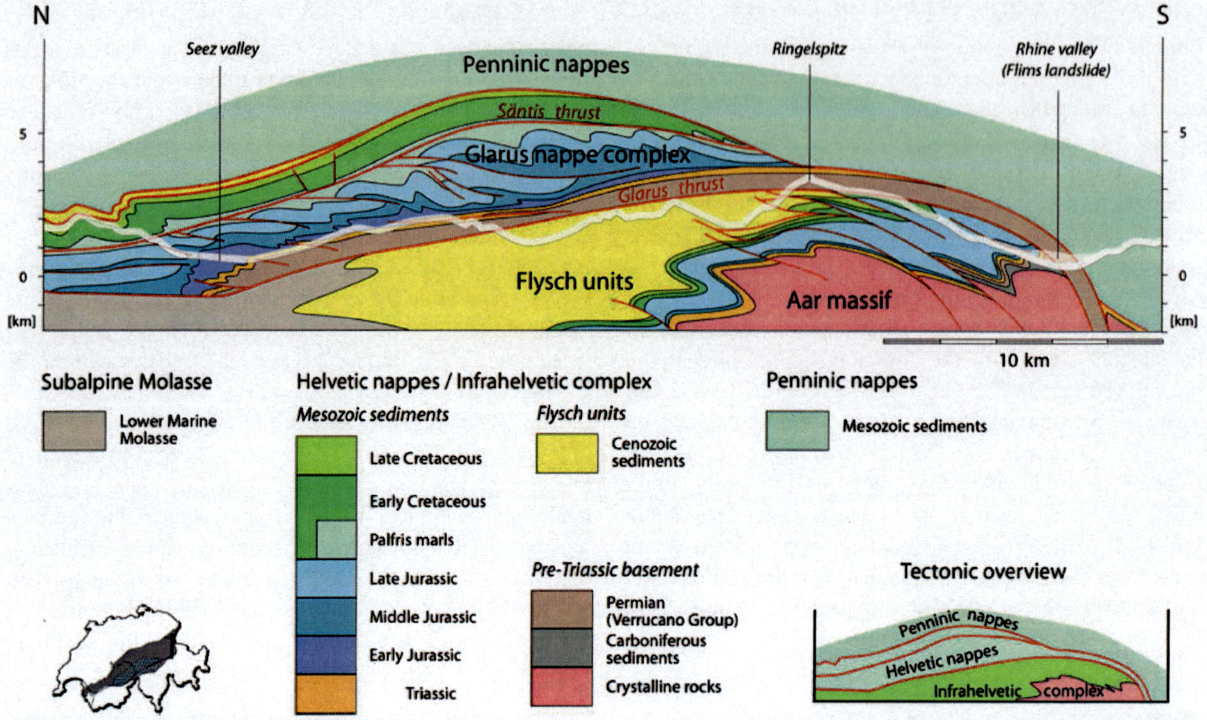

Fig. 12.3 Cross-section through the Sardona region (modified after Pfiffner 2014). The Helvetic nappes composed of Mesozoic sediments including the Verrucano Group at their base can clearly be seen thrust onto the Infrahelvetic complex, which in turn is composed of Mesozoic sediments, Flysch units and the underlying crystalline basement of the Aar massif. The trace of cross-section is also given in Fig. 12.4. White line denotes the current topographic surface

temperatures of 230 °C in the north (Lochsite) to lower greenschist facies and temperatures around 350 °C in the south (Vorab) (Mullis et al. 2002; Ebert et al. 2007). Therefore, the outcropping thrust formed as a northwards rising ramp, deep in the crust. In some instances, the age difference between hanging wall and footwall rocks across the Glarus thrust is more than 200 million years over a distance of only a few centimeters.

Besides the Glarus thrust, the World Heritage site and its surrounding Sardona region displays an extraordinary geodiversity. Erosive activity gave rise to several klippen of Permian Verrucano rocks sitting on younger rocks (see Fig. 12.1) and windows where the footwall of the Glarus thrust is exposed. Rapid uplift rates (more than 1 mm/a presently and in the past 10 million years) and ensuing fast erosion rates made the rivers cut deep valleys and gorges, some of which were further deepened and widened by the glaciers in the course of the Pleistocene. As a result outstanding landforms and landscapes developed.

The region was mapped geologically in the scale of 1:50,000 by Oberholzer (1920, 1942) and described in great detail by several hundred geology-related publications (collected in Aemisegger and Buckingham 2014). The structure of the Sardona region is best viewed on the 1:100,000 scale structural map in Pfiffner et al. (2010), while several studies on the Quaternary have been done on regional scale (e.g. Jordi 1986; Schindler 2004).

12.3 Mountain Building and the Glarus Thrust

Thrust faults are large surfaces in the Earth's crust along which kilometer-thick rock units, called nappes, are moved on top of each other over distances of tens or hundreds of kilometers. Thrust faults are therefore a direct expression of the dynamics of converging tectonic plates. During the course of the convergence of plates their margins are squashed with material escaping upward and downward. Upward escape occurs along thrust faults and creates areas of higher elevation. Thus, thrust faults are directly linked to the formation of mountain ranges like the Alps (Pfiffner et al. 2006). Downward escape results in a crustal root. The density of the crustal rocks in this root is lower than the density of the mantle rocks that were pushed aside and thus undergo buoyant rise. This in turn lifts the mountain range up and maintains it at high elevation.

The stacking of nappes by movement along thrust faults emerging from depth leads to uplift of the land surface within mountain regions. The ensuing high elevation promotes precipitation which in turn drives erosion by rivers cutting into the land surface.

Even though thrusts are very frequent structural features within collisional mountain ranges around the world, they typically are difficult to see clearly in the landscape. This often is due to a missing lithological contrast between hanging and footwall rocks. Furthermore, frail rocks along the thrust faults tend to develop shallow slopes where scree coverage is abundant.

Within the Sardona region the interactions between tectonics, metamorphism, and geochemistry have been studied in detail (e.g., Schmid 1975; Burkhard et al. 1992; Herwegh et al. 2008). From the northern to the southern limits of the Sardona region, going along the Glarus thrust one encounters rocks that represent a cross-section through the entire upper crust. The incipient Glarus thrust formed a ramp within the crust. This ramp was subsequently deformed by continued squashing of the plate margins. The southern portion was uplifted in the process to form a large-scale dome (Figs. 12.3 and 12.4) where originally deeper portions of the ramp were pushed up high into the mountains. The contours of the surface of the Glarus thrust shown in Fig. 12.4 outline this domelike structure and the depression to the north of it.

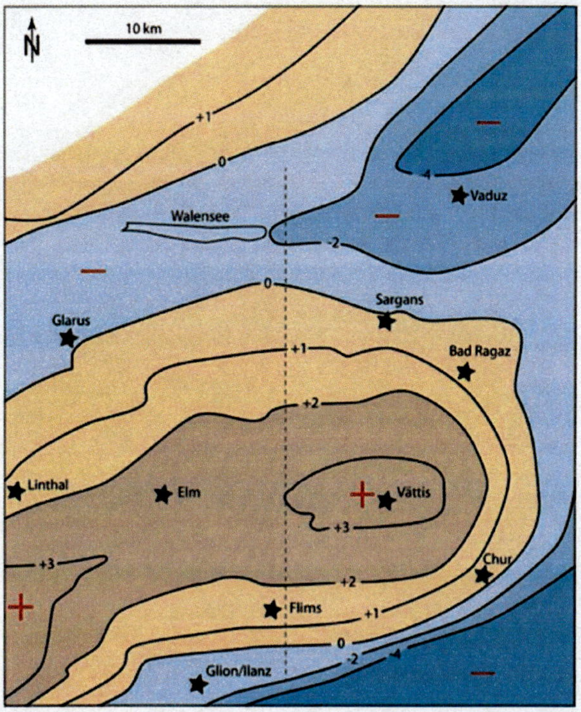

Fig. 12.4 A visualization of the basal thrust of the Helvetic nappes with structure contours (expanded from Schmid 1975). The Glarus thrust corresponds to the area south of Vaduz-Walensee-Glarus. In the south, the thrust surface forms a culmination with a dome reaching more than 3000 m a.s.l. in the air above Vättis and rising to more than 3500 m a.s.l. to the WSW. North of the culmination a WSE-ENE trending depression exists in the area of Walensee. Dashed line is the trace of the cross-section shown in Fig. 12.3

12.4 Outstanding Landforms and Landscapes

12.4.1 Morphological Expression of the Glarus Thrust

The visibility of the Glarus thrust within the Sardona region is striking. The thrust fault is visible as a distinct line in steep rock cliffs (Figs. 12.2 and 12.5). The cliffs formed because of the high resistance to erosion of the lower part of the Verrucano Group which behaved as a protecting cap rock for the underlying much more erodible Flysch units. The domelike shape of the Glarus thrust is evident from the structure contour map shown in Fig. 12.4 and the photographs shown in Fig. 12.5.

On a large scale the deeply carved valleys allow the visual tracing and observation of the Glarus thrust along several tens of kilometers along the mountain ranges. Particularly the steep and deeply carved Calfeisen Valley (Fig. 12.1) and the southern areas of the Sernf Valley (Figs. 12.1 and 12.2) display an almost 2000 m thick complex stack of sedimentary and tectonic units in the footwall of the thrust.

A number of klippen, erosional remnants of Verrucano on top of Flysch units or Mesozoic carbonates, outline the culmination of the Glarus thrust at around 3000 m a.s.l. in the southern Sardona region. It is the color contrast between the dark greenish conglomerates and shales of the Verrucano Group and the lighter gray carbonates or the grayish sandstones and shales of the Flysch units in these klippen that easily catches the eye of the observer as a "magic line" (see Figs. 12.2, 12.5 and 12.6). In the Tschingelhörner (Fig. 12.6), a naturally formed gap, Martin's Hole, exists just beneath this "magic line" (Figs. 12.2 and 12.6b). It is the scene of regular natural spectacles, when the sun or full moon shines through the opening onto the church in Elm—an event that attracts numerous visitors. The gap itself is located at the intersection of a minor thrust fault (visible by the dark Flysch slice incorporated into the Mesozoic carbonates) and a steeply dipping normal fault, which is outlined by an offset of the "magic line" (see Fig. 12.6a). Fracturing associated with these faults resulted in an area which was less resistant to erosion and allowed cliff retreat from both sides of the Tschingelhörner ridge. Ultimately a natural tunnel was formed along the intersection line of the two faults.

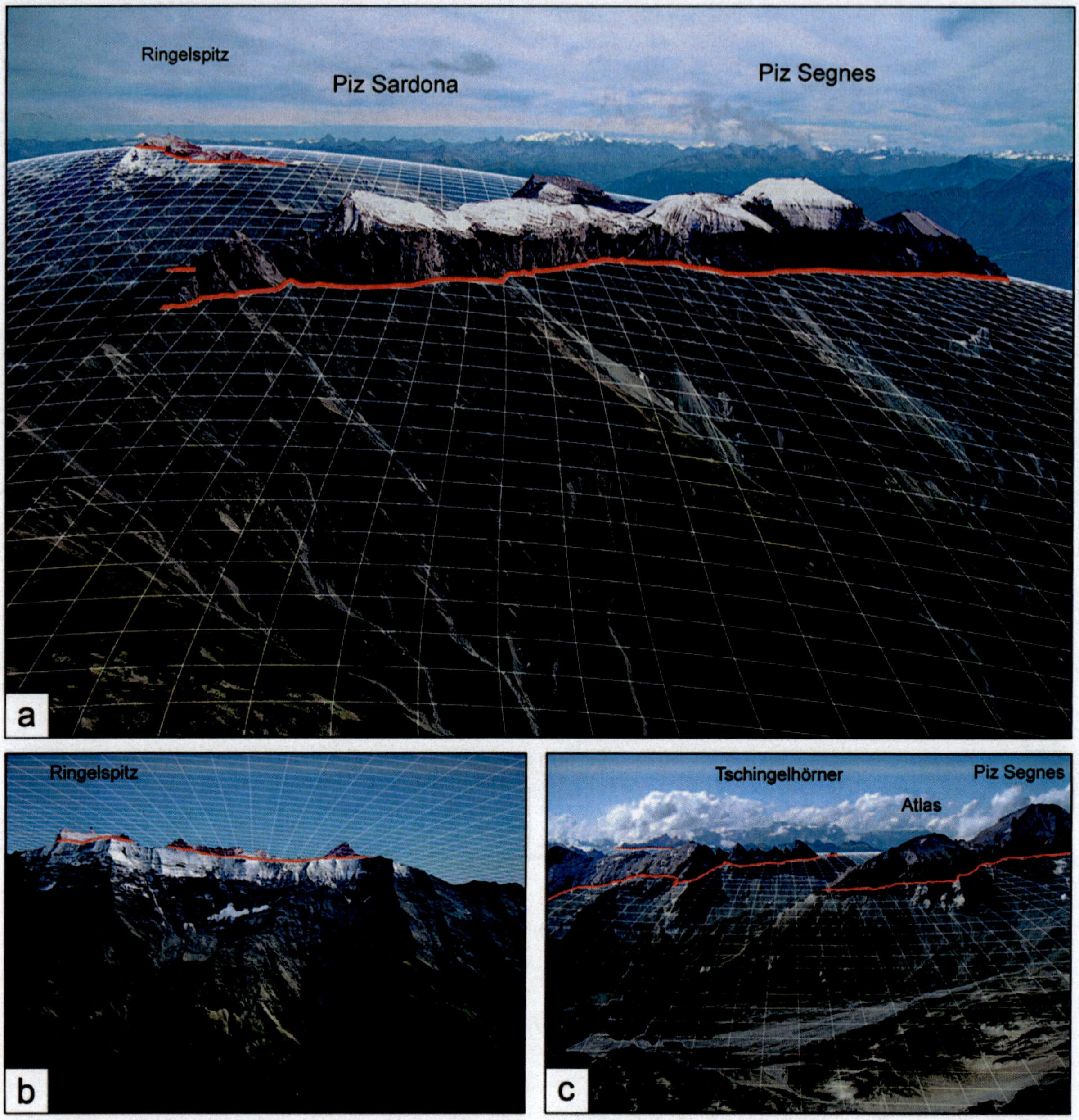

Fig. 12.5 The domelike Glarus thrust visualized as a grid structure crosscutting the mountain ranges. It is here, where the thrust unfolds its striking geomorphologic appearance. Cutting across the landscape for dozens of kilometers, it draws the eye of the visitor. Pointy peaks consisting of rocks from the Verrucano Group appear as klippen above the thrust surface; a: looking south-east; b: looking south; c: looking north-west (photo R. Homberger)

12.4.2 Landforms and Landscapes Above the Glarus Thrust

The landscapes above the thrust in the northern and southern Sardona region differ significantly. In the south, landforms are dominated by the rocks of the Verrucano Group and the overlying Mesozoic sediments. One of the key features in the landscape is the color contrast between the various sedimentary sequences: Verrucano shales (red), Triassic quartzites (white), dolostone and gypsum (yellow), and shales (red); Early Jurassic marls, sandstones and breccias (dark colored); and Late Jurassic—Cretaceous limestones (gray) (Fig. 12.7). Although these sequences are in a dominantly upward stratigraphic position, numerous decameter

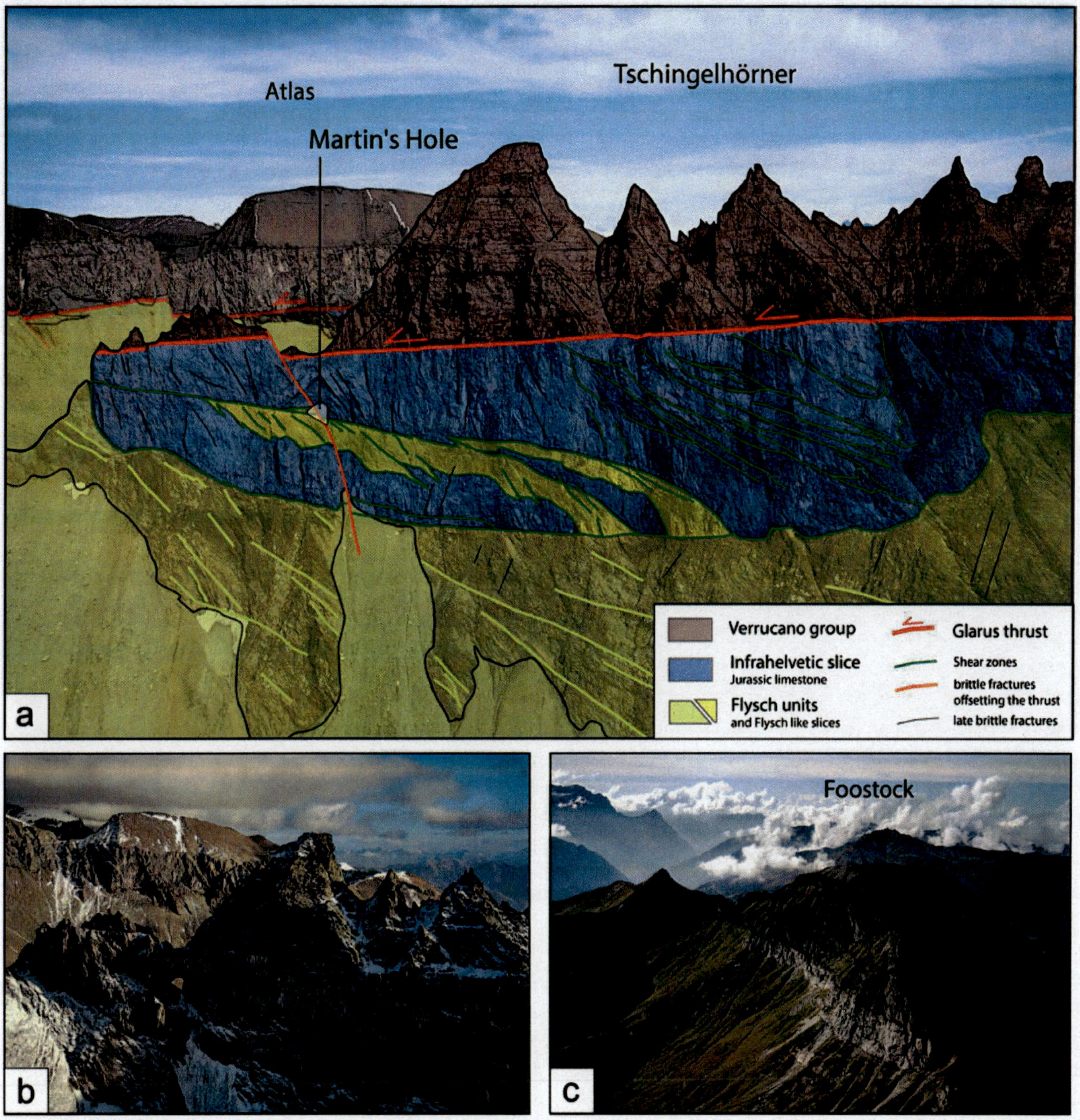

Fig. 12.6 A slice of Jurassic limestones (Infrahelvetic complex) dragged along beneath the Glarus thrust; a and b: View of Tschingelhörner looking south-east shows several shear zones, most prominent visible in the cliff below the Tschingelhörner with the Martin's Hole; c: View of Foostock looking north. The slice of Jurassic limestones (light gray) thins out toward the left (west) (photo R. Homberger)

to kilometer-scale folds and faults affect the strata and render the color pattern more complex.

In the NW part of the Sardona region, a prominent mountain, the Mürtschenstock, towers the region. The mountain is composed of Jurassic–Cretaceous carbonates that show spectacular folds in the cliffs (Fig. 12.8). Its abrupt termination in the south-west is an escarpment caused by a strike-slip fault that offsets the carbonate package. The SSE flank corresponds to a dip slope of an inverted sequence where the more easily erodible Permian strata have been removed by erosion. The northern part of the Mürtschenstock represents a fin; the prominent cliffs facing east and west mimic steeply dipping local faults and joints striking N-S.

Fig. 12.7 A view of the northern Sardona region looking south. The whitish ridge Gipsgrat ("gypsum ridge") in the foreground is comprised of light Triassic dolostone and evaporites outcropping along the crest, resting upon red shales of the Verrucano Group. The iconic mountain to the left is Spitzmeilen, built of massive brown breccias and limestones of Early Jurassic age. The variety of rock colors is nicely visible and is further pronounced by scree deposits (photo R. Homberger)

North of the Sardona region, a major valley running E-W and the Churfirsten Mountain chain to the north delimits the Sardona region. A narrow elongated lake, Walensee, occupies the valley floor and is fed by the Seez River. The valley flank to the south of Walensee has a low-angle slope shaped mainly by Permian Verrucano and Triassic—Early Jurassic strata. The overall dip of these strata is parallel to the slope. The northern flank of the Seez Valley and the lake in contrast is very steep. The slope angle is controlled by Jurassic and Cretaceous carbonates which are resistant to erosion. In the Churfirsten Mountains and in particular their summits, which mark the top of this valley flank, stratigraphic layering is clearly visible (Fig. 12.9). Several steps covered by vegetation can be detected on the lower part of this flank. They correspond to layers of marls and shales which mark thrust faults. The thrust faults used these marls and shales as décollement horizons. The most prominent one is located beneath the Churfirsten peaks and corresponds to the Säntis thrust (Figs. 12.3 and 12.9). The Säntis nappe located in the hanging wall of this thrust fault forms the Churfirsten and the Säntis massif to the north.

12.4.3 Landforms and Landscape

The morphology beneath the Glarus thrust is particularly impressive in the southern part of the Sardona region and is dominated by deep valleys carved into Flysch units and Mesozoic carbonates. The valleys were shaped by both fluvial and glacial processes. The narrow deeply carved valleys were initially interpreted by early workers as kilometer-scale fracture openings in the Earth's crust ("vallées d'écartement" of d'Omalius d'Halloy 1843). Examples include the Calfeisen Valley shown in Fig. 12.10.

The high local relief presented by these valleys is the result of the important (rock) uplift that this area has experienced in the last 10 million years (Rahn et al. 1997; Vernon et al. 2008; Pfiffner 2014), which in turn steepened river gradients and thus enhanced erosion.

In the central part of the region, Flysch units form the lower reaches of the valley flanks. Here one can observe how differences in erodibility of the individual rock types affect the slope angles. These differences are further accentuated by vegetation (Fig. 12.2). The remarkable rocky

Fig. 12.8 View of Mürtschenstock looking ENE. Mürtschenstock is built of massive carbonates that withstood erosion and now towers the surrounding mountains. The dark rocks forming a step in the cliff are marls. They separate the brownish limestones above (Early Cretaceous) from the light gray limestones below (Late Jurassic). The entire sequence forms a large-scale fold where the beds to the right (south) are vertical and even overturned. (photo R. Homberger)

spires in the summits above the Glarus thrust are mostly formed by the lower part of the Verrucano Group that consists of massive breccias. These rocks withstood erosion and stopped the headward retreat of tributaries in the underlying Flysch units. As a result a steep slope carved by numerous active gullies developed (see Piz Segnes on Fig. 12.2).

The present-day landforms within the Sardona region were also shaped by the glaciers of the Last Glacial Maximum (LGM) and the many preceding glaciations. Valley glaciers widened and deepened the main valleys surrounding the Sardona region: Linth Valley to the west, Rhine Valley to the south and east, Seez Valley and Walensee to the north. The highest-lying erratic blocks in the Sardona region indicate a minimum elevation of the glacier surface of 2050 m a.s.l. above Chur and 1750 m a.s.l. above Sargans. During the LGM, a connected network of valley glaciers existed. Even though glaciers normally merge as they descend, at two points—at Reichenau and Sargans—the Rhine Glacier was forced to split into two large side branches (Schlüchter 2009). One side branch flowed over Kunkelspass to Vättis, the other through the Seez Valley (including Walensee) joining the Linth Glacier west of Walensee.

The glacial overprint is evident in case of the wide Rhine and Linth valleys, as well as the Seez Valley. In addition a number of U-shaped side valleys show morphologies typical for glacial overprint. For instance, the Murg Valley, which drains northward into the Walensee, represents a glacially formed circus staircase (Imper 2004). Furthermore, north of Flims two alluvial plains (Segnes sut and Segnes sura) were carved out by the Segnes and Sardona glaciers and subsequently backfilled.

A special case is the torso of the valley from Kunkelspass to Vättis. As is evident in Fig. 12.11 this valley has steep flanks and a relatively flat floor. Toward the south

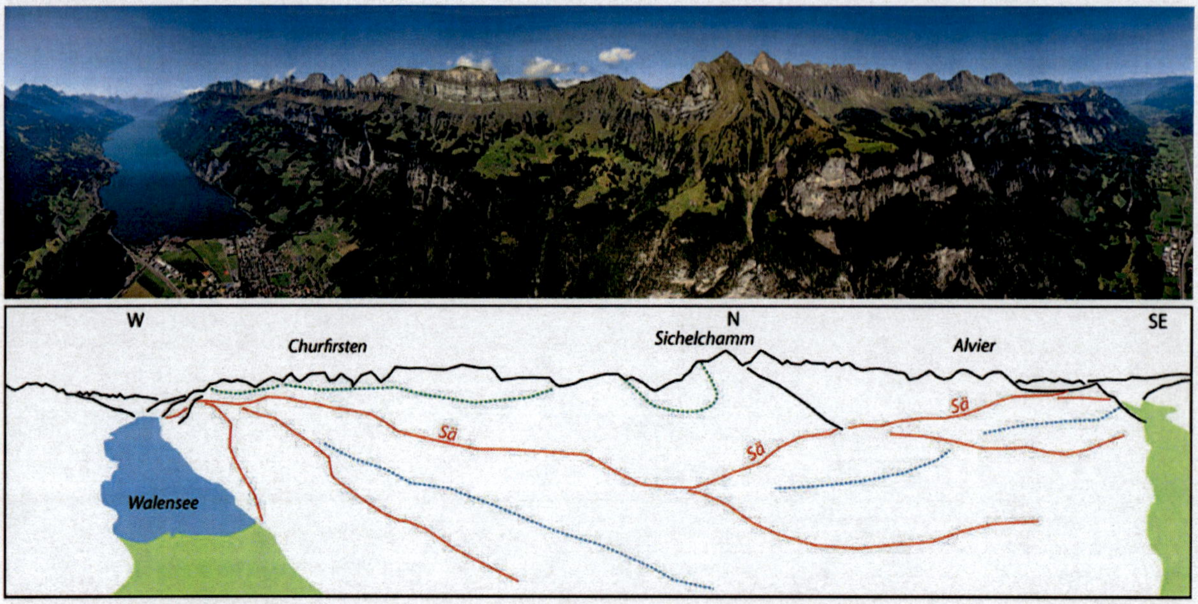

Fig. 12.9 Panorama view of Walensee—Churfirsten—Seez Valley. The geological structure is underlined by the relative resistance to erosion of massive limestones and marly or shaly strata now forming steps in the cliffs. The gray limestones beneath the Säntis thrust (Sä) are of Late Jurassic age (Quinten Limestone), the brownish limestones in the peaks of the Churfirsten of Cretaceous age. The dotted green line marks the base of the cliff-forming Schrattenkalk Formation, the dotted blue line follows the "Mergelband" Member within the Quinten Limestone. Sä: Säntis thrust (photo R. Homberger)

(Kunkelspass) several gullies end blind on top of the cliff Säsagit–Garschlichopf, which is the cutoff edge of the Tamins rock avalanche that went down into the Rhine Valley. It is positioned east of the famous Flims rockslide, the largest of its kind in the Alps (see von Poschinger et al., this volume). As evident from the shape of this torso and the erratic blocks that can be found on both flanks, this valley was overprinted by the Rhine glacier branch descending toward Vättis. The torso represents a major valley that was initially carved by a river; the question then arises from where this river came from. The area of Kunkelspass–Reichenau is presently undergoing a surface uplift of 1.5 mm/a. Apatite fission-track data suggest that this uplift maximum around Chur has persisted for the last 10 million years (Persaud and Pfiffner 2004). With this uplift rate it follows that about 750,000 years ago the present-day topographic surface of Kunkelspass was located at an elevation of around 600 m a.s.l. This altitude corresponds to the present-day altitude of the Rhine River in Reichenau and would yield a similar river gradient as is present now. It is thus concluded that an ancestral "West-Rhine" drained the area west and south of Reichenau as precursor of the modern Vorderrhine River and Hinterrhine River, respectively, but at a higher elevation (Fig. 12.12). In addition, it is interpreted that the wind gap of Lenzerheide is a remnant of an ancestral "East-Rhine," which drained the area south of Chur (Fig. 12.12). In this scenario, two passes must be assumed between these two ancestral Rhine rivers: one between "Reichenau" and "Chur" and one along the "Schin gorge" farther south (note that these locations were buried deep underground at that time). It is argued that headward erosion of tributaries to the two ancestral Rhine rivers on both sides of these passes ultimately led to the deflection of the rivers into the present-day geometry. In both cases, the deflection was aided by the rocky substrate that consists of a sandstone-shale sequence (Bündnerschiefer Group from the Penninic nappes) known to be easily erodible (Kühni and Pfiffner 2001).

Once the river capture had taken place, continued uplift in the Sardona region provoked headward erosion of the Tamina River. As a consequence the Tamina River carved a deep gorge into Flysch units south of Bad Ragaz, the Tamina Gorge. A thermal spring with water at 36.5 °C was discovered in that gorge as early as 1240 AD and it is still used in the bath resort of Bad Ragaz.

Further upstream, around the village of Vättis and at the confluence of the Tamina River that drains the Calfeisen Valley and a small creek that drains the torso, the crystalline basement rocks of the Aar massif outcrop in the so-called Vättis window (Trümpy 1980; Hesske and Imper 2010).

Fig. 12.10 View of Calfeisen Valley looking west. The deep and narrow valley flanked by massive carbonates was formerly interpreted as having formed as fracture opening in the Earth's crust (photo R. Homberger)

This confluence is towered by cliffs more than 1000 meters high, which are composed of massive Jurassic and Cretaceous carbonates (Fig. 12.10). In the cliffs of the Drachenberg northwest of Vättis (Fig. 12.11), cave bear bones were found in a cave at 2480 m a.s.l. The site is interpreted as cult place, and preliminary age dating of the bones suggests that the cave was visited by humans during the warm period preceding the LGM, some 40,000–50,000 years ago.

12.5 Geoeducation and Geotourism

As a consequence of the inclusion into the list of World Heritage sites, the region has become a driving force for geological public relations and geology-related events for amateurs within Switzerland. Awareness for geologic, tectonic, and geomorphological landscapes and landforms is being continually raised and educational projects for all target groups are being conducted (IG Tektonikarena 2010, 2013; Swisstopo 2013a, b; Erlebnis Geologie 2015). While focusing on long-term relations of regional citizens with the World Heritage and its values, a focus on projects for amateurs and support of local multipliers, so-called Geo-Guides are key actions (Zuber and Buckingham 2015). Close collaboration with all stakeholders is vital in order to preserve the outstanding geologic, tectonic, and geomorphological values, landscapes, and landforms and to have the general public appreciate the region in the long run.

12.6 Conclusions

Thrust faults represent unique features in collisional orogens, demonstrating how the collision of continental plates thickens the crust and uplifts entire slices to form mountain ranges. Structures such as the Glarus thrust are witnesses of processes that take place at great depths within the Earth's crust. The stacking of nappes along thrusts is one of the main mechanisms responsible for thickening of the upper crust, uplift, and formation of a mountain relief. While the large

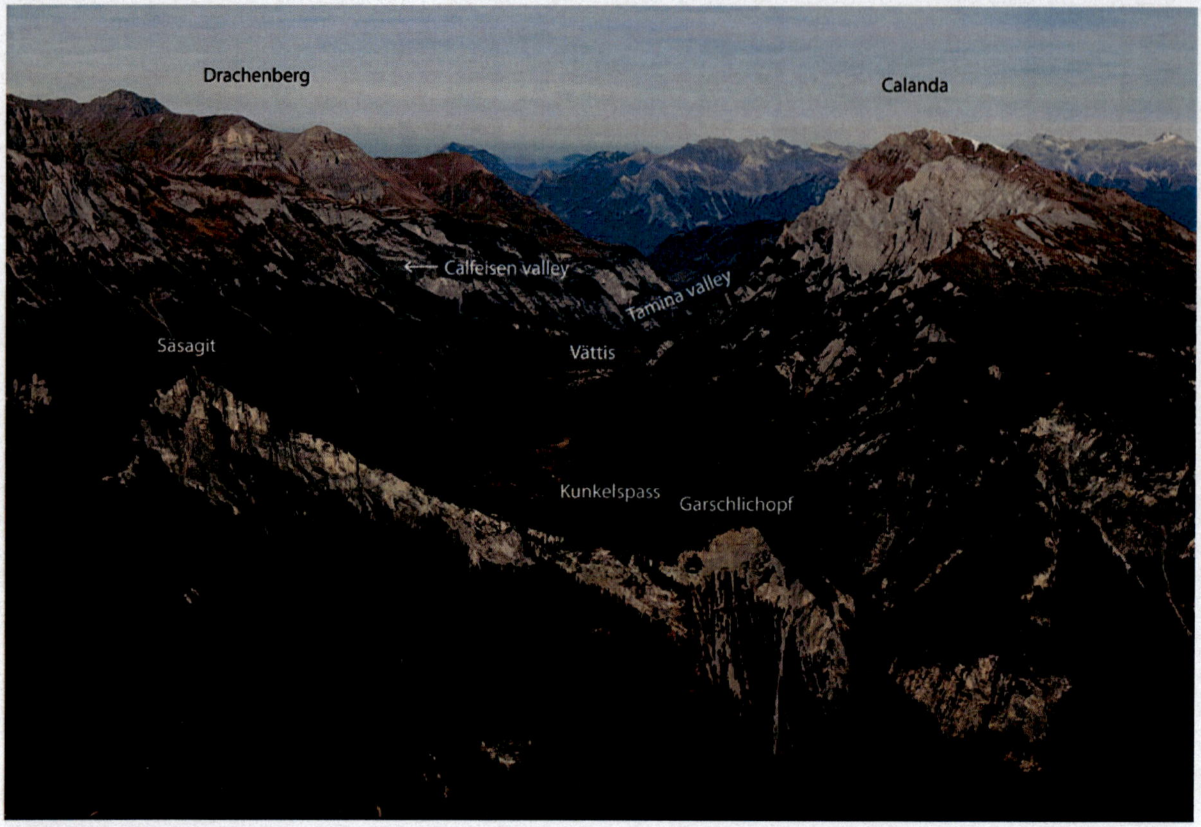

Fig. 12.11 The torso of Kunkelspass–Vättis looking toward the north. The torso ends southward at the cliffs in the foreground, which are the cutoff edge of the Tamins rock avalanche (photo © Schweizer Luftwaffe)

Glarus thrust surface once formed as a ramp some 10–15 km deep in the upper crust, differential uplift is responsible for the domelike shape it has nowadays making it exceptionally well visible from great distance over tens of kilometers.

Owing to the impressive relief, where rivers and glaciers carved out deep valleys, a vast amount of rock types, folds, and thrust faults above and beneath the Glarus thrust are exposed in steep cliffs. While the region is experiencing uplift rates of 1–1.5 mm/a, erosive activity is ongoing, leading to a large diversity of glacial, fluvial, and gravitational phenomena such as large rock avalanches. A prominent torso between Vättis and Kunkelspass indicates the course of the ancestral "West-Rhine" River some 750,000 years ago. The subsequent deflections of the ancestral rivers highlight the dynamic behavior of river networks in a geologically short time span of less than a million years.

Throughout the region remarkable pointy peaks top off the landscape as isolated klippen. The striking expression of geological structures is further owed to contrasts in rock types and colors as well as differing resistance to weathering and therefore slope steepness. A number of narrow deeply carved valleys provide a high local relief; the varying orientation of these valleys and containing landforms give an arena-like perspective and allow for a three-dimensional insight into a wide range of geological and tectonic phenomena. Or, to state it very clearly: the mountains nowadays exist because valleys have formed, and the valleys make it possible to observe how the mountain range as a whole had formed. For these reasons the region is of extraordinary scientific value and due to the easy access and visibility, is of even further importance for geological public relations, education, awareness raising, and tourism.

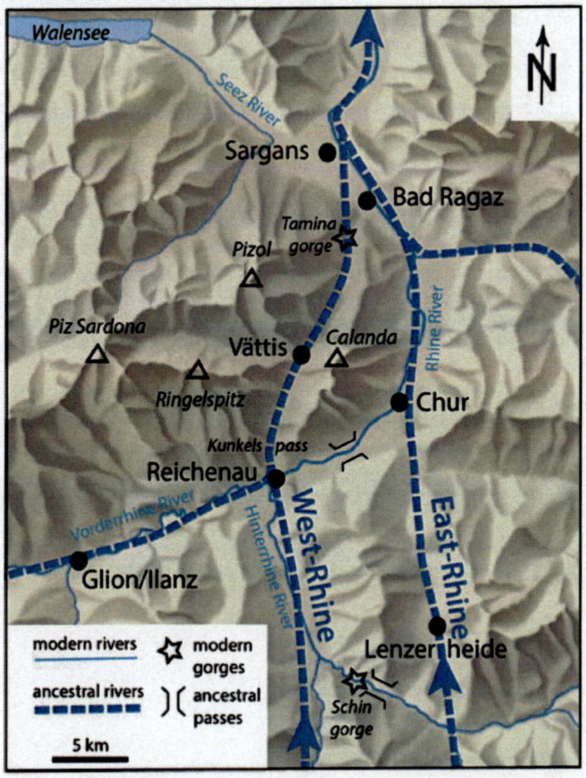

Fig. 12.12 Deflection of the Rhine River in the past 750,000 years. The ancestral "West-Rhine" River was most likely responsible for carving the valley now preserved as torso between Kunkelspass and Vättis. Two passes must be assumed between the "West-Rhine" and "East-Rhine" rivers. These were lowered by headward incision of tributaries, which was facilitated by the easy erodible sandstone-shale sequence of the Bündnerschiefer Group

Acknowledgements We thank reviewers Jürg Meyer and Emmanuel Reynard for their comments that helped to improve the manuscript.

References

Aemisegger B, Buckingham T (2014) Literaturdatenbank Geopark und Tektonikarena Sardona, https://data.unesco-sardona.ch/forschen/literaturlisten/

Burkhard M, Kerrich R, Maas R, Fyfe WS (1992) Stable and Sr-isotope evidence for fluid advection during thrusting of the Glarus nappe (SwissAlps). Contributions Mineralogy Petrology 112:293–311

D'Omalius D'Halloy JBJ (1843) Précis élémentaire de géologie. Arthus Bertrand, Paris, 742 pp

Herwegh M, Hürzeler JP, Pfiffner OA, Schmid S, Abart R, Ebert A (2008) The Glarus thrust: excursion guide and report of a field trip of the Swiss Tectonic Studies Group (Swiss geological Society, 14-16.09.2006). Swiss J Geosci 101:323–340

Hesske S, Imper D (2010) Das geologische Fenster von Vättis. Die ältesten Gesteine des Unesco-Welterbes der Tektonikarena Sardona im Taminatal. Terra Plana 3:13–17

IG Tektonikarena Sardona (2010) Alte Schichten – neue Sichten. Den geologischen Phänomenen auf der Spur. https://data.unesco-sardona.ch/info/flyer-broschueren/

IG Tektonikarena Sardona (2013) Wie Berge sich erheben. https://data.unesco-sardona.ch/info/flyer-broschueren/

Imper D (2004) Der Geopark Sarganserland-Walensee-Glarnerland. Jahrb St Gall Naturw Ges 90:101–136

Jordi U (1986) Glazialmorphologische und gletschergeschichtliche Untersuchungen im Taminatal und Rheintalabschnitt zwischen Flims und Feldkirch. Geographica Bernensia G27, Bern, 168 pp

Kühni A, Pfiffner OA (2001) The relief of the Swiss Alps and adjacent areas and its relation to lithology and structure: topographic analysis from a 250 m DEM. Geomorphology 41:285–307

Mullis J, Rahn MK, Schwer P, De Capitani C, Stern WB, Frey M (2002) Correlation of fluid inclusion temperatures with illite "crystallinity" data and clay mineral chemistry in sedimentary rocks from the external part of the Central Alps. Schweiz Mine Petrogr Mitt 82(2):325–340

Oberholzer J (1920) Geologische Karte der Alpen zwischen Linthgebiet und Rhein, 1:50 000. Geol. Spezialkarte Schweiz 63. Schweiz geol Komm, Bern

Oberholzer J (1942) Geologische Karte des Kantons Glarus, 1:50 000. Geol. Spezialkarte Schweiz 117. Schweiz geol Komm, Bern

Persaud M, Pfiffner OA (2004) Active deformation in the eastern Swiss Alps: post-glacial faults, seismicity and surface uplift. Tectonophysics 385:59–84. https://doi.org/10.1016/j.tecto.2004.04.020

Pfiffner OA (1977) Tektonische Untersuchungen im Infrahelvetikum der Ostschweiz. PhD thesis ETH Zürich and Mitteilungen Geologisches Institut ETH und Universität Zürich N.F. 217, 432 pp

Pfiffner OA, Burkard M, Schmid S (2006) Comparative study on thrust faults. IG UNESCO Weltnaturerbe Tektonikarena Sardona, https://data.unesco-sardona.ch/forschen/wissenschaftliche-publikationen/

Pfiffner, OA, Schmid SM (2008) Supplement to the Comparative study on thrust faults: The Glarus overthrust. IG UNESCO Weltnaturerbe Tektonikarena Sardona, https://data.unesco-sardona.ch/forschen/wissenschaftliche-publikationen/

Pfiffner OA, Burkhard M, Hänni R, Kammer A, Kligfield R, Mancktelow NS, Menkveld JW, Ramsay JG, Schmid SM, Zurbriggen R (2010) Structural map of the Helvetic zone of the Swiss Alps, including Vorarlberg (Austria) and Haute Savoie (France). 7 map sheets. Geological Special Map 128/1-7. Swisstopo, Bern

Pfiffner OA (2011) Explanatory Notes to the Structural map of the Helvetic zone of the Swiss Alps, including Vorarlberg (Austria) and Haute Savoie (France). Explanatory notes. Geological Special Map 128. Swisstopo, Bern

Pfiffner OA (2014) Geology of the Alps. Wiley-Blackwell, Chichester, 376 pp

Poulet T, Veveakis M, Herwegh M, Buckingham T, Regenauer-Lieb K (2014) Modeling episodic fluid-release events in the ductile carbonates of the Glarus thrust. Geophysical Research Letters 41 (207121–7128

Poschinger AV, Clague JJ, Calhoun N (this volume) The Flims and Tamins rockslide landscape

Rahn MK, Hurford AJ, Frey M (1997) Rotation and exhumation of a thrust plane: Apatite fission-track data from the Glarus thrust. Switzerland. Geology 25(7):599–602

Schmid SM (1975) The Glarus overthrust: field evidence and mechanical model. Eclogae Geol Helv 68(2):247–280

Schindler C (2004) Zum Quartär des Linthgebiets zwischen Luchsingen, dem Walensee und dem Zürcher Obersee. Beitr geol Karte Schweiz, N.F. 169, Swisstopo/Landesgeologie, Wabern

Schlüchter C (compil) (2009) Die Schweiz während des letzteiszeitlichen Maximums (LGM), 1:500 000. GeoKarten 500. Bundesamt für Landestopografie swisstopo, Wabern

Swisstopo (2013a) Minikarte ViaGeoAlpina. Swisstopo, Wabern

Swisstopo (2013b) 5080T Tektonikarena Sardona. Swisstopo, Wabern

Trümpy R (1980) Geology of Switzerland – a guidebook. Wepf, Basel, 2 vol, 104 + 334 pp

UNESCO (2008) World Heritage List. Swiss Tectonic Arena Sardona, http://whc.unesco.org/en/list/1179

Vernon AJ, van der Beck PA, Sinclair HD, Rahn MK (2008) Increase in late Neogene denudation of the European Alps confirmed by analysis of a fission-track thermochronology database. Earth and Planetary Science Letters 270:316–329

Internet References

Erlebnis Geologie (2015) Association offering an online database for geology related events and offers, www.erlebnis-geologie.ch

IG Tektonikarena Sardona (2015) UNESCO-World Heritage Swiss Tectonic Arena Sardona Website www.unesco-sardona.ch

Thomas Buckingham is a M.Sc. in Earth Sciences working in education, awareness raising and public relations for the World Heritage Swiss Tectonic Arena Sardona in eastern Switzerland. He is Co-president of the Swiss Working Group on Geotopes and Geoparks and Co-president of the Association Erlebnis Geologie. His main scientific interests are the interaction between tectonics and erosion and the links to geomorphology and natural hazards.

O. Adrian Pfiffner is Professor Emeritus of the University of Bern, president of the Scientific Council of the UNESCO World Heritage site "Tectonic Arena Sardona" and member of the Federal Commission of Geology. His main scientific interests are tectonics, structural geology, geophysics, geomorphology and mountain building processes. He has worked in the Alps and the Andes of Peru.

An Outstanding Mountain: The Matterhorn

13

Michel Marthaler and Henri Rougier

Abstract

Unique for its shape and history, the Matterhorn is an iconic mountain of the Zermatt region, in the canton of Valais, Switzerland. The rocky pyramid is a superb isolated landform in the middle of the twenty-nine "4000" summits of the Valais Alps. From a geological point of view the rocks of the Matterhorn, and those of the neighbouring mountains, have an African origin: the Dent Blanche klippe is no more than a small continental raft stranded on the remains of the disappeared Tethys Sea. The rocks of the Zermatt area, which form the base of the Matterhorn, bear witness to this Mesozoic marine origin. The pyramid itself is constituted by a gneissic and schistose Paleozoic complex, less solid than it seems, while the summit part is formed of very resistant gneisses of the Valpelline series. Geomorphologically, the Matterhorn is a karling, i.e. an angular peak with steep walls and sharp ridges, carved away by glacial erosion. Most of the pyramid is continuously frozen, especially the northern face. Gelifraction and permafrost melting are very active today, causing rockfalls dangerous for climbers. Because of its incomparable shape, the Matterhorn has acquired a very special landscape notoriety within the geomorphosites of the Alps.

Keywords

Matterhorn • Penninic nappes • Tethys Sea • Structural landforms • Glacial landforms • Permafrost

M. Marthaler (✉)
Institute of Geography and Sustainability, University of Lausanne, Géopolis, 1015 Lausanne, Switzerland
e-mail: michel.marthaler@unil.ch

H. Rougier
University Lyon 3, rue Chevreul 18, Lyon Cedex 07, 69362 Lyon, France
e-mail: henri.rougier@editionslep.ch

13.1 Introduction

The Matterhorn (45°58'21" N, 7°39'18" E; 4478 m a.s.l.) is situated at the Swiss–Italian border, in the upper part of the Vispa Valley *(Mattertal)*, and it dominates the famous tourist resorts of Zermatt in the north and Breuil-Cervinia in the south (Fig. 13.1). It is located in the heart of a massif which gathers the highest concentration of "4000 s" in the Alps: 29 summits. The Matterhorn is far from being the highest peak in the Alps but it is certainly the most iconic. Its shape resembling a colossal pyramid is extraordinary, as is the geological adventure that the landscape and rocks tell: it allows exploring the four eras of Earth's history, from the Paleozoic to the Quaternary, i.e. more than 400 million years. What makes the Matterhorn so unique is its isolation: seen from an airplane, it seems that it has managed to create a kind of vacuum around it in order to assert its dominant position (Fig. 13.2).

Zermatt is one of the most famous tourist resorts in Switzerland and in the world, nestled at the bottom of an umbilicus in the upper part of a "lost valley" (Taugwalder 1979). In 2017, it had 6,823 permanent inhabitants and 30,210 tourist beds. With more than 2 million tourist nights per year (2,087,944 nights in 2017, 55% in winter, 45% in summer; Zermatt Tourismus 2017), Zermatt is one of the most famous and active tourist destinations in Switzerland.

The Matterhorn (called *ds Horu* by the Zermatt inhabitants and *La Becca* in the Italian side) is part of the legendary mountains which would probably be nothing without the town or village for which they are the "domestic mountain" (*Hausberg* in German) (Rougier 2002, 2010). From this point of view, it is logical to assimilate Zermatt and *its* Matterhorn to Interlaken and *its* Jungfrau or Chamonix and *its* Mont Blanc. However, the aesthetics and beauty of the landscape are not enough to make the Matterhorn an exceptional mountain. Remarkably, the mountain takes on a very different look depending on the side by which it is approached. From Valtournenche or Breuil/Cervinia, seen

E. Reynard (ed.), *Landscapes and Landforms of Switzerland*, World Geomorphological Landscapes,
https://doi.org/10.1007/978-3-030-43203-4_13

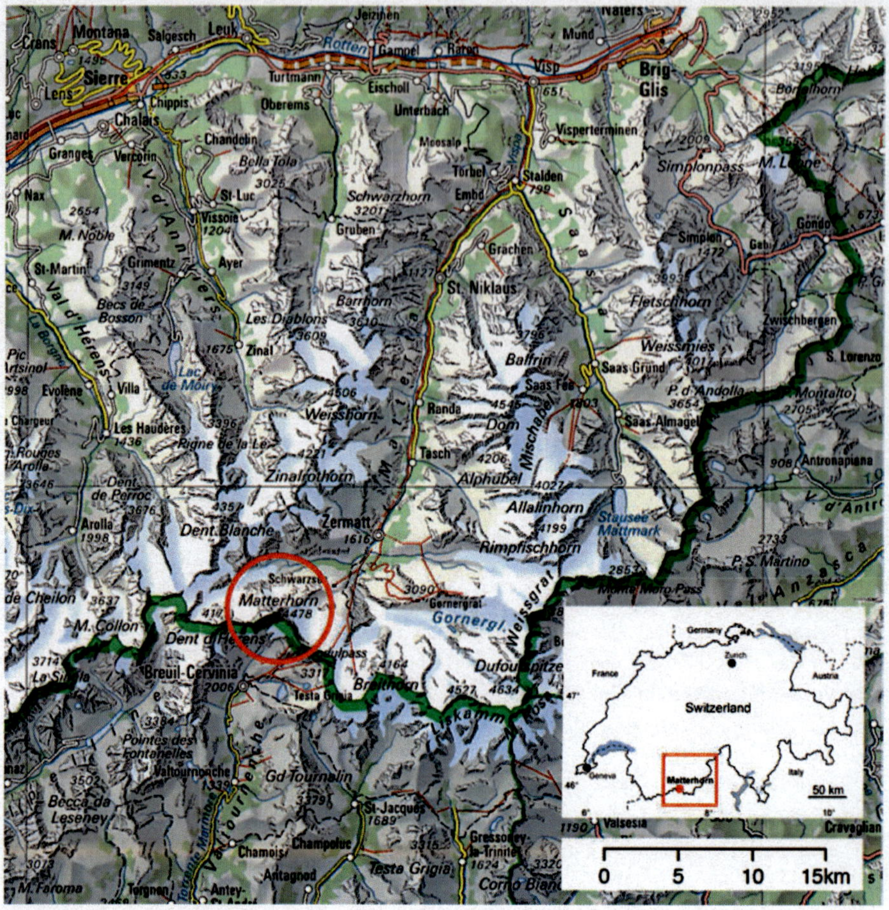

Fig. 13.1 Location of the Matterhorn and Zermatt in South-west Switzerland. The Matterhorn (red circle) is situated at the border with Italy (Valtournenche Valley), in the upper part of the *Mattertal*, a valley drained by the Vispa River, a tributary of the Rhone River (swisstopo)

from the south, it does not have the majesty that one recognizes when it is observed from Zermatt (Fig. 13.3).

A mountain unique by its shape and geological history (Marthaler 2001, 2004), the Matterhorn is also unique by the fascination it has always exerted on mankind: "*face au Cervin, face à l'œuvre accomplie depuis des millénaires, comment ne pas rêver*?" (facing the Matterhorn, facing the work accomplished over millennia, how can we not dream?) said mountaineer Gaston Rébuffat (1965). Considered as the "last problem of the Alps" by the nineteenth-century climbers, it was indeed one of the last summits to be climbed: it was not until 14 July 1865 that it was conquered. Edward Whymper (1871) wrote: *"At 1.40 P.M. the world was at our feet, and the Matterhorn was conquered"*. Unfortunately, the descent was fatal to four of the seven members of the expedition mounted by Whymper, which accentuated Matterhorn myth.

Thus a trilogy was created very quickly: a high mountain environment, a tourist resort, a name. This trilogy is based on geological and geographical contexts. At the beginning of the twentieth century, Emile Argand (1908) noticed a geological enigma, visible on both sides of the Italian–Swiss border: how is it that Paleozoic continental rocks (those of the pyramid) are stacked, in reverse of the logic of time, on Mesozoic rocks of oceanic origin? This enigma was solved and explained by the presence of a thrust (Figs. 13.4 and 13.5) at the base of the Dent Blanche nappe, which played the role of a "crushing sledge" moved over the remains of the disappeared Tethys Sea. Argand had discovered evidence of continental drift, fossilized and visible in the mountains. In direct connection with the tectonic movement, the action of geomorphological processes took place at the same time as the mountain was being established: orogenesis, tectogenesis and morphogenetic processes worked together. The Matterhorn experienced successive glaciations during the Quaternary: one reads perfectly the legacies of this grandiose sculpture, pursued by the strong action of contemporary processes.

Fig. 13.2 Aircraft view towards southeast showing how the Matterhorn is in isolated position, in front of the Monte Rosa massif in the background. The pyramid is surrounded by glaciers that emphasize its isolation. In front of the west face in the shade, the Tiefmattengletscher flows to the left, to join the Zmuttgletscher, in direction of the Zermatt Valley (photo H. Rougier)

13.2 A Long Geological History

13.2.1 The African Origin of the Pyramid's Rocks

The rocks of the Matterhorn date from the Paleozoic (Fig. 13.4). The oldest, probably pre-Carboniferous in age, are those of the darker summit (Fig. 13.6). These are hard and resistant rocks of the Valpelline series (Bücher et al. 2003): kingizitic gneisses accompanied by amphibolites and silicate marbles. The great mass of the pyramid is carved in granodioritic orthogneiss (Arolla series) dated 289 Ma by Bussy et al. (1998). These metagranites are intruded by masses of metagabbros, dated 280 Ma by Monjoie (2004). At that time, during the Permian, all continents were united in the Pangea. At the beginning of the Triassic, the Pangea broke into two mega-continents, Laurasia to the north and Gondwana to the south, separated by the Tethys Sea. The rocks of the southern continental margin of the Tethys Sea, rich in Lias breccias and witnesses of the Tethysian rifting, are well visible in the region of Mt Dolin and Arolla, 13 km to the WNW of the Matterhorn (Fig. 13.1). At the height of the Matterhorn shoulder (Fig. 13.6) only a thin band of Mesozoic calcschists is visible (Bücher et al. 2003).

13.2.2 The Oceanic Origin of the Rocks Forming the Base of the Pyramid

The rocks that bear witness to the Tethys Sea form the broad base of the Matterhorn (Figs. 13.4, and 13.5), as do all the rocks in the area between Zermatt and Breuil-Cervinia, passing through the Gornergrat and Breithorn (Bearth 1953; Bücher et al. 2003; Fig. 13.7). The Tethys Sea expanded during the Jurassic and then closed during the Cretaceous (Marthaler 2005). The main witnesses of this ocean are ophiolites (Bearth 1953, 1967): serpentinites, metagabbros and metabasaltes. Pillow lavas can be seen here, the shape of which was preserved despite the eclogitic high-pressure alpine metamorphism. A block with pillow lavas is visible at the entrance to the Zermatt Museum, near the church.

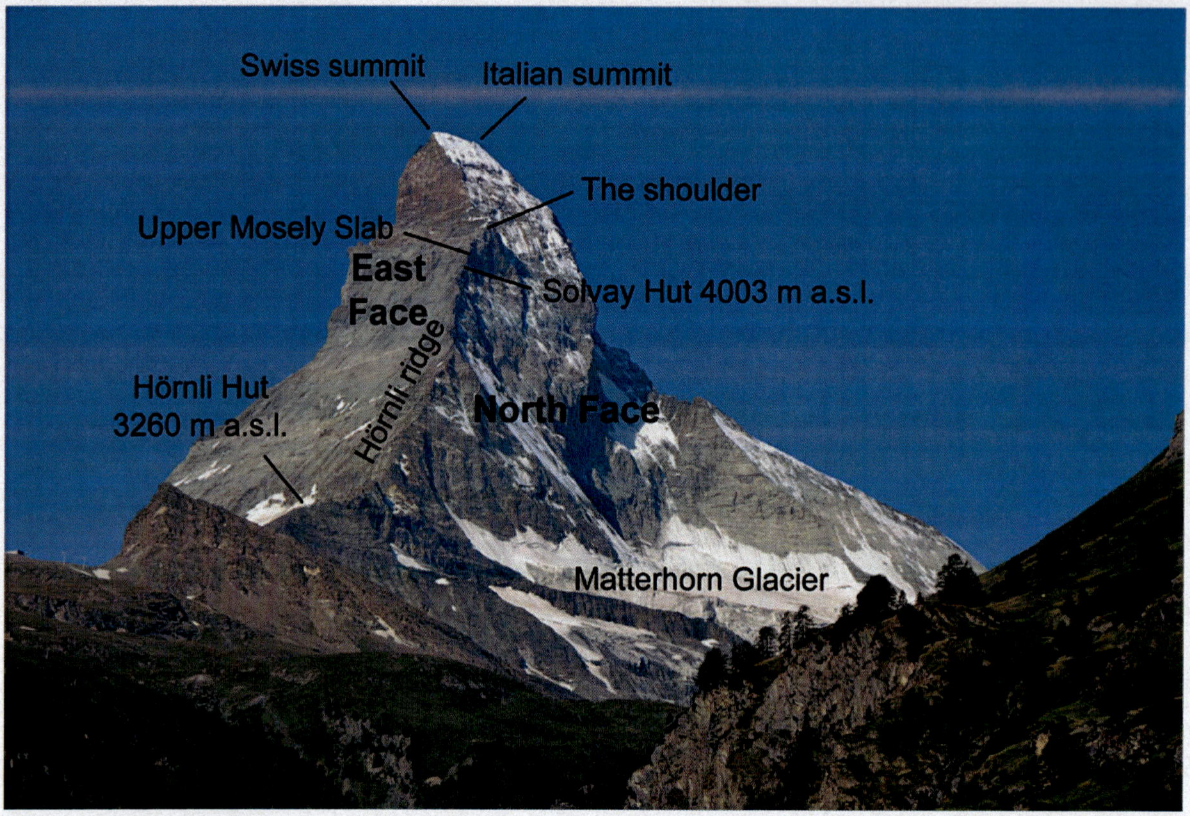

Fig. 13.3 The Matterhorn as viewed from Zermatt, with various toponyms used in the text (photo E. Reynard)

The village of Zermatt itself is built on dark and dense ophiolitic rocks, mainly metagabbros and serpentinites (Zermatt-Saas Unit; Fig. 13.7).

The sedimentary cover of this ocean floor begins with radiolarites and middle Jurassic marbles (De Wever and Caby 1981). The result is a thick series of shales and calc-schists, called lustrous shales *("Schistes lustrés)* (Deville et al. 1992), which dates from the Lower Cretaceous to the Upper Cretaceous (Marthaler 1984). These shales are visible on the path to the Hörnli hut, between 2700 and 3000 m a.s.l.; then the slope rises to cross a bar of prasinites just below the flat sector, due to the presence of serpentinites, on which the hut is built (Fig. 13.3).

13.2.3 The Rocks of the European Margin

Today, the rocks of the European margin are distributed in several basements and cover nappes (Escher et al. 1997; Fig. 13.8). These nappes all come from the Briançonnais paleogeographic domain, located south of the European plate. During the Jurassic and Cretaceous, the Briançonnais broke away from the large Eurasian plate to become a micro-continent, probably attached to the Iberian plate (Fig. 13.9).

The Basement Nappes: Siviez-Mischabel and Monte Rosa Nappes

The Siviez-Mischabel nappe, which outcrops in the Mattervispa Valley up to the entrance to Zermatt (Fig. 13.7), is formed of varied gneisses, the oldest ones, rich in amphibolites, being probably Cambrian in age (Thélin et al. 1993; Sartori et al. 2006). The large mass of Paleozoic gneisses of the Monte Rosa nappe further south has undergone high-pressure alpine metamorphism. The major part is formed of augen orthogneiss of granitic origin (Bearth 1953).

The Cover Nappes: Siviez-Mischabel, Cimes Blanches and Frilihorn nappes

The Barrhorn unit (Sartori 1990), covering the Siviez-Mischabel nappe, shows a succession of dolomitic limestone (Triassic), massive limestone (Jurassic) and

Fig. 13.4 The long geological history "told" by the Matterhorn. The rocks of the pyramid date from the Paleozoic, those of the base are Mesozoic in age. The thrust between them dates back to the Cenozoic. The shape of the mountain is even younger; it was carved by the Quaternary glaciers and related weathering (photo H. Rougier)

phyllitous limestone (Upper Cretaceous). It can be up to 300 m thick above Täsch. On the other hand, the Cimes Blanches and the Frilihorn nappes (Sartori et al. 2006) often present a stratigraphic series reduced (a few tens of metres thick) and separated from their original basement (Mont Fort nappe). They are mainly Triassic dolomitic marbles, Jurassic-laminated marbles and breccias that outcrop well in Arben, on the left bank of the Zmuttbach, a left tributary of the Mattervispa River, upstream of Zermatt (Sartori 1987).

13.2.4 Plate and Nappe Tectonics: The Matterhorn Becomes European

From a structural point of view, rocks of oceanic origin belong to two units, the Zermatt-Saas Fee Zone (Bearth 1953), rich in ophiolites, and the Tsaté nappe (Escher et al. 1997), rich in ocean sediments (Figs. 13.7 and 13.8). The latter was structured early (during Cretaceous sedimentation and subduction) as an orogenic prism (Stampfli and Marthaler 1990). The original stratigraphy is no longer preserved: Upper Cretaceous series are located at the base of the nappe and many ophiolite thrust slices are inserted in Lower to Middle Cretaceous shales in the upper part of the Tsaté nappe, near the contact with the Dent Blanche nappe. The Zermatt area is rich in high-pressure metamorphic ophiolites (Bearth 1967): it descended deeper into the ocean subduction, carrying with it the continental Monte Rosa nappe, which also had high eclogitic metamorphism.

Between the end of the Cretaceous and the beginning of the Cenozoic, a large and slow underground pile-up gave birth to the primitive Alps. On the surface, the margin of the African plate thrusted on the remains of the Tethys Sea; in depth the margin of the European plate embedded under the orogenic prism.

13.2.5 The Importance of the Matterhorn for the History of Geology

The Matterhorn rocks belong to the Dent Blanche nappe (Lower Austroalpine). Argand (1911) was the first geologist to affirm that the rocks of these more than 4000 m high mountains were of African origin (Marthaler 2001, 2005).

Fig. 13.5 View of the Zermatt Valley dominated by the Dent Blanche nappe (above the yellow thrust) which groups together the Paleozoic rocks of continental origin of the Matterhorn, Dent Blanche and Ober Gabelhorn. Dominating the village, the sedimentary strata deposited in the Tethys Sea during the Triassic, Jurassic and Cretaceous periods are clearly visible (after Rougier 2002, modified)

This nappe forms a large klippe (which also includes other famous summits as the Dent Blanche, the Weisshorn or the Ober Gabelhorn; Fig. 13.5), which floats like a raft on the remains of the disappeared ocean. Argand compared the Dent Blanche nappe to a "sledge crusher" (*traîneau écraseur* in French), but we know today that it is a small piece of the upper (African) continental plate lying on the remains of the Tethys Sea, which came to be embedded under the plate.

13.3 The Matterhorn, a Geomorphological Monument

13.3.1 The Structure: The Building Materials

The contact between the Tsaté nappe and the Dent Blanche nappe is revealed by a remarkable morphological contrast. The thrust passes at the base of the Matterhorn pyramid (Figs. 13.4 and 13.5), a hundred metres above the Hörnli hut, 3400 m a.s.l. (Figure 13.3). From this boundary between calcschists and gneiss, no more hiking or skiing is possible: it is the domain of climbers.

At first glance, the geomorphological analysis is rather simple: everything depends on the selective erosion of rocks of varied nature that form the mountain. The major part of the pyramid is carved in gneiss of the Arolla unit, in places more schistose; the innumerable slope changes of the faces and ridges (Fig. 13.3) are an expression of these selective erosion processes. On the other hand, the summit part, called "the roof", is part of the hard gneiss of the Valpelline series (Fig. 13.6). This upper part is nearly vertical, contrasting with the relatively less inclined lower sectors (especially the Hörnli ridge; Fig. 13.3) where differential dissection dominates. Between these two sections of the mountain is a flat area called "the shoulder" (Fig. 13.6). Other flat areas, often very small, are superimposed, e.g. the one that allowed the construction of the Solvay hut, very close to the vertical "Moseley Slab" (Fig. 13.3).

The four-face pyramid does not end however by a point: one is not in Giza! The Matterhorn's summit is a 110 m long ridge gently inclined to the west, so that the Italian summit is slightly lower (one metre!) than its Swiss counterpart (Fig. 13.3). In reality, everything is shared on this ridge, since the border between the two countries faithfully follows

Fig. 13.6 Close view of the Matterhorn summit, showing the boundary between the greenish orthogneiss of the Arolla series and the coloured paragneiss (pink and dark green) intermingled with silicate marble (yellowish) of the Valpelline series. The flat shoulder is due to a thin band of Mesozoic calcschists folded into the Arolla series. The base of the Valpelline series is a hundred metres higher (photo H. Rougier)

it. This almost horizontal spine, rather sharp, is just a little raised at its ends, a hollow having allowed the guides of Valtournenche to put a cross there.

13.3.2 A Highly Eroded Mountain

At the Matterhorn, instabilities are favoured by lithology. The mass of the Matterhorn's schistose gneisses, very heterogeneous in detail, provokes numerous rockfalls and rockslides. Indeed, on the east face, the most sensitive to temperature variations, freeze-thaw alternation and gelifraction are a daily process. The Matterhorn pyramid is thus less coherent that one could imagine where the uninterrupted, tireless work of erosion depending on weathering is in action. Cryoclasty is the fundamental agent of mountain dismantling. It causes the rock to burst, freeing blocks of any size. Anyone climbing the Hörnli ridge can see how unstable the rock mass is, with boulders of any size in balance: the impression is that of a dry stone wall that has collapsed. The pyramidal shape of the mountain and its isolation are two essential elements to explain remarkable aerological phenomena. The Matterhorn is not a barrier blocking the movement of atmospheric disturbances. On the contrary, there is a conjunction of aspirations on all four sides, which generate particularly violent storms provoking, on the east side mainly, runoff that wash away the wall. At the same time, the high temperature gradient increases gelifraction of fractured rocks and favours the presence of snow on all sides, giving the mountain a very aesthetic "plastered" appearance. In this sense one can consider that the Matterhorn is a very lively mountain in all seasons, with the effects of cryoclasty predominating in summer and avalanches in winter. At the foot of the walls, the rimayes of the glaciers, —that are crevasses located at the upper limit of a glacier between the moving ice and the non-moving environment, usually rock, ice attached to the rock or neve—, which hemt the pyramid (e.g. Figure 13.3) receive a high quantity of stones which feed their contents in morainic material.

Selective erosion is the major factor in the contemporary evolution of the Matterhorn pyramid, with three clear

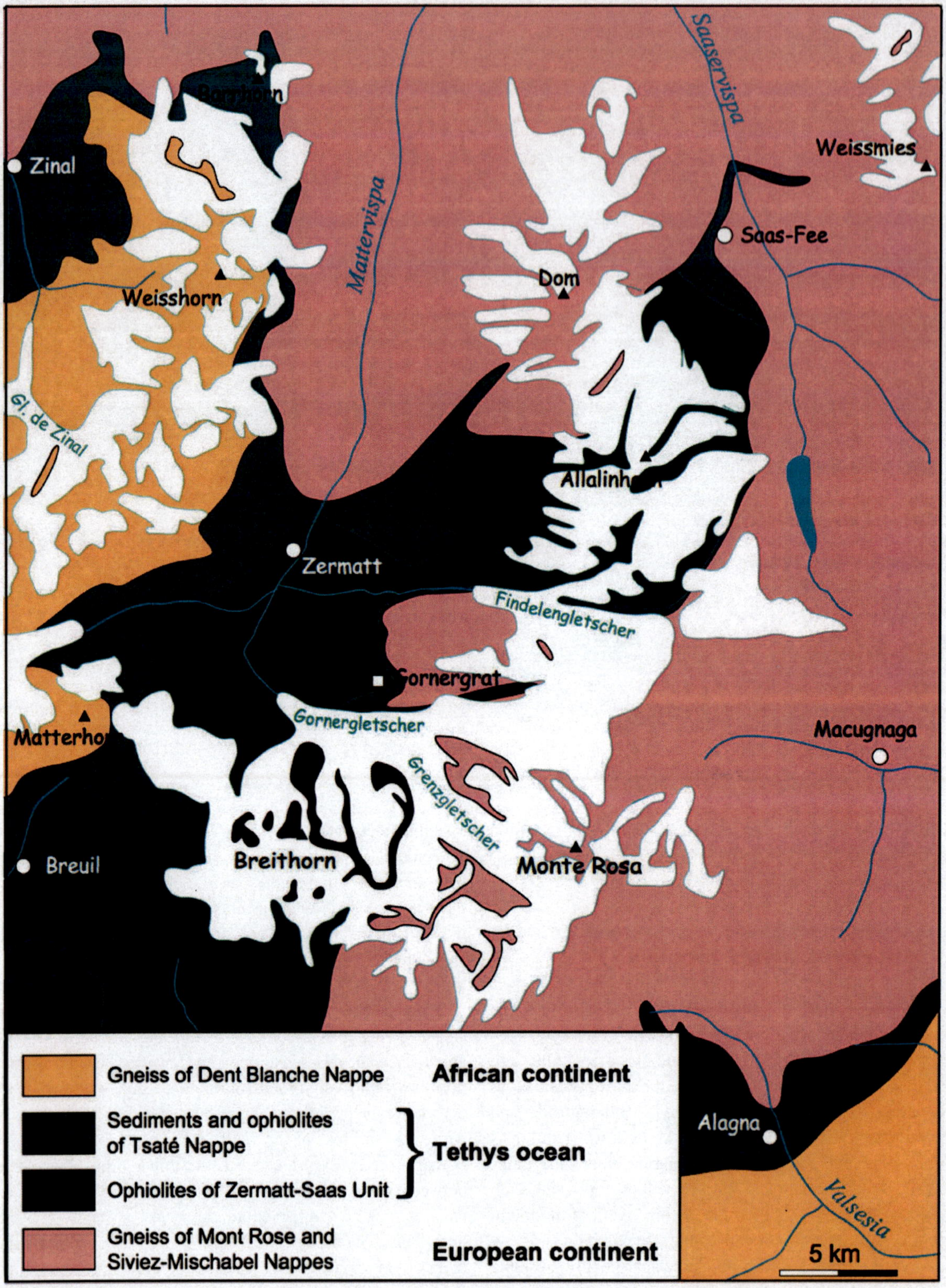

Fig. 13.7 Simplified geological map of the Zermatt region (drawing: M. Schlup)

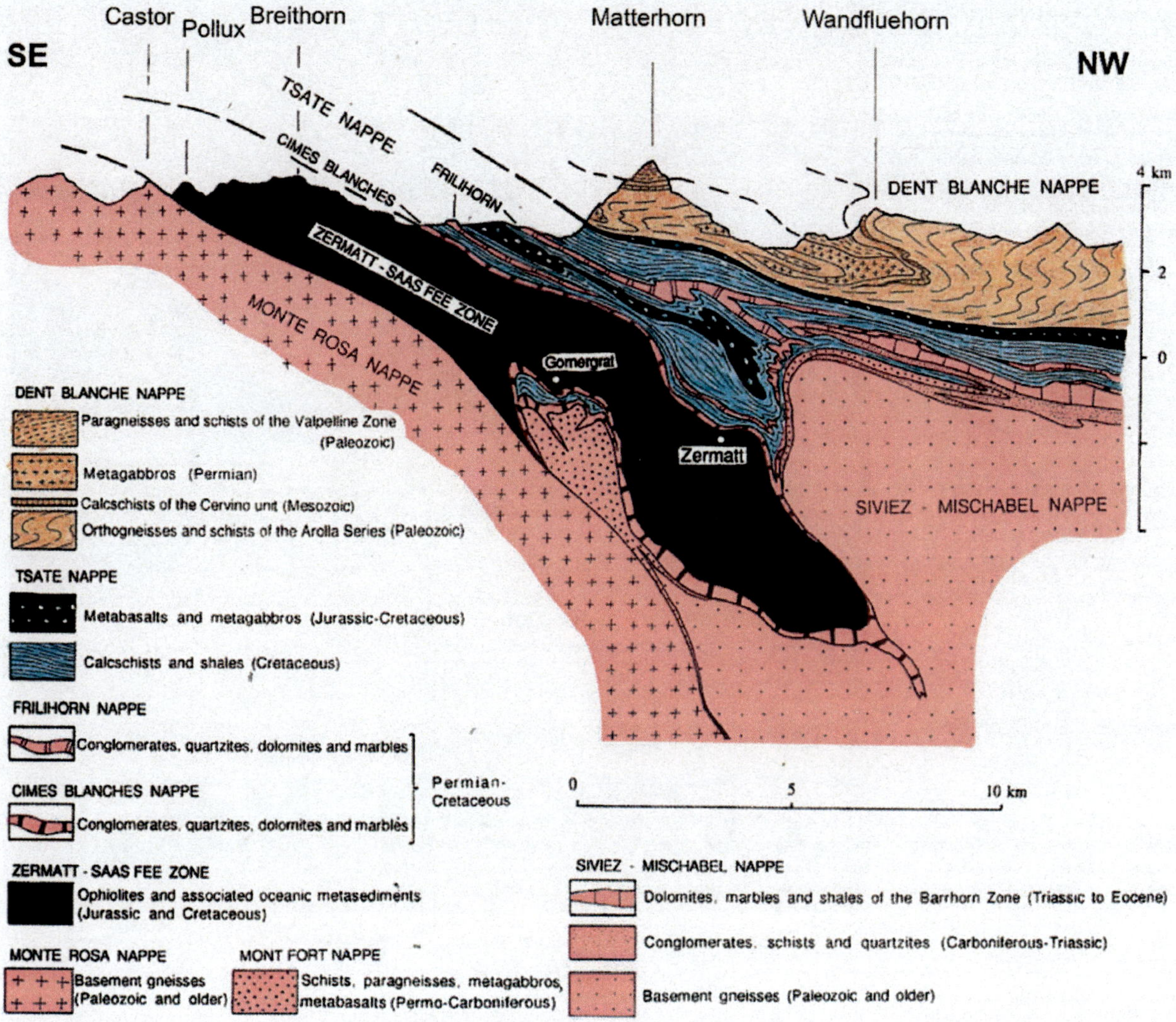

Fig. 13.8 Geological section showing the relations between the middle Penninic nappes (Monte Rosa, Mont Fort, Siviez-Mischabel, Cimes Blanches and Frilihorn, of European origin), those of the upper Penninic (Zermatt-Saas and Tsaté, of oceanic origin) and the Dent Blanche nappe (Austroalpine), of African origin (after Escher et al. 1997, modified)

oppositions between the "roof", the pyramid itself and its base, i.e. what is below the continent–ocean thrust. The three ridges visible from Zermatt are not only the ones that give the mountain its unique configuration; they also allow understanding the morphodynamic processes, generated by gravity, that govern its evolution.

13.3.3 The Role of Glaciers, in the Past and Today

On the four sides, the rocky mass dominates over many glaciers (Fig. 13.10): Tiefenmatten-, Furgg-, Matterhorngletscher and the Ghiacciao inferiore del Cervino. While still providing an incomparable ornament to the mountain, they are only a pale reflection of what they were until recently on a geological timescale.

During the Quaternary cold episodes, the upper parts of the Swiss Alps were covered by several ice domes following the water divide; one of these large icefields—also called Southern Valais icefield—was situated in the Mattertal (Kelly et al. 2004; Schlüchter et al. this volume). During the Last Glacial Maximum (LGM) the top of the icefield surface filled the valley to at least 3010 m elevation (Kelly et al. 2004). The Matterhorn was, therefore, a nunatak, affected by mechanical weathering processes. Nevertheless, if compared, for example with the Grimsel granitic area (see Schlüchter this volume), trimelines are not well developed in

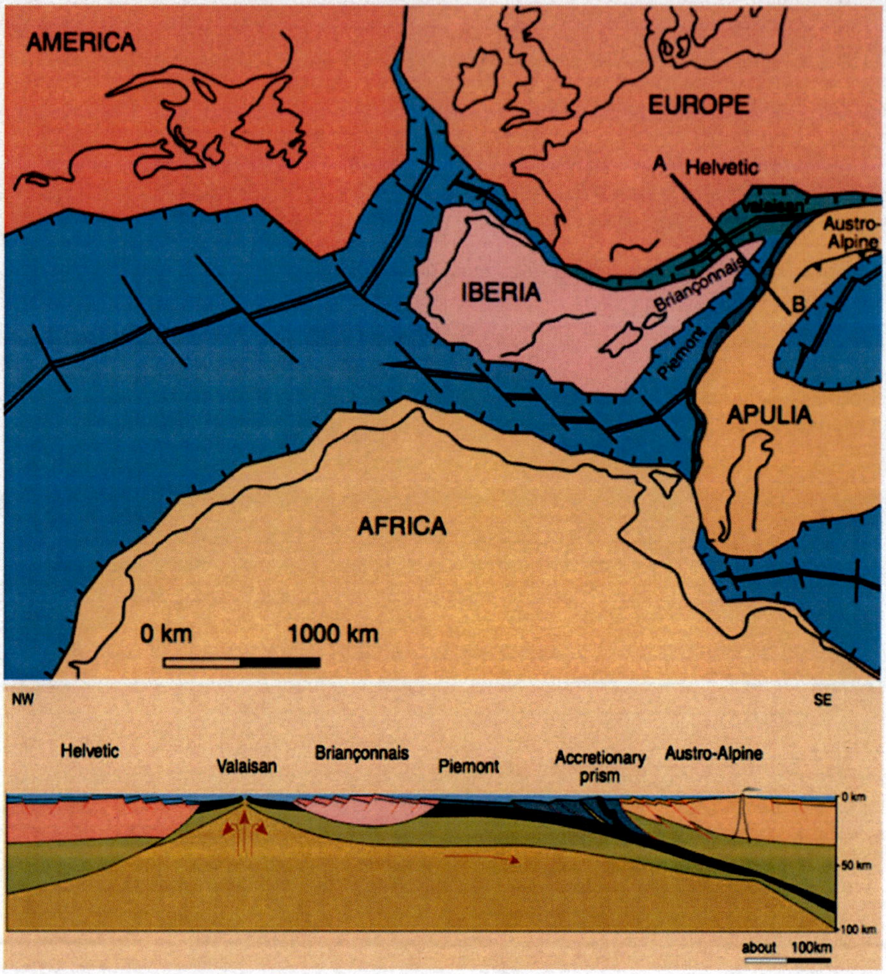

Fig. 13.9 Palaeogeographic map and cross-section, during the middle Cretaceous period, showing how the opening of the North Atlantic caused Iberia to drift off to the Southwest. Separated from Europe, Iberia became a micro-continent whose northeastern part was formed by the Briançonnais peninsula (Marthaler 2005)

the Matterhorn area because of postglacial weathering and gravity processes that have destroyed most of erosional glacial landforms such as *roches moutonnées*. Subglacial erosion by meltwater has been quite important, excavating glacial pots. For example, beautiful examples exist in the *Gletschergarten* (Glacier Garden) downstream of the Gorner glacier (Fig. 13.1), where the subglacial gorge also remains in perfect condition today. These episodes of strong glaciation were interspersed with warm interglacials during which the erosive action was more or less similar as today.

A specificity of glacial erosion at the Matterhorn is the contact of several glacial cirques forming a horn. The nature of the bedrock is important in the genesis of cirques, their extent being correlated with the density of fractures (Galibert 1965). This is exactly what happens in the Matterhorn: the "roof" provides blocks from the gelifraction of compact gneiss, but the masses of fragile schistose gneiss underneath are sensitive to glacial abrasion. Thus, through glacial abrasion during glacial periods and through freeze-thaw alternation during interglacial episodes, cirques gradually formed. A perfect example is the gigantic cirque that occupies the entire eastern face of the Matterhorn (Figs. 13.3 and 13.9).

The situation is similar on the other sides: for this reason the Matterhorn is defined as a "pyramid of cirque intersections". René Godefroy (1948) wrote: *"Les cirques entrent en lutte, leurs arêtes séparatives s'aiguillent en même temps qu'elles s'abaissent. Des pointes s'isolent, culminantes, surtout aux intersections des crêtes, là où se nouent les têtes de vallées glaciaires. Elles ont la forme de corne (all.* Horn*), comme le Cervin (Matterhorn). Les sommets correspondent aux points culminants de la surface initiale, débarrassée de neige et par conséquent attaquée en dernier lieu par la dissection des cirques"* (translation: "Cirques are fighting, their separating edges become sharper at the same time as they lower themselves. Peaks are isolated, culminating, especially at the intersections of the ridges, where the heads of glacial valleys are knotted. They have the shape of horn (German: *Horn*), like the Matterhorn. The peaks correspond to the highest points of the initial surface, cleared of snow

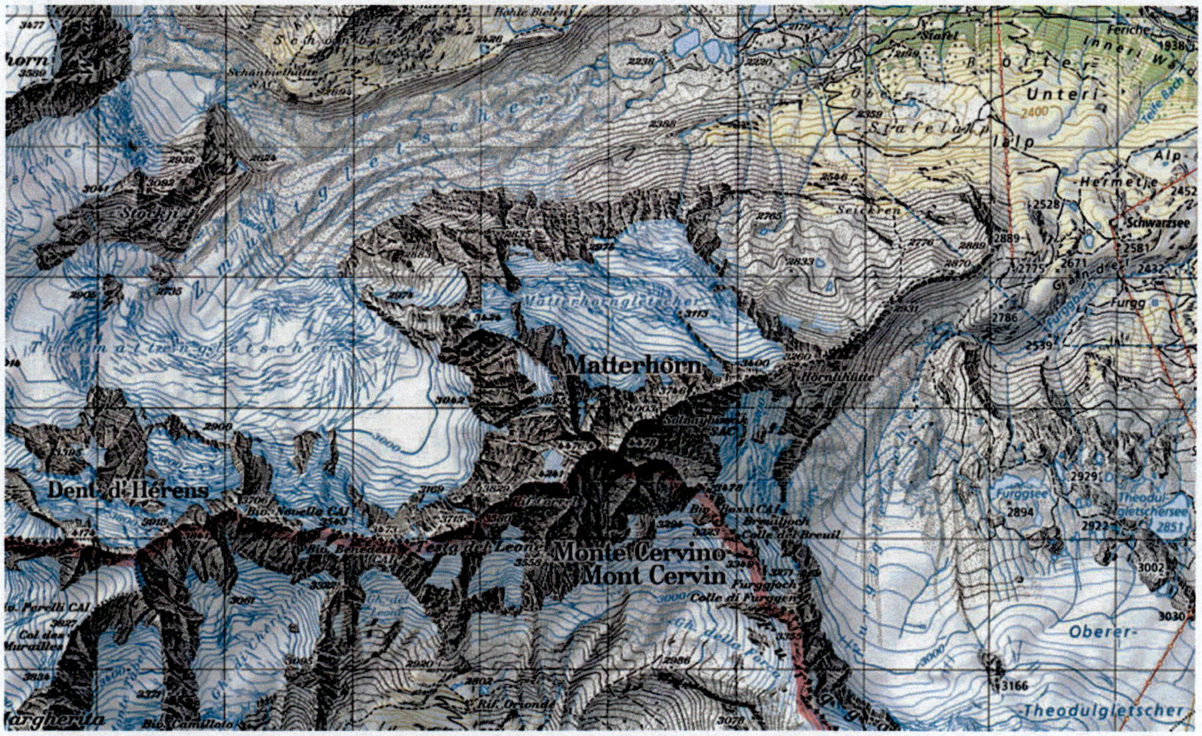

Fig. 13.10 Topographic map (scale: 1 km grid) of the Matterhorn area, showing the location of glaciers. From the glacial cirque on the eastern side of the Matterhorn an ice tongue emerged, which, on several occasions, invaded the area occupied by the Furgggletscher. Despite its melting, the Furgggletscher is still today an authentic cirque glacier (swisstopo)

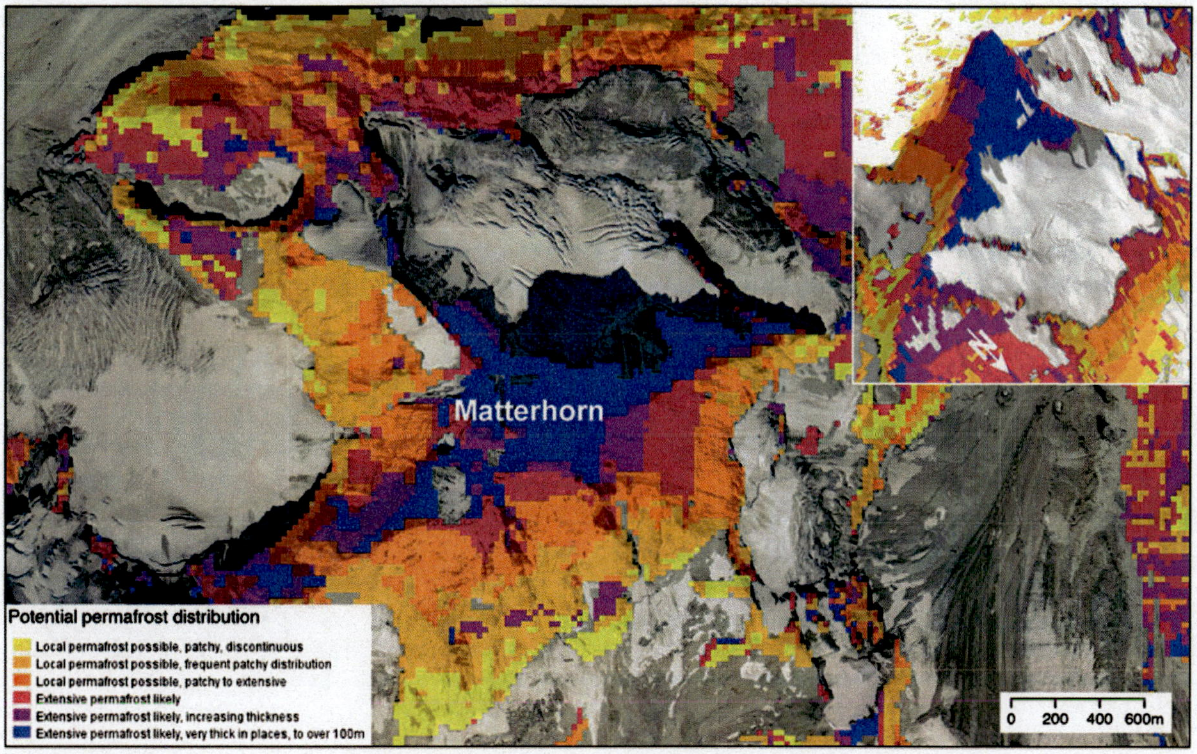

Fig. 13.11 Potential distribution of permafrost in the Matterhorn area (swisstopo)

and consequently attacked last by the dissection of the cirques"). Thus, the Matterhorn has become the archetype of a karling, i.e. an angular peak with steep walls and sharp ridges, undercut from all sides by glaciers (Haeberli et al. 2015), and the term "horn" has passed into the current language of geomorphologists.

13.4 Conclusions

Because it is both very original and representative of Alpine nature, the Matterhorn occupies a prominent place among geomorphosites. It is so attached to Zermatt that one cannot imagine separating them. From this binomial was born an authentic reference, to the point that the planet adds many "Matterhorns". Only a few kilometres away, the Dent Blanche and the Weisshorn are almost identical copies. Much further, Mount Sir Donald in the Selkirk Mountains (British Columbia, Canada) is considered "a single Matterhom-like summit towered in lonely splendor" (Palmer 1914); Mount Assiniboine, also in the Canadian Rocky Mountains, is called the "Matterhorn of the Rockies" (Sandford 2010), and the granite inselberg of Spitzkoppe in Namibia is considered the "Matterhorn of Africa" because of its conical shape (Migoń 2010). Even Berlin, far from being in a mountainous area, has its "Matterhornstrasse"!

"Il n'existe aucune autre montagne au monde qui puisse rivaliser avec le Cervin, si l'on en juge par l'attractivité exercée sur les hommes. Tous les qualificatifs ont été utilisés: unique, mythique, magique… Que ce soient les alpinistes, les géographes, les géologues ou plus prosaïquement les visiteurs contemplatifs qui se trouvent en sa présence, personne ne reste indifférent face à la "Corne" sans qui Zermatt n'existerait pas en tant que telle" (Rougier 2002) (translation: "There is no other mountain in the world that can compete with the Matterhorn, judging by the attractiveness exerted on men. All the qualifiers have been used: unique, mythical, magical… Whether mountaineers, geographers, geologists or more prosaically contemplative visitors who are in its presence, no one remains indifferent to the "Horn" without which Zermatt would not exist as such). Because it stands out totally from other mountains, thanks to its topography and its shape unlike any other, pure, alone and solitary, the Matterhorn has acquired a landscape notoriety that makes it radically unclassifiable. Whereas the oceanic sediments correspond roughly to the mountains of medium altitude and to the space favourable for hiking, when one arrives at the level of the gneissic series of the African plate, one enters without transition in the field of mountaineering. The Matterhorn, a prestigious summit, received the most flattering tributes on the occasion of the commemoration of the 150th anniversary of its first ascent on 14 July 2015. Fifty bulbs retraced the itinerary followed in 1865 on this dream mountain, at the perfect summit.

Discussion—The Matterhorn, an example of fractured bedrock permafrost

By Emmanuel Reynard, University of Lausanne

The Matterhorn is an example of fractured bedrock permafrost (Gruber and Haeberli 2007; Haeberli et al. 2015, Weber et al. 2017). Fracture kinematics in this context are influenced by several factors such as perennial ice in the permafrost body, seasonal water/ice pressure changes in the active layer, changes in seasonal snow cover, expansion/contraction in microfractures related to freeze/thaw cycles (Weber et al. 2017). Some displacements are reversible; they are due to thermoelastic strains of the material and are occurring all year round. Irreversible displacements, which can lead to rock slope failure, occur mostly during periods with temperature above 0 °C.

Based on modelling of potential permafrost distribution in Switzerland prepared in 2006 on behalf of the Federal Office for the Environment FOEN (http://map.geo.admin.ch/, accessed 20.03.2020) extensive permafrost with a thin active layer is expected in the northern face and more local permafrost is possible in the southern face with a thick active layer; Weber et al. 2017; Fig. 13.11). At the Hörnli Hut (3260 m a.s.l.) 100 m deep permafrost is present with 3 m thick active layer and at the summit, with approximately −10 to −15°C mean annual temperature, permafrost is likely to be several hundred metres thick (Haeberli et al. 2015). Fractures are filled with massive ice, as was evidenced on the detachment surface of the 15 July 2003 rockslide event (Gruber and Haeberli 2007; Haeberli et al. 2015). Permafrost and temperature variations induce instabilities at three scales (Haeberli et al. 2015): daily freeze-thaw cycles cause small rockfalls; the annual variability of temperatures influences the active layer (subsurface layer which melts every summer) and can cause large rockslides as was the case during the hot summer 2003; finally, permafrost degradation itself occurs on a secular scale. "Permafrost degradation in steep bedrock takes place by heat conduction, and by advection of heat by percolating water in fractures" (Gruber and Haeberli 2007).

References

Argand E (1908) Carte géologique du massif de la Dent Blanche. Matériaux pour la Carte géologique de la Suisse 56, Berne

Argand E (1911) Les nappes de recouvrement des Alpes Pennines et leurs prolongements structuraux. Matériaux pour la Carte géologique de la Suisse N.S. 31:1–26

Bearth P (1953) Geologische Atlas der Schweiz 1:25 000. Blatt 1348 Zermatt. Swisstopo, Wabern

Bearth P (1967) Die Ophiolithe der Zone von Zermatt-Saas Fee. Beiträge zur geologische Karte der Schweiz 132:1–130

Bücher K, Dal Piaz GV, Oberhänsli R, Gouffon Y, Martinotti G, Polino R (2003) Geologische Atlas der Schweiz 1:25 000, Blatt 1347 Matterhorn. Swisstopo, Wabern

Bussy F, Venturini G, Hunziker J, Martinotti G (1998) U-Pb ages of magmatic rocks of the western Austroalpine Dent Blanche-Sesia Unit. Schweiz Mineral Petrogr Mitt 78(1):163–168

Deville E, Fudral S, Lagabrielle Y, Marthaler M, Sartori M (1992) From oceanic closure to continental collision: a synthesis of the "Schistes Lustrés" metamorphic complex of the Western Alps. Geol Soc Am Bull 104:127–139

De Wever P, Caby R (1981) Datation de la base des schistes lustrés postophiolitiques par des radiolaires (Oxfordien supérieur—Kimméridgien moyen) dans les Alpes Cottiennes (Saint-Véran, France). Comptes Rendus de l'Académie des sciences II 292:467–472

Escher A, Hunziker J, Marthaler M, Masson H, Sartori M, Steck A (1997) Geologic framework and structural evolution of the Western Swiss-Italian Alps. In: Heitzmann P, Lehner P, Pfiffner OA, Muller S, Steck A (eds) Deep structure of the Swiss Alps. Birkhauser, Basel, pp 205–222

Galibert G (1965) La haute montagne alpine. Université de Toulouse, Thèse d'Etat, 405 pp

Godefroy R (1948) La nature alpine, exposé de géographie physique. PUF, Paris, 483 pp

Gruber S, Haeberli W (2007) Permafrost in steep bedrock slopes and its temperature-related destabilization following climate change. J Geophys Res 112:F02S18

Haeberli W, Nötzli J, Springman S (2015) Matterhorn "for ever"? In: Anker D (ed) Matterhorn—Berg der Berge. AS Verlag, Zürich, pp 294–301

Kelly MA, Buoncristiani C, Schlüchter C (2004) A reconstruction of the last glacial maximum (LGM) ice surface geometry in the western Swiss Alps and contiguous Alpine regions in Italy and France. Eclogae Geol Helv 97:57–75

Marthaler M (1984) Géologie des unités penniques entre le Val d'Anniviers et le Val de Tourtemagne (Valais, Suisse). Eclogae Geol Helv 77(2):395–448

Marthaler M (2001) Le Cervin est-il africain? Une histoire géologique entre les Alpes et notre planète. Editions LEP, Le Mont sur Lausanne (4th edn 2014), 112 pp

Marthaler M (2004) Lecture et analyse d'un paysage: Zermatt et le Cervin. Un exemple de la mémoire de la Terre révélée par les panoramas. In: Reynard E, Pralong JP (eds) Paysages géomorphologiques. Institut de Géographie, Lausanne. Travaux et Recherches 27, pp 51–66

Marthaler M (2005) The Alps and our Planet. the African Matterhorn: a geological story. Editions LEP, Le Mont-sur-Lausanne, 96 pp

Migoń P (2010) Spitzkoppe: the world of granite landforms. In: Migoń P (ed) Geomorphological Landscapes of the World. Springer, Berlin, pp 155–162

Monjoie P (2004) The Mont Collon mafic complex (Austroalpine Dent Blanche nappe): Permian evolution of the western European mantle. PhD thesis, Universities of Lausanne and Grenoble, 163 pp

Rébuffat G (1965) Le Cervin, cime exemplaire. Hachette, Paris, 215 pp

Palmer H (1914) Mountaineering and exploration in the Selkirks: a record of pioneer work among the Canadian Alps. Putnam, New York and London, 616 pp

Rougier H (2002) Au Pays de Zermatt. Editions LEP, Le Mont-sur-Lausanne, 207 pp

Rougier H (2010) Zermatt: nature, man and landscapes. Editions LEP, Le Mont-sur-Lausanne, 84 pp

Sandford RW (2010) Ecology and wonder in the Canadian Rocky Mountain Parks World Heritage Site. AU Press, Athabasca University, 379 pp

Sartori M (1987) Structure de la zone du Combin entre les Diablons et Zermatt. Eclogae Geol Helv 80(3):789–814

Sartori M (1990) L'unité du Barrhorn (zone pennique, Valais, Suisse). PhD thesis, University of Lausanne (Mémoires de Géologie, Lausanne, 6), 156 pp

Sartori M, Gouffon Y, Marthaler M (2006) Harmonisation et définition des unités lithostratigraphiques briançonnaises dans les nappes penniques du Valais. Swiss J Geosciences 99:363–407

Schlüchter C, Akçar N, Ivy-Ochs S (this volume) The Quaternary Period in Switzerland

Stampfli G, Marthaler M (1990) Divergent and convergent margins in the Western Alps. Confrontation to actualistic model. Geodinamica Acta 4(3):159–184

Taugwalder H (1979) Das verlorene Tal. Glendyn, Aarau, 312 pp

Thélin P, Sartori M, Burri M, Gouffon Y, Chessex R (1993) The Pre-Alpine basement of the Briançonnais (Wallis, Switzerland). In: Von Raumer JF, Neubauer F (eds) Pre-Mesozoic Geology in the Alps. Springer, Berlin, pp 297–315

Weber S, Beutel J, Faillettaz J, Hasler A, Krautblatter M, Vieli A (2017) Quantifying irreversible movement in steep, fractured bedrock permafrost on Matterhorn (CH). The Cryosphere 11:567–583

Whymper E (1871) Scrambles among the Alps in the years 1860–1869. Murray, London, 432 pp

Zermatt Tourismus (2017) Jahresbericht 2017. Zermatt

Michel Marthaler is an Honorary Professor of Geology at the University of Lausanne. He is working for the Swiss Geological Survey to complete the Geological Atlas of Switzerland at 1:25,000. His research and interests are focused on Alpine geology, mainly the high Valais Alps, the Pyrenees, mountain building and its links with geodynamics, as well as Earth Sciences didactics. He has published several popular scientific books including *The Alps and our Planet.*

Henri Rougier is Professor Emeritus of Physical Geography at the University of Lyon. He teaches as part of Erasmus exchanges at the two universities in Berlin and at the University of Halle-Wittenberg. He founded and chairs in Chamoson (Valais, Switzerland) the association Géoterrain, whose aim is to promote geology and geography. He has published six books on the geography of Zermatt.

14 The Aletsch Region with the Majestic Grosser Aletschgletscher

Hanspeter Holzhauser

Abstract

The Aletsch region is situated in the UNESCO World Heritage Swiss Alps Jungfrau-Aletsch and must be reckoned with right to the classic and splendid high-mountain regions in the world. The region stretches from the rocky steppes with a Mediterranean character to the glaciers. It is a perfect example of the mountain and glacier's formation. The basement of the region was formed by orogenic processes in the last approximately 500 million years. During the Ice Ages, the glaciers gave the shape of the contemporary landscape with characteristic forms, for example *roches moutonnées* and moraine deposits. Permafrost, frost weathering and rock mass movements are well presented and designed the landscape too. This rich diversity of glacial landscapes is a perfect example of the actual climatic change and geomorphological processes in high-mountain regions. The famous *Grosser Aletschgletscher*—the largest and longest glacier of the Alps—with the three world-famous peaks of Eiger, Mönch and Jungfrau in the catchment lies in the core of the World Heritage site. Its history is well documented over the last 3500 years. The once famous *Märjelensee*, an ice-dammed lake at the edge of the Grosser Aletschgletscher, has shrunk as the glacier is retreating dramatically since the end of the Little Ice Age around 1860.

Keywords

Bernese Alps • Aar massif • High-mountain geomorphology • Glaciers • Holocene glacier fluctuations • Glacial lake

14.1 Introduction

The Aletsch region (about 7°53'– 8°8'E/46°19'– 46°24'N) is enclosed in the water tower of Europe. It extends between Naters (678 m a.s.l.) and Fiesch (1049 m a.s.l.) in the Rhone River valley to the highest peaks, the Mittaghorn (3982 m a.s.l.), the Mönch (4107 m a.s.l.) and the Jungfrau (4158 m) in the north, where the Aletsch region is well developed with the Jungfrau Railway, built between 1896 and 1912 (Fig. 14.1). Up to 1,000,000 tourists drive up with this train every summer to the icy, splendid Jungfraujoch. From the south, the Aletsch region is well accessible via Blatten near Naters, Riederalp, Bettmeralp or Fiesch. Since 2001, the Aletsch region forms with the famous *Grosser Aletschgletscher* the heart of the UNESCO World Heritage Swiss Alps Jungfrau-Aletsch which covers 824 km^2 and contains a large part of the Bernese Alps.

A large part of the Aletsch region is located in the rain shadow of the Bernese Alps and the Pennine Alps. For this reason, in a relatively short distance, the annual rainfall rises from 737 mm in Brig in the south (697 m a.s.l.) to 3600 mm in the Jungfrau–Mittaghorn region in the north (3892–4158 m a.s.l.). The mean annual temperature is lowered along the same elevation profile from 8.5°C to –7.2°C. On the southern slopes this Alpine scenery is gradually transformed, via different altitudinal vegetation and climatic zones, from a sub-Mediterranean rocky steppe into a glacierised zone with arctic character. Only a narrow strip along the southern edge is a settlement and farming region. Further to the north, most of the Aletsch region is covered by glaciers. Due to the relatively dry conditions, the southern slopes of the Aletsch region need to be irrigated. Therefore miles of irrigation channels—called *Suonen*—had to be built along some steep slopes.

The following chapter contains a brief overview of the geological conditions in the Aletsch region, and the most important geological and geomorphological processes that have shaped actually the landscape are discussed in

H. Holzhauser (✉)
Ahornstrasse 38, 8051 Zürich, Switzerland
e-mail: h.holzhauser@bluewin.ch

E. Reynard (ed.), *Landscapes and Landforms of Switzerland*, World Geomorphological Landscapes,
https://doi.org/10.1007/978-3-030-43203-4_14

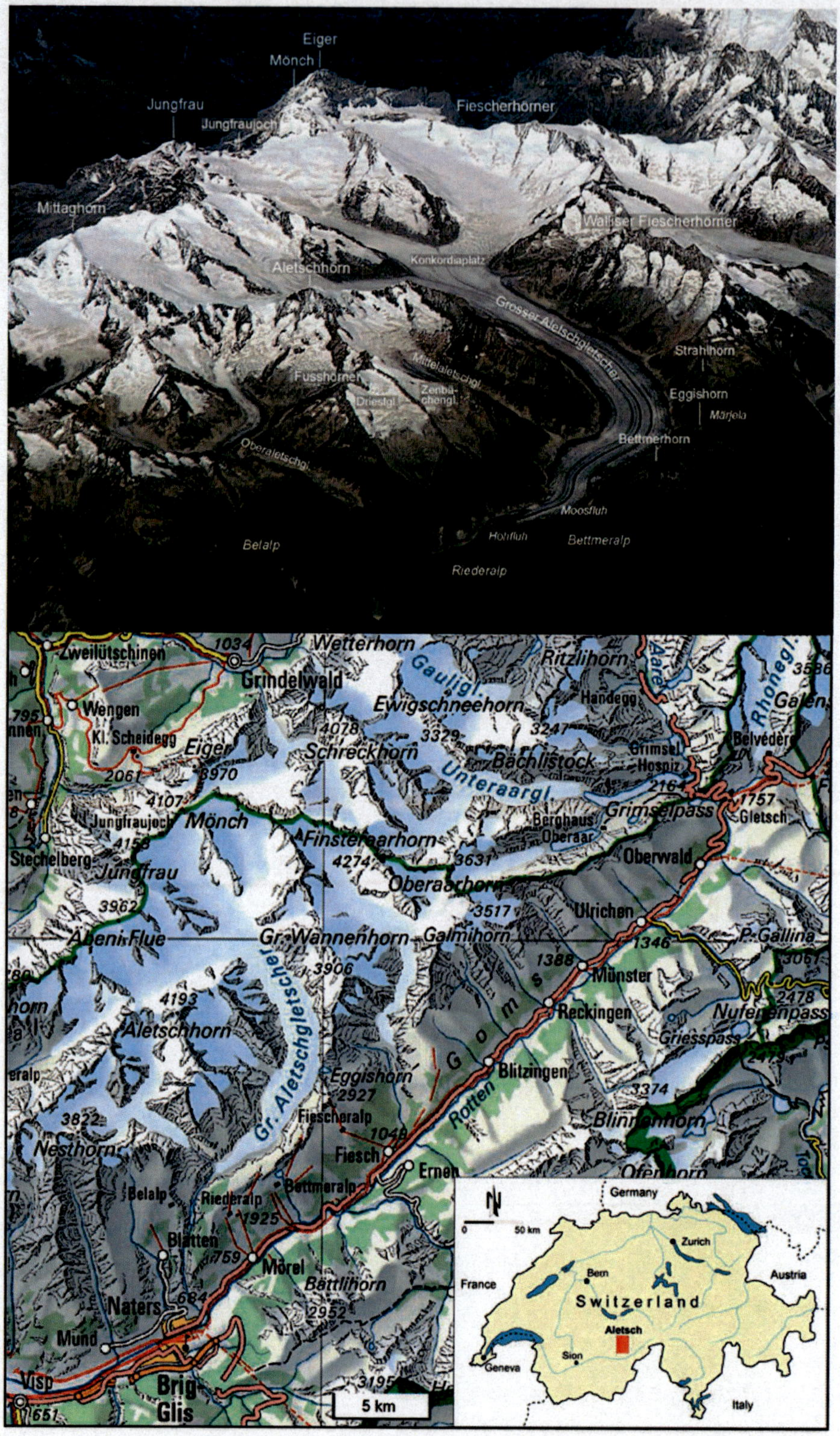

Fig. 14.1 The Aletsch region shown from space. The Grosser Aletschgletscher is clearly marked by dark medial moraines extending along the glacier's length parallel to the valley axis (photo: NASA, ISS013-E-77377, 5 September 2006; map: swisstopo)

Sect. 14.3. The major part of this contribution (Sects. 14.4–14.6) encompasses the Grosser Aletschgletscher as well as the Märjelensee (Sect. 14.7), once a piece of arctic jewellery in the Alps.

14.2 The Rock Basement of the Aletsch Region—Tectonic and Geological Overview

The Aletsch region passes from north to south across the Aar Massif, which in terms of geology is part of the European basement (Fig. 14.2). With a length of 115 km and a width of 35 km it is the largest Central Massif beside the Gotthard Massif eastwards and the Aiguilles Rouges/Arpille Massif and Mont Blanc Massif westwards. In the Mesozoic, the Aar Massif was the Helvetic and Ultrahelvetic depositional environment in the Tethys Sea that spread in the east of the supercontinent Pangea. More than half of the Aar Massif consists of gneiss and granite and owes its origin to two main orogeneses. The high-grade metamorphic gneisses and amphibolithes of the crystalline basement (*Altkristallin*) originated during the Caledonian orogeny about 450 million years ago. It occupies approximately 50 % of the Aar Massif. In the Carboniferous, about 300 million years ago, during the Variscan orogeny a tremendous amount of granitic magma invaded forming the Central Aare Granite (Labhart 2007). With a surface area of 550 km^2, the Central Aare Granite is the largest contiguous granite complex of the Swiss Alps. The contact with the crystalline basement (*Altkristallin*) is razor sharp and can be observed at various points in the Aletsch region (Fig. 14.3). During the Alpine orogeny at the beginning of the Tertiary when the African plate pushed against the European plate, the *Altkristallin* was raised and compressed. The sediments deposited on the northern edge of the Tethys were moved, slanted or folded. On the northern edge of the Aletsch region, a narrow band of the Mesozoic sedimentary cover (Autochtonous) is well visible and builds most of Wetterhorn and Eiger. Locally, for example at Jungfrau, it is embedded in the crystalline basement forming wedges (see Zumbühl et al., this volume). The north and south of the Central Aare Granite, enclosing the pre-Variscan crystalline basement, are referred to as northern and southern Gneiss Zone (Labhart 2007).

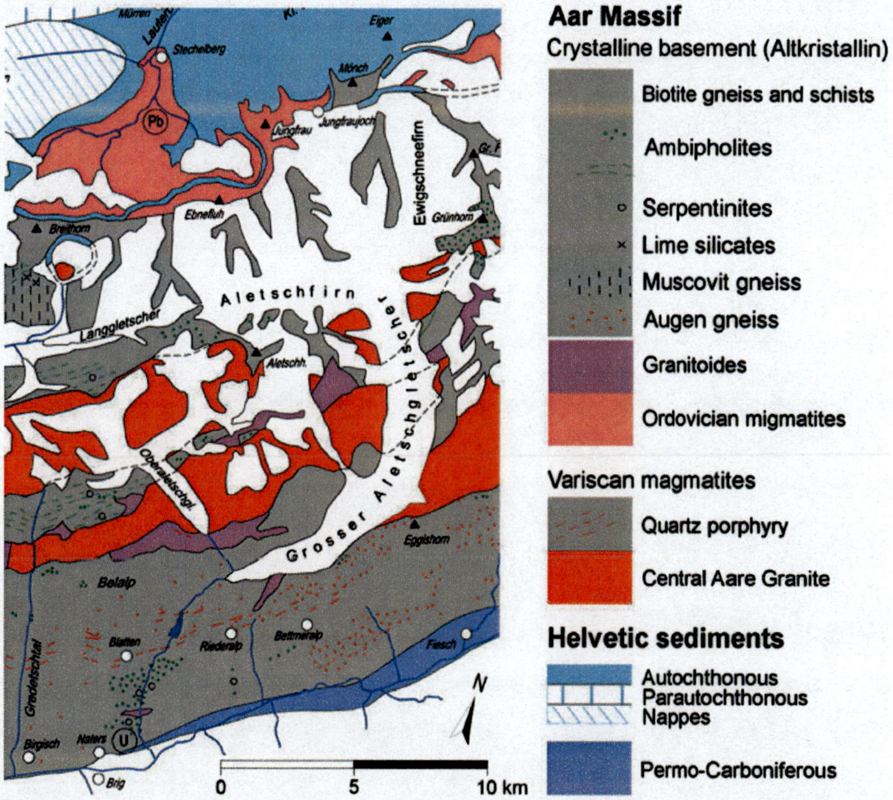

Fig. 14.2 Geological-tectonic map of the Aletsch region (modified after Labhart 2007)

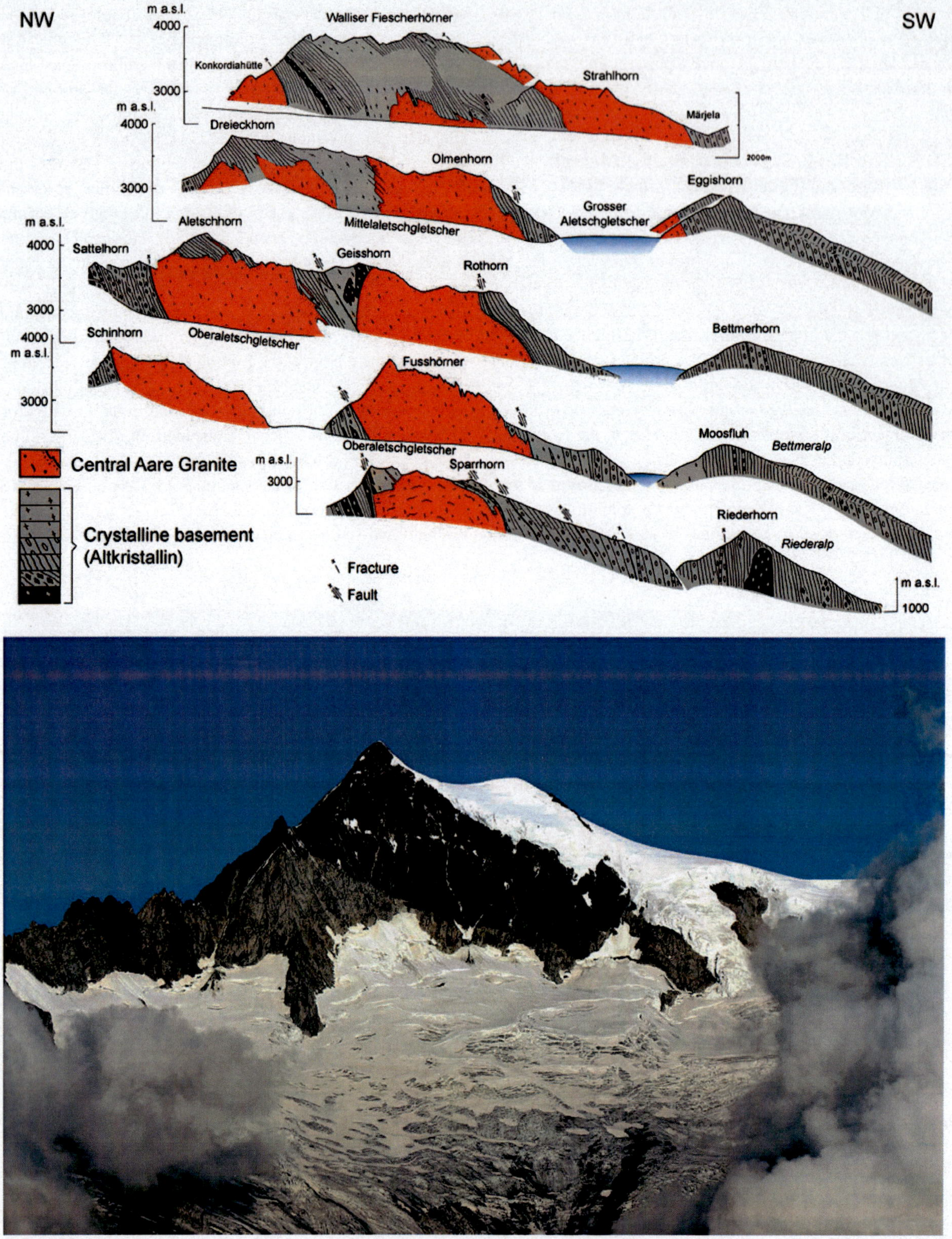

Fig. 14.3 Geological profiles and Aletschhorn (4193 m a.s.l.), the highest peak in the Aletsch region, view from Eggishorn. The dark and older crystalline basement (*Altkristallin*) lies above the younger Central Aare Granite (profile after Steck 1983; photo H. Holzhauser 2011)

14.3 The Ravages of Time Gnaws on the Rock Basement—Geomorphological Processes and Landforms

Within the last approximately 500 million years, geological processes formed the basement of the Aletsch region. The Ice Age glaciers gave it the present appearance, with characteristic surface forms. In the last 11,700 years, during the Holocene, the glaciers eroded and accumulated comparatively only in a small region and created the Little Ice Age moraine ramparts that are clearly visible in many places and limit the proglacial area. Geomorphological activities are observed today too, and increasingly due to the current global warming. They result in a complex interplay of different processes (glacial erosion and accumulation, frost weathering, gravitational movements, release of rocks due to shrinkage of glaciers and permafrost).

14.3.1 Faults and Rock Mass Movements

Various large and small fracture systems and faults that are considered mainly in the context of large-scale uplift and subsidence during the waning Alpine orogeny are visible within the Aletsch region (Fig. 14.4a; Steck 1968). The schists of the crystalline basement in the Aletsch region generally have a NW-SE strike and are slanted. Flexural toppling tilts the steeply to vertically stratified rock as a result of gravity (uphill-facing scarps, *Hakenwurf*, Fig. 14.4b).

Since the retreat of the Grosser Aletschgletscher after the Ice Age and also after the maximum Little Ice Age extent around 1860, the ice was lost as an abutment. Because of the geological and terrain conditions found at the slopes adjacent to the tongue of the Grosser Aletschgletscher, the steep slopes became unstable, and a number of slumps in the lateral moraines as well as rock mass movements (*Sackungen*, slope saggings) developed and are currently active. In addition, the rock once covered by the glacier is shattered by frost weathering what accelerated this process. A small sagging took place between 1966 and 1975 at the tongue of the Grosser Aletschgetscher near Tälli (Fig. 14.4c). First, smaller rock falls occurred and finally, the entire rock body dropped rapidly and was riddled with cracks and crevasses (Kasser et al. 1982). Since at least the 1970s, on the right side of the valley below Driest, a displacement of an unstable slope mass of about 450,000 m^2 moved downhill up to 20 cm/year (Kääb et al. 2000). By movement, unweathered bedrock was exposed as a bright stripe of about 250 m (Fig. 14.4c). A larger and still active slope movement formed after the melt back of the Grosser Aletschgletscher at the end of the last Ice Age at Moosfluh. The region is about 1.5 km wide and extends down to the maximum moraine rampart of 1860 from the Grosser Aletschgletscher. Between 1990 and 2008, horizontal movement of the rock mass in direction to the glacier amounted from 4 to 30 cm/year (Strozzi et al. 2010).

The instability on the right flank of the Moosfluh became active in autumn 2015. In 2016 and 2017, a strong acceleration of the landslide has been observed involving an estimated volume of 2.5 km^3. There were numerous cracks and rockfalls on some steep slopes. The rock movements were in the range of several centimetres and decimetres per day. In 2016, 2.5 million cubic metres of rock broke away in a single event (Kos et al. 2016).

14.3.2 Glacial Erosion and Glacial Deposits

Impressive is the environment shaped by the glacier. During the last Ice Age (called Würm in the Alps) the Grosser Aletschgletscher confluenced with the Rhonegletscher and only the highest peaks of the surrounding mountains projected above the ice. The maximum height of the Ice Age glaciation is clearly visible at the upper limit of the vertically sloping rock wall at Geissgrat, Zenbächenhorn and Rothorn, for example (Fig. 14.4d). The trimline is just below these vertical rock walls. Also the gently rounded ridge from Hohbalm below Bettmerhorn to Moosfluh and Riederfurka (Fig. 14.4e) and the depression of Märjela (see Sect. 14.7) are due to the landscape-shaping force of the ice. Numerous *roches moutonnées* were formed there by the glacier, at some place even crowds of *roches moutonnées* (*Rundhöckerflur*; Fig. 14.4f) occur.

Around 12,300 years ago, the Grosser Aletschgletscher advanced for the last time during the Lateglacial (Egesen advance during the Younger Dryas). Its tongue ended in Naters near Brig (Fig. 14.1). A moraine rampart from this former glacial extent is well preserved on the left slope of the Aletsch valley which stretches with interruptions below the Eggis- and Bettmerhorn to Riederfurka (*Moränenweg*, Fig. 14.4g). On the opposite side of the valley traces of this glacial extent can be seen at Alp Driest where the Drietgletscher confluenced with the Grosser Aletschgletscher (Fig. 14.4d).

14.3.3 Frost Weathering

Frost weathering is one of the most important processes of physical weathering in the high mountains and is strongly dependent on the nature of the rocks (lithology) and the weak zones (such as fractures, foliation surfaces). This ultimately determines the size and shape of weathering products. In the Aletsch region with schist-dominated

a
b
Sparrhorn
Driest
Tälli
M
M
c
Geissgrat
Zenbächenhorn
Rothorn
K
d
Moosfluh
e
f
Bettmerhorn
Eggishorn
g
Eggishorn
h

◀ **Fig. 14.4** Examples of geological and geomorphological landforms: **a** Faults at Belalp (arrows); **b** flexural toppling ("Hakenwurf", Moosfluh above Riederalp) in the crystalline basement; **c** slumps and rock mass movements (M) below Tälli and Driest. The black arrow indicates unweathered bedrock; **d** White arrows: Lateglacial moraine ramparts (Egesen advance, Younger Dryas, around 12,300 before present) from the Driestgletscher and former confluence with the Grosser Aletschgletscher (K). The dark arrow indicates the Little Ice Age moraine rampart. Dashed lines: Upper limit of glaciation during the Last Glacial Maximum (Würm); **e** Glacier shaped ridge between Bettmerhorn and Moosfluh; **f** Clusters of *roches moutonnées* at the Üssers Aletschji; **g** Lateglacial moraine rampart (arrows) from the Grosser Aletschgletscher (Egesen); **h** Frost weathering on Eggishorn (photos H. Holzhauser 2011)

Fig. 14.5 The Rock glacier "Grosses Gufer" (G) situated between Eggis- and Bettmerhorn (photo H. Holzhauser 2011)

crystalline basement it is well expressed. For example, the summit and the southern slope of Eggishorn referred to by the historian Mark Lutz in his book of 1827 disparaged as a "terrible heap of rubble" consists of pieces of laminated gneiss of the *Altkristallin* (Fig. 14.4h). The lower part of the northern face of Eggishorn, however, is composed of compact Central Aare Granite and mainly shows forms due to glacial processes.

14.3.4 Rock Glaciers

Rock glaciers are important indicators of permafrost in the high mountains and are located near the lower limit of discontinuous permafrost. Between Bettmer- and Eggishorn numerous rock glaciers have developed, some of which are still active. The largest rock glacier in the area is the *Grosses Gufer* (Fig. 14.5). It begins below Elselicka at 2600 m a.s.l. where the steep rock walls provide the material due to frost weathering over a distance of about 600 m. The volume was estimated at around 3 million m^3 in 1962 (Messerli and Zurbuchen 1968). The *Grosses Gufer* flows from there against the Grosser Aletschgletscher and its steep front ends today on the moraine rampart of 1860 about 2360 m a.s.l., where the trail leads to Märjela. The Lateglacial moraine rampart of the Younger Dryas (approximately. 12,300 years ago) was overran by this rock glacier long time ago. Initial studies in the 1960s showed that in 1950–1962 the rock glacier moved with the speed of 8–16 cm/year at the top and around 16–33 cm/year in the middle part. The highest flow velocities were measured in the front part, amounting to 58–75 cm/year (Messerli and Zurbuchen 1968). New measurements showed flow velocities of 30–100 cm/year since 2007/08 (Strozzi et al 2009; PERMOS 2013). In midsummer, crossing the *Grosses Gufer* allows hearing subterranean streams and in some places the melt water escapes from the rock glacier with icy cold temperatures of around 1°C. This water was captured in the sixteenth century (probably even earlier) and led in the irrigation channel "Riederfurka" to the pastures on Riederalp and Oberried (Holzhauser 1984).

14.4 The Famous Grosser Aletschgletscher—an Icy Superlative in the Alps

In his book *The glaciers of the Alps*, published in 1860, the Irish physicist and mountaineer John Tyndall (1820–1893) wrote: *"The Aletsch is the grandest glacier in the Alps: over it we now stood, while the bounding mountains poured vast feeders into the noble stream"*. For the geologist Edouard Desor (1811–1882), longtime friend and companion of Louis Agassiz (1870–1873), the Grosser Aletschgletscher was *"[...] the largest and most beautiful of the Swiss glaciers"* (Fig. 14.6a–c). The Grosser Aletschgletscher is a typical valley glacier. With a length of 20.7 km from Jungfraujoch and a surface of about 79 km^2, the Grosser Aletschgletscher is the largest and the longest glacier in the Alps. Its volume is estimated to be about 13.4 km^3. (Dr. A Bauder, Institute of Glaciology VAW-ETH Zurich, personal communication 2015). The accumulation area is bordered to the north by famous mountains such as Jungfrau (4158 m a.s.l.) and Mönch (4099 m a.s.l.). It includes the firn basins Grosser Aletschfirn, Jungfraufirn, Ewigschneefäld and Grüneggfirn (from west to east), which confluence at *Konkordiaplatz* (Fig. 14.7). They form large medial moraines such as the Kranzberg and Trugberg moraines. Their swinged shape and dark colour give the characteristic appearance to the Grosser Aletschgletscher. On Konkordiaplatz, where the glacier bed is considerably overdeepend, ice thickness is about 800–900 m as measured by a hot water drilling in the years 1990 and 1991 (Aellen and Herren 1994; Hock et al. 1999; Farinotti et al. 2009; Jouvet et al. 2011). The bedrock was not reached, as a mix of ice-rock debris made further drilling impossible. The lowest point is limited only to a very small area. At the outlet of Konkordiaplatz the ice is 550–600 m and near Märjela about 400–500 m thick. In a sweeping arc, the Grosser Aletschgletscher flows from Konkordiaplatz in southeast direction and its tongue ends deep into the coniferous forest amid abraded and fractured rocks at about 1620 m a.s.l. (Figure 14.7).

Below Konkordiaplatz the greatest flow velocities of ice were measured on the glacier surface at around 180–200 m/year and in the Aletschwald area the ice flows only about 60–80 m/year down the valley (Aellen and Röthlisberger 1981; Jouvet et al. 2011). In the ablation area the Grosser Aletschgletscher is steadily losing ice through melting. At the height of the Aletschwald the annual ice loss is comprised between 9.8 and 12.7 m (summer and winter ablation; measurements by Pro Natura Center Aletsch). This ice loss is not quite compensated by continued flow of ice from the accumulation area. Currently, the surface in the lower part of the glacier tongue is reduced by 2.25–3.75 m/year. The annual changes in thickness at Konkordiaplatz amount around 1 m/year since 1990 (Bauder et al. 2007).

During the Holocene maximum glacial extents, for the last time around 1860, the Oberaletsch- and the Mittelaletschgletscher confluenced with the Grosser Aletschgletscher. The Oberaletschgletscher, the second largest glacier in the Aletsch region with an area of 17.4 km^2 and a length of 9.1 km, was divided by a massive rock plateau in two glacier tongues. The left tongue pushed through the Oberaletsch gorge, filled it with ice and flowed together at the Tälli with the Grosser Aletschgletscher. The right glacier tongue was wider and flatter and went close to the huts of Üssers Aletschji (Fig. 14.6c). Sharp-edged moraine ramparts testify to this last maximal glacial extension in the mid-nineteenth century. A few years later, the left glacier tongue was separated from the Grosser Aletschgletscher and has retreated about 1.5 km since then. The right glacier tongue melted completely away.

On the eastern flank of Aletschhorn (4193 m a.s.l.), the highest peak of the Aletsch region, the Mittelaletschgletscher with an area of 6.8 km^2 and a length of 5.3 km emanates. It hung up until the early 1970s together with the Grosser Aletschgletscher (Fig. 14.6a).

Above the Grosser Aletschgletscher, between Fusshorn and Zenbächenhorn two other glaciers occur, the Driestgletscher with area of 2.02 km^2 and a length of 2.08 km and the Zenbächengletscher with an area of 0.85 km^2 and a length of 1.2 km (Fig. 14.6b). They end up in a broad front and are not very thick. Since the last glacial extension around 1860, they have lost about half of their areas. Both the Driest- and the Zenbächengletscher show moraine ramparts of the Little Ice Age maximum extent. They still wear little vegetation in contrast to the upstream Lateglacial moraine ramparts which are completely overgrown (Fig. 14.4d).

14.5 Research Activities on the Grosser Aletschgletscher

First measurements of ablation were taken by Arnold Escher von der Linth in 1841 when he visited the Märjelensee. Later, in 1869, Charles Grad measured also the ablation and

Fig. 14.6 Panoramic views on the Grosser Aletschgletscher with **a** the Mittelaletschgletscher (M); **b** the Driestgletscher (D) and the Zenbächengletscher (Z); **c** the Oberaletschgletscher, the Driestgletscher (D) and the Zenbächengletscher (Z); R clusters of *roches moutonnées* in the Üssers Aletschji, S rock mass movements. The solid white lines indicate the last Little Ice Age Maximum around 1860 (photos H. Holzhauser 2011)

for the first time took measurements of the velocity rate. First measurements in the firn region were carried out in 1918; first depth and velocity measurements in 1937 (Holzhauser 2009).

Comprehensive and systematic glaciological studies started when the High Alpine Research Station Jungfraujoch was established in 1931. Studies were originally limited to Jungfraufirn and Ewigschneefäld and were extended in the 1950s also to the tongue of the Grosser Aletschgletscher. Since 1922, the mass balance change has been detected first using hydrological and from the 1950s—glaciological method (Aellen and Funk 1990). Recently, radar measurements have been carried out by the Institute of Glaciology of ETH Zurich (Dr. A. Bauder, Versuchsanstalt für Wasserbau,

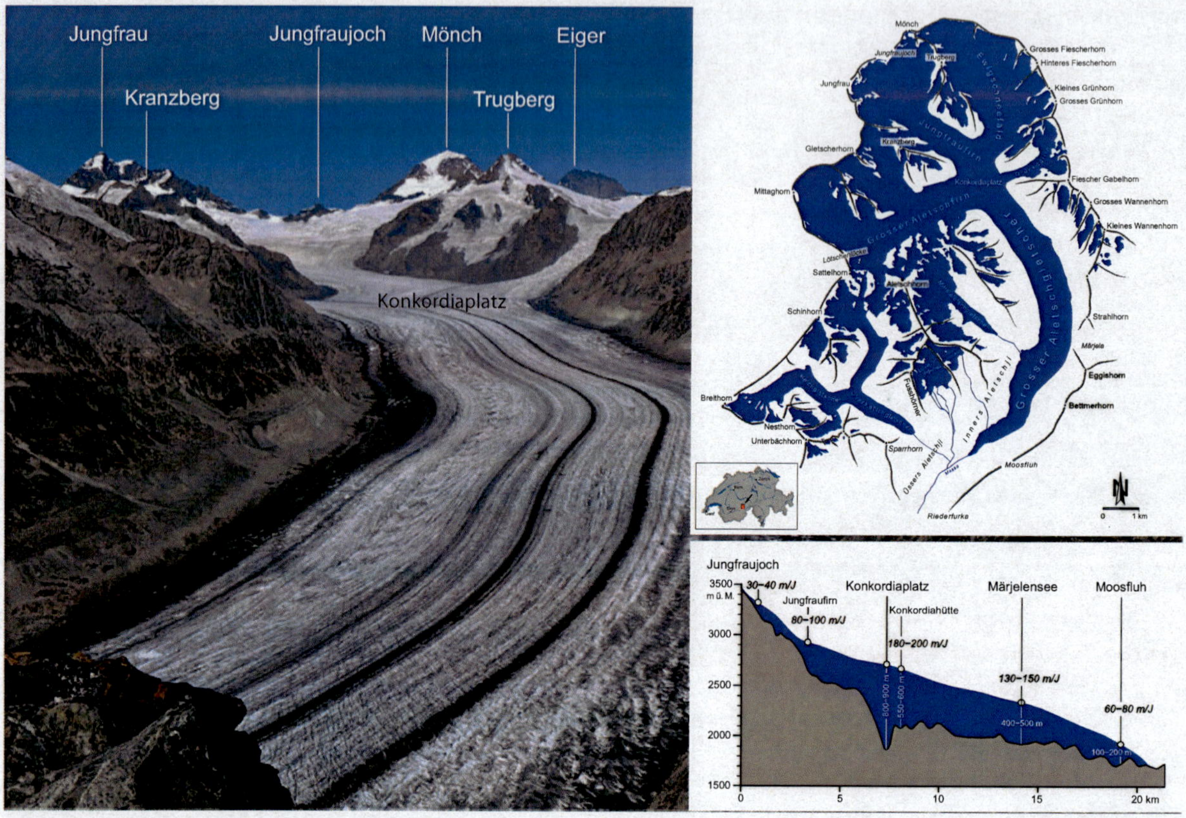

Fig. 14.7 View from Eggishorn to the Konkordiaplatz. The dark medial Kranzberg- and Trugberg moraines are clearly visible. The map shows the glaciated area and the longitudinal profile shows the overdeepening at the Konkordiaplatz as well as the thickness and flow velocities (in metres per year) of the ice (map GLAMOS.CH; profile after Jouvet et al. 2011 and Bauder A. VAW/ETHZ; photo H. Holzhauser 2011)

Hydrologie und Glaziologie (VAW), ETH Zürich, personal communication 2015). As a contribution to the “International Geophysical Year 1956” a map of the Grosser Aletschgletscher in the scale of 1:10,000 was created in 1957 by the Swiss Federal Office of Topography and VAW-ETHZ. This cartography in five map sheets published between 1960 and 1964 can be seen as a highlight of glaciological research on the Grosser Aletschgletscher (Kasser 1961).

14.6 Fluctuations of the Grosser Aletschgletscher Since 3500 BC

The tongue of the Grosser Aletschgletscher has, in the past, reached deep into the pine forest, sometimes coming very close to inhabited regions. During recent maxima of the Little Ice Age, mountain huts have been destroyed, an irrigation channel has been made unusable and pine forest and arable land have been covered. Using various methods to reconstruct Holocene glacial fluctuations—glaciological, historical, archaeological and glaciomorphological ones (Zumbühl and Holzhauser 1988, 2007), it was possible to reconstruct changes in length of the glacier tongue of the Grosser Aletschgletscher over the past 3500 years (Fig. 14.8, Holzhauser 1984, 2009; Holzhauser et al. 2005). The most recent segment of the curve presented in Fig. 14.8, from the twelfth century AD onwards, is based on dendrochronological and archaeological evidence as well as visual and written historical sources. The period from the twelfth century AD backwards is reconstructed only by means of dendrochronologically dated fossil wood found within the proglacial area, some of them actually still in situ.

From the late Bronze Age to the Middle Ages, evidence from dendrochronologically dated trees is obtained not only of growth phases of glaciers, some quite marked, but also of

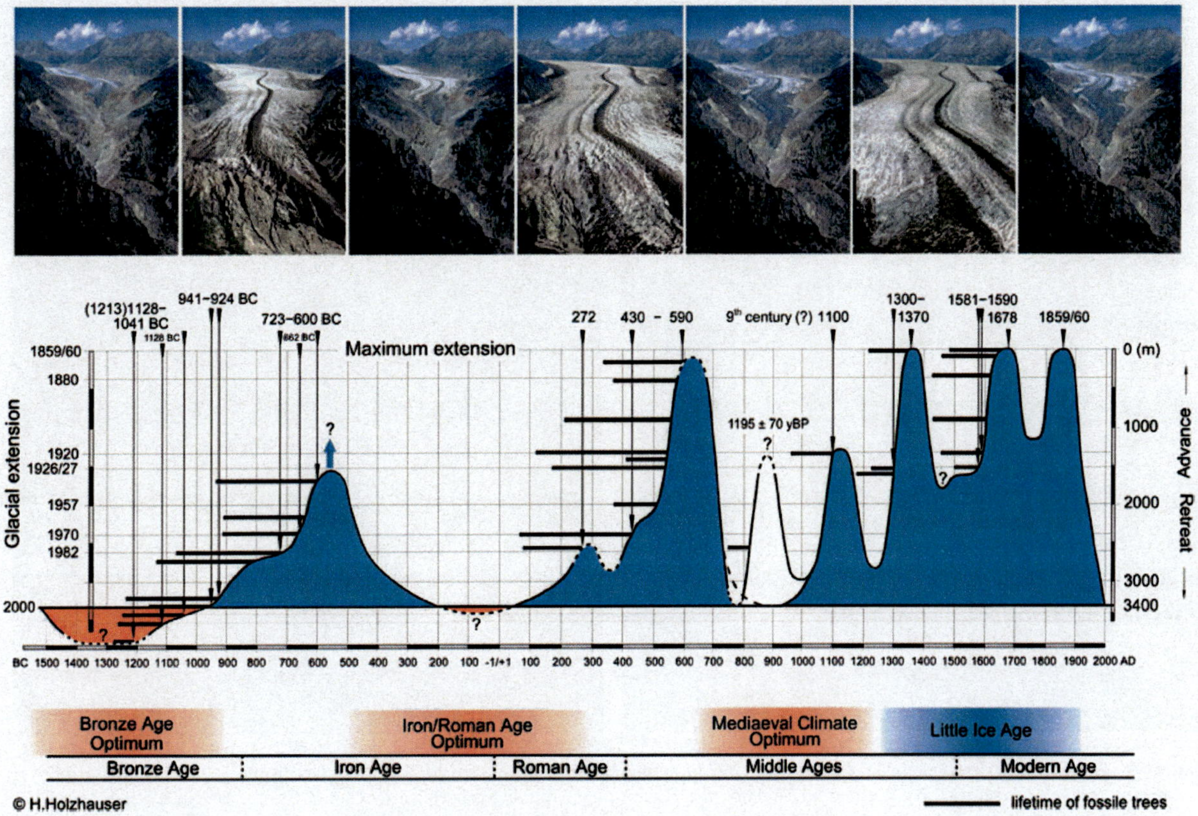

Fig. 14.8 Fluctuations of the Grosser Aletschgletscher over the last 3500 years. The above photographs show the supposed extent of the Grosser Aletschgletscher within seven periods: Bronze Age Optimum, Iron Age advance, Iron/Roman Age Optimum, early Mediaeval advance (Migration Period), Mediaeval Climate Optimum, around 1856 and in 2000 (photomontage H. Holzhauser. Original photograph from 1856: Frédéric Martens, Alpine Club London; photo H. J. Zumbühl)

periods when glacier size was similar to or smaller than today. During the late Bronze Age Optimum, the Grosser Aletschgletscher was from 1350 to 1250 BC about 900 to 1000 m shorter than it is today. During the Iron/Roman Age Optimum between 200 BC and AD 50 and in the early Middle Ages around AD 750, the glacier reached about today's extent or was even somewhat shorter than today. The Mediaeval Warm Period, from around AD 800 to the onset of the Little Ice Age around AD 1300 (Holzhauser et al. 2005), was interrupted by two weak advances in the ninth century (not certain) and around AD 1100. Precise dendrochronological dating shows a powerful advance around AD 1300. Three successive peaks characterize the Little Ice Age: a first maximum in the 1370s, a second between 1670 and 1680 and a third in 1859/60 (Fig. 14.9). Written documents indicate that in the fifteenth century the glacier was of a size similar to that of the 1930s. A very small advance around 1500 has been dated by both dendrochronology and archaeology (destruction of an irrigation channel).

The tongue length of the Grosser Aletschgletscher has been measured yearly since 1892, making the glacier behaviour in the following period extremely well documented. The glacier tongue is now approaching the previous minimum recorded length, as documented between 200 BC and AD 50. In this context, it must be borne in mind that the dynamics of the Grosser Aletschgletscher tongue not only constitute a smoothed but also a slightly delayed function of direct climate and mass balance forcing. In this case, the corresponding time lag is estimated at a few decades, which

Fig. 14.9 The Grosser Aletschgletscher viewed from Belalp around 1856 (left) and in 2018 (F. Martens, Alpine Club London; Archiv H. J. Zumbühl and H. Holzhauser). We can see the debris on the glacier tongue, produced by local rockfall due to the landslide at the Moosfluh

means that the glacier tongue would have to be hundreds of metres shorter than now in order to adjust to actual conditions. In view of the rapid warming during the past two decades, it is therefore highly probable that, in the near future, the previous minimum extent of late Bronze Age may be reached or even markedly exceeded.

14.7 The Märjelensee—Arctic Surroundings at the Foot of the Eggishorn

The Märjelensee on the edge of the Grosser Aletschgletscher was well known until the late twentieth century as one of the most beautiful and typical glacial lakes of the Alps. At its maximum extent known in 1878, the lake filled the depression of Märjela and was about 1.64 km long, 460 m wide and 78.5 m deep at its deepest point. With a surface area of 0.46 km^2 the lake contained at that time 10.7 million cubic metres of water (Lütschg 1915). The once deep blue lake with floating ice blocks together with the barren and rocky environment resembled an arctic landscape (Fig. 14.10). However, from that former "arctic jewel" is not much to see today. As a result of the profound and prolonged lowering of the surface of the Grosser Aletschgletscher the former ice barrier continuously lost thickness. Today in the spring and summertime only a small lake is formed at the edge of the ice (Fig. 14.11).

Centuries ago when the Grosser Aletschgletscher was even mightier than today the Märjelensee was a constant threat to the inhabitants in the valley below. It was infamous for his unpredictable and frequent water outburst floods. Through a fast opening basic crevasse in the icy barrier, the lake could empty suddenly (Lütschg 1915). The water masses were flowing then under the glacier as well as at the edge of the glacier and seriously damaged parts of the town of Naters as between 1813 and 1915. The Märjelensee would also flood parts of the Fieschertal, wreaking havoc on the village and the livestock. To prevent further flooding from the Märjelensee outbursts the cantonal engineer Ignaz Venetz designed in 1828 and 1829 two canals: a small canal that diverted the lateral inflows from Strahlhorn against the Fieschertal directly and a large canal that would lower the

Fig. 14.10 Historical views of the Märjelensee. **a** J. R. Bühlmann, 1835 (© Graphische Sammlung ETH Zürich); **b** H. Hogard, 1849 (the lake is empty due to a water outburst; Hogard and Dollfus-Ausset 1854; © H. Holzhauser); **c** around 1890 (postcard, private property); **d** around 1900 (postcard, private property); **e** around 1920 (postcard, private property); **f** in 1976 (photo © H. J. Zumbühl)

water level of the Märjelensee. The expected success was not achieved and the water flowing through the canals caused increased flooding in Fieschertal. Another attempt to lower the lake level and prevent lake outbursts was undertaken from 1889 to 1894, however, at a much deeper lake level. In direction to the Fieschertal a 583 m long gallery was built in the bedrock. This gallery was only flooded once, in 1896. During six weeks water flowed through the underground canal. Since then the water level of the Märjelensee has never reached the height of the gallery

Fig. 14.11 The Märjelensee in 2007 (photo H. Holzhauser)

again. In order to detect changes in the water level accurately, level constructions were installed in the years 1908 and 1909 (Lütschg 1915).

The increasing shrinking of the Grosser Aletschgletscher towards the end of the nineteenth century led to the decline in the level of the Märjelensee and a terrain threshold divided the lake into the larger and deeper lake at the edge of the glacier, the Gletschersee or Hintersee, and the shallower Vordersee, located further to the huts of the Märjelenalp (Fig. 14.12a). The Vordersee has increased since 1988 as part of the "Märjelen Project" by construction of an artificial dam and it now collects the spring water from the *Galtjinenquellen* (Galkina) below Strahlhorn (Fig. 14.12b). The water is passed from the barrier lake through the newly built, one kilometre long Tälligrat Tunnel into a reservoir. From this reservoir a pipeline leads water for irrigation to Lax and Martisberg, Goppisberg, Riederalp and Oberried. Another pipeline leads water into a reservoir on the Laxeralp where drinking water is treated. The drinking water is conducted from there to other reservoirs and is allocated to the aforementioned municipalities.

14.8 Conclusion

The Aletsch region located in the heart of the UNESCO World Heritage Swiss Alps Jungfrau-Aletsch is an eldorado not only for hikers and mountaineers, but also for glaciologists, geologists and geomorphologists. Because of its location—away from the former main tourist routes in the Rhone River valley—it was opened to tourism only late. Today, the well-accessible viewpoints at Jungfraujoch, Eggishorn and Belalp offer incomparably panoramic views onto the Pennine and Bernese Alps as well as onto the majestic Grosser Aletschgletscher, the most powerful glacier of the Alps. Its history could be completely reconstructed for the past 3500 years with the help of historical documents and analysis of fossil trees found in the glacier forefield. Once the Märjelensee, fed by glacial meltwater, was very famous because of its seemingly arctic character. This typical glacial lake was feared because of its irregular outburst floods. Today, only a small lake is formed sporadically since the surface of the Grosser Aletschgletscher has strongly decreased within the last 150 years.

Fig. 14.12 Panoramic view over the Märjela. **a** Around 1900 with the Hinter- oder Gletschersee at the edge of the Grosser Aletschgletscher and the Vordersee (on the right); **b** In 2011. The Vordersee increased since 1988 as part of the "Märjelen Project" by an artificial dam. The white line indicates the lake level in 1878 (photo H. Holzhauser 2011)

Acknowledgements Many thanks are due to Dr. A. Bauder (Versuchsanstalt für Wasserbau, Hydrologie und Glaziologie (VAW), ETH Zürich) and Dr. M. Fischer (Department of Geosciences, University of Fribourg) for the newest data from the glaciers in the Aletsch region. I dedicate this contribution to my youngest son Roberto who died much too early in October 2013 at the age of only 22. He accompanied me to the Grosser Aletschgletscher in 2011, when most of the photos for this chapter have been taken.

References

Aellen M, Herren E (1994) Die Gletscher der Schweizer Alpen 1989/1990 und 1990/1991, 111. und 112. Bericht. Glaziologisches Jahrbuch der Gletscherkommission der Schweizerischen Naturforschenden Gesellschaft (SNG), Zürich, 127 pp

Aellen M, Röthlisberger H (1981) Gletschermessungen auf Jungfraujoch. In: 50 Jahre Hoch-alpine Forschungsstation Jungfraujoch. Kantonalbank von Bern, Bern, Bulletin 23, 1981, pp 82–92

Aellen M, Funk M (1990) Bilan hydrologique du bassin versant de la Massa et bilan de masse des glaciers d'Aletsch (Alpes bernoises, Suisse). International Association of Hydrological Sciences (IAHS) Publication 193:89–98

Bauder A, Funk M, Huss M (2007) Ice volume changes of selected glaciers in the Swiss Alps since the end of the 19th century. Ann Glaciol 46:145–150

Farinotti D, Huss M, Bauder A, Funk M (2009) An estimate of the glacier ice volume in the Swiss Alps. Glob Planet Chang 68:225–231

Hock R, Iken A, Wangler A (1999) Experiments and borehole observations in the overdeepening of Aletschgletscher, Switzerland. Ann Glaciol 28:253–260

Hogard H, Dollfus-Ausset D (1854) Principaux Glaciers de la Suisse. Imprimés en Lavis-Aquarelle d'après les originaux dessinés et peints d'après nature par Henri Hogard. Matériaux pour servir à l'étude des Glaciers, Strasbourg

Holzhauser H (1984) Zur Geschichte der Aletsch- und des Fieschergletschers. PhD Thesis, University of Zurich. Zurich, Physische Geographie 13, 448 pp

Holzhauser H (2009) Die bewegte Vergangenheit des Grossen Aletschgletschers. Blätter aus der Walliser Geschichte 41:103–121

Holzhauser H, Magny M, Zumbühl HJ (2005) Glacier and lake-level variations in west-central Europe over the last 3500 years. The Holocene 15(6):789–801

Jouvet G, Huss M, Funk M, Blatter H (2011) Modelling the retreat of Grosser Aletschgletscher, Switzerland, in a changing climate. J Glaciol 57:1033–1045

Kasser P, Aellen M, Siegenthaler H (1982) Die Gletscher der Schweizer Alpen 1973/74 und 1974/75, 95. und 96. Bericht. Glaziologisches Jahrbuch der Gletscherkommission der SNG, Zürich, 163 pp

Kasser P (1961) Glaziologischer Kommentar zur neuen im Herbst 1957 aufgenom-menen Karte 1:10'000 des Grossen Aletschgletschers. IAHS Publ 54:216–223

Kääb A, Huggel C, Paul F (2000) Früherkennung hochalpiner Naturgefahren mittels Fernerkundung. In: Interpraevent 2000, Tagungspublikation, Band 1, pp 49–60

Kos A, Amann F, Strozzi T, Delaloye R, von Ruette J, Springman S (2016) Contemporary glacier retreat triggers a rapid landslide response, Great Aletsch Glacier, Switzerland. Geophys Res Lett 43:12466–12474

Labhart T (2007) Geologie—500 Millionen Jahre Erdgeschichte. In: Wallner A, Bäschlin E, Grosjean M, Labhart T, Schüpbach U, Wiesmann U (eds) Welt der Alpen—Erbe der Welt. UNESCO Welterbe-Region Jungfrau–Aletsch–Bietschhorn. Haupt, Bern, pp 23–46

Lütschg O (1915) Der Märjelensee und seine Abflussverhältnisse. Eine hydrologische Studie unter Mitberücksichtigung hydrographischer Erscheinungen in anderen Flussgebieten. Annalen der Schweizerischen Landeshydrographie, Bd. 1, Zürich, 358 pp

Messerli B, Zurbuchen M (1968) Blockgletscher im Weissmies und Aletsch und ihre photogrammetrische Kartierung. Die Alpen 3:1–13

PERMOS (2013) Permafrost in Switzerland 2008/2009 and 2009/2010. Glaciological Report (Permafrost) No. 10/11 Department of Geography, University of Zurich, 80 pp

Steck A (1968) Junge Bruchsysteme in den Zentralalpen. Eclogae Geol Helv 61(2):387–393

Steck A (1983) Geologie der Aletschregion (VS). Bull Murithienne 101:135–154

Strozzi T, Delaloye R, Raetzo H, Wegmüller U (2009) Radar interferometric observations of destabilized rockglaciers. In: Proceeding of the FRINGE 2009 Workshop, 30 November–4 December 2009, Frascati, Italy, 5 pp

Strozzi T, Delaloye R, Kääb A, Ambrosi Ch, Perruchoud E, Urs Wegmüller (2010) Combined observations of rock mass movements using satellite SAR interferometry, differential GPS, airborne digital photogrammetry, and airborne photography interpretation. J Geophys Res 115:F01014. http://dx.doi.org/10.1029/2009JF001311

Zumbühl HJ, Holzhauser H (1988) Alpengletscher in der Kleinen Eiszeit. Sonderheft zum 125jährigen Jubiläum des SAC. Die Alpen 67(3):129–322

Zumbühl HJ, Holzhauser H (2007) Annäherung an 3500 Jahre Gletschergeschichte. In: Wallner A, Bäschlin E, Grosjean M, Labhart T, Schüpbach U und Wiesmann U (eds) Welt der Alpen—Erbe der Welt. UNESCO Welterbe-Region Jungfrau–Aletsch–Bietschhorn. Haupt, Bern, pp 47–72

Zumbühl HJ, Nussbaumer SU, Wipf A (this volume) Top of Europe: the Finsteraarhorn-Jungfrau glacier landscape

Hanspeter Holzhauser was a Research Assistant and Lecturer at the Department of Geography of the University of Zurich, Research Unit Physical Geography. His main scientific interests are the Holocene history of selected Alpine glaciers (Great Aletsch Glacier, Gorner Glacier, Rhone Glacier and Lower Grindelwald Glacier). To reconstruct Holocene glacier fluctuations, he used the historical, glacio-archaeological, glacio-morphological methods including dendrochronology. He collaborates with the UNESCO World Heritage Swiss Alps Jungfrau-Aletsch and the World Nature Forum (WNF) in Naters, leads excursions and wrote scientific texts for brochures and the internet platform myswissalps.ch. He also wrote a book on the history of the Gorner Glacier and is co-author and editor of the book *Die Grindelwaldgletscher—Kunst und Wissenschaft.*

15 Top of Europe: The Finsteraarhorn–Jungfrau Glacier Landscape

Heinz J. Zumbühl, Samuel U. Nussbaumer, and Andreas Wipf

Abstract

The Finsteraarhorn, the highest peak in the Bernese Alps, and the Jungfrau, renowned for its cog railway that attracts a high number of tourists each year, are together in the heart of a high mountain glacier landscape. The Unteraar Glacier with an east-oriented and extensively debris-covered tongue has, since the eighteen/nineteenth century, been the cradle of glacier research (e.g. L. Agassiz). In turn, Lower Grindelwald Glacier became historically the best-documented Swiss valley glacier, thanks to its accessible, low-altitude ice-front position. A wealth of high-quality depictions by top artists (e.g. C. Wolf and S. Birmann) have allowed the reconstruction of the Little Ice Age (LIA) glacier fluctuations in a uniquely precise way. The Upper Lauterbrunnen Valley, dominated on both sides by huge steep rock walls with a great number of waterfalls, hosts smaller glaciers and a collection of moraines in the valley bottom. Since the end of the LIA, all the glaciers have been melting back, with a dramatic increase in recent years. The Lower Grindelwald Glacier, for instance, shows a reduction of the ice volume by 50 % since the end of the LIA. By the end of the twenty-first century, the Finsteraarhorn–Jungfrau landscape will no longer exist in the form it has been renowned for over the last centuries.

Keywords

Bernese Oberland • Glaciers • Glacier evolution • High-mountain morphodynamics • Holocene • Little Ice Age (LIA)

15.1 Introduction

The Bernese Alps, located in the western part of Switzerland, are a subrange of the Alps. On the northern slope of the Bernese Alps and situated between the lakes of Thun and Brienz, the tourist renowned town Interlaken (46°41'N, 7° 52'E) welcomes visitors from all over the world to discover gentle pastures, high mountain areas and precipitous peaks which constitute the highest of the Bernese Alps, reaching up to 4274 m a.s.l. (Finsteraarhorn). The Finsteraarhorn is the top of the catchment of Unteraar Glacier, the source of the Aare River, which drains into the lakes of Brienz and Thun. Interlaken is also the gate to the Jungfrau region, an emblematic Alpine landscape surmounted by the famous mountain peaks of Eiger (3970 m a.s.l.), Mönch (4107 m a.s.l.) and Jungfrau (4158 m a.s.l.). Together with Breithorn (3780 m a.s.l), these peaks form the edge of Lauterbrunnen Valley through which the Weisse Lütschine River flows. Further to the east, the Eiger, Schreckhorn and Wetterhorn delimit the valley of Grindelwald, headwater of the Schwarze Lütschine River. The high-Alpine areas of the Bernese Alps are accredited UNESCO World Heritage site status since 2001, the first site of the Alps given this title.

In this chapter, we aim to highlight remarkable geomorphological forms and their evolution over the last millennia, with a particular focus on glaciated landscapes. We selected three catchment areas with exceptionally rich geomorphological and historical documentation: (1) the basin of the Unteraar Glacier, (2) the basin of the Lower Grindelwald Glacier and (3) the Upper Lauterbrunnen Valley featuring

H. J. Zumbühl (✉)
Institute of Geography, University of Bern, Hallerstrasse 12, 3012 Bern, Switzerland
e-mail: zumbuehl@giub.unibe.ch

S. U. Nussbaumer
Department of Geography, University of Zurich, Winterthurerstrasse 190, 8057 Zurich, Switzerland
e-mail: samuel.nussbaumer@geo.uzh.ch

A. Wipf
Swiss Hiking Association, Monbijoustrasse 61, 3000 Bern, Switzerland
e-mail: info@andreaswipf.ch

E. Reynard (ed.), *Landscapes and Landforms of Switzerland*, World Geomorphological Landscapes,
https://doi.org/10.1007/978-3-030-43203-4_15

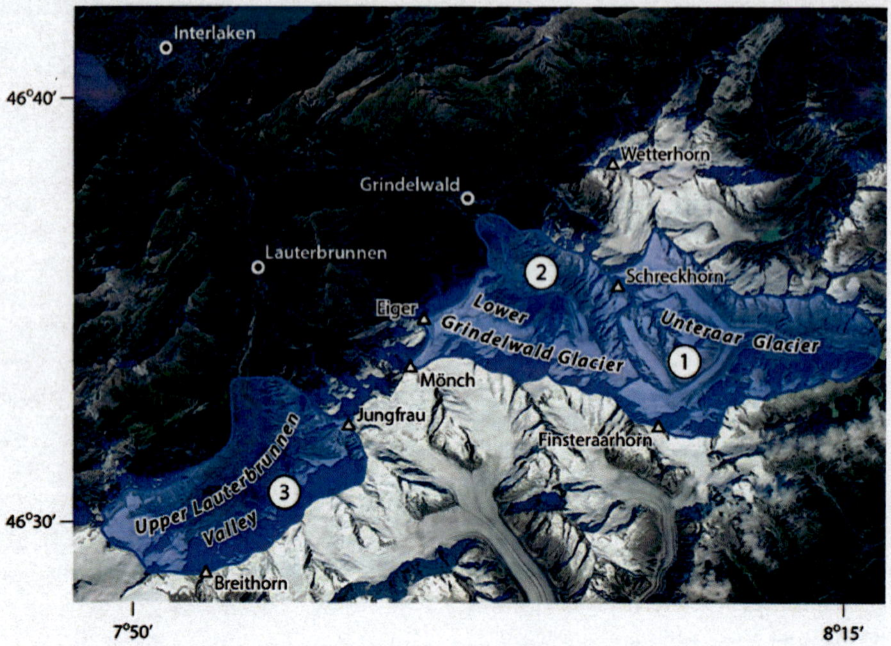

Fig. 15.1 Overview of the Finsteraarhorn–Jungfrau glacier landscape with the three catchments covered in this chapter (satellite image: Landsat 5, taken on 25 September 2013, resolution 15 m; image from USGS)

Tschingelfirn and Breithorn/Schmadri Glacier. All glaciers are located within a few kilometres of each other (Fig. 15.1).

15.2 Geological and Climatic Setting

15.2.1 Geology

The catchment areas of the three glacial regions are mainly situated in the Aar Massif (Fig. 15.2). It consists of the crystalline basement, the Innertkirchner–Lauterbrunner crystalline and the Central Aare granite (Variscan intrusion from 300 million years ago). The crystalline basement consists of gneisses and amphibolites which are highly metamorphosed, incurred during the Caledonian orogeny from older rocks. They constitute the Finsteraarhorn, the highest peak of the Bernese Alps, and the summit area of Mönch (Labhart 2007). The topmost catchment area of the Unteraar Glacier with the peaks of Schreckhorn and Lauteraarhorn consists of crystalline basement too; the glacier tongue is in the region of the Central Aare granite, a bright, granular and mostly massive rock. The glacially rounded granite forms stand out in the Grimsel region and are quite different from the overlying splintery, weathered and exposed ridges (above the trimlines from the last Ice Age). During the Alpine orogeny, fissures along faults were exposed in which the world-famous quartz crystals at Zinggenstock were found.

The Innertkirchner–Lauterbrunner crystalline is a complex intrusion and migmatite zone, which consists of a granitic matrix with inclusions of older rocks (by partial melting of the crystalline basement), and forms the northern edge of the Aar Massif (Labhart 2007). Particularly beautiful outcrops (with gneisses, marbles and amphibolites) are found in the Upper Lauterbrunnen Valley in the forefield of Tschingelfirn and Wetterlücken Glacier.

In the Upper Lauterbrunnen Valley, the Innertkirchner–Lauterbrunner crystalline rises from the valley floor to the south. Above it are the autochthonous sediments, with a prominent ocher coloured Triassic dolomite band (Röti dolomite) which can be traced through the north faces of Gletscherhorn (south of Jungfrau) to Breithorn, and north of the edge of Wetterlücken Glacier to the north side of the valley, until it re-emerges under the valley floor.

Above the autochthonous sediments, on the south side of the valley, the Doldenhorn nappe with crystalline and sedimentary rocks follows. The crystalline core of the Doldenhorn nappe forms the highest peaks from Jungfrau (Fig. 15.2) to Mutthorn. Together with the Helvetic sediments, this crystalline unit was moved to the north and partially folded during the Alpine orogeny in the Tertiary. In the region of Grindelwald, the Autochthonous was tilted and erected, forming the steep walls of Wetterhorn (Pfiffner 2014). Also, Eiger and Engelhörner north of Wetterhorn (popular with rock climbers) consist of steeply inclined

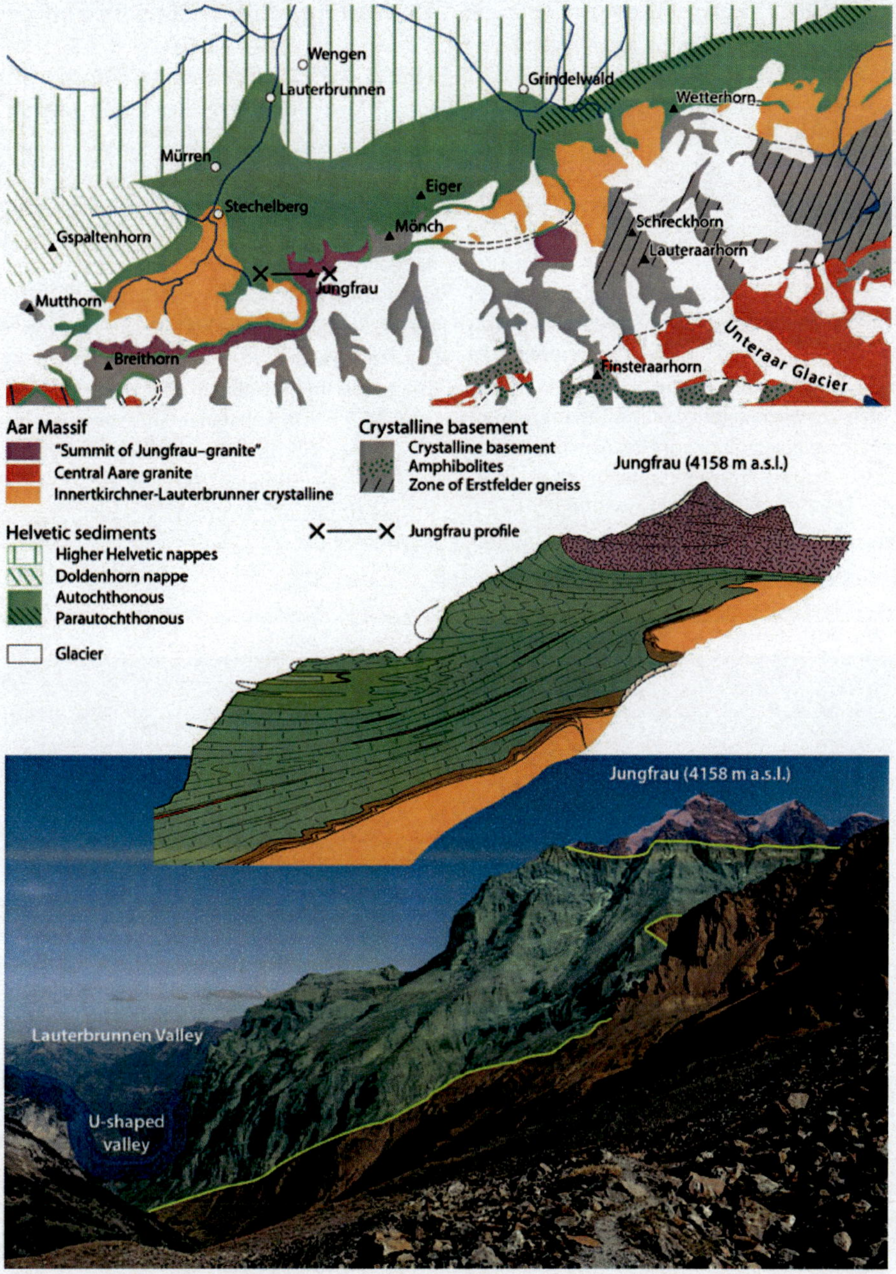

Fig. 15.2 Geological-tectonic map of the study area and profile through the Upper Lauterbrunnen Valley showing the U-shaped valley (modified after Collet and Paréjas 1931, and Labhart 2007; photo A. Wipf)

Malm limestone. As a result of the mostly steep north walls, no larger ice bodies could form except for the two Grindelwald glaciers. A great number of steep, often jagged mountain glaciers and ice aprons that periodically show ice avalanches and icefall have evolved (e.g. Eiger hanging glacier, Gutz Glacier and Hochfirn).

15.2.2 Climate

The Bernese Alps are one of the main European catchments, separating the catchment area of the Aare River (draining into the North Sea via the Rhine River) from that of the Rhone River (which flows into the Mediterranean Sea).

Situated between 46° and 47°N, the climate of the Bernese Alps is of temperate character, typical for the southern side of the extratropical westerlies. It constitutes a border between the Mediterranean and the North Atlantic climate zone (Wanner et al. 1997). The northern part of the Bernese Alps, exposed to the westerlies, shows maximum precipitation during summer, with a quite low variability, and cloudiness is higher in the north on average. This results in an increase in the height of the mean glacier elevation from about 2500 m a.s.l. in the northern part of the Bernese Alps to 2900 m a.s.l. in the south. Because of the high precipitation (locally exceeding 4000 mm per year), the altitude of the glacier equilibrium line (ELA) is relatively low in the Bernese Alps, which are a mountain range showing one of the highest degrees of glacierization in the Alps. Both the glacier with the lowest front (Lower Grindelwald Glacier) and the largest glacier of the European Alps (Great Aletsch Glacier; see Holzhauser this volume) are situated within this region.

15.3 Remarkable Glaciers and Landscapes

15.3.1 Unteraar Glacier: The Beginning of Modern Glacier Research

The Unteraar Glacier (46°35’N, 8°15’E) is the third largest valley glacier in the Bernese Alps, 11.8 km long and covering 22.5 km^2 (data from 2009; Fischer et al. 2014). A prominent feature is its large ice-cored medial moraine and extensive debris cover, typically 5–15 cm thick (Bauder 2001; Schuler et al. 2004). The tongue of the Unteraar Glacier, oriented to the east, is formed by the two main tributaries, the Lauteraar Glacier and the Finsteraar Glacier (Fig. 15.3). Mass balance measurements indicate that the present equilibrium line altitude of Unteraar Glacier is at 2850 m a.s.l. (Bauder 2001). The present glacier terminus, 1.5 km from Lake Grimsel, is at an elevation of 1930 m a.s.l. (2009).

Fig. 15.3 View of Unteraar Glacier (foreground) with its tributaries Lauteraar Glacier (right) and Finsteraar Glacier (left), with the Finsteraarhorn towering over the surrounding peaks named after famous researchers (photo A. Wipf, 29 July 2009)

Glacier Evolution During the Holocene

The Holocene history of the Unteraar Glacier involves twelve phases of retreat of different durations (100–900 years) which have been reconstructed with the help of fossil trees in the time period from 9850 cal. yBP to 1200 cal. yBP. In the last 3500 years, there are four phases of reduced ice extension as follows: 4300–3400 cal. yBP; 2800–2700 cal. yBP; 2150–1850 cal. yBP and 1400–1200 cal. yBP (cal. yBP = calibrated radiocarbon years before "today", i.e. before 1950; Joerin et al. 2006).

Glacier and Research History in the Eighteen/Nineteenth Century

The "wahre Aarequelle" (true source of the Aare River) and the ice of the Unteraar Glacier were first mentioned in a written document by the Bernese town-doctor Thomas Schoepf in connection to his excellent Map of the Canton of Berne published in 1577/78, but precise views were first painted by the famous artist of the Alps Caspar Wolf (1735–1783, see below). In the years 1774–1777 Wolf visited the Unteraar Glacier where he created several oil sketches and three oil paintings in the area of the "Abschwung", which were not only very precise, but also considered artistically as the most impactful views which were ever created from this glacier. On his travels, the painter was accompanied by either Abraham Wagner, his patron, or Jakob Samuel Wyttenbach (1748–1830), a parson and one of the most important naturalists of that time in Berne. These field companions had a direct impact on the content of the paintings of Wolf. In the time when Wolf was making drawings or paintings, Wyttenbach was conducting topographical, geomorphological and meteorological observations. On the painting "Finsteraar Glacier with view to the Finsteraarhorn", the painter's perspective from the middle moraine in the area of the "Abschwung" gives us a fantastic view of the high mountain landscape. The sense of scale is provided impressively by three tiny human figures who are standing, gesturing, and appearing nearly lost in the middle of this great, white glacier landscape (Fig. 15.4).

Fig. 15.4 C. Wolf: 1774 Finsteraar Glacier with view to the Finsteraarhorn; oil on canvas, 54 x 76 cm, signed "CWolf.1774", Kunstmuseum Basel (providing the photo)

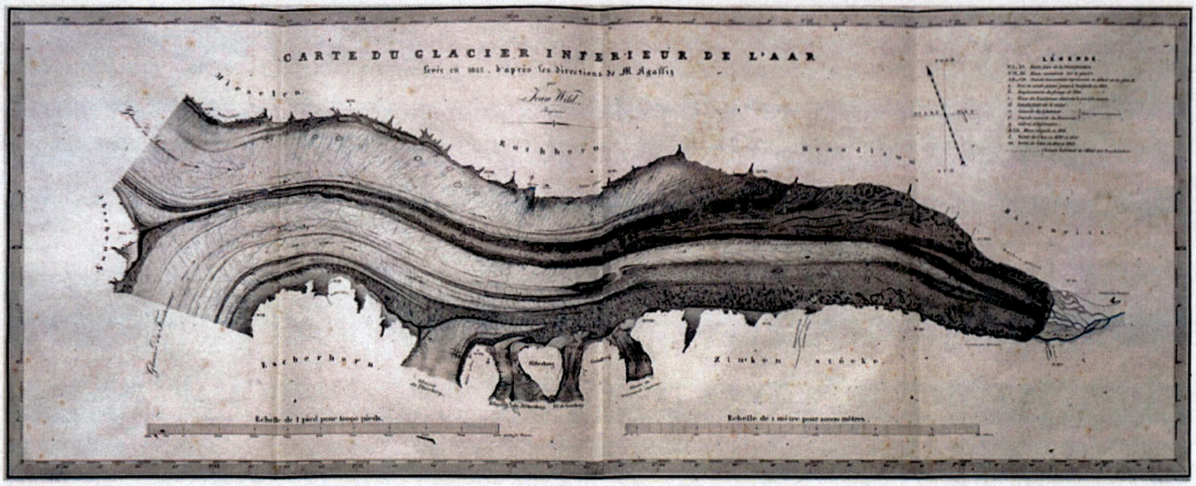

Fig. 15.5 J. Wild: 1842 map of the Unteraar Glacier, scale 1:10 000; lithograph, published by L. Agassiz 1847; private collection (photo H. J. Zumbühl)

Systematic observations on the Unteraar Glacier began with the fieldwork of the naturalist Franz Josef Hugi (1796–1855) between 1827 and 1831. He made the first observations on the surface velocity of the glacier (Hugi 1830). When his successor, Louis Agassiz (1807–1873), visited the glacier in 1839, he found to his surprise that Hugi's hut on the medial moraine (it was the same place which Wolf and Wyttenbach had visited 65 years earlier) had moved since 1827, an important indication of glacier movement (Agassiz 1847). Agassiz's observations were aimed at supporting his theory of Ice Ages. Between 1840 and 1845, Agassiz conducted a research programme on the glacier that constituted the beginning of modern experimental glaciology. He initially acted as a leader and programme manager of his interdisciplinary team and was also responsible for climatic data. His secretary and friend, the naturalist Jean Édouard Desor, conducted glaciological and geomorphological research. The results of their numerous observations were comprehensively documented by Agassiz (1847). Besides these extensive glaciological studies, the artists and engineers on the team produced the first outstanding topographically accurate representations of the glacier, among them are the panorama of Jacques Bourckhardt and the glacier maps of Johannes Wild and Johann Rudolf Stengel. A huge metamorphic boulder, the "Hôtel des Neuchâtelois" and a tent (20 m long and 5 m high) served Agassiz and his team as accommodation and as a shelter used for sleeping, drawing and studying during the summer field seasons.

The Unteraar Glacier was the subject of the first topographic map of a glacier with scientific value, generated by Wild in 1842. The lithography (scale: 1:10 000), published by Agassiz (1847), shows the tongue of the glacier, which was more than 8 km long, east of the confluence area and designated by a system of hachures (Fig. 15.5). At that time, the glacier ended in a steep, partially debris-covered ice front (Zumbühl and Holzhauser 1988; Glaciological Reports 1881–2014).

Glacier History in the Nineteenth/Twentieth Century until Today

A summary of the Unteraar Glacier history of the last 300 years is relatively simple: from 1720 to 1871 the ice tongue was advancing nearly continuously until a maximum of length of 14.49 km was reached. Since this time a continuous retreat of the front of the tongue has occurred until today, totalling 2.61 km (Glaciological Reports 1881–2014; Zumbühl and Holzhauser 1988; Steiner et al. 2008).

The Unteraar Glacier was also the focus of important studies in the twentieth/twenty-first century. In the years 1936–1950, several seismic field campaigns were executed. Since 1969 systematic measurements of ice motion, including wintertime observations, have been done with an automatic camera. In this context, it was possible to observe the peak ice speed during the melt season, as well as an upward-movement of 0.6 m (Iken et al. 1983). More recently, novel satellite-observation techniques and flow model analysis have been used to determine glacier mass balance. Results from 15 years of work on glacier-bed mapping by radio-echo soundings on Finsteraar, Lauteraar and Unteraar Glaciers resulted in new and greatly improved maps of the ice-thickness distribution. In contrast to the tongue of Unteraar Glacier, its two main tributaries, Lauteraar and Finsteraar Glaciers, are both thick and narrow (Bauder 2001).

Fig. 15.6 Overview of the basin of Grindelwald with the Upper (left) and Lower Grindelwald Glacier (right). Peaks in the background are from left to right: Wetterhorn, Schreckhorn, Finsteraarhorn, Fiescherhörner (photo R. Bösch, 2 August 2013)

15.3.2 Lower Grindelwald Glacier: The Historically Best-Documented Swiss Glacier

The Lower Grindelwald Glacier (46°35'N, 8°05'E) is a valley glacier 8.2 km long and covering 17.8 km^2 (data from 2012). Upper and Lower Ischmeer in the east and the Bernese Fiescher Glacier in the west join to form its tongue (Fig. 15.6). The geometry of Lower Grindelwald Glacier (as well as Unteraar Glacier) differs from that of the "model" glacier in having a wide variety of surface slopes and many basins that deliver ice to the mainstream.

Holocene Glacier History: Fossil Trees and Soils from the Lateral Moraines and the Forefield

From the lateral moraines near Stieregg and Zäsenberg, it was possible to collect samples of eleven fossil soils and 51 fossil trees, which allowed the reconstruction of the last 3000 years of glacier history (by means of radiocarbon dating and dendrochronology). Summarizing we can separate seven phases of advance (Holzhauser in Zumbühl et al. 2016): 1. 973–700 BC (end of Bronze Age); 2. 421–178 BC (first part of Iron Age); 3. 527–592 AD (from now on dendrochronology dates); 4. 823–836 AD; 5. 1088–1153 AD (confirmed also by written documents); 6. 1338 AD (wide extension confirmed by *Pinus cembra*); 7. 1600 until approximately 1640 AD, long lasting (confirmed by several written documents).

Little Ice Age Glacier Landscape

Since the eighteenth century, the Lower Grindelwald Glacier had an exceptionally low ice front position and was therefore highly accessible, allowing it to become a unique travel destination and object of study. In the Little Ice Age (LIA) the Lower Grindelwald Glacier ended in the area of the "Schopffelsen" (landmark rock terraces) and advanced several times into the valley bottom, ending quite near the village. As a result, the number of historical documents, including written texts and pictorial representations from villagers, is substantial (more than 400 images before 1900). Together with the high quality of the documents, it is possible to precisely reconstruct length fluctuations of the

Fig. 15.7 The last three advances of Lower Grindelwald Glacier in the LIA (all cut-outs of documents). Left: C. Wolf, 1777: the advancing ice front (oil on canvas, Kunstmuseum Bern); Middle: S. Birmann, 9.9.1826: the "Schweif" in the valley bottom (pencil/watercolour, Kunstmuseum Basel); Right: Bisson brothers in the summer/autumn of 1855/56: impressive ice tongue (photograph, Alpine Club Library London) (all photos H. J. Zumbühl)

Lower Grindelwald Glacier as far back as 1535, a feat that is not possible for most of the other Alpine glaciers. The Lower Grindelwald Glacier is probably the best-documented glacier in the Swiss Alps (Zumbühl 1980; Zumbühl et al. 1983; Holzhauser and Zumbühl 1996, 2003; Fig. 15.7).

The result of the LIA maximum extension (in the beginning of the seventeenth century) is a system of moraine ridges in the valley that are approximately 50 m further away from the moraines of the nineteenth century. From this LIA maximum advance, only a few relicts of moraine ridges survived the erosion in the valley bottom. Radiocarbon dating together with historical documents and further pollen analysis allowed dating the LIA maximum advance in the time of 1593–1606 (Zumbühl 1980).

The visual representations of the glacier can be summarized into three time windows: more isolated sources in the seventeenth/eighteenth century; the artistic peak in paintings and drawings between 1770 and 1835; and the development into the age of photography after 1850, which offered high quality topographic coverage.

Of the artistic efforts from 1770 to 1835, there are two names that were a lucky chance for the glacier history. Caspar Wolf (1735–1783) produced 62 glacier views (a quarter of it from the two Grindelwald ice streams) created in the years 1774–1778 (Brinkmann and Georgi 2014). Wolf was one of the courageous first artists to go into the high Alps where he painted the rocky and icy terrain on site, even during wintertime. One of the most dramatic and also most fascinating glacier views shows the steep blue-green frontal zone of the Lower Grindelwald Glacier with sculpted *séracs* (ice peaks) in the background, a typical and distinctive signature of advancing glaciers (Fig. 15.7).

A second artistic highlight in glacier iconography is made of the 100 glacier drawings (again, one quarter showing the Grindelwald glaciers) done by Samuel Birmann (1793–1847; Zumbühl 1997). Wolf's glacier paintings captured the morphological forms, the atmosphere and tended towards romanticism; on the other hand, Birmann's works strived for realism, with detailed and topographically accurate drawings. After the rapid advance of the Lower Grindelwald Glacier from 1814/15 to 1820/22 the "Schweif" (tail) reached the valley bottom (first time in 19th century). This setting is well captured in a drawing from the "Späten Boden" on 9 September 1826 with the Fiescherhorn in the background (Fig. 15.7). The tongue of the glacier in the valley bottom is dominated by the Fiescher Glacier part (right side), whilst the Ischmeer part (separated by a middle moraine) is pressed on the side of the Mettenberg (left side of the view).

In 1855/56 the Lower Grindelwald Glacier reached its greatest length in the nineteenth century for a second time. Thanks to the newly developed technique of photography, this event is well documented (Fig. 15.7).

During the LIA between 1560 and 1860 the history of the fluctuations of the Lower Grindelwald Glacier can be summarized as follows:

Fig. 15.8 Frontal area of the former Lower Grindelwald Glacier with Challi and "Schlossfelsplatte" (right), Stieregg (left), Zäsenberg and "Heisse Platte" (behind, middle); with dead ice, moraines, debris flows (due to permafrost degradation at Mettenberg), rock slips. Photocomposite of a remotely controlled light drone from 4 October 2014 (P. C. Jörg/S. U. Nussbaumer, University of Zurich)

- The long-lasting advance of approximately 1000 m, starting in 1575 and culminating in 1600, ended in the 1640s with the greatest ice extent in the LIA.
- At least six times relatively brief advances of 400–600 m occurred, which saw the formation of the "Schweif" (tail), a remarkable arm of ice in the valley bottom. Three times we discovered peak extensions (in 1778/79, 1820/22, 1855/56), and three times the advances were minor (e.g. in 1669).
- For several substantial periods (e.g. in 1748/49, 1762, 1794, 1808) the glacier ended in the area of the "Schopffelsen". Today, the glacier terminus (dead ice) has retreated approximately 1600 m back from the "Schopffelsen" at 1325 m altitude.

Glacier Evolution since the LIA

Since the end of the LIA, within approximately 160 years, the former Lower Grindelwald Glacier (1855/56: 10.89 km long) has lost 35–45 % of its length. More recently, from 2004 to 2014, the changes in the landscape have accelerated, particularly with dramatic melting since 2007. A photocomposite of a remotely controlled light drone from October 2014 gives a very detailed view of the actual state of the area of the glacier tongue (Fig. 15.8):

- Below Challi—Stieregg—Zäsenberg in the valley bottom the former ice tongue disintegrates in a chaotic mass of moraines, ice dams and dead ice (vertical ice loss since the LIA: 350 m). The dead ice is 1200 m long and ends at 1325 m a.s.l. Since 2014, there is a fracture between the active tongue below the "Heisse Platte" and the dead ice zone.
- Depending on the season one or two potentially high-risk glacier lakes of different size is formed. This required the construction of a costly emergency-tunnel (flood-outbreaks!) of controversial use.
- Below the Eiger east side (Challi), the "Schlossfelsplatte" was, over a longer time, disintegrated into several rock-slides (Oppikofer 2008).
- The huge moraine–debris cone of the Stieregg, an Alpine pasture, was strongly changed in its dimensions by

erosion and debris flows, hence the slopes became steeper and were sliding down; the Stieregg hut had to be abandoned.

In 2014 there was no former Lower Grindelwald Glacier but only two parts, the Fiescher Glacier (6.5–7.0 km long) and the Ischmeer (approximately 6.0 km long) remain.

15.3.3 Upper Lauterbrunnen Valley: Monumental Waterfalls and Moraine Cluster

Steep Walls and High Waterfalls

The Lauterbrunnen Valley has a characteristic U-shape due to the glacial imprint. The flat valley floor is flanked on both sides by limestone cliffs several hundred metres high (Fig. 15.2), followed above by the shoulders of Mürren and Wengen. During the Ice Ages, the glaciers eroded into a pre-existing, tectonically controlled valley. After the late glacial retreat landslides went down from both valley sides north of Lauterbrunnen, leading to the damming of the Weisse Lütschine River, forming a lake which subsequently filled with sediments that built up the valley floor.

The extreme differences in relief are exemplified by the gain in elevation of 3250 m in a 5 km horizontal distance between Stechelberg (910 m a.s.l.) and the summit of Jungfrau (4158 m a.s.l.). The vertical precipices of the Lauterbrunnen Valley make it a mecca for base jumpers.

Some of the highest waterfalls in Europe plunge over this impressive rock walls. The Mürrenbach Fall, as the highest waterfall in Switzerland, overcomes 417 m. The most spectacular and highest free-falling waterfall in Switzerland is the almost 300-m-high Staubbach Fall that has already been recorded pictorially in earlier centuries by artists and by Johann Wolfgang Goethe in his 1779 poem "Song of the spirits over the waters". Further south lie the Trümmelbach Falls that are deeply ingrained in the Jurassic limestone, waterfalls that are flowing completely inside the rock through several stages. They are fed mainly by glacier meltwater from the steep northern slopes below Eiger, Mönch and Jungfrau. At the far end of the Lauterbrunnen Valley, the Schmadribach Fall (Fig. 15.9) and the spraying Holdri Falls downstream are also impressive, the first being a known study object for drawings in the eighteenth and nineteenth century. A special feature is the hidden Talbach Fall. Its water falls through a hole in the rock into a water-filled tub. The Lauterbrunnen Valley is known as the Valley of the Falls due to its 72 waterfalls, which is reflected in the naming expressed ("Lauter" = a lot or clear, "brunnen" = brooks, springs).

The Upper Lauterbrunnen Valley (46°30'N, 7°52'E) is also dominated by steep cliffs, extending from Jungfrau to Breithorn (3780 m a.s.l.), with several spectacular hanging glaciers and some smaller mountain glaciers (Tschingelfirn: 5.23 km^2, Wetterlücken and Breithorn Glacier: 2.71 km^2, Vordra and Hindra Schmadri Glaciers: 0.95 and 0.6 km^2; Fischer et al. 2014; Fig. 15.10).

Holocene Glacier Advances—Variety of Moraine Deposits

During glacier advance periods, a unique moraine landscape has been created in the region of Lake Oberhorn (Fig. 15.11). On one hand, the plain between Lake Oberhorn and Oberhorn Alp favoured a lateral outflow of the glacier with deposition of moraines; on the other hand, the moraine ridges have been exceptionally well preserved in spite of the overprint of solifluction lobes. All vegetation-covered moraines document glacier extension larger than during the last maximum of 1850. Outside these moraines, a number of peat bogs have developed in troughs polished by the Ice Age glacier and are framed by *roches moutonnées*.

The basal date from a peat bog located only around 200 m outside of the outermost moraines is 10390 yBP (Wipf 2001; Fig. 15.11). The moraines in the plain of Lake Oberhorn are thus all of Holocene age. North of the Oberhorn Alp a podzol has developed on a late glacial moraine.

The age of all moraines could be determined using ^{14}C dating of fossil soils. The small, outermost Holocene moraine has an age of 4475 yBP. The two immediately subsequent moraines were formed during the Löbben cold phase (3340 yBP). These walls are not only a razor-sharp boundary morphologically, but also in terms of geology and botany: while the moraine deposits of calcareous rocks consist of an acidophobic vegetation, there are migmatites of the Lauterbrunnen crystalline outside, with an acidophilic vegetation.

At the second moraine west of Lake Oberhorn (3330 yBP) and at Schafläger (3530 yBP and 3180 yBP), the multipart Löbben cold phase could be detected twice. The first date is also a minimum age for a landslide that broke loose from the Oberhorn moraine down to Lake Oberhorn. The dimension of the landslide can be well traced, based on the crystalline deposits that are clearly different from the calcareous moraine. Further glacier maxima (or similar extents) occurred at about 2500 yBP, 2300 yBP, 1750 yBP, 1300 yBP, 1000 yBP and 750 yBP (Wipf 2001).

Glacier Evolution Since the Eighteenth Century

Since the eighteenth century painters frequented the Lauterbrunnen Valley. Among others, Caspar Wolf travelled across the Breithorn Glacier to Lake Oberhorn with Jakob

Fig. 15.9 Schmadribach Fall in the Lauterbrunnen Valley with Grosshorn (3754 m a.s.l.) (photo A. Wipf, 22 June 2008)

Samuel Wyttenbach. Samuel Birmann climbed for his paintings to altitudes exceeding 2500 m a.s.l., and even Johann Wolfgang Goethe visited Lake Oberhorn and mentioned glaciers in his letters.

Glacier extents close to LIA maximum can be recognized on the oil paintings by Caspar Wolf around 1774, and on the watercoloured pencil drawings by Samuel Birmann (1821, 1822 and 1827). In the period separating these artists' works

Fig. 15.10 The Upper Lauterbrunnen Valley from Jungfrau (left) to Breithorn (middle), with Breithorn Glacier (photo A. Wipf, 27 September 2014)

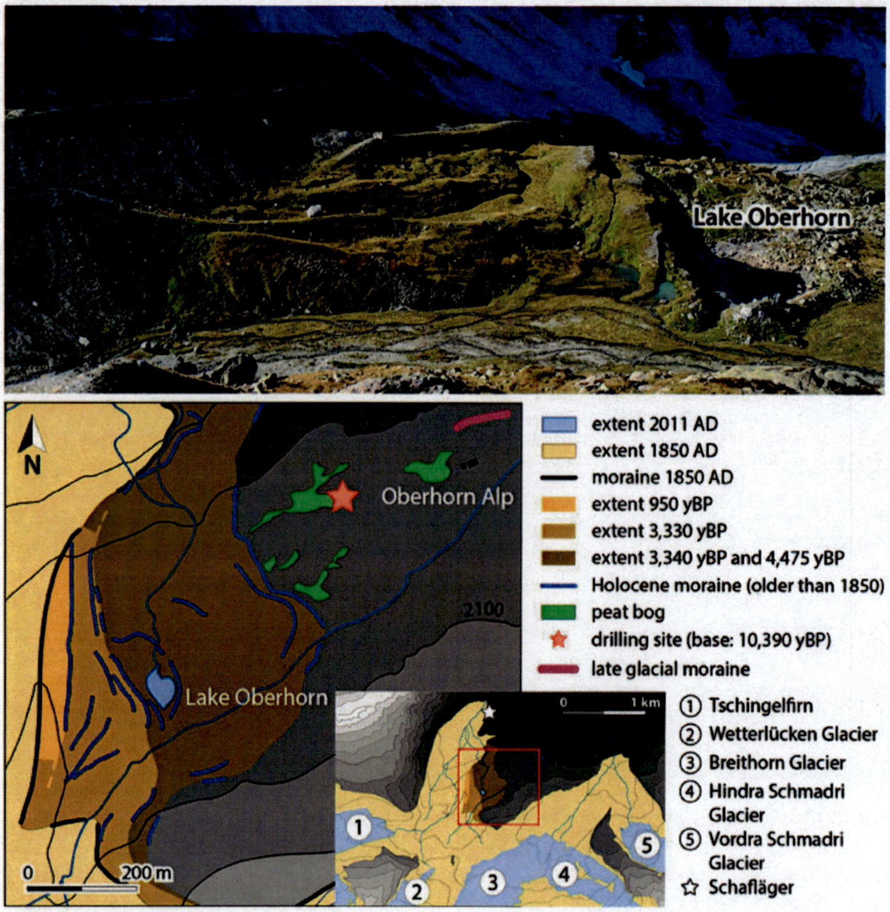

Fig. 15.11 Top: Holocene moraines at Lake Oberhorn (photo A. Wipf, 21 September 2006): 1850 ridge (left), moraines to the very right with an age of 3330 yBP; Bottom: Dated glacier maxima at Lake Oberhorn, with the largest Holocene expansion in 4475 yBP

Fig. 15.12 Evolution of the tongue of Wetterlücken Glacier from 2010 to 2018 (photos A. Wipf)

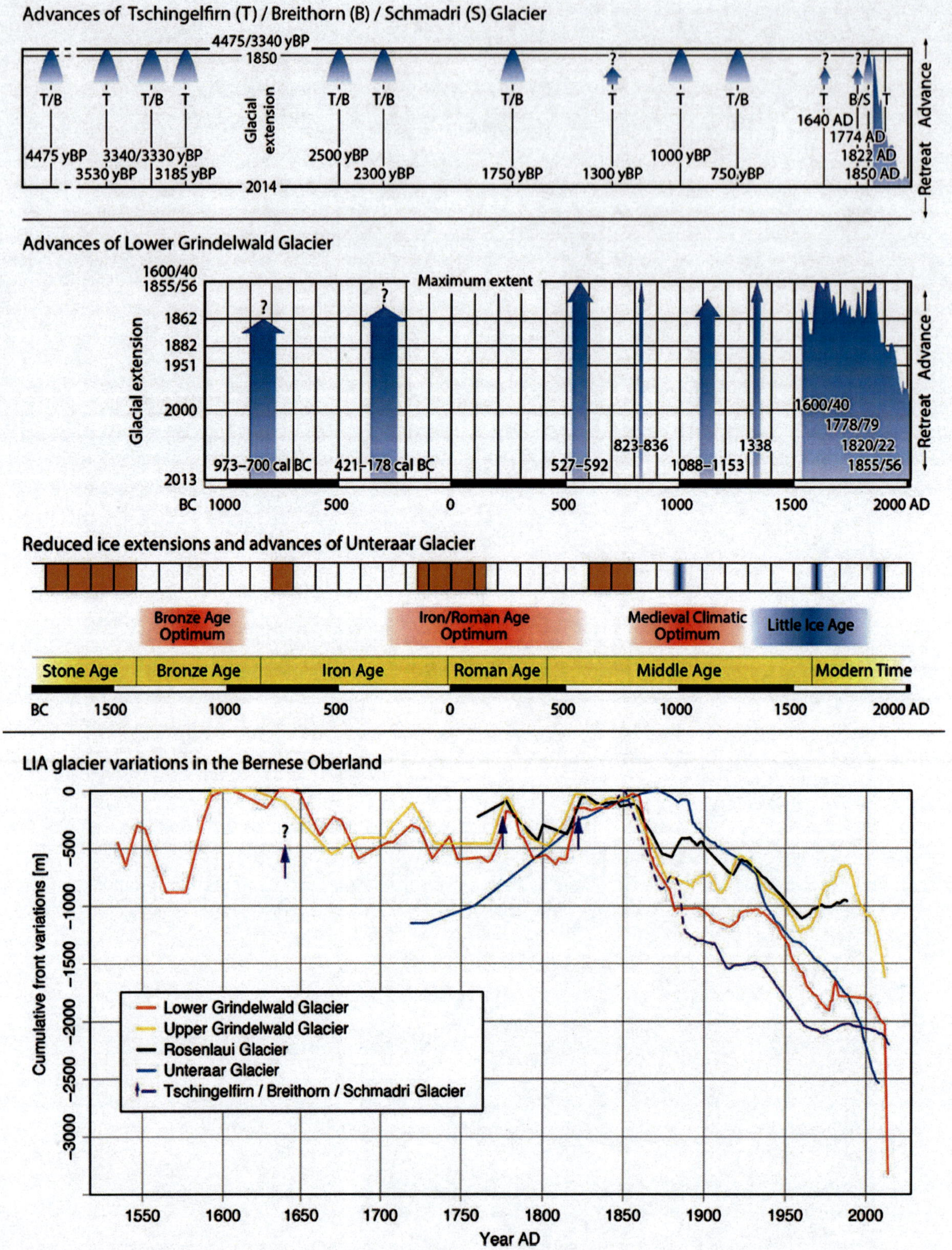

Fig. 15.13 Overview of Holocene glacier fluctuations in the three study areas: Tschingelfirn (radiocarbon dating of fossil soils; Wipf 2001); Lower Grindelwald Glacier (mainly dendrochronological evidence; Holzhauser et al. 2005); Unteraar Glacier (calibrated radiocarbon dating of fossil wood samples; periods of reduced glacier extension are given in brown; Joerin et al. 2006). Detailed reconstructions based on historical evidence are available for the glaciers shown, including nearby Upper Grindelwald and Rosenlaui glaciers (Zumbühl 1980; Zumbühl and Holzhauser 1988; Wipf 2001; Glaciological Reports 1881–2014; Zumbühl et al. 2016)

the glaciers retreated, as can be seen from renditions by Joseph Anton Koch (1794) and the Lory brothers (shortly before 1816).

The last advance of the LIA is captured on the "Messtischblatt" of Johann Rudolf Stengel from 1850/51. Moraines hardly covered by vegetation make this former glacier extent easily distinguishable today. At that time, the glaciers mentioned above were still connected to each other and covered approximately 15 km^2 (Wipf 1999). Since 1850, 5.4 km^2 or about 36 % of the glacial ice has melted away (Wipf 1999; Fischer et al. 2014). The overall loss was interrupted by three short re-advances around 1880/90, 1920/30 and in the 1980s, during which up to one to three moraine crests were deposited in all glacier forefields. The last two short advance periods are also reflected in the surveys that have been conducted on Tschingelfirn since 1893.

Since the late 1980s, Vordra Schmadri Glacier has lost the foremost part of its tongue through the melt-out of a rock step, the tongues of Breithorn and Hindra Schmadri Glaciers are highly down-wasted and covered by a dense supraglacial moraine, and length change measurements on Tschingelfirn constantly show negative values. Today, walls prone to rockfall remain where ice-crusted cliffs were encountered years ago. The retreat of Wetterlücken Glacier is given in Fig. 15.12.

In the former confluence zone of Wetterlücken Glacier and Tschingelfirn, there are beautiful exposures of glacial striations and *roches moutonnées*. They show instructive rock inclusions (marbles, etc.) and even a fossil coral reef can be admired.

15.4 Conclusions and Outlook

The glaciers discussed in this chapter dominate the study area not only because of their dimensions, but they are also prominent for the role they played in the history of science and physical geography. The discovery of the landscape goes hand in hand with the (beginning of) geomorphological and glaciological research. Pioneering studies in the eighteenth century by Jakob Samuel Wyttenbach, accompanied by painter Caspar Wolf, have been undertaken in all three study areas, and give us a holistic view of the glacier landscape at that time.

The summary of the (late) Holocene glacier history shows a coherent picture with glacier advances documented in all three study sites (Fig. 15.13). Due to its large debris cover, Unteraar Glacier reacts slowly to climatic changes and only shows low-frequency frontal variations. A wealth of fossil soils and dendrochronologically dated fossil trees in the lateral moraines document several pre-LIA advances of Lower Grindelwald Glacier. Its fluctuations over the past centuries are exceptionally well documented thanks to historical accounts. The forefield of Tschingelfirn and Breithorn Glacier reveals Holocene glacier variations as documented by a cluster of lateral moraines of various ages (outside the LIA margin).

In the nineteenth century, the ice and snow landscape with its impressive summits attracted interest of the first alpinists, resulting in the first ascents of Jungfrau in 1811, and Finsteraarhorn, probably in 1812. At that time it was not only a mountaineering hobby, but also scientific curiosity that justified the arduous travels in the high mountains. Several mountain peaks named after important scientists are proof of the pioneering spirit at that time (e.g. Agassizhorn, Hugihorn/Hugisattel, Scheuchzerhorn, Escherhorn, Grunerhorn, Studerhorn, Desorstock, Altmann; cf. Figure 15.3). A century later, the inauguration of the Jungfraujoch railway station, Europe's highest railway station at 3454 m a.s.l., opened the mountains for the broader public. Offering a gateway to the entire region, this tourist highlight has not lost its attractiveness despite the general glacier retreat. The easy access to a high-alpine site was the prerequisite for the foundation of the High Altitude Research Station Jungfraujoch, which allows a wealth of scientific studies and projects profiting from this unique spot (high elevation, clear atmospheric conditions, low disturbances).

The region is intensively used by tourism, and the expected further rise in temperature and climatic conditions will force the glacier landscape to change at a rapid pace. As a follow-up, natural hazards are expected to increase, as the example of the formation of a glacial lake at Lower Grindelwald Glacier has shown. An integrative hazard analysis is required to cope with the various geomorphological hazards.

The future evolution of Unteraar Glacier has been assessed for the period 2005–2050 using a flowline model by Huss et al. (2007). The model predicts a retreat of the glacier terminus of 800–1025 m by 2035, and of 1250–2300 m by 2050. The debris cover of the glacier tongue reduces the retreat rate by a factor of three. The thinning rate would increase by 50–183 % by 2050 depending on the scenario applied, compared to the period 1997–2005.

Modelled retreat of the Lower Grindelwald Glacier (Huss in Zumbühl et al. 2016) shows a reduction of the current ice volume by 50 % since the end of the LIA. The glacier tongue is expected to retreat further, and both tributaries, the Fiescher Glacier and the Ischmeer, will retreat into their respective firn areas in 2050. More dramatic will be the situation towards the end of the twenty-first century, when the model predicts complete disintegration of the glacier. Only in the uppermost locations, some remnants of ice are expected to survive, especially below Mönch next to the station "Eismeer" of the Jungfrau railway. The Lower Grindelwald Glacier and the Finsteraarhorn–Jungfrau glacier landscape will not exist anymore in the form for which it has been known over the past 500 years (or even longer!).

Acknowledgments Our big thanks go to Laura Thomson who carefully checked the English. Comments by Philip Deline, Emmanuel Reynard and Piotr Migoń gave valuable input and are very much appreciated. We also thank Philipp Rastner for processing and providing the satellite image of the study area. Philip C. Jörg and Johann Kaufmann are acknowledged for great help with fieldwork at Lower Grindelwald Glacier, and Thomas Burri, Naturhistorisches Museum Bern, for the geological consulting.

References

Agassiz L (Guyot, A/Desor E) (1847) Système glaciaire ou recherches sur les glaciers, leur mécanisme, leur ancienne extension et le rôle qu'ils ont joué dans l'histoire de la terre. Première partie: Nouvelles études et expériences sur les glaciers actuels, leur structure, leur progression et leur action physique sur le sol. V. Masson, Paris, L. Voss, Leipzig, 2 vol

Bauder A (2001) Bestimmung der Massenbilanz von Gletschern mit Fernerkundungsmethoden und Fliessmodellierungen. Eine Sensitivitätsstudie auf dem Unteraargletscher. Versuchsanstalt für Wasserbau, Hydrologie und Glaziologie (VAW) der Eidgenössi-schen Technischen Hochschule Zürich (ETHZ), Mitteilungen No. 169, 171 pp

Brinkmann B, Georgi K (2014) Caspar Wolf und die ästhetische Eroberung der Natur. Katalog. Ausstellung Kunstmuseum Basel, 19.10.2014–01.02.2015. Hatje Cantz, Ostfildern, 231 pp

Collet LW, Paréjas E (1931) Chaîne de la Jungfrau. Matériaux pour la carte géologique de la Suisse, no 63. Francke, Bern, 64 pp

Fischer M, Huss M, Barboux C, Hoelzle M (2014) The new Swiss Glacier Inventory SGI2010: relevance of using high-resolution source data in areas dominated by very small glaciers. Arct Antarct Alp Res 46:933–945

Glaciological Reports (1881–2014) The Swiss Glaciers. Yearbooks of the Cryospheric Commission of the Swiss Academy of Sciences (SCNAT), published since 1964 by the Laboratory of Hydraulics, Hydrology and Glaciology (VAW) of ETH Zürich, No. 1–130, https://www.glamos.ch/

Holzhauser H (this volume) The Aletsch region with the majestic Grosser Aletschgletscher

Holzhauser H, Magny M, Zumbühl HJ (2005) Glacier and lake-level variations in west-central Europe over the last 3500 years. The Holocene 15:789–801

Holzhauser H, Zumbühl HJ (1996) To the history of the Lower Grindelwald Glacier during the last 2800 years – palaeosols, fossil wood and historical pictorial records – new results. Zeitschrift für Geomorphologie N. F. Suppl. 104:95–127

Holzhauser H, Zumbühl HJ (2003) Nacheiszeitliche Gletscherschwankungen. In: Weingartner R, Spreafico M (eds) Hydrologi-scher Atlas der Schweiz. Special edition for the 54th "Deutscher Geographentag" in Berne. Bundesamt für Landestopographie, Bern-Wabern, Tafel 3.8

Hugi FJ (1830) Naturhistorische Alpenreise. Amiet-Lutiger, Solothurn, 378 pp

Huss M, Sugiyama S, Bauder A, Funk M (2007) Retreat scenarios of Unteraargletscher, Switzerland, using a combined ice-flow mass-balance model. Arct Antarct Alp Res 39:422–431

Iken A, Röthlisberger H, Flotron A, Haeberli W (1983) The uplift of Unteraargletscher at the beginning of the melt season – a consequence of water storage at the bed? J Glaciol 29:28–47

Joerin UE, Stocker TF, Schlüchter C (2006) Multicentury glacier fluctuations in the Swiss Alps during the Holocene. The Holocene 16:697–704

Labhart T (2007) Geologie – 500 Millionen Jahre Erdgeschichte. In: Wallner A, Bäschlin E, Grosjean M, Labhart T, Schüpbach U, Wiesmann U (eds) Welt der Alpen – Erbe der Welt. UNESCO Welterbe-Region Jungfrau–Aletsch–Bietschhorn. Jahrbuch der Geographischen Gesellschaft Bern, Band 62. Haupt, Bern/Stuttgart/Wien, pp 23–46

Oppikofer T, Jaboyedoff M, Keusen H-R (2008) Collapse at the eastern Eiger flank in the Swiss Alps. Nat Geosci 1:531–535

Pfiffner OA (2014) Geology of the Alps. Wiley-Blackwell, Chichester, 376 pp

Schuler T, Fischer UH, Gudmundsson GH (2004) Diurnal variability of subglacial drainage conditions as revealed by tracer experiments. J Geophys Res 109:F02008. https://doi.org/10.1029/2003JF000082

Steiner D, Zumbühl HJ, Bauder A (2008) Two Alpine glaciers over the past two centuries: a scientific view based on pictorial sources. In: Orlove B, Wiegandt E, Luckman BH (eds) Darkening peaks: glacier retreat, science, and society. University of California Press, Berkeley, pp 83–99

Wanner H, Rickli R, Salvisberg E, Schmutz C, Schüepp M (1997) Global climate change and variability and its influence on Alpine climate – concepts and observations. Theoret Appl Climatol 58:221–243

Wipf A (1999) Die Gletscher der Berner, Waadtländer und nördlichen Walliser Alpen. Eine regionale Studie über die Vergletscherung im Zeitraum "Vergangenheit" (Hochstand von 1850), "Gegenwart" (Ausdehnung im Jahr 1973) und "Zukunft" (Gletscherschwund-Szenarien, 21. Jhdt.). Physische Geographie, Vol. 40. Geographi-sches Institut der Universität Zürich, 295 pp

Wipf A (2001) Gletschergeschichtliche Untersuchungen im spät- und postglazialen Bereich des Hinteren Lauterbrunnentals (Berner Oberland, Schweiz). Geographica Helvetica 56:133–144

Zumbühl HJ (1980) Die Schwankungen der Grindelwaldgletscher in den historischen Bild- und Schriftquellen des 12. bis 19. Jahrhunderts. Ein Beitrag zur Gletschergeschichte und Erforschung des Alpenraumes. Denkschriften der Schweizerischen Naturforschenden Gesellschaft (SNG), vol 92. Birkhäuser, Basel/Boston/Stuttgart, 279 pp

Zumbühl HJ (1997) Die Hochgebirgszeichnungen von Samuel Birmann – ihre Bedeutung für die Gletscher- und Klimageschichte. In: Katalog zur Ausstellung "Peter und Samuel Birmann – Künstler, Sammler, Händler, Stifter" des Kunstmuseums Basel vom 27.09.1997–11.01.1998. Schwabe, Basel, pp 59–74

Zumbühl HJ, Holzhauser H (1988) Alpengletscher in der Kleinen Eiszeit. Sonderheft zum 125jährigen Jubiläum des SAC. Die Alpen 64:129–322

Zumbühl HJ, Messerli B, Pfister C (1983) Die Kleine Eiszeit: Gletschergeschichte im Spiegel der Kunst. Katalog zur Sonderausstellung des Schweizerischen Alpinen Museums Bern und des Gletschergarten-Museums Luzern vom 09.06.–14.08.1983 (Luzern), 24.08.–16.10.1983 (Bern), 60 pp

Zumbühl HJ, Nussbaumer SU, Holzhauser H, Wolf R (eds) (2016) Die Grindelwaldgletscher – Kunst und Wissenschaft. Haupt, Bern, 255 pp

Heinz J. Zumbühl is Emeritus Professor of Geography at the University of Bern. The dominant part of his research interest is concentrated on history and art history of glaciers, landscapes and climate especially in the Alps during the Little Ice Age, mostly in a team with other colleagues. Another part of his interest is (the teaching of) Urban Geography, especially history and public space in New Towns of Europe and Asia. To support this he visited intensively the places of his interests, e.g. Arctic (Greenland, Svalbard, North Polar Sea, Alaska), Patagonia and Antarctic Islands, in Asia especially India, Nepal/Himalaya, Tibet and New Zealand/Southern Alps.

Samuel U. Nussbaumer studied Geography, Earth Sciences and Physics at the University of Bern and finished with a PhD thesis on historical glacier fluctuations in the Alps and Scandinavia. With a fellowship for prospective researchers by the Swiss National Science Foundation, he conducted a research project on glacier variations in southern South America, with research stays in Mendoza (Argentina) and London (Canada). He is currently a Science Officer with the World Glacier Monitoring Service and a Senior Researcher at the Department of Geography of the University of Zurich.

Andreas Wipf finished his studies in Geography with a Diploma thesis about the glacier history in the Upper Lauterbrunnen Valley followed by a PhD thesis about the impact of climate change on the glaciers in the Bernese Alps since 1850 (Swiss glacier inventory) at the University of Zurich. He worked for the inventory of the glacier forefields and alpine floodplains to integrate them into the Swiss Federal Inventory of alluvial zones. He is the author of hiking books leading to the glaciers in the Swiss Alps and is an enthusiastic (panorama) photographer of high mountain areas. Today, he is a GIS specialist at the Swiss Hiking Association.

16 Rockglaciers of the Engadine

Isabelle Gärtner-Roer and Martin Hoelzle

Abstract

The Engadine is one of the rockglacier hot spots in the European Alps. Many rockglaciers in all states of activity (active, inactive, relict) are found, indicating the former and present occurrence of permafrost. This is due to continental climate conditions, high elevation, and high weathering rates. Rockglaciers are developed in valley bottoms, such as the Val Sassa and the Val da l'Acqua rockglacier, or in formerly glaciated cirques, such as the Muragl or the Murtèl rockglacier. Hence, the Engadine is the home of research on rockglaciers in Europe with the first studies on rockglaciers of the Swiss National Park one century ago. Engadine was also the first place in the world where boreholes in rockglaciers were drilled in 1987. Nowadays, several Engadine rockglaciers are monitored within the Permafrost Monitoring Network Switzerland (PERMOS).

Keywords

Rockglaciers • Mountain permafrost • PERMOS • Engadine

16.1 Introduction

Active rockglaciers are defined as "lobate or tongue-shaped bodies of perennially frozen unconsolidated material supersaturated with interstitial ice and ice lenses that move downslope or downvalley by creep" (Haeberli 1985). Due to their ice content and flow behavior, rockglaciers are classified into the following types: active, inactive, and relict. Often they form a sequence of active rockglaciers in higher altitudes to relict landforms in lower altitudes. The Engadine (Eastern Switzerland, 46°35' N, 9°58' E; Fig. 16.1) is one of the rockglacier hot spots in the European Alps. A large number of rockglaciers in all states of activity (active, inactive, relict) are found, indicating the former and present occurrence of permafrost. Continental climate conditions, high elevation (most of the valley bottoms in the Engadine are above 1800 m a.s.l.), and weathering rates allowing the build-up of large sediment deposits are prerequisites for the occurrence of rockglaciers. In the Alps, these landforms are between 200 and 800 m long and about 20-100 m thick. Typically, they develop in valley bottoms, such as the Val Sassa and the Val da l'Acqua rockglacier, or in formerly glaciated cirques, such as the Muragl or the Murtèl/Corvatsch rockglacier (see location map, Fig. 16.1).

The history of rockglacier research in Europe had its origin in the Engadine, in the—at that time—newly founded Swiss National Park (see Schlüchter et al., this volume). In 1918, Emile Chaix was the first one to pay special attention to the peculiar features he observed. He started geomorphological investigations, which were continued by André Chaix. By measuring stone lines on the rockglaciers of Val Sassa and Val da l'Acqua, André Chaix delivered the first evidence of the movement of these landforms and compared them to similar landforms in California (Chaix 1923; 1943).

A new and very important chapter of rockglacier science began in 1987, with the first deep borehole ever drilled through a rockglacier (e.g., Vonder Mühll and Haeberli 1990); this was performed on Murtèl/Corvatsch, where Dietrich Barsch had already started measurements in the 1970s (Barsch and Hell 1975). Several boreholes followed on rockglaciers in the Engadine and elsewhere and allowed for a more detailed description of permafrost temperatures (Fig. 16.2), landform composition (ice content, sediment

I. Gärtner-Roer (✉)
Department of Geography, University of Zurich, Winterthurerstrasse 190, 8057 Zurich, Switzerland
e-mail: isabelle.roer@geo.uzh.ch

M. Hoelzle
Department of Geosciences, University of Fribourg, Chemin du Musée 4, 1700 Fribourg, Switzerland
e-mail: martin.hoelzle@unifr.ch

E. Reynard (ed.), *Landscapes and Landforms of Switzerland*, World Geomorphological Landscapes,
https://doi.org/10.1007/978-3-030-43203-4_16

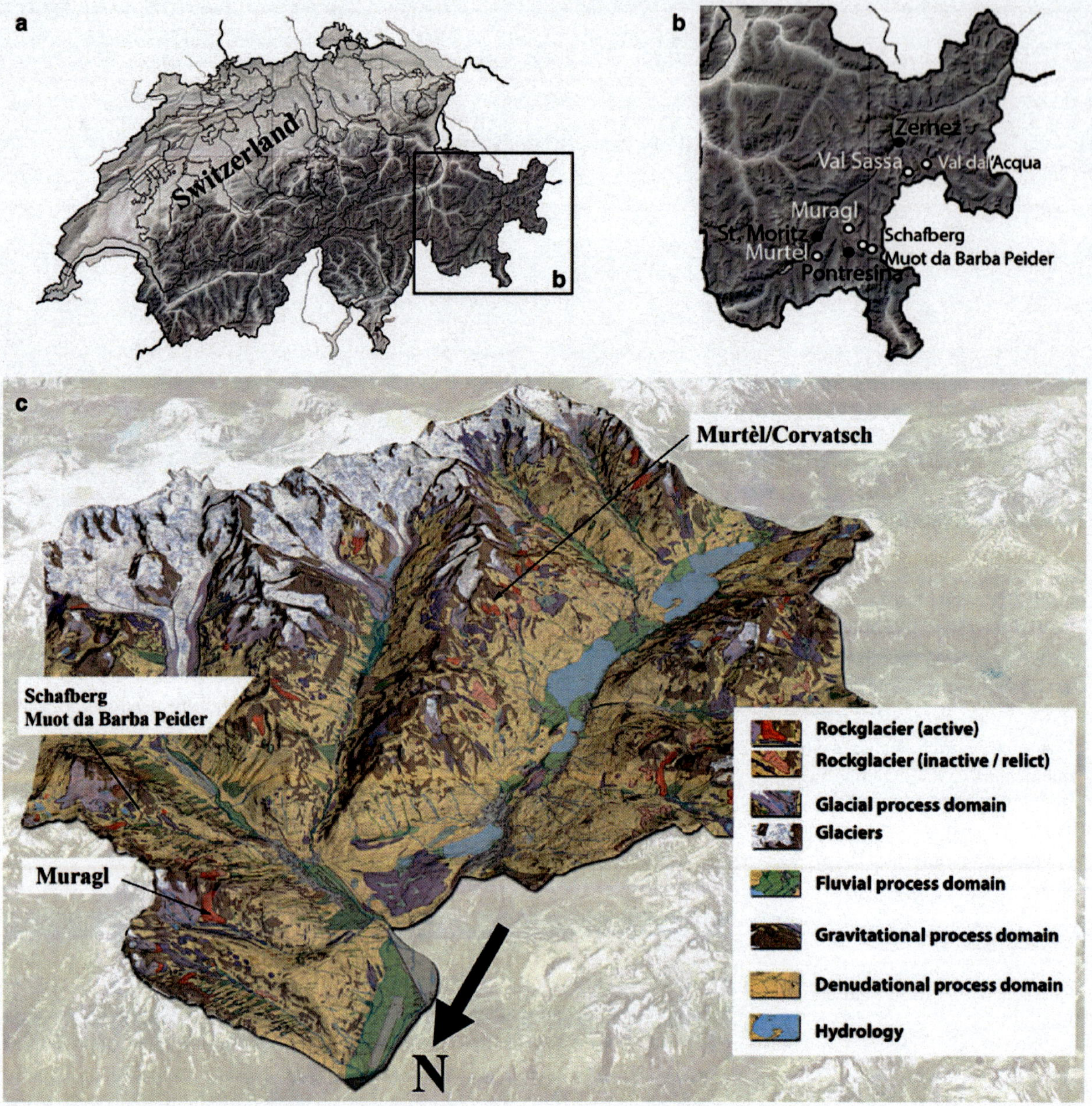

Fig. 16.1 **a** Location of Engadine region in Switzerland; **b** Main rockglacier sites (white dots) and towns (black dots); **C** Geomorphological map of the Upper Engadine with rockglaciers in different states of activity given in red colors (*source* M. Maisch and C. Levy, unpublished)

characteristics), as well as its deformation in depth (e.g. Arenson et al. 2002).

Nowadays, several rockglaciers in the Engadine are investigated and monitored by different research groups. At least three rockglaciers are part of the Permafrost Monitoring Network Switzerland (PERMOS), which was founded in 2000 and involves long-term and systematic measurements of permafrost temperature, active layer thickness, deformation and other characteristics, such as ice content. These are Murtèl/Corvatsch, Muragl, and Schafberg (see Phillips and Kenner, this volume). The borehole temperatures at Murtèl/Corvatsch and Muot da Barba Peider, a frozen talus slope,

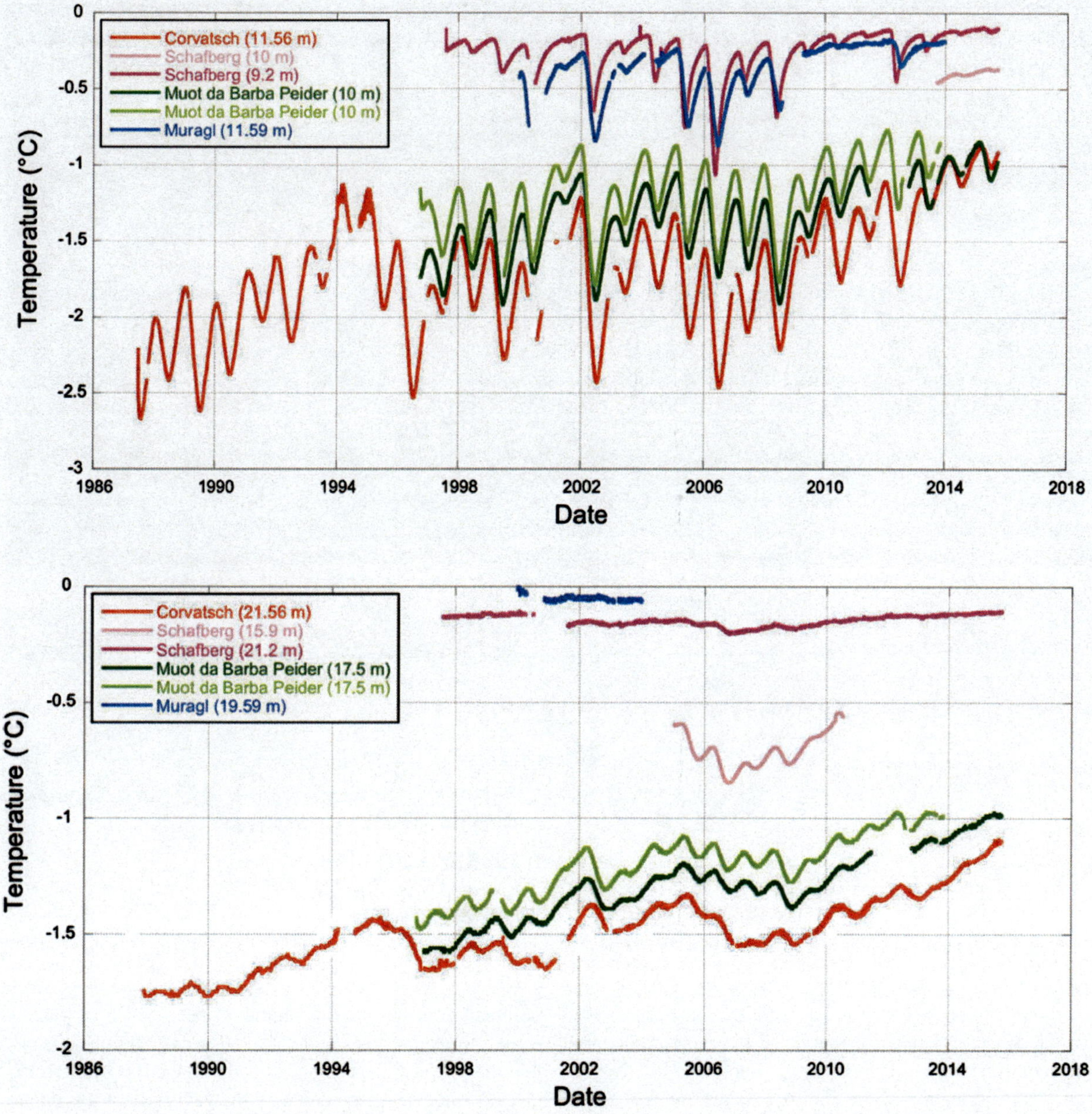

Fig. 16.2 Borehole temperatures at different depths monitored at four rockglacier sites in the Engadine (*source* PERMOS 2016,www.permos.ch)

are between −1 and −2 °C at about 20 m depth and indicate an increase of about 0.5 °C during the last 15 years (Fig. 16.2). For the sites Muragl and Schafberg, the temperatures are much higher and cooling periods are less pronounced in the last 8 years (nicely visible at about 10 m depth; Fig. 16.2). This long-term monitoring of permafrost temperatures allows for the documentation of warming trends and the assessment of landform-specific sensitivities to climatic changes.

16.2 Rockglaciers Val da l'Acqua and Val Sassa—Where It Began

The rockglaciers Val da l'Acqua and Val Sassa, situated in the core zone of the Swiss National Park (see Schlüchter et al., this volume), were the first landforms described and investigated systematically. Here, movement rates—quantified by repeated measurements of painted stone lines—were

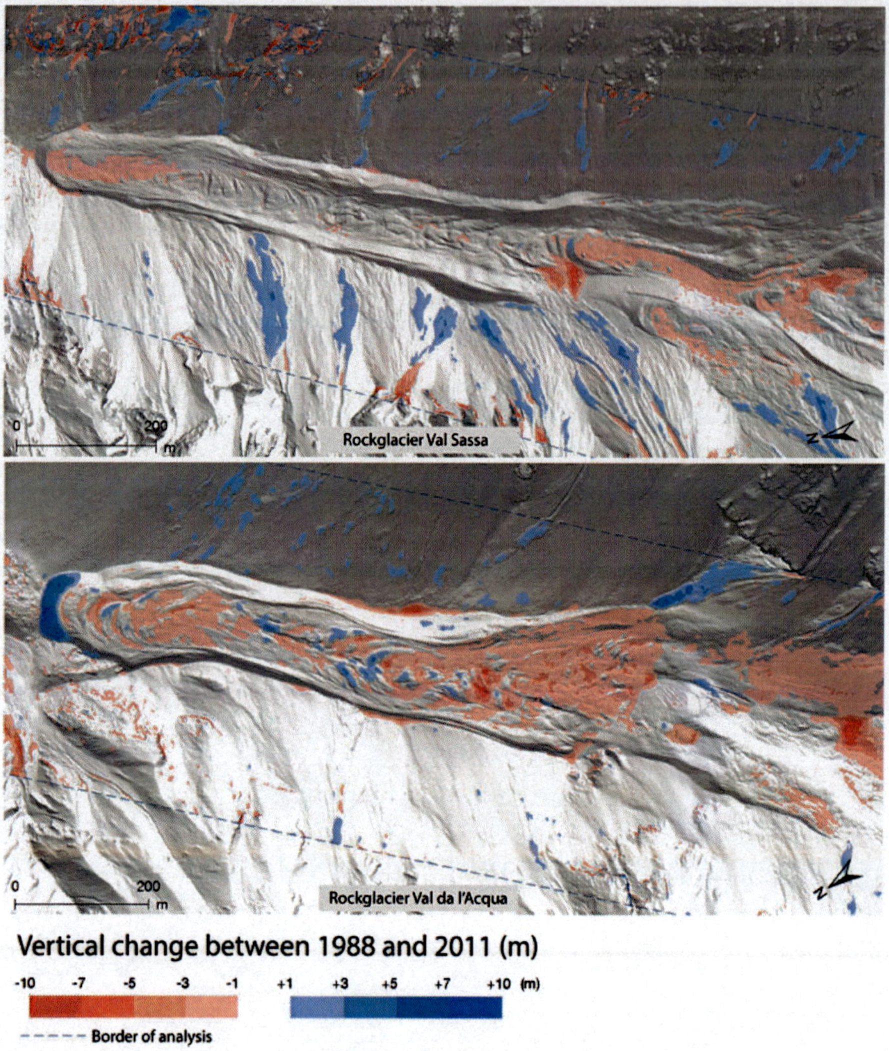

Fig. 16.3 Comparison of digital elevation models of Val Sassa and Val da l'Acqua rockglaciers from the years 1988 and 2011, indicating vertical changes of the rock glacier surfaces (*source* Baur and Scheurer 2014)

described for the first time (Chaix 1923). Since the 1950s, the Val Sassa rockglacier is surveyed by photogrammetric and geophysical investigations (e.g., Barsch 1969). Nowadays, the deformation of the Val da l'Acqua and Val Sassa rockglaciers is measured by photogrammetry and differential GPS and a new measurement network was established in 2007 (Haller et al. 2013). Val da l'Aqua rockglacier shows distinct permafrost creep near the front (Fig. 16.3), illustrated by a distinct pattern of gains and losses due to the advance of the furrow-and-ridge topography (Haller et al. 2013). Val Sassa rockglacier (Figs. 16.3 and 16.4) seems to be inactive in large parts; a small advancing front is visible in the rooting zone (Fig. 16.3). Latest findings show general thinning in large parts of the rockglaciers, which may indicate changes in the input of sediment and ice and thus a dynamic inactivation. The comparison of digital elevation models from the years 1988 and 2011 (Fig. 16.3) shows sediment transfer on the lateral slopes, which is mostly related to debris flows and rock fall events. It is nicely visible that the sediment is not directly connected to the rockglacier. Another reason could be the vanishing of ice, due to a relatively thin debris cover; this would indicate permafrost degradation and climate-induced inactivation of the landform.

16.3 Rockglacier Murtèl/Corvatsch—The Outdoor Laboratory

The Corvatsch–Furtschellas area is one of the best-investigated cryospheric sites of the Alps. Several rockglaciers, glaciers, and glacier forefields dominate the landscape (Fig. 16.1). In the past, the eastern slope of the Corvatsch mountain crest was dominated by glaciers, which have strongly retreated since the

Fig. 16.4 Aerial view of the Val Sassa rockglacier (photo C. Levy, 20.08.2011)

Little Ice Age and today they are covering only small parts. The whole area is used as a ski resort, with ski tracks running from the top of Corvatsch (3400 m a.s.l.) down to the valley bottom (1800 m a.s.l.). Large parts of the ski tracks are prepared in early winter each year by artificial snow production. The rockglacier Murtèl (Fig. 16.5) spreads down from about 2750 to 2600 m a.s.l., with an approximate surface area of 0.5 km^2, in the middle of the ski area. It is therefore easily accessible.

16.3.1 Short Historical Outline of Past Research Activities

First research activities on this rockglacier started in the mid-1970s with photogrammetrical investigations (Barsch and Hell 1975). The first shallow borehole was drilled by Barsch (1977) to a depth of 11 m. It constituted the first step to have direct access to the interior of these unique creeping permafrost features. In 1987, two boreholes (called 1/87 and 2/87) were drilled down to a depth of 58 m and 21 m, respectively (Haeberli et al. 1988; Haeberli 1990; Vonder Mühll and Haeberli 1990). These drillings allowed for answering questions related to (i) the thermal conditions within the rockglacier, (ii) the creep behavior along a vertical depth profile, and (iii) the analysis of the physical–chemical properties of ice/rock mixtures within the core (Haeberli et al. 1998). Temperatures have been measured since 1987 in the 58 m deep borehole (Fig. 16.2) and have been continued until today, although considerable shearing has occurred at a depth of around 30 m. Between 1987 and 2014, the thickness of the active layer has changed only a little (Fig. 16.6), probably because of the massive ice core occurring between around 3.5 and 25 m depth. However, since 2011 the active layer depth is continuously increasing, pointing to a probable change caused by increasing melt at the ice-rich permafrost table. Such changes are particularly important in relation to natural hazards, as in recent years several rockglaciers within the Alpine belt have become unstable and could therefore be a remarkable hazard for infrastructure and human beings in

Fig. 16.5 Rockglacier Murtèl with the distinctive ridge-and-furrow topography in the frontal part, as seen from the air (photo C. Rothenbühler, 23.10.2004)

the Alpine valleys (Kääb et al. 2007; Roer et al. 2008; Delaloye et al. 2010). Rockglaciers seem to have strong heterogeneity, which is confirmed at Murtèl rockglacier, where results from the borehole 2/00 drilled in the year 2000 show strong seasonal changes down to 25 m indicating seasonal groundwater flow across the rockglacier body. Springman et al. (2012) report that the development of a sub-permafrost talik is measured at a depth of about 57 m, corresponding to a similar depth of 52 m in the borehole 2/87, where positive temperatures were reported by Vonder Mühll (1992).

16.3.2 New Borehole Drilled in 2015

Nearly 30 years later, in late autumn 2015, a new borehole 3/15 was drilled to a depth of 60 m, financed by the Federal Office for Environment (FOEN) and organized by the University of Zurich, responsible for the monitoring at this site within PERMOS. The drilling was performed with the intention to replace the old borehole 2/87, where the equipment reached its end of life mainly because of the aging of the material and as well as the break of the wires along the shearing zone at a depth of around 30 m. First results show that temperatures measured in both boreholes, 2/87 and 3/15, are similar over the whole profile allowing for continued and uninterrupted monitoring into the next decades (Fig. 16.7; Bast et al. 2016).

16.3.3 Measurements Performed in the Boreholes Drilled in 1987, 2000, and 2015

In the 2/87 borehole, as well as in the more recently drilled holes, slope indicator measurements (to quantify deformation in the borehole) were performed showing the existence of an ice-supersaturated layer in the upper 30 m indicating

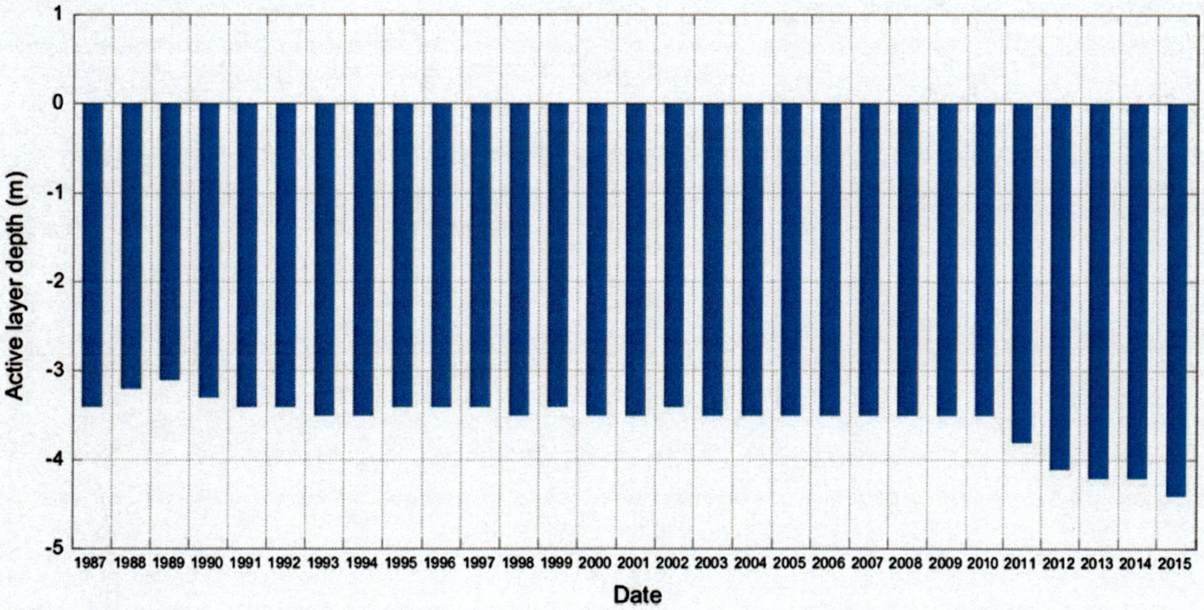

Fig. 16.6 Active layer depth at the Murtèl rockglacier (2/87) from 1987 to 2016 (*source* PERMOS 2016, www.permos.ch)

steady-state creep. According to the borehole deformation and corresponding photogrammetric measurements at the surface, the velocity at the drill site is about 6 cm a^{-1}. The main part of the deformation (about 66 %) takes place at a depth of 28–30 m, i.e., within the transition zone between the two main permafrost layers, with the upper (supersaturated) layer obviously undergoing steady-state creep and overriding the non-deforming (structured) lower layer (Wagner 1992). The slope indicator measurements stopped in 1994, when shearing prevented further application of the sensor. Vertical deformation mainly occurs between 25 and 30 m depth: the layer immediately above the shear horizon is being shortened by roughly 1 cm a^{-1}. The inclination of the flow vectors resulting from the horizontal and vertical displacements is about 10° and is comparable to the surface inclination of the compressing flow part of the rockglacier. Wagner (1992) and Leysinger Vieli and Gudmundsson (2003) tried to model this type of rockglacier creep. However, assumptions that vertical strain rates are linear and the volume is constant appear to be inappropriate. This is very important for rockglacier modeling approaches (Arenson et al. 2010). A reduction in volume can be caused by compressive flow due to the presence of air voids within the permafrost body. The flow velocities depend on temperatures, water content, slope, and composition of the ground material. Within degrading permafrost, the ice content decreases, the creep velocity increases due to increasing temperature and thus influence the rate factor in the flow law (Müller et al. 2016) and increase the unfrozen water content and water flow. With time the shear zone moves closer to the surface where seasonal temperature changes. All these information lead to a general model of the evolution of rockglaciers as described by Haeberli et al. (1998).

In addition to the study of ongoing processes, the physical–chemical properties have been investigated in order to explain the genesis of the landform. ^{14}C-dating of organic material found in one of the cores confirmed an earlier assumption that the rockglacier was formed during a time interval of several thousands of years (Haeberli et al. 1999). This finding is consistent with age estimates from photogrammetrically determined flow fields (Kääb et al. 1998), implying that the formation of the rockglacier started after the area became deglaciated at the beginning of the Holocene. Photogrammetric surveys from the 1970s to 2006 show that the whole perennially frozen body is creeping with horizontal velocities in the range of 0.05–0.15 m a^{-1} (Barsch and Hell 1975; Kääb 1998; Kääb et al. 1998; Kääb and Vollmer 2000). In the lower and flatter part a compression zone is found, whereas upstream of the rockglacier more extension has been observed. The compressive zone is characterized by distinct ogive-like ridges and furrows (Fig. 16.5), which were recently explained by an analytical buckle-folding expression including a relationship between observed wavelength, layer thickness and effective viscosity ratio between the folded layer and the underlying ice (Frehner et al. 2014). The authors conclude that their model, based on published surface flow velocities, uses about 1000–1500 years to produce the observed furrow-and-ridge morphology.

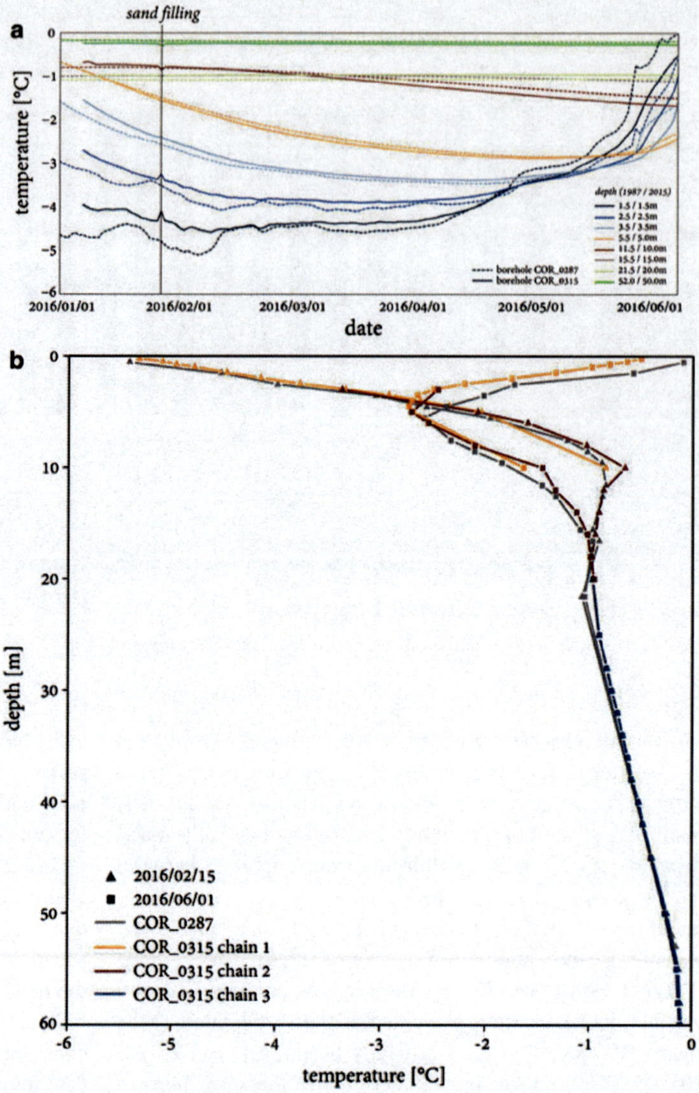

Fig. 16.7 Comparison of temperatures of the old (2/87) and the new (3/15) borehole at Murtèl rockglacier. **a** Temperature development at different depths between January 1 and June 1 2016; **b** Temperature/depth distribution (snapshots of February and June 2016) (*source* Bast et al. 2016)

16.3.4 Geophysical Investigations

Several geophysical investigations using methods like Refraction Seismic Tomography (RST), Electric Resistivity Tomography (ERT), Ground-Penetrating Radar (GPR), gravimetry and Electromagnetics (EM) have provided data for better interpretation of subsurface structures (Vonder Mühll 1993; Vonder Mühll and Klingelé 1994; Lehmann et al. 1998; Hauck et al. 2001, 2011; Hilbich et al. 2009). The results were validated with direct borehole information. The main goals of these investigations are the determination of: (i) the active layer thickness; (ii) the depth of the shear horizon forming the base of the creeping part of permafrost; (iii) the amount of ice; and (iv) the depth to bedrock. In saturated and supersaturated debris-rich permafrost, the combination of ERT and RST is the best method for investigating the permafrost table (Hauck 2013).

An additional geophysical method, mainly used in the early 1990s, was the so-called Bottom Temperature of the winter Snow cover (BTS) method. Several BTS measurements were carried out constituting a database to test the permafrost distribution models in the Corvatsch–Furtschellas area (Hoelzle et al. 1993; Hoelzle and Haeberli 1995). Continuous near-surface ground temperature measurements with miniature loggers were introduced at the end of the 1990s to better understand and interpret BTS mapping results (Hoelzle et al. 1999; Hoelzle et al. 2003) and in general, to investigate the temporal and spatial changes of the atmosphere-ground interactions (Bernhard et al. 1998; Hanson and Hoelzle 2004, 2005; Hoelzle and Gruber 2008;

Gubler et al. 2011; Schneider et al. 2012, 2013; Schmid et al. 2012). During the past years, several studies revealed that snow cover seems to be one of the most important factors influencing ground temperatures through its high insulating effects (Vonder Mühll et al. 1998; Hoelzle et al. 1999, 2003; Hanson and Hoelzle 2004; Gruber and Hoelzle 2008; Schmid et al. 2012; Schneider et al. 2012). A snow cover with a thickness of more than about 0.6 m acts as insulation on the rockglacier that has a high surface roughness (Hanson and Hoelzle 2004). It conserves the heat introduced in summer and protects the permafrost from cold winter temperatures. In contrast, a thin (5–15 cm) snow cover in late autumn is most efficient in allowing cooling of the ground (Keller and Gubler 1993). Local effects such as variations of snow-cover distribution as a function of boulder size or local climate cause particular conditions for every site. Since 1997, microclimatological measurements have been performed. These measurements allow the calculation of heat and energy fluxes between the atmosphere and the surface and between the active layer and the underlying permafrost (Mittaz et al. 2000; Etzelmüller et al. 2001; Mittaz et al. 2002; Hoelzle et al. 2001, 2010; Hanson and Hoelzle 2004; Hoelzle and Gruber 2008; Scherler et al. 2014). These energy fluxes, together with the heat flow from the earth interior, govern mainly the thermal regime of the subsurface, which is important for the whole spatial and temporal landform development.

16.3.5 Modeling Permafrost Distribution

Since research started at Murtèl rockglacier considerable efforts have been undertaken to model permafrost at this site. One can distinguish more process-oriented and more empirical-statistical models in 1D or 2D. Empirical-statistical models directly relate documented permafrost occurrences to topoclimatic factors (altitude, slope and aspect, mean air temperature, solar radiation), which can easily be measured or computed (e.g., Keller 1992; Keller et al. 1998). In the Corvatsch–Furtschellas area, several of these model types were developed and applied during the past decades (Keller 1992; Hoelzle and Haeberli 1995; Hoelzle 1996; Hoelzle et al. 2001). In these models, complex energy exchange processes at the surface and within the active layer are not treated explicitly. The model results represent therefore yes/no functions about the presence or absence of permafrost, primarily applicable to certain areas and assume steady-state conditions. Such models can be used carefully in well-calibrated areas for palaeo-reconstructions and simple future simulations (Hoelzle and Haeberli 1995; Frauenfelder and Kääb 2000; Frauenfelder et al. 2001).

Model approaches allowing for spatio-temporal extrapolations should be process based. Such models focus on more detailed understanding of the energy fluxes between the atmosphere and the permafrost and were developed also for the Murtèl/Corvatsch site (Etzelmüller et al. 2001; Marchenko 2001; Stocker-Mittaz et al. 2002; Salzmann et al. 2007a, b; Gruber and Hoelzle 2008; Dall'Amico et al. 2011; Fiddes and Gruber 2012, 2014). They explicitly parameterize solar radiation, sensible heat, surface albedo, and heat conduction. They are often complex and need a correspondingly large amount of precisely measured or computed data and are especially well suited for sensitivity studies with respect to interactions and feedbacks involved with climate change scenarios.

16.3.6 Sediment Storage and Transfer

Besides the more climate-related approaches, Murtèl rockglacier has also been analyzed to quantify sediment storage and to assess sediment transfer rates (Gärtner-Roer 2012; Müller et al. 2016). The energy-related approach by Müller et al. (2016) includes also the headwall above the rockglacier and helps in understanding the complex dynamic behavior of a periglacial mountain slope. While the highest amount of energy is released within the headwall areas, the rockglacier is mobilizing the highest amount of sediment. This pattern emphasizes that rockglaciers can be considered as long-term sediment sinks.

16.4 Rockglacier Muragl

The rockglacier Muragl (Figs. 16.8 and 16.9) originates from the La Laungia cirque, north of Piz Las Sours, and extends over 700 m from 2700 to 2480 m a.s.l.; its width varies between 100 and 300 m. The average inclination is at 15°, with steepest slopes (up to 25°) in the rooting zone (Springman et al. 2012). With its rectangular form resulting from the drastic change of its flow direction, the Muragl rockglacier is one of the most noticeable landforms and perfectly situated at the touristic "climate trail" between Muottas Muragl and Alp Languard (https://www.myswitzerland.com/en-ch/muottas-muragl-alpine-peak-languard.html, accessed 20.03.2020). Diverse lobes on top of each other as well as the outbreak lobe to the north reflect the complex genesis and history of the landform. The single lobes show different dynamics, surface structures, as well as vegetation cover. The rockglacier is part of a very distinctive landscape of the upper Muragl valley showing complex structures of different landforms related to cryospheric processes, which have been shaped during the Holocene and Lateglacial periods (Maisch et al. 2003).

Fig. 16.8 Rockglacier Muragl (view to the east) (photo I. Gärtner-Roer, 18.08.2009)

After intensive geophysical soundings to describe the small-scale structure of the landform as well as the occurrence of ice, four deep boreholes were drilled in 1999 in the lower part of the rockglacier, in order to investigate the rockglacier structure, its thermal properties as well as the deformation characteristics in the compression zone (Arenson et al. 2002). The multidisciplinary approach included monitoring of thermal conditions and deformation (Arenson et al. 2002; PERMOS 2013), geophysical investigations (Vonder Mühll 1993), pressuremeter testing and sampling of rockglacier material for laboratory tests (Arenson 2002), and allowed for much better process understanding (Springman et al. 2012).

One of the boreholes (1/1999) is located outside the active part of the rockglacier. No permafrost was recorded, but cold air penetrates during winter due to natural convection, indicating coarse material with high permeability (Springman et al. 2012). The other boreholes (2–4/1999) were all drilled into permafrost. They depicted a frozen layer of about 15 m, which is overlying a gravelly and blocky layer before bedrock was reached at a depth of 30 m (Springman et al. 2012).

Borehole temperatures were recorded by in situ thermistor chains, which sheared off in most of the boreholes in the meantime. Thermistors in the borehole 2/1999 kept recording until May 2006, whereas those in the borehole 3/1999 worked until June 2005. In the borehole 4/1999 temperatures are still monitored in the top 13.6 m, but the lower part sheared off in January 2004 (Springman et al. 2012). Nevertheless, the Muragl rockglacier contributed valuable data and information within PERMOS and for the Engadine in particular.

Borehole temperatures at the borehole 4/1999 at a depth of 11.59 m typically indicate seasonal fluctuations between –0.2 and –0.8 °C during the last decade and are very much in parallel with the temperatures in the adjacent borehole at Schafberg (see Fig. 16.2 and Phillips and Kenner, this volume). The reported active layer thickness between 2006 and 2013 was constantly at a depth of 4.5 m. In general, thickness of the active layer in the period 2001–2005 was smaller than in subsequent years (PERMOS 2013).

Borehole deformations were derived from inclinometers installed in two boreholes and for the first time seasonal variations were described, with maximum deformation rates during winter and slower rates during summer (Springman et al. 2012). This is explained by the time needed for warmer (and also colder) temperatures to penetrate to the depth of the shear zone, at about 16 m (Arenson et al. 2002).

Photogrammetric analyses as well as in situ investigations have been conducted on the Muragl rockglacier in order to describe kinematics and dynamics (Kääb 1998). As depicted in the complex surface topography, the landform showed a complex flow pattern with maximum deformations of 0.5 m a^{-1} in the 1990s (Kääb 1998). An in situ kinematic

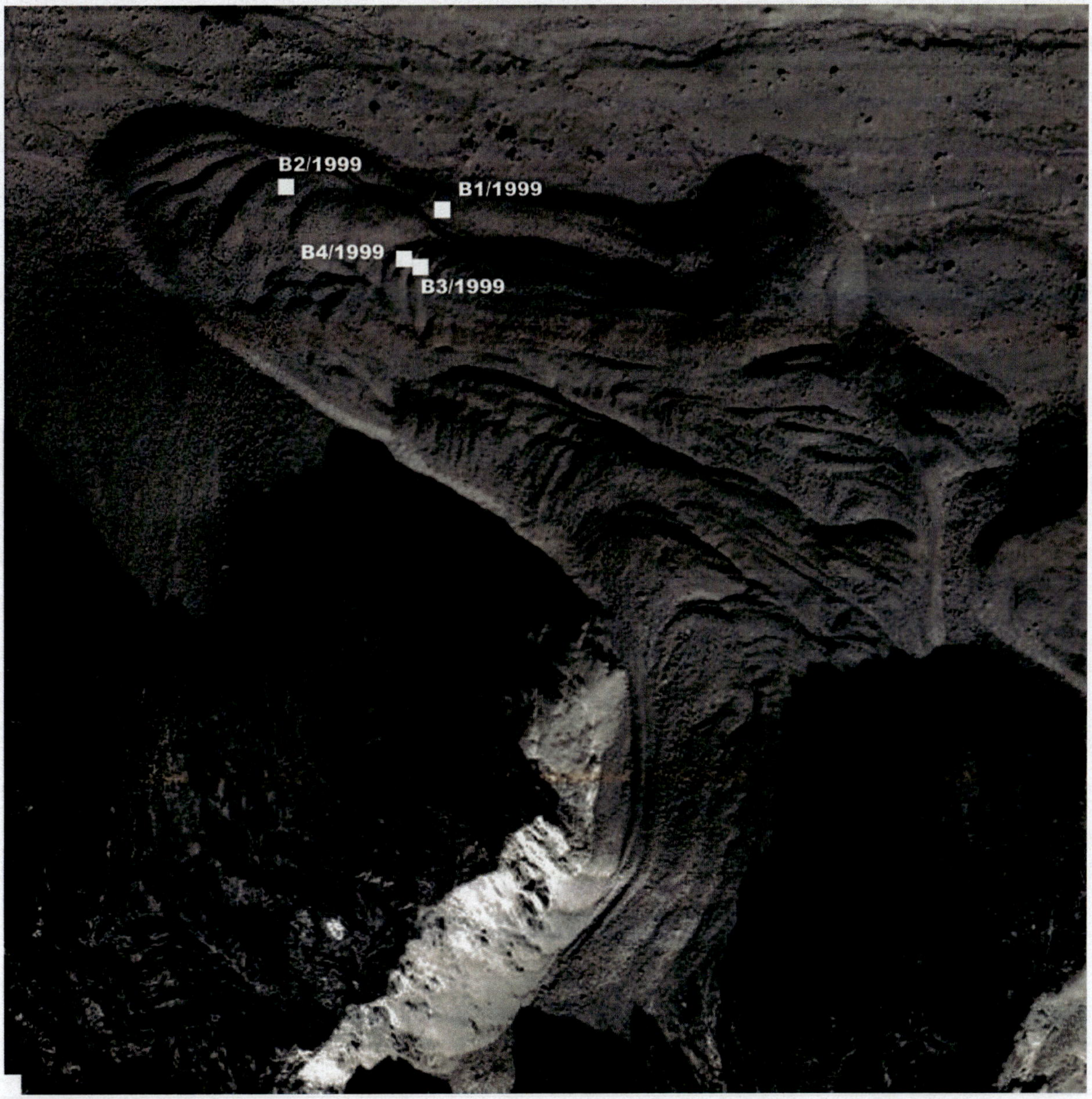

Fig. 16.9 Aerial image (Airborne Digital Sensor) of the Muragl rockglacier showing the surface structure of ridges and furrows as well as the outbreaking lobe to the north (*source* Swisstopo, PERMOS, August 2012). The approximate locations of the four boreholes are marked with white squares

monitoring was re-established in 2009 and now shows much higher horizontal velocities in the central part of the rockglacier, up to 1.5 m a^{-1} (PERMOS 2013).

16.5 Conclusion

The Engadine is a rockglacier hot spot, but also a hot spot of rockglacier research, from its very beginning in the 1910s to the most recent findings. Research on these creeping permafrost bodies progressed from simple mapping to elaborated borehole loggings and complex modeling approaches. Most of the approaches are motivated by climate-related research, whereas others focus on geomorphological investigations, including hazard assessment, or testing of sophisticated methodology. Besides the research-driven investigations, long-term monitoring programs of subsurface temperatures, ground surface temperatures, geophysics, and kinematics run on several rockglaciers in the Engadine.

The findings from rockglacier research in the Engadine indicate warming permafrost conditions and related process changes, such as increased horizontal velocities on the Muragl rockglacier, as well as degradation of certain landforms, such as the Val Sassa rockglacier. These observations are consistent with findings from other parts in Switzerland and the entire European Alps, confirming that the observed changes do not reflect local conditions but rather regional trends. Such trends can be important for the assessment of natural hazards as in recent years several rockglaciers within the Alpine belt have become unstable and represent—depending on the topographic position of the landform—a remarkable hazard for infrastructure and human beings in the Alpine valleys. Therefore, continuous monitoring of permafrost evolution in general and rockglacier development in particular is of great importance.

References

Arenson LU (2002) Unstable alpine permafrost: a potentially important natural hazard–variations of geotechnical behaviour with time and temperature. PhD thesis, Swiss Federal Institute for Technology, Zurich

Arenson LU, Hoelzle M, Springman S (2002) Borehole deformation measurements and internal structure of some rock glaciers in Switzerland. Permafrost Periglac Process 13:117–135

Arenson L, Hauck C, Hilbich C, Seward L, Yamamoto Y, Springman S (2010) Sub-surface heteorogenities in the Murtèl-Corvatsch rock glacier, Switzerland. 6th Canadian Permafrost Conference. CNC-IPA/NRC, Calgary, pp 1494–1500

Barsch D (1969) Studien und Messungen an Blockgletschern in Macun, Unterengadin. Zeitschrift für Geomorphologie, Suppl. 8:11–30

Barsch D (1977) Ein Permafrostprofil aus Graubünden, Schweizer Alpen. Zeitschrift für Geomorphologie 21:79–86

Barsch D, Hell G (1975) Photogrammetrische Bewegungsmessungen am Blockgletscher Murtèl I, Oberengadin, Schweizer Alpen. Z Gletscherk Glazialgeol 11:111–142

Bast A, Noetzli J, Gärtner-Roer I, Hoelzle M, Hauck C. Vonder Muehll D, Müller J, Phillips M, Wasser D, Vieli A (2016) And it continues: A new borehole extends the longest mountain permafrost temperature record on rock glacier Murtèl, Swiss Alps. In: Abstracts of the 11th International Conference on Permafrost, Potsdam, Germany

Baur B, Scheurer T (eds) (2014) Wissen schaffen. 100 Jahre Forschung im Schweizerischen Nationalpark. Haupt, Bern, 384 pp

Bernhard L, Sutter F, Haeberli W, Keller F (1998) Processes of snow/permafrost-interactions at a high-mountain site, Murtèl/Corvatsch, Eastern Swiss Alps. In: Lewkowicz AG, Allard M (eds) 7th International Conference on Permafrost, Yellowknife. Proceedings. Centre d'Etudes Nordiques, Université Laval, pp 35-41

Boeckli L, Brenning A, Gruber S, Noetzli J (2012a) Permafrost distribution in the European Alps: calculation and evaluation of an index map and summary statistics. Cryosphere 6:807–820

Boeckli L, Brenning A, Gruber S, Noetzli J (2012b) A statistical permafrost distribution model for the European Alps. Cryosphere 6:125–140

Chaix A (1923) Les coulées de blocs du Parc National Suisse d'Engadine (Note préliminaire). Le Globe 62:1–35

Chaix A (1943) Les coulées de blocs du Parc National Suisse: nouvelles mesures et comparaison avec les « rock stream » de la Sierra Nevada de Californie. Le Globe 82:121–128

Dall'Amico M, Endrizzi S, Gruber S, Rignon R (2011) A robust and energy-conserving model of freezing variably-saturated soil. Cryosphere 5:469–484

Delaloye R, Lambiel C, Gärtner-Roer I (2010) Overview of rock glacier kinematics research in the Swiss Alps. Geogr Helv 65:135–145

Etzelmüller B, Hoelzle M, Heggem ESF, Isaksen K, Mittaz C, Vonder Mühll D, Ødegård RS, Haeberli W, Sollid JL (2001) Mapping and modelling the occurrence and distribution of mountain permafrost. Norwegian J Geography 55:186–194

Fiddes J, Gruber S (2012) TopoSUB: a tool for efficient large area numerical modelling in complex topography at sub-grid scales. Geosci Model Develop 5:1245–1257

Fiddes J, Gruber S (2014) TopoSCALE v. 1.0: downscale gridded climate data in complex terrain. Geosci Model Develop 7:387–405

Frauenfelder R, Haeberli W, Hoelzle M, Maisch M (2001) Using relict rockglaciers in GIS-based modelling to reconstruct younger Dryas permafrost distribution patterns in the Err-Julier area, Swiss Alps. Norwegian J Geography 55:195–202

Frauenfelder R, Kääb A (2000) Towards a palaeoclimatic model of rock glacier formation in the Swiss Alps. Ann Glaciol 31:281–286

Frehner M, Ling AHM, Gärtner-Roer I (2014) Furrow-and-ridge morphology on rockglaciers explained by gravity-driven buckle folding: a case study from the Murtèl rockglacier (Switzerland). Permafrost Periglac Process 26:57–66

Gärtner-Roer I (2012) Sediment transfer rates of two active rockglaciers in the Swiss Alps. Geomorphology 167–168:45–50

Gruber S, Hoelzle M, (2008) The cooling effect of coarse blocks revisited: a modelling study of a purely conductive mechanism: In: Kane DL, Hinkel KM (eds) 9th International Conference on Permafrost. Proceedings. Institute of Northern Engineering, University of Alaska, Fairbanks, pp 557–561

Gubler S, Fiddes J, Keller M, Gruber S (2011) Scale-dependent measurement and analysis of ground surface temperature variability in alpine terrain. Cryosphere 5:431–443

Haeberli W (1985) Creep of mountain permafrost: internal structure and flow of alpine rock glaciers. Mitteilungen der VAW/ETH Zürich 77:1–119

Haeberli W (1990) Pilot analysis of permafrost cores from the active rock glacier Murtèl I, Piz Corvatsch, Eastern Swiss Alps. Versuchsanstalt für Wasserbau, Hydrologie und Glaziologie, ETH Zürich

Haeberli W, Hoelzle M, Kääb A, Keller F, Vonder Mühll D, Wagner S (1998) Ten years after drilling through the permafrost of the active rock glacier Murtèl, Eastern Swiss Alps: answered questions and new perspectives. In: Lewkowicz AG, Allard M (eds) 7th International Conference on Permafrost, Yellowknife. Proceedings. Centre d'Etudes Nordiques, Université Laval, pp 403–410

Haeberli W, Huder J, Keusen HR, Pika J, Röthlisberger H (1988) Core drilling through rock glacier-permafrost. In: Senneset K (ed) 5th International Conference on Permafrost, Yellowknife. Proceedings. Tapir Publishers, Trondheim, pp 937–942

Haeberli W, Kääb A, Wagner S, Vonder Mühll D, Geissler P, Haas JN, Glatzel-Mattheier H, Wagenbach D (1999) Pollen analysis and ^{14}C-age of moss remains recovered from a permafrost core of the active rock glacier Murtèl/Corvatsch (Swiss Alps): geomorphological and glaciological implications. J Glaciol 45:1–8

Haller H, Eisenhut A, Haller R (2013) Atlas des Schweizerischen Nationalparks. Die ersten 100 Jahre. Haupt, Bern

Hanson S, Hoelzle M (2004) The thermal regime of the active layer at the Murtèl rock glacier based on data from 2002. Permafrost Periglac Process 15:273–282

Hanson S, Hoelzle M (2005) Installation of a shallow borehole network and monitoring of the ground thermal regime of a high alpine discontinuous permafrost environment, Eastern Swiss Alps. Norwegian J Geography 59:84–93

Hauck C (2013) New concepts in geophysical surveying and data interpretation for permafrost. Permafrost Periglac Process 24:131–137

Hauck C, Böttcher M, Maurer H (2011) A new model for estimating subsurface ice content based on combined electrical and seismic data sets. The Cryosphere 5:453–468

Hauck C, Guglielmin M, Isaksen K, Vonder Mühll D (2001) Applicability of frequency-domain and time-domain electromagnetic methods for mountain permafrost studies. Permafrost Periglac Process 12:39–52

Hilbich C, Marescot L, Hauck C, Loke MH, Mäusbacher R (2009) Applicability of Electrical Resistivity Tomography monitoring to coarse blocky and ice-rich permafrost landforms. Permafrost Periglac Process 20:269–284

Hoelzle M (1996) Mapping and modelling of mountain permafrost distribution in the Alps. Norwegian J Geography 50:11–15

Hoelzle M, Gruber S (2008) Borehole and ground surface temperatures and their relationship to meteorological conditions in the Swiss Alps. In: Kane DL, Hinkel KM (eds) 9th International Conference on Permafrost. Proceedings. Institute of Northern Engineering, University of Alaska, Fairbanks, pp 723–728

Hoelzle M, Haeberli W (1995) Simulating the effects of mean annual air temperature changes on permafrost distribution and glacier size. An example from the Upper Engadin. Swiss Alps. Ann Glaciology 21:400–405

Hoelzle M, Haeberli W, Keller F (1993) Application of BTS-measurements for modelling permafrost distribution in the Swiss Alps. In: Guodong C (ed) 6th International Conference on Permafrost. Proceedings. South China University Technology Press, Beijing, pp 272–277

Hoelzle M, Haeberli W, Stocker-Mittaz C (2003) Miniature ground temperature data logger measurements 2000-2002 in the Murtèl--Corvatsch area. In: Phillips M, Springman S, Arenson L (eds) 8th International Conference on Permafrost, Zurich. Proceedings. Swets & Zeitlinger, Lisse, pp 419–424

Hoelzle M, Hauck C, Scherler M, Schneider S, Bircher S, Panz M (2010) Active layer processes on rock glacier Murtèl-Corvatsch. In: Mertes JR, Christiansen HH, Etzelmüller B (eds) 3rd European Conference on Permafrost. University Center Svalbard, Longyearbyen, p 326

Hoelzle M, Mittaz C, Etzelmüller B, Haeberli W (2001) Surface energy fluxes and distribution models of permafrost in European mountain areas: an overview of current developments. Permafrost Periglac Process 12:53–68

Hoelzle M, Wegmann M, Krummenacher B (1999) Miniature temperature dataloggers for mapping and monitoring of permafrost in high mountain areas: First experience from the Swiss Alps. Permafrost Periglac Process 10:113–124

Kääb A (1998) Oberflächenkinematik ausgewählter Blockgletscher des Oberengadins. In: Vonder Mühll D (ed) Beiträge aus der Gebirgs-Geomorphologie. VAW/ETH Zürich, Samedan, pp 121–140

Kääb A, Frauenfelder R, Roer I (2007) On the response of rockglacier creep to surface temperature increase. Global Planet Change 56:172–187

Kääb A, Gudmundsson GH, Hoelzle M (1998) Surface deformation of creeping mountain permafrost. Photogrammetric investigations on Murtèl rock glacier, Swiss Alps. In: Lewkowicz AG, Allard M. (eds) 7th International Conference on Permafrost, Yellowknife. Proceedings. Centre d'Etudes Nordiques, Université Laval, pp 531–537

Kääb A, Vollmer M (2000) Surface geometry, thickness changes and flow fields on permafrost streams: automatic extraction by digital image analysis. Permafrost Periglac Process 11:315–326

Keller F (1992) Automated mapping of mountain permafrost using the program PERMAKART within the geographical information system ARC/INFO. Permafrost Periglac Process 3:133–138

Keller F, Frauenfelder R, Gardaz JM, Hoelzle M, Kneisel C, Lugon R, Phillips M, Reynard E, Wenker L (1998) Permafrost map of Switzerland. In: Lewkowicz AG, Allard M (eds) 7th International Conference on Permafrost, Yellowknife. Proceedings. Centre d'Etudes Nordiques, Université Laval, pp 557–562

Keller F, Gubler HU (1993) Interaction between snow cover and high mountain permafrost, Murtèl-Corvatsch, Swiss Alps. In: Guodong C (ed) 6th International Conference on Permafrost. Proceedings. South China University of Technology Press, Beijng, pp 332–337

Lehmann F, Vonder Mühll D, van der Veen M, Wild P, Green A (1998) True topographic 2-D migration of georadar data. In: Proceedings of the symposium on the application of geophysics to environmental and engineering problems (SAGEEP), Chicago, pp 107–114

Leysinger Vieli GJ-MC, Gudmundsson GH (2003) Evolution of rock glaciers and alpine glaciers: a model-model approach. In: Phillips M, Springman S, Arenson L (eds) 8th International Conference on Permafrost, Zurich. Proceedings. Swets & Zeitlinger, Lisse, pp 673–677

Maisch M, Haeberli W, Frauenfelder R, Kääb A (2003) Lateglacial and Holocene evolution of glaciers and permafrost in the Val Muragl, Upper Engadine, Swiss Alps. In: Phillips M, Springman S, Arenson L (eds) 8th International Conference on Permafrost, Zurich. Proceedings. Swets & Zeitlinger, Lisse, pp 717–722

Marchenko SS (2001) A model of permafrost formation and occurrences in the intracontinental mountains. Norwegian J Geography 55:230–234

Mittaz C, Hoelzle M, Haeberli W (2000) First results and interpretation of energy-flux measurements of Alpine permafrost. Ann Glaciol 31:275–280

Mittaz C, Imhof M, Hoelzle M, Haeberli W (2002) Snowmelt evolution mapping using an energy balance approach over an Alpine terrain. Arct Antarct Alp Res 34:274–281

Müller J, Vieli A, Gärtner-Roer I (2016) Rockglacier on the run–Understanding rockglacier landform evolution and recent changes from numerical flow modeling. Cryosphere Discuss 35:1–40

PERMOS (2013) Permafrost in Switzerland 2008/2009 and 2009/2010. Noetzli J (ed). Glaciological Report (Permafrost) No. 10/11 of the Cryospheric Commission of the Swiss Academy of Sciences

PERMOS (2016) Permafrost in Switzerland 2010/2011 to 2013/2014. Noetzli J, Luethi R, Staub B. (eds). Glaciological Report (Permafrost) No. 12-15 of the Cryospheric Commission of the Swiss Academy of Sciences

Phillips M, Kenner R (this volume) Periglacial landscapes and protection measures above Pontresina

Roer I, Haeberli W, Avian M, Kaufmann V, Delaloye R, Lambiel C, Kääb A, (2008) Observations and considerations on destabilizing active rock glaciers in the European Alps. In: Kane DL, Hinkel KM (eds) 9th International Conference on Permafrost, Fairbanks. Proceedings. Institute of Northern Engineering, University of Alaska, Fairbanks, pp 1505–1510

Salzmann N, Frei C, Vidale PL, Hoelzle M (2007a) The application of Regional Climate Model output for the simulation of high-mountain permafrost scenarios. Global and Planetary Change 56:188–202

Salzmann N, Notzli J, Hauck C, Gruber S, Hoelzle M., Haeberli W (2007b) Ground surface temperature scenarios in complex high-mountain topography based on regional climate model results. J Geophys Res-Earth Surf 112:F02S12

Scherler M, Schneider S, Hoelzle M, Hauck C (2014) A two sided approach to estimate heat transfer processes within the active layer of rock glacier Murtèl-Corvatsch. Earth Surf Dyn 2:141–154

Schlüchter C, Lozza H, Haller R (this volume) Geomorphology and landscapes of the Swiss National Park

Schmid TO, Gubler S, Fiddes J, Gruber S (2012) Inferring snowpack ripening and melt-out from distributed measurements of near-surface ground temperatures. The Cryosphere 6:1127–1139

Schneider S, Daengeli S, Hauck C, Hoelzle M (2013) A spatial temporal analysis of different periglacial materials by using geoelectric, seismic and borehole temperature data at Murtèl-Corvatsch, Upper Engadin, Swiss Alps. Geogr Helv 68:265–280

Schneider S, Hoelzle M, Hauck C (2012) Influence of surface heterogeneity on observed borehole temperatures at a mountain permafrost site in the Upper Engadine, Switzerland. The Cryosphere 6:517–531

Springman S, Arenson L, Yamamoto Y, Maurer H, Kos A, Buchli T, Derungs G (2012) Multidisciplinary investigations on three rock glaciers in the Swiss Alps: legacies and future perspectives. Geografiska Annaler A 94:215–243

Stocker-Mittaz C, Hoelzle M, Haeberli W (2002) Permafrost distribution modeling based on energy-balance data: a first step. Permafrost Periglac Process 13:271–282

Vonder Mühll D (1992) Evidence of intrapermafrost groundwater flow beneath an active rock glacier in the Swiss Alps. Permafrost Periglac Process 3:169–173

Vonder Mühll D (1993) Geophysikalische Untersuchungen im Permafrost des Oberengadins. Versuchsanstalt für Wasserbau, Hydrologie und Glaziologie der ETH Zürich, Zürich

Vonder Mühll D, Haeberli W (1990) Thermal characteristics of the permafrost within an active rock glacier (Murtèl/Corvatsch, Grisons, Swiss Alps). J Glaciol 36:151–158

Vonder Mühll D, Klingelé E (1994) Gravimetrical investigation of ice rich permafrost within the rock glacier Murtèl-Corvatsch. Permafrost Periglac Process 5:13–24

Vonder Mühll D, Stucki T, Haeberli W (1998) Borehole-temperatures in alpine permafrost: a ten year series. In: Lewkowicz AG, Allard M (eds) 7th International Conference on Permafrost, Yellowknife. Proceedings. Centre d'Etudes Nordiques, Université Laval, pp 1089–1095

Wagner S (1992) Creep of Alpine permafrost, investigated on the Murtèl-rock glacier. Permafrost Periglac Process 3:157–162

Isabelle Gärtner-Roer is Senior Researcher at the Department of Geography at the University of Zurich and Science Officer at the World Glacier Monitoring Service (WMGS). She is active in the Executive Committee of the International Permafrost Association (IPA) and member of the scientific committee of ICSU World Data System (WDS). Between 2011 and 2013 she was President of the Swiss Geomorphological Society (SGS). Her research focuses on mountain geomorphology, cryospheric processes, and related sediment transfer processes.

Martin Hoelzle is Professor of Physical Geography at the University of Fribourg, Switzerland. He is Vice-President of the Swiss Cryospheric Commission, member of the Advisory Body on Climate Change (OcCC) of the Federal Department of the Environment, Transport, Energy and Communications (DETEC), scientific advisor of the World Glacier Monitoring Service (WMGS) and former president of the Swiss Snow, Ice and Permafrost Society between 2006 and 2009. He has been involved for 25 years in the development of modern, strongly interdisciplinary cryospheric monitoring strategies and has a broad knowledge about the complex processes chains related to interactions between climate and mountain environment.

17 Geomorphology and Landscapes of the Swiss National Park

Christian Schlüchter, Hans Lozza, and Ruedi Haller

Abstract

The landscape of the Swiss National Park (SNP) is an area of unique beauty within the morphological systems of the Swiss Alps. This is due to bedrock lithology and its deformation history: the Austroalpine facies and structures are restricted to the southeasternmost corner of the country. The morphologically dominant and most visible rock type in the landscape is the Triassic "*Hauptdolomit*." It forms high cliffs, which supply extensive scree slopes. The Park is split into two areas: the larger heartland of the Engadine Dolomites and the glacially sculptured pass areas in the south and the much smaller mountain area of Macun to the north of the Engadine Line in the crystalline Austroalpine basement rocks. The high mountain areas of the Park display various types of periglacial features, including the famous and long-researched rock glaciers.

Keywords

Swiss National Park • Engadine dolomites • Engadine Line • Glaciation • Human impacts • Engadine

17.1 Introduction

The Swiss National Park (SNP) is a Wilderness Area of Category 1a according to the classification scheme of the International Union for Conservation of Nature (IUCN), implying strict preservation of all natural processes. The evolution of the landscape has therefore not been influenced by anthropogenic landscape design and protective geo-engineering structures for one hundred years, with the exception of the hydropower schemes in the Spöl Valley (*Val dal Spöl* in Romansh language). The SNP is situated in the southeastern corner of Switzerland, between the Engadine Valley (*Engiadina*) to the north and the Val Müstair to the southeast, extending into adjoining valleys to the south towards the border with Italy (Fig. 17.1). It has a surface area of 170.3 km^2, split into two parts with an area of 167 km^2 mainly along the Pass dal Fuorn and a 3.6 km^2 exclave at Macun (Fig. 17.1). The "entrance doors" to the Park are the village of Zernez in the Engadine Valley at 46°42′ N/10°06′ E and the Pass dal Fuorn when the Park is approached from Val Müstair (Fig. 17.1). The highest peak in the Park area is Piz Pisoc at 3,173 m a.s.l. The lowest point is in Val S-charl along the River Clemgia at 1,380 m a.s.l. The alpine topography in this area is unique for the Swiss sector of the Alps, due to the Eastern Alpine Facies bedrock with the Engadine Dolomites (*Engadiner Dolomiten* in German), a substantial sedimentary rock sequence of predominantly Lower Mesozoic age (in yellow and blue in Fig. 17.2), which is structurally surrounded by crystalline units.

This chapter gives a brief outline of the relationships between bedrock lithologies, structural deformations, and the current landscape in an alpine environment of exceptional beauty. Various publications have been produced for the Park's 100-year jubilee celebrations (e.g., Kupper 2012; Haller et al. 2013; Baur and Scheurer 2014). These publications are a rich source of information and also provide the background for this manuscript.

C. Schlüchter (✉)
Institute of Geological Sciences, University of Bern, Baltzerstrasse 1 + 3, 3012 Bern, Switzerland
e-mail: christian.schluechter@geo.unibe.ch

H. Lozza · R. Haller
Swiss National Park, Chastè Planta-Wildenberg, 7530 Zernez, Switzerland
e-mail: hans.lozza@nationalpark.ch

R. Haller
e-mail: rhaller@nationalpark.ch

E. Reynard (ed.), *Landscapes and Landforms of Switzerland*, World Geomorphological Landscapes,
https://doi.org/10.1007/978-3-030-43203-4_17

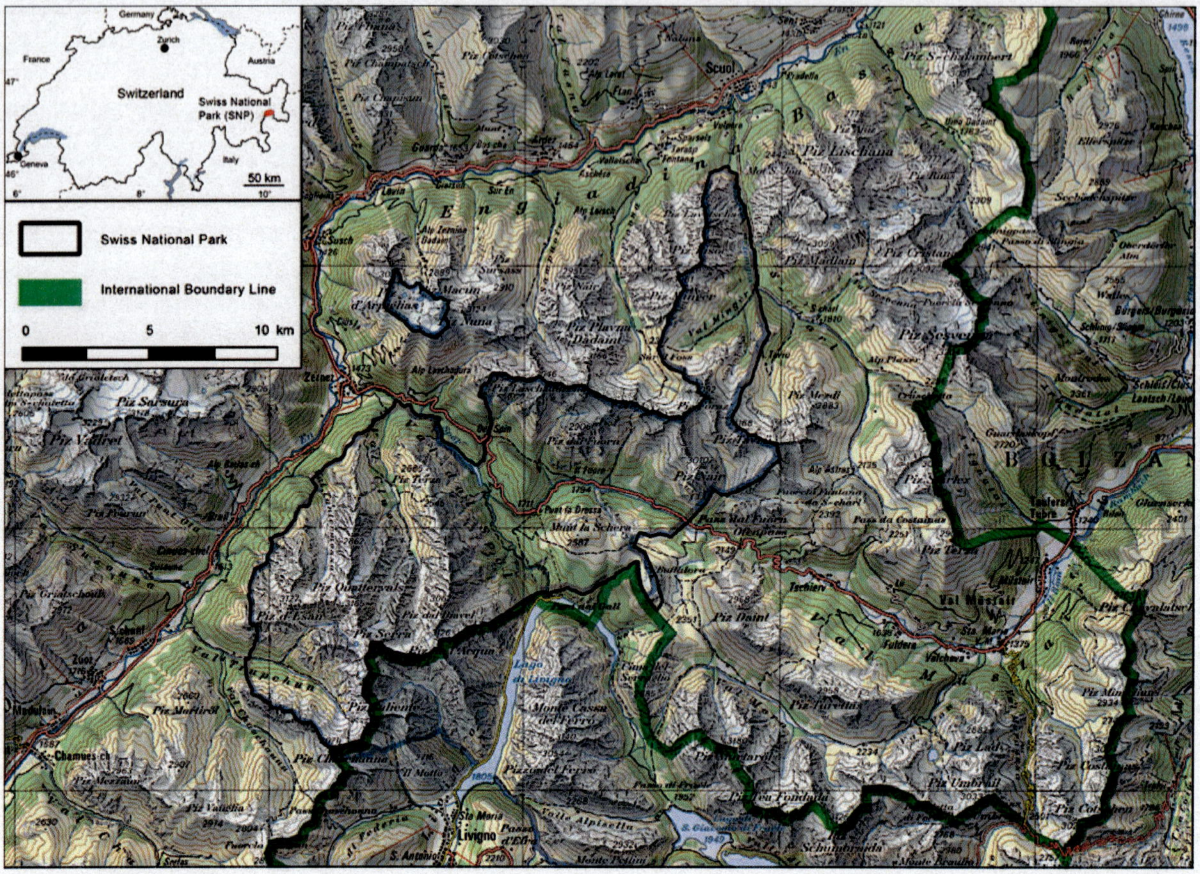

Fig. 17.1 Situation and topographic map of the Swiss National Park and the surrounding area (*Source* Swisstopo)

17.2 Geographical and Geological Setting

17.2.1 Geography

The SNP lies entirely in the Canton of Graubünden and, from a cultural point of view, in the realm of the Romansh language, the fourth official language in Switzerland. The official name of the Park is therefore "*Parc Naziunal Svizzer.*"

The boundaries of the SNP are essentially political as it is restricted to the higher parts of the mountains with less productive agricultural and forest areas (Fig. 17.1). These boundaries are the result of intense discussions between all parties involved, which finally led to the establishment of a protected area, which became effective under Federal law on August 1st, 1914. Some boundaries changed over the years: Val Tavrü (SW of Val S-charl) was given back to agricultural use in 1936 and, in contrast, the exclave of Macun was added to the Park in the year 2000, substantially supplementing the geomorphological diversity of the Park (Fig. 17.2).

The founding members of the Park are the Swiss Confederation, Pro Natura (the largest organization for nature conservation in Switzerland) and the Swiss Academy of Sciences. The Park is managed by the central park administration under supervision of the Federal National Park Commission; scientific activities are commissioned to the Research Commission of the Swiss National Park along with the Swiss Academy of Sciences. Today, the Swiss National Park is embedded within a set of neighbouring areas of varying protection status: the UNESCO Biosphere Val Müstair Parc Naziunal, the National Park Stilfserjoch (in Italy), the regional parks Adamello Brenta and Brenta (in Italy) and the Naturpark Kaunergrat (in Austria). In combination these parks form an almost uninterrupted protected area in the border zone of Switzerland, Austria, and Italy from the Central to the Southern Alps.

Five local communities contribute property to the SNP: Zernez (68.4%), S-chanf (13.5%), Scuol (13.4%), Val Müstair (4.7%), and Lavin (2.1%). Ownership of the parkland is both public and private and is compensated by a yearly interest paid by the Federal government. 50.9% of the surface area consists of primary soils devoid of vegetation

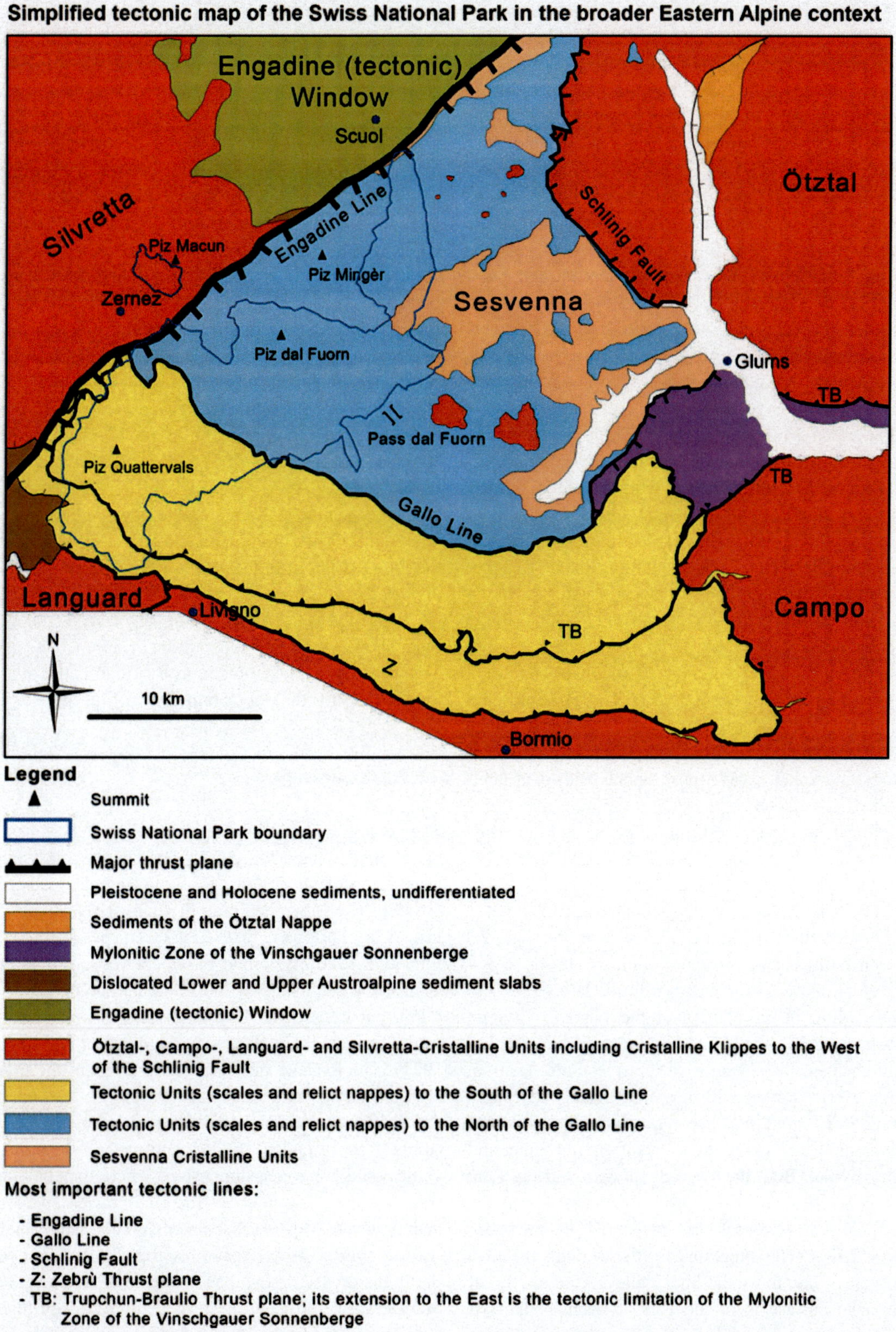

Fig. 17.2 Simplified tectonic map of the Swiss National Park in the broader Eastern Alpine context (modified after Zimmermann et al. 2014)

cover (*Rohböden* in German), 17.1% are grassland and alpine meadows, 31.4% are forested and 0.6% are rivers and lakes. Infrastructure accounts for 0.1% of the total area. The tree limit is between 2,200 and 2,300 m a.s.l. All surface waters naturally drain toward the Inn River (*En*, Fig. 17.1), which joins the Danube River on its course to the Black Sea.

The morphological heart of the Park is the broad valley of the Pass dal Fuorn, with the river Spöl in the lower and the river Il Fuorn in the upper sector. In the western part of the Park there are valleys draining directly to the Engadine Valley: Val Tantermozza and further to the west Val Trupchun. Entering the northernmost sector of the Park from the Lower Engadine Valley in Scuol the Val S-charl and its tributary Val Mingèr and Foraz can be followed. The park exclave of Macun is a unique plateau with about 40 small lakes and ponds.

The Park area has a very long tradition of economic activities, the most important of which are mining and forestry. The Engadine Valley was probably inhabited as early as 3,600 years BC. Around 1000 AD agricultural activity was widespread and around 1300 AD mining started to develop, with peaks of activity in the fourteenth and fifteenth centuries, and then again in the seventeenth century. Mining continued sporadically and was reconsidered as late as in the early twentieth century. The target ore in the Pass dal Fuorn area was iron. In the Val S-charl (at Piz Madlain) silver and lead have been mined for over 300 years. There is a museum worth visiting on the medieval mining industry in the village of S-charl. Local small scale mining produced anthropogenic morphology and small excavations, tunnels and local landfills are still visible. The most extensive mining operations with resulting landscape modifications were on Munt Buffalora just outside the borders of the Park. A comprehensive report on the history of the mining activities in the Engadine–Val Müstair area is given by Schläpfer (2013).

The evolution of the natural environment in the Park did not experience any major disruption until the 1950s. However, this quiet phase was interrupted in the late 1950s and early 1960s with the installation of the Lago di Livigno hydropower scheme (Fig. 17.1). This major impact has drastically changed the character of the river Spöl, from an alpine high-energy torrent with maximum yearly discharge of about 12 m^3s^{-1} until 1962 to the reduced, legally contracted remaining flow of 0.5 m^3s^{-1} resulting in a quiet and low-energy rivulet. Both the Lago di Livigno and the Punt dal Gall concrete dam are outside the Park, but the geo-ecological consequences are registered in the Park (Schlüchter 2014). The situation is different with the Ova Spin reservoir further down the Spöl Valley. Around 30% of the upper lake surface is in the Park as well as the left orographic border of the lake and half of the foundation of the dam wall. The hydropower scheme in the Spöl Valley causes a unique situation where waters from outside are brought into a natural park where they are fully managed for hydropower production. The Ova Spin reservoir is a central installation of the Engadine Hydropower Scheme (*Engadiner Kraftwerke*) as it is a reservoir for temporary water storage. From there, water is either pumped to the Lago di Livigno or sent to turbines further down the Inn Valley. This intermediate position makes the Lake of Ova Spin a sedimentation laboratory because the water remains on average only for around two days, thus causing an extremely high sedimentation rate of >60 cm/y. This sedimentation rate is among the highest known. Another aspect of interest is that some of the water brought to the Ova Spin reservoir is diverted further up valley from the Inn River, which is, to some degree at least, contaminated by the Upper Engadine wastewater treatment plant. The sediments in the Ova Spin lake therefore contain the wastewater signal from one of the most developed tourist resorts in the world: St. Moritz. The National Park is thus a sort of monitoring site of the human tourist ecology and its impacts on a largely unspoiled high alpine environment. A much more obvious human impact to the regular visitor of the Park is the road from Zernez across the Pass dal Fuorn to the Val Müstair (Fig. 17.1). This road divides the nature reserve in two parts, both acoustically and mechanically as seen in the number of accidents along the road, above all by motorbike traffic in summer and early fall. This is a significant problem, as this road is the only access to the north for the population of Val Müstair.

17.2.2 Geology

The geological setting for the area of the Park is best shown on a simple map (Fig. 17.2). One of the most important tectonic accidents of the Alps—the Engadine Line, a sinistral (left-lateral) wrench fault (Trümpy 1977)—cuts the area just to the northwest of the Park. To the northwest of the Engadine Line crystalline basement rocks are brought up against the sediments of the Engadine Dolomites by about 4 km, but in fact, the Engadine Line as a mechanical discontinuity cuts the rocks down to a depth of 10 km (Trümpy 1977). The exclave of Macun is situated in the crystalline rocks of the Silvretta to the northwest of the tectonic line. To the southeast the Engadine Dolomites are a tectonically complex set of faulted, folded, and pushed slabs of varying vertical and horizontal dimensions where they belong to the typical eastern alpine (Austroalpine) facies succession of mainly lower Mesozoic sediments. This sedimentary succession (Furrer 1993) spans roughly 210 million years of Earth history with the basal terrestrial sediments of the Münstertaler Verrucano (Permian), followed by thick shallow marine sandstones and dolomites (Triassic) with characteristic dinosaur tracks overlain by deeper marine marls, limestones, and radiolarian cherts (Jurassic to Lower

Cretaceous). The full sedimentary succession records the terrestrial environment of the old supercontinent Pangea during the Permian, its break-up phase with the shallow marine dolomites reflecting slightly changing sea levels (Triassic) and the full break-up and rifting of the northern from the southern continental margin with extensional faulting around 202 million years ago, with deeper marine sediments (Jurassic to Early Cretaceous) where the sedimentation ended (Furrer et al. 2013). The mountain building processes were initiated by compressive deformations involving folding and thrusting. The main lithologies, which not only control the deformation pattern of the Austroalpine sequence but also the morphology of today's landscape are: (i) the basal continental sediments (Permian); (ii) the main Triassic 1500 m thick unit of the "*Hauptdolomit*" (Main Dolomite); (iii) the Lower Middle Triassic Raibl-Group with evaporites; and (iv) the predominantly marly limestones and marls of the Jurassic and Cretaceous. The Main Dolomite dominates the attractive morphology of the mountains (Zimmermann et al. 2014).

17.3 High Diversity of Landforms

The landscape of the SNP is a scenery of contrasting landforms: the rolling lower hilly areas of the Spöl Valley and the upper reaches of the Fuorn Valley contrast sharply with the high rock walls of Piz Laschadurella, dal Fuorn and Nair to the north or with Piz Terza, Murter, da l'Acqua and Quattervals to the south. These peaks are the prominent mountains of the Engadine Dolomites in the heart of the SNP (Fig. 17.3). They all consist of Triassic Main Dolomite. Piz Laschadurella, Piz Fuorn, and Piz Nair are located to the north, Piz Terza, Piz da l'Acqua, and Piz Quattervals (Fig. 17.1) to the south of the Gallo Line, a major thrust with northerly/northwesterly compression onto deformable units of the Raibler-Beds (Fig. 17.2). The high peaks are tectonically speaking solitary scales or slabs, or more complex stacks of dislocated slabs. This means that the peak-and-wall morphology is preconditioned by tectonics. This is given also by the lithology of the high peaks: the Main Dolomite is a hard rock resistant to folding and mechanical thinning.

Fig. 17.3 A typical view of the Engadine Dolomites with Piz Quattervals and Piz d'Esan (photo H. Lozza)

Fig. 17.4 Northward thrust of the Piz Nair Slab marked by the highly deformed (brownish) Raibler-Beds in the rocky slopes to the east of Fuorcla da Val Botsch. The deformed and easily eroded Raibler-Beds are responsible for rounded morphologies in the Engadine Dolomites (photo C. Schlüchter)

However, it is sensitive to brittle faulting and a number of substantial faults are visible along the road between Zernez and the Pass dal Fuorn. The high peaks of the Engadine Dolomites are spread over a substantial area between the Engadine Valley and the Val Müstair. The northernmost peaks Piz Lavetscha and Piz Pisoc toward Val S-charl (Fig. 17.1) within the SNP are also prominent members of the Engadine Dolomites. They are a key example of the broken-up and thrusted sequence of scales consisting of the Main Dolomite. A morphological textbook example is the northward thrust of the Piz Nair slab toward the Piz Murters at the Fuorcla da Val Botsch (Fig. 17.4), which the hiking trail between Val dal Botsch (south) and Val Plavna/Val Mingèr (north) crosses and where the Raibler-Beds display the mechanical deformation resulting in the soft morphology of the Fuorcla.

The mountains of southeastern Switzerland, namely, the Engadine Dolomites, are characterized by simple geometry, in sharp contrast to the Central and Western Swiss Alps (see for example Pfiffner, this volume): they are of a uniform elevation of 3,100 m. This situation is called "*Ostalpine Gipfelflur*" or Eastern Alps Summit Level (Staub 1934; Trümpy et al. 1997) and is considered to be a remnant of an orogenetically older part of the Alpine landscape than the Penninic and Helvetic central and western sectors of the Alps. This also implies that the Engadine Dolomites are an erosional relict of a formerly much more comprehensive mountain range or of a vast plateau dissected by important faults along which early drainage evolved (possibly as early as the Late Cretaceous). The triangle of the Engadine Dolomites—consisting of sedimentary rocks of mainly Triassic age on continental units of Permian age, overlain with some Jurassic to Lower Cretaceous sediments and surrounded by Austroalpine crystalline basement rocks (Fig. 17.2)—tectonically forms a vast arch-like A-form, oriented SW–NE. The soft rounded morphology of the central Val dal Fuorn and further to the SE with the plain at Buffalora and the Val Müstair lies in the basal Permian sandstones and fanglomerates of the Münstertaler Verrucano, where the Triassic sequence has mainly been removed. This erosion has formed the wide-open Val dal Fuorn.

Landscape elements of exotic beauty can be found in the upper reaches of Val da Stabelchod and at Chaschabella

Fig. 17.5 Unique pillars in the National Park (photo H. Lozza)

(south of Munt la Schera; Fig. 17.1). They are individual, free standing pillars which are up to 10 meters high and several meters in diameter (Fig. 17.5). They consist of angular, secondary cemented polymictic rock fragments of sand, gravel and boulder size and they occur in the gypsum-bearing Raibler-Beds. They should be referred to as "secondary karst features" as they constitute refilled karst openings and tunnels by scree from slopes situated above washed into the system through flood events. The sediments of the pillars are a relict and breccia-type facies of what is called *Rauhwacke* (cornieule). The cement of these sediments is secondary calcite and is therefore more resistant to current erosion than the original Raibler Beds (containing gypsum) where the karst channels originally formed. Specific karst landforms are of restricted importance within the Park and are not shown on maps.

A striking morphological element of the Engadine Dolomites are the extended, voluminous, and characteristic scree slopes below the high walls of the peaks. Such vast scree slopes do not form elsewhere in the Swiss Alps and there are no other dolomite sequences of comparable thickness forming high peaks in Switzerland. Physical weathering of the dolomite produces characteristic gravelly to bouldery scree sediments which form steep slope surfaces determined by the internal angle of friction (Fig. 17.6). These scree slopes can be potential hazard areas as the material is easily mobilized by running water and thus sensitive to locally concentrated rainfall during summer storms. Intense erosion and scree remobilization occurred in Val dal Botsch, in the Piz Nair area, and Val Cluozza (Fig. 17.1) in summer 2014, with substantial accumulations in the depositional fans below the transport channels. In Val dal Botsch aggradation locally exceeded 2 m in a single event (Fig. 17.7).

In the current climatic conditions, dolomitic rocks, especially when interbedded with gypsum or *Rauhwacke* in tectonically stacked and therefore highly deformed situations, produce spectacularly collapsing slopes caused by intense physical erosion. Typical examples are the uppermost part of Val Vallatscha at the easternmost limit of the Park and Piz Daint to the south of Pass dal Fuorn. In both cases, the lithology of the rocks involved are the Raibler-Beds and the geometry of the strata (dip slopes on Piz Daint) is crucial.

The position of the Engadine Valley is set by the Engadine Line, which cuts the mountains to a depth of over

Fig. 17.6 Extensive scree slopes, typical morphological element of the Engadine Dolomites. Physical weathering of dolomitic carbonates favors their formation (photo H. Lozza)

Fig. 17.7 Voluminous scree formations easily remobilized during heavy rainfall, producing considerable deposits further downvalley (Val da Botsch July 2014 flooding) (photo C. Schlüchter, 2015)

Fig. 17.8 The high Macun plateau, view to the north, with lakes and rock glaciers (photo H. Lozza)

10 km. The Silvretta crystalline unit to the north of this lineament is part of the Austroalpine basement rocks—a complex succession of granites, gneisses, amphibolites, and ultrabasic rocks penetrated to the east by the Engadine Window of Penninic rocks (Fig. 17.2). At Zernez, the Engadine Valley turns to the north (Fig. 17.1) and cuts a multistep open gorge into the crystalline basement rocks of the Silvretta nappe. The Engadine Line and the present valley merge again at Scuol. The morphogenetic penetration of the river Inn to the basement rocks, thus abandoning the Engadine Line, is a major morphological dilemma in the SNP area, which remains unsolved. It is not merely coincidence that postglacial displacement along the Engadine Line has been mapped only four kilometers to the east of Zernez in the Val Laschadura (Schlüchter et al. 2013). The evidence for postglacial movement is given by "parasitic structures" with dextral orientation. Such observations make a tectonically controlled course of the northward bend of the river Inn the most likely explanation.

The SNP Macun exclave is in the crystalline Austroalpine basement rocks and is therefore not part of the Engadine Dolomites; as a consequence, it displays a different morphology. Macun has a complex multi-cirque morphology containing a wide plateau in the center with about 40 shallow small lakes and ponds (the so-called Macun "*Seenplatte,*" Fig. 17.8) at an average elevation of about 2,625 m a.s.l. The surrounding mountain crests are pointed and sharp, with an unnamed peak to the south of Piz Macun being the highest point with 3,046 m a.s.l. The high Macun plateau is open to the north where surface waters drain directly to the Engadine Valley.

At the Last Glacial Maximum (LGM), the area of the SNP was almost completely covered by glacier ice. The main valley glacier descending the Engadine Valley from the Engadine Ice Dome was joined by a second ice lobe through the Valley of Livigno. At Piz Mezdi in the Macun mountains the Engadine Valley Glacier reached an elevation of 2,820 m a.s.l. Glacial striae at Il Jalet (directly south of the Pass dal Fuorn) at 2,300 m a.s.l. clearly indicate ice flow across the pass toward the Val Müstair. The cirques developed around the high peaks of the Engadine Dolomites show that local ice joined the main ice streams above 2,800 m a.s.l. (Florineth 1998). Glacial morphological terms have an interesting significance in the Romansh language: **Munt** la Schera (=Mountain la Schera), **Mot** Tavrü, or **Mont** in some dialects mean an "easy to climb" mountain top, which has been moulded by glacial erosion, producing a rounded topography. In contrast a **Piz** (d'Arpiglias) is a pointed peak: its morphology is exclusively controlled by frost weathering and it has not been overridden by a glacier. The exact position of the trimline as the morphological marker of maximum elevation of glacial erosion is difficult to determine in the Engadine Dolomites. Fortunately, in the Macun mountains made of crystalline rocks glacial erosional features are better preserved. The glacially scoured lower valleys and mountain slopes are a prominent contrast to the "unpolished" rock walls above around 2,800 m of the Engadine Dolomites. Constructive glacial landforms are rare in the area of the SNP, most likely due to active morphological shaping of the mountains during the retreat of the LGM glaciers. In a few high mountain cirques moraines from unknown Lateglacial or Early Postglacial local

Fig. 17.9 A periglacial specialty at Macun: the "salami boulders," which move slowly downslope, depending on the moisture content of the soil matrix (photo C. Schlüchter)

readvances are present. Despite the fact that they are rare, they document multiple local glacier advances. The most characteristic examples are in Val dal Botsch and Val da Stabelchod, west of Piz Nair, between 2,300 and 2,600 m a.s.l.

The high mountain areas above the forests and continuous grasslands of the SNP are world famous for periglacial landscape elements as pioneering work was carried out here on such features and processes (see Gärtner-Roer and Hoelzle, this volume).

After the discovery of rock glaciers in the Alps, André Chaix started a general survey of this landscape element in the SNP in 1917 (Chaix 1923). It was also here that displacement rates of rock glaciers were measured over decades (Eugster 1973) and the rock glacier in Val Sassa inspired Jäckli (1978) for a model-case in soil mechanics (for a review of the importance of the SNP for research on permafrost and rock glaciers in Switzerland, see also Gärtner-Roer and Hoelzle, this volume). Measurements on creep in rock glaciers over the past 30 years show a marked reduction in displacement rates, which is of interest in the context of climate warming. Rock glaciers have been mapped in most of the high valleys in the Park. Exceptionally well-developed successions of inactive and active examples are around the peaks of Macun (Fig. 17.8). The high Macun plateau displays unique periglacial features, the so-called "salami boulders." Boulders of several cubic meters in original size decompose into slab-shaped slices through differential soil creep as a function of mainly soil moisture and slope angle. A simple precondition for this peculiarity to occur is the gneiss lithology of the boulders (Fig. 17.9).

An example of well-developed and widespread periglacial features are the earth lobes (Fig. 17.10), for example on Munt la Schera, and the delicate slopes with terracettes on Margunet, east of Piz dal Fuorn (Fig. 17.1). The earth lobes on Munt la Schera are monitored continuously. Furrer (1954) did a first study on the processes involved. The solifluction landforms are favored by bedrock lithology (Engadine Dolomites), which produces large quantities of scree and by the inner-Alpine continental climate with strong annual temperature and humidity gradients. Recently, Keller (2013) produced a map of the periglacial facies associations in the SNP based on climate modeling and field surveys.

Fig. 17.10 Example of solifluction landforms to the north of Pass dal Fuorn (western slope of Munt de la Bescha, view north toward Chaschlot) (photo C. Schlüchter)

Snow avalanches and their geomorphic effects in gullies and in the corresponding run-out areas are the high-energy version of periglacial processes. Repeated avalanche activity has caused sustainable interaction between surface geology and vegetation. An example is still visible in Val da Barcli, just west of the SNP boundary, where a destructive avalanche moved downhill from the southern slopes of Macun in 1999.

Deep valleys constitute another striking landscape element in the SNP. Some of them are deeply cut gorges, mainly around the western park border: Spöl River below Lake Ova Spin or Lower Cluozza and Fuorn up to Punt la Drossa. The northeastern boundary of the SNP to the south of Scuol follows the river Clemgia which flows for several kilometers in a morphologically active gorge in the dolomite rocks. The upper reaches of the rivers are largely characterized by open, braided channel systems (Fig. 17.11) with substantial sediment input from lateral tributaries. These rivers are an important landscape element enjoyed by visitors, who recognize them as unspoiled wilderness (Fig. 17.12).

17.4 Conclusions

The SNP is characterized by landscapes of contrasting beauty: the high rock walls and the pointed peaks (Piz) of the Engadine Dolomites versus the rolling hills of the Fuorn Valley leading to the Pass, the deeply cut valleys and gorges of the lower parts versus the open and broad upper valley floors, the high mountain periglacial and bare rock areas versus the forest of the lower slopes and the Mots, Munts, and Monts, the mountains smoothly eroded by the glaciers of the last ice age. The Macun exclave is a landscape of its own in the crystalline rocks of the Silvretta, contrasting sharply with the sedimentary bedrock landscape in the heartland of the SNP. In addition, it represents a unique cirque plateau of lakes and ponds, with exceptional periglacial features. To explain the course of the Engadine Valley bending around the mountains of Macun and leaving the tectonically predetermined direction of the valley remains a challenging and open geomorphological question in the larger area of the SNP.

Fig. 17.11 Braided river (in low-energy mode) in the Swiss National Park, Upper Val dal Botsch (photo C. Schlüchter, 2016)

Fig. 17.12 The rivers in the National Park are an important element of unspoiled wilderness (photo H. Lozza)

Acknowledgments We formally and cordially thank the Park staff for supporting and understanding the wishes and needs of scientists and the students from University of Bern and ETH Zurich who contributed to stimulating research in the SNP over many years. Marcia Phillips (SLF Davos) and Mélanie Clivaz (University of Lausanne) are acknowledged for useful comments on the manuscript and cartography of Figs. 17.1 and 17.2, respectively.

References

Baur B, Scheurer T (eds) (2014) Wissen schaffen. 100 Jahre Forschung im Schweizerischen Nationalpark. Haupt, Bern, 384 pp [exists also in French]

Chaix A (1923) Les coulées de blocs du Parc national suisse d'Engadine (Note préliminaire). Le Globe 62(1):1–35

Eugster H (1973) Bericht über die Untersuchungen des Blockstroms in der Val Sass im Schweizerischen Nationalpark von 1917–1971. Ergebnisse der wissenschaftlichen Untersuchungen im Schweizerischen Nationalpark 29, Zernez

Florineth D (1998) Surface geometry of the last glacial maximum (LGM) in the southeastern Swiss Alps (Graubünden) and its paleoclimatological significance. Eiszeit Gegenw 48:23–37

Furrer G (1954) Solifluktionsformen im Schweizerischen Nationalpark. Ergebnisse der wissenschaftlichen Untersuchungen im Schweizerischen Nationalpark 68, Zernez

Furrer H (1993) Stratigraphie und Facies der Trias/Jura-Grenzschichten in den oberostalpinen Decken Graubündens. PhD thesis, University of Zurich, 99 pp

Furrer H, Schlüchter C, Inderbitzin L, Pointner E, Margreth A, Gaar D, Frei J, Hanitzsch J (2013) Geologie und Erdgeschichte. Lithologisches und zeitliches Fundament des SNP. In: Haller H, Eisenhut A, Haller R (eds) Atlas des Schweizerischen Nationalparks. Die ersten 100 Jahre. Haupt, Bern, pp 16–17

Gärtner-Roer I, Hoelzle M (this volume) Rockglaciers in the Engadine

Haller H, Eisenhut A, Haller R (2013) Atlas des Schweizerischen Nationalparks. Die ersten 100 Jahre. Haupt, Bern, 247 pp [exists also in French]

Jäckli H (1978) Der Blockstrom in der Val da l'Acqua im Schweizerischen Nationalpark. In: Kasser P (ed) Gletscher und Klima. Jahrbuch der Schweizerischen Naturforschenden Gesellschaft 1978, pp 213–221

Keller F (2013) Permafrost. Klimasignale aus dem Untergrund. In: Haller H, Eisenhut A, Haller R (eds) Atlas des Schweizerischen Nationalparks. Die ersten 100 Jahre. Haupt, Bern, pp 32–33

Kupper P (2012) Wildnis schaffen. Eine transnationale Geschichte des Schweizerischen Nationalparks. Haupt, Bern, 371 pp

Pfiffner OA (this volume) The geology of Switzerland

Schläpfer D (2013) Die Eisenberge am Ofenpass. Haupt, Bern, 174 pp

Schlüchter C (2014) Dynamik im regulierten Spöl dank Hochwasser. In: Backhaus N et al. (eds) Wieviel Schutz(gebiete) braucht die Natur? Jahreskongress SCNAT 2014. Reihe Nationalpark-Forschung in der Schweiz 103:1–5

Schlüchter C, Clausen M, Stadelmann F, Kissling E (2013) Tektonik. Das bewegte und gestapelte Gebirge. In: Haller H, Eisenhut A, Haller R (eds) Atlas des Schweizerischen Nationalparks. Die ersten 100 Jahre. Haupt, Bern, pp 18–19

Staub R (1934) Grundzüge und Probleme alpiner Morphologie. Denkschriften der Schweizerischen Naturforschenden Gesellschaft 69:1–183

Trümpy R (1977) The Engadine Line: a sinistral wrench fault in the Central Alps. Mem Geol Soc China 2:1–12

Trümpy R, Schmid SM, Conti P, Froitzheim N (1997) Erläuterungen zur Geologischen Karte 1:50,000 des Schweizerischen Nationalparks (Geologische Spezialkarte Nr 122). Nationalparkforschung in der Schweiz 87, Zernez

Zimmermann S, Furrer H, Stoffel M, Schlüchter C (2014) Das Gesicht der Erde. In: Baur B, Scheurer T (2014) (eds) Wissen schaffen. 100 Jahre Forschung im Schweizerischen Nationalpark. Haupt, Bern, pp 114–147

Christian Schlüchter is Professor Emeritus at the University of Bern. He is an active member of several professional organizations and of editorial boards of international peer review journals. He has been a member of the Executive Committee of the International Union for Quaternary Research (INQUA) and actively engaged in its commission work. The XVIII International Congress of INQUA was held in Bern in 2011 under his presidency. He is a former president of the German Quaternary Association (DEUQUA) and recipient of the Albrecht Penck Medal. His research activities in glacial geology, environmental, and engineering geology have brought him to mountains on all continents. He has been working for many years with students in the Swiss National Park and has been the president of the science commission.

Hans Lozza studied Earth Sciences at the ETH Zurich. His Master thesis dealt with the stratigraphy and tectonics of the Lower Austroalpine Err-Carungas zone in the Val d'Err (Graubünden). After his studies he worked for five years as a geology-project manager for Electrowatt Engineering AG in Zurich. Since 1995 he is working for the Swiss National Park (SNP) and he concluded in parallel his studies as a teacher in Geography at the Didactic Unit of ETH Zurich. Since 1997 he is a member of the Executive Board of the SNP and acts as head of Communication and Public Relations.

Ruedi Haller studied Geography with focus on GIS and Remote Sensing at the University of Zurich (Ph.D.). He started his professional career as a programmer in forest information systems (1993–1996) before he came to the Swiss National Park (SNP) in 1997 and started to build up a GIS division. From 2000 to 2011 he was the head of the GIS and ICT division of the SNP and member of the Executive Board of the SNP. From 2012 to 2019 he has been leading the Research and Geoinformation Division of the SNP. Since 2019, he is the Director of the SNP. In 2013 he published a well recognized Atlas of the Swiss National Park, incorporating over 90 relevant topics of 100 years of research and management of the Swiss National Park. The current research activity includes the use of GIS in wildlife ecology, ecological connectivity, and the impact of spatially inaccurate data on results in wildlife management. His interests also cover location-based services on protected areas and the international habitat definitions, including Natura 2000.

18 Glacial and Periglacial Landscapes in the Hérens Valley

Christophe Lambiel

Abstract

The Hérens Valley is one of the main tributary catchments of the upper Rhone River valley. Despite the relative dryness of the climate, many large glaciers are present, thanks to altitudes rising up to 4,357 m a.s.l., reached at the top the Dent Blanche. Many early Lateglacial moraine deposits occur in the northern part of the valley (Euseigne Pyramids, La Luette moraines), whereas moraines from the Younger Dryas are mainly present on its orographic left side. Little Ice Age moraines present a large diversity of facies, from long moraine ridges to thick morainic bastions. The Ferpècle and Arolla catchments contain large glacier forefields, where strong paraglacial activity occurs. Periglacial landforms are also well developed in the valley. In its northern part, large relict rock glaciers are present, among which many are densely vegetated thanks to the schist lithology. Active rock glaciers are concentrated in the southern half of the catchment, where altitudes are higher. The valley presents also a large diversity of solifluction landforms.

Keywords

Hérens Valley • Glacial and periglacial landforms • Moraines • Rock glaciers • Solifluction

18.1 Introduction

The Hérens Valley is located on the orographic left side of the Rhone River in the Valais Alps (46°07′ N, 7°29′ E, Fig. 18.1). Since the 1960s, this valley has attracted a very large number of national and international scientists from the geological, glaciological and geomorphological communities. If the high glaciological and geo(morpho)logical diversity might be one of the main reasons explaining this interest, the long tourist tradition of the valley is probably another explanation. Indeed, since the nineteenth century, many travellers, climbers and painters, especially from the UK, came for the spectacular mountain landscapes and for mountaineering (Fig. 18.2). Another reason is the easy access to the glaciers, in particular thanks to the roads built for hydroelectric infrastructures. Accordingly, the glaciers of the valley, in particular the Haut Glacier d'Arolla, have been subject of a huge number of scientific publications (e.g. Gomez and Small 1985; Nienow et al. 1998; Goodsell et al. 2002; Gabbud et al. 2015; Capt et al. 2016). In addition to glaciological studies, several research programmes focused on glacial geomorphology, including glacial stadial reconstruction (Winistörfer 1977), moraine dating (Schneebeli and Röthlisberger 1976) and paraglacial evolution of Little Ice Age moraines (Curry et al. 2006). Another research topic is periglacial geomorphology, which has been mainly investigated in the Arolla Valley, with a special focus on the permafrost distribution and internal characteristics (e.g. Gardaz 1997; Lambiel and Pieracci 2008; Scapozza et al. 2011). Other recent studies focused on sediment transfer and slope movements (e.g. Barboux et al. 2014; Micheletti et al. 2015; Vivero and Lambiel 2019). Finally, an integrated analysis of a mountain system was recently proposed by Reynard et al. (2012a) and a complete geomorphological map of the Hérens Valley was published by Lambiel et al. (2016).

This large number of publications dedicated to glaciology and high mountain geomorphology of the Hérens Valley denotes its high interest to the scientific community. This chapter aims thus at showing the diversity of glacial and periglacial landscapes of the valley, by highlighting the geomorphic characteristics of various sites of the catchment.

C. Lambiel (✉)
Institute of Earth Surface Dynamics and Interdisciplinary Centre for Mountain Research, University of Lausanne, Géopolis, 1015 Lausanne, Switzerland
e-mail: christophe.lambiel@unil.ch

E. Reynard (ed.), *Landscapes and Landforms of Switzerland*, World Geomorphological Landscapes,
https://doi.org/10.1007/978-3-030-43203-4_18

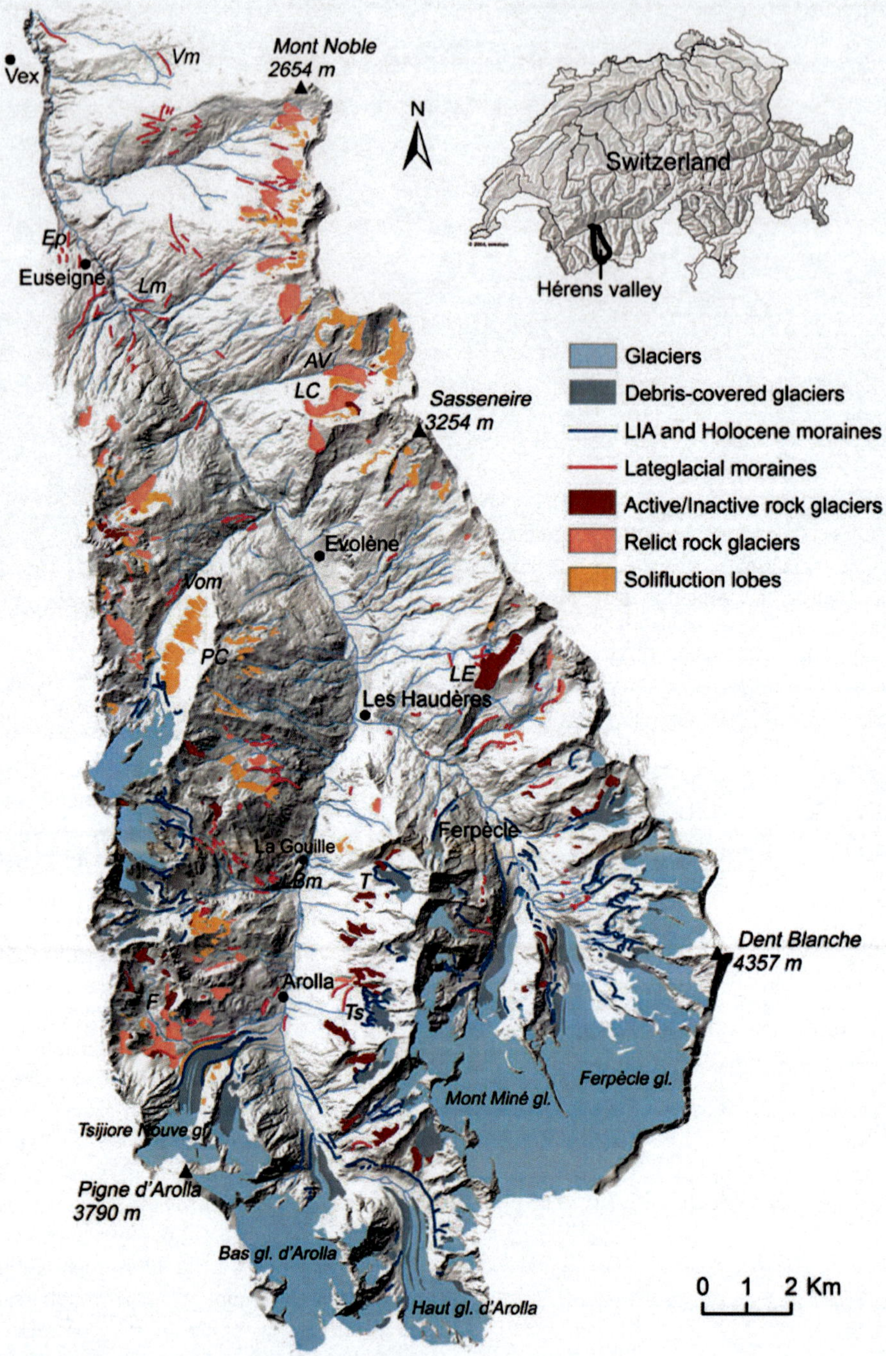

Fig. 18.1 The Hérens Valley, with the glaciers and the main glacial and periglacial landforms described in the text. *Vm* = Vernamiège moraine, *Ep* = Euseigne pyramids, *Lm* = La Luette moraines, *LBm* = Lac Bleu moraines, *Vom* = Vouasson moraines, *F* = Fontanesses, *T* = Tsarmine, *Ts* = La Tsa, *AV* = L'A Vieille, *LC* = Les Cliosses, *LE* = Liapey d'Enfer, *PC* = Palanche de la Cretta (hillshade relief: swisstopo)

Fig. 18.2 The Ferpècle Valley, with the coalescing Mont Miné (right) and Ferpècle (left) glaciers, painted by Johann Rudolf Bühlmann in July 1835 (Graphische Sammlung ETH Zürich)

18.2 Geographical, Geological and Climatic Setting

With a length of 30 km and an area of 270 km^2, the Hérens Valley is one of the largest tributary catchments of the upper Rhone River valley (Fig. 18.1). This south-north catchment divides into two main valleys just north of the village of Euseigne, at 710 m a.s.l.: the Hérémence Valley to the west and the Hérens Valley strictly speaking to the east. Only the latter is presented here. A second division occurs in Les Haudères, at 1,420 m a.s.l., with the Ferpècle Valley to the east and the Arolla Valley to the west. Altitudes range between 470 m a.s.l. at the outlet near Sion and 4,357 m a.s.l. at the top of the Dent Blanche.

The valley is located in the Pennine Alps. Five main tectonic units cut the valley transversally from north to south (Steck et al. 2001) (Fig. 18.3). The northern sector belongs to the Siviez-Mischabel and Mont Fort nappes (Middle Penninic). These units are made of quartzite, micaschist, gneiss and amphibolite. Further south is the Upper Penninic unit, with the Cimes Blanches nappe and the Tsaté nappe, which is mainly formed of oceanic sediments. Main rocks are calcschists, but prasinite (metabasalt) and serpentinite of the oceanic crust are locally also present. The southern part of the valley belongs to the Dent Blanche nappe (Austroalpine unit) and is composed mainly of orthogneiss of the Arolla series (Arolla gneiss). However, metagabbro (at the Mont Collon) as well as dolomite and breccia (Mont Dolin series) are also present in the Arolla Valley. The topography is very different between the Tsaté nappe sector and the Dent Blanche one. In the first sector, the slopes are rather smooth and rock walls are generally of reduced dimensions. In the second, the slopes are much steeper and rock walls can be up

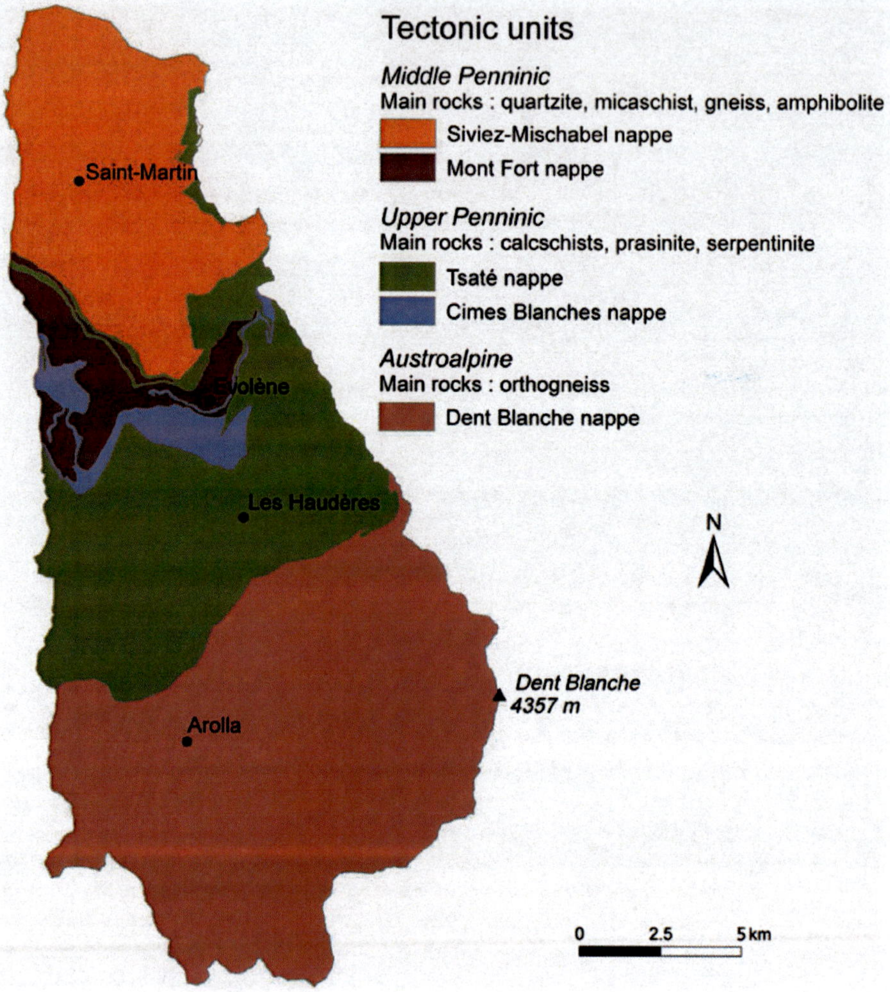

Fig. 18.3 Map of the main rocks and tectonic units of the valley

to 700 m high. Accordingly, sediment accumulations, such as talus slopes, moraine deposits or rock glaciers are generally thicker in the southern side of the valley.

Due to its position in the central part of the Alps, the Hérens Valley is sheltered from the main atmospheric disturbances. As a consequence, the climate of the valley is relatively dry, with the mean annual precipitation reaching 720 mm (mean 1987–2012) at the Evolène-Villa climate station (1825 m a.s.l.). 0 °C isotherm is around 2600 m a.s.l. Despite relatively low precipitation, around 40 km^2 of the valley is covered with glaciers. Five main glaciers occupy the head of the valleys (Fig. 18.1): Ferpècle, Mont Miné, Haut Glacier d'Arolla, Bas Glacier d'Arolla and Tsijiore Nouve, and four of them are integrated in the Swiss glacier monitoring network (https://www.glamos.ch). Outside the main glaciated areas more than 20 smaller glaciers are present, of which many are cirque glaciers located at the foot of high rock walls. The most significant are the Pièce, Aiguilles Rouges, Vouasson, Bricola, Dent Blanche and Manzettes glaciers. The landscapes of the Hérens Valley have been deeply influenced by the history of the past glacial fluctuations. In particular, large Little Ice Age (LIA) glacier forefields are present in the front of the main glaciers and thick moraine deposits are found at the front of smaller cirque glaciers.

18.3 Landforms and Landscapes

This section is divided into six subsections that present different sectors or landform categories present in the valley. Both WGS84 and Swiss coordinates are indicated after the site names, so that the sites can be either visualized on Google Earth or on https://www.map.geo.admin.ch, the

Swiss geoportal website, which contains topographical maps, orthophotos, DEM and geological maps. The location of the sites described can also be found in Fig. 18.1.

18.3.1 Lateglacial Moraine Deposits

Despite the generally steep valley sides, which locally prevented good preservation of the deposits, and the intense paraglacial and gravitational processes, which partially reworked the deposits, several Lateglacial moraines from different glacial stages are present in the valley. The oldest moraines are those of Vernamiège (46°12′44″ N, 7°26′54″ E; 2′600′800/1′117′900; Fig. 18.1), located at 1750 m on the right side of the valley, and those of Les Collons on the left side (46°10′54″ N, 7°23′08″ E; 2′595′900/1′114′500; not visible on Fig. 18.1), where three lateral crests 2 km long have been well preserved between 1800 and 1940 m a.s.l. These moraines correspond to the right, respectively, the left lateral moraines of an early Lateglacial stadial of the Hérens glacier.

Located more to the south, the Euseigne pyramids (46°10′24″ N, 7°25′02″ E; 2′598′350/1′113′520; Fig. 18.4a, b) are hoodoos built by gullying and gravitational erosion (Bollati et al. 2017) of till deposited on the right side of the Hérémence glacier. They are remarkable signatures of past glacier extension and have consecutively been listed in the inventory of Swiss geosites (Reynard et al. 2012b). Deltaic deposits with foreset dip angle of 20° and south-north direction are present at the base of the pyramids, indicating the former presence of a lake in the north part of the valley. Similar deposits are found 3 km to the north, south-east of Vex, with foresets oriented towards the south (Fig. 18.4b). They correspond to the northern border of the lake, whose presence would have been caused by the obstruction of the valley by the Rhone glacier (Sylvain Coutterand, unpublished data). La Luette moraines (46°09′50″ N, 7°26′30″ E; 2′600′200/1′112′500; Fig. 18.4c) are located 1.5 km

Fig. 18.4 Lateglacial moraines: **a** The Euseigne pyramids (photo C. Lambiel); **b** The Euseigne pyramids and, in the background, the Vex delta (arrow) (photo E. Reynard); **c** La Luette moraines (photo B. Maillard); **d** The Lac Bleu moraines (photo M. Kummert)

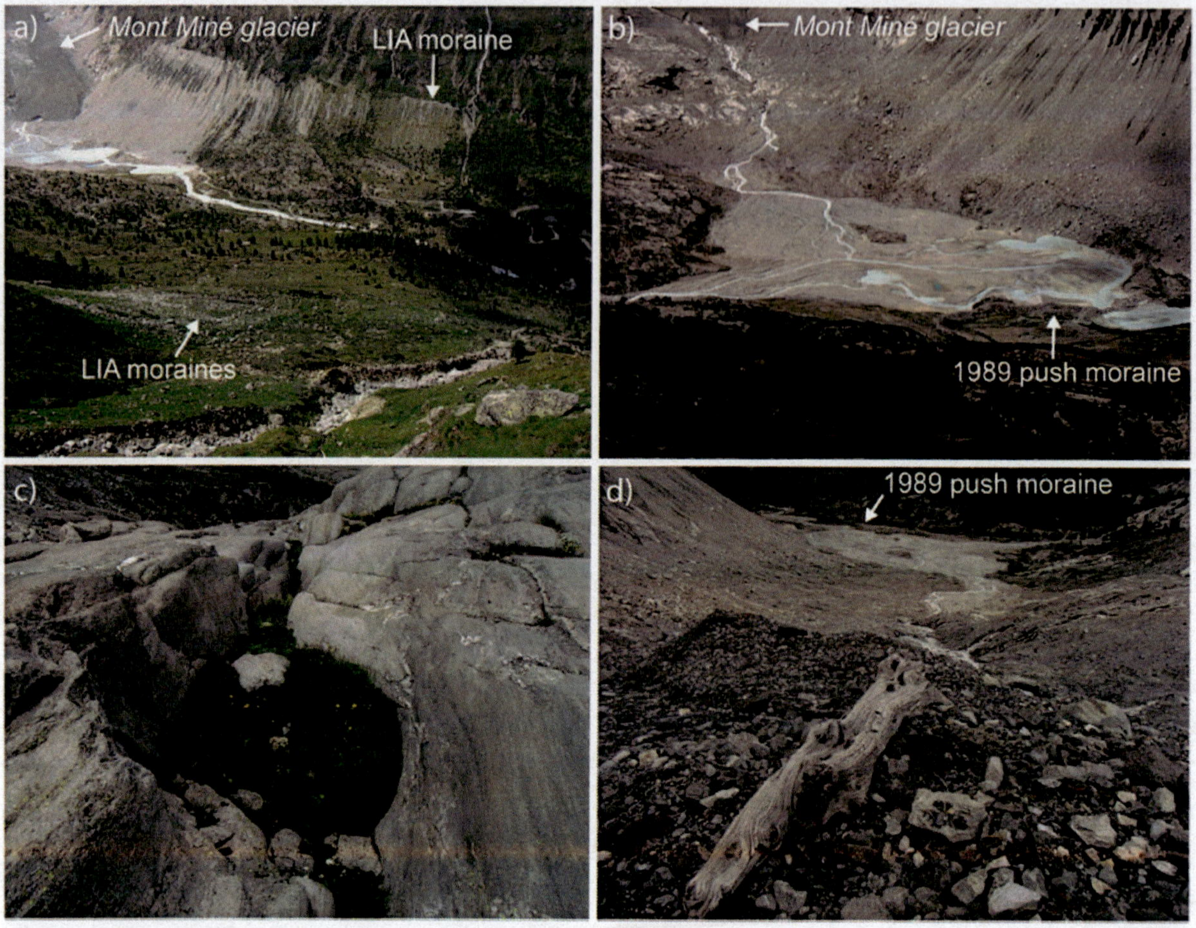

Fig. 18.5 Glacial landforms in the Ferpècle area. **a** General view of the glacier forefield, with the left lateral moraine of Mont Miné glacier; the spatio-temporal evolution of the vegetation colonization is visible (photo C. Lambiel 2004); **b** The sandur and the 1989 push moraine (photo C. Lambiel 2014); **c** Nye channels in metagranodiorites at the front of Ferpècle glacier (photo C. Lambiel); **d** Subfossil trunk (Swiss pine or larch) found at the surface of the Mont Miné glacier in July 2014 (photo C. Lambiel)

south-east of the Euseigne pyramids. They were built by the Hérens glacier and constitute the most prominent Lateglacial landforms of the valley.

Several Lateglacial moraines of adjacent glaciers can be identified higher up in the valley. Those of the Lac Bleu in the Arolla Valley are among the most remarkable ones (46° 03′02″ N, 7°28′48″ E; 2′603′200/1′099′900; Fig. 18.4d). This complex made of four crests was built by the confluence of the Aiguilles Rouges and the Arolla glaciers. The Lac Bleu (*Blue lake*) occupies a depression formed by the morainic complex. It is supplied by a spring located directly above the lake, which gives the lake its shining blue colour. The most significant Younger Dryas moraines are found on the orographic left side of the Hérens Valley, as for instance the Vouasson moraines (46°06′26″ N, 7°27′04″ E; 2′601′ 000/1′106′250; Fig. 18.1).

18.3.2 Ferpècle and Arolla Little Ice Age Glacier Forefields

Glacier retreat that followed the Little Ice Age (LIA) is at the origin of wide areas of unconsolidated sediments, the so-called glacier forefields or proglacial margins. The largest are naturally found at the front of the largest glaciers, that is the Ferpècle and Mont Miné glaciers in the Ferpècle catchment and the Haut and Bas Glaciers d'Arolla, in the Arolla Valley (Fig. 18.1).

During the LIA, the Ferpècle and Mont Miné glaciers were coalescing and occupied the entire Ferpècle area up to around 1800 m a.s.l (46°03″ N, 7°33′15″ E; 2′608′900/1′ 100′000; Fig. 18.2). The left limit of this position is nowadays clearly visible all along the valley flank (Fig. 18.5a). From the current position of the Mont Miné glacier to the

Fig. 18.6 The frontal part of the Mont Miné glacier during its last advancing phase in 1984 (photo M. Fort). We clearly see the frontal push moraine in formation and gullying of the steep LIA moraine

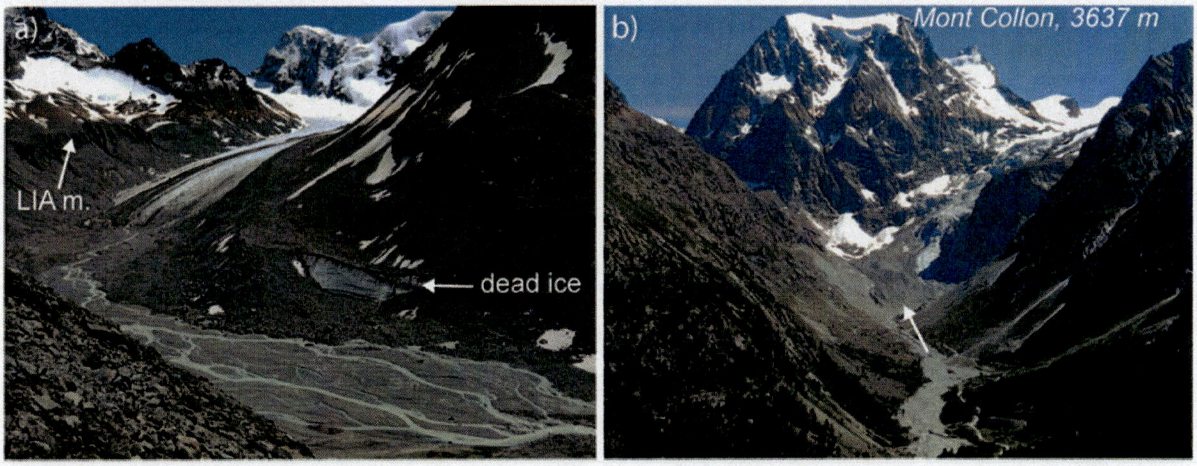

Fig. 18.7 **a** The Haut Glacier d'Arolla, with its proglacial margin (photo C. Gabbud 2013) and **b** The Bas Glacier d'Arolla, with its steep, crevassed section and the debris-covered frontal part (photo C. Gabbud 2013); the arrow indicates the glacier front; braided river system in the foreground

former frontal LIA position, that is, from south to north, the moraine shows evidence of typical spatio-temporal evolution of paraglacial reworking and plant colonization (Fig. 18.5a). Near the glacier, the slope gradient of the higher parts of the moraine is significantly greater than the repose angle. Gully systems cut through the moraine, whereas the material transported by debris flows has accumulated in the lower levels (Fig. 18.6; Curry et al. 2006). Towards the north, the slope has progressively adjusted and vegetation colonized the moraine. On the right side of the system, the slope gradient is less pronounced and thus the glacier built several adjacent moraines (Fig. 18.5a). As there is almost no headwall at the top of the accumulation zone of the Ferpècle glacier, the sediment load transport is low and thus the moraines are small.

The Ferpècle and Mont Miné glaciers separated in the early 1960s and the twentieth-century glacier retreat was then interrupted by a glacier advance in the 1970s and 1980s, due to a positive mass balance of both glaciers for about 20 years. The Mont Miné glacier excavated fluvio-glacial deposits and built a push moraine at its maximal extent in 1989 (Figs. 18.5b and 18.6). The depression formed upslope the moraine is nowadays occupied by a lake, which is progressively being filled by the new fluvio-glacial deposits. A sandur has consecutively developed in this sector. Between 1989 and 2014, the glacier retreat was 700 m.

Glacial erosion landforms are well present in Ferpècle, in particular *roches moutonnées* shaped by Lateglacial and Holocene glacier advances on gneissic rocks. At the front of the Ferpècle glacier a wide sector of post-LIA outcrops of *roches moutonnées* is present, with several Nye channels and glacial potholes (Fig. 18.5c).

Another specificity of the area is the presence of subfossil trunks expelled by the Mont Miné glacier (Fig. 18.5d). As no trees are present on the slopes overlooking the glacier nowadays, this indicates that the treeline was higher than today in some periods of the Holocene. Radiocarbon dating of samples, including peat layers collected in the lateral moraines, showed that the glacier was smaller than today several times during the Holocene (Schneebeli and Röthlisberger 1976; Joerin et al. 2006).

Two other large glacier forefields are present in the front of the Arolla glaciers. The Haut Glacier d'Arolla is a valley glacier with a particular low slope angle (45°58′ N, 7°31′25″ E; 2′606′600/1′92′000; Fig. 18.7a). About a hundred of scientific papers have been published on this glacier, which makes it one of the most investigated glaciers in the World (e.g. Gomez and Small 1985; Nienow et al. 1998; Gabbud et al. 2015). The glacier retreat has been especially significant since the end of the LIA. The internal flank of the lateral LIA moraines is currently being remobilized by very active gullying and gravitational processes, whereas strong fluvio-glacial activity occurs in the flat proglacial margin. Downslope, the Bas Glacier d'Arolla presents a spectacular, steep crevassed slope at the foot of the Mont Collon (45°59′ N, 7°29′40″ E; 2′604′400/1′92′500; Fig. 18.7b). The glacier forefield is also well developed and is marked by a very dynamic braided river system (Lane et al. 2014). Due to lower elevations, the proglacial margin here has been much more colonized with pioneer vegetation.

18.3.3 La Montagne d'Arolla: The Tsijiore Nouve Glacier and the Periglacial Catchment of Les Fontanesses

Located south-west of Arolla, the Tsijiore Nouve Glacier and its well-preserved moraines is another system that has been intensively studied by glaciologists and geomorphologists (Fig. 18.8). It is also included in the inventory of Swiss geosites (Reynard et al. 2012b). The glacier tongue is located at the base of a 500 m high, steep and strongly crevassed glacier wall (Fig. 18.8a). Today the tongue is largely covered with debris and is delimited by two voluminous moraines. The left lateral one is locally subdivided into four juxtaposed crests (Fig. 18.8b). Radiocarbon dating by Schneebeli and Röthlisberger (1976) showed that this complex formed during the last 3000 years and that the LIA deposits correspond only to a small part of the moraine. Another smaller moraine, also locally subdivided into four adjacent crests, is located outside this system on the left orographic side of the glacier (Fig. 18.8b, c). According to Scapozza (2013) it could date back to the end of the Younger Dryas and would then correspond to the Egesen stadial. The Holocene moraine is currently affected by intense paraglacial activity, as shown by the recent 500 m long slide of the internal part of the left lateral moraine (Fig. 18.8d).

A high concentration of active and relict rock glaciers made up of different lithologies typifies the Fontanesses catchment, northwest of the Tsijiore Nouve glacier (Figs. 18.1 and 18.9). In the lower part of the sector, rock glaciers either composed of gneiss, dolomite or breccia face each other, coming from the large Tsena Refien talus slope in the south, the small Petit Mont Rouge talus slope in the west and the Fontanesses catchment in the north (Fig. 18.9 a). At higher elevations, rock glaciers are still active, as for instance, the Mont Dolin rock glacier, with well-delimitated compression ridges (Fig. 18.9b).

Permafrost is not only present in rock glaciers, but also in the numerous talus slopes located at the base of the rock walls of the Fontanesses valley (Lambiel and Pieracci 2008). A thorough investigation of the Petit Mont Rouge talus slope by Scapozza et al. (2011) and Scapozza (2013) evidenced contrasted ground characteristics, with the presence of ice-cemented debris in the lower half of the slope and the absence of ground ice at the top of the slope. Delaloye and Lambiel (2005) showed that air advection in the debris is the main factor explaining this situation, by provoking cold winter air aspiration in the lower part of the slope.

Fig. 18.8 The Tsijiore Nouve glacier. **a** General view (photo C. Lambiel 2003); the letter "b" and the arrow indicate from where picture "b)" was taken; **b** Left lateral moraines (photo C. Lambiel 2015); the plain arrows indicate three different Holocene positions; the dotted arrow indicates a Lateglacial position (Younger Dryas); **c** The debris-covered glacier tongue and the two Holocene and Lateglacial moraine complexes (photo C. Lambiel 2003); the letter "d" and the arrow indicate from where picture "d)" was taken; **d** Slump in the left Little Ice Age lateral moraine, activated between 2012 and 2013 (photo C. Lambiel 2015)

Fig. 18.9 Rock glaciers in the Fontanesses catchment. **a** Relict rock glaciers, protalus ramparts and Lateglacial moraine in the lower part of the catchment; **b** The active Mont Dolin west rock glacier (photos C. Lambiel)

18.3.4 The Right Side of the Arolla Catchment: A Complex and Dynamic Valley Side

The right side of the Arolla catchment, between La Gouille and 1 km south of Arolla, is a steep valley side with altitudes roughly comprised between 1800 m a.s.l. and 3700 m a.s.l. (46°02′ N, 7°30′ E; 2′605′000/1′98′000; Fig. 18.1). This slope is affected by strong glacial, periglacial, gravitational and torrential activity.

During the Last Glacial Maximum, the main glacier thickness was 800 m in the Arolla Valley. The glacier surface was thus at an elevation of around 2600 m a.s.l. (Kelly et al. 2004). Glacial shoulders just above this altitude can be observed locally on the valley side and the impact of glacier erosion is evidenced by large sectors of *roches moutonnées*. Glacier retreat led to strong post-glacial decompression and consecutive large rock slope deformations occurred.

Due to steep slopes, the Lateglacial moraine deposits of the local glaciers were largely remobilized and are hardly visible today. On the other hand, thick moraine deposits were built during the Holocene above the glacial shoulders, as is best illustrated by Tsarmine (Fig. 18.10a) and La Tsa morainic bastions (Fig. 18.10b), located at the base of 500 m high rock walls. Nowadays, bare ice in the Tsarmine and La Tsa glaciers is restricted to the foot of the head walls, but actually debris-covered ice still occupies a large part of the sectors located directly upslope from the LIA frontal moraines, as shown by field observations, geophysical data and kinematic data (Barboux et al. 2014; Bosson and Lambiel 2016; Capt et al. 2016).

The valley side also houses several rock glaciers, the most active being the Tsarmine and La Roussette rock glaciers (Fig. 18.10a, b) (Barboux et al. 2014; Micheletti et al. 2015).

18.3.5 The Rock Glaciers of the Right Side of the Hérens Valley

The northern part of the orographic right side of the valley hosts several large relict rock glaciers between 2300 m and 2500 m a.s.l. Most of them are made of calcschist blocks, which have promoted the development of soil and, accordingly, herbaceous vegetation. According to Scapozza (2013), these large landforms may have started to form during the Bølling, that is, during a warming period between the colder Older and Younger Dryas periods over which the erosion rates were significantly higher than today. The L'A Vieille rock glacier is one of the most prominent landforms in the sector (46°09′ N, 7°30′ E; 2′605′000/1′111′000; Fig. 18.11 a). Its upper half presents angular ridges, suggesting that a glacier partially overlapped the rock glacier, probably during the Younger Dryas. This illustrates the complex glacier-permafrost interactions that probably occurred during the Lateglacial, not only in L'A Vieille, but also on the other large relict rock glaciers of the sector. Another similar landform is located just south of L'A Vieille, in Les Cliosses (Fig. 18.11b). Here the roots have been overlaid by an active rock glacier. According to its topoclimatic situation (exposure west, front at 2460 m a.s.l. and 0 °C isotherm at around 2600 m a.s.l.), this rock glacier is located at the very lower limit of discontinuous permafrost in this part of the Alps.

In this area, some rock glaciers made of quartzitic blocks can also be observed (Fig. 18.11c). Here the coarse blocks and the lack of fine-grained sediments prevented the development of soil and vegetation. On the other hand, the blocks are almost completely covered with lichens (rhizocarpons). These blocky landforms contrast strongly with the much smoother surfaces of schistose rock glaciers.

Fig. 18.10 The right side of the Arolla Valley in the **a** Tsarmine (photo R. Delaloye) and **b** La Tsa (photo C. Lambiel) sectors. Both sectors host a small debris-covered glacier above a thick morainic bastion, as well as active rock glaciers

Fig. 18.11 Rock glaciers on the right side of the Hérens Valley. **a** The L'A Vieille relict rock glacier (photo C. Lambiel); **b** Roots of a relict rock glacier overlaid by an active rock glacier at Les Cliosses (photo S. Rüttimann); **c** The quartzitic relict rock glacier at Les Rechasses (photo B. Maillard); **d** The Liapey d'Enfer rock glacier (photo U. Raz)

South of the Sasseneire and up to the Pointe du Tsaté, steep headwalls are uncommon due to the combination of tectonic and lithological factors. The sediment production has thus been reduced and, accordingly, rock glaciers or thick moraine deposits are rare. Nevertheless, one large rock glacier is present south of Pointe du Tsaté at the foot of a serpentinite rock wall. The place is named *"Liapey d'Enfer"*, which means "Hell's block field", probably because it was a "hell" for the cattle adventuring at its surface (46°05′30″ N, 7°32′50″ E; 2′608′200/1′104′400; Fig. 18.11d). In addition to its large size, the interest of this rock glacier resides in the orange colour of the blocks, which is due to the alteration of serpentinite.

18.3.6 Solifluction Landforms

In temperate mountain regions, seasonal frost, steep slopes and fine-textured soils are necessary conditions for the occurrence of solifluction (Matsuoka 2001). This explains the high concentration of solifluction landforms on the schistose debris slopes of the Tsaté nappe. Good examples are found on the western slope of Palanche de la Crettaz (46° 05′50″ N, 7°27′20″ E; 2′601′300/1′105′400; Fig. 18.12a) or to the north of Mont-Noble (46°12′45″ N, 7°29′09″ E; 2′603′650/1′117′900; Fig. 18.12b). In the latter, despite the overall coarse grain size, solifluction occurs thanks to the supply of fine sediments by occasional debris flows. As fine matrix is necessary to retain moisture, solifluction also affects lodgement till deposits or morainic crests, like the LIA moraines of the Vouasson (46°05′20″ N, 7°26′50″ E; 2′600′700/1′104′150; Fig. 18.12c) and the Tsijiore Nouve glaciers (46°00′56″ N, 7°27′528″ E; 2′601′500/1′96′000; Fig. 18.12d). Figure 18.12 also illustrates considerable diversity of solifluction landforms, from densely vegetated lobes to ones completely free of vegetation. This is partly due to contrasts in the activity of mass transport processes.

Fig. 18.12 Solifluction landforms on debris slopes, **a** west of Palanche de la Crettaz (photo B. Maillard) and **b** north of Mont Noble (photo C. Lambiel) and on the LIA moraines of **c** the Vouasson glacier (photo B. Maillard) and **d** the Tsijiore Nouve glacier (photo C. Lambiel)

18.4 Conclusion

This chapter gives an overview of the diversity of glacial and periglacial landscapes in the Hérens Valley. Glacial accumulations from several Lateglacial and Holocene stages, in particular the Little Ice Age, are numerous all along the catchment. At the front of the main glaciers, large glacier forefields have developed, whereas thick morainic bastions are present near the small cirque glaciers located at the foot of high gneissic rock walls in the southern part of the valley. Landscape diversity is particularly well expressed by different types of rock glaciers, from large and smooth relict landforms in the northern part of the valley to the coarse steep active ones in the southern part. This diversity is explained mainly by lithological variability and resultant various topographies. The Hérens Valley is thus an ideal area to observe glacial and periglacial processes, as well as landforms derived from these processes. Due to its geographical characteristics, especially size, altitudinal range and climate, it is representative of the glacial and periglacial geomorphology of other valleys located in the Valais Alps and in the Graubünden. Regarding current climate change, this landscape is, however, experiencing rapid changes, in particular an accelerated glacier shrinkage.

References

Barboux C, Delaloye R, Lambiel C (2014) Inventorying slope movements in Alpine environment using DInSAR. Earth Surf Proc Land 39(15):2087–2099

Bollati I, Pellegrini M, Reynard E, Pelfini M (2017) Water driven processes and landforms evolution rates in mountain geomorphosites: examples from Swiss Alps. Catena 158:321–339

Bosson JB, Lambiel C (2016) Internal structure and current evolution of very small debris-covered glacier systems located in Alpine permafrost environments. Front Earth Sci. https://doi.org/10.3389/feart.2016.00039

Capt M, Bosson JB, Fischer M, Micheletti N, Lambiel C (2016) Decadal evolution of a very small heavily debris-covered glacier in Alpine permafrost environment. J Glaciol 62:535–552

Curry AM, Cleasby V, Zukowskyj P (2006) Paraglacial response of steep, sediment-mantled slopes to post-'Little Ice Age' glacier recession in the central Swiss Alps. J Quat Sci 21:211–225

Delaloye R, Lambiel C (2005) Evidences of winter ascending air circulation in talus slopes situated near the lower limit of alpine

discontinuous permafrost (Swiss Alps). Nor Geogr Tidsskr 59:194–203
Gabbud C, Micheletti N, Lane SN (2015) Lidar measurement of surface melt for a temperate Alpine glacier at the seasonal and hourly scales. J Glaciol 61(229):963–974
Gardaz JM (1997) Distribution of mountain permafrost, Fontanesses Basin, Valaisian Alps, Switzerland. Permafr Periglac Process 8:101–105
Gomez B, Small RJ (1985) Medial moraines of the Haut Glacier d'Arolla, Valais, Switzerland: debris supply and implications for moraine formation. J Glaciol 31:303–307
Goodsell B, Hambrey MJ, Glasser NF (2002) Formation of band ogives and associated structures at Bas Glacier d'Arolla, Valais Switzerland. J Glaciol 48(161):287–300
Joerin UE, Stocker TF, Schlüchter C (2006) Multicentury glacier fluctuations in the Swiss Alps during the Holocene. Holocene 16:697–704
Kelly MA, Buoncristiani JF, Schlüchter C (2004) A reconstruction of the last glacial maximum (LGM) ice-surface geometry in the western Swiss Alps and contiguous Alpine regions in Italy and France. Eclogae Geol Helv 97(1):57–75
Lambiel C, Pieracci K (2008) Permafrost distribution in talus slopes located within the alpine periglacial belt (Swiss Alps). Permafr Periglac Process 19:293–304
Lambiel C, Maillard B, Kummert M, Reynard E (2016) Geomorphology of the Hérens valley (Swiss Alps). J Maps 12(1):160–172
Lane SN, Bakker M, Balin D, Lovis B, Regamey B (2014) Climate and human forcing of Alpine river flow. In: Schleiss AJ, de Cesare G, Franca MJ, Pfister M (eds) River Flow 2014. Taylor and Francis, London, pp 7–15
Matsuoka N (2001) Solifluction rates, processes and landforms: a global review. Earth Sci Rev 55:107–133
Micheletti N, Lambiel C, Lane SN (2015) Investigating decadal scale geomorphic dynamics in an Alpine mountain setting. J Geophys Res: Earth Surf 120:2155–2175
Nienow P, Sharp M, Willis I (1998) Seasonal changes in the morphology of the subglacial drainage system, Haut Glacier d'Arolla, Switzerland. Earth Surf Proc Land 23:825–843
Reynard E, Lambiel C, Lane SN (2012a) Climate change and integrated analysis of mountain geomorphological systems. Geogr Helv 67(1–2):5–14
Reynard E, Berger JP, Constandache M, Felber M, Grangier L, Häuselmann P, Jeannin PY, Martin S (2012b) Révision de l'inventaire des géotopes suisses: rapport final. Groupe de travail pour les géotopes en Suisse, Lausanne
Scapozza C (2013) Stratigraphie, morphodynamique, paléoenvironments des terrains sédimentaires meubles à forte déclivité du domaine périglacaire alpin. PhD thesis, University of Lausanne (Géovisions n°40)
Scapozza C, Lambiel C, Baron L, Marescot L, Reynard E (2011) Internal structure and permafrost distribution in two alpine periglacial talus slopes, Valais, Swiss Alps. Geomorphology 132:208–221
Schneebeli W, Röthlisberger F (1976) 8000 Jahre Gletschergeschichte im Wallis. Die Alpen (SAC) 52(3/4):5–134
Steck A, Epard JL, Escher A, Gouffon Y, Masson H (2001) Carte tectonique des Alpes de Suisse occidentale. Notice explicative. Office fédéral des eaux et de la géologie, Berne (Carte géologique spéciale n° 123)
Vivero S, Lambiel C (2019) Monitoring the crisis of a rock glacier with repeated UAV surveys. Geogr Helv 74(1):59–69. https://doi.org/10.5194/gh-74-59-2019
Winistörfer J (1977) Paléogéographie des stades glaciaires des vallées de la rive gauche du Rhône entre Viège et Aproz. Bull Murithienne 94:3–65

Christophe Lambiel is Senior Lecturer in Geomorphology at the Institute of Earth Surface Dynamics at the University of Lausanne. His main scientific interest is in the field of mountain geomorphology, with a special focus on mountain permafrost (spatial distribution, ground characterisation, rock glacier dynamics, permafrost monitoring, glacier-permafrost relationships). He mainly works in the Swiss Alps and he is a member of the Scientific Committee of the Swiss Permafrost Monitoring Network (PERMOS) and of the Committee of the Swiss Geomorphological Society (SGS).

19 The Glacial Landscape at Wangen an der Aare

Susan Ivy-Ochs, Kristina Hippe, and Christian Schlüchter

Abstract

The glacial landforms at Wangen an der Aare provide a footprint of the extent of the Valais (Rhone) glacier during the Last Glacial Maximum. The early studies of the region played a key role in development of the glacial theory during the early nineteenth century. The Valais glacier advanced into the region several times leaving evidence of its presence in deposits of moraines and outwash. The former are recognized as the broad band of hills that extends from Oberbipp to Bützberg and around to the south at Steinhof–Steinenberg. Direct dating of erratic boulders with cosmogenic ^{10}Be suggests withdrawal of the Valais glacier from the area starting at about 24,000 years ago. Coarse gravels carried by meltwater streams emanating from the glacier were deposited as outwash plains to the northeast of the moraines. As the glacier withdrew stepwise incision into the plains left a flight of terraces extending down to the present flood plain and path of the Aare River.

Keywords

Valais (Rhone) glacier • Alps • Last Glacial Maximum (LGM) • Moraines • Erratics • Outwash terraces

S. Ivy-Ochs (✉) · K. Hippe
Laboratory for Ion Beam Physics, ETH Zurich, 8093 Zurich, Switzerland
e-mail: ivy@phys.ethz.ch

K. Hippe
e-mail: kristina.hippe@fu-berlin.de

K. Hippe
Department of Earth Sciences, Freie Universität Berlin, Malteserstrasse 74-100, 12249 Berlin, Germany

C. Schlüchter
University of Bern, Institute of Geological Sciences, Baltzerstrasse 1 + 3, 3012 Bern, Switzerland
e-mail: christian.schluechter@geo.unibe.ch

19.1 Introduction

Wangen an der Aare (47°14′ N, 7°40′ E) lies in the westernmost part of the Swiss Alpine foreland (Fig. 19.1) in the region known as Oberaargau, which includes parts of the cantons of Bern, Solothurn, and Aargau. The dominant element is the Aare River (here at about 450 m a.s.l.), which flows in a northeasterly direction to join the Reuss and Limmat rivers and eventually the Rhine River. Following the trace of the Aare are the main highway and railway thoroughfares connecting Bern, Basle, and Zurich. Looking out over the area one sees a broad valley filled with low relief tree-covered hills. It is bound to the west by the steeply rising Weissenstein ridge of the Jura Mountains (Fig. 19.2), which attains an elevation of nearly 1300 m a.s.l. To the east rounded Molasse hills rise gently in the direction of the Napf highlands (1400 m a.s.l.). In the background the high Alps of the Berner Oberland can be seen. The mean annual temperature is about 8–9 °C (1961–1990). Annual precipitation sums are in the range of 1000–1400 mm per year with the higher values associated with the Jura Mountains.

The landforms and deposits in the Wangen an der Aare region and the adjacent Jura Mountains played a key role in the birth of the glacial theory in the early nineteenth century. The scientific dilemma was how did blocks of granite, gneiss, and gabbro from the Alps end up on top of the limestones of the Jura Mountains? The blocks, which were also scattered across the Swiss lowlands, were called *Findlinge*, *Irrblöcke*, which means erratic blocks (Binggeli 1983). Already in 1825 Bernhard Studer provided a detailed description of the landforms of the Wangen area, although he attributed the deposits and boulders to massive floods. In 1834, Jean de Charpentier gave a lecture entitled "Notice sur la cause probable du transport des blocs erratiques de la Suisse" (de Charpentier 1841). He presented concepts he had discussed formerly with Jean-Pierre Perraudin and Ignaz Venetz in the upper Rhone Valley about glacial deposits in the Alps and the relevance of such concepts to the deposits

E. Reynard (ed.), *Landscapes and Landforms of Switzerland*, World Geomorphological Landscapes,
https://doi.org/10.1007/978-3-030-43203-4_19

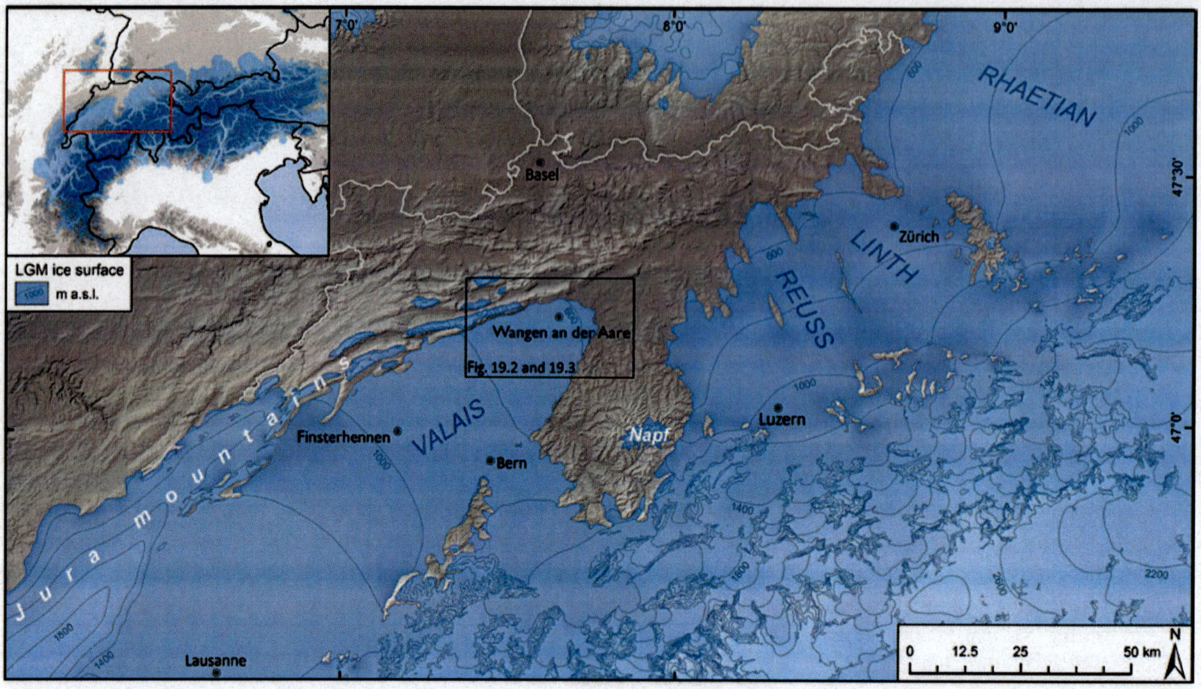

Fig. 19.1 Overview map of the Swiss Alpine foreland showing the piedmont glacier lobes of the Swiss Alpine foreland during the Last Glacial Maximum (modified after Schlüchter 2009)

and landforms of the Wangen area. The huge blocks in the Jura and on the forelands must have been brought from the Alps by glaciers. These concepts were slow to gain acceptance as the flood theory held sway at the time. Nevertheless, after strong promoting by Louis Agassiz the theory of the Ice Ages was finally accepted (Krüger 2008). The culmination of the understanding of the remarkable expansion of glaciers from the Alps in the past is displayed in the map of Schlüchter (2009). It shows the Valais, Reuss, Linth, and Rhaetian (Rhine) glaciers on the northern Swiss forelands as they must have looked 24,000 years ago[1] (Fig. 19.1).

In this paper we discuss the landscape elements of the Wangen an der Aare region with special focus on those formed through glacial processes. Based on the geological record it has been concluded that during the last 2.5 Ma, corresponding to the Quaternary period of Earth's history, glaciers have extended from the Alps onto the Swiss forelands at least 15 times (Schlüchter et al. this volume). The glacial and glaciofluvial deposits of past glaciations in Switzerland have been subdivided into high *(höhere)* and low *(tiefere) Deckenschotter*, high terraces (*Hochterrasse*, in classical terms "Riss") and low terraces (*Niederterrasse*, classical "Würm"). These correspond to the early (2.58–0.788 Ma), middle (788–120 ka), and late Pleistocene (the last 120 ka), respectively (Graf and Burkhalter 2016). In the Wangen region, deposits assigned to the *Hochterrasse* and *Niederterrasse* categories are present. Evidence suggests that in the Wangen region several glaciations must have occurred prior to the last glaciation although ages for all but the last glaciation have not been determined. Direct dating suggests that the Valais glacier constructed the moraines at Wangen an der Aare synchronous with the global Last Glacial Maximum (LGM) during marine isotope stage 2 (Ivy-Ochs 2015; Seguinot et al. 2018 and references therein). In this paper we use the terms Pre-LGM and LGM for the several landform assemblages (cf. Bitterli et al. 2011), with distinct focus on the latter.

19.2 Geological Setting

Underlying the Quaternary deposits in the Wangen region are Molasse sandstones and conglomerates and below them Triassic, Jurassic, and Cretaceous calcareous rocks (Ledermann 1978; Gerber and Wanner 1984; Bitterli et al. 2011) (Fig. 19.2). The Jura Mountains are made up of Triassic (230–200 Ma) evaporites, including anhydrite and rock-salt layers, overlain by Jurassic and Early Cretaceous (200–130 Ma) limestones and marls. The structure of the Jura is dominated by west and northwest directed thrusting with a suggested basal décollement along the Triassic evaporites

[1]Age based on cosmogenic ^{10}Be exposure dating (see also below).

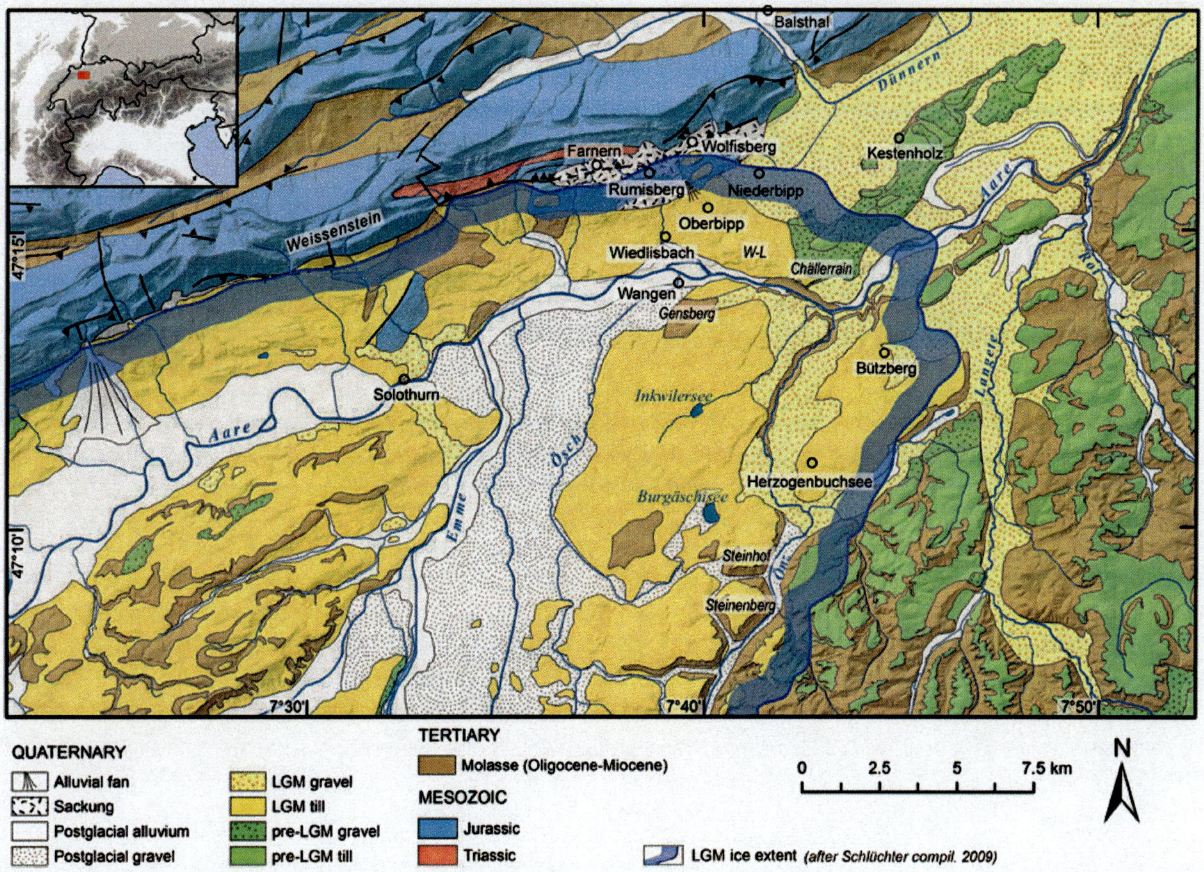

Fig. 19.2 Geological map of the region of Wangen an der Aare (based on Ledermann 1978; Gerber and Wanner 1984; Bitterli et al. 2011) superimposed on a hillshade image based on the Swiss ALTI3D DEM (swisstopo). W-L stands for Walliswil-Längswald. The location of the Walliswil-Bipp gravel pit (Fig. 19.4) is just south of the W-L on the map

(Pfiffner 2015). Thrusting occurred in response to Alpine orogeny during the Miocene, starting around 10 Ma and continued at least into the late Pliocene. The present ridge and valley topography characterized by sharp crests of anticlines and intervening synclines in part reflects this fold and thrust activity.

The northeast trending parallel ridges of the Jura Mountains are perpendicularly breached at several points along their length by steep-walled gorges called *Klus* (local word for an opening or break that runs perpendicular to the general trend of the ridges). The Balsthal Klus (Fig. 19.3) crosses the Weissenstein anticlinal ridge (Fig. 19.2), with spectacular views of the internal structure of the anticline. The Dünnern stream flows through the Balsthal Klus following a path that may also in part be tectonically controlled.

In the Aare Valley itself the bedrock underlying the Quaternary deposits comprises the Oligocene to Miocene Molasse units of the *Untere Süsswassermolasse* (USM) and *Obere Meeresmolasse* (OMM). Beige to gold, at times green glauconitic, sandstones, and marls comprise the OMM. The USM encompasses brown, yellow, red, and violet sandstones, marls, and interbedded conglomerate layers (Bitterli et al. 2011). USM sandstones with concretions form marked benches in the topography. The isolated Molasse hills that are exposed below the glacial deposits along the orographic right side are the morphological extension of the Napf area (see Fig. 19.1 for location). The latter comprises a gentle topography that was dissected by meltwater streams over several glacial periods. Repeated periods of Valais glacier advance led to carving of the broad trough between the hilly Molasse topography to the SE and the Jura to the NW. The top surface of the bedrock is irregular and shows the existence of several buried channels, reflecting the patterns of erosion by the Aare River and the Valais glacier during middle Pleistocene glaciations (Dürst Stucki and Schlunegger 2013). Their data compilation reveals the presence of two main Aare River (paleo-) channels. They are separated by the SW to NE trending Kestenholz Molasse bedrock high, which is mantled with pre-LGM tills and gravels (Fig. 19.2). The broader, deeper bedrock channel lies

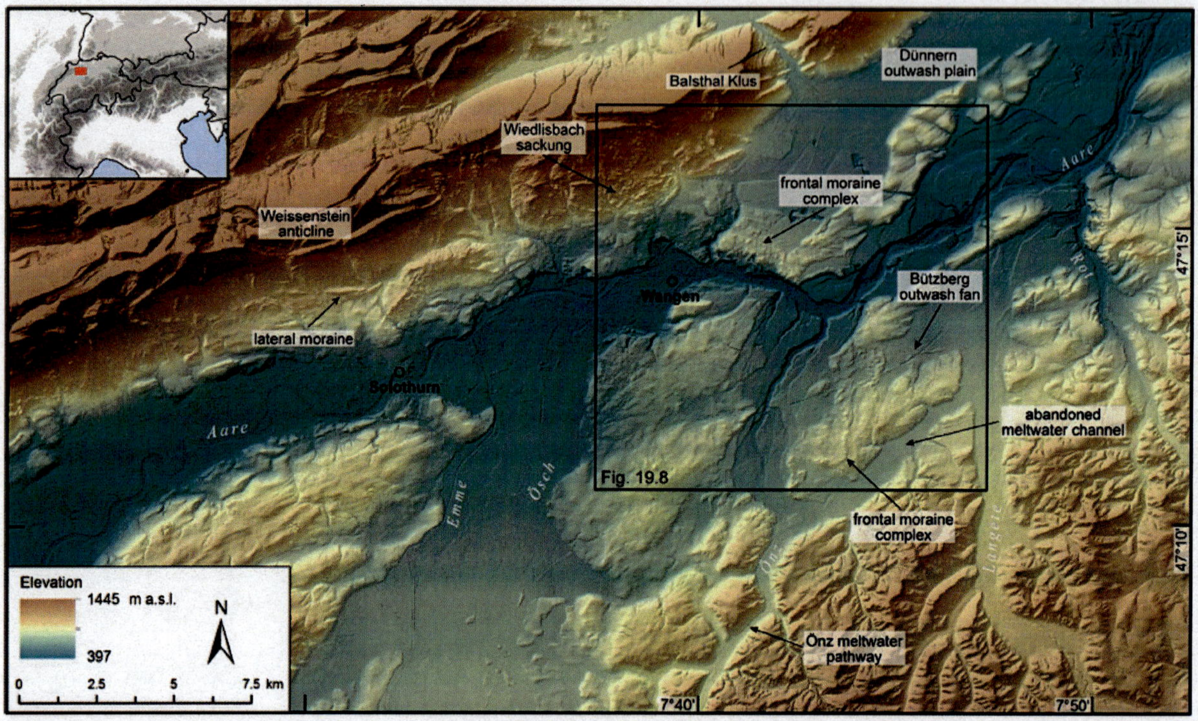

Fig. 19.3 Shaded relief image of the Wangen an der Aare region superimposed by a color-coded DEM (swissALTI3D, swisstopo). Altitude is given by color code

Fig. 19.4 View of the east wall of the Walliswil-Bipp gravel pit (location shown in Fig. 19.2). Alternating layers of outwash gravel, thin-bedded sands, and silty-sandy till originally deposited in front of and beneath the advancing glacier, were deformed into a repeating sequence due to subglacial thrusting from right to left (to the north-northeast) (see also Fig. 19.5). Height of the section is about 5 m (photo C. Schlüchter, 2015)

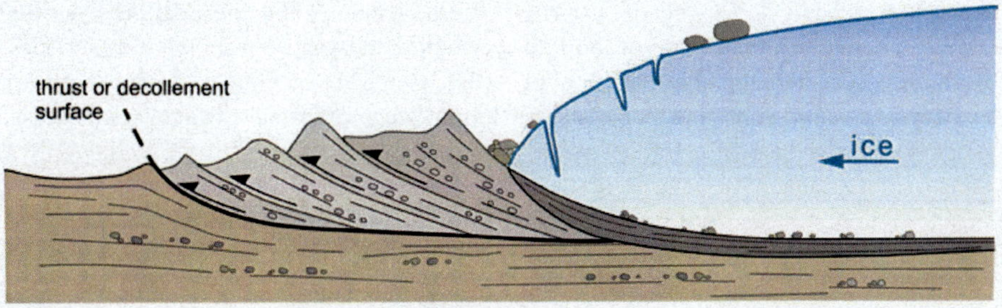

Fig. 19.5 Schematic cross-section through an advancing glacier. Slices of former subglacial and proglacial outwash sediments are stacked, while being pressed upwards and forwards by the compressional forces at the front of the glacier (modified after Binggeli 1983; Aber et al. 1989)

to the N along the foot of the Jura and is filled with sediments; therefore, its presence is not discernible from the surface morphology. The shallower channel presently houses the Aare River.

19.3 Landforms and Landscapes

Past glaciations have left their mark not only in the shape of the underlying bedrock, but also in the huge volumes of glacial and related sediments, till and outwash that fill the bedrock topography (Ledermann 1978; Gerber and Wanner 1984; Bitterli et al. 2011). Till is sediment that has been deposited solely from a glacier with little to no involvement of water or gravity. It is unsorted (matrix-supported glacial diamicton) containing all grain size fractions, ranging up to even many meter diameter blocks, embedded in a matrix of clayey silt. Associated with the purely glacial sediments are the outwash gravels, which were transported away from the ice margin by meltwater streams. Outwash is better sorted and poorer in the fine-grained components than till. Outwash is often found as crudely bedded to well-stratified, clast-supported gravels, interbedded with thin-bedded to massive sand layers. The abundance of gravel-rich outwash deposits in this region has played an important economic role as shown by the many gravel quarries.

19.3.1 Pre-LGM Glacial Deposits

Deposits related to older pre-LGM glaciations are found outside of and beyond the well-defined footprint of the LGM Valais glacier (Fig. 19.2). In comparison to the gravels of the last glacial cycle, the older gravels are often better cemented and contain slightly different pebble lithologies. During the earlier glaciations, ice from the Alps may have covered much of the Jura and reached higher along the Napf peak than ice during the LGM (Schlüchter et al. this volume). In addition, gravel deposits from the pre-LGM glaciations underlie the gravels of the last glaciation on the Dünnern plain. The former are exposed in the deeper levels of many of the gravel pits in the region. The older outwash gravels and associated till mantle the Kestenholz ridge (Fig. 19.2) where they outcrop at elevations 30–40 m higher than the adjacent LGM outwash surface. The deposits show that during earlier glaciations huge outwash fans were built up in front of the expanded Valais glacier. During subsequent interglacials and glacials the fans were dissected. The outwash fans of the last glaciation are inset into this incised fan configuration (Zimmermann 1963; Bitterli et al. 2011). As the landscape of the region formed under multiple repeated glaciations, several of the meltwater pathways used during the LGM likely existed already prior to the last glaciation (Haefeli 1981).

19.3.2 The Last Glaciation: Deposits and Landforms

19.3.2.1 Moraine Complexes

A broad band of ice marginal landforms forms an arc that can be followed along the foot of the Jura. The line of hills curves around to the east and crosses the Aare Valley toward Bützberg, and then extends southwestward to Steinhof and Steinenberg (Fig. 19.2). Here lies the footprint of the LGM Valais glacier. The glacier remained in the region of Wangen for several thousand years, oscillating at various positions. The upstream retreat stadial positions are not discussed further here (see Bitterli et al. 2011). Based on sediment characteristics, especially beds with concentrations of huge blocks, observed in construction sites, gravel pits, and sediment cores, it has been suggested that the actual furthest extent of the Valais glacier during the last glaciation may have been a kilometer or two beyond the margin defined by the surface landforms (Haefeli 1981). This is the LGM ice extent shown in Fig. 19.2 (Schlüchter 2009). Geomorphological evidence of this ice extent is lacking.

The moraine complex that extends from Oberbipp to Bützberg and dominates the landscape (Fig. 19.3) marks an important stabilization position of the Valais glacier and corresponds to the extent mapped by de Charpentier (1841). The LGM left-lateral ice margin along the slope of the Jura Mountains can be followed by tracking the elevation of erratics and till, although in some spots the boundary is masked by talus from the steep slopes. The former glacier boundary attains an elevation of about 700 m a.s.l. north of Solothurn, and drops down to 600 m a.s.l. at Oberbipp. Just north of Solothurn a steep-walled left-lateral moraine (crest elevation 580 m a.s.l.), which is just internal to the outermost ice margin, is found (Zimmermann 1963). In the region of Wiedlisbach, the past ice margin is obscured by gravitationally collapsed bedrock blocks (see also below).

The left side of the Valais glacier was constrained by the Jura slope, and the glacier end was located in the broad Aare River valley below. A 1.5 km wide (measured in the direction of palaeo-glacier flow) band of hummocky terrain from Oberbipp to Bützberg including the Walliswil-Längswald hills marks the frontal position of the LGM Valais glacier (Figs. 19.2 and 19.3). The complex gives a faint impression of subparallel ridges. The 20-m high hills attain elevations of up to 510 m a.s.l. Exposures in gravel pits show that these ridges are the topographic expression of a series of pressed-up slices of outwash gravels (Zimmermann 1963) (Fig. 19.4). Compressional forces at the front and base of the glacier led to squeezing-up of previously deposited gravel beds (Fig. 19.5). The several ridges formed due to glaciotectonic activity (Aber et al. 1989) and do not individually record varying ice margin positions.

Along the southern margin of the Valais glacier the right latero-frontal position between Bützberg and Herzogenbuchsee is marked by sets of moraines (Nussbaum 1911). The moraines rise 30–40 m high above the Molasse bedrock (Haefeli 1981). Outcrops show the presence of ice marginal till with abundant erratics and glacially striated clasts. The clasts were striated and abraded into a flat-iron shape during their time at the base of the moving glacier. An especially well-developed moraine extends from Bützberg to Herzogenbuchsee (Fig. 19.3).

One can also map the footprint of the LGM Valais glacier based on the distribution of erratic blocks. Only a small fraction of these huge blocks remains today (Akçar et al. 2011); most were used for construction purposes during the nineteenth century. The lack of crystalline building material in the region led to the use of the erratics for railway beds, retaining walls, as well as for decorative purposes. It was recognized early that an erratic's rock type provides a unique fingerprint of where it came from in the Alps. This fact has been used since the beginning of Quaternary research in the Alps to track the origin and path of past glaciers from their source areas to the forelands (Favre 1884). Several lithologies present in the deposits of the Wangen area point to homes in the southern valleys of Wallis (Jouvet et al. 2017). These include Arolla gneiss of the Dent Blanche nappe and eclogites and gabbro from the Zermatt-Saas ophiolite zones (Fig. 19.6). The latter, the Allalin gabbro, with its distinctive bright green pyroxenes, is an especially useful index erratic as it comes from an outcrop covering only a few km^2 in the bed of the Allalin glacier near Saas-Fee. The presence of Mont Blanc granite, as well as Vallorcine conglomerate, attests to significant contribution of ice from the Mont Blanc massif into the Valais glacier catchment. Finally, granitic gneiss blocks originating from the Aar massif in the upper Rhone Valley are also present (Nussbaum 1911; Schmalz 1966).

One place where a high concentration of erratics is still found is the Steinhof–Steinenberg area on the southeast side of the former glacier tongue (Fig. 19.2). Steinhof and Steinenberg correspond to topographic Molasse bedrock highs with several meters thick till cover. One of the largest erratic blocks (1200 m^3) is the *Grosse Fluh* at Steinhof (Fig. 19.7). It bears drill holes and was partly split on the side with explosives. The lithology of the block is green hornblende granitic gneiss, also termed Arkesine, a subdivision of the Arolla gneiss (Schmalz 1966). The fact that more than 50 blocks at Steinhof–Steinenberg are all of very similar and distinctive rock type suggests that they originated from a massive rockfall the glacier in one of the southern Valais valleys (Bagnes, Hérens, Anniviers) during the LGM (Nussbaum 1911; Zimmermann 1963).

The timing of construction of the moraines by the Valais glacier has been established by exposure dating of boulders at Steinhof–Steinenberg with cosmogenic ^{10}Be (Ivy-Ochs et al. 2004). As ^{10}Be is produced inside minerals exposed to cosmic rays, measurement of the ^{10}Be concentration in the surface of an erratic boulder allows for determination of how long it has been exposed at that location. This allows for dating of the time elapsed since stabilization of an ice margin. Withdrawal from the Wangen position began no later than 24,000 years ago (Ivy-Ochs 2015). Similar ages were obtained from erratic boulders at the maximum position of the Reuss glacier (Fig. 19.1) (Reber et al. 2014). At the Finsterhennen site, 45 km upstream from Wangen (Fig. 19.1), a mammoth tusk recovered from glaciofluvial gravels deposited in front of the advancing glacier shows that the Valais glacier had advanced past that point already by 30,000–29,000 years ago (Schlüchter 2004). Luminescence dating of sediment from the site yielded similar results (Preusser et al. 2007). Taken together, we may conclude that the Valais glacier reached Wangen an der Aare not long after 30,000 years ago and remained there until no later than 24,000 years ago (Ivy-Ochs 2015).

19.3.2.2 LGM Outwash Terraces

Fluvial terraces comprise an exceptional landscape element in the Wangen an der Aare region. The relative age of

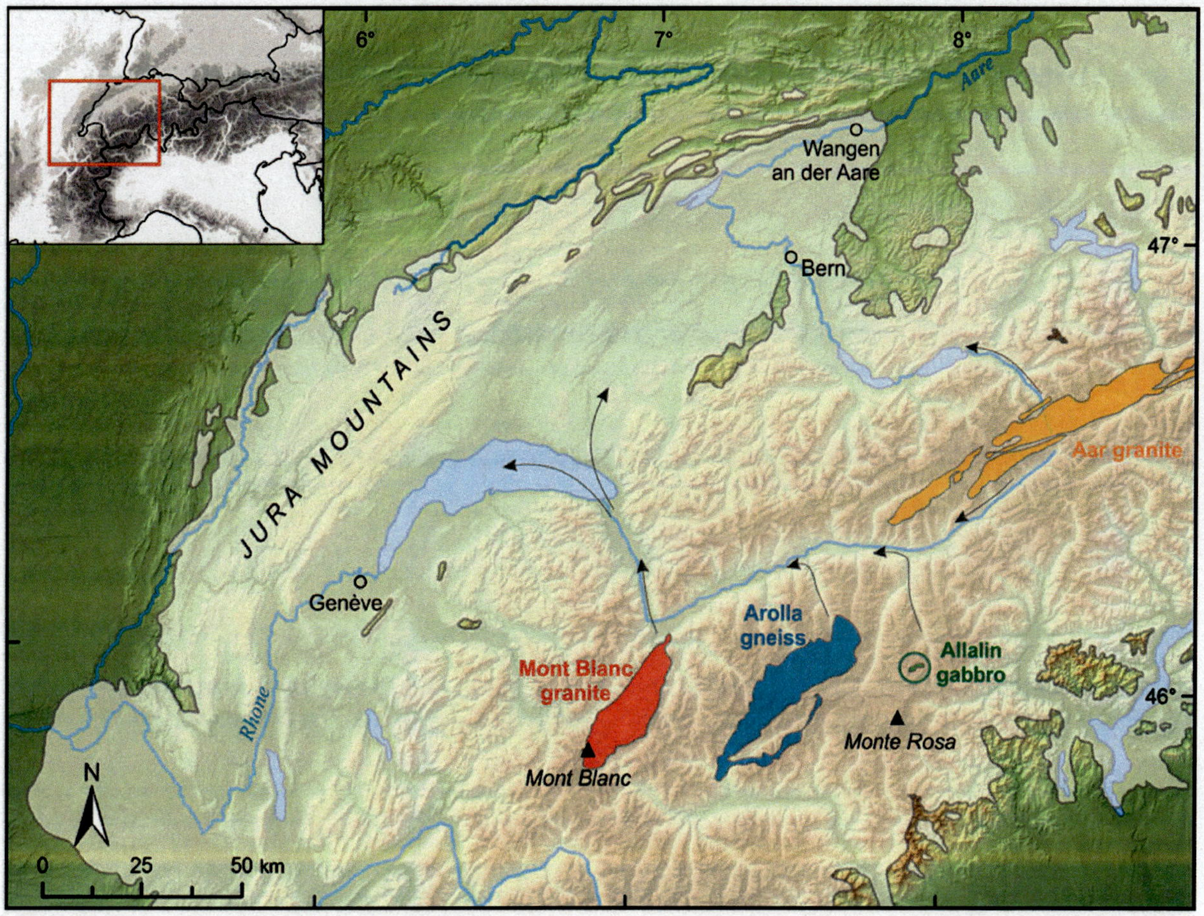

Fig. 19.6 Map of the northwestern Alps showing the extent of the LGM Alpine ice cap (after Ehlers et al. 2011). The arrows indicate the direction of the former ice flow directions. Bedrock outcrops of characteristic lithologies (Mont Blanc granite, Arolla gneiss, Allalin gabbro, and Aar granite) are from Swisstopo (2005). Figure modified from Jouvet et al. (2017)

Fig. 19.7 *Grosse Fluh* at Steinhof (for location see Fig. 19.2), one of the largest erratics ever recorded on the Swiss forelands (photo M. Ochs, 1994)

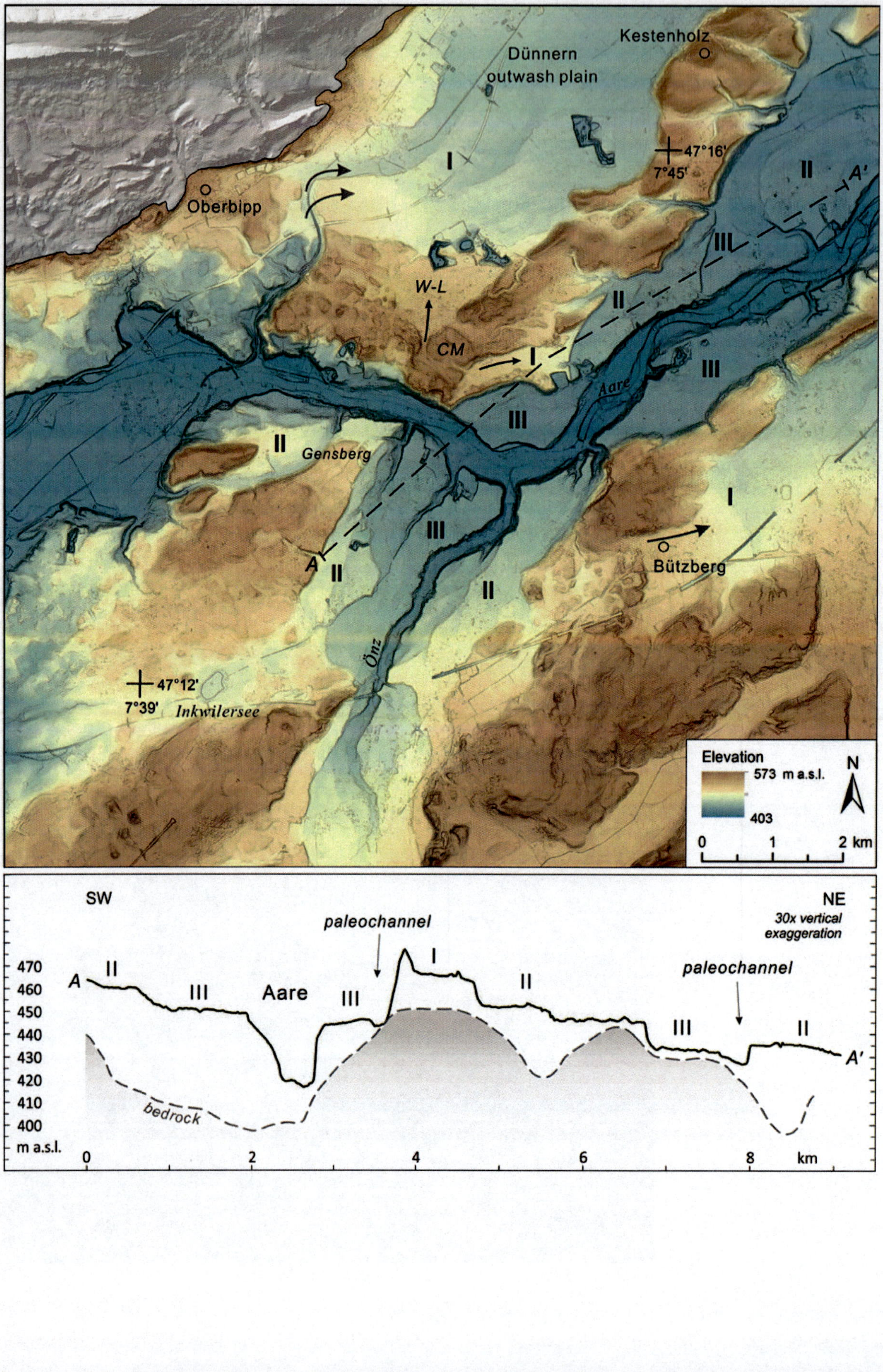
Kestenholz
Dünnern
outwash plain
I
47°16'
7°45'
II
A'
Oberbipp
III
W-L
II
CM
I
III
Aare
III
II
Gensberg
I
III
A
II
Bützberg
II
Önz
47°12'
7°39'
Inkwilersee
Elevation
573 m a.s.l.
403
N
0 1 2 km
SW
NE
30x vertical exaggeration
paleochannel
paleochannel
A
II
III
Aare
III
I
II
III
II
A'
bedrock
470
460
450
440
430
420
410
400
m a.s.l.
0 2 4 6 8 km

◀ **Fig. 19.8** Detail of the ice marginal landforms and outwash terraces of the Wangen an der Aare region. Terrace relative ages from oldest to youngest are I, II, III. Arrows show the meltwater paths. The pattern of meltwater channels suggests that the glacier was stabilized at the Chälerrain moraine (CM) as the (Walliswil-Längswald) (W-L) terrace formed (phase I). Stepwise retreat of the Valais glacier led to incision of each lower terrace level (II, III) whose elevations are shown in the cross-section in the lower panel. Slope raster superimposed by a color-coded DEM (swissALTI3D, swisstopo). Altitude is given by color code. Lower panel topographic profile based on DEM (swissALTI3D, swisstopo). Elevation of bedrock surface from the *Felsreliefkarte Kanton Bern* (Geodaten des Kantons Bern) (see also Dürst Stucki and Schlunegger 2013)

various terraces as well as correlation between terraces can be established based on spatial relationships and terrace top surface elevations (Fig. 19.8). When its snout was in the Wangen region, the Valais glacier was actively carrying sediment from the Alps to the foreland. Each terrace level represents a period of time when sediment-laden meltwater emanating from the glacier front led to build up of tens of meters thick outwash deposits. In that sense, each terrace level can be connected to a certain position of the ice margin, although the associated former ice margin may or may not be marked by preserved or recognizable moraine ridges. Apart from the lowest terrace level and the present channels, the deposits discussed in the following paragraphs are attributed to the LGM. In the Wangen region, three main terrace levels (I, II, III) can be recognized as depicted in Fig. 19.8, which is a modified version of the terraces subdivision of Zimmermann (1963). By dividing the terraces into relative time slots it is possible to visualize the evolving glacial phases and processes of deglaciation that led to aggradation versus incision of the outwash plains.

The most elevated, thus the oldest, outwash fans were active as the meltwater streams exited the glacier while it stood at the frontal moraine complex at Walliswil–Längswald (W-L on Figs. 19.2 and 19.8). These are indicated as I in Fig. 19.8. The three main meltwater paths during this phase were from west to east: (i) at Oberbipp out toward the Dünnernplain; (ii) along the present Aare channel; and (iii) at Bützberg out to the northeast (see arrows on Fig. 19.8). During the earliest phase of glacier advance, huge volumes of glacial outwash poured out in the direction of the Dünnern plain. There, the elevation of the point where the outwash is connected to the moraines is 470 m a.s.l. (Fig. 19.3). Meltwater exiting the glacier formed broad ephemeral channels on a braided river plain. As still visible in the landscape today, channel configuration changed constantly as the discharge and sediment load varied (Figs. 19.3 and 19.8). The gently sloping surfaces of the group I terraces mirror the outwash plain topography during the LGM. The present-day Aare River path was also likely a meltwater channel but much of the evidence for the passage of meltwater during this early phase was destroyed by incision during subsequent phases. Two paleochannels extend from the Chälerrain moraine (Figs. 19.2 and 19.3); one to the ENE and one directly to the N. These appear to have breached the moraine ridges (Fig. 19.8), which suggests that the glacier was still in contact with the moraines there. The onset points are at around 470 m a.s.l. These two small meltwater channels have the character of erosional surfaces cut into older gravel bodies.

During a slightly later phase of the maximum advance period, a meltwater channel was located at the north side of the glacier right along the foot of the Jura. This channel, indicated by the northernmost arrow in Fig. 19.8, is inset 5–10 m into the earlier deposits of terrace I. Postglacial alluvial fan deposits from the Jura slope obscure the upstream extension of this channel. Blocking of the northernmost meltwater channel was not only due to the moraines themselves, but may as well have been related to activity at the deep-seated gravitational displacement above Wiedlisbach (see also below). The abandoned channel (note two arrows in Fig. 19.8), now used by the railroad between Niederbipp and Wangen, may also have been an active meltwater channel at this time. In order to breach the barrier created by the moraine complex the portal must have been close to the moraines. As the Valais glacier melted back from its position along the outermost moraine front, the locus of meltwater and outwash deposition became topographically constrained behind the landforms of the outermost glacier margin position. The phase I outwash plains were abandoned and the present-day Aare channel became the main meltwater system. Whether it formed this breakthrough the Kestenholz bedrock high epigenetically at this time or found a break that it had used earlier is not clear. Bedrock exposed along the left bank near Chällerrain as well as further downstream testify to how shallow the bedrock is here, suggesting that this was a relatively young path for the Aare River to follow (Figs. 19.2 and 19.8, profile).

After reaching the maximum position, the Valais glacier stagnated and melted back several kilometers to the region between Gensberg and Inkwillersee (Figs. 19.2 and 19.3). The ice margin location is suggested by the hummocky ground there, which may reflect ice stagnation at the end of this phase. The onset points of the meltwater streams and outwash fans (Figs. 19.3 and 19.8) at about 465 m a.s.l. provide a rough approximation of the position of the ice margin (Zimmermann 1963). We refer to the two meltwater portals marked by II in Fig. 19.8 near Gensberg and just NE of Inkwilersee (compare Fig. 19.2). Based on topographic

Fig. 19.9 View to the southwest from Wiedlisbach to the region of Wangen an der Aare. In the foreground are the stepped terraces of the Wiedlsibach deep-seated gravitational slope deformation. In the center of the photograph the moraine complex is seen onlapping onto the Kestenholz bedrock high (photo K. Hippe, 2015)

relationships, the latter terrace can be connected to terrace remnants at the same elevation across the Önz Valley and along the northern side of the Aare River, both marked with II (Fig. 19.8 and profile). Construction of these II terraces shows that the ice margin had retreated back far enough that the meltwater pathway in the present Önz Valley (making a left turn just after Steinhof) was now unobstructed. During the maximum extent phase the glacier blocked this pathway. At that time, the main meltwater avenue on the right-lateral side was along the Önz channel just south of Steinhof–Steinenberg, with flow continuing toward the Langete River valley (compare Fig. 19.2). As the present path of the Önz River became free the latter channel was abandoned (see also Fig. 19.3). The third terrace level is inset into the previous one in the Önz Valley and can be followed especially well along the north side of the Aare River. This lower level (shown as III in Fig. 19.8) indicates that the Valais glacier had retreated even further back and out of the Wangen area.

19.3.3 Post-LGM Landscape Evolution

After the Valais glacier had withdrawn from the Wangen region the source of gravels was cutoff. This allowed tributary streams and the Aare River to incise into the outwash deposits.

Along the Jura slope just above Wiedlisbach a section of bedrock collapsed along the northern limb of the Weissenstein anticline and moved downwards several hundred meters (Sackung, Figs. 19.2 and 19.3). The detachment scar can be seen along the ridgeline. Several kilometer-long packages of limestone and marl bedrock broke up in the deep-seated collapse. The movement had a rotational component resulting in the formation of gentle horizontal terraces perfect for agriculture near the towns of Farnern and Wolfisberg (Fig. 19.9). The resulting exposure of the stratigraphically deeper Triassic layers on the slope above Rumisberg brought gypsum beds to light; they were periodically mined in the past (Binggeli 1983). Because of the degree of weathering of the slumped rock and the possibility that glacial deposits overlie it, the age has been interpreted as pre-LGM. Nevertheless, the till deposits reported to drape the slumped bedrock slabs may at least in part be LGM till that was displaced with the unstable mass. The feature may also be interpreted as post-glacial in age. The deep-seated structure may have formed during several phases of slow, non-catastrophic gravitational movement. The blocking of the far northwestern meltwater channel (shown as I on Fig. 19.8) during the earliest phase of ice withdrawal from the Oberbipp moraine position may be related to blockage by the displaced mass (Haefeli 1981). Final modification of the Jura slopes related to rapid denudation at the end of the glacial phase moved loose sediment downslope and several alluvial fans formed along the Jura foot (Fig. 19.2).

19.3.4 Archaeological Landscape Elements

Oberaargau is rich in archaeological rests (Hodel et al. 2011). Several erratic blocks at Steinenberg bear man-made bowl-shaped depressions called *Schalen*. For this reason they are called *Schalensteine* (blocks with man-made shallow depressions or cup marks). The depressions range in diameter from 3.5 to 9 cm (Schmalz 1966), and one block has 38 of these depressions,

Near Oberbipp, in 2011, a Neolithic dolmen was discovered as attempts were made to remove a granite plate from a field, which had hampered moving. The dolmen consists of a granite cover plate resting on several granite and gneiss feet. All of the boulders are erratics. The dolmen was not previously recognized as it had been nearly completely buried by alluvial deposits (Ramstein et al. 2013). The dolmen was reconstructed, with the original blocks, near the church in Oberbipp.

Evidence of *Pfahlbau* (stilt-house) settlements has been found at Burgäschisee and Inkwilersee (Fig. 19.2); both are kettle holes located a few kilometers southeast of Wangen. It is thought that the lakes formed after huge blocks of ice left by the downwasting Valais glacier were buried by sediment. The ice blocks subsequently melted leaving the holes that filled with water. Evidence of extent of stilt dwellings became clear in the nineteenth century as the water table and lake levels were lowered during drainage operations to allow surrounding areas to be used for farming (Müller-Beck 2008). It is thought that the dwellings were situated on the lakeshore rather than on the lake itself but were on stilts because of the swampy underground and the threat of flooding. The sites eventually became inundated as lake levels rose. The archaeological materials suggest attribution to the Cortaillod culture, 4th millennium BC, and the *Schnurkeramik* culture (corded pottery), 3rd millennium BC (Harb et al. 2010). In 2011, both sites became part of the Prehistoric Pile Dwellings around the Alps UNESCO World Heritage site (http://whc.unesco.org/en/list/1363, accessed 20.08.2019).

19.4 Conclusions

Landforms in the region of Wangen an der Aare are unique in their remarkable depiction of the shape of a piedmont glacier lobe, the Solothurn lobe of the Valais (Rhone) glacier. During the LGM the Valais glacier reached the Wangen an der Aare region and oscillated at that position for several thousand years. The frontal moraine complex is made up of parallel ridges that, based on sediment exposed in gravel pits, formed due to glaciotectonic activity rather than as nested frontal moraines. Scattered erratic boulders provide spectacular evidence of the shape of the former ice margin and provide clues to the source of the ice and sediment in the Valais. Huge volumes of sediment-choked meltwater poured outwards from the glacier. The northernmost pathway toward Dünnern was eventually blocked and abandoned. After minor retreat the glacier stabilized at a position just a kilometer or so upstream of the maximum extent moraine complex. Meltwater then followed the present-day path of the Aare River. As the Valais glacier melted back and withdrew from the Wangen region the outwash gravels were deposited at successively lower elevations. Stepwise incision into the outwash deposits left terraces that record several glacier stabilization phases during retreat. After abandonment of the area by the Valais glacier, the rate of landscape change slowed down markedly. Reworking of glacial and related sediments through hillslope processes, notably along the Jura slope, ensued.

Acknowledgments We sincerely thank E. Reynard for organizing this volume and for his patience. J. Seguinot and G. Jouvet kindly provided Fig. 19.6 for our modification. We acknowledge the Amt für Geoinformation des Kantons Bern for access to the bedrock surface Digital Elevation Model (DEM) of the Wangen region and swisstopo for authorizing use of the ALTI3D DEM (JA100120).

References

Aber JS, Croot DG, Fenton MM (1989) Glaciotectonic landforms and structures. Kluwer, Dordrecht, 200 pp

Akçar N, Ivy-Ochs S, Kubik PW, Schlüchter C (2011) Post-depositional impacts on 'Findlinge' (erratic boulders) and their implications for surface-exposure dating. Swiss J Geosci 104:445–453

Binggeli V (1983) Geographie des Oberaargaus. Jahrbuch des Oberaargaus, Sonderband 3. Herzogenbuchsee, 278 pp

Bitterli T, Jordi HA, Gerber ME, Gnägi C, Graf HR (2011) Geologischer Atlas der Schweiz 1:25,000, Blatt 1108, Murgenthal; mit Erläuterungen. Schweiz Geol Kommission, Basel

de Charpentier J (1841) Essai sur les glaciers et sur le terrain erratique du bassin du Rhône. Ducloux, Lausanne, 363 pp

Dürst Stucki M, Schlunegger F (2013) Identification of erosional mechanisms during past glaciations based on a bedrock surface model of the central European Alps. Earth Planet Sci Lett 384:57–70

Ehlers J, Ehlers J, Gibbard PL, Hughes PD (2011) Quaternary glaciations-extent and chronology: a closer look. Elsevier, Amsterdam, 1108 pp

Gerber ME, Wanner J (1984) Geologischer Atlas der Schweiz 1:25,000, Blatt 1128, Langenthal; mit Erläuterungen. Schweiz Geol Kommission, Basel

Graf HR, Burkhalter R (2016) Quaternary deposits: concept for a stratigraphic classification and nomenclature—an example from northern Switzerland. Swiss J Geosci 109:137–147

Favre A (1884) Carte du phénomène erratique et des anciens glaciers du versant nord des Alpes suisses et de la chaîne du Mont-Blanc, 1:250,000. Commission géologique de la Société Helvétique des Sciences Naturelles, Berne

Haefeli C (1981) Grundlagen für die siedlungswasserwirtschaftliche Planung des Kantons Bern. Hydrogeologie Oberaargau. Wasser- und Energiewirtschaftsamt des Kantons Bern, Bern, 114 pp

Harb C, Hafner A, Harb P (2010) Die UNESCO-Welterbe-Kandidatur "Prähistorische Pfahlbauten um die Alpen". Die Fundstellen im Oberaargau und im Wasseramt. Jahrbuch des Oberaargaus 53:123–148

Hodel C, von Burg A, Marti R, Bachmann A (2011) Archäologie des Oberaargaus. Jahrbuch-Vereinigung des Oberaargaus, Riedtwil, 210 pp

Ivy-Ochs S (2015) Glacier variations in the European Alps at the end of the last glaciation. Cuadernos de Investigación Geográfica 41:295–315

Ivy-Ochs S, Schäfer J, Kubik PW, Synal HA, Schlüchter C (2004) Timing of deglaciation on the northern Alpine foreland (Switzerland). Eclogae Geol Helv 97(1):47–55

Jouvet G, Seguinot J, Ivy-Ochs S, Funk M (2017) Modelling the diversion of erratic boulders by the Valais Glacier during the last glacial maximum. J Glaciol 63(239):487–498

Krüger T (2008) Die Entdeckung der Eiszeiten. Internationale Rezeption und Konsequenzen für das Verständnis der Klimageschichte. Schwabe Verlag, Basel, 619 pp

Ledermann H (1978) Geologische Atlas der Schweiz 1:25,000, Blatt 1127, Solothurn; mit Erläuterungen. Schweiz Geol Kommission, Basel

Müller-Beck H (2008) Seeberg, Burgäschisee-Süd: Beiträge zur prähistorischen, klassischen und jüngeren Archäologie. Teil 2, Bauten und Siedlungsgeschichte. Stämpfli, Bern, pp 178

Nussbaum F (1911) Das Endmoränengebiet des Rhonegletschers von Wangen a. A. Mitteilungen der Naturforschenden Gesellschaft Bern 1910:141–168

Pfiffner OA (2015) Geology of the Alps. Wiley, Chichester, pp 376

Preusser F, Blei A, Graf H, Schluechter C (2007) Luminescence dating of Würmian (Weichselian) proglacial sediments from Switzerland: methodological aspects and stratigraphical conclusions. Boreas 36 (2):130–142

Ramstein M, Rüttimann D, Lösch S (2013) Oberbipp, Steingasse. Ein neolithischer Dolmen. Archéologie Bernoise 2013:103–106

Reber R, Akçar N, Ivy-Ochs S, Tikhomirov D, Burkhalter R, Zahno C, Lüthold A, Kubik PW, Vockenhuber C, Schlüchter C (2014) Timing of retreat of the Reuss Glacier (Switzerland) at the end of the Last Glacial Maximum. Swiss J Geosci 107(2/3):293–307

Schlüchter C (2004) The Swiss glacial record: a schematic summary. In: Ehlers J, Gibbard PL (eds) Quaternary glaciations: extent and chronology. Part I: Europe. Elsevier, London, pp 413–418

Schlüchter C (compil) (2009) Die Schweiz während des letzteiszeitlichen Maximums (LGM), 1:500,000. GeoKarten 500. Bundesamt für Landestopografie swisstopo, Wabern

Schlüchter C, Akçar N, Ivy-Ochs S (this volume) The Quaternary period in Switzerland

Schmalz KL (1966) Steinhof und Steinenberg. Jahrbuch des Oberaargaus 12–58

Seguinot J, Ivy-Ochs S, Jouvet G, Huss M, Funk M, Preusser F (2018) Modelling last glacial cycle ice dynamics in the Alps. The Cryosphère 12(10):3265–3285

Studer B (1825) Beyträge zu einer Monographie der Molasse. Jenni, Bern, 427 pp

Swisstopo (2005) Geological map of Switzerland 1:500,000. Compiled by: Geological Institute, University of Bern, and Federal Office for Water und Geology

Zimmermann HW (1963) Die Eiszeit im westlichen und zentralen Mittelland (Schweiz). Mitteilungen der Naturforschenden Gesellschaft Solothurn 21:11–146

Susan Ivy-Ochs is a Professor at the Earth Sciences Department of ETH Zurich and in the Laboratory of Ion Beam Physics of ETH Zurich. She is a Quaternary geologist who studies landscape change, especially in the European Alps. She and her students use field mapping, remote imagery interpretation, and dating with cosmogenic nuclides to understand the evolution of the shape of the Alps. Her BSc, MSc, and Ph.D. degrees are all in Geology. In 2010 her habilitation was accepted at the Earth Science Department of ETH Zurich, with an emphasis on Quaternary Geology and Geomorphology.

Kristina Hippe is a Dr. in Geology and is currently a Guest Professor in the Sedimentary Systems Group at the Department of Earth Sciences, Freie Universität Berlin, Germany. Her research focusses on understanding the processes that shape the surface of the Earth and developing new methods to quantify them by cosmogenic radionuclide techniques. She has been studying Quaternary glacier fluctuations in the Swiss Alps as well as sediment transfer processes in the Eastern Andes.

Christian Schlüchter is Professor Emeritus at the University of Bern. He is an active member of several professional organizations and of editorial boards of international peer review journals. He has been a member of the Executive Committee of the International Union for Quaternary Research (INQUA) and actively engaged in its commission work. The XVIII International Congress of INQUA was held in Bern in 2011 under his presidency. He is a former president of the German Quaternary Association (DEUQUA) and recipient of the Albrecht Penck Medal. His research activities in glacial geology, environmental and engineering geology have brought him to mountains on all continents. He has been working for many years with students in the Swiss National Park and has been the president of its science commission.

20 The Landscape of the Rhine Glacier in the Lake Constance Area

Oskar Keller

Abstract

In the northeast of Switzerland, the region between Lake Constance and the Hörnli molasse mountains consists of a landscape of rolling hills seamed by broad valleys. The *Deckenschotter* on top of the isolated elevations are relicts of Early Pleistocene glaciations. The redirection of the Alpine Rhine from the River Danube to the low-lying Upper Rhine Graben caused a complete restructuring of the drainage system. By means of abrasion, the ice streams of the following Middle Pleistocene basin glaciations formed overdeepened basins and troughs in the new valleys. The Rhine Foreland Glacier of the latest Würm glaciation covered the entire region with a thick dendritic glacier. Distinctive push moraines were formed during meltback. Most of the—originally numerous—lakes in the glacial basin silted up in late and post glacial periods, except the Untersee (Lower Lake) and the Bodensee (Lake Constance).

Keywords

Glaciations • *Deckenschotter* glaciation • Overdeepened troughs • Rolling hills

20.1 Introduction

With Lake Constance (*Bodensee*) at its centre, the deep Lake Constance basin stretches from the Alps to the Jura (Fig. 20.1). It reaches to the Alpine front in the south and almost up to the River Danube in the north. The outer border of the foreland basin is marked in the west by the Hörnli mountains at 1 133 m a.s.l and the Randen highlands at 912 m a.s.l., while in the east, accordingly, by the Adelegg mountains at 1 124 m a.s.l. This extended basin formed the core area of the Rhine Piedmont Glacier in the two last glacial cycles (see also Schlüchter et al. this volume). Today it is shared by Switzerland, Austria and Germany, the latter country having the northeastern part of the basin.

The subject of this chapter is the southwestern part of the basin, between the High Rhine (or Alpine Rhine), Lake Constance and Hörnli. From east to west, it stretches out along around 60 km and from north to south around 35 km. This amounts to an area of around 1000 km^2. The landscape is characterized by single-standing elevations rising around 300–500 m above the current level of Lake Constance at 395 m a.s.l. They are considered to be relicts of a high-altitude old landscape. Curiously, the entire area drains towards the river Thur (Fig. 20.2), with the exception of a narrow strip of land along Lake Constance and the High Rhine. The source of the river Thur is located near the Alpine border and the river flows through the region in a wide arc. The entire landscape is covered with glacial sediments, much like a carpet. The Molasse bedrock only becomes visible in a few places. Current topography is the result of glacial structuring through glaciers in the late glacial periods. In the following, the geological and particularly geomorphological processes that lead to today's landscape are illustrated.

O. Keller (✉)
Falzigenweg 1, 9450, Lüchingen, Switzerland
e-mail: o.keller@paus.ch

O. Keller
University of Teacher Education, St. Gallen, Switzerland

O. Keller
Department of Geography, University of Zurich, Zurich, Switzerland

E. Reynard (ed.), *Landscapes and Landforms of Switzerland*, World Geomorphological Landscapes,
https://doi.org/10.1007/978-3-030-43203-4_20

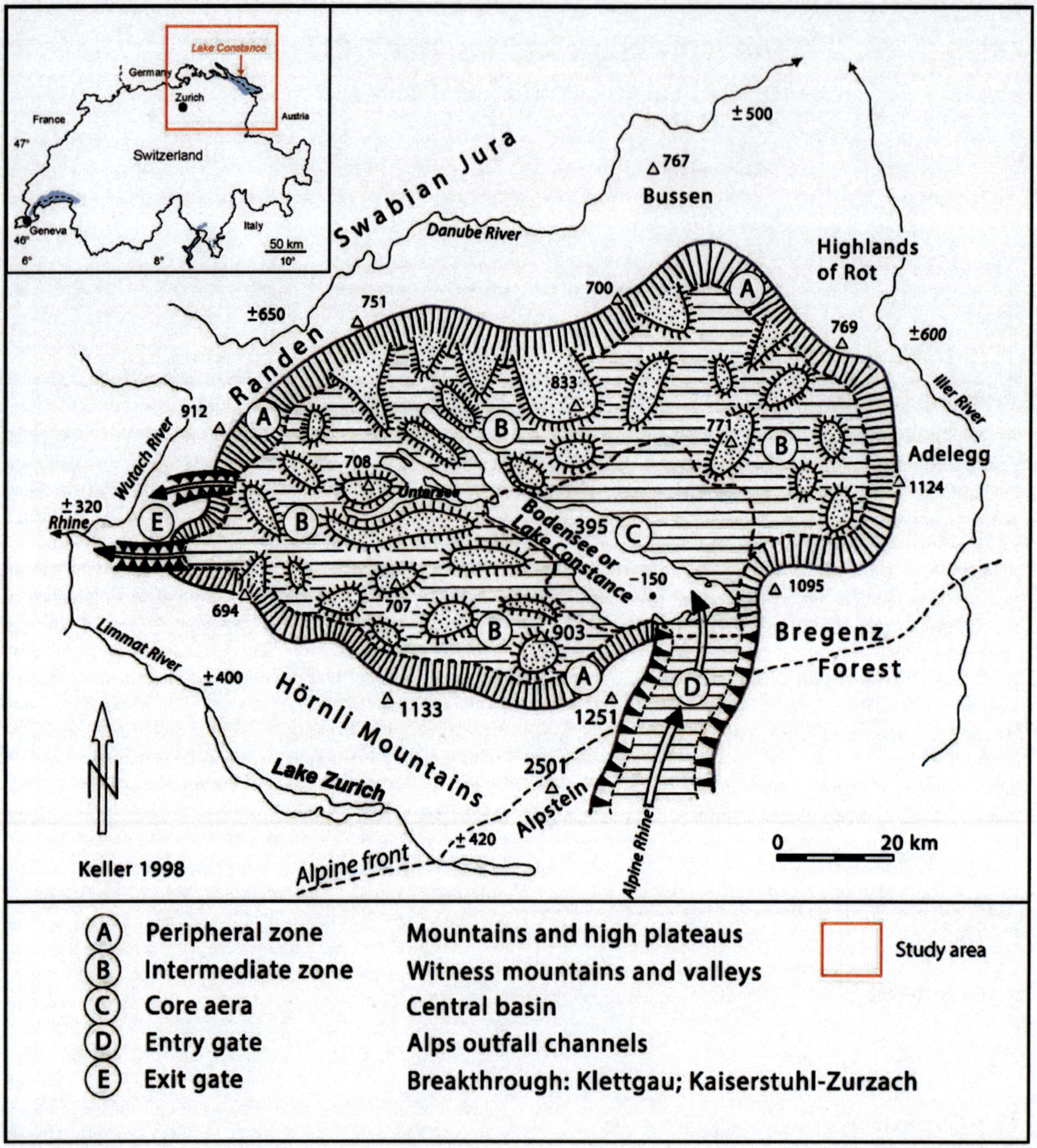

Fig. 20.1 Localization of the Lake Constance basin, which stretches across over 130 km between the Alps in the southeast and the Jura in the northwest. The portion belonging to northeastern Switzerland encompasses the area southwest of Lake Constance (modified after Keller)

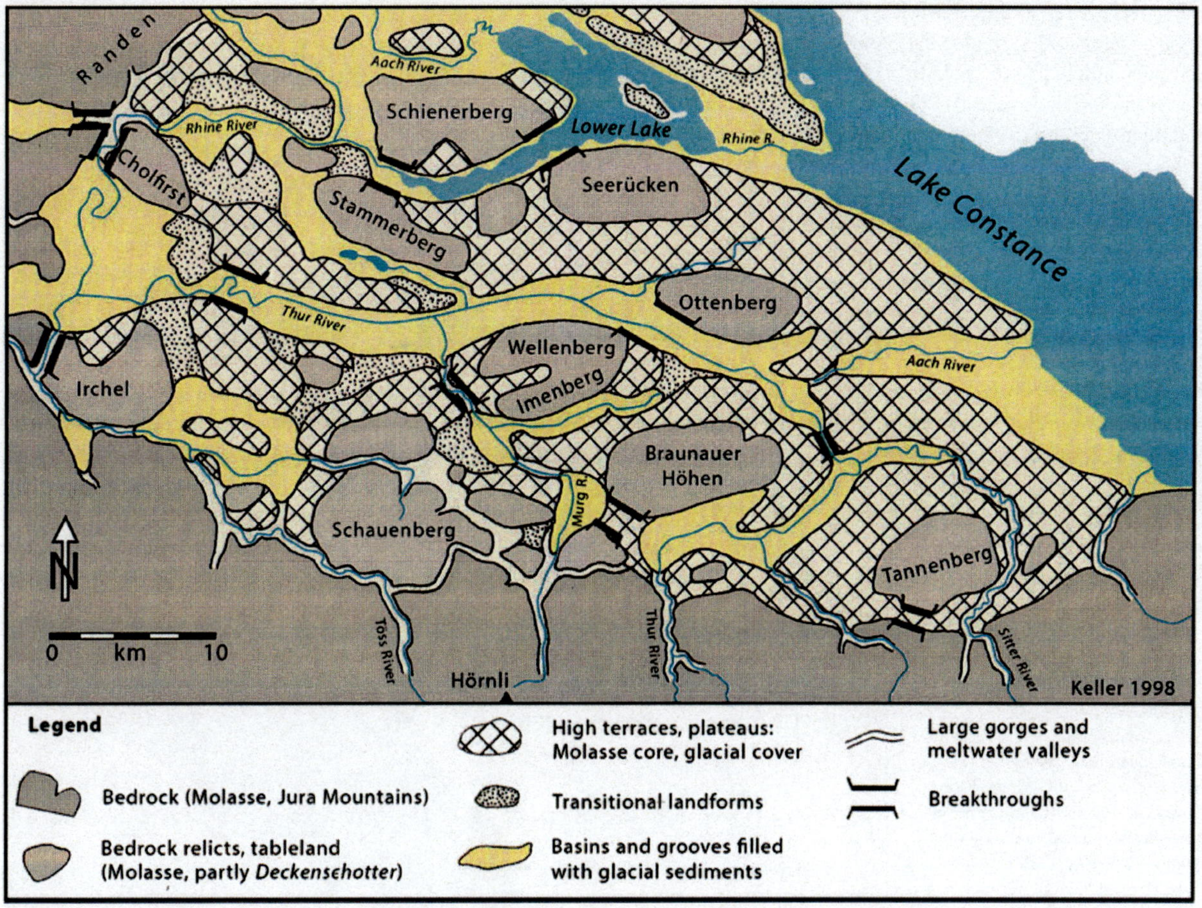

Fig. 20.2 Landscape structures of Canton Thurgau between Lake Constance and Hörnli dominated by isolated mesas (bedrock relicts) and a network of low-lying wide valleys and narrow breakthroughs (modified after Keller)

20.2 Main Geographic and Geologic Features

20.2.1 Geographic Features

The Swiss Canton of Thurgau covers the largest part of the region south of Lake Constance and the High Rhine by far. In the southwest, certain areas are located in the Canton of Zurich, while some areas in the south are in the Canton of St. Gallen. The entire region is interspersed by numerous mesa-type hills with high plateaus and plateaus of lower altitude in between (Fig. 20.2). They are separated from each other by deep valleys and long, drawn-out basins, with the middle Thur Valley constituting its most significant deep zone. Breakthroughs and tunnel valleys act as intermediaries between the basins. Both Lake Constance (475 km^2) and Untersee (Lower Lake) (64 km^2) are glacial basins, connected by the 4 km-long Seerhein ("Lake-Rhine"). The Obersee (Upper Lake) fills the central part of the Lake Constance basin. A string of molasse mountains forms the border of the region in the south, while the northwest is contained by the Randen range (Fig. 20.1) as a part of the Jura (Keller and Krayss 1999; Wyss and Hofmann 1999).

The overall nature of the landscape is smooth and thus allows for agricultural use, mainly farming and orcharding. Viticulture is practised on the southern slopes of the Thur Valley and in some areas of the Untersee. Forests are limited to the steeper slopes and ravines. The main areas of settlement are the large linear valleys, whereby some of the plateaus of middle altitude also feature numerous villages.

20.2.2 Geology and Tectonics

The bedrock of the entire region is almost exclusively made up of Tertiary Upper Freshwater Molasse (*Obere Süsswassermolasse*, OSM), which consists mainly of sandstone and marls. Upper Marine Molasse (*Obere Meeresmolasse*, OMM) is visible in the lowest point of the Thurgau Canton in the municipality of Schlatt (Amt für Raumentwicklung des Kantons Thurgau 2007; Geosite TG128, Schlatt; https://raumentwicklung.tg.ch/themen/natur/geotope.html/4222, accessed 04.01.2019). Conglomerates (*Nagelfluh*) are predominant in the southwest, in the former embankment centre of the primeval Rhine River. As the Molasse is part of the Swiss Plateau, it exhibits a more or less horizontal position. Some cover tuffs are enclosed to the south of the Untersee; they are related to the Late Tertiary volcanism in Hegau (Germany), northwest of the Untersee (Wyss and Hofmann 1999). There are known to be some fault systems with low displacement levels in the land strip close to the lake. As revealed by sparse drilling, the crystalline basement is present in depths of 2–3.5 km (Naef 1999). During the general uplift in the Pliocene, the entire Lake Constance foreland was elevated from the lowlands slightly above sea level by 800–1000 m to the "Lake Constance highland" (Keller and Krayss 1999).

20.3 Landscapes and Landforms

20.3.1 Old Landscapes and Redirection of the Alpine Rhine

During the Middle and Late Miocene around 15–5 million years ago, the latest significant period of Alpine orogeny led to a thrust of the foreland mountains to the south of the Lake Constance region (Hofmann 1982). Additionally, a general rebound occurred in the Late Miocene in the Alpine border mountains and the foreland, likely creating a rolling highland landscape. The Alpine Rhine, emerging from the inner Alps of what today is Graubünden, was a significant tributary to the river Danube (Keller 1994). Its altitude can be estimated at around 700 m a.s.l. in the upper Lake Constance area. There, the Alpine Rhine River—among others—gathered the rivers from the regions of the today Appenzell Prealps, the northeastern slopes of the Hörnli Mountains and the Randen-Hegau region. This river system thus ran in a general eastern direction, in reverse to its current drainage

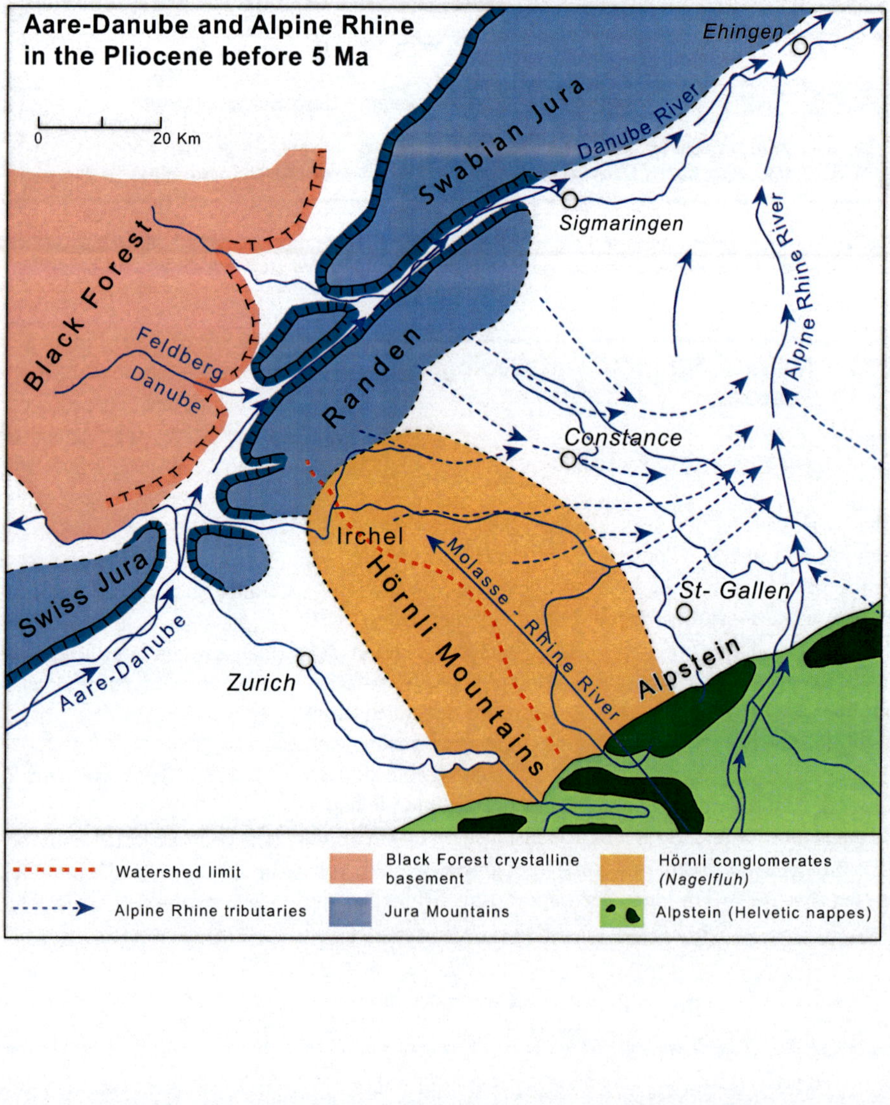

Fig. 20.3 Pliocene palaeo-hydrography of the Lake Constance area. During the Molasse period, the Alpine Rhine flowed northwest. Later in the Pliocene, it drained to the Danube River. Reversely to their present flow, the rivers from Alpstein over the Hörnli Mountains to the Randen flowed in a generally eastern direction towards the Alpine Rhine. These valleys are still visible today in the arrangement of mountains and hills (modified after Keller)

at a substantially higher altitude towards the Alpine Rhine (Fig. 20.3; Keller and Krayss 1999).

Whenever the glacial advance reached the foreland during the *Deckenschotter* period (>2 million years—450 000 years BP; see Schlüchter et al. this volume), a foreland glacier was formed over the Lake Constance highland. On its outer border, meltwater streams deposited melt-out sediments in the valleys which can be consolidated under the term "*Deckenschotter*". According to the studies by Graf (2009), at least four, maybe eight, *Deckenschotter* glacial periods occurred. In the western Lake Constance region, *Deckenschotter* relicts can be found exclusively in the form of high-lying, often cemented deposits at the top of the mesas. Being glaciofluvial, and partially glacial, sediments, they hint at the outer margin of the former foreland glacier (Fig. 20.4; Amt für Raumentwicklung des Kantons Thurgau 2007; Geosite TG107, Groswies; https://raumentwicklung.tg.ch/themen/natur/geotope.html/4222, accessed 04.01.2019). Apart from draining to the river Danube, the westward-flowing outlet of the river Aare towards the low-lying Upper Rhine Graben is particularly striking. The foreland glaciers always closed off the northern axis to the River Danube, causing drainage to the west over the former Hörnli-Randen watershed due to impoundment (Keller and Krayss 1999). In the interglacial periods, drainage readjusted towards the Alpine Rhine and, thus, to the river Danube (Fig. 20.5).

After the latest *Deckenschotter* glacial period (around 450,000 y BP), the river gradient westwards to the Upper Rhine Graben was steeper than the one facing north to the river Danube. Evidence of its altitude can be deduced from the firmly cemented Felsenholz gravels ("*Schotter*") at Bischofzell (Hipp 1986), which according to their false bedding could be considered as a delta of the Ur-Lake Constance following the latest *Deckenschotter* glacial period at ±600 m a.s.l. Therewith, the Alpine Rhine was permanently—including interglacial periods—redirected away from the river Danube and towards the river Aare and the Upper Rhine.

Fig. 20.4 Cemented *Deckenschotter* forming the plateau at the top of the Seerücken, around 700 m a.s.l., south of the Untersee (see Fig. 20.2 for location). The delta-stratification proves the existence of a body of standing water in a valley basin of the former Lake Constance highland (photo J. Früh, around 1910). See also Amt für Raumentwicklung des Kantons Thurgau 2007; Geosite TG107, Groswies; https://raumentwicklung.tg.ch/themen/natur/geotope.html/4222, accessed 04.01.2019)

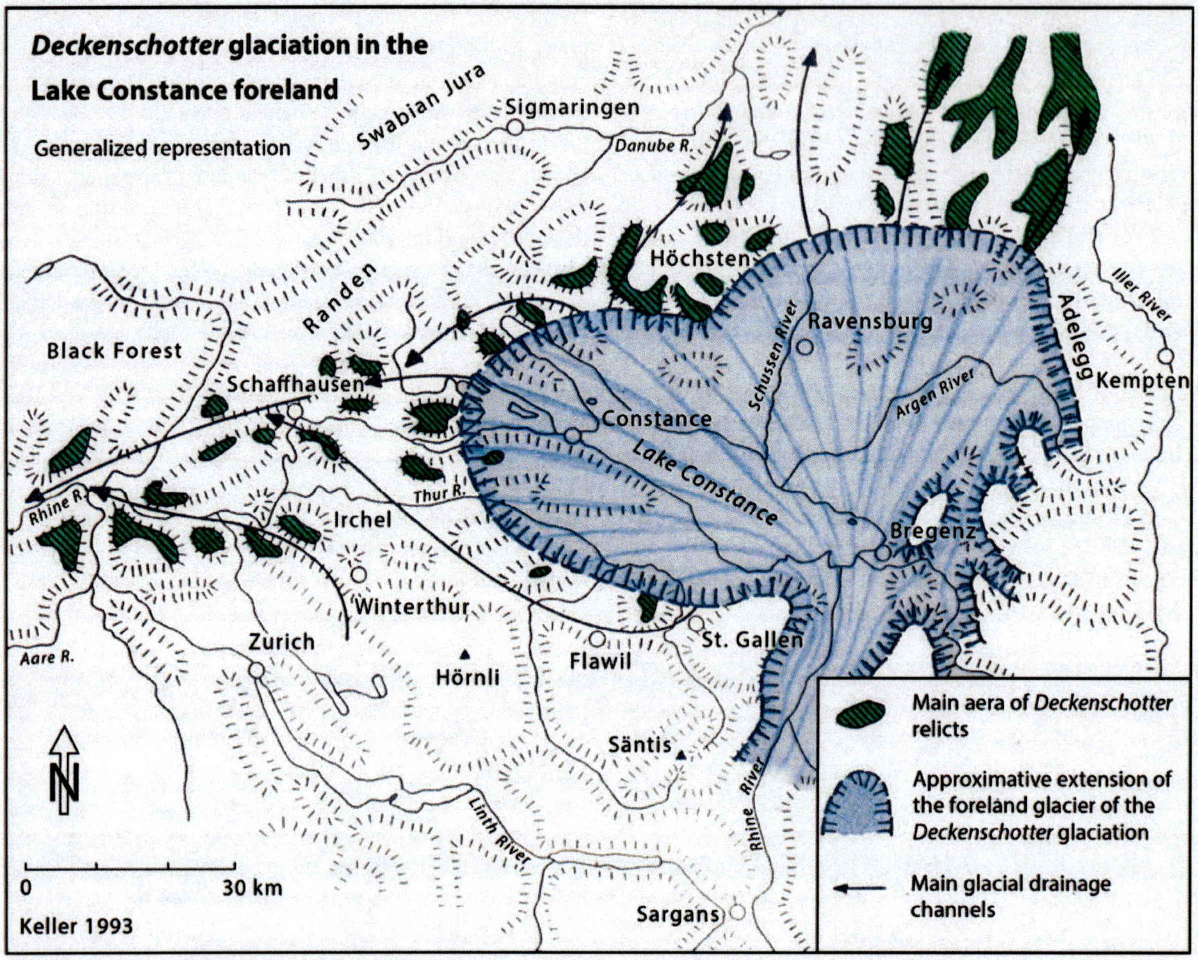

Fig. 20.5 Medium extension of the foreland glacier of the *Deckenschotter* glaciation. The *Deckenschotter* areas are uniformly marked. Drainage occurred in the north towards the river Danube, and in the west due to impounding over the Hörnli-Irchel-Randen ridge towards the Upper Rhine Graben (modified after Keller)

20.3.2 Restructuring of the Landscape in the Late Glacial Periods

The westward redirection of the Alpine Rhine goes along with the beginning of fluvial erosion reaching into deeper levels, from the Lake Constance region back up into the Alps, which caused the river courses to overdeepen by up to 300 m. This pronounced emergence of valleys affected the river Rhine and its larger tributaries (Hipp 1986).

The glaciers of the following Möhlin glacial period (Preusser et al. 2011), also known as the Greatest Rhine Glacial, formed the Most Extensive Glaciation (MEG) (Schlüchter et al. this volume). During their advance, they followed the new drainage network and began restructuring it. In particular, they created extensive, heavily deepened basins and troughs in the foreland and in the drainage valleys from the Alps. The Lake Constance basin was eroded to more than -100 m a.s.l., the Thur Valley to almost 0 m a.s.l. (Fig. 20.6). Evidence for the Holstein interglacial was detected in the basal sediments of some basins (Ellwanger et al. 2011). The Möhlin glacial is thus older and can be dated at ±350,000 years BP. The restructuring of the landscape over a wide area must have occurred during this glacial period. Further restructurings can be attributed to the following glacial periods (Keller and Krayss 2010). The glaciations typical for such deep basins are referred to as "basin glaciations" (Keller and Krayss 1999).

Several trough valleys and deep basins exhibit a characteristic succession of glacial sediments (Fig. 20.7). The higher mesas accompanying the valleys are crowned by

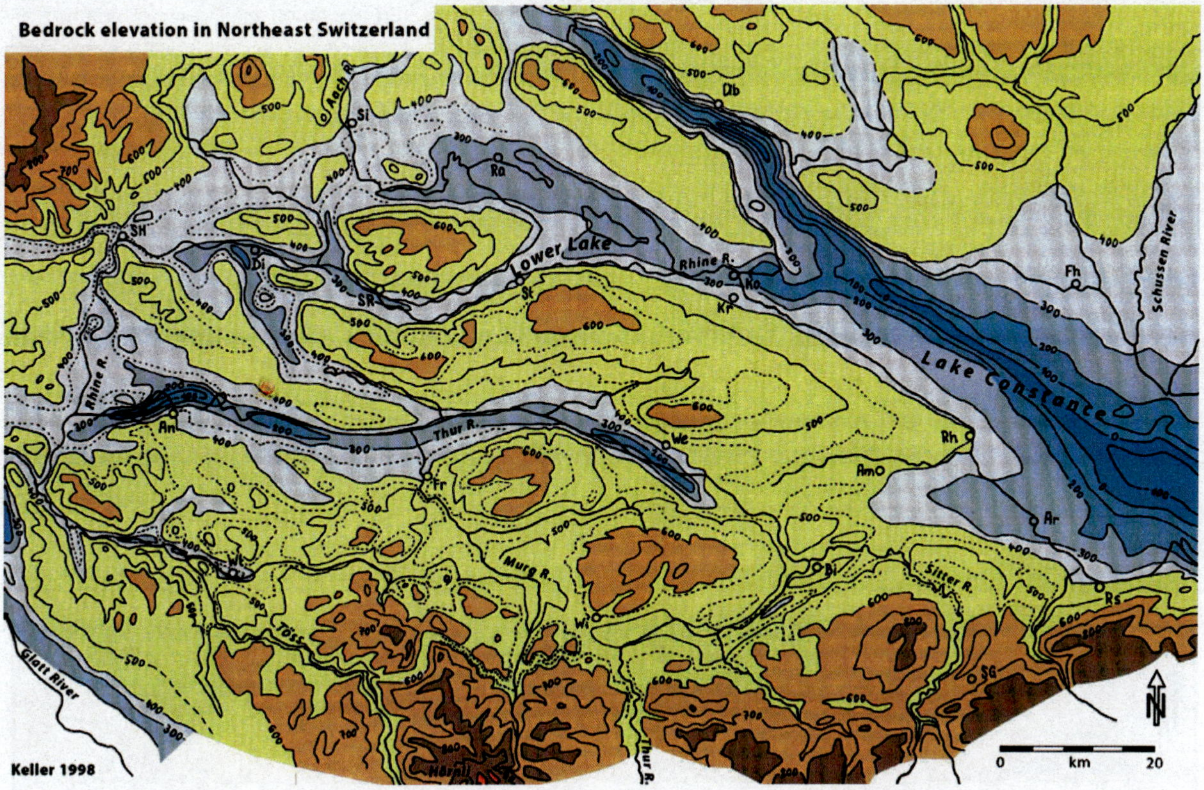

Fig. 20.6 Bedrock elevation in Northeast Switzerland. In the blue-tinted areas, the bedrock lies deeper than the mean elevation of Lake Constance and Thur Valley at ±400 m a.s.l. Glacial abrasion of troughs and basins is clearly visible (modified after Keller)

Fig. 20.7 Quaternary sediments in Thurgau. The composite profile shows the high-lying *Deckenschotter*, the late formation of the trough valleys and the different sediments therein from the basin glaciations. In general, everything is covered by the ground moraine of the Last Glacial (modified after Keller)

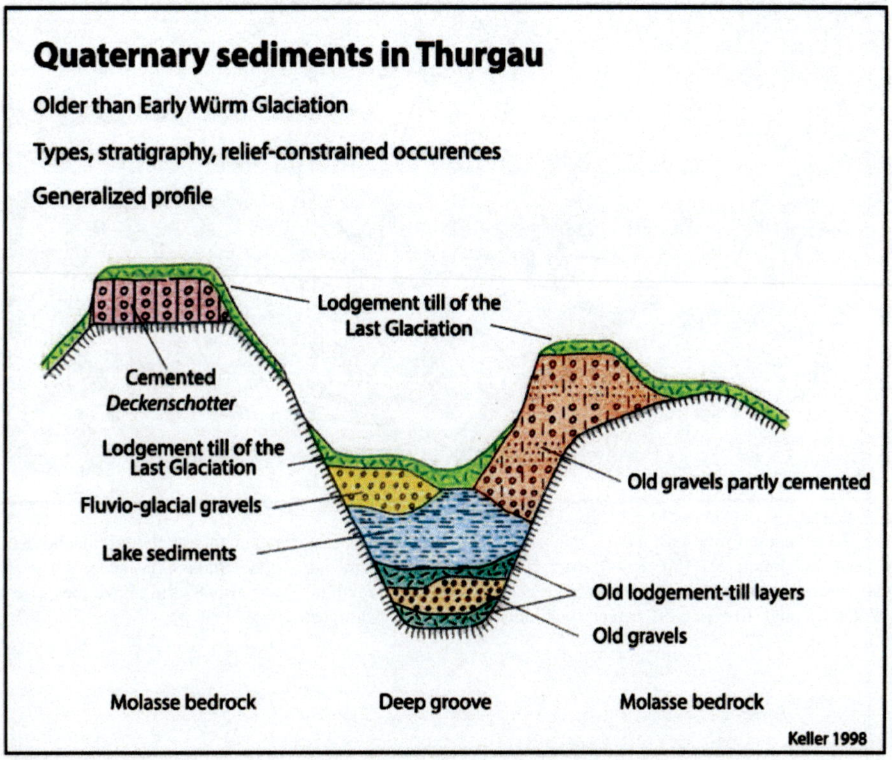

sediments from the *Deckenschotter* glaciation. The deeper valleys were created by means of glacial abrasion in the early basin glaciations (Müller 1996). In some regions, they feature sediments from draws from the older basin glaciations and/or gravels at the edge of the beds of valleys. In the hanging wall, the slopes are covered by the ground moraine of the Last Glaciation, with regional gravels present.

20.3.3 The Legacy of the Last Glaciation

After the Eemian Interglacial, the climate began to cool, causing a new glaciation within the Alps. Following numerous observations, it can be assumed that a first glacial advance into the foreland occurred around 70–60,000 years BP, reaching into the region of the Untersee, the western part of Lake Constance (Hipp 1986; Keller and Krayss 1998). In the following, milder Middle Würm, the glaciers melted back into the Alps. In the period between 60,000 and 30,000 years BP, the Lake Constance foreland was free of ice, which can be derived from datings made at the Mörschwil lignite, south of Arbon (Keller and Krayss 1998).

After the end of the Middle Würm and the climate turning towards high glacial, the Rhine Glacier re-advanced from the region of Chur (Graubünden). Around 22,100 ± 4.3 years BP (Schlüchter et al. this volume), the foreland glacier reached its maximal extension, advancing in the north until

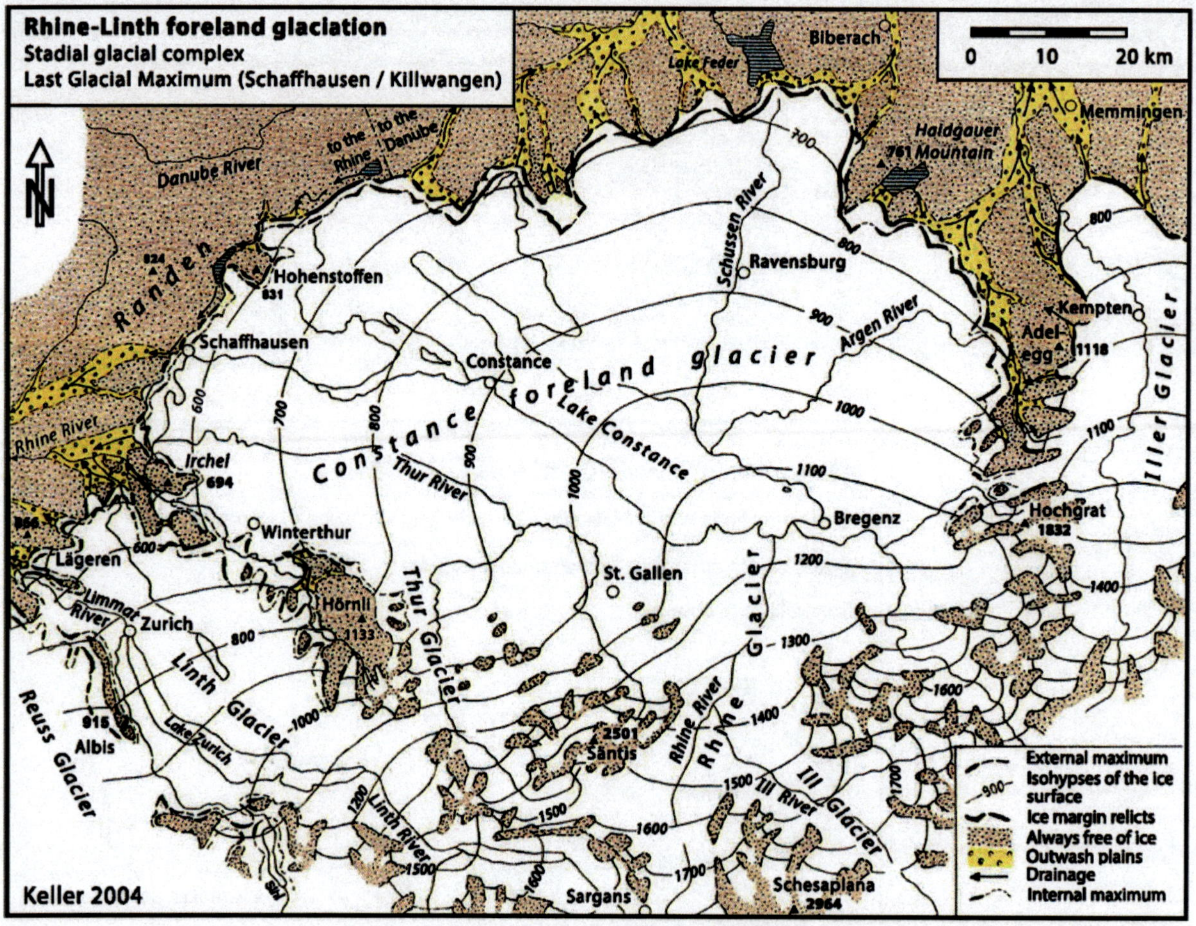

Fig. 20.8 Rhine-Linth foreland glaciation. At the peak of the Last Glacial Maximum (22,100 ± 4.3 years BP), the massive foreland glacier stretched across the entire Lake Constance basin. In the south of St. Gallen and Bregenz, it merged with the ice streams from the mountains at the Alpine borders into a dendritic glacier. Above the upper Lake Constance, the ice surface was at around 1200 m a.s.l., which is 800 m above the current lake level (modified after Keller)

Lake Federsee slightly south of the river Danube, and in the west until Schaffhausen (Hantke 1980; Keller and Krayss 1993, 2005; Fig. 20.8). The entire area south of Lake Constance, including all mesa hills, was thereby engulfed by the ice masses (Fig. 20.8).

After the maximum, the general meltdown began to set in, occurring in stages with limited re-advances (Müller 1979). In Fig. 20.9, the terminal positions of the glacier deciphered from the landscape are illustrated. These narrow belts of witnesses from the glacier front form the "Stadial glacial boundary complexes". Particularly distinctive is the glacial boundary complex visible at Stein am Rhein, also known as "Internal Younger Terminal Moraine" ("*Innere Jungendmoränen*"), corresponding to Nr 7 on Fig. 20.9. It distinguishes itself by its particularly pronounced glacial basin, moraine ramparts, gravel field and meltwater gullies (Figs. 20.10 and 20.11). The moraines at Stein am Rhein are responsible for the creation of Lake Constance (Fig. 20.9). At around 19,000 years BP, the Rhine Glacier melted back from the Stein am Rhein complex and continuously uncovered ever-larger growing areas of the glacial basin. Since the meltwater drainage through the river Rhine occurred at a higher level than the floor of the glacial basin, the water was impounded into a lake in front of the melting ice border. This caused the "birth of Lake Constance" (Fig. 20.12). In its course, the continuously growing lake followed the retreating glacier, with the result of the Untersee first becoming free of ice (Fig. 20.9).

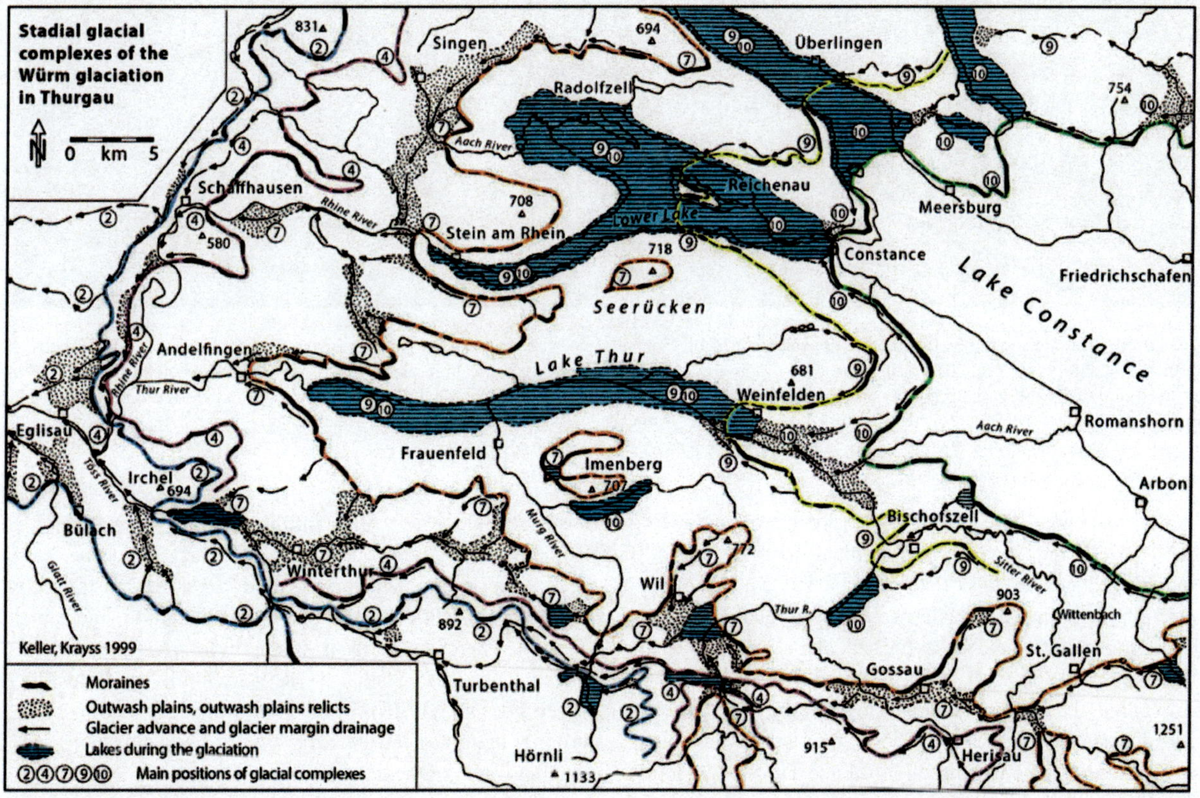

Fig. 20.9 Glacial boundary complexes around and south of Lake Constance from the moraine ramparts of the Last Glacial Maximum (22,100 ± 4.3 years BP) to the Constance Stadial (around 18,000 years ago). The coloured strands indicate the main positions of the individual complexes: (2) blue = Last Glacial Maximum, Schaffhausen; (4) violet = Feuerthalen Stadial; (7) red = Stein am Rhein Stadial; (9) yellow = Reichenau Stadial; (10) green = Constance Stadial (modified after Keller and Krayss 1999)

Fig. 20.10 View to the northwest into the glacial basin of the Seetal north of Frauenfeld (see Fig. 20.9 for location). Here, a few small lakes survived, the main one in the foreground being the Hüttwiler Lake. In the background to the left, covered by the forest belt, the frontal moraine is sprawled out. The Stammerberg, covered with woods, a *Deckenschotter* high plateau, is on the right. This area is geosite Nr 98 of the Thurgau geosite inventory (photo Archäologie Thurgau)

Simultaneously, the 30 km-long Lake Thur developed in the Thur Valley beyond the terminal moraine ridge of Andelfingen, eventually reaching as far as Weinfelden (Fig. 20.9). During the meltdown of the foreland glacier in the region of Gossau-Wil, a further lake developed in the Thur Valley to the west of Bischofszell due to the ice still barricading the old eastern overflow of the river Thur to Lake Constance (Fig. 20.9). The accumulated water bursts out to the north of Bischofszell into Lake Thur (Hipp 1982). The thereby created water gap became the new path of the Thur Valley, which still flows through today. The debris of the Thur Valley Wil-Bischofszell and from the breakthrough was essential for the quick infilling of Lake Thur, which is estimated to have disappeared by the Early Holocene.

The ice in the Untersee melted back to Constance by 18,000 years BP, where a phase of stagnation interrupted ablation. Thereby, the far-traceable moraine rampant ridges of the Constance Stadial W/K were built up (Figs. 20.9 (green line) and 20.13). In the area of the city of Constance, terminal moraines caused a division of Lake Constance into Untersee and Obersee (current Bodensee). Impressive drumlin fields were created during ablation and oscillation of the glacier fronts on the high plateaus directly outside of the moraine ramparts of Constance (Fig. 20.14). They are

Fig. 20.11 Isolated mesa hills in the southern areas at the border of the Hörnli Mountains (view toward southeast). Above the centre of the image, to the left, are visible the gravel fields of Aadorf (Stein am Rhein Stadial). Drainage occurred through the Eulach Valley in the lower foreground. This area is geosite Nr 2 of the Thurgau geosite inventory (photo N. Wächter)

partially streaked by meltwater draws which are still clearly visible today (Fig. 20.15).

The Rhine Glacier had already freed the Obersee basin of Lake Constance at around 17,500 years BP and retreated into the Rhine Valley (Keller and Krayss 2005). In doing so, the entire region south of Lake Constance had also become free of ice, signalling the end of the glacial period in that part. Since then, most of the remaining lakes have silted up, and only Lake Constance and Untersee have remained prominent lakes until today, whereby especially the Untersee features constantly growing, local deltas (Fig. 20.16). Apart from partial infilling of steeply sloped meltwater valleys, only a few instances of restructuring occurred in the Thur region during the Holocene. A tentative reforestation began in the Allerød period around 14,000 years ago (Keller and Krayss 1999).

20.4 Conclusions

The region from south of Lake Constance to the molasse Hörnli Mountains proves to be formed exclusively by glaciation. The isolated elevations of the molasse mesas are witnesses of the Early Quaternary Lake Constance highlands;

Fig. 20.12 Outflow of the river Rhine from the Untersee (foreground) with Stein am Rhein. The frontal moraine ramparts are the two hill ridges covered by woods at the top right. This area is geosite Nr 103 of the Thurgau geosite inventory (photo Archäologie Thurgau)

their cover mainly consists of strongly cemented gravels. These so-called *Deckenschotter* sediments accumulated in former meltwater valleys as glaciofluvial and sometimes glacial sediments during the *Deckenschotter* glaciations. Due to their distribution, a rough estimation of the extension of the corresponding foreland glaciers can be made.

The redistribution of the Alpine Rhine from the river Danube to the Upper Rhine Graben caused a complete change of the river system and pervasive fluvial erosion. The ice streams of the following basin glaciations, particularly those of the early and, coincidentally, most extensive of all foreland glaciations, formed heavily overdeepened trough

Fig. 20.13 The Rhine Glacier during the Constance Stadial around 18,000 years ago. The glacier extended to the area of the historic district of Constance. Untersee (left) and Lake Überlingen (upper left) were both already free of ice, while the Obersee was still covered by the Rhine Glacier. This ice border caused the division of Lake Constance. Part of the area is geosite Nr 76 of the Thurgau geosite inventory

valleys and basins with bedrock sometimes as deep as below sea level by means of abrasion.

In the Last Glacial Maximum, the entire region south of Lake Constance and the Rhine Valley was overtaken and mostly engulfed by ice. During the stage-by-stage ablation of the foreland glacier, combined with limited re-advancements, a series of arranged moraine ramparts with corresponding glacial basins, meltwater valleys and gravel fields, as well as numerous lakes and drumlin fields, were created. When the glacial basin limited by the Stein am Rhein complex became free of ice, an initially small body of water impounded as the beginning of Lake Constance. The Untersee began to grow in

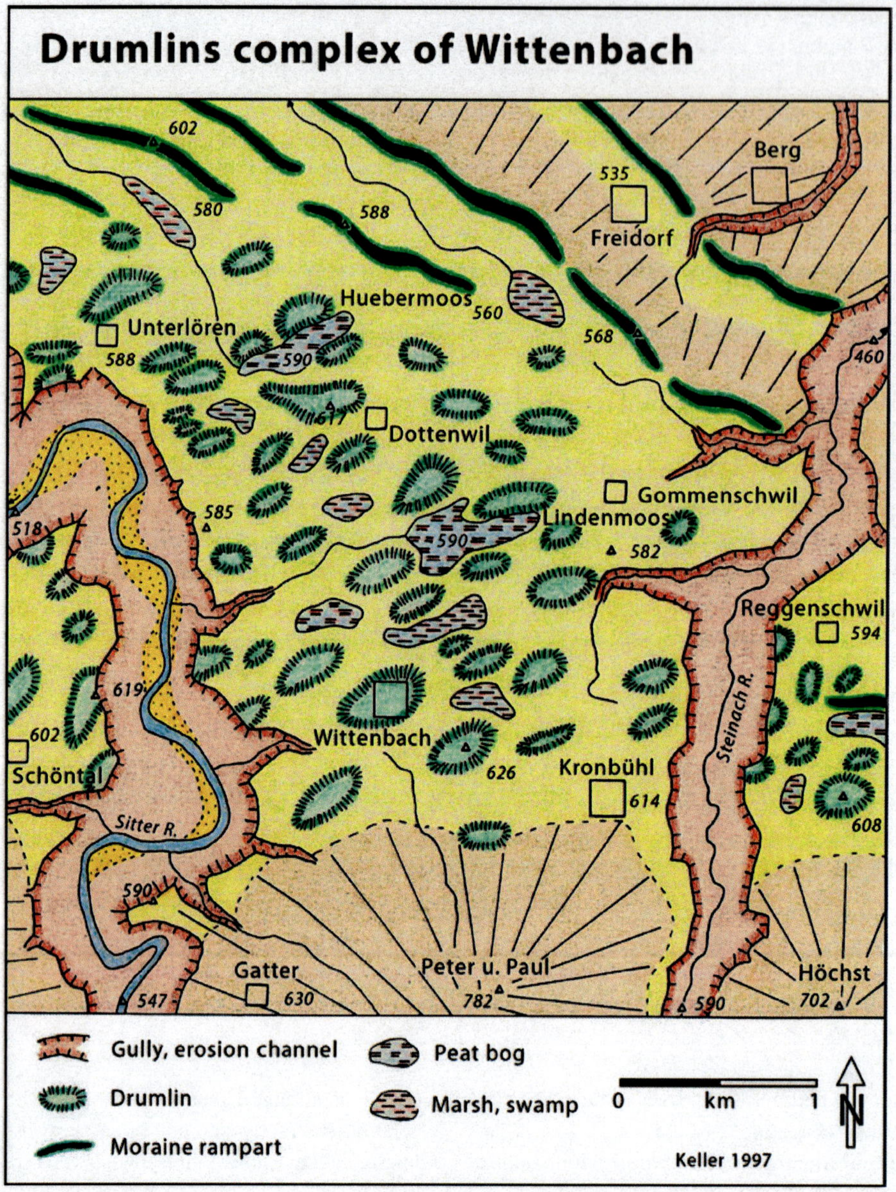

Fig. 20.14 Highland of Wittenbach north of St. Gallen (location on Fig. 20.9), covered with drumlins, arranged in the direction of the ice flow from east to west. On the slopes towards Lake Constance, glacial boundary ramparts (in black) trace the Constance Stadial (modified after Keller)

Fig. 20.15 Drumlin field of Gottshaus and meltwater valley near Bischofszell (photo Archäologie Thurgau)

Fig. 20.16 The Untersee (Lower Lake) filling a glacial basin. Downwards from the Seerücken (left), postglacial deltas continue to advance further into the lake (photo Archäologie Thurgau)

step with the further retreat of the Rhine Glacier, only to be followed by the Lake Constance (Obersee). The present morphology of the landscape at and south of Lake Constance was mainly shaped by glaciation during the Last Glacial.

References

Amt für Raumentwicklung des Kantons Thurgau (2007) Geotop-Inventar Thurgau. Kurzbericht mit vollständiger Objekt-Liste und Literaturverzeichnis, 33 pp. https://raumentwicklung.tg.ch/themen/natur/geotope.html/4222

Ellwanger D, Wielandt-Schuster U, Franz M, Simon T (2011) The quaternary of the southwest German Alpine Foreland (Bodensee, Oberschwaben, Baden-Württemberg, Southwest Germany). E&G Quat Sci J 60(2):306–328

Graf HR (2009) Stratigraphie und Morphogenese von frühpleistozänen Ablagerungen zwischen Bodensee und Klettgau. E&G Quat Sci J 58(1):12–53

Hantke R (1980) Eiszeitalter, vol 2. Ott Verlag, Thun, 703 pp

Hipp R (1982) Der Stand Bischofszell, eine Eisrandlage des würmzeitlichen Bodensee-Rhein-Gletschers. In: Gamper M (ed) Physische Geographie 1. Geogr Inst Univ Zürich, pp 215–220

Hipp R (1986) Zur Landschaftsgeschichte der Region Bischofszell. Eine Glazialmorphologische Arbeit. Mitteilungen der Thurgauischen Naturforschenden Gesellschaft 47:3–117

Hofmann F (1982) Die geologische Vorgeschichte der Bodenseelandschaft. Schriftenreihe VG Bodensee 99(100):35–67

Keller O (1994) Entstehung und Entwicklung des Bodensees – ein geologischer Lebenslauf. In: Maurer H (ed) Umweltwandel am Bodensee. UVK, St. Gallen, pp 33–92

Keller O, Krayss E (1993) The Rhine-Linth-Glacier in the Upper Würm: a model of the last alpine glaciation. Quat Int 18:15–27

Keller O, Krayss E (1998) Datenlage und Modell einer Rhein-Linth-Vorlandvergletscherung zwischen Eem-Interglazial und Hochwürm. GeoArcheoRhein 2:121–138

Keller O, Krayss E (1999) Quartär und Landschaftsgeschichte. Erläuterungen zur Geologischen Übersichtskarte des Kantons Thurgau. Mitteilungen der Thurgauischen Naturforschenden Gesellschaft 55:39–67

Keller O, Krayss E (2005) Der Rhein-Linth-Gletscher im letzten Hochglazial. Vierteljahresschrift der Naturforschenden Gesellschaft in Zürich 150(1–2):19–32, 150(3–4):69–85

Keller O, Krayss E (2010) Mittel- und spätpleistozäne Stratigraphie und Morphogenese in Schlüsselregionen der Nordschweiz. E&G Quat Sci J 59(1–2):88–119

Müller E (1979) Die Vergletscherung des Kantons Thurgau während der wichtigsten Phasen der letzten Eiszeit. Mitteilungen der Thurgauischen Naturforschenden Gesellschaft 43:47–72

Müller E (1996) Die Ittinger Schotter und ihr morphologisches Umfeld. Eclogae Geol Helv 89(3):1077–1092

Naef H (1999) Übersicht über die Erdgeschichte des Thurgaus seit dem Paläozoikum (Erdaltertum). Erläuterungen zur Geologischen Übersichtskarte des Kantons Thurgau. Mitteilungen der Thurgauischen Naturforschenden Gesellschaft 55:17–26

Preusser F, Graf HR, Keller O, Krayss E, Schlüchter C (2011) Quaternary glaciation history of northern Switzerland. E&G Quat Sci J 60(2–3):281–305

Schlüchter C, Akçar N, Ivy-Ochs S (this volume) The Quaternary Period in Switzerland

Wyss R, Hofmann F (1999) Molasse im Thurgau. Erläuterungen zur Geologischen Übersichtskarte des Kantons Thurgau. Mitteilungen der Thurgauischen Naturforschenden Gesellschaft 55:27–38

Oskar Keller was a Professor of Geography and Prorector at the University of Teacher Education St. Gallen (PHSG) and an External Lecturer for Physical Geography at the University of Zurich. His scientific research has focused on the Würm glacial period in the extended area around Lake Constance (Rhine foreland glacier), on the late glacial period in the Rhine marginal Alp areas and on the sequence of glacial periods in the Middle and Late Pleistocene in Northern Switzerland and the Lake Constance region. He retired in 2003.

21 Lake Lucerne and Its Spectacular Landscape

Beat Keller

Abstract

The complex shape of Lake Lucerne, and the surrounding historically significant landscape offer a unique insight into the alpine mountain structure and geomorphological development of the region. The geology and geomorphology have fascinated generations of natural scientists and have led to fundamental geoscientific discoveries. The irregular shape of Lake Lucerne comprising multiple arms, bays and basins is related to the underlying geologic-tectonic architecture. Those geomorphological elements were subsequently carved out during the Ice Age by glaciers that transitioned from alpine valley to piedmont type in the region of the lake. The genesis of the Lake Lucerne region can be reconstructed from the remarkable preservation of macro- to microscale geomorphologic features. These include giant stoss and lee forms, nunataks, areal scouring, and spectacular glacial potholes. The glacial potholes were formed by the subglacial drainage systems, and are the main attraction of the Glacier Garden in Lucerne. Flooding events, and frequent gravity mass movement processes, such as massive rock slope failures, and the subsequent rise of the lake level, illustrate the active morphodynamics of the region until today.

Keywords

Alpine lakes • Glacial erosion • Glacial potholes • Glacier Garden • Natural hazards

21.1 Introduction

The foundation of the Swiss Confederation, the precursor of present-day Switzerland, took place in 1291 within the spectacular region of Lake Lucerne (German: *Vierwaldstättersee*, lit. Four Forested-Cantons Lake). Because of its extraordinary ensemble, the scenery of Lake Lucerne is listed in the Federal Inventory of Landscapes and Natural Monuments of National Importance (ILNM, objects 1605/1606).

Hence, already in the 1820s, the well-connected region around the city of Lucerne (47°3′0″ N, 8°18′0″ E; Fig. 21.1) belonged to the main touristic attractions in the Alpine region. In the pioneering era of tourism, this highly attractive landscape was often difficult to access. It was made reachable by innovative civil engineering structures like the Mt. Rigi railway established in 1871 as the first mountain railway in Europe or through the Mt. Pilatus railway in 1889 which is the steepest cogwheel railway in the world to date.

As the shortest trans-European transportation corridor, the Gotthard Route is one of the most important Alpine crossings. Therefore, the region around Lake Lucerne has had an important role as a trade corridor linking southern and northern Europe since the prehistoric times. This route follows the Gotthard Pass (2106 m a.s.l., 46°33′33″ N, 8°33′42″ E) northwards through the Urner Reuss Valley directly to Lake Lucerne. Up until the mid-nineteenth century, goods were transported by ship on Lake Lucerne and northwards along the Reuss River to the Aare River, and eventually along the Rhine River to the North Sea.

Along with the modernisation of transportation, the Gotthard Route has been continuously extended. This started in 1882 when the Gotthard Railway Company commissioned a 15 km-long railway tunnel. With a length of 57 km, the new Gotthard Base Tunnel (GBT), which was put in operation in 2016, is the longest railway tunnel in the world.

Since the sixteenth century, the easily accessible Alpine geology and geomorphology have fascinated generations of

B. Keller (✉)
Keller+Lorenz AG, Tribschenstrasse 61, CH–6005 Luzern, Switzerland
e-mail: beat.keller@keller-lorenz.ch

E. Reynard (ed.), *Landscapes and Landforms of Switzerland*, World Geomorphological Landscapes,
https://doi.org/10.1007/978-3-030-43203-4_21

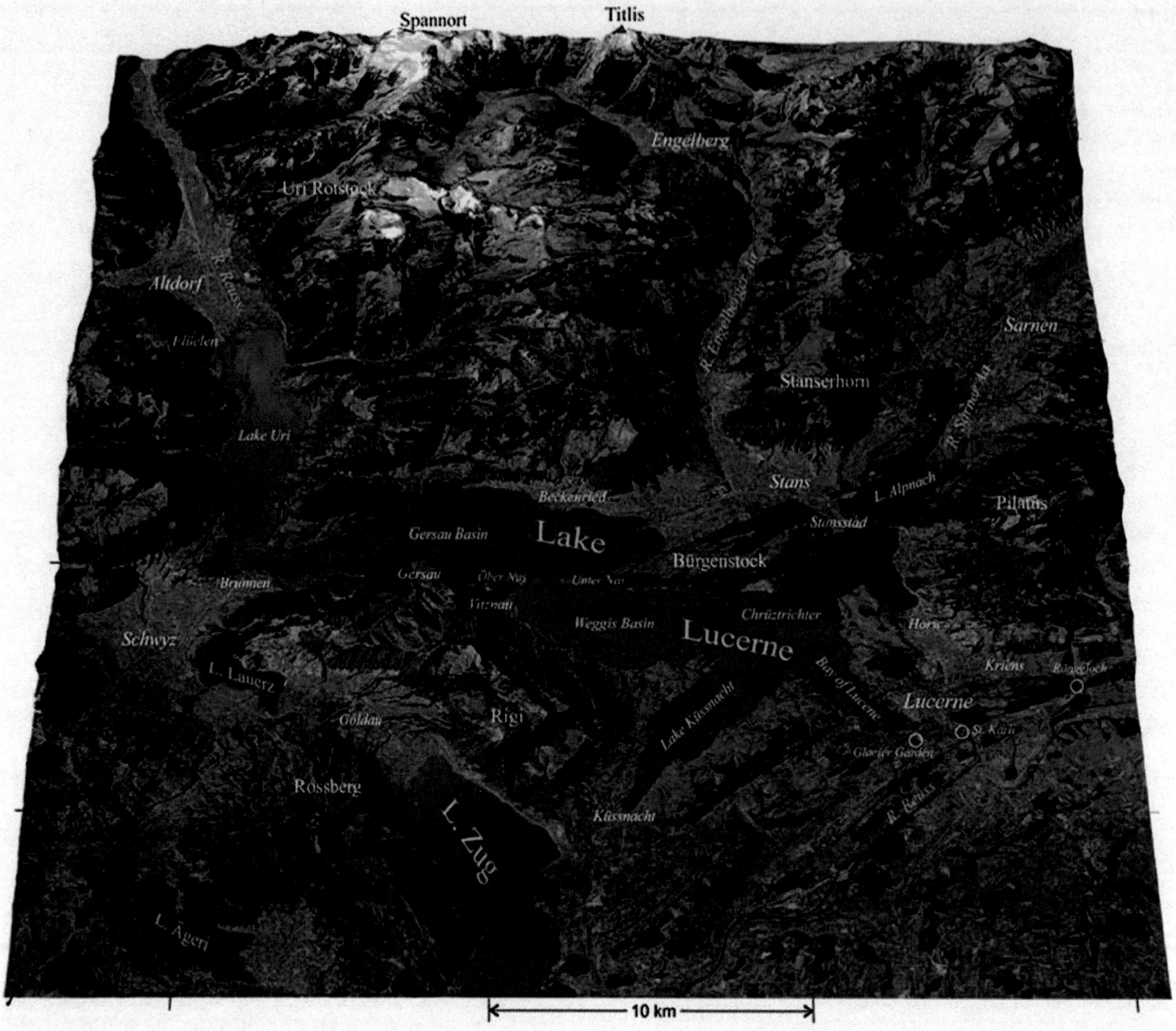

Fig. 21.1 Topographic block model illustrating the multifaceted Lake Lucerne with its arms, bays and multiple basins. See labels for localities mentioned in the text (orthophoto by Swissimage 25 © swisstopo)

well-known natural scientists. Conrad Gessner (1516–1565) described for the first time an altitudinal zonation, based upon the different types of vegetation. Moritz A. Kappeler (1685–1769) applied as the first one the concept of actualism in his 1767 paper *Pilati Montis Historia*, a century before Charles Lyell, usually credited for this (Trümpy 2003). He linked ripple marks in sandstones of the Upper Marine Molasse to shallow marine structures he had observed along the Mediterranean shore whilst working as an engineer with the Imperial troops in Naples. Finally, the Lucerne-based geologist, Franz J. Kaufmann (1825–1892) proved that the Cretaceous Helvetic nappes are overlain by remnants of older Jurassic nappes based on the fossil record. He interpreted the Jurassic remnants as outliers of the Pre-alpine Penninic nappes (Fig. 21.2) that were part of a former mountain chain. His 1876 paper, titled 'Five new Jurassic mountains', provides the first account of the tectonic term 'Klippen' (Kaufmann 1876).

This chapter attempts to facilitate understanding the geomorphological complexity of the Lake Lucerne region and tries to explain its genesis by interlinking the geologic-tectonic architecture with glacial processes during the Ice age and erosional as well as sedimentary processes, mainly beginning with deglaciation, that are ongoing.

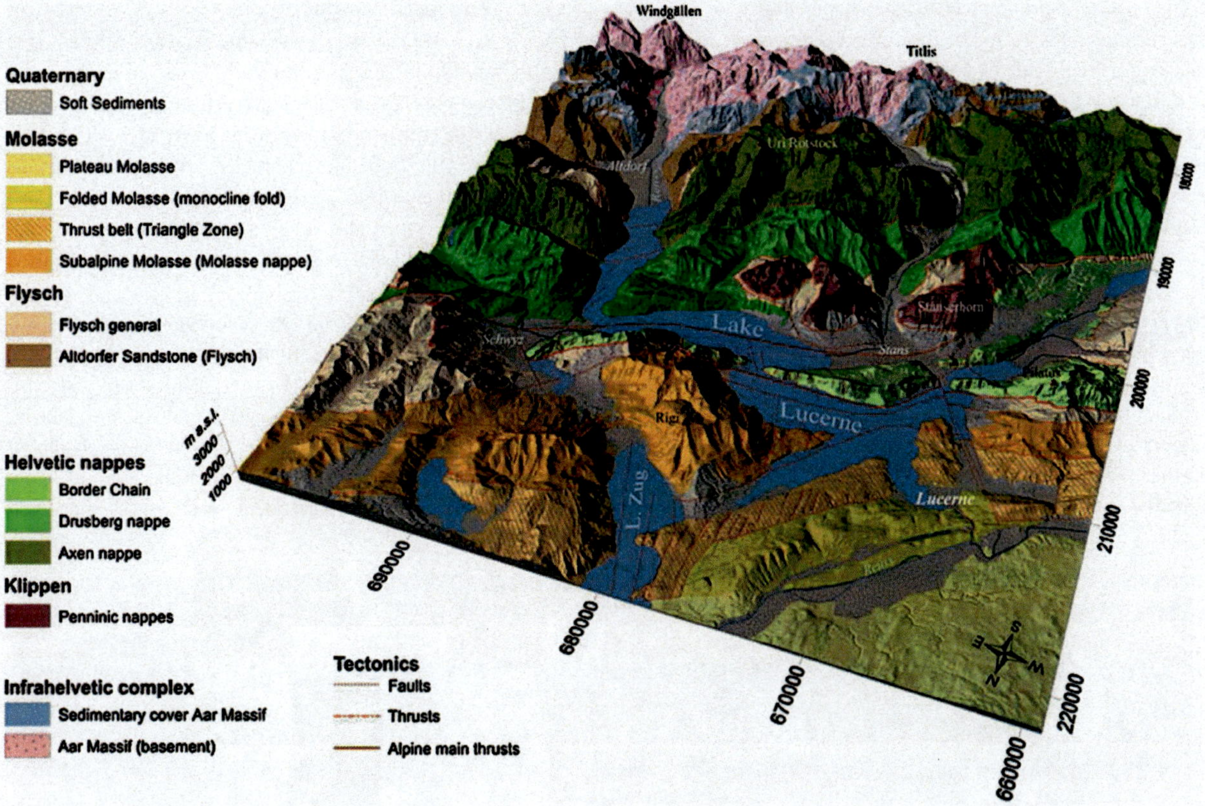

Fig. 21.2 Geological block model illustrating the unique cross section through the northern Alps from the crystalline basement in the south to the foreland basin to the north (vertical exaggeration factor of 1.5; DTM by swissALTI3D © swisstopo)

21.2 Geological and Geomorphological Setting of Lake Lucerne

21.2.1 Geotectonic Overview

The region of Lake Lucerne represents a complete geological section from the crystalline basement in the south through the Alpine nappe pile to the foreland basin in the north (Fig. 21.2; see also Piffner this volume). Four major tectonic units form the mountains in this region, described from south to north:

- The stacked European upper crust forms the dome-shaped ridge of the Aar Massif, which consists of gneisses, granite bodies, and volcanic rocks.
- Overlying the Aar Massif is the northern Helvetic Flysch that is overlain itself by the Helvetic nappes. The Helvetic nappes consist mostly of Mesozoic limestone and thin Eocene siliciclastic sediments and are divided into three distinct nappes. These are, from south to north, the Axen nappe, the Drusberg nappe and the Border Chain that were thrusted along steeply dipping thrust faults onto the Subalpine Molasse. The Klippen nappes, a distinct feature of this region, and the basal Flysch series were preserved within the structural depressions of the Helvetic nappes as the next higher tectonic level of the Penninic nappes.
- The Subalpine Molasse tectonic complex represents the part of the southern basin of the Alpine foredeep that was subjected to alpine deformation. It comprises a nappe, the overthrusted, southward-dipping, Oligocene Rigi-Rossberg Megafan. This prograding alluvial fan is marked by a ~3 km thick accumulation of coarsening and thickening upwards conglomerate sequence.
- North of the main thrust fault of the Subalpine Molasse is a broad, complex fold-and-thrust belt that contains a northward-dipping monocline, which forms a transverse ridge as topographic ending of Lake Lucerne. It consists mainly of shallow marine sandstones of the Upper Marine Molasse and to a lesser extent the Lower Freshwater Molasse.

During Alpine orogenesis, the Rigi-Rossberg Megafan played an important tectonic role acting as a rigid indenter, which the nappes were thrusted onto. This led to the formation of a pronounced dome-shaped bulge of the displaced nappe stack around the former depocenter of the fan and to a regional extension. This extension feature is recorded by the plunge of the fold axes of the Axen nappe towards the WSW and ENE, west and east of Lake Uri, respectively. Therefore, Lake Uri (Fig. 21.3), as the southernmost basin of Lake Lucerne, subsided along a graben-like transverse fault flanked by a series of N–S-striking horst-graben structures bound by normal faults. In the western depression of the Drusberg nappe (Arbenz 1912), the Penninic Klippen of the Buochserhorn and Stanserhorn were preserved. West of the Rigi-Rossberg Megafan, the Border Chain was gradually thrusted further north along wrench faults onto the foreland basin. As a result, the Pilatus region of the Border Chain was thrusted at least 10 km further north than the Hochflue region of the Border Chain directly at the southern edge of the fan.

21.2.2 Physical Geography

From the morphological point of view, Lake Lucerne shows the most complex shape of all lakes throughout the Alps, which will be further explored in Sect. 21.3. The lake (113.6 km^2) with fjord-like character within the montane zone can be divided into a chain of main basins (i.e. Lake Uri, basin of Gersau, Lake Weggis and the bay of Lucerne) and side arms (i.e. Lake Alpnach, Lake Küssnacht and the cove of Horw) (Fig. 21.1).

The catchment area for Lake Lucerne is 2140 km^2 of which ~75% occurs in the montane (1400–2200 m a.s.l.) and alpine (2200–2700 m a.s.l) zones. With an overall volume of 11’800 M m^3, the average depth of the lake is 104 m, and the maximum depth is 214 m in the basin of Gersau. During the flood event of August 2005, a maximum inflow for the whole catchment area was measured

Fig. 21.3 *Top*: Panorama of the bay of Lucerne with Mt. Rigi (Subalpine Molasse), the Border Chain with Vitznauerstock, Bürgenstock, Bürgenstock and Pilatus, the Drusberg nappe with Niderbauen, the Walenstöcke (Axen Nappe), and Stanserhorn (Penninic nappes, Klippe). The wooded hills in front belong to the fold-and-thrust belt of the Molasse. *Middle*: Panorama of the eastern part of Lake Uri showing the impressive cross section from the basement to the Helvetic nappes as well as the Klippen of higher tectonic levels (Penninic and Eastern Alpine units). To the right Mt. Bristen (crystalline basement), to the left follow Windgällen (Autochtonous), Mt. Rophainen (Axen Nappe), Fronalpstock (Drusberg Nappe) as well as the two limestone pyramid-shaped Grosse and Kleine Mythen (Penninic nappes, Klippe) and Hochstuckli (Flysch). *Beneath*: View of left bank of Lake Uri from Flüelen with Scharti (Axen Nappe), Ober- and Niderbauen (Drusberg Nappe), and to the right in the background Mt. Rigi Hochflue (Border Chain) (photos B. Keller)

at ~1400 m^3/s, and the average outflow of the Reuss River in the city of Lucerne was 473 m^3/s.

Around Lake Lucerne, the distribution of precipitation is mainly influenced by altitude zoning. The average annual precipitation ranges between 1000 and 1200 mm in the foothill zone, 1200–1400 mm in the submontane zone, 1400–2500 mm in the montane zone and is >2500 mm in the alpine zone.

Enormous differences in the precipitation rate at the same altitudinal level around Lake Lucerne can be attributed to the different topographic exposition and the proximity to the Alpine border. This was especially evident during the flood event of 2005. The cumulative rainfall between 19th and 25th August 2005 for the stations Lucerne (located at the southern margin of the Swiss Plateau), Weggis (in a sheltered spot at the foot of Mt. Rigi) and Vitznau (in a funnel-shaped terrain concavity at the foot of the Border Chain) was 185 mm, 176 mm and 312 mm, respectively. Another characteristic of regional climate is the transalpine wind called 'Föhn' (Swiss German: hairdryer). This katabatic wind causes the lee side of the Alps to have mild and dry conditions. This is especially pronounced during a 'Südföhn' event, which originates south of the Alps and moves northward towards the southern area of Lake Lucerne where wind speeds can reach up to hurricane strength (>118 km/h).

21.3 Landforms and Landscapes

21.3.1 The Geomorphological Evolution of the Lake Lucerne

To date there have only been rudimentary investigations on the extremely variable geomorphological landscape of the Lake Lucerne region (Keller 2007). However, there is an excellent record of the enormous variety of erosional patterns formed by the Reuss glacier during the last Ice Age. Several lake basins in Switzerland (e.g. Lake Constance, Lake Zurich, Lake Geneva), which are elongated and parallel to the glacier flow direction, were clearly formed by glacial erosion. The many arms of Lake Lucerne extending in different directions make a similar interpretation problematic and have caused many controversies in interpretation. Penck and Brückner (1909) attributed the complex shape of Lake Lucerne solely to glacial erosion. In contrast, Heim (1919) interpreted Lake Lucerne and its arms as paleo-river valleys, which were drowned due to subsidence during the late Alpine phase of mountain building. Since Arbenz (1912), it is clear that the arrangement of the fluvial valleys in the Pliocene and therefore the present-day geomorphological shape of the lake and its sub-basins are controlled by the mountain structure and the spatial arrangement of geological-tectonic features (Fig. 21.3; see also Pfiffner this volume).

Since the end of Miocene Molasse deposition, the Alpine paleo-rivers instigated ~5 Ma of fluvial erosion during the Pliocene and Pleistocene, creating a dendritic river system with deep V-shaped valleys. However, there is minimal information preserved in the rock record about the river network prior to the onset of the last Ice Age (Keller 2007). Presumably, the paleo-Reuss River flowed out of the Urner Reuss Valley directly into Lake Zug (Fig. 21.1), and from there through a subterranean bedrock valley that reached down to 240 m a.s.l. towards the lower course of the modern Reuss River. The paleo-pathways of the Engelberger and Sarner Aa rivers likely merged in the area of Stansstaad and flowed northwards through the Lucerne Basin or the Bay of Horw.

During the Ice Age, and in particular during the Last Glacial Maximum (LGM, between 26.5 ka cal BP and 19–20 ka cal BP), the Reuss Glacier, which originated from the Gotthard region, dominated the glacial network in the region of Lake Lucerne (Fig. 21.4). Also in the region were the smaller Engelberg and Brünig glaciers, with the latter as a result of glacial transfluence from the Aar Glacier. The confluence of these three valley glaciers occurred in the large central basin of the 'Chrüztrichter' (lit. *cross funnel*, 47°01′09.3″N 8°21′43.1″E, Fig. 21.1), and is the transition zone into the spreading piedmont glacier at the northern end of Lake Lucerne. The surface of the Reuss valley glacier at northern Lake Uri reached up to 1260 m a.s.l. during the LGM, as documented at Mt. Niderbauen. The height of the glacier surface in the transition zone is well-documented by lateral moraines at the foot of Mt. Pilatus at Krienseregg (47°00′40.2″N 8°16′02.3″E), and at the foot of Mt. Rigi at 1026 and 1027 m a.s.l., respectively. The average grade for the ice surface is calculated to be 1.5% during the LGM. The equilibrium line altitude (ELA) during the LGM oscillated around 1000 m a.s.l. and reached up to 1300 m a.s.l. (Keller and Krayss 2005). The glaciers in the region of Lake Lucerne showed the transition from high velocity (>1 km·a^{-1}), and high basal shear stress (150–200 kPa) from the inner Alpine valley glaciers to low velocity (<0.1 km·a^{-1}) and low basal shear stress (<50 distal to 100 kPa proximal) at the piedmont glacier in the Swiss Plateau (Haeberli and Penz 1985).

The geomechanical perturbation in more strongly tectonised zones is expressed as discontinuities such as joints, fractures, faults and schistosity, which are more susceptible to erosion. Also easily erodible are the weak shaly Flysch and Lower Marine Molasse units (Fig. 21.2). Glacial erosion in this environment was facilitated by the following key factors:

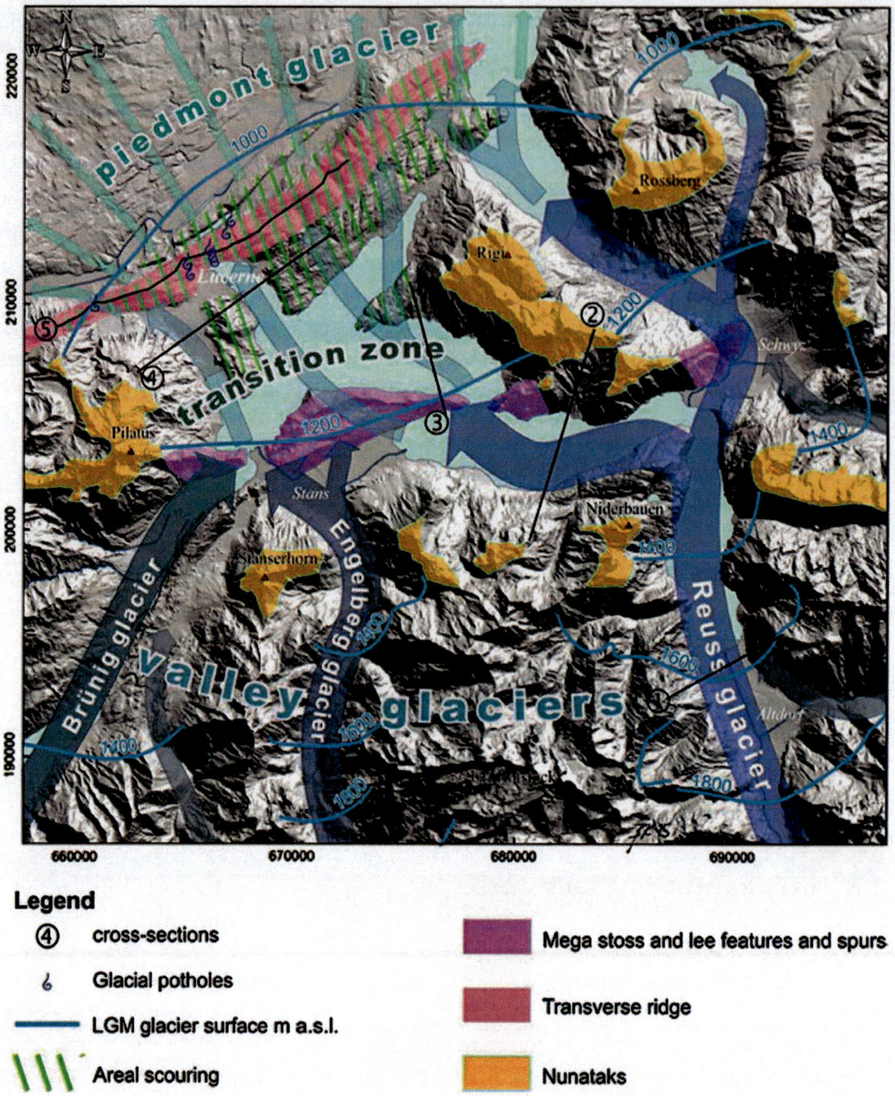

Fig. 21.4 Geomorphological overview of Lake Lucerne with the different Ice Age glaciers as well as the explained geomorphological features (shaded relief based on DTM swissALTI3D, swisstopo)

- The over-consolidation of bedrock. Compaction of the Subalpine Molasse by the overlying nappe units with an overburden of >3.5 km of sediments resulted in an overburden pressure of more than 90 MPa. Glacial erosion from the overlying ice of varying thickness created a progressive unloading effect. In the elastoplastic continuum, this effect caused ongoing deeper reaching stress-release joints that resulted in migration of weathering front to greater depths, up to 150 m below the bedrock surface in the area of Lucerne. Hence, each glacial advance took place in a strongly weakened stress relief, leading to excessive initial depth erosion.
- High porewater pressures at the base of the glacier, which led to progressive reduction in the rock strength, in particular in over-consolidated shales and mudstones of Flysch and Molasse. These two factors are highly dependent on lithology and are strongly self-reinforcing (positive feedback). The influence of weakening increases with growing ice thickness, and increasing basal shear stress towards the proximal part of Lake Lucerne in a disproportional manner.

These factors also played an important role in creating the lake's shape and remarkable macroscale characteristics, including along the talweg. Elevations of the bedrock surface along the talweg from the distal (NW) to the proximal (SE) part of the basin are as follows: 408.8 m a.s.l. within the tunnel valley at the St. Karli bridge in Lucerne (Fig. 21.1), 335 m a.s.l. at the Lucerne railway station (Keller 2013), 190 m a.s.l. at the Weggis Basin, 30 m a.s.l. at Gersau (Finckh et al. 1984) and about 160 m below sea level at the Reuss Delta located at the south end of Lake

Lucerne (Nitsche et al. 2002) (Fig. 21.5). The elevation difference is 570 m over a distance of 38 km, thus the average grade of the bedrock surface slope along the talweg towards the Alps is 1.5%. Therefore, the bedrock surface slope is inverse to the slope of the glacial surface during the Ice Age and is directly proportional to the basal shear stress.

The variety of bedrock valley shapes along Lake Lucerne can be attributed to several factors. These are the variability and extent of the Ice Age glaciers, their variable thickness, and the associated differences in basal shear stress, as well as the glacier flow direction, the deflection of the glaciers by the fluvial paleo-valleys, geologic-tectonic architecture and erosion by subglacial meltwater streams. Examples of localities along Lake Lucerne with different bedrock valley shapes are as follows (Fig. 21.5; lines of cross sections are indicated on Fig. 21.4):

– *Reuss delta* (Fig. 21.1-1): Lake Uri, and the Urner Reuss, Engelberger and Stansstaad valleys are excavated along graben-like transverse faults, with the former two in the

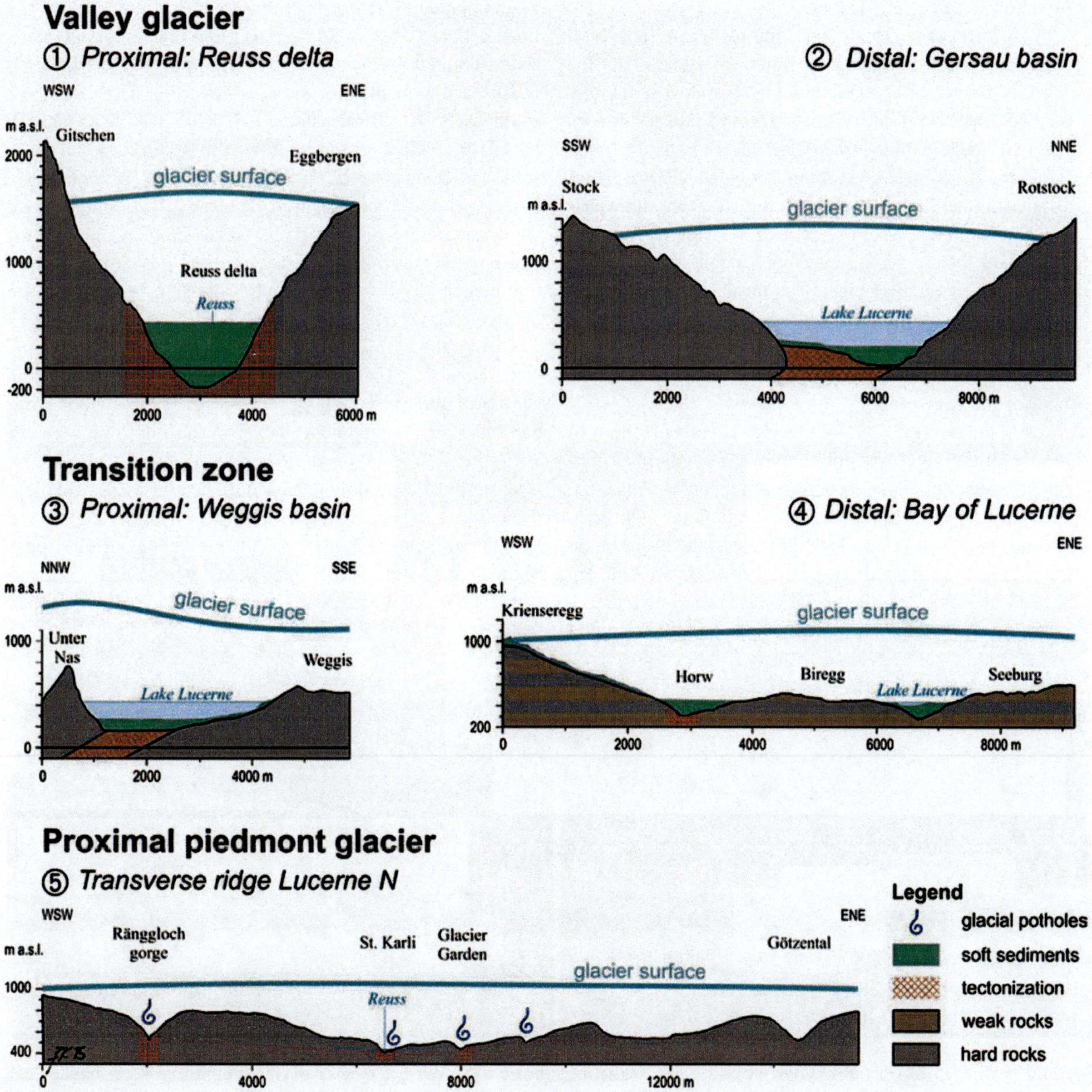

Fig. 21.5 Changes in valley cross sections around Lake Lucerne ranging from inner alpine valley glacier to proximal piedmont glacier type (vertical exaggeration factor of 2.0; for location of topographical profiles see Fig. 21.4)

Axen and Drusberg nappes that extend to Lake Zug. The trough of Lake Uri (Fig. 21.3) has a typical U-shape that was formed under valley glacier conditions with ice thicknesses up to 1800 m.

- *Gersau basin* (Fig. 21.5-2): The shape of the wider Schwyz–Brunnen–Gersau–Stans transversal trough is a result of the geologic-tectonic architecture. The trough is embedded in a syncline made out of soft Flysch rocks that comes along with a nappe overthrust (Pfiffner this volume). The basin of Lake Alpnach is also located in a syncline consisting of soft marly shales with a partial nappe overthrust. The thickness of the ice sheet reached around 1300 m at this location.
- *Weggis basin* (Fig. 21.3): The high towering transversal ridge of the Border Chain separates the Gersau and the Weggis basins. The breakthroughs are located in tectonised zones with prominent strike-slip faults, between the two spurs of the Nasen (lit. *noses*, Figs. 21.1 and 21.6) and the Lopper crest. In the Weggis—Chrüztrichter Basin, in the inner transition zone, the ice thickness reached around 1000 m. The valley shape is strongly controlled by the geologic-tectonic architecture and the confluence of three glaciers. The basin is located in the dominantly weak shales (Flysch and Lower Marine Molasse) with three thrust faults lying on top of each other (Fig. 21.2). Lake Küssnacht follows the basal thrust fault of the Rigi nappe.
- *Bay of Lucerne* (Fig. 21.5-4): Due to lateral spreading of the ice in the outer transition zone, ice overburden continued to decline down to 800 m. As a consequence, the bedrock basin is significantly wider and flatter with twin valleys in the bedrock surface: in the east the basin of the bay of Lucerne and in the west the bay of Horw. Both are located on wrench faults of the thrust belt. Towards the north, these valleys are increasingly carved as tunnel valleys into the bedrock, due to the increasing influence of subglacial meltwater erosion. The lake basin finally terminates at the transverse ridge north of Lucerne which is cut through by canyon-like tunnel valleys (Fig. 21.5-5).

21.3.2 Mesoscale Glacial Landforms

There is an enormous variety of mesoscale, warm-based glacial erosion landforms in the region of Lake Lucerne due to local variations in the rock mass strength of the bedrock. The mountains comprising limestone that are part of the Border Chain (Fig. 21.2) were resistant to erosion and acted as a prominent barrier to ice flow resulting in the formation of spectacular giant stoss and lee forms (Fig. 21.4). The best example is Mt. Bürgenstock (47°00′03.3″N 8°23′43.8″E, Fig. 21.6) with a gentle southward-dipping, smooth stoss side, and a very steep, north-facing, lee side marked by a cliff up to 900 m high. The rock faces at Mt. Bürgenstock were intensely quarried by glacial erosion, and since deglaciation, the highly unstable slopes are very prone to fragmental rock fall and massive rock slope failures (Fig. 21.15).

During the LGM, Mt. Rigi, Mt. Rossberg, and Mt. Pilatus were nunataks (Fig. 21.4). At Mt. Pilatus, the trimline is geomorphologically highly pronounced. The transition from the valley to the piedmont glacier in the hill zone between Weggis and the transversal ridge north of the city of Lucerne is strongly embossed by areal scouring (Fig. 21.4). Thereby, the harder sandstone channel belts of the thrust belt of the Lower Freshwater Molasse and the sandstones of the Upper Marine Molasse were modelled by glacial erosion (Fig. 21.7). The mesoscale geomorphological whaleback features are distinctly aligned with the bedding strike. In

Fig. 21.6 *Left*: View of Ober Nas to the left and Unter Nas to the right. In the strait between those two spurs is a prominent terminal moraine from the stage of 'Vitznau' (16.8 ka cal BP). *Right*: View of the border chain with Mt. Bürgenstock and Mt. Pilatus in the background with the Weggis basin in the front. The steep and shattered and unstable rock cliffs of Mt. Bürgenstock represent the mega stoss and lee features (photos B. Keller)

Fig. 21.7 Shaded topographic relief of areal scouring around Lake Küssnacht. In the fold-and-thrust belt, the sandstone channel belts of the Lower Freshwater Molasse as well as the transgressive units of the Upper Marine Molasse were carved out by the temperate-based glacier as a streamlined bedrock eminence ('whalebacks'). Those whalebacks mostly follow the strike of bedding. Apart from U-shaped outcrop of a fold structure (down left), the traces of the wrench faults are recognisable (shaded relief based on DTM swissALTI3D, swisstopo)

contrast, the rock drumlins of the Weggis Peninsula are aligned consistently with the glacier flow direction from south to north.

The transverse ridge near the city of Lucerne, made out of resistant sandstone of the Upper Marine Molasse, was eroded by the proximal zone of the piedmont glaciers and forms a mega-whaleback that is incised by several perpendicular valleys. Along the whaleback strike, a pronounced, wide depression can be observed that reflects an increasing paleo-ice thickness (Fig. 21.5). The deepest incisions in the centre of the transverse ridge are increasingly V-shaped, which clearly indicates the orientation of subglacial tunnel valleys. The deepest one is located in the Reuss breakthrough at St. Karli (Fig. 21.5-5), and was part of the main drainage system. It has a lowermost elevation of 408.8 m a.s.l., which marks the base level of the entire catchment area of Lake Lucerne. The canyon is flanked by the spectacular Ränggloch Gorge (N 47°2′19.28″ E 8°14′22.85″, Fig. 21.5-5) to the west and by the hanging tunnel valley at the Glacier Garden to the east with a southward-dipping thalweg.

The transverse ridge (Fig. 21.8) acted as a significant barrier against the subglacial meltwater flow originating from the Ice Age Reuss Glacier. The amount of meltwater generated from the ~1900 km^2 glacier ice and firn/snow areas in the conditions of an average discharge of the melting period would have reached between 2000 and 2200 m^3/s. In comparison, the discharge during flooding of the present-day Reuss River in the city of Lucerne reaches up to 473 m^3/s for an event with a recurrence interval of >150 years. This indicates that all subglacial tunnel valleys (Fig. 21.4) could have been active around the same time. Another plausible possibility is that they represent several successive incisions from earlier glaciation periods, as discussed in Sect. 21.3.3.

Near the bedrock riegel of the Helvetic Border Chain, and the transversal ridge near the city of Lucerne, multiple sets of crevasses opened at the glacial surface, which transformed into temporary moulins. Through moulins, the supraglacial channels supplied water through englacial siphons and conduits into the subglacial drainage system (Fig. 21.8). The subglacial meltwater, which was subjected to high hydrostatic pressures, moved through a labyrinth of subglacial channels from Chrüztricher towards the north, and was increasingly channelised towards the transversal tunnel valleys. Originating at the present-day railway station in Lucerne, a large tunnel valley crosses the city of Lucerne generally following the present-day Reuss River. Its genesis as a tunnel valley is inferred from the presence of local esker gravels above the bedrock surface and clusters of glacial potholes. The deepest drilling in the alluvial basin of Lucerne was at the railway station where the paleo-valley floor could be localised at 332.5 m a.s.l., 103.5 m below the surface. From here, the bottom of the tunnel valley rises towards up to 30 m deeply incised bedrock canyon at St. Karli situated at 408.8 m a.s.l. The canyons situated at the Glacier Garden and the Ränggloch Gorge (Figs. 21.4 and 21.5-5) are subordinate tunnel valleys at higher levels.

21.3.3 Glacier Garden of Lucerne

The Glacier Garden of Lucerne (Fig. 21.9) is a unique natural monument. In 1872, J.W. Amrein (1842–1881) purchased the land adjacent to the Lion Monument where he intended to excavate a wine cellar. At the beginning of the excavation, the Lucerne-based geologist F.J. Kaufmann discovered a remarkable, basin-shaped depression in the bedrock, and correctly identified it as a feature produced during the last Ice Age. Together with the well-known geologist A. Heim, he was able to convince J.W. Amrein to continue further excavations in order to preserve this natural monument as an impressive record of glaciation in the region. The natural history museum of the Glacier Garden includes the famous topographic relief of Central Switzerland by Pfyffer de Wyher (1716–1802), as well as tourist attractions.

The excavated rock surface is made of sandstone of the Upper Marine Molasse (Miocene), which covers an area of 5000 m^2 and exhibits a variety of small-scale erosional

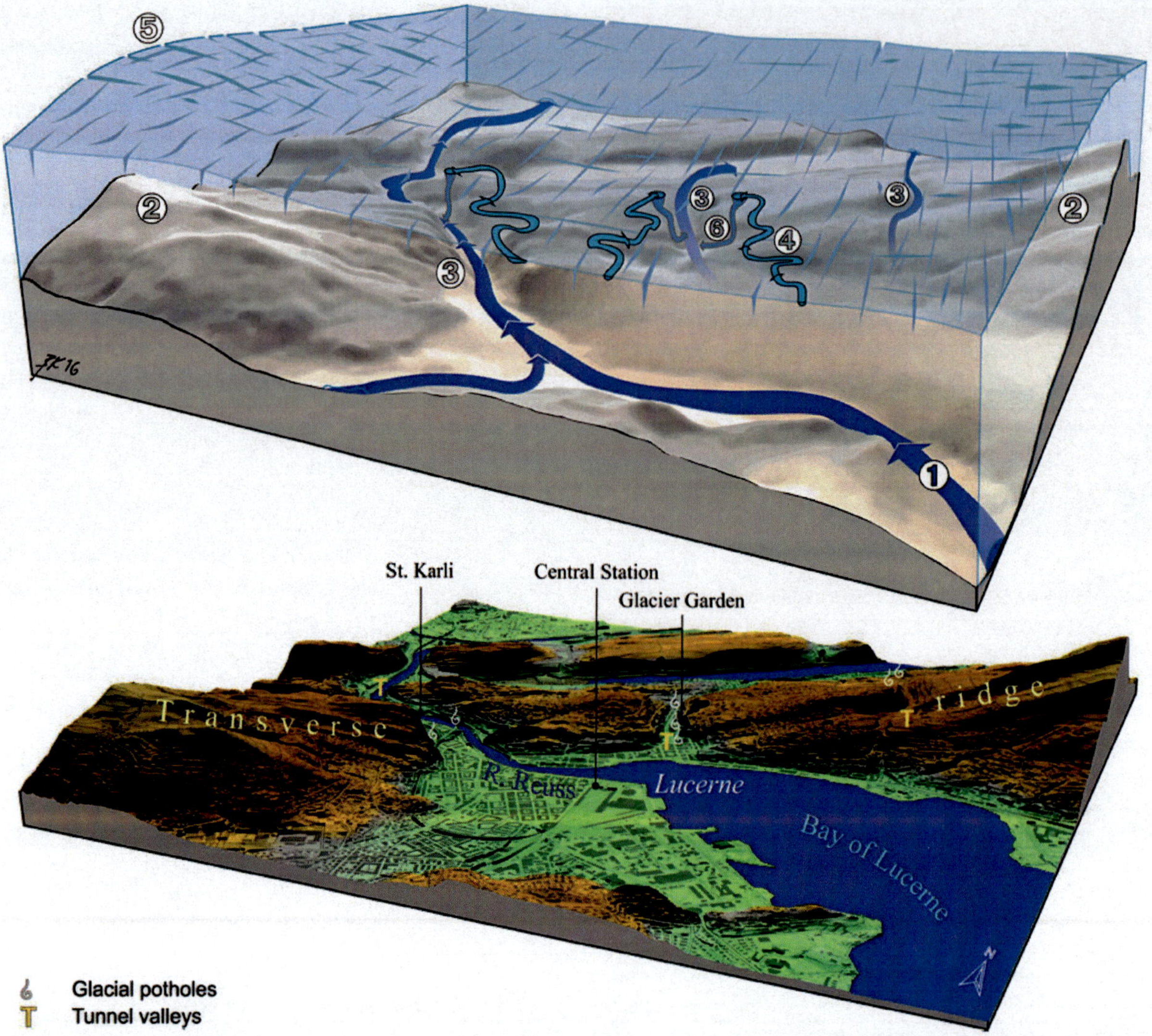

Fig. 21.8 *Top:* Block model of Lucerne during the ice decay, with glacier surface of about 850 m a.s.l. (1) Subglacial meltwater tunnels. Through the transverse ridge (2), the runoff is dammed and cannelized into the narrow oblique tunnel valleys (3). Supraglacial meltwater channels (4). Crossing the transversal ridge, crevasses break open (5), and temporary moulins (6) form. The supraglacial meltwater vanishes in the moulins and flow through tube-like conduits down to the base of the glacier. In pale brown, the bedrock valleys below present-day lake level is indicated. *Below:* Corresponding topographic block model illustrating the transverse ridge delimiting the bay of Lucerne. In green are the Quaternary basin fills. Vertical exaggeration factor of 2

features, such as rat tails, crescentic gouges, lunate fractures and glacial striae (Fig. 21.9). These features indicate flow direction of the Reuss Glacier from south to north. Also present are p-forms that formed by abrasion below the plastically deformed ice from glacial and fluvioglacial processes. Apart from the undulating, mammillated bedrock surfaces, these features occur together with 32 potholes—the largest having the diameter of 8 m, and the depth of 9.5 m (Fig. 21.9).

There have been several different theories proposed over the last hundred years explaining the formation of the potholes. At the time of the Glacier Garden excavation, between 1872 and 1876, the prevailing 'glacial mill theory' held that the potholes were erosional features formed by waterfalls plunging vertically down through glacial crevasses or moulins onto the glacier bed (Fig. 21.10-1). The descending water caused boulders to rotate and thereby excavated holes into the bedrock. Thus, according to this theory, every pothole should contain a well-rounded boulder that would have acted as a 'grinder'. To demonstrate this erosional process, a model of a 'glacial mill' was built into a rock cavern at the Glacier Garden by the Swiss Federal

Fig. 21.9 *Left top*: Historical view of the freshly excavated bedrock surface with some erratic boulders in the Glacier Garden Lucerne. Glacial striae, crescentic features like gouges indicate former glacial motion from south to north (from the front to the rear of the picture) (photo Glacier Garden Lucerne). *Left beneath*: Largest and most beautiful pothole (9.5 m in depth, 8 m in diameter) in the Glacier Garden Lucerne. The depth of the pothole gives an idea at what high pressure and velocity the glacial meltwater hollowed out the bedrock surface whilst the complicated path of the water jet is reflected in the great variety of forms occurring in the pothole. The eddie in the giant pothole seems to have spiralled down clockwise into the depths. Somewhat more solid sandstone in the middle of the pothole was preserved as a pointed 'rock-nose' (photo Glacier Garden Lucerne). *Right*: The legendary Ränggloch Gorge that has not, as often thought, been carved by man. It rather represents a subglacial meltwater channel that has been carved into the bedrock (see Fig. 21.5) (photo B. Keller)

Institute of Technology (ETH) in 1896. The model comprises an artificial block that is hollowed on the inside, and a water jet crashing down from above causing the block to rotate.

Roesli (1957) continued with the glacial mill theory but no longer considered the rotating blocks as an important factor for erosion (Fig. 21.10-2). The observation of a glacial pothole on the summit of a *roche moutonnée* in the Glacier Garden indicated that the hole could not have originated from a subglacial stream, but by waterjet action through *moulins*, as illustrated by the moulin-waterjet theory of Roesli and Wick (1979). However, Streiff-Becker (1951) interpreted this to indicate that the potholes were formed by subglacial meltwater flow prior to the formation of the *roche moutonnée*. This interpretation is supported by the recent discovery of the deeply scoured Reuss tunnel valley (Fig. 21.8), more than 100 m lower than the southern end of the 'hanging' tunnel valley of Glacier Garden. It is therefore plausible that this tunnel valley and the potholes of the Glacier Garden were formed during earlier glacial advances before the LGM, presumably during the first advance of the Birrfeld glaciation ('Early Würmian'), some 110 ka cal BP.

The moulin-waterjet theory contradicts the fact that temperate glaciers during the last Ice Age flowed several meters per day, thus the plunging waterjet would have moved over the rock surface and created groove-like channels. It has been established that U-shaped gouge holes or plunge pools are also eroded with their long axis perpendicular to the flow direction. Hence, the erosional capacity of

Fig. 21.10 The development of interpretation of the potholes: (1) 'Glacial mill' theory with plunging waterfall (late nineteenth-century postcard; *left* closed, *right* open with insight); (2) Waterfall into a crevasse with grinding stone (Roesli, 1957); (3) 'Moulin-waterjet' theory with only the minor role of the 'grinding stone' (Roesli and Wick 1979); (4) Current interpretation: various phases of the formation of a glacial pothole in the bedrock under the base of the glacier. The sediment-loaded meltwater currents flow at an enormous pressure and high speeds through irregular bedrock of the subglacial meltwater tunnel. Irregularities cause turbulences, flow separation and complicated vortex motions, which, in short time, carve out glacial potholes (from Keller 2007)

stationary eddies in flowing currents is a viable explanation for the formation of glacial potholes, which has been supported by experiments of Alexander (1932). According to this theory, which is analogous to river pothole formation, water spirals in an separated eddy downward along the pothole walls to its base, and then rises finally in a very tight spiralling movement in the center axis of the eddy. Keller and Wick (1985) adapted this theory to explain the formation of the potholes of the Glacier Garden (Fig. 21.10-3). The glacier potholes are interpreted to represent holes formed by eddies that occur in subglacial meltwater streams that are associated with flow separation. Their formation is also comparable to the recently formed river potholes observed at the entry of the Ränggloch Gorge near the city of Lucerne. Contrary to surface waters, subglacial meltwater is subjected to very high pressures: during the LGM the ice thickness at the current-day location of Lucerne was between 500 and 600 m, which would correspond to water pressures of up to 55 kPa.

The most recent explanation (Keller 2013) for the formation of glacial potholes takes into account glacial activity in the region of Lucerne during the last glaciations. As shown in the preceding section, the Glacier Garden is situated in one of three major tunnel valleys that are incised into the transverse ridge (Fig. 21.5-5). Subglacial meltwater typically flowed through most of the glacial network at velocities of only a few km/h except where it was channelled in bottlenecks at valley crossings, resulting in accelerated flow velocities. This was especially the case for tunnel valleys located at the Glacier Garden, the River Reuss and at the

Ränggloch Gorge (Fig. 21.9). Comparable subglacial esker channels with a width of ~20 m and a height of ~10 m have a flow capacity of ~500 m^3/s (Shreve 1972). These channels offered ideal conditions for the torrential, subglacial meltwater streams to scour the rock beneath by stationary eddy currents that formed potholes, and less frequently Sichelwannen (curved grooves formed by water under immense pressure at the base of the glacier). The sedimentary load that was transported by the turbid, turbulent meltwater stream (i.e. carrying silt, sand, gravel, stones and blocks) scoured the bedrock and formed the potholes. According to the experiments of Alexander (1932), rock fragments circulating in eddies are thrown against the walls of the potholes at velocities of up to 200 km/h and erode out the potholes in a short amount of time (one up to a few meltwater periods), comparable to a sandblaster. Evidence for the spiralling water movement is also indicated by the helical erosional gouges along the pothole walls (Figs. 21.9 and 21.10-4).

In the tunnel valleys at the outer transition zone of the Reuss Glacier (Fig. 21.4), glacial potholes are expected to have formed in clusters. Until now they have been observed almost exclusively in the presumably pre-LGM tunnel valley at the Glacial Garden. In the same valley, numerous other potholes were discovered during underground excavations. The biggest one had a diameter of over 10 m and was found at the northern exit of the valley near Schlossberg. A very deep glacial pothole has been detected through seismic investigations and pile drilling work south of the Glacial Garden (N 47°3′27.4″ E 8°18′37.2″). It has the diameter of 10 m and the depth of 12 m. In 2008, first glacial potholes were discovered at the flanks of the Reuss breakthrough (Keller 2013).

21.3.3.1 Landscape Evolution from the Ice Decay Period to the Present

The maximum ice levels during the LGM occurred at ~24 ka cal BP, and the ice decay of the piedmont glacier (Fig. 21.11) started at 23 ka cal BP (Keller and Krayss 2005). The glacier retreat (Fig. 21.12) was interrupted by several climatically controlled, but also rheologically induced readvances. At 18 ka cal BP (Stage 'Gisikon'), the glaciers had already receded from the midland Sempach, Hallwil and Baldegg lakes located north of city of Lucerne. The Reuss Glacier stagnated around 17.5 ka cal BP south of the transversal ridge, and dammed the topographic depressions by kame terraces or end moraines to form temporary lakes. One of these temporary lakes was Lake Kriens, which dewatered via the Ränggloch gorge into Lake Littau situated ~10 km away in the valley of the Kleine Emme River. After a stagnant phase at Lucerne, ice decay, which was interrupted by two stagnant ice stages, continued into the transition zone. Around 16.8 ka cal BP, an initial, distinctive subaqueous terminal moraine with frontal debris flows and flow tills was deposited between the two spurs of Nasen at Vitznau (Fig. 21.1). In addition, a second, subaqueous terminal moraine was formed at ~16.5 ka cal BP in the Gersau Basin (Hilbe et al. 2011). Prior to 16 ka cal BP, Lake Lucerne was completely ice free with fjord-like lake arms in the proximal U-shaped valleys, and Mt. Bürgenstock protruding initially as an island. At the exit of the tunnel valley of the Reuss River north of the city of Lucerne (47°03′59.0″N 8°17′05.0″E), another terminal moraine is situated below the valley floor. This moraine dammed Lake Lucerne, which successively grew in size through progressive ice decay. The crown height of this terminal moraine reaches up to 432 m a.s.l., corresponding to the temporal storage level of Lake Lucerne.

Fig. 21.11 The wall painting 'Lucerne during the Ice Age' (by E. Hodel, 1926/27, Glacier Garden Lucerne) shows the view from a virtual hill in the front towards the Alps at the beginning of the ice decay, some 20 ka cal BP ago (compare with Fig. 21.3 top)

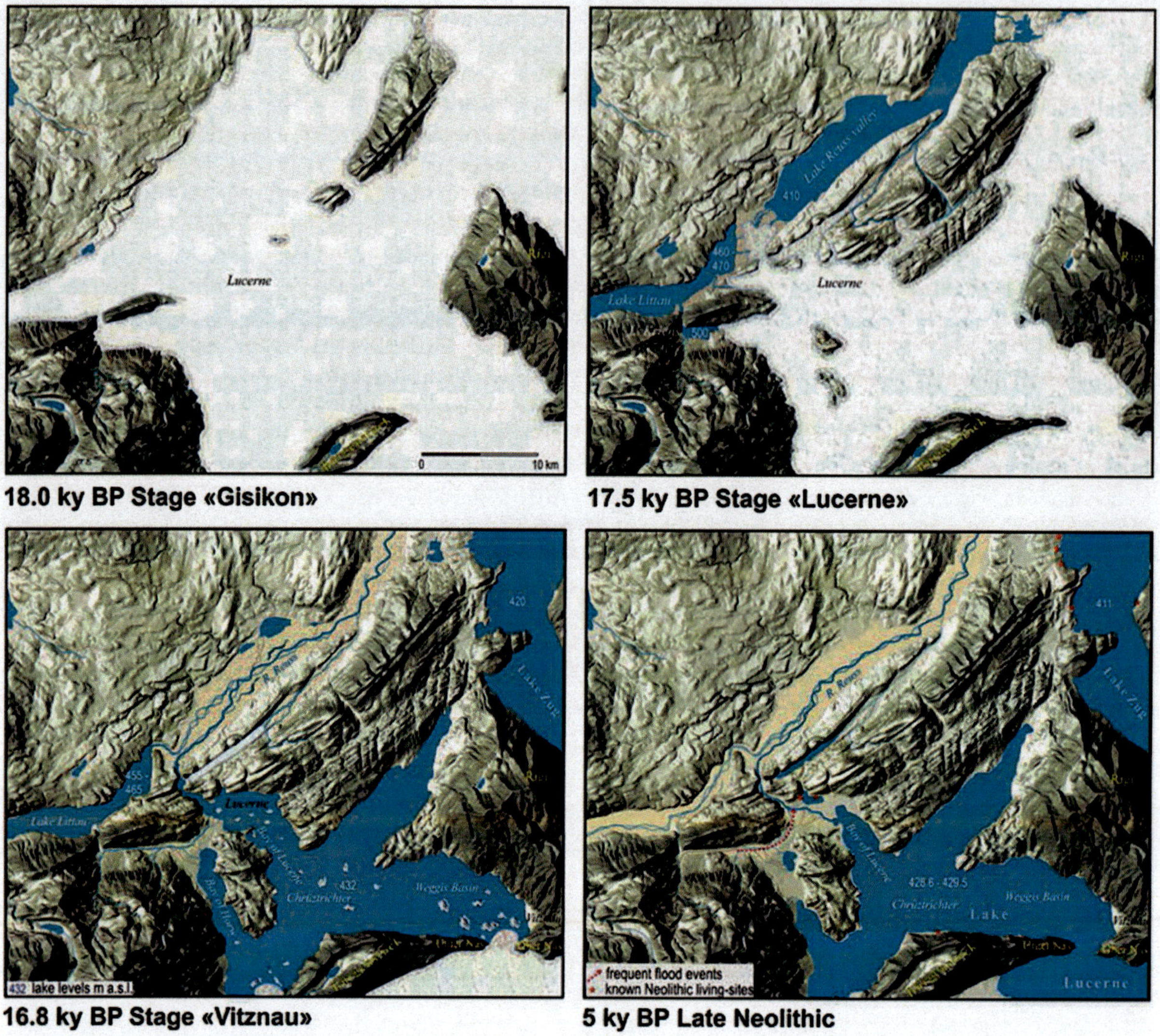

Fig. 21.12 Reconstruction of the geomorphological evolution around the city of Lucerne during the Late glacial ice decay between 18.0 ka cal BP and 5 ka cal BP (after Keller 2013)

The glaciers receded from the valley flanks 18 ka cal BP ago. Disappearance of the ice cover caused sedimentary gravity flows that transported the remaining unconsolidated tills from the valley flanks to their floors. The sediments were subsequently transported and distributed by young streams and torrents towards Lake Lucerne. The same erosional processes are still active today. Unlike the bay of Lucerne described above, the sedimentary facies of the infillings of the fjord-like proximal valleys consist of prograding, mainly Gilbert-type deltas, and alluvial plains, as observed in the Urner Reuss Valley.

Several deep drillings around the city of Lucerne provide information on the geomorphological evolution in the region. They indicate that the distal tunnel valleys of Lake Lucerne were almost completely incised into bedrock (Fig. 21.8). Only a few deposits of esker gravel and basal melt-out tills remained on the bedrock surface. Accompanied by the ice decay, the gradually growing lake basin increasingly acted as a large-scale sediment trap which received meltwater discharge with pronounced suspended sediment load from the subglacial conduits, deposited as heterolithic sandy-silty turbidites. In addition, subaqueous ice-terminal fans were locally deposited into the distal bedrock valley of Lake Lucerne. Dropstones within these glacio-lacustrine facies constrained the position of the ice front that calved into the young Lake Lucerne between 17.5 and 16 ka cal BP. Drifting icebergs reached the present-day city of Lucerne (Fig. 21.12) as recorded by a 2.5 ton boulder of Aar Granite, which was dug out of a slotted wall in 1993 at the Lucerne railway station. As the ice decay continued, the

Fig. 21.13 *Left*: Paleogeographical reconstruction of the Bay of Lucerne in the Neolithic Age (5 ka cal BP) with a low lake level between 428.6 and 429.5 m a.s.l. showing that the lake shore was situated further south and that the River Reuss meandered through a swampy alluvial plain. The rounded bight is a probable footprint of the glacial tongue at the time of the turbiditic infill of the former lake basin. *Right*: The same area with the present-day extent of the lake at a lake level of 434 m a.s.l. The water tower is located in the water (see text)

glacial sediment supply gradually decreased, resulting in the deposition of glacial sediment in the more proximal sediment traps of the Gersau Basin, and into Lake Uri. By 14.7 ka cal BP (Bølling period), based on palynological constraints, the bay of Lucerne was already filled by sediments up to a level of 422 m a.s.l. in the area of the present-day railway station and up to 426 m a.s.l. at the northern lake shore.

The history of the prehistoric lake level was reconstructed in detail from drill cores, palynology data, and radiocarbon dating of wood (Keller 2013). The lake level during the Late glacial stage was 432 m a.s.l, and subsequently lowered to an average level of 428.6 and 429.5 m a.s.l. in the Neolithic Age (between 7 and 4 ka cal BP), with an amplitude of 1.5 m from seasonal variations. At this time, the shoreline of the bay of Lucerne was situated 1.6 km further south-east, and only a small pond remained (Fig. 21.13). In the area between the former shoreline and Lucerne a swampy flood plain existed. Along its margins, near the hill flanks, at depths of 5–8 m below the recent terrain surface an Upper Paleolithic artefact and numerous traces of Neolithic settlements have been discovered.

Since the end of the last Ice Age, the Ränggbach and Krienbach torrent systems have dominated the geomorphological arrangement of the flood plain on the western bank of the bay of Lucerne. The catchment area of these torrents is situated on the western flank of Mt. Pilatus, which is covered by a thick series of easily erodible tills. During intense rainfall events, these torrents have led to catastrophic flooding (Figs. 21.12 and 21.13). Such events are recorded by the dominantly gravelly alluvial fan deposits and sandy crevasse and overbank sands, which cover most of the flood plain situated in the Neustadt (lit. *new village*) right bank of the city of Lucerne. The remaining water and coarse-grained bedload from the floods were gradually fed into the slow flowing Reuss River, and dominantly deposited as gravelly lobes. The low energy Reuss River was only partially able to erode away these coarse-grained deposits that were supplied by the torrents. This remaining sediment accumulated and increasingly dammed the lake. The successive damming of the lake can explain the numerous transgressive-regressive sedimentary cycles from shallow lake to alluvial plain and peat-rich swamp deposits. Since the deglaciation, at least two catastrophic debris flows entered the Reuss River, documented in the sedimentary record by up to 2 m thick beds of diamictons. The younger debris flow was deposited at the end of or after the Neolithic and may have further dammed the lake resulting in an average lake level of 429.5 m a.s.l during the Roman period.

During the foundation of the city, after 1200 AD, the average lake level was between 432.5 and 433.0 m a.s.l. Therefore, the lake basin became submerged making the city of Lucerne accessible by boat. The Wasserturm (lit. *water tower*) as Lucerne's landmark symbol was built in the fourteeth century. Based on its grounding ledge at 432.9 m a.s.l. and the corresponding lake levels it becomes clear that this tower was able to be built on land. Since the Middle Ages, the lake level has been artificially controlled (see Keller 2013).

The flooding and overbank sedimentation from the Krienbach torrent system down to the western bank of the Reuss River and the Kleinstadt neighbourhood have caused

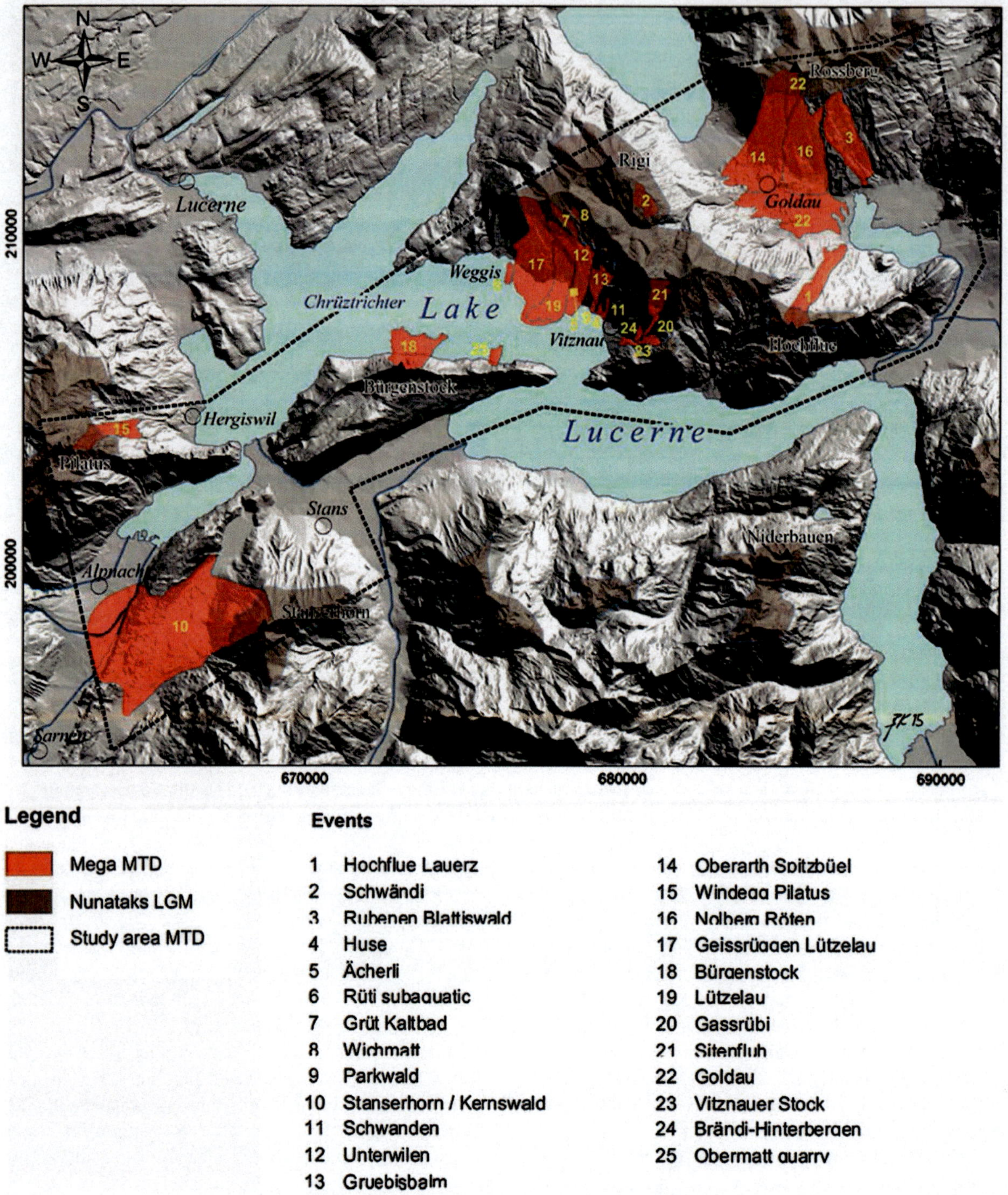

Fig. 21.14 Shaded relief inventory of 25 known prehistoric and historic massive rock slope failures (MRSF) in the area of the northern Lake Lucerne (from Keller 2017; DTM based on SwissALTI3D, swisstopo)

significant problems for the city of Lucerne since the Middle Ages. Flooding events are documented for the first time in 1333, subsequently 4 events in the fifteenth century, 14 in the sixteenth century, 7 in the seventeenth century and 2 in eighteenth and nineteenth centuries each. Only in the 1890s preventive watershed and flood prevention operations were

Fig. 21.15 In 2003, a high-energy fragmental rock fall broke through three rows of steel rope nets, causing major damage on the main road south of the town of Vitznau, and some fragments eventually arrested at the rockfall protection dam (N 47°0′11.95″ E 8°29′9.74″). The source area was located within the steep, north-facing and intensely quarried bluffs of Mt. Vitznauer Stock (photo B. Keller)

put into place. These included reforestation, construction of check dams, drop structures, riprapping and an extension of the straight, paved water race downstream to the Ränggloch Gorge. The flood abatement measures were completed in 1992 with the building of a bypass tunnel that diverts floodwater from the municipality of Kriens directly to the Reuss River.

21.3.4 Contemporary Catastrophic Mass Movements

Although the unique landscape of the region of Lake Lucerne is beautiful, inhabitants are often exposed to risks of catastrophic gravity mass movements, due to the accumulation of unstable Quaternary glacial deposits on steep slopes. A recent inventory of mass-transport deposits (MTDs) in the area of northern Lake Lucerne (Keller 2017) has yielded an astonishing number of identifiable 25 massive rock slope failures (MRSF) since deglaciation (Fig. 21.14 and 21.15), corresponding to calculated return periods of 121 ± 47 y for the historic, and 215 ± 75 y for the datable MTDs. This can be explained by the very unfavourable geologic-geomorphological predisposition, mainly caused by sedimentary successions, tectonic framework and glacial erosion processes. Rock irregular slides like the Bürgenstock event in 1601 (20 M m^3) or the Vitznauer Stock event in 1879 (>1 M m^3) are mainly bound to the Border Chain. In contrast, rock planar slides typically occurred on the broad dip-slopes of the Subalpine Molasse. Examples are Lützelau south of Vitznau in 1661 or the remarkable Goldau rock planar slide—rock avalanche in 1806, which resulted from the slope failure of Mt. Rossberg. This massive rock slope failure (3–4 M m^3) completely buried the village of Goldau and resulted in 457 casualties (Heim 1932). Some of the catastrophic MRSF were triggered by earthquakes, such as the Weggis debris-landslide, the previously mentioned Bürgenstock mega rock fall, as well as 11 coeval mass-flow deposits detected on the lake floor. These MRSF were triggered by an earthquake in 1601 with a moment magnitude of ~6.2 that caused a tsunami wave in Lake Lucerne and some damage to the city of Lucerne (Schnellmann et al. 2006). In addition, steep, unstable cliffs that exhibit megascale stoss and lee features are sources of permanent fragmental rock falls. Although earlier settlements and transportation routes avoided the gravitational natural hazard zones, this has not been the case in the twentieth century with development increasingly extending into these critical zones (Keller 2007). Because of risk exposure, extensive protective and monitoring measurements have been put in place.

21.4 Conclusions

The extraordinarily diverse landscape of Lake Lucerne is geologically and geomorphologically fascinating. Since the Middle Ages, this region has been part of the important transalpine corridor over the Gotthard Pass, which was developed at the end of the nineteenth century by the pioneer mountain Gotthard Railway. This region has inspired natural scientists for centuries and its easy accessibility due to transportation has facilitated early scientific investigations and has led to the development of several fundamental concepts in geosciences.

The Lake Lucerne region offers the shortest geological cross section in the Alps from the crystalline basement through the Mesozoic nappe pile to the Tertiary foreland basin. The complex shape of Lake Lucerne, with its bays, basins and lake arms, is unequivocally a product of the macroscale geologic-tectonic architecture, which was subsequently carved by glacial erosion during the Ice Age.

In the proximal areas of the lake, the valley glaciers Brünig, Engelberger, as well as the larger Reuss Glacier, formed a dendritic glacier network and eroded deep U-shaped valleys. At the confluence of the three glaciers in the transition zone that was confined by the two prominent nunataks of Mt. Rigi and Mt. Pilatus. The ice spread outwards towards the Swiss Plateau in the north, as a piedmont glacier.

These valley and piedmont glaciers left impressive examples of meso- and microscale forms. These well-preserved features include: (i) the spectacular stoss and lee forms, that include spurs from the Border Chain; (ii) the nunataks, such as Mt. Pilatus and Mt. Rigi; (iii) the elongate whalebacks and *roches moutonnées*, which are part of areal scouring from the transition zone glaciers; (iv) the glacial striae on the bedrock surface exhibited at the Glacier Garden; (v) the famous giant potholes of the Glacier Garden, which formed in subglacial tunnel valleys at high water pressures due to water jets carrying very abrasive, suspended sediment load that circulated in stationary eddies.

During deglaciation, as the glaciers decayed, Lake Lucerne increasingly became a large-scale sedimentary trap. At the onset, the bay of Lucerne became ice free, and within a short time period, was filled by a thick series of dropstone-bearing, glacio-turbiditic sediments. During the Neolithic, the lake level was approximately 5 m lower than today, and the bay of Lucerne was swampland. Along the swamp margins there were several Neolithic living sites, which were flooded when coarse-grained fan deposits from neighbouring torrents caused a rise in the lake level. During the Middle Ages, the first weir was constructed that facilitated a further increase in the lake level. Not only did glaciation produce a beautiful landscape, but also a geomorphologically active one. The region is prone to widespread natural hazards that include frequent fragmental rock falls and massive rock slope failures.

Acknowledgments The author gratefully acknowledges Nathalie Lefebvre and Marco Billia for the accurate translation of the manuscript as well as Peter Heitzmann, an anonymous reviewer and Emmanuel Reynard for the careful review.

References

Alexander HS (1932) Pothole erosion. J Geol 40:305–337

Arbenz P (1912) Der Gebirgsbau der Zentralschweiz. Verh Schweiz Natf Ges 95:95–120

Finckh P, Kelts K, Lambert A (1984) Seismic stratigraphy and bedrock forms in perialpine lakes. Geol Soc Am Bull 95(9):1118–1128

Haeberli W, Penz U (1985) An attempt to reconstruct glaciological and climatological characteristics of 18 ka cal BP Ice Age glaciers in and around the Swiss Alps. Z Gletscherkunde Glazialgeologie 21:351–361

Heim A (1919) Geologie der Schweiz. Band I: Molasseland und Juragebirge. Tauchnitz, Leipzig, 704 pp

Heim A (1932) Bergsturz und Menschenleben. Fretz & Wasmuth Verlag, Zürich, 218 pp

Hilbe M, Anselmetti FS, Eilertsen RS, Hansen L, Wildi W (2011) Subaqueous morphology of Lake Lucerne (Central Switzerland): implications for mass movements and glacial history. Swiss J Geosci 104:425–443. https://doi.org/10.1007/s00015-011-0083-z

Kaufmann FJ (1876) Fünf neue Jurassier: Mythen, Buochserhorn und Stanzerhorn, Enzimattberg und Rothspitz. Nebst Bemerkungen über den Giswilerstock. Jahrb Schweiz Alpenclub 11:45–74

Keller B (2007) So entstand der Vierwaldstättersee. In: Stadelmann P (ed) Vierwaldstättersee. Brunner Verlag, Kriens, pp 30–53

Keller B (2013) Geologische Geschichte der Luzerner Reuss. In: Paravicini G (ed) Gestautes Wasser—Regulierter See. Kantonaler Lehrmittelverlag, Luzern, pp 103–122

Keller B (2017) Massive rock slope failure in Central Switzerland. History, geologic-geomorphological predisposition, types and processes, triggers, and resulting risks. Landslides 14(5):1–21. https://doi.org/10.1007/s10346-017-0803-1

Keller B, Wick P (1985) Glacier Garden Lucerne. Gletschergarten Luzern, Luzern

Keller O, Krayss E (2005) Der Rhein-Linth-Gletscher im letzten Hochglazial. Vjschr Natf Ges Zürich 150(1-2):19–32 and 150 (3-4):69–85
Nitsche FO, Green AG, Horstmeyer H, Bücker F (2002) Late Quaternary depositional history of the Reuss delta, Switzerland: constrains from high-resolution seismic reflection and georadar surveys. J Quaternary Sci 17:131–143
Penck A, Brückner E (1909) Die Alpen im Eiszeitalter. Tauchnitz, Leipzig, 3 vol
Pfiffner OA (this volume) The structural landscapes of Central Switzerland
Roesli FJ (1957) Der Gletschergarten von Luzern. Luzern i. Wandel d. Zeiten 7. Verlag Eugen Haag, Luzern, 48 pp
Roesli FJ, Wick P (1979) Glacier Garden Lucerne. Gletschergarten Luzern, Luzern
Schnellmann M, Anselmetti FS, Giardini D, McKenzie JA (2006) 15,000 years of mass-movement history in Lake Lucerne: Implications for seismic and tsunami hazards. Eclogae Geol Helv 99:409–428
Shreve RL (1972) Movement of water in glaciers. J Glaciol 62:205–214
Streiff-Becker R (1951) Potholes and glacier mills. J Glaciol 1:488–490
Trümpy R (2003) Trying to understand Alpine sediments—before 1950. Earth-Sci Rev 61:19–42

Beat Keller is CEO of the consulting company Keller+Lorenz AG in Lucerne, member of the International Association of Sedimentologists, the Society for Sedimentary Geology, and the Swiss Geological Society. His main scientific interests are the sedimentology, hydrogeology and petrophysics of the Molasse basin and the geotechnical properties and genesis of Quaternary sediments. His practical experience encompasses amongst other disciplines the engineering geology of large buildings and tunnels as well as natural hazards. He was freelance scientific employee of the Glacier Garden Lucerne (1985 and 2010), is lecturer of special courses at ETH Zurich, and author of the geology and geomorphology assessment for the Federal Inventory of Landscapes and Natural Monuments of National Importance (ILNM).

22 Between Glaciers, Rivers and Lakes: The Geomorphological Landscapes of Ticino

Cristian Scapozza and Christian Ambrosi

Abstract

The upper valleys of the Canton Ticino have been the theatre of important descriptions and interpretations about the morphogenesis of Alpine glacial valleys since the mid-nineteenth century. William Morris Davis, for example, visited this region at the end of the nineteenth century. Numerous sites host remarkable associations of glacial, periglacial and/or fluvial landforms. We present here, as examples, (a) the Greina plateau, a symbol in the preservation of natural heritage in Switzerland, which is considered a fragment of arctic tundra in the Alps; (b) the rock glaciers of the southern side of Adula, the most emblematic periglacial site in the Southern Swiss Alps; and (c) the Bolle di Magadino, corresponding to the youngest part of the Ticino River delta in Lake Maggiore. The ideas and landscape evolution related to the origin of the southern Alpine valleys are finally discussed.

Keywords

Southern Swiss Alps • Glacial erosion • Permafrost • Floodplain • Historical geomorphology

22.1 Introduction

The upper valleys of the Canton Ticino (Leventina, Blenio and Riviera) constitute the southern pillar of the St. Gotthard massif, which is considered one of the mountain symbols of the Swiss Alps, cradle of springs of the most important Swiss Alpine rivers (Reuss, Rhine, Ticino and Rhone rivers). Considering also the Ticino valley south and west of Bellinzona, this region extends from Lake Maggiore (193 m a.s.l., 46° 9′ N 8° 50′ E, the lowest area in Switzerland) to the summit of Adula/Rheinwaldhorn (3,402 m a.s.l., 46° 29′ 38″ N 9° 2′ 25″ E) (Fig. 22.1).

Since the mid-nineteenth century, these valleys have been the theatre of important descriptions and interpretations about the morphogenesis of Alpine glacial valleys. These deeply incised valley troughs have been inhabited since the Early Neolithic (ca. 7.0 ka cal BP). Since the Bronze Age, they have acted as an important corridor of transit from the north to the south of the Alps, and vice versa. This characteristic was reinforced in the Contemporary Epoch by the opening of the Gotthard rail tunnel in 1882, the Gotthard road tunnel in 1980, and the Alp Transit base tunnel, which was opened the 1st June 2016. The Gotthard rail tunnel in particular, made it possible not only the circulation of people and goods, but also the circulation of ideas. The region was also visited at the end of the nineteenth century by one of the fathers of modern geomorphology, William Morris Davis (Davis 1900a, b), who partially developed here the theory about "the sculpture of mountains by glaciers" (Davis 1906).

After an overview of the region, three sites are described, making it possible to present the glacial, periglacial and fluvial landscapes of Upper Ticino valleys. A fourth section focuses on the evolution of ideas concerning the genesis of relief of this region. Valleys' origin and evolution are presented focusing on landforms genesis, valley infilling and slope sedimentary dynamics.

22.2 Geographical and Geological Setting

For its geographical location and its particular morphological configuration, the Upper Ticino is located between the harsh Alpine climate and the more temperate Mediterranean climate. The relief of the upper valleys is characterised by

C. Scapozza (✉) · C. Ambrosi
Institute of Earth Sciences, University of Applied Sciences and Arts of Southern Switzerland (SUPSI), Campus Trevano, 6952 Canobbio, Switzerland
e-mail: cristian.scapozza@supsi.ch

C. Ambrosi
e-mail: christian.ambrosi@supsi.ch

E. Reynard (ed.), *Landscapes and Landforms of Switzerland*, World Geomorphological Landscapes,
https://doi.org/10.1007/978-3-030-43203-4_22

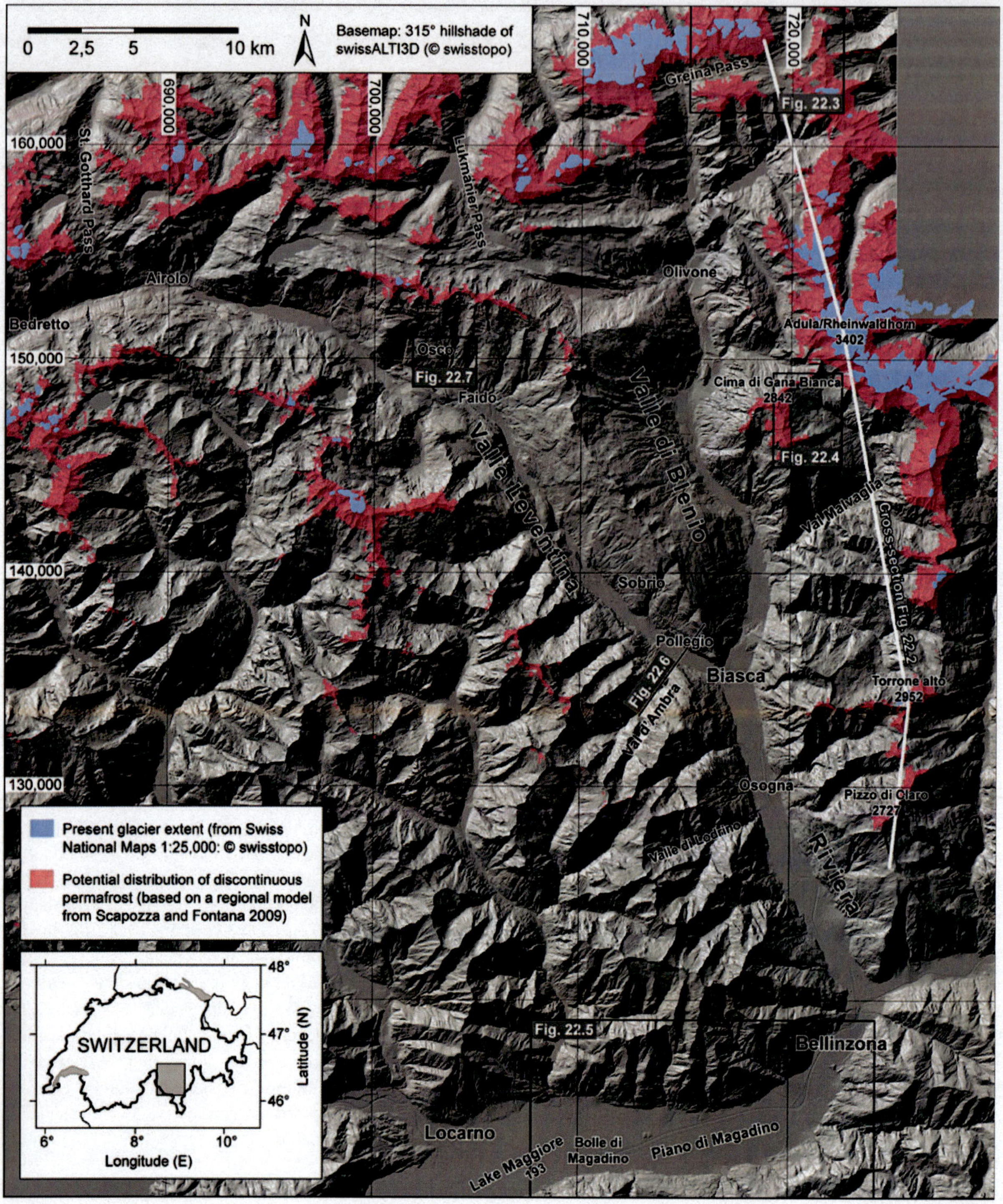

Fig. 22.1 The upper valleys of the Canton Ticino, with location of the localities cited in the text. Coordinates: Swiss Grid system CH1903/LV03 (swisstopo)

steep slopes, where significant rock walls, peaks exceeding 3,000 m a.s.l., partial lack of vegetation above 2,000 m a.s.l., and the presence of perennial snow and glaciers above 2,800 and 3,000 m a.s.l. (altitude of the equilibrium line of the glaciers, Scapozza and Fontana 2009) indicate typical Alpine geomorphological conditions. Downstream of Bellinzona, the valleys widen, extensive chestnut forests cover the valley slopes, and climate is more temperate and becomes almost Mediterranean in proximity of the Lake Maggiore.

The Southern Swiss Alps, and in particular the Ticino Canton, are known in Switzerland as the "sun living room" (*Sonnenstube*) of the country. This is not really the case in the upper valleys, where the shadow effect of the rock walls reduces the number of sunny hours to 1,400–1,500 per year. From the right side of the Piano di Magadino to the Lake Maggiore, this value overcomes 2,000 h per year, with a maximum value of 2,155 h per year in Locarno–Monti (366 m a.s.l. on the Lake Maggiore right side; 1961–1990 mean). This significant sunny spot contrasts with the relatively humid climate, with mean annual precipitations (1961–1990 period) of 1,500–1,600 mm, in the upper valleys and in the region of Bellinzona, and even more than 1,700 mm on the Lake Maggiore shoreline (with a maximum peak of 2,019 mm in Brissago). However, precipitations are concentrated in more or less 100 days per year (97 days of Biasca, at the crossroad between Leventina, Blenio and Riviera valleys; 103 days between Bellinzona and Lake Maggiore). Mean annual air temperatures in the valley bottoms are generally between 9.6 °C (Comprovasco, 575 m a.s.l. in the mid-Blenio Valley) and 11.5 °C (Locarno–Monti). At 2,500 m a.s.l., the mean annual precipitation is 2,300 mm/a and the mean annual air temperature is −1 °C (Scapozza and Fontana 2009).

The Ticino Alps are located between two zones characterised by completely different petrography, tectonics and palaeogeography: the Central Lepontine Alps to the north, and the Southern Alps to the south. These two zones are separated by the Insubric Line, also called Tonale Line, crossing the Morobbia Valley and the Piano di Magadino in East–West direction.

From a tectonic point of view, north of the Insubric Line the Gotthard massif (composed by granites and metagranitoids) with its autochthonous and para-autochthonous cover of Triassic calcareous and dolomite marbles and Jurassic calcschists and clay schists, and several nappes and zones belonging to the Penninic domain are present, whereas south of it, units belonging to the Sudalpine domain are located. In the Penninic domain, nappes belonging to the Infrapenninic crystalline basement (Leventina/Lucomagno, Simano/Antigorio, and Adula nappes), the Lower Penninic crystalline basement (Southern Steep Belt [SSB] zones and nappes, corresponding with the former *Wurzelzone*), and the Middle Penninic crystalline basement (Maggia nappe), are present (Fig. 22.2). These nappes are mainly composed of gneisses and micaschists. From a palaeogeographical point of view the Infrapenninic nappes represent the European continental margin and the transitions between the Helvetic and the Valais domains, and are the deepest structures of the entire Alps. The Simano and Antigorio nappes are two deepest structures over the Ticino culmination, represented by the Leventina/Lucomagno gneisses (called also Lepontine basement), and are considered as the deepest and southernmost portions of continental crust of the Helvetic domain. The Lower Penninic nappes belonging to the SSB are comprised in the Paleogene Tectonic Accretion Channel (PTAC) and represent the continental crust forming the basement of the Valais Basin. The Middle Penninic, finally, represents the Briançonnais continental crust. The Mesozoic cover of the Penninic nappes was almost completely dislocated against the Gotthard massif.

In the Sudalpine domain, the Ivrea–Verbano (on the orographic right side of Lake Maggiore) and Strona–Ceneri zones, as the Upper and Lower Orobic nappes, are present. These Sudalpine units are composed by gneisses and micaschists and belong to the Adriatic continental margin. Directly across the Insubric Line, intrusive Alpine rocks such as diorites, tonalites and granites of the Masino-Bregaglia massif are present.

22.3 Landforms and Landscapes

Three sites illustrating the diversity of landscapes and processes in the Upper Ticino valleys were selected to present, respectively, the glacial, periglacial and fluvial landforms: (a) the Greina plateau, a symbol in the preservation of natural heritage in Switzerland, which is considered a fragment of arctic tundra in the Alps; (b) rock glaciers of the southern side of Adula, the most emblematic periglacial site in the Southern Swiss Alps and (c) the Bolle di Magadino, corresponding to the youngest part of the Ticino River delta in Lake Maggiore.

22.3.1 The Greina Plateau: A Slice of Arctic in the Alps

The Greina region is located geographically between the cantons of Ticino and Graubünden (see location in Fig. 22.1) and tectonically between the Gotthard massif and its autochthonous and para-autochthonous cover-rocks (South-Helvetic domain) and the Mesozoic cover of the Lower Penninic nappes (Fig. 22.2). This tectonic and geological diversity favours considerable geomorphological diversification of the landscape. The region is well-preserved

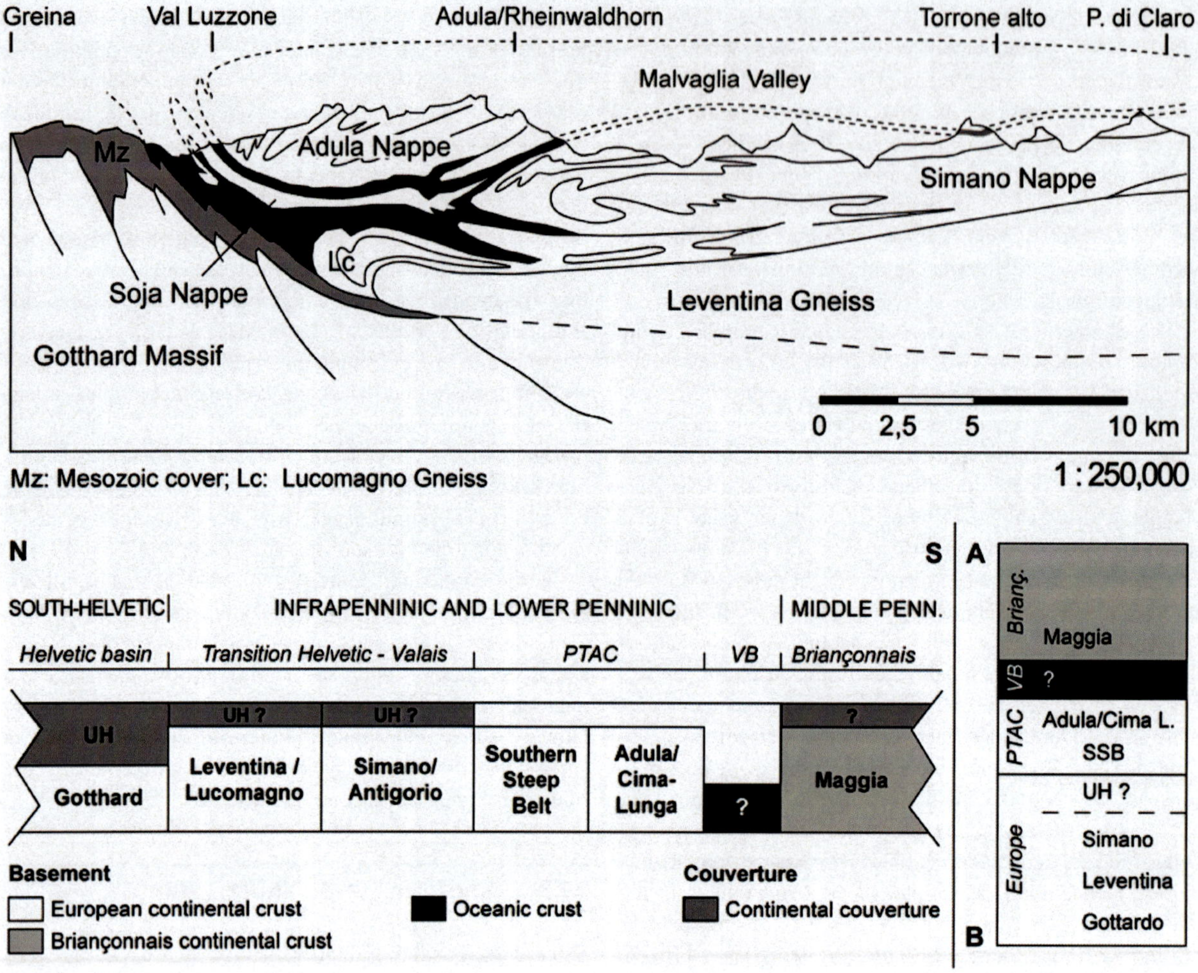

Fig. 22.2 Geology and palaeogeography of the Ticino Alps. Above: Tectonic cross-sections of the Penninic nappes of the Eastern Ticino Alps (see location on Fig. 22.1). Below: Palinspastic cross-section of the Ticino Alps during the Early Cretaceous (ca. 130 Ma ago) (at left) and actual tectonostratigraphy (A: Above; B: below). PTAC: Paleogene Tectonic Accretion Channel; SSB: Southern Steep Belt; UH: Ultra-Helvetic; VB: Valais Basin (Modified after Scapozza and Fontana 2009)

from human influences and is characterised by a large and relatively hanging valley bottom situated between 2,200 and 2,350 m a.s.l., outlined by peaks passing 3,000 m a.s.l. Considering the lack of trees, the valley bottom—called *Piano della Greina* in Italian or *Plaun la Greina* in Romansh (which means Greina plateau)—presents aspects recalling the Arctic tundra. The Greina plateau is well known in Switzerland as a symbolic locality for natural heritage protection because it was the subject, between the 1950s and 1970s, of strong debates related to the project of creation of a dam for hydropower production. The project was finally rejected on 11.11.1986 due to economic and ecological issues.

Thanks to systematic description and analysis of glacial erosion features, such as trimlines, *roches moutonnées*, striation directions, etc., it was possible to postulate that the Greina region was occupied by an ice flow coming from the north of the Alps during the Last Glacial Maximum (LGM). This ice flow marked the landscape until 2,600 m a.s.l (Fig. 22.3). Coming from the northeast, the ice flow presented a diffluence in *Crap la Crusch*, where part of the ice transgressed to the south in direction of the *Alpe di Motterascio*, whereas the main part transgressed to the West over the *Passo della Greina*. This ice flow was also fed by local glaciers coming from the southern side of Piz Gaglianera and the northern side of Pizzo Coroi (Scapozza et al. 2011). As postulated by Florineth and Schlüchter (1998), the important ice flow coming from the north of the Alps in the Greina region may be explained only by the presence of an ice dome located on the Upper Surselva (Rhine ice dome).

After the collapse of the ice dome and local glacier regression during the Lateglacial (see Scapozza et al. 2014a),

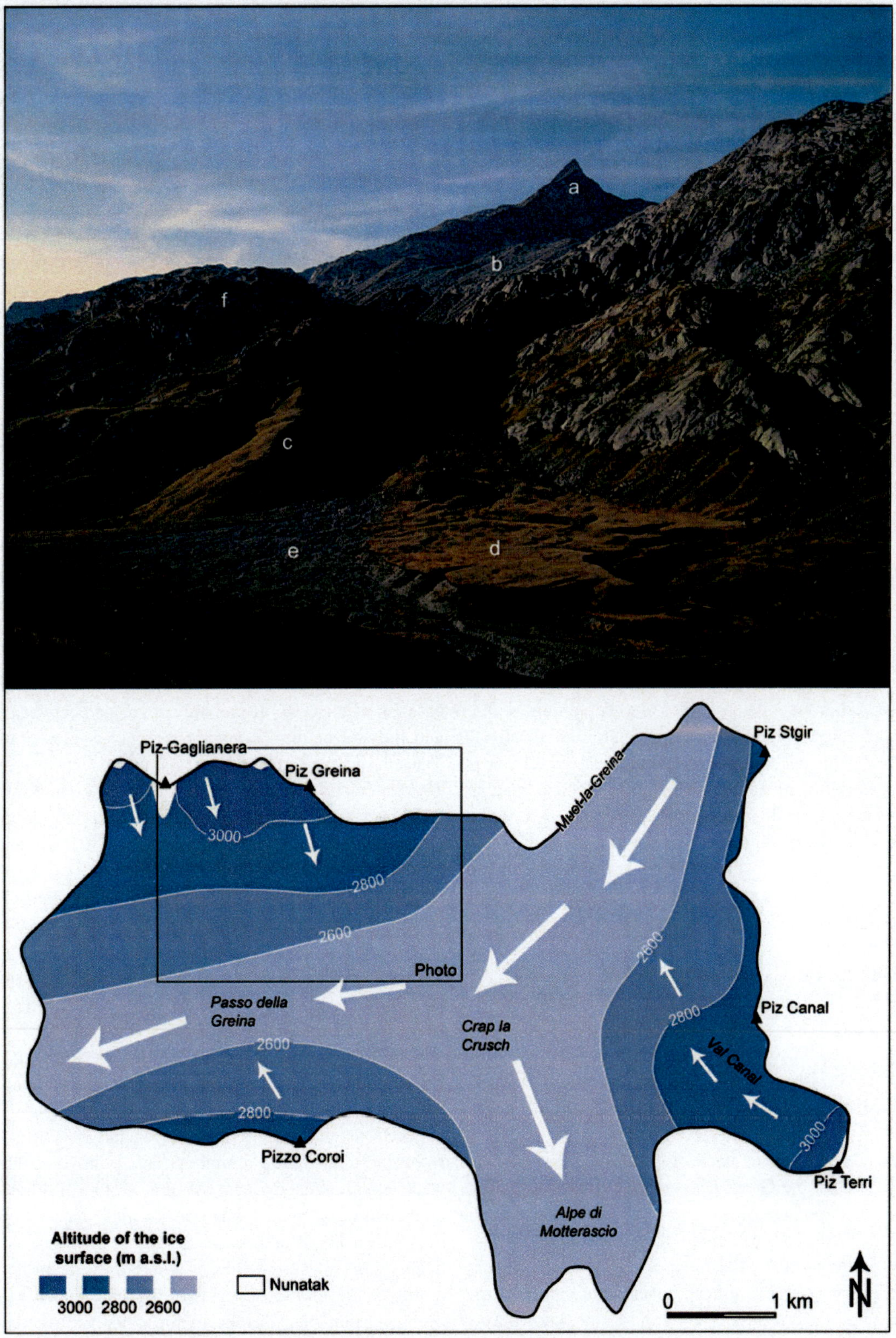
a
b
f
c
e
d
Piz Gaglianera
Piz Greina
Piz Stgir
Muot la Greina
3000
2800
2600
Photo
Passo della Greina
Crap la Crusch
2600
2800
Pizzo Coroi
2600
2800
Piz Canal
Val Canal
3000
Piz Terri
Alpe di Motterascio
Altitude of the ice surface (m a.s.l.)
3000 2800 2600
Nunatak
0 1 km
N

◀ **Fig. 22.3** Glacial and glaciofluvial landscape in the Greina region. Above: A frontal view in direction of the Piz Gaglianera (3,121 m a.s.l.) horn (**a**). The Little Ice Age moraines are visible in grey just in front of the peak (**b**). In the foreground, a Younger Dryas right lateral moraine ridge (vegetated) (**c**), an Early Holocene glaciofluvial fan (terraced and vegetated) (**d**), and the present alluvial fan (**e**), are visible. Note the morphology of the *roches moutonnées* in the middle of the photograph (**f**), indicating the direction of the ice flow from East to West (from right to left on the photograph) (photo G. Scapozza). Below: Origin and direction of the ice flow and elevation of the ice surface during the LGM (modified after Scapozza et al. 2011)

as evidenced on the Greina plateau by moraine ridges attributed to the end of the Oldest Dryas and to the Younger Dryas, the valley bottom was strongly influenced by glaciofluvial and fluvial reworking of glacial deposits. A very educational alluvial fan located at the foot of Piz Gaglianera presents the traces of this history (Fig. 22.3). It is composed of a relict part, constituted by glaciofluvial deposits covered by Alpine grassland and incised by several palaeo-channels, and an active part, incised in the relict one, characterised by unvegetated coarse torrential deposits. The relict glaciofluvial fan is terraced four meters above the active alluvial fan. This is probably the result of a paraglacial reworking of glacial deposits related to the Younger Dryas stadials, taking place before the Early-to-Mid-Holocene temperature optimum (9.5–6.3 ka cal BP in Central Europe). During this period, as a consequence of the exhaustion of the sediment stock in the upper part of the catchment and of climate optimum, the glaciofluvial fan was probably incised by fluvial erosion, with resedimentation in the *Plaun la Greina* alluvial plain, and further modelled by pedogenesis in the part preserved from erosion. Since the beginning of the Meghalayan Stage, a significant increase in torrential activity related to the ice melting after repeated glacier advances caused reworking of almost all the Holocene glacial deposits. This was particularly marked after the end of the Little Ice Age, when fluvial erosion completely reworked glacial deposits related to the 1820 advance. It is probable that sediment wasting from the upper part of the catchment took place only several decades after glacier recession. For the recent period, this is proved by the fact that the moraine bastion built by the Gaglianera glacier between 1900 and 1940 is still intact (Fig. 22.3).

22.3.2 Rock Glaciers of the Adula Massif: From Strabo to Global Warming

The Adula massif takes its name from *Mons Aduelas*. This was the name that the Greek geographer Strabo, in 20 AD, gave to all the Lepontine Alps comprised between the source of the Rhine River and the source of the Ticino River. The Adula/Rheinwaldhorn (3,402 m a.s.l.; see Fig. 22.1) is characterised by several significant glaciers, in particular on its northern and eastern sides. The southern side, known also as Cima di Gana Bianca massif (Fig. 22.4), is not high enough to have significant glacier accumulation (the highest point, the Cima di Gana Bianca, reaches 2,842 m a.s.l.), but is sufficiently high for well-established periglacial conditions, with the lower limit of discontinuous permafrost between 2,350 and 2,700 m a.s.l. according to the slope aspect (Scapozza and Fontana 2009; see also Fig. 22.1). This region is characterised by a typical Alpine glacial and periglacial geomorphology, presenting various rock glaciers of different degrees of activity, well-developed talus slopes and significant Lateglacial moraine ridges.

Thanks to geomorphological and geophysical study of periglacial and glacial landforms in the area, the activity of rock glaciers, permafrost distribution and the palaeoenvironmental history of the Cima di Gana Bianca massif were documented in detail (for a summary, see Scapozza 2013; Scapozza et al. 2014b). Thanks to palaeogeographical and palaeoclimatical reconstructions, as well as to Schmidt hammer exposure-age dating (SHD), three generations of rock glaciers were identified (RG III, II and I). Relict rock glaciers compose RG III and RG II. RG III started its development at the end of the Oldest Dryas (after 16.0 ka cal BP), whereas RG II probably developed between Bølling and the end of the Younger Dryas (14.5–11.6 ka cal BP). The minimal age of RG III is comprised between 15.7 and 10.4 ka cal BP, which indicates that these landforms were probably already inactive or even relict during the Younger Dryas cold period. The minimal age of RG II relict rock glaciers is comprised between 12.8 and 7.2 ka cal BP. This may indicate their inactivation during the Early-to-Mid-Holocene thermal optimum. RG I generation, which consists of currently active and inactive rock glaciers, started its development during or just after the Early-to-Mid-Holocene temperature optimum, as indicated by the SHD comprised between 5.8 and 3.2 ka cal BP.

SHD of periglacial talus slopes between 10.5 and 6.0 ka cal BP indicates that these landforms may have developed rapidly after the end of the last glacial cycle, between the end of the Oldest Dryas and the Early-to-Mid-Holocene temperature optimum. This points out that rockwall retreat after the end of the Early-to-Mid-Holocene temperature optimum was weak, and that the interval between maximal and minimal ages is in most cases relatively short (4–6 millennia). Therefore, rockwall retreat during the development period of the talus slopes must have been considerable. Thanks to the calculation of rockwall erosion rates based on the volume of talus accumulations, it was possible to find evidence of the existence of

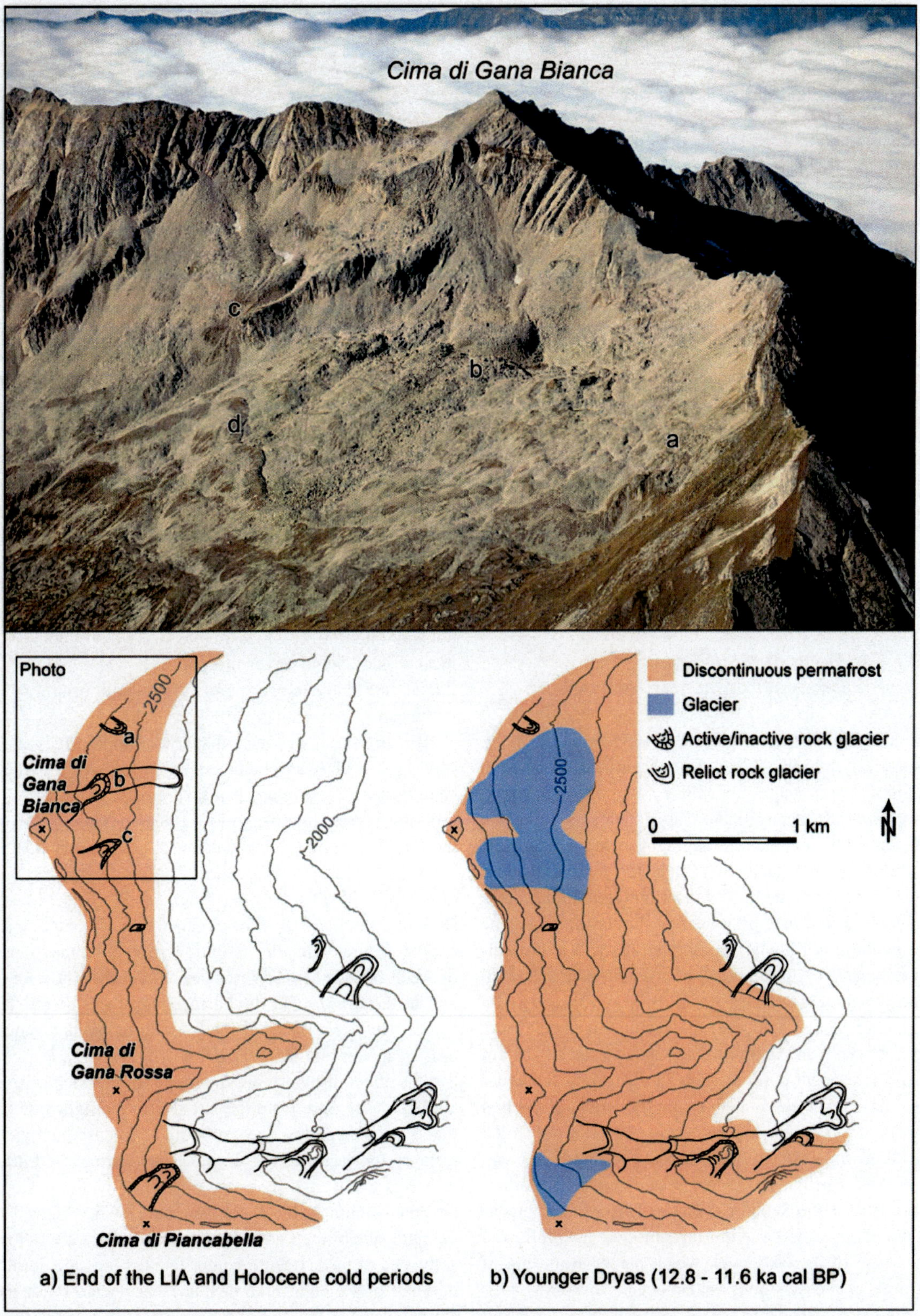

Fig. 22.4 Glacial and periglacial landscape in the Adula massif. Above: The eastern side of the Cima di Gana Bianca (2,842 m a.s.l.), with three active rock glaciers (**a**, **b** and **c**) and the moraine ridge related with the Younger Dryas glacier advance (**d**) (photo M. Buzzi). Below: Reconstruction of the cryosphere evolution in the Cima di Gana Bianca massif between the end of the Pleistocene and the Holocene (modified after Scapozza and Fontana 2009)

"paraperiglacial rockfall phases" related to permafrost degradation in rockwalls (Scapozza 2013, 2016). These phases correspond to rapid climate warming periods, as the beginning of Bølling, during the Greenlandian Stage and, maybe, the recent period after the 1980s.

The thermal regime and kinematics of rock glaciers of the Cima di Gana Bianca massif have also been studied since 2006 to provide data on the permafrost evolution in the southern Alpine morphoclimatic context. Ground surface temperatures (GST) are monitored using autonomous mini-loggers and annual to pluriannual surface displacements are measured by differential GPS. The movements of the Stabbio di Largario rock glacier (2300–2500 m a.s.l., 46° 28′ 40″ N 8° 59′ 10″ E, see Scapozza et al. 2014b) over three different (annual, decadal and millennial) time scales were assessed, showing a link between the periods of increase in mean air temperature on different time scales and variations in rock glacier kinematics. The significant link between changes in temperature and rock glacier behaviour may indicate that its kinematics is probably influenced principally by the first order (i.e. decadal) climatic variation at the super regional (entire Alps) or continental scale (north Atlantic and Western Europe).

22.3.3 The Bolle di Magadino: A Palaeoenvironmental Archive

A joint analysis of several boreholes in the Bolle di Magadino (see location in Fig. 22.1) at the locality Castellaccio in Magadino di Sopra and at the Military Airport of Locarno, cartographic and literary historical documents, and the comparison with the pollen stratigraphy determined in the Origlio and Muzzano lakes located in the Southern Swiss Alps, was carried out. It allowed determining the main steps of the Ticino River delta progradation during the Holocene and the evolution of fluvial geomorphology and hydrosedimentary dynamics of the Ticino River in the Piano di Magadino (Scapozza and Oppizzi 2013; Scapozza 2016). At the beginning of the Holocene, the Ticino River delta was located between San Antonino and Cadenazzo, ca. 10 km upslope of its current position (Fig. 22.5). During the Older Atlantic phase, in correspondence with the arrival of the first inhabitants of the Southern Swiss Alps, the delta prograded several kilometres, and was located between Cadenazzo and Gudo. A third position was reconstructed between Quartino and Riazzino for the Subboreal phase. The delta progradation in direction of the Bolle di Magadino was also followed by a decrease of the lake level, which passed from 204 to 207 m a.s.l. at the beginning of Holocene, to 200 m a.s.l. during the Older Atlantic phase, to 193–197 m a.s.l. during the Subboreal phase (Fig. 22.5). The present lake level (193 m a.s.l.) was reached during the Subatlantic phase.

During the Roman Period (15 BC–400 AD in Southern Switzerland), the Ticino River delta front was located somewhere between Riazzino and Gordola on the right side, and between Quartino and Magadino di Sopra on the left side of the Piano di Magadino (Scapozza and Oppizzi 2013). During this period, the Ticino River hydrosedimentary dynamics was relatively calm, as the fine sediments deposited in the Bolle di Magadino show. This interpretation is supported by natural climatic—Roman Warm Period—and anthropogenic events, with the end of the artificial forest fires and the anthropogenic introduction of chestnut (*Castanea*) in the Southern Swiss Alps ca. 200 AD. The morphosedimentary dynamics became more intense during the Late Antiquity and the Early Middle Ages (400–1000 AD), and was particularly powerful during the period comprised between 700 and 1000 AD, mainly because of intensification of anthropogenic activity (cereal culture and deforestation). The Ticino River delta front made an important progression and attained Gordola around 1100 AD, where at that time there was the most important lacustrine port of all the upper basin of the Lake Maggiore (Scapozza and Oppizzi 2013). The morphosedimentary dynamics during the High and Late Middle Ages (1000–1500 AD) were relatively calm because of the Medieval Warm Period and decrease of the anthropogenic pressure. It is possible to make the hypothesis of a meandering Ticino River morphology, which indicates that fluvial environments were probably in equilibrium with this multisecular period of hydrological calm.

The transition between the Late Middle Ages and the Modern Epoch was characterised by a large flood event caused by the emptying of a lake created in 1513 by a 50 million m^3 rockslide in the lower Blenio Valley. Known as the *Buzza di Biasca*, which occurred on 20th May 1515, this event modified in an important way the fluvial morphology of the Ticino Valley between Biasca and the Lake Maggiore (Scapozza et al. 2015). The event, which was superimposed on the natural climatic evolution related to the transition between the Medieval Warm Period and the Little Ice Age, was at the origin of a fluvial metamorphosis of the Ticino River, which passes from meandering to braided morphology (Scapozza and Oppizzi 2013). The Buzza di Biasca also had the effect of displacing the mouth of the Ticino River in the Lake Maggiore from the left side to the right side of the Piano di Magadino. The particularly intense morphosedimentary dynamics of the Ticino River during the Little Ice Age was manifested by a new migration of the position of the river mouth in the Lake Maggiore, which moved from the right side of the Piano di Magadino, where it was located at the end of the sixteenth century, to the left side during the first half of the eighteenth century. The Ticino River mouth moved again at the end of the eighteenthh century towards the central part of the Piano di Magadino, to return completely to the left side of the floodplain during the first half of

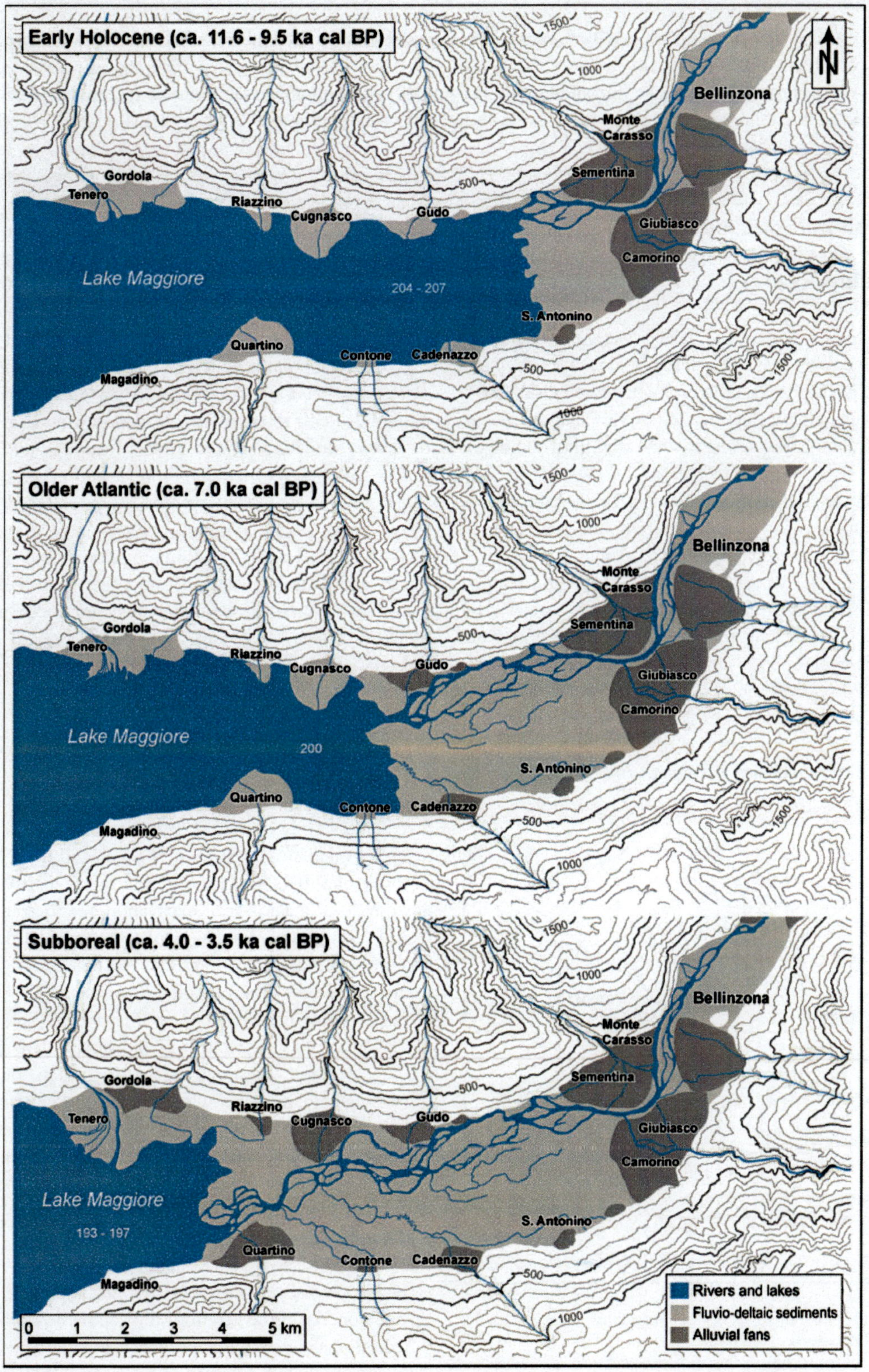

Fig. 22.5 Palaeogeography of the Ticino River fluvio-deltaic plain in the Piano di Magadino at three moments of the Holocene (modified after Scapozza 2016)

the nineteenth century. The braided morphology was maintained until construction of the Ticino River embankment during the second half of the nineteenth century when a new metamorphosis occurred, this time entirely due to anthropogenic reasons, with the river passing from braided to rectilinear single channel morphology.

22.4 Evolution of Landscape and Ideas: On the Origin of the Southern Alpine Valleys

Southern Alpine valleys (and lakes) were considered for more than one century as of exclusively glacial origin. Since the introduction of the glacial erosion model (Ramsay 1864), and in particular after the publication by Penck and Brückner (1909), this theory was strongly supported until the beginning of the 1980s (Nangeroni 1980).

William Morris Davis himself was in Upper Ticino at the end of the nineteenth century, describing and analysing the morphology of these deeply incised valley troughs. He noted, for example, the discordant altitude of the hanging valleys with respect to the main valley floor between Val d'Ambra and Leventina in Pollegio (Fig. 22.6a), and between Valle di Lodrino and Riviera in Osogna (Davis 1900b). Based on these observations, Davis (1900b) developed a general model for over-deepened valleys (Fig. 22.6 b). ABC represents a cross-section of the older Ticino valley, into which the side stream DB enters at accordance grade.

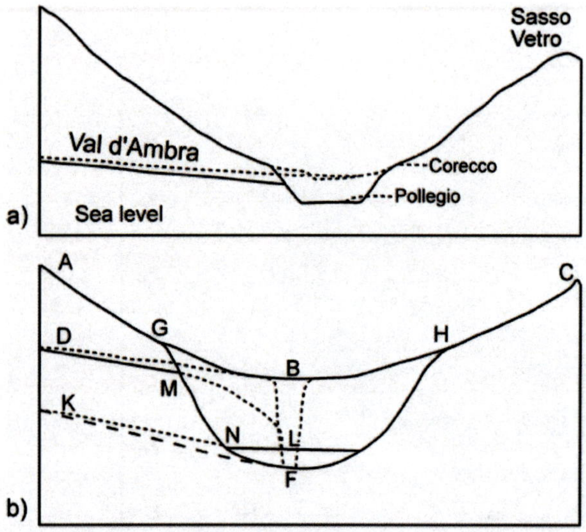

Fig. 22.6 William Morris Davis's cross-sections inspired by his fieldwork in Leventina, Blenio and Riviera valleys. **a** True-scale cross-section of the Ticino valley in the lower part of Leventina, illustrating the discordant altitude of the hanging valleys with respect to the main valley floor. **b** Theoretical diagram of a glaciated valley. See text for details (re-drawn after Davis 1900b)

An uplift of the region initiates a new erosion cycle and causes incision of the river and then creation of the canyon BF. In brief time, the side stream may cut the lateral canyon DML. Over long times, the young canyon BF will evolve into the "mature" open valley GFH with, as a consequence, incision by the side river reaching a slope such as KF. The infilling of the valley floor (described in the Piano di Magadino, between Bellinzona and the Lake Maggiore; for an historical overview, see Scapozza 2016), will provoke aggradation of the valley floor to L, followed by the aggradation of the side valley to a slope KL.

William Morris Davis can also be considered as a precursor of geotourism. In his work about the glacial erosion in the valley of Ticino, he wrote that «*it is my hope that this article may lead a few of the many thousand travellers who annually invade Italy by the St. Gotthard route, to stop a day or two at the village of Biasca for a walk up and down the Ticino valley in that neighbourhood, and perhaps for some short climbs up to one or more the discordant lateral valleys*» (Davis 1900b: 144).

On the tracks left by William Morris Davis, Albrecht Penck and Eduard Brückner, the upper Ticino became an important theatre of research on erosion terraces, interpreted as the remnants of ancient valley floors created by fluvial and glacial erosion from the end of the Miocene to the Holocene (Lautensach 1910; Annaheim 1946). Four systems of terraces were described. Between the end of Miocene and the beginning of Pliocene, incision of the original relief by fluvial erosion, as a consequence of the rise of the Alps, led to the origin of the first system of terraces, called "Pettanetto level". A second system, the "Bedretto level", was formed in a similar manner between the Pliocene and the Pleistocene. A further uplift of the Alps during the Pleistocene caused new phase of glaciofluvial excavation due to both tectonics (Alpine rising) and climate (glaciation), originating the third system of terraces, the "Sobrio level" (Fig. 22.7). According to this model, the present valley floor was finally incised by glacial erosion during the Late Pleistocene and filled by glacial, glaciofluvial, lacustrine and fluvial deposits during the deglaciation, forming a fourth system of terraces. According to current knowledge about the genesis of these terraces, most of them are the result of deep-seated landslides, as it is the case for the "Bedretto level" in Bedretto valley (Ambrosi and Scapozza 2014) or for the "Sobrio level" in Osco (Strozzi et al. 2013). Other terrace levels are also in correspondence with lithological and/or tectonic changes in the geological structure.

In the second half of the 1970s, an important paradigm change concerned the origin of the southern Alpine valleys. Based on seismic reflection profiles below the Southern Alps lakes and on erosion surfaces on the *Gonfolite lombarda* (the Southern Alpine Oligocene and Miocene molasse), the crypto-depressions occupied by the lakes and then the entire

Fig. 22.7 The Osco terrace, interpreted by Lautensach (1910) as belonging to the "Sobrio level" system (photo C. Scapozza). In reality, this terrace is probably related to the deep-seated landslide of Osco, which concerns the main part of the left side of the Leventina valley around Osco (Strozzi et al. 2013)

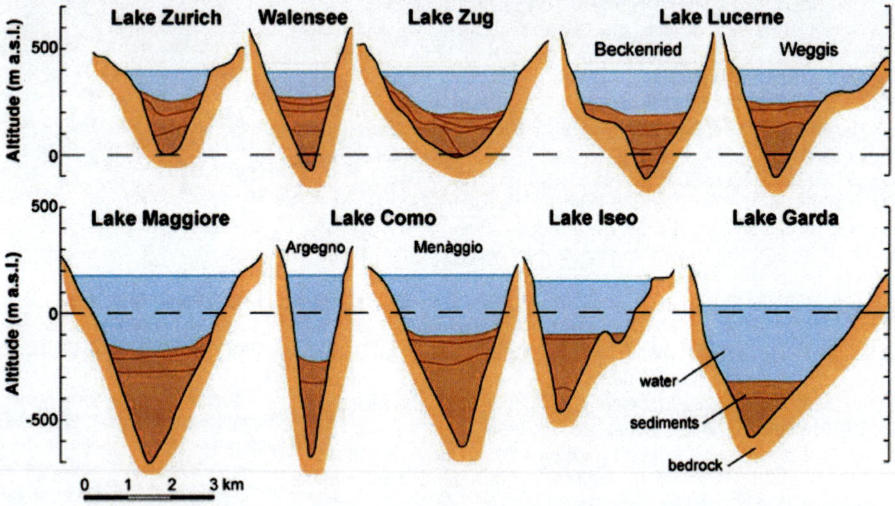

Fig. 22.8 Cross-sections of the deepest part of some northern (above) and southern (below) Alpine lakes based on seismic reflection and refraction prospecting (modified after Finckh 1978)

southern Alpine valleys were interpreted as over-deepened Paleogene and Neogene palaeo-valleys (Bini et al. 1978; Finckh 1978). In all the southern Alpine lakes, the lake bottoms are located below the present sea level, with depths reaching – 200 m a.s.l. and altitudes of mean water levels are on average 266 m lower than in the Northern Alps. Despite this, the southern Alpine lakes are on average 130 m deeper than northern Alpine lakes. Taking also into account their particularly embedded morphology ("V" profile) and the depth of the bedrock that can reach 600–700 m below the present-day sea level (Fig. 22.8), the southern Alpine valleys are now interpreted as the result of intense fluvial erosion during the Miocene. This incision was causally related to the "Messinian salinity crisis", created by the repeated closure of the Strait of Gibraltar for tectonic reasons, which resulted in repeated desiccation events of the Mediterranean Sea. This event caused significant lowering of the base level of fluvial erosion.

22.5 Conclusions

Upper Ticino is well known for the UNESCO World Heritage of the Three Castles, Defensive Wall and Ramparts of the Market-Town of Bellinzona, the ancient door between the Alpine valleys leading to the main Central Alpine passes (St. Gotthard, Lukmanier and San Bernardino passes) and the Insubrian region marked by the southern Alpine lakes and Mediterranean climate. Less known is the importance of this region for the history of science, in particular for the development of the cycle of erosion model by William Morris Davis and for the theories about the origin of southern Alpine valleys and lakes. The importance of the landscape is, therefore, not only related to the specific particularly impressive, interesting or remarkable landforms, which is the case of the Greina plateau or the international natural reserve of the Bolle di Magadino, but also to their cultural importance for the history of geomorphology, particularly the development of ideas on glacial, periglacial and fluvial landscape evolution.

Acknowledgements A special thank to the editor, Emmanuel Reynard, and to an anonymous reviewer for their useful feedback.

References

Ambrosi C, Scapozza C (2014) Improvement of 3D digital mapping in geomorphological and Quaternary geological cartography. Geogr Helv 70:121–133. https://doi.org/10.5194/gh-70-121-2015

Annaheim H (1946) Studien zur Geomorphogenese der Südalpen zwischen St Gotthard und Alpenrand. Geogr Helv 1:65–149. https://doi.org/10.5194/gh-1-65-1946

Bini A, Cita MB, Gaetani M (1978) Southern Alpine lakes—Hypothesis of an erosional origin related to the Messinian entrenchment. Mar Geol 27(3–4):271–288. https://doi.org/10.1016/0025-3227(78)90035-x

Davis WM (1900a) Glacial erosion in France, Switzerland and Norway. Proc Boston Soc Nat Hist 29(14):273–309

Davis WM (1900b) Glacial erosion in the Valley of the Ticino. Appalachia 9:136–156

Davis WM (1906) The sculpture of mountains by glaciers. Scott Geogr Mag 22(2):76–89. https://doi.org/10.1080/00369220608733647

Finckh PG (1978) Are southern Alpine lakes former Messinian canyons?—Geophysical evidence for preglacial erosion in the southern Alpine lakes. Mar Geol 27(3–4):289–302. https://doi.org/10.1016/0025-3227(78)90036-1

Florineth D, Schlüchter C (1998) Reconstructing the Last Glacial Maximum (LGM) ice surface geometry and flowlines in the Central Swiss Alps. Eclogae Geol Helv 91:391–407

Lautensach H (1910) Glazialmorphologische Studien im Tessingebiet. Dissertation, Friedrich-Wilhelms-Universität zu Berlin, Teubner, Dresden, 70 pp

Nangeroni G (1980) È ancora valida l'ipotesi dell'origine glaciale dei grandi laghi prealpini italiani da escavazione glaciale? Geografia 3:127–130

Penck A, Brückner E (1909) Die Alpen in Eiszeitalter. Tauchnitz, Leipzig, 3 vol

Ramsay AC (1864) On the erosion of valleys and lakes. Taylor & Francis, London, 34 pp

Scapozza C (2013) Stratigraphie, morphodynamique, paléoenvironnements des terrains sédimentaires meubles à forte déclivité du domaine périglaciaire alpin. PhD thesis, University of Lausanne (Géovisions 40), 551 pp

Scapozza C (2016) Evidence of paraglacial and paraperiglacial crisis in alpine sediment transfer since the Last Glaciation (Ticino, Switzerland). Quaternaire 27(2):139–155. https://doi.org/10.4000/quaternaire.7805

Scapozza C, Fontana G (2009) Le Alpi Bleniesi: storia glaciale e periglaciale e patrimonio geomorfologico. Memorie della Società ticinese di scienze naturali e del Museo cantonale di storia naturale, Lugano 10:1–111

Scapozza C, Oppizzi P (2013) Evolution morpho-sédimentaire et paléo-environnementale de la plaine fluvio-deltaïque du Ticino pendant l'Holocène récent (Canton du Tessin, Suisse). Géomorphol Relief Proces Environ 19(3):265–286. https://doi.org/10.4000/geomorphologie.10289

Scapozza C, Castelletti C, Soma L, Dall'Agnolo S, Ambrosi C (2014a) Timing of LGM and deglaciation in the Southern Swiss Alps. Géomorphol Relief Proces Environ 20(4):307–322. https://doi.org/10.4000/geomorphologie.10753

Scapozza C, Lambiel C, Bozzini C, Mari S, Conedera M (2014b) Assessing the rock glacier kinematics on three different time scales: a case study from the Southern Swiss Alps. Earth Surf Proc Land 39:2056–2069. https://doi.org/10.1002/esp.3599

Scapozza C, Tognacca C, Ambrosi C, Seno S (2015) 20 maggio 1515: la "Buzza" che impressionò l'Europa. Bollettino della Società ticinese di scienze naturali 103:71–80. http://repository.supsi.ch/7111/

Scapozza G, Scapozza C, Reynard E (2011) Morphogenèse de la région de la Greina depuis le Dernier Maximum Glaciaire. In: Lambiel C, Reynard E, Scapozza C (eds) La géomorphologie alpine: entre patrimoine et contrainte. Actes du colloque de la Société Suisse de Géomorphologie, 3–5 Septembre 2009, Olivone (Géovisions 36). Université, Institut de géographie, Lausanne, pp 99–111. http://repository.supsi.ch/3150/

Strozzi T, Ambrosi C, Raetzo H (2013) Interpretation of aerial photographs and satellite SAR interferometry for the inventory of landslides. Remote Sens 5:2554–2570. https://doi.org/10.3390/rs5052554

Cristian Scapozza is Senior Researcher and Lecturer in Physical Geography and Geomorphology at the University of Applied Sciences and Arts of Southern Switzerland (SUPSI), where he leads the Geology Research Group, and Director of the Valle di Blenio Museum. He his the President of the Swiss Geomorphological Society (SGmS) and member of the Scientific Committee of the Swiss Permafrost Monitoring Network PERMOS. His main scientific interests are in the field of geomorphological evolution of Alpine valleys, with a particular focus on Quaternary geological mapping, glacial and periglacial processes, hillslope processes and dynamics, natural and historical evolution of Alpine floodplains and geoheritage promotion.

Christian Ambrosi is Professor of Applied Geology at the University of Applied Sciences and Arts of Southern Switzerland (SUPSI), where he directs the Institute of Earth Sciences. He is a member of the Quaternary Working Group of the HARMOS project led by swisstopo. His main scientific interests are in the field of geological mapping, natural hazards assessment, rock mechanics, numerical modelling of hillslope processes and dynamics. He has worked also in Romania (Carpathian Region), Nepal, Buthan and South Tyrol (Italy).

The Rhine Falls

23

Peter Heitzmann

Am Rheinfall

Halte dein Herz, o Wanderer, fest in gewaltigen Haenden!
Mir entstuerzte vor Lust zitternd das meinige fast.
Rastlos donnernde Massen auf donnernde Massen geworfen,
Ohr und Auge wohin retten sie sich im Tumult?
Wahrlich den eigenen Wutschrei hoerete nicht der Gigant hier,
Laeg ervom Himmel gestuerzt unten am Felsen gekruemmt!
Rosse der Götter, im Schwung, eins ueber dem Ruecken des andern,
Stuermen herunter und streun silberne Maehnen umher;
Herrliche Leiber unzaelbare, folgen sich, nimmer dieselben,
Ewig dieselbigen—wer wartet das Ende wohl aus?
Angst umzieht dir den Busen mit einsund, wie du es denkest,
Über das Haupt stürzt dir krachend das Himmelsgewölb!

Eduard Moerike 1846

At Rhine Falls

Hold your heart firmwayfarer, in stalwart hands!
My hands trembling with emotion almost let go.
Relentless thundering masses thrown onto thundering masses,
Whither shall eye and ear seek shelter in such tumult?
Truly, the giant would not hear his own cry of fury here,
Were he lyingfallen from the sky cramped under the rock!
Vaulting steeds of gods, one up over the other's back
Plunge downward scattering silver manes about,
Gorgeous bodies, countless, in pursuit never the same,
Ever the same—who will ever see this to the end?
Suddenly fear pervades your bosom, and, while you ponder.
Upon your head the heavenly vault crashes down in thunder!

Translation: Charles L. Cingolani
http://www.cingolani.com/60em.html

P. Heitzmann (✉)
Geowissenschaftliche Öffentlichkeitsarbeit und Geologische Beratungen, Niederriedweg 110, 3019 Bern, Switzerland
e-mail: heitzmann@bluewin.ch

E. Reynard (ed.), *Landscapes and Landforms of Switzerland*, World Geomorphological Landscapes,
https://doi.org/10.1007/978-3-030-43203-4_23

Abstract

The Rhine Falls lie in northern Switzerland at the border between the Swiss Plateau and the Tabular Jura. During the glaciations, the course of the Rhine River changed several times. When the river channel was filled with gravel during the last glaciation, the river changed its course allowing it to drain the melting water. The present-day riverbed intersects a riverbed from the Riss glaciation and this is the place where the Rhine Falls are located. The river eroded the gravel in the former valley creating a cliff between the higher limestone plateau and the deeply eroded old valley. The Rhine Falls are 23 m high and 150 m wide and present outstanding geomorphological features of stream erosion and accumulation. Since the Middle Ages, the power of the falls has been used for industrial applications. The Rhine Falls also have a long tradition in tourism, literature and art.

Keywords

Rhine Falls • Waterfall • Fluvial erosion and accumulation • Geoheritage • Geotourism

23.1 Introduction

The Rhine Falls (*Rheinfall* in German), described in the folder of the Rhine Falls Coordination Office as "Europe's biggest waterfall", lie in northern Switzerland (47°40′41″ N, 8°36′54″ E) near the town of Schaffhausen (Fig. 23.1). In this region, the Rhine River forms the border between the Swiss cantons of Zurich and Schaffhausen; therefore, the northwestern side of the Rhine Falls is in the Canton of Schaffhausen and the southeastern side in the Canton of Zurich. The normal level of the Rhine River at Schaffhausen is about 283 m a.s.l. The Rhine Falls represent an outstanding geological and geomorphological feature (Fig. 23.2), because it is the only really big waterfall along the Rhine River between the Lake of Constance and the North Sea. In other Swiss rivers outside of the Alps, only smaller waterfalls are found.

A geological overview and the explanation of the origin of the waterfall will follow the geographical description of the waterfall. A specific section is dedicated to the geomorphological features. The Rhine Falls have been described in the literature as a special natural site since the sixteenth century. Since 1544, pictures and paintings are available and for long time the Rhine Falls have played an important role for tourism and water exploitation. All these aspects are treated in specific sections.

23.2 Geographical and Geological Setting

Between the Lake of Constance and Basle where the Rhine is called *Hochrhein* (High Rhine; Preusser 2008), the Rhine Falls, also called *"Grosser Laufen"* (big waterfall), are not the only waterfall along this river. There were two other but smaller falls that have to be mentioned:

- About 2 km upstream from the Rhine Falls, at the place where today an electric power station is located (Fig. 23.1, indication EW), irregularities in the limestone underground are responsible for some water rapids, called "*Oberer Laufen*" (upper waterfall).
- About 15 km east of Basle, the Rhine River runs on the crystalline rocks of the Black Forest Massif through the old town of Laufenburg (which means "castle at the waterfall"). Until the beginning of the twentieth century, a deep gorge with a splendid waterfall called "*Kleiner Laufen*" (small waterfall) crossed the town (Heitzmann 2014). The gorge was destroyed in favour of an electric power station.

As part of the Swiss Plateau, the climate of the Rhine Falls region is highly influenced by the westerly winds from the Atlantic. Mean annual air temperature is 9.5 °C and total rainfall brings about 900 mm precipitations per year. The hydrology of the Rhine River at Schaffhausen is characterised by the fact that the Alpine Rhine flows through the Lake of Constance. Consequently, all the clastic load is deposited and the Rhine River upstream of the Rhine Falls has nearly no solid load. The catchment area of the Rhine River at Schaffhausen is 11,890 km^2.

Compared to other big waterfalls, the dimensions of the Rhine Falls are rather humble: the total width is 150 m and the total height is 23 m. The depth of the kolk basin is 13 m, eroded in the Riss-age gravel filling. The average water flow is about 600 m^3/s during summer and about 250 m^3/s in wintertime. The maximal discharge was measured in 1965 with 1250 m^3/s and on 11 June 1999 1180 m^3/s was attained. The minimum was on 3 June 1921 with 95 m^3/s, indicated on the information table at the Castle of Laufen.

Since 1983, the Rhine Falls and surroundings are included in the Federal Inventory of Landscapes and Natural Monuments of National Importance (ILNM, site n° 1412; Beutler and Gerth 2015) and in 1999 they were integrated in the Inventory of Swiss Geosites (Working Group 1999; Reynard et al. 2012; Heitzmann 2017; http://www.naturalsciences.ch/organisations/geosciences/projects/wg_geotopes, accessed 31.08.2019; geosite n° 39).

The Rhine Falls lie at the border between the Tabular Jura in the north and the Swiss Plateau with the Molasse Basin in the south (Fig. 23.3). Both tectonic units in this region are

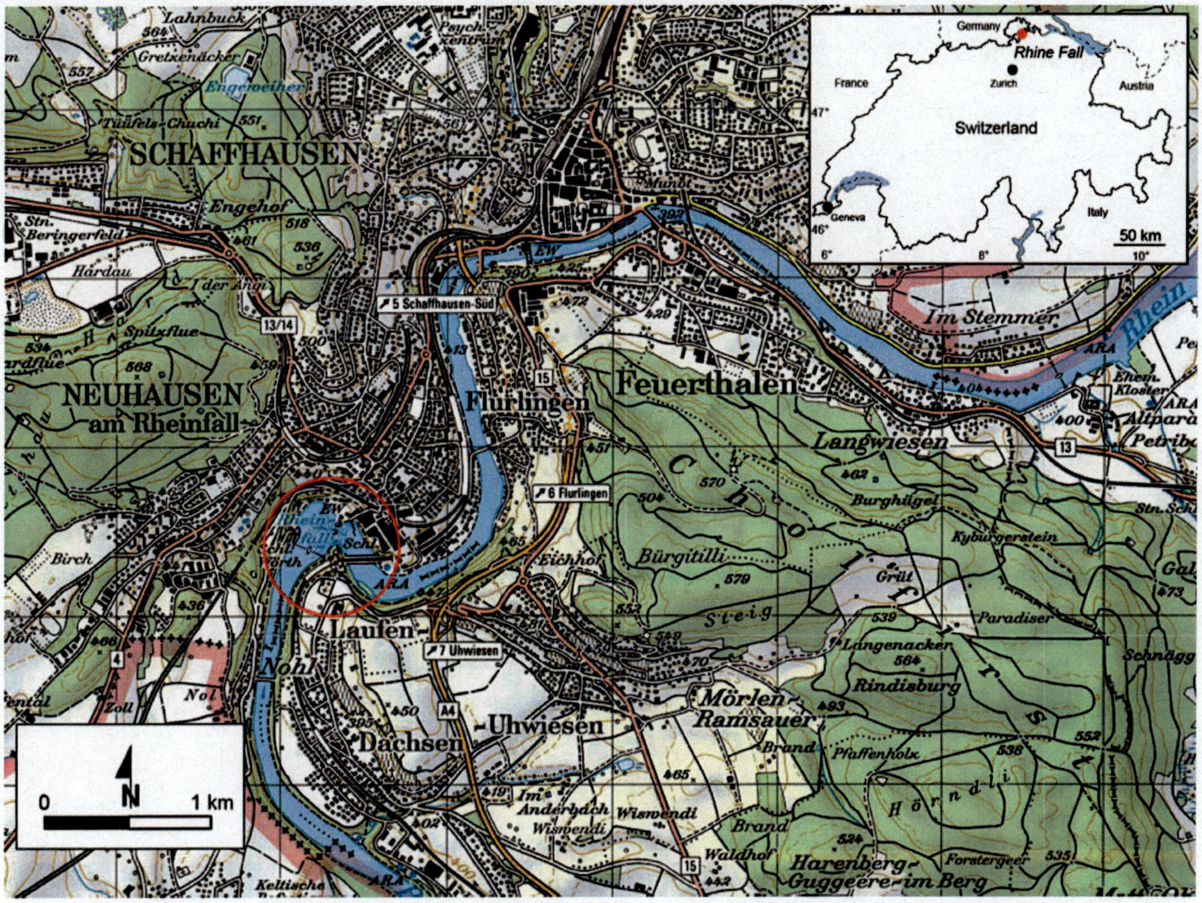

Fig. 23.1 Topographic map of the Rhine Falls area. The Rhine Falls are in the red circle (swisstopo)

Fig. 23.2 Panorama of the Rhine Falls, viewed towards east. In the centre of the picture, the waterfall is divided in several flows by the relics of the former limestone cliff. On the right (southern) side, belonging to the Canton of Zurich, is the historic Castle of Laufen, which is today a restaurant with special events and a youth hostel. Below the castle is the access way to the waterfall and to the magnificent Känzeli viewpoint. On the left side, in the Canton of Schaffhausen, the industrial buildings of the SIG Company dominate the site (*Source* Any1s, 24.08.2010, http://commons.wikimedia.org/wiki/File:Rheinfall_Panorama.jpg?uselang=de-ch, accessed 31.08.2019)

almost completely covered with Quaternary deposits (glacial and fluvial). The two geological map sheets covering the Rhine Falls and Schaffhausen (Geologischer Atlas der Schweiz, 1:25,000, sheet 1031 Neunkirch 1981 and sheet 1032 Diessenhofen 1964) give a general introduction to the regional and local geological situation.

The Tabular Jura is made of thick-bedded Younger Jurassic limestone. The bedding is dipping about 5 degrees

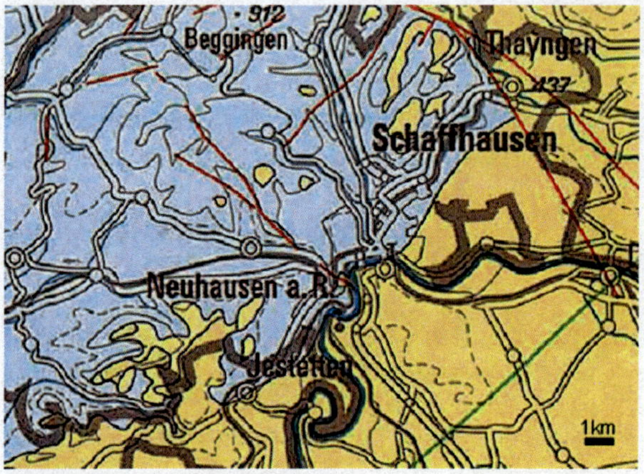

Fig. 23.3 Tectonic Map of Switzerland (original scale 1:500,000) of the Rhine Falls region. In the northwest, the Tabular Jura (blue), in the southeast, the Molasse Basin. The Rhine Falls are just below the "R" of "Neuhausen a. Rhein." (swisstopo)

towards southeast (Fig. 23.4). We can distinguish a lower part with massive, fossiliferous limestone, forming the cliffs of the Rhine Falls themselves, and an upper part with more bedded limestone and marly interlayers, which can be observed upstream of the railway bridge. At its top, kaolinitic and ferruginous, Paleocene/Eocene clays ("*Bohnerz-Formation*"), often in karst pockets and caves, overlay the limestone. There is a time gap, because sediments of youngest Jurassic and Cretaceous age were never deposited or eroded before the Paleocene. The Rhine River upstream of the Rhine Falls runs over the limestone beds. On the southern side of the Rhine River, Molasse deposits overlying the Tabular Jura are of continental origin. The Lower Freshwater Molasse ("*Untere Süsswassermolasse*"), which forms the southern riverside, is composed of sandstones, especially sands derived from erosion of crystalline rocks, and of red marls in the upper part.

Quaternary deposits can often be related to glaciation periods. The hills southeast of Flurlingen (Cholfirst, Fig. 23.1) are built up by gravel of the Younger *Deckenschotter* Formation (see Schlüchter et al., this volume, for a discussion on the age of glaciations in Switzerland). The Riss glaciation (or Most Extensive Glaciation; Trümpy 1980) reached Möhlin near Basle and covered the whole region of the Rhine Falls during its maximum expansion. During the last glaciation, the Würm Ice Age, the Rhine Glacier reached with its maximum expansion (Last Glacial Maximum or LGM) the Rhine Fall region (Keller this volume). The outermost moraines are found just west of the present-day Rhine bed and the moraines of the Würm retreat stages are situated on the left side further east.

Before the Riss glaciation, the Rhine River was flowing north of the present Rhine Falls, from Schaffhausen directly westwards into the wide Klettgau valley (Fig. 23.5; see also Preusser 2008). During the following glaciation that valley was filled up with gravel ("Klettgau gravel"), which blocked the way towards the west (Fig. 23.5a). Therefore, the Rhine River was deviated towards the south and created a new, deeply incised valley (the "Rhine Falls valley"; Fig. 23.5b). During or after the Riss Ice Age this second Rhine valley was again filled ("Riss gravel"). During the LGM, the whole region including the gravel-filled Rhine valley was covered with moraine materials. After the glacial retreat, the Rhine River searched for a new bed, which did not follow exactly the old "Rhine Falls valley" (Fig. 23.5b). The river created a new bed in the moraine deposits, made a new bed at Schaffhausen south of the older one and reached the old bed again, which it left south of Flurlingen before returning to it at the location of the Rhine Falls. After the erosion of the Würm morainic material, the Rhine River bed upstream of the future Rhine Falls reached the Jurassic limestone beds and from here downstream the "Riss gravel". The weaker, loose "Riss gravel" was easily eroded and washed away, but the harder Jurassic limestone upstream was much more resistant to erosion. Consequently, the old "Rhine Falls valley" was formed again and at the border of the Jurassic limestone to this new valley, a cliff more than 20 m high developed between the two parts of the riverbed where the water dropped down in the older "Rhine Falls valley", again eroded. Upstream of the falls where the river flows on the Jurassic limestones two wide channels were eroded leading to the two main parts of the falls, the "Zürich and the Schaffhausen Falls" (Figs. 23.2 and 23.6). On the northernmost side of the falls, a third, very small fall formed (the "Mill Fall"), which is important for the economic use of the waterfall (see also Sect. 23.5).

The backward erosion during the last 15,000 years softened the originally subvertical cliff and created the present waterfall. The two towers in the centre of the falls (Fig. 23.7) are witnesses of the former cliff. Below the falls, an over-deepened basin with a depth of 13 m was eroded in the "Riss gravel" during flooding periods, especially in the beginning of summertime due to snowmelt in the Alps.

This interpretation of the Rhine Falls genesis was first described in 1871 by L. Würtenberger. An extended research about the waterfall was then published by the famous Swiss geologist Albert Heim, together with a detailed geological map at 1:10,000 scale (Heim 1931, see also Hofmann 1987, Heitzmann 2017).

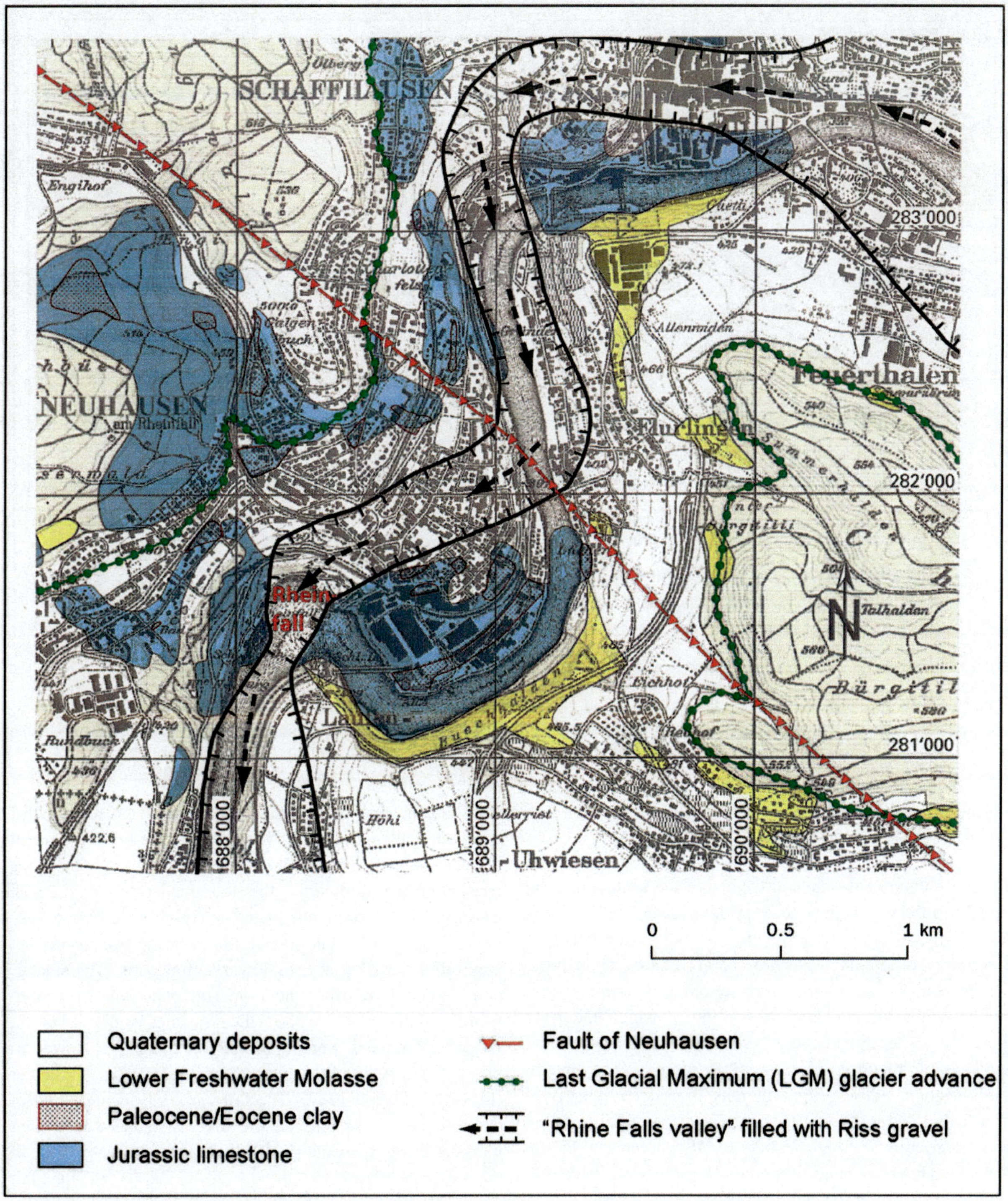

Fig. 23.4 Geological map of the Rhine Falls region. The older Rhine Falls valley of Riss age can clearly be distinguished from the present Rhine River course (modified after Rick 2006, courtesy of Museum zu Allerheiligen, Schaffhausen, Hier und Jetzt, Baden)

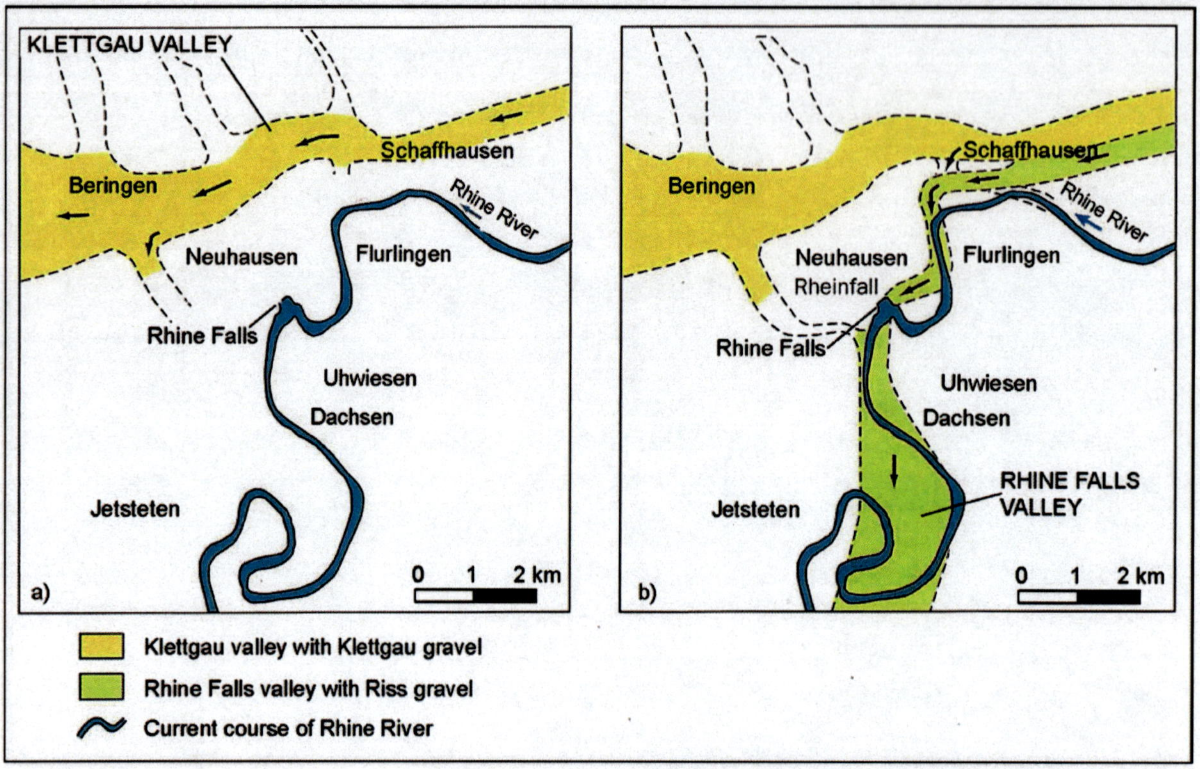

Fig. 23.5 Deviation of the Rhine River course during the Quaternary: **a** Before the Riss glaciation, the Rhine passed north of Schaffhausen and followed towards west into the Klettgau valley. During or after the Riss glaciation this valley was filled with gravel ("Klettgau gravel", yellow); **b** The Rhine River found a new course towards south. Also this valley was filled with gravel ("Rhine Falls gravel", green). The post-Würm Rhine River course did not follow exactly the Riss Rhine River course, but found it again at the site where the contemporary Rhine Falls are located (modified after Rick 2006, courtesy of Museum zu Allerheiligen, Schaffhausen, Hier und Jetzt, Baden)

23.3 Landforms and Landscapes

The morphological features of the bed of the Rhine Falls can best be observed at low to very low water level (below 200 m^3/s). At higher water levels, after all during the summer, there is too much water for geomorphological analysis.

Upstream of the waterfalls, morphology is dominated by the flat-lying Jurassic limestone beds on which the river flows very silently. In these beds, SSE to NNW-oriented channels can be observed; they follow more or less joints in the limestone bedrock. From the railway bridge on, the river itself is divided into two general flows corresponding to the two main falls ("Zurich Fall" and "Schaffhausen Fall"). These two main channels deepen much more towards the fall than in the more horizontal central part (Fig. 23.6).

The waterfalls themselves can be divided into a flatter upper part in the uppermost layers of the massive limestone formation and a steeper lower part in the massive limestone itself. Everywhere erosion of resistant rocks can be detected. Between the two main falls, two relics of the former cliff dominate the central part. The bigger of the two towers can be reached with touristic boats and the top is a marvellous terrace in the middle of the foaming water (Fig. 23.7). Because of low sediment load, the Rhine River has a very low erosion power, but despite this fact, the two towers are attacked by erosion at their lower, softer parts. The smaller one required reinforcement at the end of the nineteenth century (1879–1880). During the winter 1984/85, substantial repairs were executed on the bigger tower to prevent the voluminous top from falling in the river (Fig. 23.8; Deppe 1987). The flood of 1999, which did not damage the towers, proved the success of these important works. During low water in 1921, a karst cave was visible in the massive limestone of the lower slope of the fall (Fig. 23.9), a witness to the widely distributed karst phenomena in this geological formation.

Below the waterfalls, water accumulates in a wide circular basin, caused by erosion in the Riss gravel found in the former Rhine course. Near the fall, the basin is more than

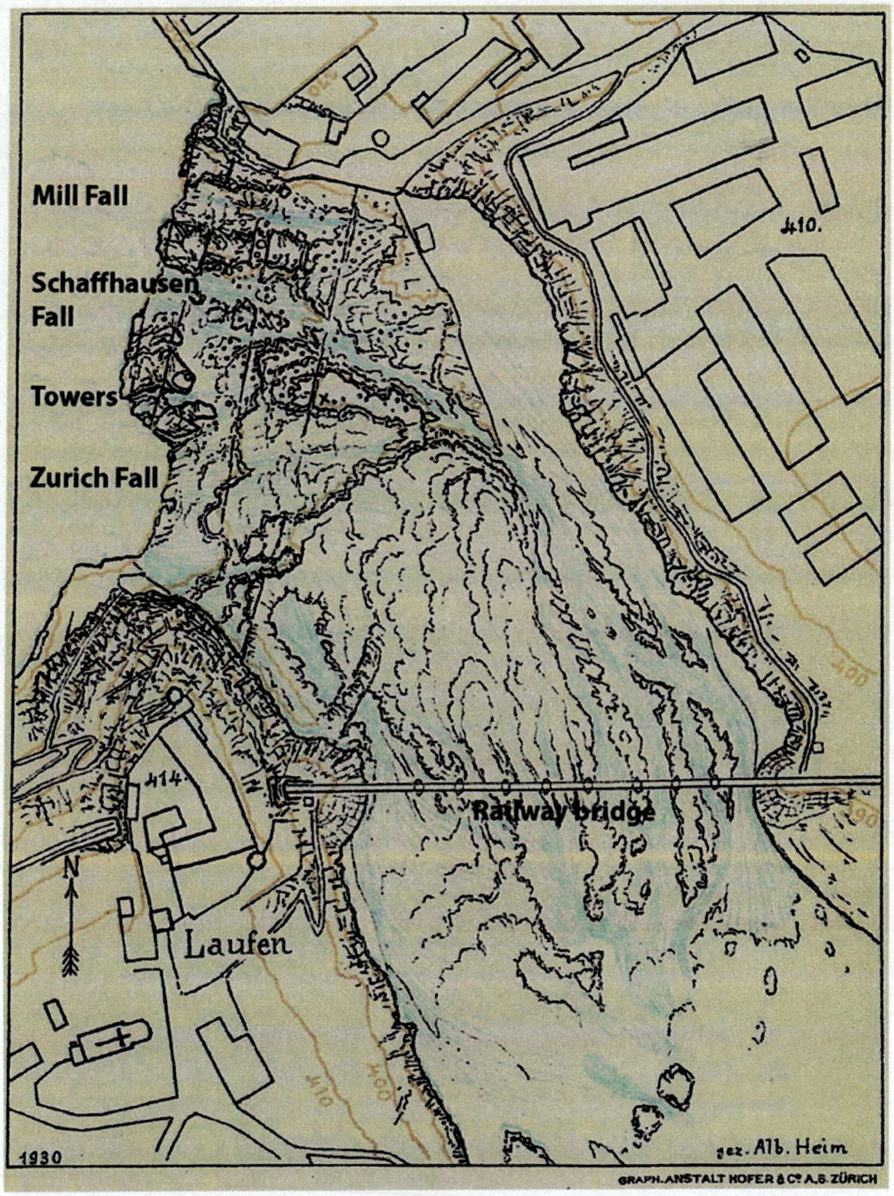

Fig. 23.6 The Rhine bed upstream of the waterfall. We can clearly distinguish the two main courses leading to the "Zurich Fall" (south) and the "Schaffhausen Fall" (north). At the northern end, there is a small fall, called "Mill Fall" ("*Mühlefall*"), where water was used for energy production (modified after Heim 1931)

10 m deep (deepest point is 13 m), carved out by the falling and rotating water. Both sides of the Rhine Falls are dominated by high limestone cliffs upon which are situated the Castle of Laufen and the industrial buildings of SIG Company. Towards the south, the new Rhine bed lies in the Riss gravels. We can conclude that the older Riss valley was much larger than the present one. On the western side of the valley, the Riss gravels are found in five terraces, corresponding to the lowering of the late Würm erosion levels during the glacier retreat (Keller this volume).

23.4 The Economic Use of Waterfalls

Since the Middle Ages two main activities have developed in the Rhine Falls area:

- In the oxygen rich water below the waterfall, fishermen found a rich fish fauna to be recovered. Today people still own a fishing right, which can be handed to the descendants;

Fig. 23.7 The Rhine Falls, viewed from the castle of Laufen (southern side). In the centre, two towers are visible as relics of the former cliff. They separate the southern "Zürich Fall" from the northern "Schaffhausen Fall". The bigger, northern of the two towers can be reached by tourist boats and people can climb on its top. The two different slopes of the falls can clearly be distinguished, a flatter upper part and steeper lower part. In the background at the left side, the more silent water in kolk basin can be observed (photo Nikater, https://commons.wikimedia.org/w/index.php?curid=36383350, accessed 31.08.2019)

- On the Schaffhausen side, the small so-called "Mill Fall" was used for energy supply. Several professions installed mills at the Schaffhausen side along the small waterfall: grain millers, oilseed millers and blacksmiths.

In the fifteenth century, the water rights passed from the Monastery of Allerheiligen to the State of Schaffhausen, where they are still now. In the sixteenth century, the exploited iron ore of the Eocene clays from the Laufenberg north of the Rhine River were brought to the Mill Fall and smelted; an iron foundry and a furnace were installed. At the beginning of the nineteenth century, the family Neher bought the iron industry complex and enlarged the production enormously. In 1888 the Neher Company was sold and the new owners changed from iron to aluminium production. Instead of waterpower, electric power served for energy supply. The first aluminium production in Europe by Alusuisse lasted at the Rhine Falls until 1945. It was allowed to use 20 m^3 water per second for electric power supply (a demand for more water to increase the production was several times rejected). Today the follower of Alusuisse, Rio Tinto Alcan, and then 3ATM, concentrate on research and development in the field of composites. The industrial buildings near the Rhine Falls of Neher/Alusuisse were destroyed in 1944 and today the terrain is integrated in the wide tourist park along the northern side of the Rhine River as part of the ILNM site 1412.

In 1853, a new company for the production of railway wagons was founded and installed on the northern promontory of the Rhine Falls (Figs. 23.2 and 23.10). This company, today known as SIG (Combibloc), enlarged in 1860 its range of products by manufacturing also firearms.

Fig. 23.8 Reinforcement works at the two towers separating the two main falls. The smaller was repaired in winter 1879/80, the bigger during low water in winter 1984/85 (*Source* Stadtarchiv, J 02.14.01.05/22, Schaffhausen)

Fig. 23.9 The Rhine Falls during the 1921 low water period. In the lower part of the fall, a large karst cave was visible in the massive limestone of Upper Jurassic age (Stadtarchiv, J 02.14.01.05/32, Schaffhausen)

Fig. 23.10 Industrial complex on the northern, right side of the Rhine Falls. In the foreground is visible the iron company of Neher, above the falls the wagon company. From the Rhine River, a 70 m long drive shaft supplied the wagon company with mechanical energy (detail of a lithography "Rhine Falls Panorama" by Theophil Beck; courtesy of Museum zu Allerheiligen, Schaffhausen)

Since the beginning the twentieth century, also packing machines and packing materials are produced (Pfaff 1987; Grütter 2006).

For the aluminium production an electric power station was established, which also served SIG. Several times, demands from the industry companies to increase the production were rejected by the Canton of Schaffhausen. In the middle of the twentieth century, at Rheinau about 10 km downstream from the Rhine Falls, a new electric power station was built; the level of the Rhine River was elevated by 6 m (instead of 8 m in the original project). That influenced the water level in the basin below the waterfall, but there is no important impact on the water regime. On 18 May 2014, a proposal by the Canton itself to change the water policy law allowing the building of a new electric power station functioning only during the night was rejected in a referendum.

Another problem, which was discussed for nearly 200 years, was the opening of the Rhine Falls for boats. As long as the Rhine River served as transport way between the Lake of Constance and Basle, the wares had to be transported on land at Schaffhausen. In 1817, a Bavarian inspector of mines proposed to destroy totally the Rhine Falls and to build a channel for boats. In 1839, a new project was proposed to build ten locks on the Zurich side of the fall. Finally, during the twentieth century, several propositions to overtake the barrier of the Rhine Falls by a tunnel were planned. None of them were executed.

23.5 The Tourism Value of Rhine Falls

The Rhine Falls are also known for their tourist interest. Since the Middle Ages, persons who visited the Rhine Falls described them as a big mass of water falling down from 20, 50 or 100 m and creating a horrible noise. With the beginning of the Enlightenment, the number of visitors increased. The period between the sixteenth and the nineteenth century was the time of the classical visit to the Rhine Falls during educational travels in Europe. Probably the most famous among these visitors was J. W. von Goethe.

In the middle of the nineteenth century, a new kind of tourism began and more and more people visited the Rhine Falls, not only during an educational travel, but also for

Fig. 23.11 Poster for the Hotels Schweizerhof and Bellevue at the Rhine Falls by Melchior Annen (87 × 58 cm). The lower picture was taken from the terrace of Grand Hotel Schweizerhof, the upper picture from the Castle of Laufen, seen in the lower picture (courtesy of Museum zu Allerheiligen, Schaffhausen, Hier und Jetzt, Baden)

holidays. On the north side of the big water basin, the Hotel Weber was built first in 1842–44 and in 1862–63 it was transformed into a bigger hotel, the Grand Hotel Schweizerhof. Later, a second hotel, the Hotel Bellevue was built nearby (Fig. 23.11). The great period of the Rhine Falls lasted until the First World War. The Hotel Schweizerhof was then transformed into a school, and the Hotel Bellevue had to overcome many difficulties to survive. Both hotels have since been destroyed and the ground was integrated into the large touristic park.

A period of mass tourism began after the Second World War and up to 3 million visitors per year were counted in the 1960s. Since then, the number of tourists has been decreasing: in the 1990s it dropped to about 2 million annually and in 2009 only 1.1 million tourists came to the Rhine Falls. In 2014 the number of visitors was about 1.3 million. Today the Rhine Falls see mostly day visitors rather than tourists coming for a longer stay. Since 2008, both cantons have made new investments in the infrastructure (Castle of Laufen, Visitor Centres, access to the falls) and founded the Rhine Falls Coordination Office, which includes all partners on both sides of the Rhine Falls. A unique website informs the public about all aspects of the Rhine Falls (http://www.rheinfall.ch/, accessed 31.08.2019).

23.6 The Rhine Falls in the Fine Arts and Literature

Many early visitors left descriptions (Butz 2009), drawings or paintings of the Rhine Falls. The first mention with a description and a representation of the Rhine Falls was made by Sebastian Münster in 1544 in his *Cosmographia universalis* (Fig. 23.12). J.W. Goethe visited the Rhine Falls four times: in 1775, 1779 and twice in 1797. He mentioned his visits in different texts and used his emotional experience for poems. The poem by E. Moerike (at the beginning of the chapter) is probably the most famous of all. Many paintings exist from the Rhine Falls (Fig. 23.13; Steiner 1958; Fayet 2006; Heitmann 2015). Often the representation of the Rhine Falls is exaggerated; the natural and idyllic aspects are in the centre and the economic facts are neglected. In relation to tourism, many posters were also produced, often with a rich artistic aspect (Fig. 23.11).

Fig. 23.12 The first graphic representation of the Rhine Falls: "The Rhine Falls" by Sebastian Münster, 1544. Xylography from *Cosmographia. Beschreibung aller Lender* (courtesy of Falke Verlag, Bern)

23.7 Conclusions

The Rhine Falls are an outstanding natural and cultural site, which presents a high diversity of landforms. They are considered as the waterfall in Europe with the biggest flow rate. Their origin is explained by the deviation of the Rhine River during the last glaciation, where the river found its older bed and eroded the former infilling. The falling water over the limestone cliff created the present waterfalls. Beside the natural facts, cultural aspects also play an important role for the Rhine Falls. For long time water resources have been used for economic purposes, beginning in the Middle Ages with mills that were developed in the nineteenth century for big industrial complexes held by innovative companies (the first aluminium company, the wagon industry). Tourism has been very important for the site too, as visitors described the beauty of the site, asked for the preservation of its natural environment and contributed significantly to the culture of the Rhine Falls through literature and the arts.

Fig. 23.13 Idyllic picture of the Rhine Falls: "Vue de la Chute du Rhin près de Schaffouse", by Johann Ludwig (Louis) Bleuler, about 1840 (guache, 32 × 50.1 cm) (courtesy of Museum zu Allerheiligen/Sturzenegger Stiftung, Schaffhausen)

Acknowledgements For the improvement of the manuscript, I would like to thank Lee Burton, my wife Anni and our son Michael for reading the manuscript, and R.T. Van Balen and E. Reynard for helpful suggestions for the final version. The following institutions helped with the illustrations by providing high-resolution digital copies and the authorisation to publish them: Federal Office of Topography, swisstopo, Bern; Museum zu Allerheiligen, Schaffhausen; Stadtarchiv, Schaffhausen; Verlag Hier und Jetzt, Baden.

References

Beutler R, Gerth A (2015) Naturerbe der Schweiz. Haupt, Bern, 385 pp

Butz HG (2009) Sie waren am Rheinfall. Der Rheinfall in der europäischen Literatur vom Mittelalter bis in die Gegenwart. Chronos, Zürich, 392 pp

Deppe J (1987) Sanierung der Rheinfallfelsen. In: Isler K (ed) Der Rheinfall. Neujahrsblatt der Naturforschenden Gesellschaft Schaffhausen 39:21–26

Fayet R (2006) Ästhetik der Entfesselung. Der Rheinfall und die Kunst. In: Fayet R (ed) Der Rheinfall. Strömungen, Tumulte, Reflexionen. Interdisziplinäre Schriftenreihe des Museums zu Allerheiligen Schaffhausen, Band 2. Hier und Jetzt, Baden, pp 196–247

Grütter D (2006) „Das ist Schaffhausens Zierde und Pracht". Historische Annäherungen an den Rheinfall. In: Fayet R (ed) Der Rheinfall. Strömungen, Tumulte, Reflexionen. Interdisziplinäre Schriftenreihe des Museums zu Allerheiligen Schaffhausen, Band 2. Hier und Jetzt, Baden, pp 30–60

Heim A (1931) Geologie des Rheinfalls. Mitteilungen der Naturforschenden Gesellschaft Schaffhausen 10:1–70

Heitmann C (2015) (ed) Der Rheinfall. Erhabene Natur und touristische Vermarktung. Schnell & Steiner, Regensburg, 184 pp

Heitzmann P (2014) Aus der Hölle fast bis in den Himmel—Valorisierung des nationalen Geotops von Laufenburg (CH). From Hell to Heaven—Valorisation of the national Geotope Laufenburg (Switzerland). Schriftenr Dt Ges Geowiss 84:49–54

Heitzmann P (2017) Der Rheinfall—Ist das überhaupt ein Geotop? In: Röhling HG, Zelmer H (eds) GeoTop 2017. Von der „Klassischen Quadratmeile" bis ins 21. Jahrhundert. Die gesellschaftliche Relevanz von Geotopen im Wandel der Zeit. Schriftenr Dt Ges Geowiss 90:22–35

Hofmann F (1987) Geologie und Entstehungsgeschichte des Rheinfalls. In: Isler K (ed) Der Rheinfall. Neujahrsblatt der Naturforschenden Gesellschaft Schaffhausen 39, pp 10–20

Keller O (this volume) The landscape of the Rhine glacier in the Lake Constance area

Pfaff R (1987) Nutzung der Wasserkräfte und industrielle Entwicklung am Rheinfall. In: Isler K (ed) Der Rheinfall. Neujahrsblatt der Naturforschenden Gesellschaft Schaffhausen 39:27–35

Preusser F (2008) Characterisation and evolution of the River Rhine system. Neth J Geosci 87:7–19

Reynard E, Berger JP, Constandache M, Felber M, Grangier L, Häuselmann P, Jeannin PY, Martin S (2012) Révision de l'inventaire des géotopes suisses: rapport final. Groupe de travail pour les géotopes en Suisse, Lausanne

Rick B (2006) Stabile Felsen im tosenden Wasser. Zur Geologie des Rheinfalls. In: Fayet R (ed) Der Rheinfall. Strömungen, Tumulte, Reflexionen. Interdisziplinäre Schriftenreihe des Museums zu Allerheiligen Schaffhausen, Band 2. Hier und Jetzt, Baden, pp 13–28

Schlüchter C, Akçar N, Ivy-Ochs S (this volume) The Quaternary Period in Switzerland

Steiner H (1958) Das Rheinfallerlebnis in Dichtung und bildender Kunst. In: Guyan WU, Steiner H (1958) (eds) Der Rheinfall. Schweizer Heimatbücher 83, Haupt, Bern, pp 7–25

Trümpy R (1980) Geology of Switzerland, a guide-book. Part a: an outline of the geology of Switzerland. Wepf, Basel, 104 pp

Working Group for the Protection of Geotopes in Switzerland, SAS/SANW/ASSN (1999) Inventory of geotopes of national significance. Geol Insubrica 4(1):25–48

Maps

Geologischer Atlas der Schweiz, Blatt 1031 Neunkirch, 1:25,000 (1981) Map and explanations by F. Hofmann. Bundesamt für Landestopografie swisstopo, Wabern

Geologischer Atlas der Schweiz, Blatt 1032 Diessenhofen, 1:25,000, (1964) Map by J. Hübscher and explanations by F. Hofmann and R. Hantke. Bundesamt für Landestopografie swisstopo, Wabern

Peter Heitzmann (Dr. sc. nat. ETH) studied geology at the ETH Zurich and worked as geologist in the field of structural geology/tectonics at different Universities in Switzerland (ETH Zurich, Neuchâtel, Bern). He was Director of the Swiss Geological Survey where projects like digital geological maps and GIS systems, storage of radioactive waste or the construction of the long transalpine tunnels were his main undertakings. Today he is retired and works in the field of geoheritage, bringing geology to a wider public. He is a member of the Swiss Working Group on Geotopes and Geoparks since its foundation in 1994.

24 The Allondon River: Decadal Planform Changes Under Changing Boundary Conditions

Nico Bätz, Ion Iorgulescu, and Stuart N. Lane

Abstract

The Allondon River (Canton of Geneva) is a historically well-documented piedmont river system that reflects interactions between an evolving climate, a changing catchment and local management practices and the suite of internal biogeomorphological responses and feedbacks that result. Over the last century, the river has been evolving from a braided to a wandering system. This appears to be driven by two key factors: (i) changes in hydrological forcing (notably in flood seasonality and flood power) coupled to a decrease in sediment availability, which has increased river incision; (ii) the angle of the shallow groundwater table, which creates drier conditions in the upper reach and wetter conditions in the lower reach. The coupling of both effects influences vegetation encroachment rates and the associated channel stabilisation. Hence the change in the rate of disturbance versus the rate of stabilisation along the river reach causes the morphological evolution to be spatially heterogeneous and variably resilient to extreme events, validating observations of the effects of vegetation made in flume studies.

N. Bätz (✉) · S. N. Lane
Institute of Earth Surface Dynamics, University of Lausanne, Géopolis, 1015 Lausanne, Switzerland
e-mail: nico.baetz@eawag.ch

S. N. Lane
e-mail: stuart.lane@unil.ch

N. Bätz
Department Surface Waters Research and Management, Swiss Federal Institute of Aquatic Science and Technology (Eawag), Seestrasse 79, 6047 Kastanienbaum, Switzerland

I. Iorgulescu
Etat de Genève, Département du Territoire, Service de l'Ecologie de l'Eau, Case Postale 206, 1211 Genève 8, Switzerland
e-mail: ion.iorgulescu@etat.ge.ch

Keywords

Braided river • Wandering river • Processes forcing • Morphological transition • Decadal timescale

24.1 Introduction

Piedmont rivers are an important element of the Swiss Alpine landscape. In general, north-west European rivers appear to be more fluvially active (aggradation and incision) during climate transition phases, such as the cooling at the start of the Little Ice Age (LIA), around 1200 AD and the warming since the end of the LIA around 1850 AD (Rumsby and Macklin 1996). Thus, climate cycles may impact river morphodynamics through time, such as by changing channel planform. However, most North-West Europe rivers have also been impacted upon by human activities, both locally (e.g. engineering for flood risk reduction) and across the wider catchment (e.g. hydroelectric exploitation, dam construction etc.). These impacts increased dramatically from the eighteenth century onwards in European Alpine piedmont rivers (e.g. Vischer 1989; Girel et al. 2003). In turn, a suite of recent new objectives for river management has developed. Rivers are recognised as both sustaining and being sustained by ecosystems, while taking on important functions for human civilisation (e.g. flood retention). Thus, what can be done to rivers and floodplains is increasingly regulated to protect or even to restore the ensemble of river and floodplain ecology.

The result of these climatic and human-induced changes is that rivers are rarely stationary. Further, there may be few examples of especially braided piedmont rivers that have been unimpacted by direct human activities. This makes it difficult to appreciate the dynamic relationship between geomorphology and climate and its impact on river bio-geomorphodynamics in such systems. In this chapter, we present an exception: the braided-wandering reach of the

E. Reynard (ed.), *Landscapes and Landforms of Switzerland*, World Geomorphological Landscapes,
https://doi.org/10.1007/978-3-030-43203-4_24

Allondon River, as a rare example of a morphologically evolving piedmont river in Switzerland that has seen a little direct human impact.

Unlike many other Swiss piedmont rivers (e.g. the Aare River, Canton Bern and the Saane/Sarine River, Canton Fribourg), and partly relating to its particular geomorphological and sedimentological setting, the Allondon River has not been straightened or embanked. The peculiarity comes from the fact that due to base level control, notably by the main Rhone River (Moscariello this volume), the Allondon River has incised into fluvioglacial and morainic deposits to create a relatively wide but incised floodplain given the catchment size. This floodplain is wide enough to support, historically, a braided river, but it is also largely uninhabited, meaning that the river generally reflects the interaction between evolving catchment processes, as influenced by factors like climate, and processes (e.g. channel change) internal to the river–floodplain system. Historical imagery reveals that its planform is evolving and given low levels of human impact, the Allondon provides a natural laboratory for quantifying and explaining morphodynamics over the decadal timescale. However, morphodynamics are influenced by different vegetation encroachment rates along the river reach due to a downstream trend in water table and so water accessibility (Harner and Stanford 2003). The associated stabilising effect of vegetation influences river system resilience to extreme events and may also lead to thresholds between different stable river morphologies (e.g. Francis et al. 2009). This makes the Allondon unique as compared with other Swiss braided piedmont rivers that have experienced much greater direct human impacts (e.g. the Maggia River, Ticino; Aare River, Bern/Thun; Saane/Sarine River, Fribourg).

24.2 Geographical Setting and Climatic Context

The Allondon River is fed by a small catchment (145 km^2) in the west of the Canton of Geneva (46°11′54″ N, 6°00′07″ E; Fig. 24.1). Its altitude ranges from 1705 m a.s.l. at Crêt de la Neige to 349 m a.s.l. at the confluence with the Rhone River (mean 650 m a.s.l.). Its slope ranges from 155 to 0.0007 % (mean 17 %). The majority of the catchment is located in the karstic French Jura Mountains where a network of small torrent systems and creeks collects water, with relatively little sediment charge, and flows towards the Swiss Plateau. Only close to the Swiss–French border the runoff combines into a single river, forming a 3 km long and 200–300 m wandering/braided floodplain (Fig. 24.2) until its confluence with the Rhone River.

Whilst the upper catchment controls flow accumulation, sediment availability and sediment dynamics relate to more local controls. As the river arrives at the base of the Jura Mountains, it encounters fluvioglacial and morainic deposits. The river has incised into these deposits (laterally and vertically). Thus, in contrast to its upstream tributary, this 3 km reach clearly distinguishes itself from the surrounding landscape. It is this river reach, which for shorthand we refer to as the "Allondon River", that is the subject of this chapter.

The entire river basin is characterised by a steep gradient in temperature and precipitation falling from an annual average of 3.9 °C and 1888 mm/year at 1669 m a.s.l. (La Dôle in the Jura) to 10.5 °C and 1005 mm/year at 420 m a.s.l. (Geneva Cointrin, to the east of the study site; data MeteoSwiss). The altitudinal range of the basin means that the regime is "pluvial jurassien" (Pflaundler et al. 2011) with floods normally occurring within 6 h of storm events (Hottinger 1998). However, the hydrology appears to be strongly non-stationary in time and we discuss this below.

Land use in the catchment is mostly forest, meadows and pasture (70 %), followed by agriculture (15 %). There is some human impact: both industry and the European Organization for Nuclear Research (CERN) research centre near Saint-Genis-Pouilly use the water of the Allondon River. Despite extensive wastewater treatment, several polluting events impacted the river ecology between 1970 and 1990 (DIM 2010). Since 2011, wastewater has been redirected into the Geneva wastewater treatment system and no longer reaches the river.

Generally, there is little physical river management within the 3 km reach considered in this study and most of the interventions (e.g. spur dykes) are located on the edges of the steep incised valley slope to prevent slope failure or protect infrastructure. Many interventions were removed in 2000 during a revitalisation and renaturation program. A major invasion of the neophyte (*buddleia*) was brought under control in the year 2004. Nowadays, the river corridor is recognised both nationally and internationally as a protected site (DIM 2010).

Only a few comparable examples exist in Switzerland, as for instance the Sense/Singine River and Ärgera/Gérine River, both in the Canton Fribourg. This makes this site very particular as its morphological changes are mainly influenced by natural changes in the external forcing (hydrology) and changes in reach internal biogeomorphic feedbacks as explained below.

24.3 Hydro-Geological Setting

The floodplain valley slope direction follows the tectonic lineament of the Vuache Fault system, indicating that the ancient river floodplain was eroded into the molassic deposits by following a major geologic fault (Fig. 24.3; Moscariello this volume). The molassic deposits are (mainly

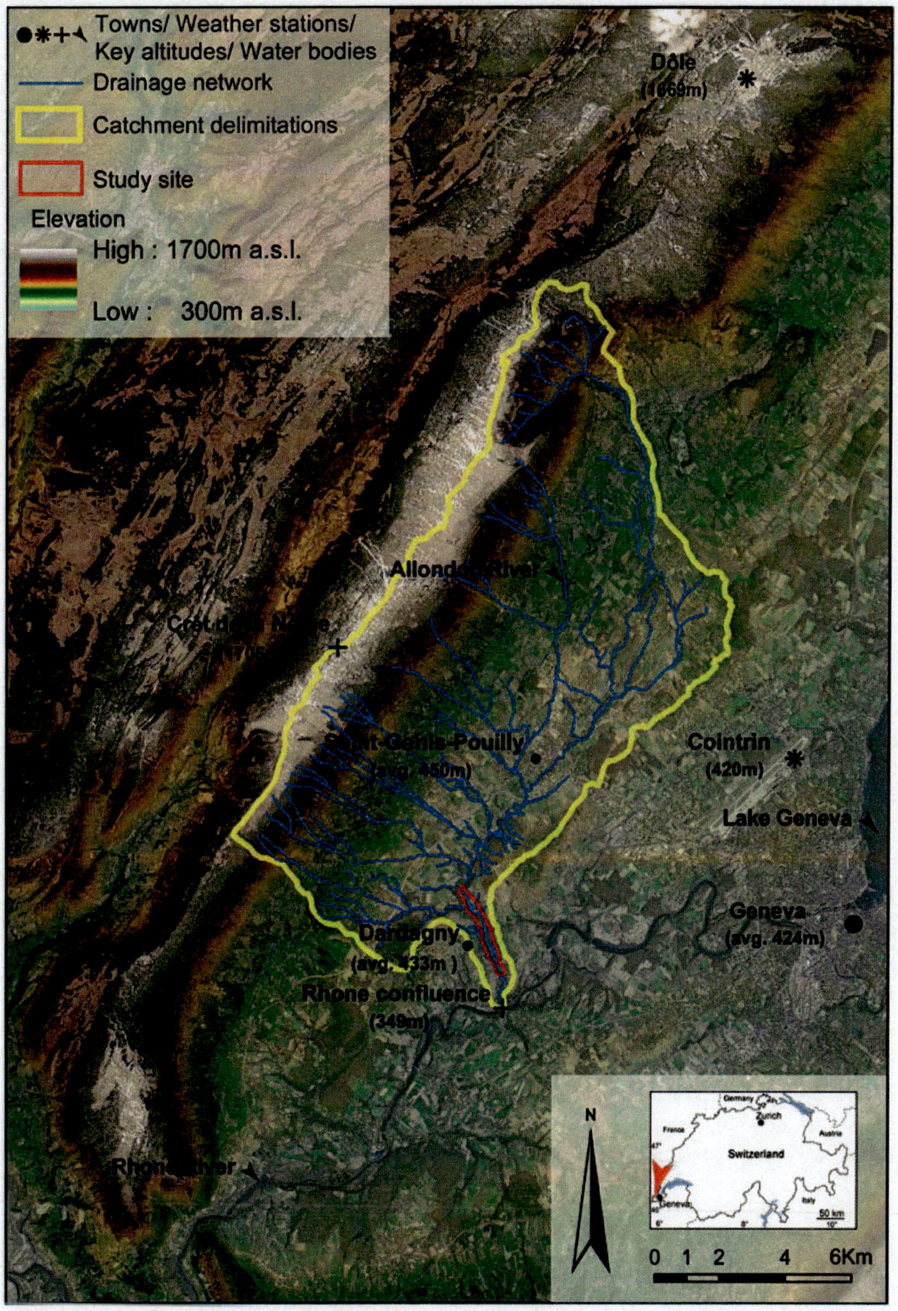

Fig. 24.1 Map of the catchment showing the topography with the superimposed aerial image. Key locations that are mentioned in the text are shown

sandstone and conglomerate) formed in the Oligocene due to massive erosion during Alpine orogenesis. The valley was then filled by a series of morainic deposits. The oldest has been mainly attributed to the Rhone glacier during the Riss (370.000–130.000 BP). However, the Jura ice cap and Arve glaciers (from the Mont Blanc massif) also affected the landscape morphology of the area. Due to the recession of Rissian glaciers, a layer of fluvioglacial sediments was deposited, which was then covered again by morainic deposits during the Last Glacial Maximum (LGM), called Würm in Alpine context, 70.000–10.000 BP (Moscariello this volume). During the last recession of the Rhone glacier a large proglacial lake formed, with water levels reaching a maximum of 470 m a.s.l., compared with the altitude of the study reach close to 450 m a.s.l. Hence, the region corresponding to the present river reach was inundated, leaving a thin layer of glaciolacustrine deposits (Coutterand 2010; Moscariello this volume). This layer has largely disappeared

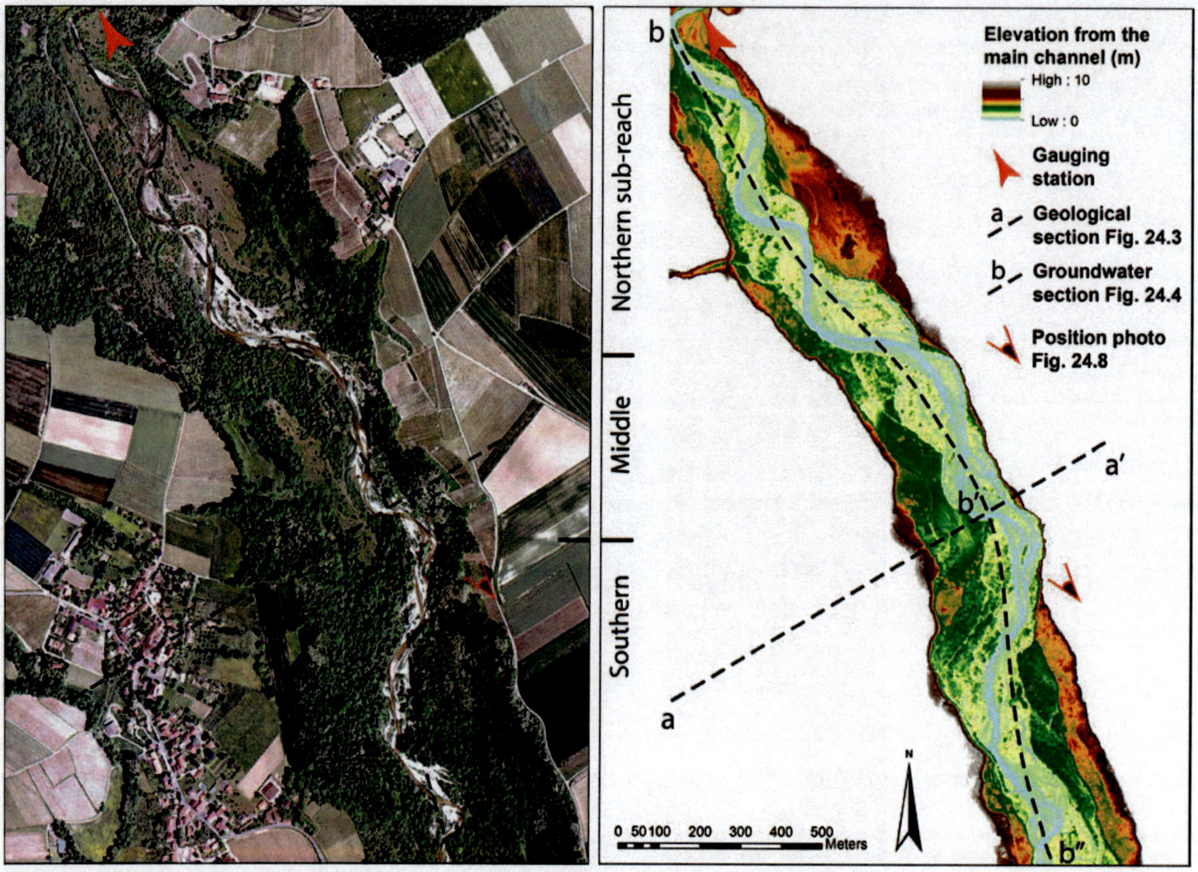

Fig. 24.2 Aerial image (left) and river slope detrended topography (right) of the Allondon River (2009) (*Source* Swisstopo). The locations of the geological section (Fig. 24.3), of the groundwater section (Fig. 24.4) and of the panorama photo (Fig. 24.8) are indicated

from the floodplain due to erosion. Between these three mainly depositional periods, the Allondon has incised into these deposits, preserving the incised floodplain along the fault (Fig. 24.3). For more details about the overall long-term geomorphological development of the Geneva Basin, we refer to Moscariello (this volume).

In terms of river morphodynamics, this geological setting is important. First, the moraine and fluvioglacial deposits create groundwater conditions that influence vegetation dynamics and so river dynamics. There is evidence of downwelling in the first kilometre of the reach (groundwater depth of about 10 m), and upwelling in the last kilometre

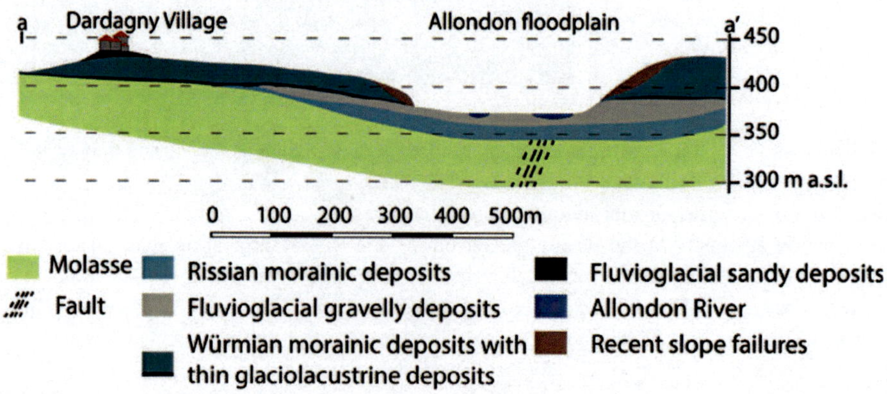

Fig. 24.3 Geological section through the Allondon floodplain (modified after Wellhauser and Géroudet 1978, and Geological Atlas of Switzerland 1:25 000—Special geological map "St-Julien en Genevois", swisstopo, 1997)

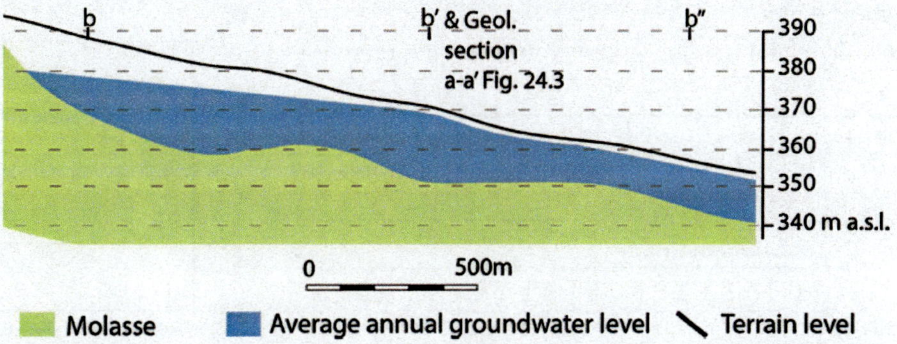

Fig. 24.4 Cross-section showing the trend of the groundwater table along the study reach (modified after Hottinger 1998)

where groundwater feeds the river flow (Fig. 24.4; Hottinger 1998). Second, the nature of the sedimentary deposits leads to a local source of erodible and transportable sand and gravel material. This is particularly important for the morphodynamics as it enables deposition of bars within floodplain as well as the formation of terrace systems.

24.4 The Hydrology of the Allondon River

The Allondon streamflow has been monitored continuously since 1985 at Dardagny-Les Granges (Fig. 24.2), 0.5 km upstream of the studied reach. The catchment area at the gauging location is 119 km^2. A stream gauge was operated at the same location between 1918 and 1935 (Fig. 24.5a). The analysis of the historical record reveals that the hydrological regime has significantly evolved over the past century. As the changes in the hydrological forcing are highly relevant for the geomorphological and ecological processes in the study area, we present the current hydrological regime and compare it with the one that prevailed almost a century ago.

The annual discharge regime is illustrated in Fig. 24.5b using the monthly Pardé coefficients: the ratios between the mean monthly and the mean annual flows. The current regime is characterised by a late winter/early spring maximum and a late summer/early autumn minimum. During winter (December to February), the coefficients at the beginning of the twentieth century were considerably lower, with maximum discharges recorded later in the spring (March to May). Minimum flows correspond in both cases to late summer, but historical values were higher (Pardé coefficients, Fig. 24.5b). However, absolute values of maximal discharge were generally lower in the past (Fig. 24.5a). The classification of the discharge regime has also changed. Presently, the Allondon is one of the seven reference basins used to define the "pluvial jurassien" regime (Pflaundler et al. 2011). However, the analysis performed by Aschwanden and Weingartner (1985) on the 1918–1935 data, and included in the 1992 edition of the Swiss Hydrological Atlas, classified the discharge regime as "pluvial-nival jurassien". Snow seems to be less important in the basin at present.

In terms of flows, a significant increase with respect to 1918–1935 can be noted. For example, the Q_{95} calculated for the 1986–2013 period (10.3 m^3/s) is 30 % higher than during the 1918–1935 period (7.9 m^3/s). The statistical analysis of maximum annual flows shows a significant increase in the mean of the distribution, but not in terms of dispersion. This translates into an upward shift in the fitted Gumbel distributions (Fig. 24.5a and c). As a consequence, a maximum relative effect is observed for frequent floods. For example, the median flood (the flow with a recurrence interval of two years) is 46 m^3/s, while during the 1918–1935 it was 31 m^3/s, which represents a 48 % increase. These changes, as discussed in the next sections, are very important for fluvial dynamics as most sediment transport laws are a nonlinear function of discharge excess over a critical value, and so it is probable that sediment transport capacity in the river is now markedly higher.

Finally a change in the seasonality of the maximum annual flows is illustrated in Fig. 24.5d. During the 1918–1935 period more than one in five, below-median snowmelt-type, annual floods occurred in May while none occurred in February. During the last 30 years, none of the annual floods occurred in May while almost one in five occurred in February. This indicates a change in the flow regime from pluvial-nival to pluvial, associated with less seasonal snow storage (Fig. 24.5b and c), and may be of great geomorphological importance. The present higher flows occur early in, if not before, the start of the growing season. Lower spring and summer flows may be making vegetation and hence fluvial dynamics more groundwater dependent (Bätz et al. 2016).

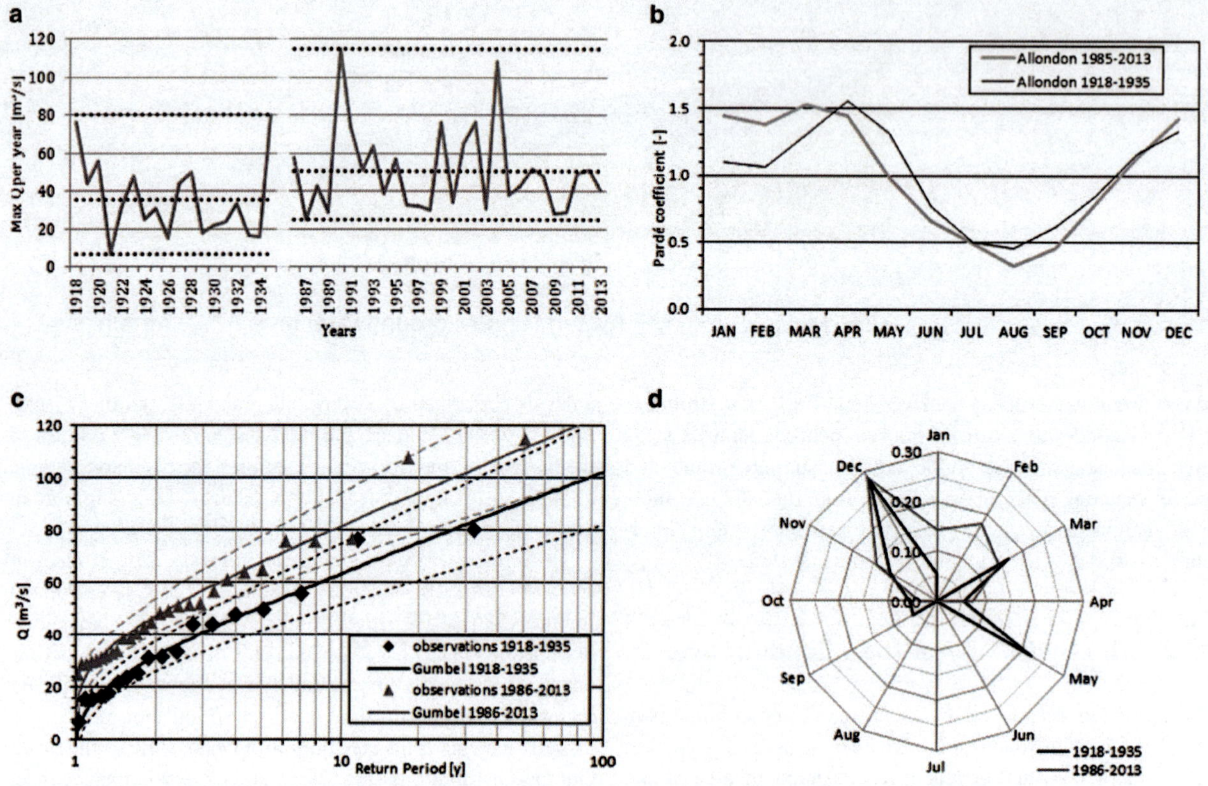

Fig. 24.5 **a**. Peak annual discharge for the periods 1918–1935 and 1986–2013; **b**. Comparison of the monthly discharge regimes (Pardé coefficients) for the periods 1918–1935 and 1985–2013; **c**. Comparison of the recurrence intervals of observed (Gringorten plotting position) and fitted Gumbel distributions (method of moments) with 80 % confidence peak annual flows for the periods 1918–1935 and 1986–2013; **d**. Comparison of the relative frequency of occurrence of maximum annual flows during each month (seasonality) for the periods 1918–1935 and 1986–2013

24.5 The Geomorphology of the Allondon River

The recent morphology of the Allondon River is characterised by a wandering river system (Fig. 24.2). Such systems can have high rates of channel change but, as compared with the continuous changes associated with river braiding, flow tends to be concentrated into a primary channel (Wang et al. 2012). The geomorphic activity of the river leads to an active zone that is sufficiently reworked to avoid vegetation colonisation and stabilisation. This zone is surrounded by a number of terrace levels, which can be up to 4–5 m above the active river channel (Fig. 24.2).

Four key elements allow understanding the geomorphology of the Allondon River. First, erosion of the river terraces, but also slope failure of the undercut fluvioglacial deposits of the lateral valley slopes, appear to be an important sediment source for braiding and bar formation along the entire reach. Observations from aerial images indicate lateral erosion rates of 1.3–2.0 m/year and up to 10 m/year for the terraces in the northern and middle sub-reaches (Fig. 24.7, images 1999–2012). Valley slope failure processes occur continuously, but are accelerated when erosion removes the slope talus and eventually undercuts the ancient deposits (Figs. 24.6a and 24.7, images 1999–2005).

Second, despite having a wandering morphology, the Allondon River can develop braiding sections (Figs. 24.2 and 24.7). Figure 24.7 shows several zones in which the flow braids around both bars and vegetated islands. It is well established that to maintain the braiding process, the river needs to be subject to a rate of perturbation that is greater than the rate of stabilisation, for example by vegetation development. Thus, maintaining braiding will require relatively frequent sediment delivery from upstream and so either active sediment sources or transport effective flow events. If these are able to lead to bar formation, reworking and migration, and these processes occur at a greater rate than the vegetation capacity to encroach the channel, that is lower than the geomorphic turnover rate (Murray et al. 2008; Tal and Paola 2007), then braiding will be maintained.

Fig. 24.6 **a**. Slope failure processes of the fluvioglacial sediments of the valley side slopes, which are considered the main sediment source; **b**. Large woody debris deposit on an alluvial bar. Pioneer plants (*Salix ssp.*) colonised the lee of the bar, enhancing landform stability; **c–d**. Cut banks of a river terrace under heavy lateral erosion of the highest terrace (**c**) and the lowest terrace (**d**). Note the difference in rooting depth, which affects river bank stability (photos: Nico Bätz)

For most of the time, the Allondon River is transport limited. Braiding and (vegetated) bar dynamics can only be observed within the 250 m downstream of the main sediment sources, often associated with major terrace or valley slope erosion. Vegetation dynamics in the Allondon River, as observed in similar environments (e.g. Corenblit et al. 2011; Gurnell 2014), actively influence the morphological development by taking on an engineering role (e.g. Fig. 24.6). Pioneer species such as *Salix spp.* and *Populus spp.* are capable of reinforcing stable river bank deposits with their roots and creating an above ground rough surface able to trap sediments at the timescale of years to decades. This promotes bar accretion and reduces lateral erosion. Only powerful events are able to reset the system (Corenblit et al. 2011; Gurnell 2014). In the studied reach, the styles of channel change (e.g. Ashmore 2013) characteristic of a fully braided system are rare. However, because of different rates of vegetation encroachment along the reach current morphology is changing from wandering (upper sub-reach) to meandering (lower reach).

Third, deposition of large woody debris on young bars of the Allondon River has been observed to facilitate and to sustain pioneer vegetation colonisation within the active zone. The resprouting woody debris, but also the establishment of pioneer vegetation in the lee of (dead) woody debris, may stabilise deposited sediment and so enhance bar aggradation (Fig. 24.6b). Such processes have been shown to influence fluvial landform development significantly (Corenblit et al. 2011; Gurnell 2014). In the Allondon River it appears to take about 20–30 years to transform part of the active zone into a vegetated island. Such islands and also older vegetated terraces may in turn become the source of woody debris.

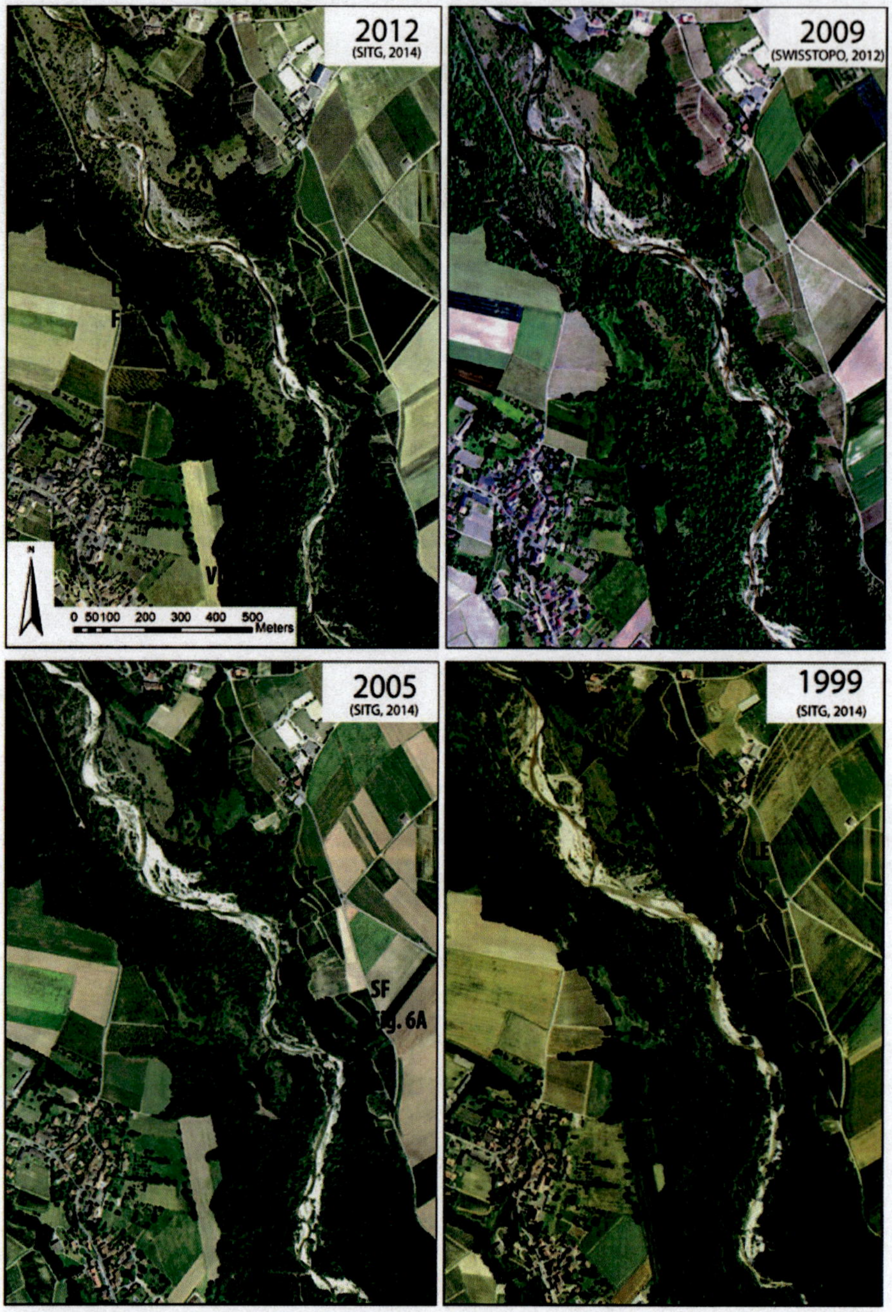

Fig. 24.7 Series of aerial pictures and plans, documenting the morphological evolution of the Allondon River. The arrows give some examples of prominent processes that can be compared to the previous image: SF = valley slope failure processes; LE: lateral terrace erosion; AV: channel avulsion processes; VI: vegetated island formation; BD: bar dynamics; TF: terrace formation *Sources* Federal Office of Topography Swisstopo; Territorial Information System of Geneva (SITG); Cabrit A. 1836. Carte du cours de la riviere "la London" et des Hauteurs qui la bordent—Plan géomètrique. Archives d'Etat de Genève (AEG) CH AEG Travaux BB 12.53

Fig. 24.7 (continued)

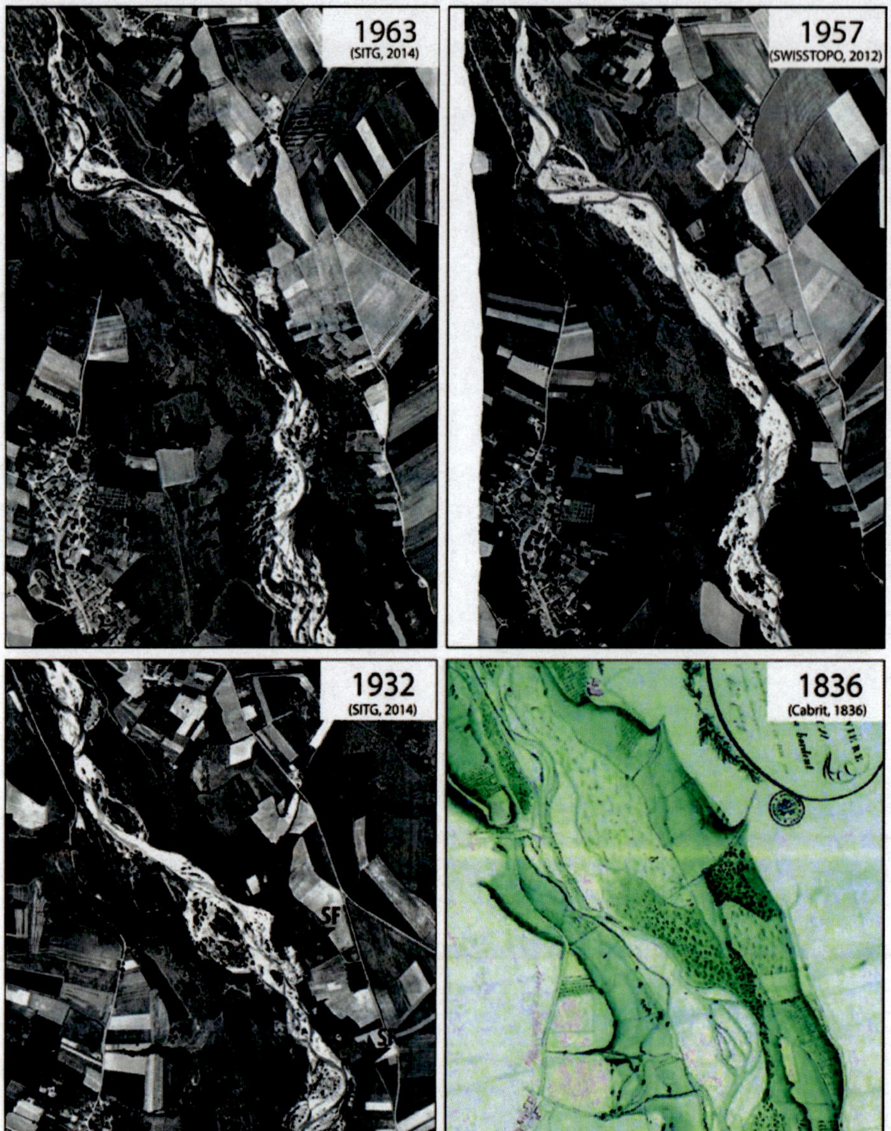

Fig. 24.7 (continued)

Fourth, despite not braiding in a more classical sense, high lateral erosion rates can lead to local increases in primary channel curvature until the flow cuts-off, causing local channel avulsion or chute cut-off (Ashmore 2013). The gravelly material of cut banks appears to be very easily eroded (Fig. 24.6c and d).

Despite being able to find elements of these four key processes throughout the reach, their importance varies from north to south down the river reach. The groundwater distal northern sub-reach (Fig. 24.2) shows a sinuosity of about 1.23 and width of the active zone ranges from 50 to 100 m. Here, vegetation colonisation and growth rates appear to be slower (Bätz et al. 2016) and there is much less resistance provided by older terrace vegetation (mostly *Fraxinus excelsior* and *Quercus robur* or dry prairies).

The groundwater proximal southern sub-reach has a much straighter channel morphology (sinuosity 1.11), and the active zone reaches a maximum width of only 60 m. Rapid scrollbar processes and small scale avulsions discontinuously rework the active zone, with periodic erosion of the vegetated landforms. The terraces associated with this sub-reach have much denser vegetation and are much more resistant to erosion. Compared to the northern sub-reach, fresh deposits appear to have faster pioneer vegetation colonisation rates, which helps to explain why the active channel is much narrower (Bätz et al. 2016).

In the intermediate transitional section, weaker bank material and a larger active zone allow much larger-scale avulsions than in the southern sub-reach. There is evidence of recent strong coupling to the valley side walls. However, compared to the northern reach, the abandoned channels show faster pioneer vegetation recolonization, which through the associated engineering effect on stable banks enhances landform stability.

24.6 Geomorphic Evolution of the Allondon River

The geomorphic evolution of the Allondon River can be considered over two distinct timescales: annual to multi-annual and decadal.

24.6.1 Annual to Multi-annual Changes

To consider the most recent evolution of the Allondon River, we consider the images 1986–2012, comparable to the second period described in Sect. 24.4: a generally higher discharge (Figs. 24.5a/c) but also greater differences between high and low flows (Fig. 24.5b). The general trend during this period was characterised by high lateral erosion rates of the outer meander belts and rapid revegetation of the active zone after two major flooding events (1990 and 2004; Fig. 24.4a). In detail, the active channel in the year 1986 (Fig. 24.7) rarely exceeded 50 m over the entire reach. However, by 1991 the active width in the northern sub-reach had doubled, removing up to 80 m of terrace material. The 1990 flood (115 m^3/s; Fig. 24.5a) appears to be the main driver for the observed morphological changes. The middle and southern reaches appear more resilient to both events.

The following period (1991–1999) shows revegetation of the formerly exposed deposits, notably in the north. However, a channel avulsion can be observed in the north, with pioneer plants rapidly colonising the site (1999; Fig. 24.7). The morphological changes in the middle reach during the same period were characterised by lateral erosion, transforming the straight channel into a more meandering flow. This suggests morphological adaptation of this sub-reach through erosion of relatively weak bank material, following the upstream responses to the 1990 flood.

The period 1999–2005 was characterised by greater activity due to wetter years and a major flood event in 2004 (108 m^3/s; Fig. 24.5a). Meander dynamics (lateral erosion) dominate in the northern and middle sub-reach. The sinuosity of this reach increased from 1.03 (1999) to 1.24 (2005), with rapid vegetation colonisation of the generated point bars. The first meander of this sub-reach undercut the lateral valley slope, activating a sediment source through slope failure (e.g. Fig. 24.6a). This caused bar formation during the following years. Lateral erosion of up to 18 m in the northern sub-reach increased sinuosity from 1.10 to 1.19. At the same time the inner meander bends (point bars) were slowly colonised by vegetation. The 2004 flood event, but also the depth of groundwater (Hottinger 1998), may have prevented rapid vegetation colonisation processes as compared to the middle sub-reach (Bätz et al. 2016). The development of the southern sub-reach during this period (1999–2005) did not involve major morphological changes (Fig. 24.7). Only the 2004 flood reactivated part of the floodplain by increasing the active width from about 35 m to 50 m.

The morphological development between 2005 and 2012 was characterised by a stabilisation of the morphology. No major flood occurred (Fig. 24.5a), allowing overall vegetation encroachment. Lateral erosion in the north and associated cut bank slope failure increased local sediment input and caused the formation of some bars (2005–2012; Fig. 24.7). Although hydrological forcing was low in this period, high meander curvature encouraged rapid channel migration (2009 and 2012; Fig. 24.7), showing how the continued occurrence of less hydrologically intense events can still lead to major system changes. Compared to the northern reach, the former channels of the middle and the active zone of the southern sub-reach witnessed rapid vegetation colonisation.

A prominent example of vegetated island formation, and thus the vegetation-induced engineering, can be seen in lower middle sub-reach (Fig. 24.7 from the year 1986 on). The landform rapidly developed a vegetation cover (1986–2005; Fig. 24.7) that stabilised the deposits. Only the major flood event of 2004 caused massive lateral erosion and a channel avulsion, partly removing the vegetated patch (2009; Fig. 24.7).

Summarising, there is some spatial variability in the annual to multi-annual changes in this river. The series of recent aerial images (1986–2012; Fig. 24.7) shows that to the north, the river remains more dynamic. To the south, it has developed a more stable meandering morphology. In gravel-bed rivers, such as the Allondon River, water is a limiting resource on vegetation growth rates during the hot and dry summer months, thus hampering vegetation establishment on fresh alluvial deposits (e.g. Wolman and Gerson 1978). However, access to groundwater may significantly change vegetation growth and hence the associated engineering role (Bätz et al. 2016). As noted above, groundwater depth of the Allondon shows a particular trend; the northern sub-reach appears as groundwater distal (10 m), while in the southern sub-reach upwelling groundwater has been observed (Hottinger 1998). This groundwater trend may change vegetation colonisation and growth rate (Harner and Stanford 2003), suggesting differences in system resilience to exceptional perturbations (Francis et al. 2009) from north to south.

24.6.2 Decadal Changes

These annual to multi-annual changes are superimposed upon the long-term evolution of the system (Fig. 24.7). The floodplains of 1836, 1932 and 1957 appear to comprise a 110–140 m wide vegetated island braided system throughout the river reach. The high braiding activity is most likely to be related to high sediment remobilisation rates over large areas of the active zone, but appears also to be associated with floodplain incision (1936–1972; Fig. 24.7). Several terrace levels that persist nowadays are formed during this period and are up to 2 m higher than the current channel.

By 2012, notably in the southern sub-reach, the active width has decreased to about 40 m and the aerial images of 1963, 1972 and 1980 (Fig. 24.7) show the associated transition from an island braided to a wandering river morphology. This transition is correlated with progressive encroachment by vegetation and reflects similar transitions from braided to wandering systems observed in the laboratory (e.g. Tal and Paola 2007). Between 1963 and 1980 there was a substantial change in system morphodynamics either because of a reduction in perturbation rate (sediment dynamics and/or hydrological forcing) or an increase in vegetation encroachment rate. The latter may well be negatively correlated to the former, such that only small changes in the perturbation rate may be needed to achieve the observed transition.

The hydrological changes illustrated above suggested a general increase in discharge and related peak discharges during the last century. One result is that the enhanced sediment transport capacity may not be balanced by sediment delivered to the reach. This may have led to an increase in floodplain incision and to the associated narrowing of the active zone during recent decades. In turn, this should have helped the southern reach to establish a better connection with groundwater whilst incision may have also made it more difficult for avulsion to perturb the older terraces. Hence, terraces have developed and have continued to maintain a mature vegetation cover and so a relative resistance to erosion. Notwithstanding, the northern reach remains relatively vegetation free because the deeper groundwater reduces vegetation growth rates (Bätz et al. 2016). Moreover, temperature and precipitation data suggest that climate may be becoming drier, something that should make vegetation development more dependent upon hyporrheic flows. This may be exacerbated by a shift in peak flow timing to the winter (Fig. 24.5d) and hence a reduction of disturbance during the growing period. The associated stabilisation by vegetation, as controlled by the water table trend along the reach, may have increased the resilience of the fluvial system to the increased perturbation observed over the last few decades (Fig. 24.5c), thus facilitating the morphological transition from a braided to a wandering river system. This is not to say that extreme events cannot still have a very big impact (cf. Fig. 24.7), but rather that the size of the event needed to have a substantial impact, notably in the southern reach, may be increasing (e.g. Francis et al. 2009).

Historical sources (Chapeaurougy 1774) document the importance of the Allondon floodplain as a resource for the two neighbouring municipalities (e.g. firewood, grazing). However, economic changes during the twentieth century have led to the abandonment of most of the floodplain activities and steep slope agriculture. Woody and shrubby vegetation now dominates the floodplain cover, stabilising the river banks and south exposed slopes. Figure 24.8 shows the Allondon floodplain around 1945. Shrubby vegetation is

Fig. 24.8 Recent (left: 9th September 2015; photo: Nico Bätz) and historical (right: around 1945; Services Industriels de Genève (SIG), Projet d'une usine hydroélectrique sur l'Allondon, 1948; Archives d'Etat de Genève (AEG) CH AEG 1995 va 37.110) morphological aspect of the northern reach of the Allondon River. Note the difference in vegetation cover and the related width of the active zone. A–E indicate the location of several places

restricted to the left bank. Moreover, the southern exposed floodplain valley slopes appear to be under agricultural use. The easily eroded, vegetated terraces and the steep and loose slopes (fluvioglacial sediments) would have provided possible sediment sources for braiding. Since then, reforestation may have disconnected the river from these sources (Figs. 24.8 and 24.7, image 1932 and 2012). Similar processes have been observed in all the tributaries of the Allondon, but also in French Alps, Pyrenees and New Zealand.

24.7 Conclusions

The historical development of the Allondon River illustrated here shows how, for a river relatively uninfluenced by human activity, climate change interacts with hydrology, hydrogeology and vegetation to determine river morphodynamics. In turn, these morphodynamics induce an autogenic response, changing the feedbacks between river morphodynamics and vegetation encroachment. The Allondon is a classic case because most other Swiss braided piedmont river systems have been highly managed, and interpreting their morphodynamics is difficult given such strong historical human influences.

The data suggest that over the last century the river is in a non-stationary state, with the evolution from an island gravel-bed braided system to a meandering/wandering gravel-bed system, and an associated vegetation encroachment of the active zone (Figs. 24.7 and 24.8). Similar observations have been made in laboratory flume models (e.g. Tal and Paola 2007) and are thought to relate to changes in the rate of disturbance processes (power and frequency) versus the rate of stabilisation processes (e.g. vegetation encroachment rate). In similar environments, vegetation has shown to take on an engineering role and hence drive river morphodynamics by colonising the sufficiently stable river banks (Tal and Paola 2007; Corenblit et al. 2011; Gurnell 2014).

Data on the hydrological forcing of the last century show a general increase of about 15 m^3/s in discharge (Q_2) and an increase in difference between high and low flows. Moreover, a change in seasonality was observed, changing the hydrological regime from bimodal "pluvial-nival jurassien" to unimodal "pluvial jurassien" (Fig. 24.5). During the same period, floodplain land use changed from an agro-pastoral system to a protected forested site, thus reducing sediment availability for transport due to the increased river bank and valley slope stability.

The observed change in hydrological forcing coupled to reforestation may have led to a considerable increase in incision rate. However, the observed change in hydrological seasonality may have a second crucial effect. The hydrological changes mean that perturbations and morphological changes are now concentrated into a single period outside the vegetation growth period. This may allow a longer growing season combined with greater riverbank stability, thus increasing the chances of surviving the next bigger winter flood. The stabilisation of riverbanks and deposits by vegetation, further forces the channel into a single flow hence promoting incision.

Despite the general trend explained above, the morphological response of the Allondon River is spatially heterogeneous. The morphological transition shows a trend from north to south with higher vegetation encroachment rates towards south. The groundwater nappe follows this trend (Fig. 24.4), suggesting that the vegetation establishment during the drier growing period is limited by the access to the groundwater table (Bätz et al. 2016). Similar observations have been made by Harner and Stanford (2003), who observed different vegetation growing rates based on upwelling and downwelling river reaches in USA. The vegetation encroachment rates influence the morphological resilience along the reach, thus the critical threshold for the transition to a different stable state (e.g. braided to single channel; Francis et al. 2009).

Future morphological response of the Allondon River will potentially be different for the upper and the lower reach. A more pluvial regime will concentrate disturbing high-energy floods into winter, and summers will have the tendency to become drier. Thus, the longer undisturbed period (April–November; Fig. 24.5d) may encourage vegetation encroachment in the groundwater upwelling southern reach. In the northern reach, due to the higher groundwater dependency of the vegetation during the warmer summer months, morphological response will largely depend on the groundwater depth (accessibility) during summer. Distance to the groundwater table has been increasing over the last 10 years (measurements available since 2003), mainly because of the reduction in summer precipitation. Vegetation establishment, and thus channel stabilisation, has become more restricted, to the low water line (images 2005–2012; Fig. 24.7). Only sufficiently erosive winter floods, which are able to break through the stabilising vegetation along the channels, will cause avulsion, keeping a more braided character of the upper reach.

These observed morphological changes and their potential triggers also raise questions about future sustainable river management. To what extent could and should high turnover rates of the active zone be preserved to maintain braided river morphology? As this example shows, braided morphodynamics are highly dependent on the balance between disturbing and stabilising processes. Catchment processes, such as land use and climate change, but also reach scale human impacts may change this relation irrevocably, making local river preservation strategies ineffective. More holistic approaches may help in defining how boundary conditions have changed and the measures needed to maintain the relationship between vegetation, form and process necessary to sustain the Allondon River.

Acknowledgments Special thanks to: the State of Geneva (Etat de Genève, Département Du Territoire), for data and access to the natural reserve for research purposes; Hans-Rudolf Pfeifer for providing valuable literature and inspiring discussions on the Allondon River; to the presenters Ion Iorgulescu, Erik Bollaert, Arielle Cordonnier, Marc Hottinger and the participants of the "Atelier Allondon" held in November 2014 who inspired this work. We also would like to thank Emmanuel Reynard and the anonymous reviewer for the precious comments that improved this manuscript.

References

Aschwanden H, Weingartner R (1985) Die Abflussregimes der Schweiz, Publikationen Gewässerkunde 65. University of Bern, Institute of Geography, 205 pp

Ashmore P (2013) Morphology and Dynamics of Braided Rivers. In: Shroder J, Wohl E (eds) Treatise on Geomorphology. Elsevier, San Diego, pp 289–312

Bätz N, Colombini P, Cherubini P, Lane SN (2016) Groundwater controls on biogeomorphic succession and river channel morphodynamics. J Geophys Res: Earth Surf 121(10):1763–1785

Chapeaurougy (1774) Mémoire manuscrit des communiers de Dardagny au Petit-Conseil concernant les droits des riverains. Archives d'Etat de Genève (AEG), CH AEG 1971 va 8

Corenblit D, Baas ACW, Bornette G, Darrozes J, Delmotte S, Francis RA, Gurnell AM, Julien F, Naiman RJ, Steiger J (2011) Feedbacks between geomorphology and biota controlling Earth surface processes and landforms: a review of foundation concepts and current understandings. Earth Sci Rev 106:307–331. https://doi.org/10.1016/j.earscirev.2011.03.002

Coutterand S (2010) Etude géomorphologique des flux glaciaires dans les Alpes nord-occidentales au Pléistocène récent—Du maximum de la dernièr glaciation aux premières étapes de la déglaciation. PhD thesis, University of Savoie, 468 pp

DIM (2010) L'Allondon—Fiche-rivière n. 1. Département de l'Intérieur et de la Mobilité, Genève, 56 pp

Francis RA, Corenblit D, Edwards P (2009) Perspectives on biogeomorphology, ecosystem engineering and self-organisation in island-braided fluvial ecosystems. Aquat Sci 71:290–304

Girel J, Vautier F, Peiry JL (2003) Biodiversity and land use history of the alpine riparian landscapes (the example of the Isère river valley, France). In: Mander U, Antrop M (eds) Multifunctional landscapes: continuity and change. International Series on Advances in Ecological Sciences. WIT-Press, Southampton, pp 167–200

Gurnell AM (2014) Plants as river system engineers. Earth Surf Proc Land 39:4–25. https://doi.org/10.1002/esp.3397

Harner MJ, Stanford JA (2003) Differences in cottonwood growth between a losing and gaining reach of an alluvial foodplain. Ecology 84:1453–1458. https://doi.org/10.1890/0012-9658(2003)084%5b1453:dicgba%5d2.0.co;2

Hottinger M (1998) Etude hydrogéologique de la nappe alluviale de l'Allondon genevoise (GE). Centre d'hydrogéologie de l'Université de Neuchâtel, Neuchâtel, 122 pp

Moscariello A (this volume) The geomorphological landscape in the Geneva Basin

Murray AB, Knaapen MAF, Tal M, Kirwan ML (2008) Biomorphodynamics: physical-biological feedbacks that shape landscapes. Water Resour Res 44:W11301. https://doi.org/10.1029/2007wr006410

Pflaundler M, Dübendorfe C, Zysset A (2011) Méthodes d'analyse et d'appréciation des cours d'eau. Hydrologie—régime d'écoulement niveau R (région). Office fédéral de l'environnement, Berne, 113 pp. http://www.bafu.admin.ch/publikationen/publikation/01603/index.html?lang=fr

Rumsby BT, Macklin MG (1996) River response to the last neoglacial (the "Little Ice Age") in northern, western and central Europe. Geol Soc Lond Spec Publ 115:217–233. https://doi.org/10.1144/gsl.sp.1996.115.01.17

Tal M, Paola C (2007) Dynamic single-thread channels maintained by the interaction of flow and vegetation. Geology 35:347–350. https://doi.org/10.1130/g23260a.1

Vischer D (1989) Impact of 18th and 19th century river training works: three cases studies from Switzerland. In: Petts GE, Möller H, Roux AL (eds) Historical change of large alluvial rivers: western Europe. Wiley, Chichester, pp 19–44

Wang Z, Lee JHW, Melching CS (2012) River dynamics and integrated river management. Springer, Berlin

Wellhauser P, Géroudet P (1978) (eds) Le Vallon de l'Allondon. Association genevoise pour la protection de la nature, Genève, 56 pp

Wolman MG, Gerson R (1978) Relative scales of time and effectiveness of climate in watershed geomorphology. Earth Surf Process Landf 3:189–208. https://doi.org/10.1002/esp.3290030207

Nico Bätz obtained his Ph.D. in Geoscience at the University of Lausanne. His research interest lies in developing the fundamental understanding of the interactions between the abiotic and the biotic environment of river systems and particularly how these affect river morphodynamics, ecology and ecosystem services. Currently, he is engaged at the Swiss Federal Institute of Aquatic Science and Technology (Eawag), where he studies the impact of hydroelectric power production (i.e. hydropeaking) on river ecosystems.

Ion Iorgulescu is a hydrologist at the Cantonal Office for Water (OCEau) of the State of Geneva (Switzerland). He is a hydraulic engineer and he obtained his Ph.D. in hydrology at the EPFL (Swiss Federal Institute of Technology in Lausanne).

Stuart N. Lane is Professor of Geomorphology at the University of Lausanne where he runs the research group Water, ice, sediment and ecology in Alpine environments.

The Illgraben Torrent System

25

Brian W. McArdell and Mario Sartori

Abstract

The Illgraben catchment is one of the most active torrents in the European Alps, with between 2 and 7 debris flows or similar debris floods occurring every year since the start of systematic observations from 2000 to the present. The ultimate cause of the debris-flow activity is the presence of a variety of rock types, which have been substantially deformed and fractured due to tectonic processes. The change in the orientation of the major Alpine Simplon fault line resulted in the development of many additional faults, which is the cause of the highly fractured bedrock. Sediment delivery to the torrent-channel system is dominated by landslides, the deposits of which are temporarily stored within the channel. Subsequent mobilization of sediment, mainly by intense rainfall, results in the formation of debris flows and debris floods, which entrain the sediment within the torrent system and potentially endanger the community.

Keywords

Debris flow • Sediment production • Erosion • Landslides

25.1 Introduction and Geographical Setting

The Illgraben catchment is one of the most active debris-flow catchments in the Alps, with between 2 and 7 debris flows occurring every year since the start of systematic observations in June 2000 (Hürlimann et al. 2003). The Illgraben catchment and fan are situated in the southern side of the valley of the Rhone River near the village of Susten (Fig. 25.1). The catchment extends from the summit of the Illhorn mountain (2716 m a.s.l.) to the confluence with the Rhone River at about 610 m a.s.l.

The Illgraben fan is a prominent landform in the Rhone River valley, with a radius of about 2 km (Fig. 25.2). The Rhone River is confined by the fan to the northern edge of the valley. Prior to its channelization, the Rhone River upstream of the fan had a meandering planform which is evident due to the presence of oxbow lakes on the floodplain, near Agarn, upstream of Susten. This was due to the very low slope of the river due to the damming by the Illgraben fan associated with comparatively low sediment charge from the Rhone River. Below the confluence with the Illgraben channel, the Rhone River has a classical braided-river morphology which can be explained by the large sediment influx from the Illgraben (Schlunegger et al. 2009).

The Illgraben catchment comprises two sub-drainage basins (Figs. 25.1 and 25.2), the Illbach Valley on the eastern side (sub-catchment area of 5.7 km^2) and the Illgraben valley (4.7 km^3), drained by the Illgraben stream, on the west. Upstream of the confluence, the Illgraben Valley is sometimes labeled Vanoschi or Fanöischi on topographic maps (e.g., Fig. 25.1). The Illbach Valley consists of a conventional small glacial cirque trending perpendicular to the main Rhone Valley. The Illbach Valley enters the Illgraben Valley as a hanging valley (Fig. 25.2), clearly indicating that the erosion rate is larger in the Illgraben Valley. Downstream of the confluence, the channel is termed the Illgraben channel. The Illgraben Valley constitutes one of the most impressive active torrential erosion features of Switzerland.

The eastern half of the Illgraben fan is largely settled and also used for agriculture purposes. The western half is largely covered by the Pfynwald, a largely continuous pine forest. The lower part of the fan is crossed by the cantonal road of Valais and by an artificial channel, part of the

B. W. McArdell (✉)
Swiss Federal Institute for Forest, Snow and Landscape Research WSL, Zürcherstrasse 111, 8903 Birmensdorf, Switzerland
e-mail: brian.mcardell@wsl.ch

M. Sartori
Department of Earth Sciences, University of Geneva, Rue des Maraîchers 13, 1205 Geneva, Switzerland
e-mail: mario.sartori@unige.ch

E. Reynard (ed.), *Landscapes and Landforms of Switzerland*, World Geomorphological Landscapes,
https://doi.org/10.1007/978-3-030-43203-4_25

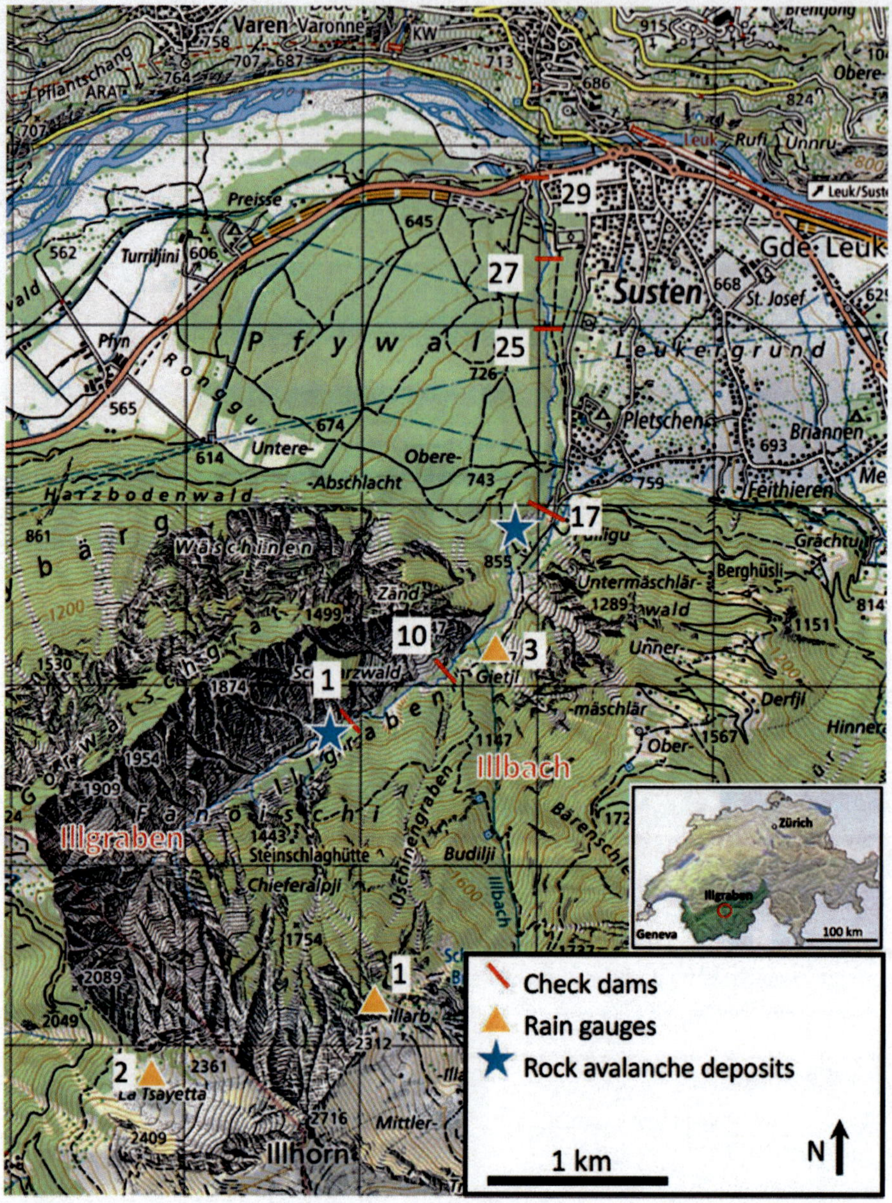

Fig. 25.1 Location map of the Illgraben catchment and torrent system illustrating the location of the Illgraben and Illbach subcatchments (red text). Features described in the text include the locations of selected dams or check dams (red lines) and their corresponding number, rock avalanche deposits (blue stars), and rain gauges (yellow triangles). The influence of the sediment supply on the Rhone River is apparent from the transition from a single-thread meandering river upstream of the confluence with the Illgraben to a multi-thread braided river downstream of the confluence

Susten–Chippis hydropower plant. In the future, the fan will be crossed by a gallery of the Rhone motorway, which is currently under construction.

Since 2005, the Illgraben is part of the Pfyn-Finges Nature Park recognized by the Canton of Valais. In 2013, the park has been recognized as Regional Nature Park of national importance by the Swiss government. The Illgraben is also a landscape of national importance (Object Nr 1716, Pfynwald—Illgraben) and a geosite of national significance (Object Nr 249 Illgraben torrential system).

In this chapter, we summarize some of the main geomorphic features of the catchment and describe the geological setting which can be used to help explain the exceptionally high frequency of debris flows.

25.2 Monitoring and Early Warning Systems at the Illgraben

The Illgraben debris flow observation station was installed by the staff of the Swiss Federal Institute for Forest, Snow and Landscape Research, WSL, in June of 2000 (Hürlimann et al. 2003). Instrumentation consists of geophones to detect vibrations of passing debris flows and to initiate recording of

Fig. 25.2 Overview of the Illgraben fan (view to the south) showing the village of Susten on the eastern side of the Illgraben channel and the Pfynwald forest on the west. The Rhone River is visible in the foreground (photo B. McArdell)

data, sensors to measure the surface elevation of the flow (laser and radar), a large force plate to measure basal forces produced by debris flows (McArdell et al. 2007), and an instrumented wall, parallel to the channel, which allows detection of flow properties at different heights within the flow (Berger et al. 2011a). The force plate was destroyed by a debris flow in 2017; however, it has been redesigned and is scheduled to be replaced in 2019. The observation station, designed for scientific purposes, is optimized to detect and record debris flows and not floods.

A separate Early Warning System (EWS) for the community was installed in 2007 (Badoux et al. 2009). The EWS also relies on geophones and radar depth sensors to detect hazardous events (debris flows and flash floods) within the channel. When a hazardous event is detected, acoustic and visual alarms situated near footpaths warn people to stay out of the channel, and local hazard managers also receive an alert. The EWS also consists of an observation program, where local hazard managers regularly inspect the catchment for new landslide activity and the channel for potential problems. Warning signs installed at regular intervals near the channel provide additional information to keep people away from the channel. Finally, information is also provided to tourists and school children.

Data from both the scientific observation station and EWS are used for research purposes. The ease of access and regularity of debris flow activity have helped transform the Illgraben catchment into a natural laboratory, where many different research groups are working to improve our understanding of the debris-flow processes, the conditions of debris-flow formation, and the development and structure of the debris-flow fan.

25.3 Geological Setting

The position and orientation of the Illgraben Valley, upstream of the confluence with the Illbach, is a deep V-shaped feature which is tectonically and lithologically controlled. The talweg follows a contact between two thrusted middle Penninic nappes, which have been deformed to a vertical orientation by late Alpine backfolding (Figs. 25.3 and 25.4).

On the eastern side of the Illgraben stream, the former hanging wall of the thrust consists of a thick sequence of folded and/or stacked Permo-Triassic quartzites forming the frontal part of the Siviez–Mischabel nappe (Escher 1988; Scheiber et al. 2013; Fig. 25.3). White to green-colored conglomeratic- and green sericitic-quartzites with thin layers of brownish ankerite alternate in a 1400 m-thick sequence providing most of the sediment feeding the debris-flow system. Some decameter-thick bands of Triassic dolomitic marble and of Palaeozoic granitic gneiss horizontally cross the Illhorn's northern side (Gabus et al. 2008).

The left side of the Illgraben stream, the former footwall of the thrust, consists of the Mesozoic cover of the Zone Houillère (Fig. 25.3). A thick middle Triassic series of grey marbles and yellowish dolomites runs parallel to the Rhone

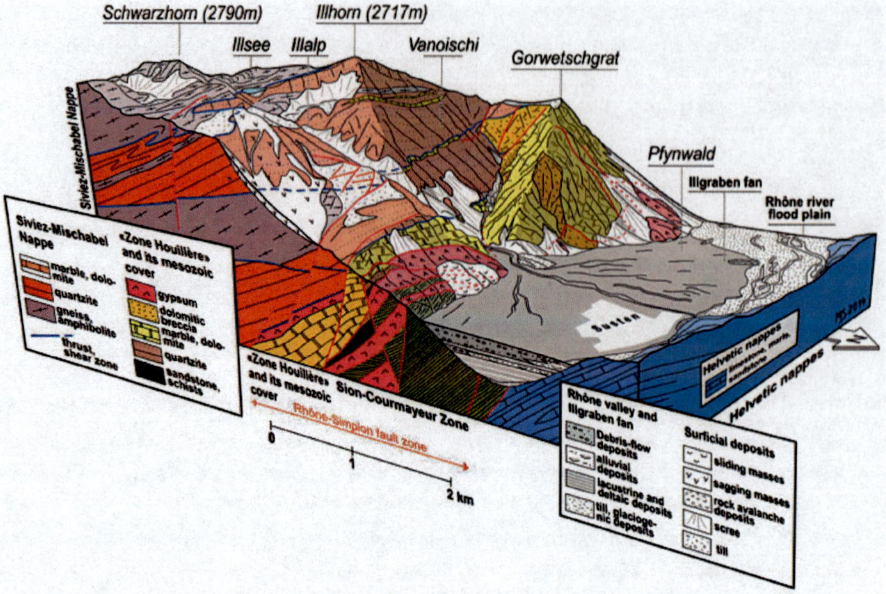

Fig. 25.3 Geological and tectonic structure of the Illgraben (illustration M. Sartori)

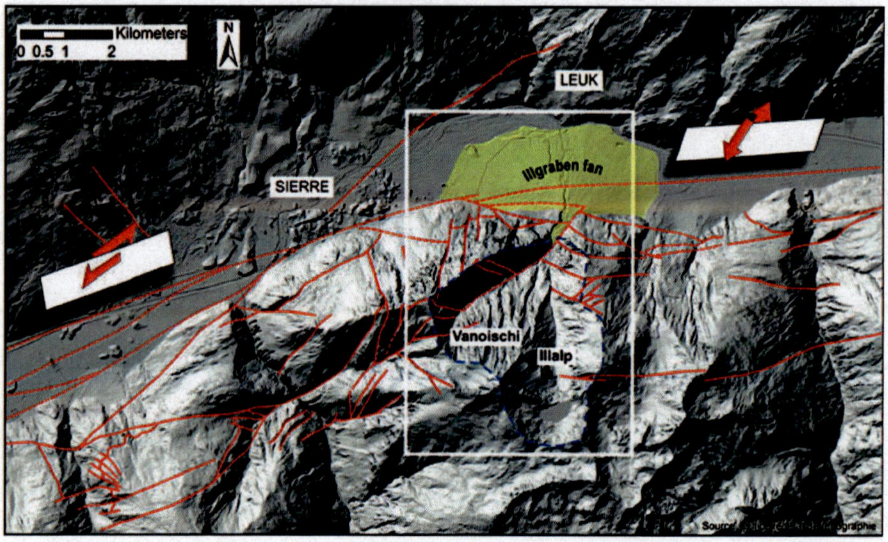

Fig. 25.4 Locations of fault lines near the Illgraben (illustration M. Sartori)

Valley, forming the Gorwatsch ridge (Fig. 25.1). On the Illgraben side of this ridge, the top of the series plunges steeply in dip slope plates toward the south. Weaker, friable upper Triassic dolomitic breccia and greywacke rest locally on top of the marble slabs and also form the largest part of the rocky wall that closes the Illgraben's valley toward west.

This wall is limited in its northern edge by a regional fault that runs parallel to the Illgraben's valley, dipping 50° toward SE and following the Illgraben's streambed toward the apex of the alluvial fan (Figs. 25.3 and 25.4). Numerous additional faults trend parallel and perpendicular to the Rhone Valley (Fig. 25.4). They belong to the Rhone–Simplon fault system, active since about 20 Ma (Campani et al. 2010). The Illgraben catchment developed where the Rhone Valley trend bends from a dextral strike-slip fault zone trending WSW-ENE along the Sion–Sierre part of the valley to a dextral-normal fault zone trending roughly W-E along the Leuk–Visp segment (Fig. 25.4). This change of direction of the main fault line results in a regional extensional bending zone, as evidenced by results on fault analyses (Champagnac et al. 2003) and on seismicity (Maurer et al. 1997).

Erosion processes in the Illgraben cirque are enhanced by the very dense joint network that has been developed in this highly deformed zone. This region is also located near the center of the higher seismic hazard zone of Switzerland

(Giardini et al. 2004) in which earthquakes of magnitude Mw ≥ 6 occur with a return period of approximatively one hundred years (the last ones in 1855 and 1946). Repeated strong seismic events could have favored erosion by triggering of landslides and by opening of the fracture network near the surface.

25.4 Landforms and Landscapes

The Illgraben torrent system contains a variety of landforms and landscapes ranging from the strongly tectonically conditioned and mechanically weathered bedrock hillslopes in the source area (Fig. 25.5), often mantled with debris and vegetation, to talus slopes and other depositional landforms within the torrent system. The large fan shows various depositional lobes and abandoned channels, which are best preserved on the fan to the west of the main Illgraben channel, in the largely undisturbed Pfynwald.

25.4.1 Landslide Activity

In the upper catchment, two main types of landslide have been identified based on the interpretation of aerial photographs (Bennett et al. 2012): shallow failures of highly fractured bedrock near the surface and deep-seated failures which may be up to several decameters thick. The small landslides also include the collapse of bedrock towers, e.g., on the north face of the Illhorn mountain (Fig. 25.5). They collapse, probably by toppling, sending significant amounts of sediment into the torrent channels.

The largest-known landslide is the rock avalanche with a volume of ~3.5 10^6 m^3 that occurred on March 26, 1961 in the Illgraben sub-catchment (Lichtenhan 1971; Gabus et al. 2008). The landslide left a deposit, up to nearly 90 m thick (Oppikofer et al. 2006), in the upper catchment (southern-most blue star in Fig. 25.1), which dammed the channel, causing the formation of a lake, which eventually drained, resulting in the largest documented debris flow on the alluvial fan (Bardou and Jaboyedoff 2008) with a volume estimated at ~500,000 m^3 based on an unpublished consulting reports (cited in Hürlimann et al. 2003). To prevent additional erosion of this deposit, a large 48-m tall dam was installed at the toe of the landslide (Fig. 25.6; see location on Fig. 25.1). Twenty-nine additional check dams (30 dams total, including the uppermost dam, labeled Check dam 1 in Fig. 25.1) were constructed in the channel to prevent vertical and lateral erosion of the channel. A large rock avalanche deposit, situated just upstream of the fan apex (Fig. 25.1), has a surface exposure age of 3220 ± 250 years BP (Schürch et al. 2016). A comparison of volume and reach angle with the 1961 event suggests that the volume of that rock avalanche event was on the order of 10 million cubic meters.

Smaller volume landslides are significant at annual timescale. Berger et al. (2011b) reported 10 landslides (volumes of hundreds to thousands of cubic meters) for several observation periods in 2007–2009 (Fig. 25.7). Using

Fig. 25.5 Summit of the Illhorn mountain. Strongly jointed bedrock features and "headless" talus slopes where the source area has been removed by erosion are visible on the left (photo B. McArdell)

Fig. 25.6 Check dam 1 which was constructed to stabilize the toe of the 1961 rock avalanche (RA) visible above and to the left of the check dam (CD1), covered by trees (photo B. McArdell)

ground-based real-aperture radar, Caduff et al. (2014) described the slow deformation of a large bedrock landslide with an estimated volume on the order of 500,000 m^3. Landslide activity documented by Bennett et al. (2012) between 1963 and 2005 confirms the importance of landslides in delivering sediment to the torrent channel system. The resulting rate of erosion is 0.39 m/y for the most active part of the catchment on the north face of the Illhorn (Fig. 25.8).

25.4.2 Debris Flows and Sediment Transfer

The torrent system within the catchment is exceptionally active. The vegetated areas of the catchment, between some of the active gullies, are much less important in terms of sediment delivery in comparison with the gullies and torrent channels (Schlunegger et al. 2009). On warm days during the snowmelt period (approximately. May and June), abundant rockfalls, very small debris flows, and dry granular flows are commonly observed upstream of the fan apex. These small mass movements generally remain in the upper catchment and are temporarily stored within the Illgraben channel (Fig. 25.7) (Berger et al. 2011b). Debris flows, debris floods (e.g., Hungr et al. 2014), and sediment-transporting flood flows are typically triggered by intense rainfall (Badoux et al. 2009), although observations by WSL staff suggest that some debris floods and small debris flows may be triggered by snowmelt alone. These small events (maximum depths $< \sim 0.3$ m) typically do not activate the debris flow observation station, or the EWS of the community (Badoux et al. 2009). In some cases, more than one tributary gully contributes a debris flow during one storm event (Fig. 25.9), resulting in debris flows with more than one main surge.

The sediment flux in the torrent system indicates a general pattern of landslide activity resulting in sediment storage in steep gullies and in the main channel above the fan apex (Berger et al. 2011b; Bennett et al. 2012, 2013; Fig. 25.7) followed by transfer from the gullies to the valley bottom either directly as debris flows (Fig. 25.9), debris floods, or as sediment transport. Sediment deposition in the gullies probably mainly occurs during the snowmelt period, which is suggested by observations of up to 4-5 m of sediment deposits in the channel upstream of check dam 1 (Fig. 25.1), the so-called sediment reservoir (Berger et al. 2011b), during the winter and spring, and subsequent erosion of these deposits during the summer debris-flow season.

The eastern side of the main Illgraben channel upstream of check dam 1 consists of rock-avalanche deposits from the 1961 event. Due to the significant rockfall hazard within the catchment upstream of check dam 1, direct observations in the field are limited. The area between the fan apex and check dam 1 provides some indication of the processes which may occur upstream of check dam 1. Rarely, debris-flow surges have been observed to stop in the channel (Fig. 25.10), which is one process by which the channel reservoir becomes filled with sediment.

The geomorphic apex of the Illgraben fan is currently situated at the downstream end of the oldest-documented rock avalanche (Schürch et al. 2016). Using field mapping and cross-checking with a digital terrain model, Schürch et al. (2016) were able to develop a plausible chronology of the alluvial fan and surficial structure of the fan (Fig. 25.11). It is clear that the active channel has changed its location on the fan many times over the past several thousand years. A detailed chronology near the current active channel on the fan was developed using dendrochronological methods on trees which were damaged but not destroyed (Stoffel et al. 2008). Debris-flow activity could be dated to events back to 1793. The chronology of events over the last few hundred years is certainly incomplete because many flows do not leave the channel, and some large historical events have not yet been observed in the tree-ring record (Schürch et al. 2016).

Currently, the main channel is confined by check dams and some lateral berms to the current location. Observations of the structure of the fan with depth are rare; however, a ground-penetrating radar measurement campaign (Franke et al. 2015) identified buried channels and levees at several locations on the fan, indicating that debris-flow activity and overbank flooding were perhaps similar to what is occurring today.

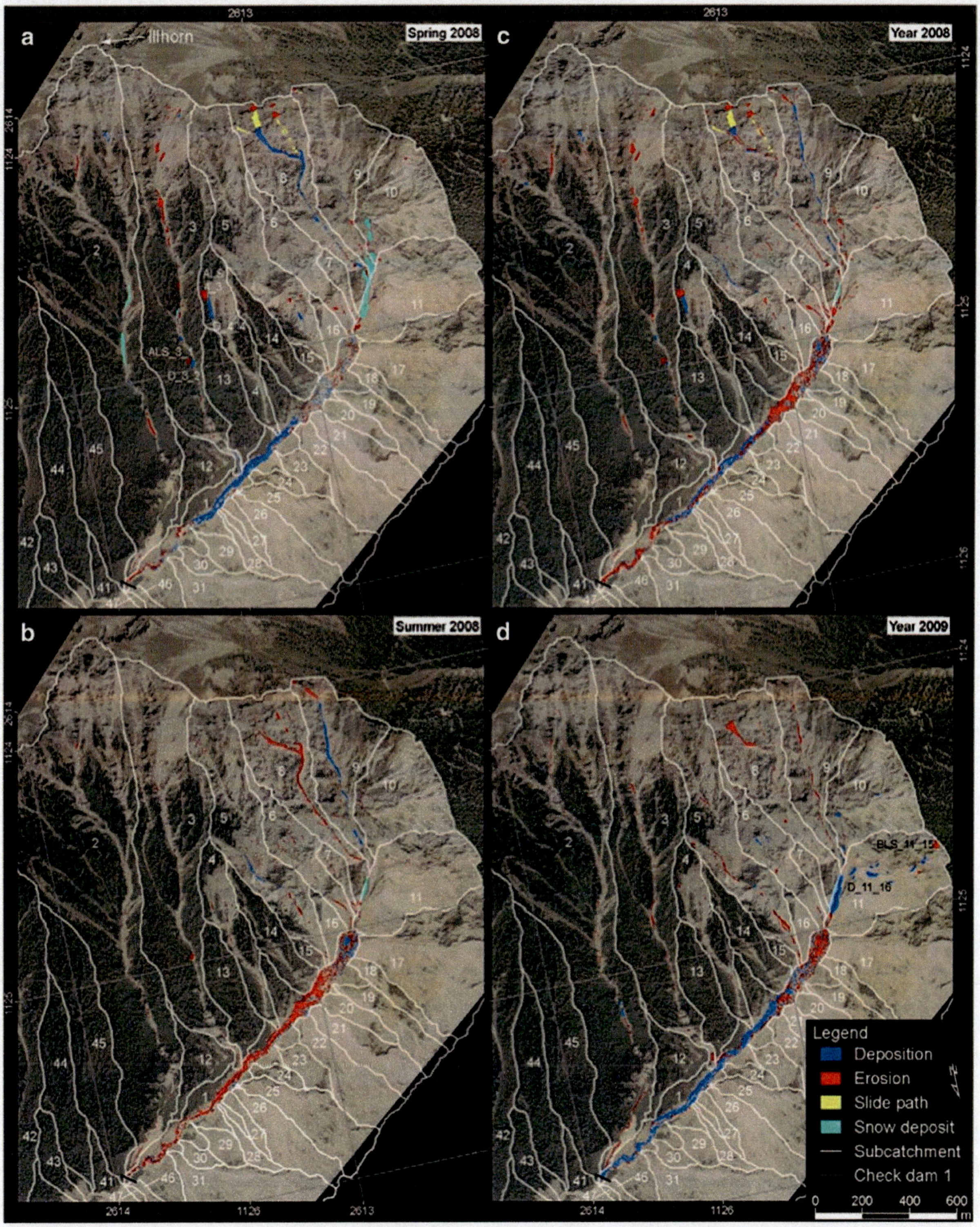

Fig. 25.7 Patterns of erosion (red) and deposition (blue) for **a**. spring 2008; **b**. summer 2008; **c**. year 2008; **d**. year 2009. The background image is the orthophoto from 2008 (from Berger et al. 2011b)

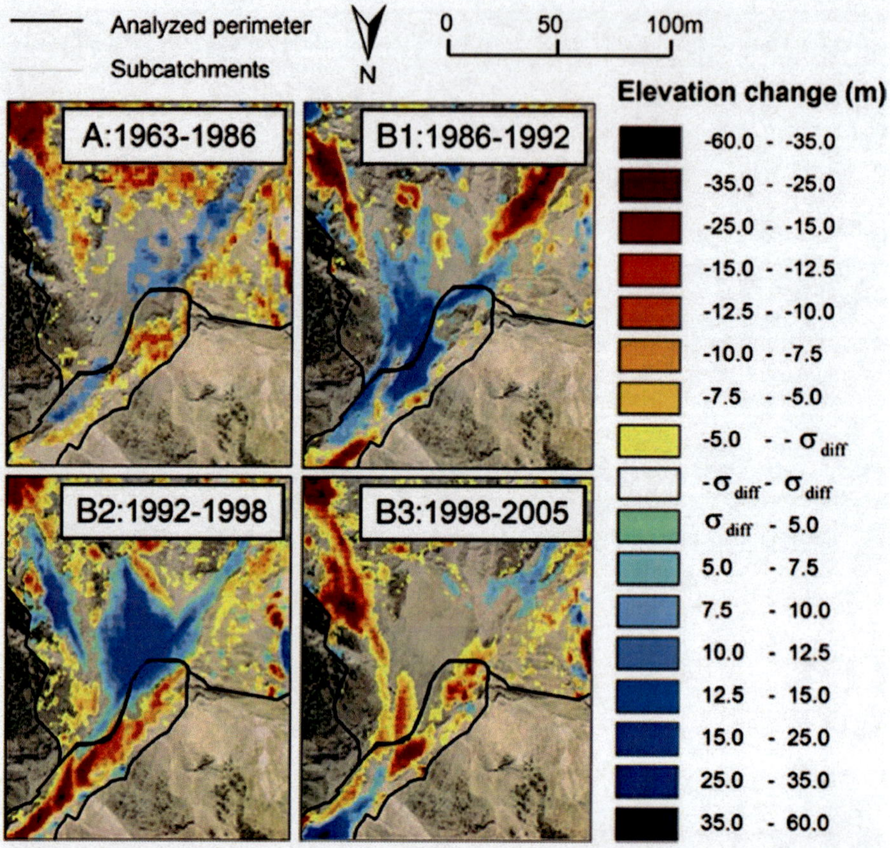

Fig. 25.8 Patterns of erosion (red) and deposition (blue) based on photogrammetric analyses (from Bennett et al. 2013, based on data from Bennett et al. 2012)

Fig. 25.9 Debris-flow levees at a tributary to the main Illgraben channel (levee spacing is ~3 m) (photo B. McArdell)

Fig. 25.10 Debris-flow lobe deposit (thickness ~2 m) in the channel upstream of the fan apex (photo B. McArdell)

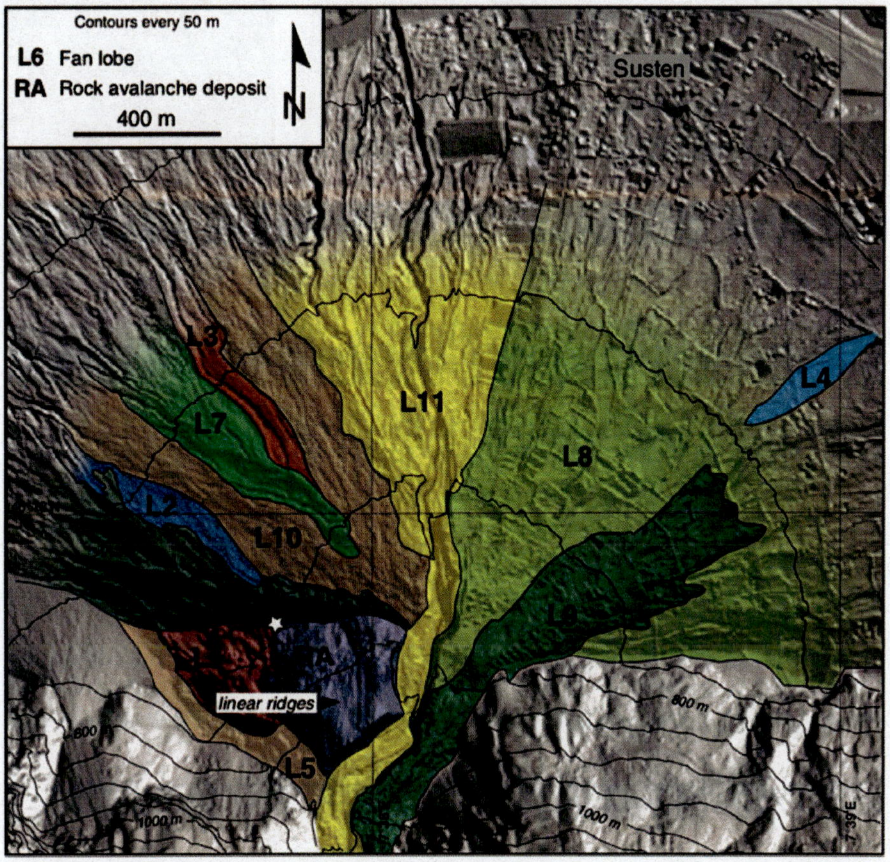

Fig. 25.11 Depositional lobes determined by field mapping and in comparison with an analysis of the digital terrain model (from Schürch et al. 2016)

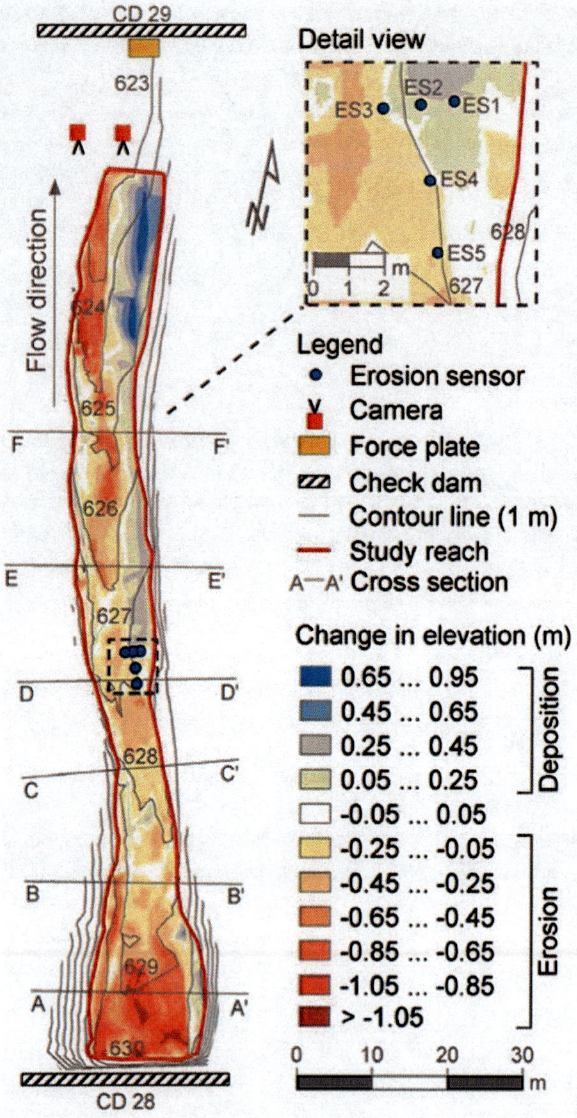

Fig. 25.12 Patterns of erosion (red) and deposition (blue) for a large debris flow on July 1, 2008 (from Berger et al. 2011a from Bennett et al. 2012)

Fig. 25.13 **a**. Warning light and siren at a footpath near check dam 27 (see location on Fig. 25.1), view downstream; **b**. View upstream at a typical Illgraben check dam (CD27; (see location on Fig. 25.1) with a footpath crossing the channel upstream of the check dam. A scale is visible on the right side; each black or white bar is 0.5 m tall (photos B. McArdell)

The elevation of the channel bed may vary significantly through time due to erosion or deposition within the channel itself (Fig. 25.12), deposition of levees at the margins of the active channel or, more rarely, cessation of a debris-flow surge within the channel.

The Illgraben debris-flow observation station has recorded approximately 75 debris flows or debris floods between 2000 and 2018. Typical debris-flow depths, at the front of the flow, are 1–2 m (McArdell et al. 2007; Berger et al. 2011a; Schürch et al. 2011; McArdell 2016) and typical front velocities are 1–5 m/s, although some surges have been observed to temporarily stop in the channel and re-activate as secondary surges reaching the front of the flow (McArdell 2016). Debris flows at the Illgraben generally have a relatively steep boulder-rich front, although events with smaller boulder concentrations are also common. Typical bulk densities of debris flows, measured using a large force plate installed in the channel bed at the distal end of the Illgraben channel (McArdell et al. 2007; Schlunegger et al. 2009) are typically 1800–2200 kg/m^3, and the typical event volume is $\sim$20,000 m^3.

25.5 Outlook: The Illgraben, a Hazardous System Under Surveillance

The large degree of debris-flow activity at the Illgraben provides an unprecedented opportunity to investigate the underlying geomorphic processes, especially as new measurement technologies are introduced and new research paradigms arise. The high frequency of debris flows has led local authorities to construct an Early Warning System (Badoux et al. 2009). The goal of the Early Warning System is to alert authorities to debris-flow occurrence and to prevent tourists from entering the active channel when conditions could be hazardous. Warning signs have been installed at regular intervals along the channel on the fan, and tourists are provided with information about the hazard. When an event is detected within the channel (using a combination of flow depth and ground-vibration intensity), a warning, consisting of a siren and a blinking light (Fig. 25.13a, is automatically issued at all major footpaths which cross the channel, e.g., in the vicinity of check dam 27 (Fig. 25.13b. During the first year of operation, the warning system successfully detected 19 hazardous floods and debris flows within the channel, with only one false alarm (Badoux et al. 2009). Recent advances in measurement technologies (e.g., seismic or infrasound) should make it possible to detect landslides and debris-flow initiation without having to install equipment in hazardous areas (e.g., Burtin et al. 2014; Walter et al. 2017).

Acknowledgments The research at the Illgraben has been greatly facilitated by the innovative ideas and competent assistance from WSL technical staff, including B. Fritschi, S. Boss, and C. Graf. Many of the ideas presented herein resulted from interesting discussions with collaborators. We regret if any such ideas have not been properly acknowledged in the citations herein. We are also grateful for the helpful comments and constructive editing by E. Reynard.

References

Badoux A, Graf C, Rhyner J, Kuntner K, McArdell BW (2009) A debris-flow alarm system for the Alpine Illgraben catchment: design and performance. Nat Hazards 49(3):517–539

Bardou E, Jaboyedoff M (2008) Debris flows as a factor of hillslope evolution controlled by a continuous or a pulse process? In: Gallagher K, Jones SJ, Wainwright J (eds) Landscape evolution: denudation, climate and tectonics over different time and space scales. Geological Society, London, Special Publications 296, pp 63–78

Bennett GL, Molnar P, Eisenbeiss H, McArdell BW (2012) Erosional power in the Swiss Alps: characterization of slope failure in the Illgraben. Earth Surf Process Landf 37(15):1627–1640

Bennett GL, Molnar P, McArdell BW, Schlunegger F, Burlando P (2013) Patterns and controls of sediment production, transfer and yield in the Illgraben. Geomorphology 188:68–82. https://doi.org/10.1016/j.geomorph.2012.11.029

Berger C, McArdell BW, Schlunegger F (2011a) Direct measurement of channel erosion by debris flows, Illgraben. Switzerland Jour Geophys Res 116:F01002. https://doi.org/10.1029/2010jf001722

Berger C, McArdell BW, Schlunegger F (2011b) Sediment transfer patterns at the Illgraben catchment, Switzerland: implications for the time scales of debris flow activities. Geomorphology 125(3):421–432

Burtin A, Hovius N, McArdell BW, Turowski JM, Vergne J (2014) Seismic constraints on dynamic links between geomorphic processes and routing of sediment in a steep mountain catchment. Earth Surf Dyn 2(1):21–33. https://doi.org/10.5194/esurf-2-21-2014

Caduff R, Kos A, Schlunegger F, McArdell B, Wiesmann A (2014) Terrestrial radar interferometric measurement of hillslope deformation and atmospheric disturbances in the Illgraben debris-flow catchment, Switzerland. Geosci Remote Sens Lett 11:434–438. https://doi.org/10.1109/lgrs.2013.2264564

Campani M, Mancktelow N, Seward D, Rolland Y, Müller W, Guerra I (2010) Geochronological evidence for continuous exhumation through the ductile-brittle transition along a crustal-scale low-angle normal fault: Simplon Fault Zone, central Alps. Tectonics 29(3):1–25

Champagnac JD, Sue C, Delacou B, Burkhard M (2003) Brittle orogen-parallel extension in the internal zones of the Swiss Alps (south Valais). Eclogae Geol Helv 96(3):325–338

Escher A (1988) Structure de la nappe du Grand Saint-Bernard entre le val de Bagnes et les Mischabel. Service hydrologique et géologique national, Berne (Rapport géologique 7)

Franke D, Hornung J, Hinderer M (2015) A combined study of radar facies, lithofacies and three-dimensional architecture of an alpine alluvial fan (Illgraben fan, Switzerland). Sedimentology 62:57–86. https://doi.org/10.1111/sed.12139

Gabus JH, Weidmann M, Burri M, Sartori M (2008) Feuille 1287 Sierre. Atlas géol Suisse 1:25,000. Notice explicative 11

Giardini D, Wiemer S, Fäh D, Deichmann N (2004) Seismic Hazard Assessment of Switzerland 2004. Swiss Seismological Service Report, Zurich

Hungr O, Leroueil S, Picarelli L (2014) The Varnes classification of landslide types, an update. Landslides 11(2):167–194

Hürlimann M, Rickenmann D, Graf C (2003) Field and monitoring data of debris-flow events in the Swiss Alps. Can Geotech J 40:161–175

Lichtenhahn C (1971) Zwei Betonmauern: die Geschieberückhaltesperre am Illgraben (Wallis) und die Staumauer des Hochwasserschutzbeckens an der Orlegna im Bergell (Graubünden). International Symposium Interpraevents. Villach, Kaernten, Austria, pp 451–463

Maurer HR, Burkhard M, Deichmann N, Green AG (1997) Active tectonism in the central Alps: contrasting stress regimes north and south of the Rhône Valley. Terra Nova 9(2):91–94

McArdell BW, Bartelt P, Kowalski J (2007) Field observations of basal forces and fluid pore pressure in a debris flow. Geophys Res Lett 34:L07406. https://doi.org/10.1029/2006gl029183

McArdell BW (2016) Field measurements of forces in debris flows at the Illgraben: implications for channel-bed erosion. Int J Eros Control Eng 9(4):194–198. https://doi.org/10.13101/ijece.9.194

Oppikofer T, Bardou E, Jaboyedoff M (2006) DTM analysis of the 1961 rock avalanche, Illgraben creek (Switzerland). European Geosciences Union, Geophysical Research Abstracts 8, 02291

Scheiber T, Pfiffner OA, Schreurs G (2013) Upper crustal deformation in continent-continent collision: a case study from the Bernard nappe complex (Valais, Switzerland). Tectonics 32(2):1320–1342

Schlunegger F, Badoux A, McArdell BW, Gwerder C, Schnydrig D, Rieke-Zapp D, Molnar P (2009) Limits of sediment transfer in an alpine debris-flow catchment, Illgraben. Switzerland. Quat Sci Rev 28(11–12):1097–1105

Schürch P, Densmore AL, Rosser NJ, McArdell BW (2011) Dynamic controls on erosion and deposition on debris-flow fans. Geology 39 (9):827–830

Schürch P, Densmore AL, Ivy-Ochs S, Rosser NJ, Kober F, Schlunegger F, McArdell B, Alfimov V (2016) Quantitative reconstruction of late Holocene surface evolution on an alpine debris-flow fan. Geomorphology 275:46–57. https://doi.org/10.1016/j.geomorph.2016.09.020

Stoffel M, Bollschweiler M, Leutwiler A, Aeby P (2008) Tree-ring reconstruction of debris-flow events leading to overbank sedimentation on the Illgraben cone (Valais Alps, Switzerland). Open Geol J 2:18–29

Walter F, Burtin A, McArdell BW, Hovius N, Weder B, Turowski JM (2017) Testing seismic amplitude source location for fast debris-flow detection at Illgraben, Switzerland. Nat Hazards Earth Syst Sci 17(6):939–955. https://doi.org/10.5194/nhess-17-939-2017

Brian W. McArdell is a Geologist and scientific staff member at the Swiss Federal Institute for Forest, Snow and Landscape Research (WSL) in Birmensdorf, Switzerland, since 2001. He is the scientific coordinator for the Illgraben debris-flow observation station. His research interests include debris-flow process research in the field and laboratory, and applied research on debris flow and landslide runout.

Mario Sartori is Senior Lecturer at the Section of Earth and Environmental Sciences of the University of Geneva, and a member of the Swiss Geological Commission. He carried out extensive geological mapping in the Alps, especially in the Valais. His research concerns mainly structural geology and natural hazards.

The Landslide of Campo Vallemaggia

26

Luca Bonzanigo

Abstract

The deep-seated Campo Vallemaggia landslide is located in the southern Swiss Alps. It covers about 6 km^2 and reaches the depth of about 300 m, incorporating up to 800 Mm^3 of strongly weathered and broken metamorphic rocks. It does not form a single body but it is divided into blocks, which follow pre-existing fault zones. A total displacement of about 30 m has been measured since the end of nineteenth century, with an average of about 5 cm/a, with a pulsing character. It means that events of rapid movement are separated by relatively quiet periods. The cause of the sliding resides in water pressures inside and under the sliding mass, reaching artesian heads. The activity has been stopped in 1995 by means of a drainage system through the stable rockmass under the slide.

Keywords

Deep-seated landslide • Artesian head • Metamorphic rocks • Viscosity • Campo Vallemaggia

26.1 Introduction

The deep-seated landslide of Campo Vallemaggia is situated in the Rovana River catchment, a tributary of the Maggia River, in the canton of Ticino, South Switzerland (Fig. 26.1). It covers a surface of about 6 km^2, from behind the crest of Grosshorn-Pizzo Bombögn (46°18′ N, 8°28′ E) at about 2100 m a.s.l. to the Rovana River (46°17′ N, 8°30′ E) at 1100 m a.s.l. (Fig. 26.2). This slope instability phenomenon is extraordinary in many ways, first of all for the great size (volume of approximately 800 Mm^3; thickness reaching 300 m; Fig. 26.3a). Its recurrent activity was of great concern not only for the two villages of Campo Vallemaggia and Cimalmotto on the slide itself, but for the whole Maggia Valley down to the town of Locarno, on the shore of Lake Maggiore (called Lake Verbano in Italy). The displacement records show a pulsing or "stick-slip" behaviour, with sudden accelerations associated with periods of intense precipitations, separated by relatively quiet periods. Interesting is that heavy precipitations do not always result in acceleration and the greatest displacements are not always related to the more severe rainfall events. The artesian hydraulic heads observed in boreholes and in some springs led to consider high pore pressure as the cause of instability. This chapter presents the geological, geomorphological, geomechanical and hydrodynamical characteristics of the landslide, based on the detailed investigation carried out between 1982 and 1995. The mitigation measures that were undertaken and their results are also described.

26.2 Geographical and Geological Setting

Two inhabited areas were affected by the damage due to the slope instabilities, Campo Vallemaggia and Cimalmotto, both typical villages of mountain farming (Fig. 26.2). They had populations of some hundreds of people until the mid-twentieth century and, as most of the villages in the region, they are nowadays permanently inhabited only by a few tens of people each. The climate is harsh in winter but very nice in summer, so the site is beloved by tourists and as a secondary home area for the region of Locarno. The Quadrella pass at 2137 m a.s.l. links Campo Valle Maggia to Bosco Gurin in the valley adjacent to the north.

The Campo Vallemaggia landslide is located within the crystalline zone of the Penninic nappes of the southern Swiss Alps. The bedrock is composed of a metamorphic series of amphibolites and micaceous schists, gneisses presenting all sorts of textures and mineralogy, ultramafic metaperidotites and metacarbonates (Grütter 1929; Hall 1972) (Fig. 26.4).

L. Bonzanigo (✉)
Consultant geologist, geolog.ch SA, 6501 Bellinzona, Switzerland
e-mail: luca.bonzanigo@geolog.ch

E. Reynard (ed.), *Landscapes and Landforms of Switzerland*, World Geomorphological Landscapes,
https://doi.org/10.1007/978-3-030-43203-4_26

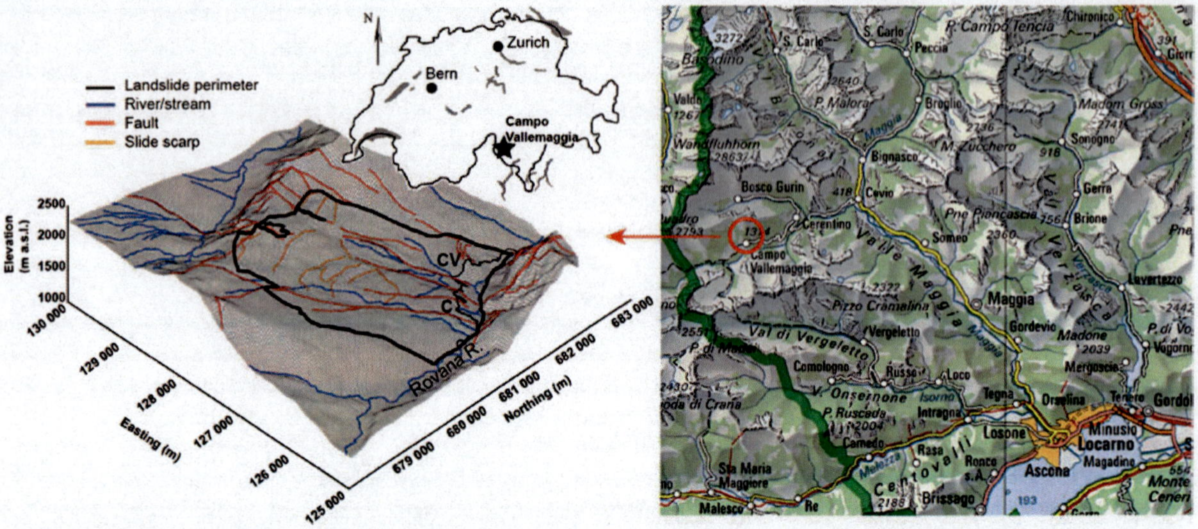

Fig. 26.1 Location and digital elevation model of the Campo Vallemaggia landslide. Reference is made to the Swiss coordinate system. The latitude and longitude of the centre point of the landslide is approximately 46°17′ N, 8°29′ E. CV = Campo Vallemaggia; C = Cimalmotto (modified after Bonzanigo et al. (2007))

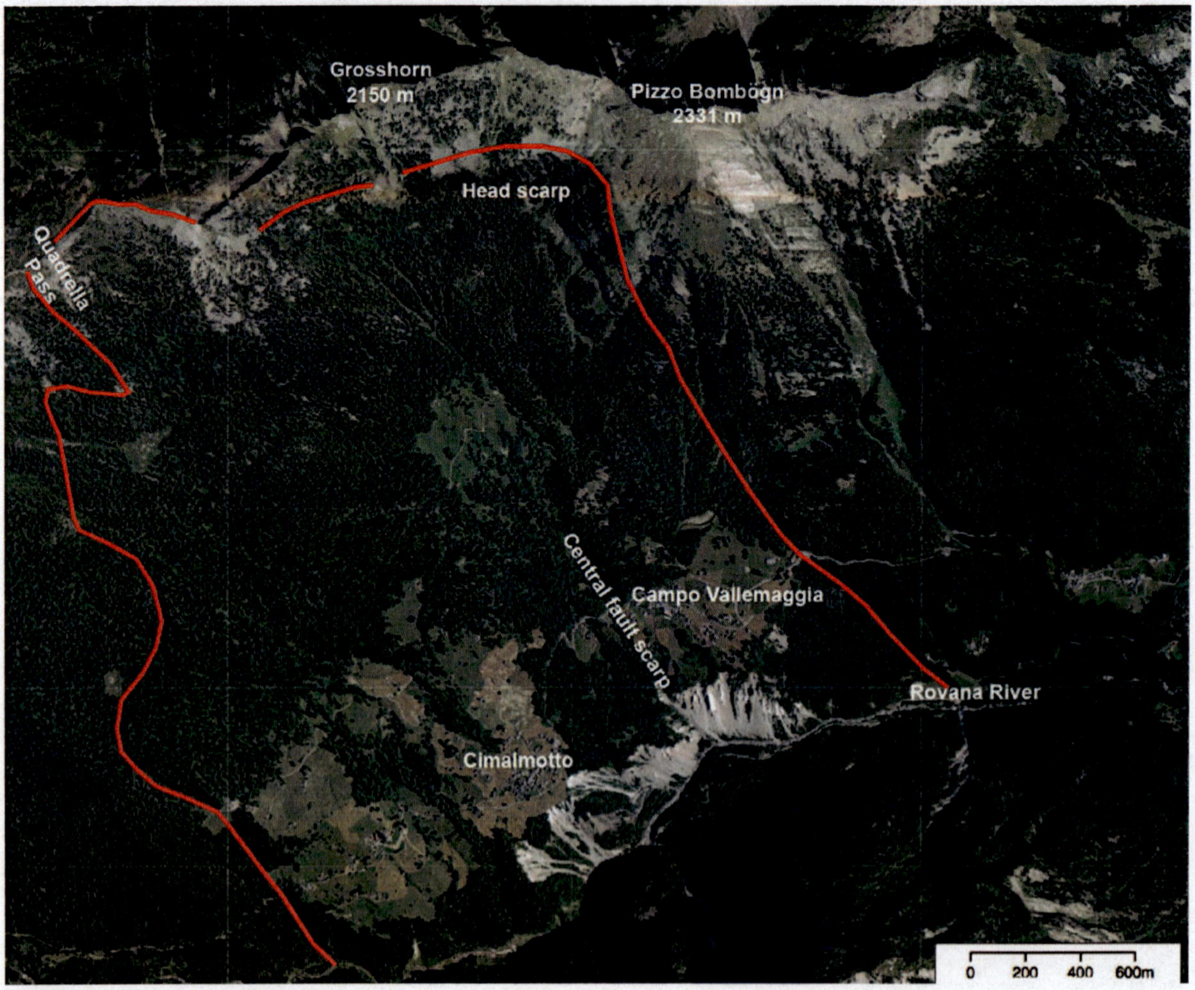

Fig. 26.2 Aerial photograph of the landslide with the 150 m high scarp on the left side of the Rovana River and the approximative limit of the landslide in red (*source* Swisstopo)

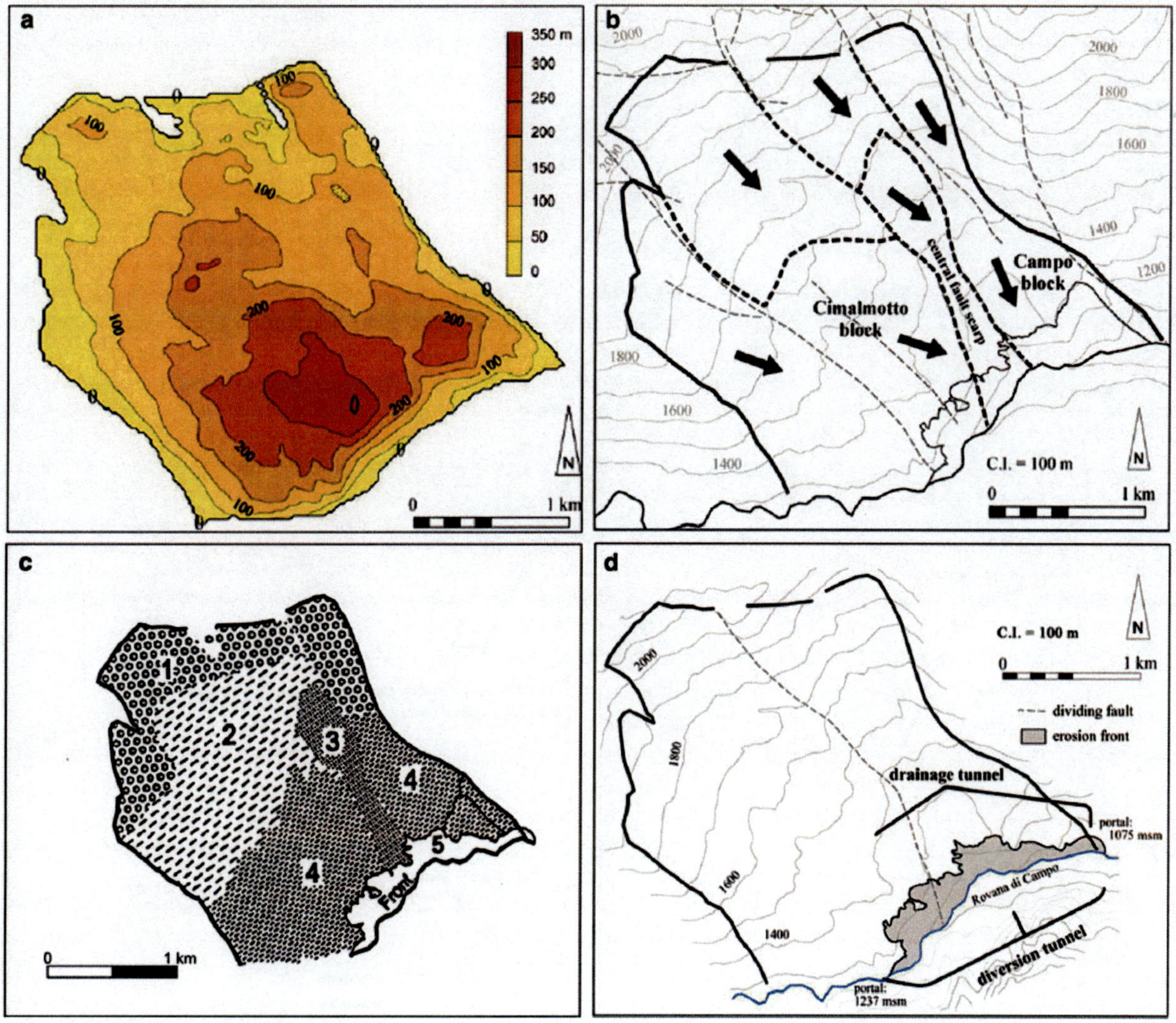

Fig. 26.3 **a** Estimated thickness of the displaced mass based on geodetic survey, boreholes and seismic prospection; **b** Geodetically independent areas. This subdivision of the sliding mass is correlated to existing tectonic features of the stable rockmass; **c** Subdivision of geomorphologically differentiated areas (see text for the numbers); **d** Location of the two adits excavated, one (diversion tunnel) to deviate the flow of the river from the toe, the other (drainage tunnel) to drain the artesian pressures observed inside and under the sliding mass

The main structures are controlled by the isoclinal folding-foliation dipping 20–30° to SSE. Intense subvertical brittle faulting divides bedrock along a NNW-SSE main alignment (Hall 1972) (Fig. 26.4). Minor joint systems are also observed, for example one dipping about 80° towards N or NE.

The slide mass itself is composed of the same lithology, but in a severely weathered form. The structures are also displaced or rotated (Fig. 26.4b). Some particular lithologies, like massive gneisses or metaperidotites, are only disturbed by intense fracturing, but remain sound in their matrix (Fig. 26.5). The sliding mass is subdivided into distinct blocks, their displacements being partially independent of one another (Fig. 26.3b). The longitudinal separation is controlled by pre-existing tectonic features (Heim 1932; Bonzanigo 1999; Bonzanigo et al. 2007; Fig. 26.4). The transversal subdivision is instead controlled by fracturing deriving from the post-glacial relaxation, and by the orientation of schistosity. The displacements are generally aligned to the direction and dip of the metamorphic foliation. This type of compound sliding masses is typical for deep-seated landslides in the crystalline environment (Cronin 1992).

At the regional level, the hydrogeological circulations are typical of crystalline metamorphic environments, the hydraulic conductivity being the result of discontinuities rather than porosity. Schistosity plays a secondary role. Already in the nineteenth century, the Swiss geologist Albert Heim suspected water pressure to be responsible for

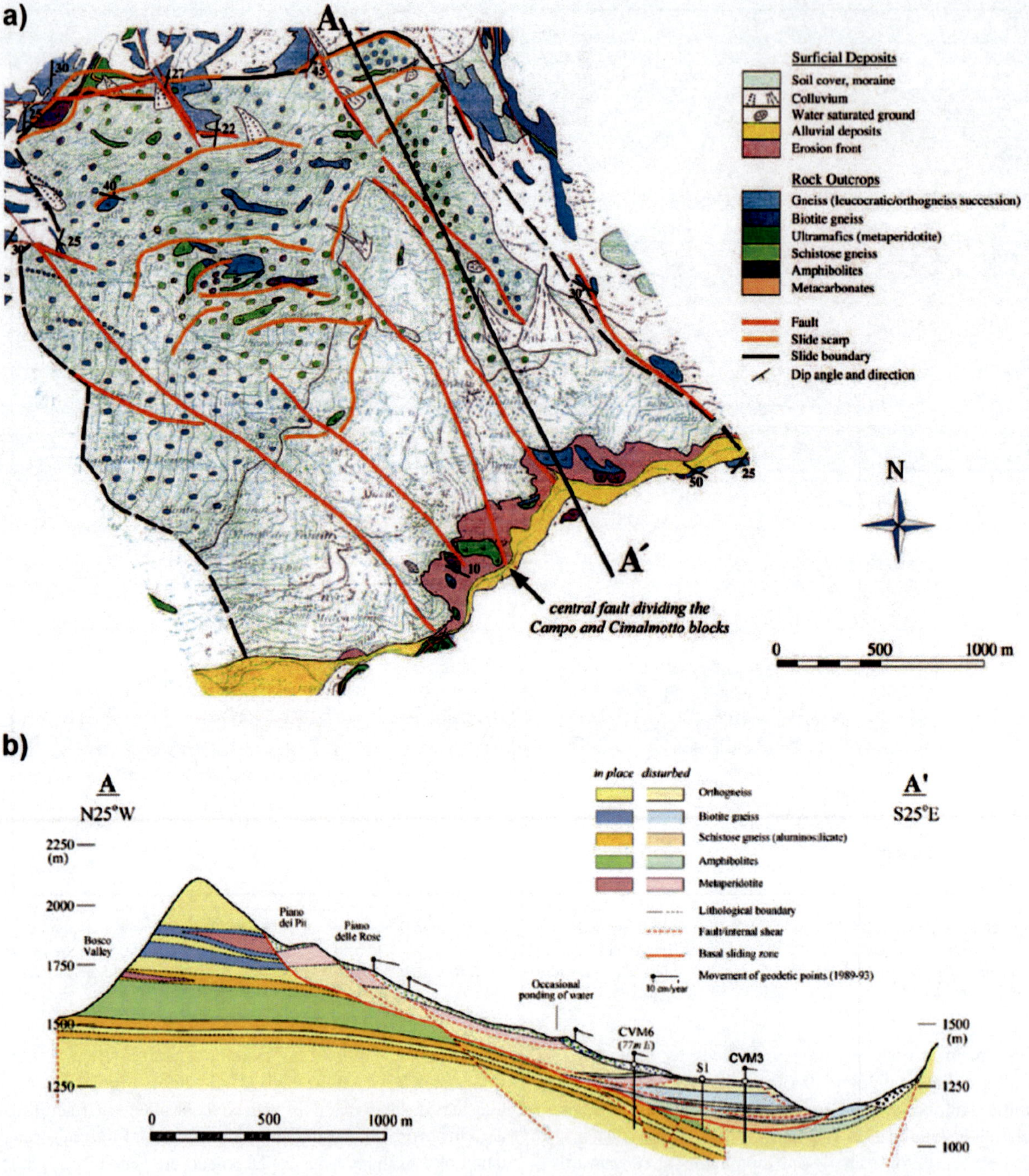

Fig. 26.4 **a** Geological map of the landslide; **b** Simplified geological section through the landslide. The sound rock formations and the displaced ones are differentiated, but the apparently loose material conserves the original organisation of the different rock types. Vectors show the vertical orientation and the amount of displacement between 1989 and 1993 (modified after Bonzanigo et al. 2007)

landslide movement and associated damages, basing on the observation of some springs showing amazing behaviour, as described with some animistic fear by the locals (Heim 1897, 1932). In 1962, a first borehole showed the presence of two zones of artesian pressure at 80 m and 120 m depth, respectively. Other boreholes drilled in 1991 (Bonzanigo 1999) also highlighted high pore pressure heads, both in the slide body and in discontinuities of the underlying bedrock.

Fig. 26.5 Front of the erosion scarp with the evidence of differential weathering between amphibolites and schistous gneisses (photo L. Bonzanigo)

Other hydrogeological investigations were also performed, for instance comparison of Oxygen and Deuterium isotopes equilibrium with Standard Mean Ocean Water (SMOW) to estimate the elevation origin of spring water. Balance of water flows with rainfall regime and hydrochemical investigations also supported the conclusion that at least part of the water of the artesian aquifers comes from the adjacent valley of Bosco Gurin. Finally, a hydrogeological model, which takes into account the strong anisotropy of hydraulic conductivity, was developed (Bonzanigo 1999; Bonzanigo et al. 2007). Numerical modelling was conducted at the British Columbia University after the construction of the drainage adit (see below) and used data collected before, during and after the drainage (Bonzanigo et al. 2007).

26.3 History

Descriptions of damage are reported since 1780, when the region was probably temporarily abandoned (Mondada 1977). Between 1834 and 1839, losses of pasture land were reported after heavy floods. Timber logging began in 1857 and appeared to be dangerous for the stability of the erosion scarp at the bottom of the landslide. Timber exploitation was therefore forbidden after a couple of years. For a long time, until about 1990, logging activity of that short period was accused of being the cause of all the damages. In 1897, after deep erosion episodes associated with losses of land and housing, and after the failure of the construction of transverse check dams to protect the base of landslide from erosion, the professor Albert Heim of the Swiss Federal Institute of Technology (ETH Zurich) was called by the federal authorities to investigate. He is the first who suspected underground water to be responsible for damage. The first geodetic measurements—only at two points, on two churches—were undertaken with heavy theodolites in 1892. The average rate of deformation was measured as being approximately 30 cm/a, with a minimum of 1 cm/a and a maximum of 5 cm/day in 1940 (Bonzanigo et al. 2000). The period 1938–1942 was the peak of activity (Fig. 26.6). The recent history is characterised by increasing investigations in the short times of acceleration, associated with great political interest, followed by forgetfulness at the end of emergency situation.

In the meanwhile, people left the place, attracted by more profitable activities in lowland than upland agriculture. Exceptional precipitations in 1978 produced some losses of land but no noticeable deformations at the surface (Fig. 26.6). The geodetic measurements (see below) did not record important movements. In 1987, after a short period of rainfall, less severe than in 1978, up to 1 m/a displacement was recorded. This induced new investigations, detailed mapping and deep borehole drilling (Bonzanigo 1990a, b, 1999).

26.4 Landforms and Landscapes

The slide is divided into five morphological elements (Fig. 26.3c; see also Fig. 26.4):

1. The head scarp is composed of large intact blocks of rock. Morphology is typical of a slow, creeping landslide or sagging. Open tension cracks can be seen on the other side of the crest.
2. In the upper and central slide mass, large intact blocks are very fractured and in part back-rotated. They are mostly covered by glacial till and dry colluvium. Open fractures between moving blocks are often filled with weathered material. The transverse fractures confer a very irregular morphology to the slope and rain can easily infiltrate. The wide open fractures guide caverns that were used in the past by the locals as cellars for food and wine. Where fractures are sealed by fine-grained material ponds and groves are formed.
3. The central fault scarp (see also Fig. 26.5), delimited by two relict faults, is the fastest part of the slide.
4. The lower part of the slide body forms a flattish terrace, where the two villages lie. It is formed by highly fractured and weathered rock and the surface is covered by a few metres of glacial deposits, slightly consolidated.
5. The toe of the slide is heavily eroded by the river and forms an impressive 150-m high and about 2.5 km long scarp (see also Fig. 26.2). Here the strongly weathered rock mass looks like soil, except some hard massive gneiss and amphibolite fragments. The shape of the scarp has changed continuously due to displacements and erosion.

26.5 Geodesy and Geomechanical Model

Geodetic measurement data are available for more than one century, the first records being from 1892 (Heim 1897). They show a horizontal displacement up to 30 m and vertical movement of one to two metres, depending on the exact location. Each time some damage was recorded, new investigations were undertaken, generally limited to the implementation of new geodetic points. This is why the quantity of geodetic information is so unusually large. In addition, in 1989, an automatic geodetic station was installed, which allowed for daily observation of the movements.

The analysis of geodetic data highlights the pulsing mechanism of the slide (Fig. 26.6). The periods of rapid displacement are not directly correlated to the amount of rainfall. This led to the search of a triggering mechanism related to the pore pressure inside and under the sliding mass rather to the intensity of precipitation.

The peculiar aspect of the Campo Vallemaggia landslide and similar phenomena resides in the fact that they are never completely still. In such cases, one has to accept that the terrain moves anyway and that the main issue is its velocity. If it moves centimetres per year or even more, the slide is destructive. If the displacements are limited to millimetres per year, nobody and nothing suffers. The proposed approach for the description of the mechanism of deep-seated landslides is based on the viscosity model (Vuillet and Hutter 1988; Vuillet and Bonnard 1996), where the varying parameters of the adopted physical laws rule the velocity and the "nervosity" of the phenomenon, namely its tendency to behave in a gentle and self-braking manner, or catastrophically (Bonzanigo 1999; Bonzanigo et al. 2007).

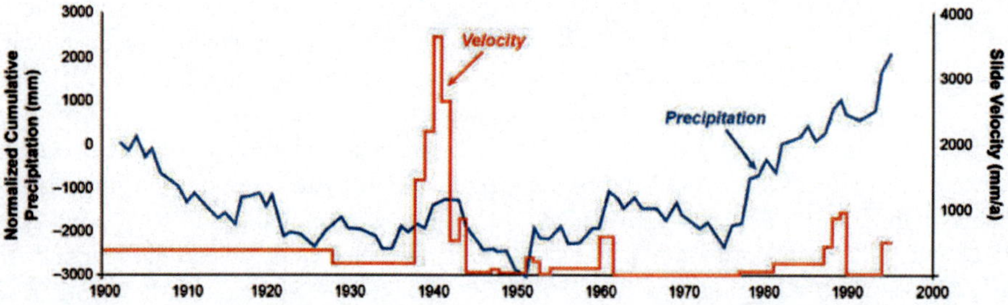

Fig. 26.6 Comparison between rainfall and landslide velocity for the period 1900–1995. The annual cumulative precipitations are normalised to a yearly average of 1825 mm/year. Displacement velocities are based on geodetic measurements of the Campo block

26.6 Mitigation

Mitigation attempts have been conducted since the end of the nineteenth century. The first measure was the construction of a series of rock-boulder weirs and check dams along the Rovana River in front of the slide mass. After one year, the check dams were completely destroyed and swept away. Intensive planting of alder trees and bush was also undertaken, as well as control of surface runoff by means of diversion wooden channels, with some local benefit to surface structure, but none to the sliding mass displacements. Finally, after extensive studies, a deep drainage solution was claimed (Bonzanigo 1990a, b; Lombardi 1996; Bonzanigo et al. 2007), as already proposed by Albert Heim ninety years earlier and by Carlo Liechtenhahn later (Liechtenhahn 1971). Yet despite a long history of no success, many still favoured a solution focused on erosion control. This led to a typically Swiss decision to implement both options. The first involved the construction of a diversion tunnel on the opposite side of the valley (Fig. 26.3d) to redirect the river away from the erosion front of the landslide. At the same time, in 1993–1995, a deep drainage adit (Fig. 26.3d) was drilled in the stable rockmass under the unstable slide, from which drainage boreholes could be drilled upwards into the base of the slide body.

The result of the drainage adit was a nearly immediate stop of the displacements that could be observed thanks to the automatic geodesic station (Fig. 26.7). The sank zone of drainage could also later be seen in InSAR picture, showing as the geodetic records, a sink of nearly one metre.

26.7 Conclusions

The Rovana Valley, a side valley of the Maggia River, represents a very spectacular landscape. Unlike the steep walls of most valleys of Ticino that could oppress the visitor unaccustomed to mountain environments, the area of Campo Vallemaggia presents more gentle slopes. But this gentleness is apparent and related to the sliding activity over centuries or even millennia that has shaped the territory in a complex manner.

The landslide is a massive deep-seated creeping slide mass of approximately 800 million m^3 of metamorphic weathered crystalline rock. It is subdivided into distinct blocks by pre-existing tectonic features like faults and fracturing. It reaches a depth of about 300 m and an average deformation of about 30 cm/a during the twentieth century, with several intense events (e.g. at the beginning of the 1940s), before mitigation measures were undertaken in the mid-1990s. The peculiarities of this landslide are its size and the particular deformation processes, ruled by viscosity and high water pore pressure inside and under the sliding mass. These features have been disturbed since 1995 for the interest of human beings, by means of a drainage adit underneath the slide body, and the damaging displacements have almost stopped since that time.

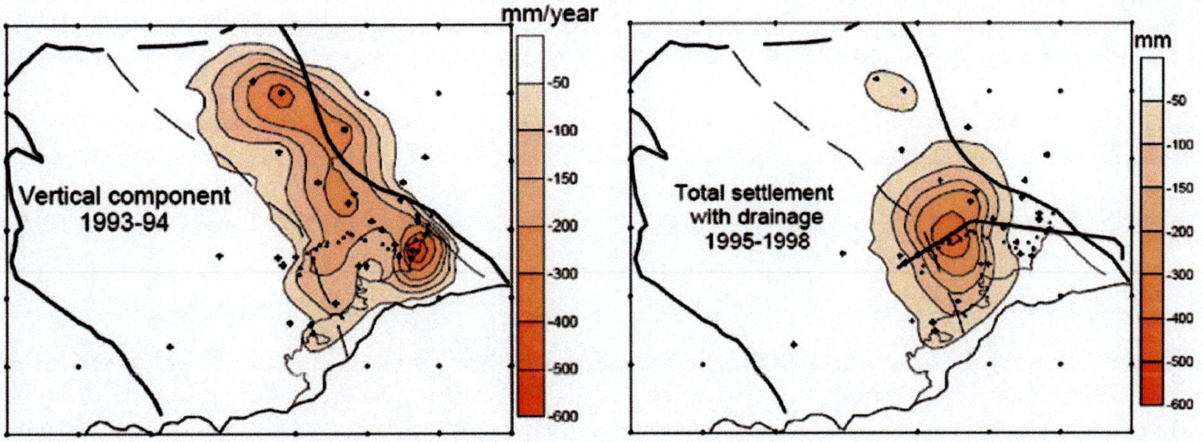

Fig. 26.7 Vertical settlements measured between 1993 and 1994, before the driving of the drainage adit, and between 1995 and 1998, with drainage adit active

Acknowledgments The author thanks Prof. Conrad Schindler and Dr. Giovanni Lombardi for supporting his Ph.D. The former ministers for the environment of the Canton Ticino, Switzerland permitted the investigations by political support. Dr. Giovanni Lombardi supported the author's views though not very fashionable at that time. Pierfrancesco Bertola, engineer at Lombardi Ltd, designed the drainage adit. Prof. Simon Löw of ETH Zurich and Prof. Erik Eberhardt of British Columbia University are acknowledged for their help during the Ph.D. thesis writing and for inviting the author to publish with them.

References

Bonzanigo L (1990a) Etude des mécanismes d'un grand glissement en terrain cristallin: Campo Vallemaggia. In: Proceedings of the 5th International Symposium on Landslides, Lausanne, 1988. Balkema, Rotterdam, pp 1313–1315

Bonzanigo L (1990b) Lo slittamento di Campo Vallemaggia; fenomeni artesiani e presenza di gas in ambiente cristallino. Bulletin der Vereinigung Schweiz. Petroleum-Geologen und -Ingenieure 57(131):65–72

Bonzanigo L (1999) Lo slittamento di Campo Vallemaggia. PhD thesis, Swiss Federal Institute of Technology (ETH Zurich), Zurich, 206 pp. http://e-collection.library.ethz.ch/view/eth:23420

Bonzanigo L, Eberhardt E, Loew S (2000) Measured response to a drainage adit in a deep creeping slide mass. In: Bromhead E, Dixon N, Ibsen ML (eds) Landslides in research. Theory and practice. Proceedings of the 8th International Symposium on Landslides, Cardiff. Thomas Telford, London, pp 151–156

Bonzanigo L, Eberhardt E, Loew S (2007) Long-term investigation of a deep-seated creeping landslide in crystalline rock—Part 1: geological and hydromechanical factors controlling the Campo Vallemaggia Landslide. Can Geotech J 44:1157–1180

Cronin VS (1992) Compound landslides: nature and hazard potential of secondary landslides within host landslides. Geol Soc Am Rev Eng Geol 9:1–9

Grütter O (1929) Petrographische und geologische Untersuchung in der Region von Bosco (Valle Maggia), Tessin. Verhandlungen der Naturforschenden Gesellschaft in Basel 40:78–152

Hall WDM (1972) The structural geology and metamorphic history of the lower penninic nappes, Valle di Bosco, Ticino, Switzerland. PhD thesis, University of London

Heim A (1897) I movimenti di terreno di Campo V.M. Report by Albert Heim (translated from German into Italian by Alberto Totanzi, Cimalmotto). Received as handwritten copy from Dr. Giovanni Lombardi, transcripted and included in PhD thesis of L. Bonzanigo as attachement. http://e-collection.library.ethz.ch/view/eth:23420

Heim A (1932) Bergsturz und Menschenleben. Fretz and Wasmuth Verlag, Zurich, 218 pp

Liechtenhahn C (1971) Zwei Stollenbauten: Stollen im Eis zur Verhinderung von Ausbrüchen eines Sees im Grubengletschergebeit (Wallis) und Stollen im Felsen zur unterirdischen Entwässerung des Rutschgebietes von Campo Vallemaggia (Tessin). Internal Report, unpublished

Lombardi G (1996) Der Drainagestollen von Campo, Rovana. Wasser, Energie, Luft 11(12):281–287

Mondada G (1977) Commerci e commercianti di Campo Valmaggia nel Settecento. Edizioni del Cantonetto, Lugano, 248 pp

Vuillet L, Bonnard C (1996) The Clöwena Landslide: Prediction with a viscous model. In: Senneset K (ed) Proceedings of the 7th International Symposium on Landslides, Trondheim. vol 1, Balkema, Rotterdam, pp 397–402

Vuillet L, Hutter K (1988) Viscous-type sliding laws for landslides. Can Geotech J 25:467–477

Luca Bonzanigo obtained his Ph.D. at ETH Zurich in 1999 and is a Consultant Geologist in Applied geology, Geotechnics, Rock mechanics and Hydrogeology. His main activities are consulting for tunnelling (for instance AlpTransit Gotthard Base Tunnel), landslide mitigation, rockslope instability, groundwater security and blasting in difficult conditions. He is also active as a court expert and president of LHO SIA 106. He is still active when possible in academic environment.

The Flims and Tamins Rockslide Landscape

27

Andreas von Poschinger, John J. Clague, and Nancy Calhoun

Abstract

The landscape of the Rhine River valley between Chur and Ilanz bears the signature of two remarkable events that happened shortly after the end of the Last Glacial Period. Two huge rockslides blocked the river and formed upstream lakes. The Flims rockslide (10 km^3) liquefied about 1 km^3 of alluvial and lacustrine sediments, and the liquefied slurry flowed down the Vorderrhein Valley, over the older Tamins rockslide barrier, and far up the Hinterrhein Valley. The deposit of this slurry, termed Bonaduz gravel, rafted huge masses of rockslide material ('tumas'). Soon after these momentous events, the Vorderrhein and Hinterrhein rivers cut down through the rockslide deposits and the Bonaduz gravel, while Lake Ilanz, impounded behind the Flims rockslide barrier, drained in a series of huge downstream floods. A legacy of these events is the Rhine gorge, which was cut by the Vorderrhein River where it flowed across the Flims rockslide barrier.

Keywords

Flims rockslide • Tamins rockslide • Bonaduz gravel • Tuma • Liquefaction

A. von Poschinger (✉)
Bavarian Environment Agency, Geological Survey, Bgm.-Ulrich-Str. 160, 86179 Augsburg, Germany
e-mail: Andreas.Poschinger@lfu.bayern.de

J. J. Clague · N. Calhoun
Department of Earth Sciences, Simon Fraser University, 8888 University Drive, Burnaby, BC V5A 1S6, Canada
e-mail: jclague@sfu.ca

N. Calhoun
e-mail: nancy.calhoun@oregon.gov

N. Calhoun
Oregon Department of Geology and Mineral Industries, Portland Office, 800 NE Oregon Street, Suite 965, Portland, Canada

27.1 Introduction

For any geoscientist and especially for a geomorphologist, the region between Chur, Ilanz and Thusis in central Grisons is a treasure trove full of fascinating landforms and outstanding outcrops. Here, the largest rockslide in the Alps, the Flims rockslide, has completely remodelled the Vorderrhein Valley. Some kilometres to the east is the deposit of the Tamins rockslide, a smaller, but still impressive rockslide (Fig. 27.1). Together, the two events and especially their interaction have puzzled geoscientists for over a century. The events played out over an area of more than 95 km^2. In view of the large area involved, it is not possible to visit all the features and understand the processes during a short trip. Visitors should take their time to get an adequate impression and appreciation of this remarkable area.

27.2 Geographical and Geological Setting

The Flims and the Tamins rockslides have reshaped the landscape of the Vorderrhein, Hinterrhein, and Rhine valleys (Fig. 27.1). The Vorderrhein and Hinterrhein rivers join at Reichenau and form the Rhine River (*Rhein* in German). The Vorderrhein Valley stretches from the Oberalp Pass on the west in an almost straight line to Reichenau on the east. The Hinterrhein River is sourced near the San Bernardino Pass and flows north to join the Vorderrhein River at Reichenau.

Between Ilanz and Reichenau, forested deposits of the-Flims rockslide obstruct the Vorderrhein Valley, and the Vorderrhein River flows through this barrier in a spectacular, 150–350 m deep gorge (Fig. 27.2). For a long time, the rockslide was an important geographic and cultural barrier, giving rise to the local name Surselva—'above the forest'—for the upper Vorderrhein Valley.

In the Flims-Reichenau area, sedimentary rocks crop out on both sides of the valley. On the south side of the valley are metamorphosed Jurassic and Cretaceous clastic rocks of

E. Reynard (ed.), *Landscapes and Landforms of Switzerland*, World Geomorphological Landscapes,
https://doi.org/10.1007/978-3-030-43203-4_27

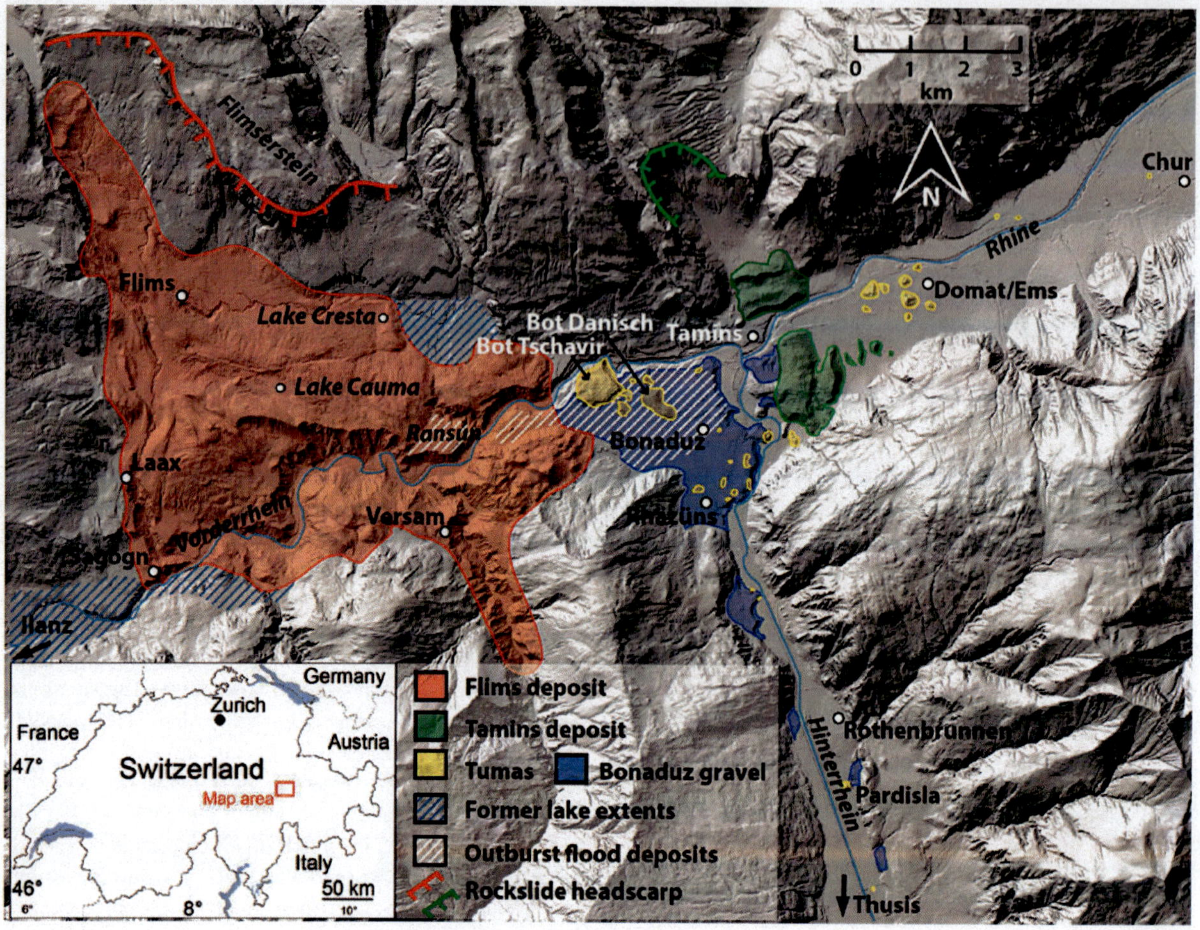

Fig. 27.1 Sketch map of the area near the confluence of the Hinterrhein and Vorderrhein rivers showing the distribution of the Flims and Tamins rockslide deposits and the Bonaduz gravel (digital elevation model: swisstopo)

the Penninic nappe system (so-called 'Bündnerschiefer'). On the north side of the valley, massive grey Late Jurassic and Early Cretaceous limestones of the Helvetic nappe system form high cliffs that dominate the landscape. The Penninic nappes were thrust onto the Helvetic nappes from the late Eocene to the Miocene, producing intensive deformation in the latter. Bedding and tectonic foliation within the Helvetic nappe system dip with about 30° to the south-southeast, parallel to the surface slopes, making these classic 'dip slopes' (see Buckingham and Pfiffner this volume). The pronounced asymmetry of the lower Hinterrhein Valley, with gentle slopes to the west and steep rock faces to the east, reflects the easterly dip in the Bündnerschiefer sequence.

The difference in lithology between Helvetic rocks to the north and Penninic rocks to the south enables a clear tracing of the rockslide material that entered the valleys. Early workers interpreted outcrops of rocks in the Hinterrhein Valley as in situ and typical of the Helvetic realm (Heim 1891; Arbenz and Staub 1910). Later, however, Cadisch (1944), Nabbolz (1954), and particularly Remenyik (1959) argued that these outcrops are isolated non-rooted hills of rockslide debris (tumas in Fig. 27.1). The largest of these Helvetic tumas in the Hinterrhein Valley is Pardisla hill, 2 km south of Rothenbrunnen. A geophysical investigation by Scheller (1970) proved that at least some of the tumas are rootless.

27.3 Landforms and Landscapes

27.3.1 The Flims Rockslide

Albert Heim in 1878 was the first to provide a detailed description of the huge, crushed rock mass in the Vorderrhein Valley near Flims. He concluded that the mass had come down in a single enormous 'stroke' (Heim 1883). Due to the giant dimensions of the slide mass, many later

Fig. 27.2 The Rhine gorge, known locally as "Ruinaulta". The Vorderrhein River has incised through the Flims rockslide barrier. The forested plain located within the meander was produced by overtopping of the barrier by outflow from Lake Ilanz (photo A. von Poschinger)

researchers could not accept this audacious theory and offered the alternative hypothesis that it is in situ bedrock (see Poschinger et al. 2006 for a summary of the controversial discussions). Today, it is accepted that the rock mass detached from the Flimserstein, a nearly 400 m high cliff in Helvetic limestones. The bedding planes dip at an angle of 20–30° towards the valley, making such an enormous failure possible. Using a Digital Elevation Model (DEM), Caprez (2008) calculated the volume of the remaining rockslide mass to be about 8.6–9.3 km^3. He also estimated the volume of the material eroded along the Rhine gorge after the event to be 1.5 km^3, giving a total volume of the failure of about 10–11 km^3. Erratic boulders and even some patches of till are present on top of the rockslide mass, which led some researchers to infer a pre- or syn-glacial age for the event. Absolute dating by the radiocarbon method, however, has shown that the rockslide occurred during the Holocene, about 9500 years ago (Poschinger and Haas 1997; Deplazes et al. 2007). The erratics and patchy till atop the slide mass obviously had been transported, as on horseback, from their former positions high on the slopes above to where they are found today. New findings indicate that some Quaternary sediments not only lie on top of, but also within, the rockslide mass. They must have been squeezed into fissures in the rockslide mass under very high pressure.

The full extent of the Flims deposit is not easy to determine (Fig. 27.1). On the west, a line from Laax to the Sagogn golf course only marks the surface outcrops. From boreholes, Flims rockslide material is known to extend much farther to the west, beyond Ilanz, but in that area it is buried under younger sediments. The eastern boundary is even more difficult to delineate. The rockslide mass is continuous between Sagogn and Trin, but to the east comprises a series of hills surrounded by younger sediments. It is not known if the hills of Bot Danisch and Bot Tschavir are connected to the main rockslide body underground or if they are tumas—isolated hills of rockslide material (Fig. 27.1). These hills, however, are clearly Flims rockslide material. Farther east, to Bonaduz and Rhäzüns, the tumas decrease in size and their connection to the Flims event is less evident. It might also be possible that these are remnants of the Tamins rockslide, remobilised and transported by the Flims-Bonaduz event (see below).

The rockslide material is extremely crushed and shattered (Fig. 27.3). Some parts are a chaotic loose assemblage of blocks set in a more-or-less fine matrix. Other parts reveal the intact former sedimentary bedding, but nevertheless are crushed to centimetre or millimetre size and form a three-dimensional jigsaw puzzle. Although not strongly cemented, the material stands in vertical faces many tens of metres high. One of the most accessible places to view the material is along the road from Bonaduz to Versam (parking between the two tunnels).

The trigger for the detachment of such an enormous volume of rock is unknown; we can only speculate. Obviously, the whole slope was unstable and approaching failure before the event; it needed only a minor impulse to collapse. Failure was possibly triggered by an earthquake, either a

Fig. 27.3 Thoroughly crushed Flims rockslide debris exposed in a gravel pit 2 km west of Bonaduz (photo N. Calhoun)

tectonic one or one induced by another large rockslide, for example the Tamins rockslide. In any case, the Flims rockslide itself certainly caused a severe earthquake. Another explanation might be climate change. The event happened during the Boreal Period, the first warmer and wetter period after the last glaciation (Gruner 2006). A loss of permafrost at higher elevations may have changed the karst-hydrological regime, perhaps producing exceptional water pressures within the bedrock mass.

The obstruction of the Vorderrhein River and its tributaries by the rockslide mass gave rise to several lakes. The city of Ilanz is located in the basin of the largest of these former lakes, so-called 'Lake Ilanz'. Evidence for this lake can be seen in delta sediments exposed in a gravel pit 1 km east of Ilanz and in the gorge of the Laaxerbach south of Laax. Silty, well-bedded lake-bottom sediments are exposed south-east of Sagogn and are visible from the road between Ilanz to Versam on the opposite slope. Similar fine sediments have also been found in boreholes under younger Rhine River fluvial deposits in Ilanz and farther west. All these sediments are below 820 m a.s.l., indicating a long-lived lake at this level, with a volume of about 1.5 km^3 and a length of about 23 km, up to the village of Darvella. Most likely, however, this was not the maximum level of the lake because downvalley outburst sediments indicate that the water was dammed higher up, probably to an elevation of about 900–920 m a.s.l. At that time, the lake had a water volume of 3 km^3 and a length of 29 km, reaching up to the village of Rabius (Poschinger 2005). After several months or a few years, the upper part of the dam breached, presumably shortly after the first overflow and the lake lowered to the 820 m level. The then-lower lake may have persisted for some hundreds of years before emptying by stepwise erosion of the Vorderrhein River to its present position. Eroded remnants of the dam are visible today in the inclined plane of Ransun and can be seen from a viewpoint on the opposite side of the river, 200 m west of the parking area between the two tunnels on the road from Versam to Bonaduz.

The land surface near Bonaduz and south of Reichenau, as well as from Reichenau to Chur shows the effects of catastrophic outbursts from Lake Ilanz. A gently inclined huge fan stretches about 3.5 km east of the breach. Near Bonaduz, the fan transforms into channels eroded into the flat Bonaduz gravel plain. The inclined surface continues to the east and passes through the breach in the Tamins rockslide deposit near Reichenau. Farther east, it continues at least down to Domat/Ems. The material is exposed in the banks of the Rhine River downstream of the dam (west of Domat/Ems) and in excavations in Bonaduz and Reichenau. The exposures show a sandy to bouldery and blocky diamicton with partly rounded components and a maximum block size up to some meters. The fact that this flood deposit overlies the Bonaduz gravel and that it eroded this plain indicates that the outburst events are younger than the Bonaduz gravel.

27.3.2 The Tamins Rockslide

The Tamins rockslide left a deposit on the floor of the Rhine valley at Reichenau. The source area of the rockslide is a deep, amphitheatre-shaped bowl in Late Jurassic limestone of the Helvetic nappe system (Figs. 27.1 and 27.4). The bowl is crowned by the 2000 m high summit of Sennenstein (Fig. 27.5). About 2 km^3 of rock is missing from this amphitheatre. In the literature this rockslide has also been named after the village of Reichenau, the Säsagit summit or the Kunkels pass entering the bowl.

A south-sloping plain separates the head scarp from the outcropping rockslide mass to the south. This plain was formed by infilling of a depression at the trailing edge of the rockslide mass by debris flow deposits that are younger than

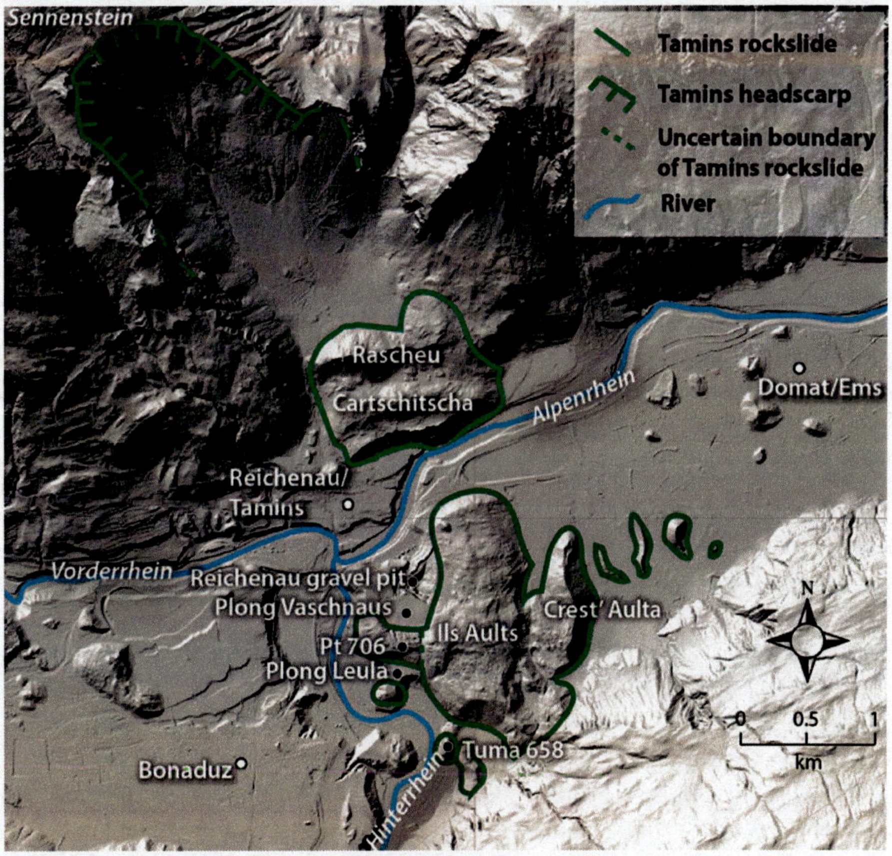

Fig. 27.4 Tamins rockslide headscarp and deposit (green areas) (digital elevation model: swisstopo)

Fig. 27.5 Tamins rockslide headscarp and deposit (forested) (photo N. Calhoun)

the rockslide. A peat bog on this plain bears witness to a former small lake. The plain is bordered on the south by an east-trending ridge (Rascheu; Fig. 27.4). Farther south is a second ridge (Cartschitscha) that is parallel to the first. Both ridges are slide blocks emplaced at the trailing edge of the Flims rockslide (Fig. 27.1). The more northerly ridge is up to 100 m high and 1 km long; the southern ridge has lower relief.

The Rhine River has eroded a 400–500-m-wide and approximately 100-m-deep valley in the Tamins rockslide mass. Nested within this valley are downriver-sloping terraces underlain by mass flow deposits of the Flims rockslide and the younger Lake Ilanz outburst event.

The main part of the Tamins rockslide mass lies south of the Rhine River. It extends for almost 2 km across the Rhine Valley and forms the hills of Ils Aults and Crest' Aulta (Fig. 27.4). The surface of debris sheet in this area is undulating and reaches up to 170 m above the Rhine River. Bonaduz gravel, more than 70 m thick, is inset into the rockslide debris near Reichenau and is also exposed at the western margin of the debris sheet beneath a terrace capped by Hinterrhein fluvial gravel.

A remarkable feature of the Tamins debris sheet south of the Rhine River is that it appears to be disarticulated—it comprises a 'central block' (Ils Aults) and attached and detached smaller blocks to the east, south, and west (Fig. 27.4). A block of fragmented and crushed limestone rockslide debris 500 m wide and 1 km long (Crest' Aulta) forms a semi-detached wing at the east edge of the central block (Fig. 27.4). It is partly covered by a secondary slide mass that connects Ils Aults to Crest' Aulta. Three small, linear tumas consisting of crushed limestone are located 100–800 m east of, and parallel to, Crest' Aulta. Two or three other tumas formed of limestone debris border Ils Aults on the south, and several others lie along the western margin of the central block, just east of the Hinterrhein River (Fig. 27.4). The latter tumas comprise rockslide debris of Helvetic lithologies—Triassic and Jurassic limestone and dolostone, and Permian metamorphosed volcanic rock (albite-chlorite schist). The proximity of these latter tumas to Ils Aults and their position suggest that they may have detached and moved away from Ils Aults due to liquefaction of the Rhine valley fill caused by one or both of the huge rockslides. It is possible, however, that they were rafted during the Bonaduz mass flow event (see below) from the main mass of the Flims debris sheet to the east and came to rest against Ils Aults.

The age of the Tamins rockslide is constrained by its temporal relation to the Flims rockslide and Bonaduz mass flow and by the time of terminal Pleistocene deglaciation. The Tamins rockslide is clearly older than 9500 years, which is the age of the Flims rockslide. The evidence for this relation is that the Tamins rockslide barrier (Ils Aults) split the Bonaduz flow into two lobes, one of which was diverted up the Hinterrhein Valley. The Tamins rockslide happened, however, after the Rhine Valley was deglaciated about 12,000 years ago.

27.3.3 Bonaduz Gravel

The massive Flims rockslide accelerated away from the Flimserstein. The fragmenting rock mass crossed the Vorderrhein River valley at high speed and slammed into the opposite valley wall. The impact raised the pore water pressure within alluvial sediments, liquefying them and sending a huge mass flow, which we refer to as the Bonaduz flow, down the valley.

Abele (1997) estimated that about 1 km^3 of valley-fill material was mobilized by the impact of the rockslide on the Vorderrhein valley floor. The Bonaduz mass flow, which was more than 100 m thick, encountered the barrier formed by the Tamins rockslide and split into the two masses, one of which turned 90° and flowed more than 12 km up the Hinterrhein Valley, rafting huge blocks of rockslide material with it. The other overtopped and burst through the barrier of the Tamins rockslide and continued east down the Rhine River valley at least as far as Chur. Good places to see the Bonaduz gravel are natural outcrops and gravel pits in the Reichenau-Realta region.

In all exposures, the Bonaduz gravel unit fines upward from cobble gravel at the base to granule-rich sand and silt at the top. The pebbles and cobbles are well rounded, yet are separated by a silt-rich matrix (Fig. 27.6). Although the gravel fines upward, rip-up clasts of stratified silt up to several metres across occur throughout the unit. The gravel is also laced with 'Pavoni pipes', which are subvertical tube-like pipes free of silt that formed when the Bonaduz gravel flow came to rest and the entrained water moved upward and out of the sediment (Fig. 27.6). All of these features create a unique deposit that is quite a sight to see!

The villages of Bonaduz and Rhäzüns are located on the Bonaduz plain, a flat to gently undulating surface underlain by Bonaduz gravel (Fig. 27.1). The Hinterrhein River between Rothenbrunnen and Reichenau has incised 40–50 m into the Bonaduz gravel. Tumas are embedded within the Bonaduz gravel along this reach (Fig. 27.1). Near the medieval church of Sogn Gieri, one can see the crushed Helvetic bedrock interior of a tuma with Bonaduz gravel on both sides. In addition, the root of one tuma farther up the Hinterrhein River is exposed in the valley wall. The

Fig. 27.6 Bonaduz gravel exposed in the Reichenau gravel pit; note darker vertical zones (Pavoni pipes) that formed by forceful extrusion of water in the wake of the impact of the rockslide (photo N. Calhoun)

boundaries between tumas and Bonaduz gravel are accentuated by agricultural practices. The former are wooded hills that rise above plowed fields of Bonaduz gravel. Channels cut by the Lake Ilanz outburst floods also incise the Bonaduz plain, and the Hinterrhein River flowed in a shallow channel across the plain after it was blocked by the Bonaduz gravel.

The genesis of the Bonaduz gravel has long been debated. Early researchers considered it to be till or glaciofluvial sediment. In 1968, Nazario Pavoni hypothesized that the Bonaduz gravel was the product of wet gravel slurry mobilized by the impact of the Tamins rockslide. He noted several unique sedimentological features, including the above-mentioned dewatering pipes, ubiquitous upward fining and Helvetic limestone clasts in the Hinterrhein valley, an area where there is only Penninic bedrock. Evidence that the Bonaduz gravel is contemporaneous with the Flims rockslide can be seen at several sites where rounded pebbles and cobbles fill clastic dykes in rockslide debris.

Although the Bonaduz mass flow overtopped and breached the Tamins rockslide barrier, there is no terrace equivalent to the Bonaduz plain down-valley of Reichenau. It is likely that the Bonaduz mass flow deposits downstream of Reichenau are covered by Rhine River alluvium, probably in part reworked from Bonaduz sediments deposited farther upstream.

There has not been liquefaction-induced mass flow on this scale identified elsewhere in the world. This is not to suggest that the event is unique, although the transport of 500 m size tumas over such long distances is, to our knowledge, unprecedented. There are several localities where a rockslide has liquefied its substrate, but not on the scale of the Flims rockslide.

27.3.4 Sequence of Events

The landscape between Ilanz, Chur, and Thusis was shaped by events that unfolded at the end of the Pleistocene and in the early Holocene. Up to now, only the Flims rockslide has been accurately dated; it happened about 9500 years ago. Ivy-Ochs (personal communication, 2014) dated the Tamins rockslide using ^{36}Cl surface exposure techniques and obtained a similar age, albeit with large uncertainties. The sequence of events described below is based on stratigraphic and geomorphologic evidence.

The Tamins rockslide was the first event. Its debris blocked the drainage upstream of Reichenau, creating a lake into which the Vorderrhein River advanced a delta. About 9500 years ago, the Flims rockslide crashed into the Vorderrhein valley and instantly mobilised the water-saturated fluvial, deltaic, and lacustrine sediments upstream of Reichenau. Part of the mobilised valley fill was deflected by the Tamins rockslide barrier and travelled up the Hinterrhein valley, depositing the Bonaduz gravel as far south as almost to Thusis. Another part of the mass flow overtopped and breached the Tamins barrier and swept down the Rhine valley as far as Chur. The flow also transported large masses of rockslide material from the Tamins rockslide, stranding them as tumas (Figs. 27.7 and 27.8). An unusual layer of silt and sand found in two cores taken from

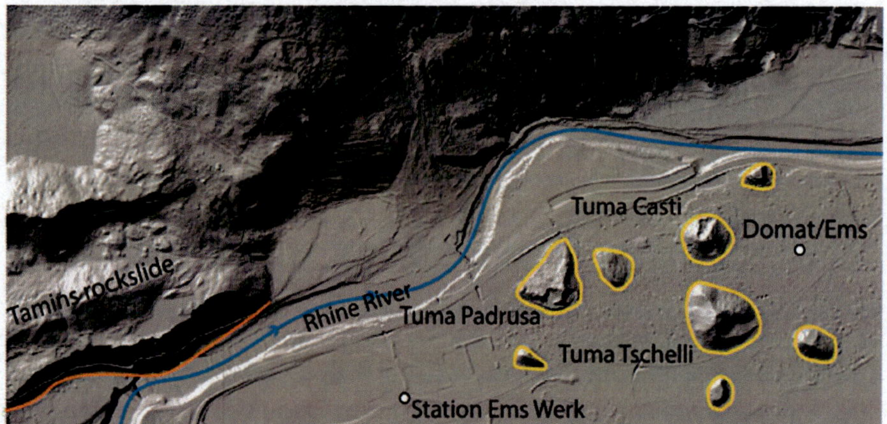

Fig. 27.7 LiDAR-derived hillshade DEM of the Rhine River valley between Tamins and Domat/Ems. Note the breach in the Tamins rockslide (in orange), occupied by the Rhine River (digital elevation model: swisstopo)

Fig. 27.8 Tuma Tschelli in Domat/Ems, seen from Tuma Casti. It is one of the largest tumas. The core is not exposed. It is assumed to be of rockslide material, outcropping in other tumas as e.g. in Tuma Casti (photo J. Clague)

Lake Constance, 100 km downvalley, probably records this event (Wessels 1998).

The Flims rockslide barrier dammed Lake Ilanz. A first outburst flood happened after Lake Ilanz overtopped the barrier. A fan of outburst flood debris accumulated at Bonaduz, and the floodwaters eroded into the Bonaduz gravel plain and coursed down the Rhine Valley to Domat/Ems and Chur. It may be responsible for a second anomalous layer of sediment in Lake Constance cores mentioned by Wessels (1998). Shortly afterward, reshaping of the fresh rockslide surface began and the Vorderrhein River incised the relief to its present level.

27.4 Conclusion

The rockslide landscape between Chur and Ilanz is the product of several huge events that are closely spaced in time. The geomorphology is dominated by the main body of the Flims rockslide mass and the erosive trench of the Rhine River gorge—the famous Ruinaulta. Less prominent, but even more interesting are the eastern margin of the Flims rockslide mass, the Bonaduz plain, the Tamins rockslide deposit, and the numerous tumas around Chur, Bonaduz, Domat/Ems, Domleschg, and Rhäzüns. After decades of debate, today there is no doubt that the tumas are large rafted fragments of one of the big rockslides. They were transported within or on a flow of fluidized alluvium that left the Bonaduz gravel. Some of the tumas probably derive directly from the Flims rockslide and some were part of the Tamins mass, mobilised during the Bonaduz flow event. A similar event, on this scale, is without known precedent, and further study will be required to establish the details of the genesis and movement of such a huge mass flow. In the meantime, visitors can marvel at the splendid scenery that was produced in a 'blink' of geologic time.

References

Abele G (1997) Rockslide movement supported by the mobilization of groundwater-saturated valley floor sediments. Zeitschrift für Geomorphologie N F 41(1):1–20

Arbenz P, Staub W (1910) Die Wurzelregion der helvetischen Decken im Hinter-rheingebiet und die Überschiebung der Bündnerschiefer südlich von Bonaduz. Vjschr Naturforsch Ges Zürich 55:23–62

Buckingham T, Pfiffner OA (this volume) Mountain building and valley formation in the UNESCO World Heritage Tectonic Arena

Cadisch J (1944) Beobachtungen im Bergsturzgebiet der Umgebung von Reichenau und Rhäzüns. Eclogae Geol Helv 37:409–412

Caprez J (2008) Das Flimser Bergturzereignis; 3D-Geländerekonstruktionen und Volumenberechnungen mit Hilfe von GIS. Unpublished diploma thesis, Universität Zürich, Zürich, 99 pp

Deplazes G, Anselmetti F, Hajdas I (2007) Lake sediments deposited on the Flims rockslide mass: the key to date the largest mass movement in the Alps. Terra Nova 19(4):252–258

Gruner U (2006) Bergstürze und Klima in den Alpen—Gibt es Zusammenhänge? Bulletin für Angewandte Geologie 11(2):25–34

Heim A (1878) Untersuchungen über den Mechanismus der Gebirgsbildung: im Anschluss an die Geologische Monographie der Tödi-Windgällen-Gruppe. Schwabe, Basel, 3 vol

Heim A (1883) Der alte Bergsturz von Flims. Jb des Schweizer Alpenclubs 18:295–309

Heim A (1891) Geologie der Hochalpen zwischen Reuss und Rhein. Francke, Bern (Beitr z Geol Karte der Schweiz 25), 503 pp

Nabholz W (1954) Neue Beobachtungen im Bergsturzgebiet südlich von Reichenau-Tamins (Graubünden). Verh naturf Ges Basel 65 (1):67–81

Pavoni N (1968) Über die Entstehung der Kiesmassen im Bergsturzgebiet von Bonaduz-Reichenau (Graubünden). Eclogae Geol Helv 61 (2):494–500

Poschinger Av (2005) Der Flimser Bergsturz als Staudamm. Bulletin für angewandte Geologie 10(1):33–47

Poschinger Av, Haas U (1997) Der Flimser Bergsturz, doch ein warmzeitliches Ereignis? Bulletin für angewandte Geologie 2 (1):35–46

Poschinger Av, Wassmer P, Maisch M (2006) The Flims rockslide; history of interpretation and new insights. In: Evans SG, Scarascia-Mugnozza G, Strom A, Hermanns RL (eds) Massive Rock Slope Failure. Springer, Amsterdam, pp 329–356

Remenyik T (1959) Geologische Untersuchungen der Bergsturzlandschaft zwischen Chur und Rodels. Eclogae Geol Helv 52(1):177–235

Scheller E (1970) Geophysikalische Untersuchungen zum Problem des Taminser Bergsturzes. ETH Zürich, Dissertation Nr 4560, 91 pp

Wessels M (1998) Late-Glacial and postglacial sediments in Lake Constance (Germany) and their paleolimnological implications. Archive für Hydrobiogie 53:411–449

Andreas von Poschinger is a geologist working for the Bavarian Geological Survey, Germany. His focus is on geological hazards in the Bavarian Alps. Beside this, he studies the Flims and Tamins rockslides since the 1980s by field work. Based on his field knowledge he encourages other researchers to work on these rockslides.

John J. Clague is Emeritus Professor of Earth Sciences at the Simon Fraser University, Vancouver, Canada. He is a former President of the Geological Association of Canada and former President of Engineers and Geoscientists of British Columbia. His principal research interests are Quaternary geology, geomorphology, and natural hazards. His other professional interest is making science accessible to, and appreciated by, the public.

Nancy Calhoun is a natural resources specialist for the Oregon Department of Geology and Mineral Industries (DOGAMI) in Portland, Oregon. She received a B.A. degree in Geology from Colorado College and a M.Sc. degree in Earth Sciences from Simon Fraser University, British Columbia, Canada. She conducts landslide hazard mapping and risk assessments, and contributes to a variety of projects within the Natural Hazards and Earth Sciences Group of DOGAMI.

28 Periglacial Landscapes and Protection Measures Above Pontresina

Marcia Phillips and Robert Kenner

Abstract

The landscape above Pontresina in the Eastern Swiss Alps is characterized by mountain peaks reaching around 3,000 m a.s.l, steep rock walls, ice-bearing scree slopes and dynamic rock glaciers in a corrie—all located above a precipitous gully leading down to the village. A strong visual impact is caused by the presence of over 15 km of avalanche defence structures and a large retention dam. These protection measures were progressively built over the past 130 years. The most recent ones are specially adapted for unstable permafrost conditions and to provide protection against both avalanches and debris flows triggered in over-steepened ice-bearing terrain. The Foura da l'Amd Ursina site (also known as Schafberg) above Pontresina displays the ideal implementation of research results in the development of practical solutions against natural hazards in a mountain environment undergoing climate-related changes.

Keywords

Mountain permafrost • Ice-bearing substrates • Rock glaciers • Debris flows • Natural hazards • Protection measures • Avalanche defence structures

28.1 Introduction

The site *Foura da l'Amd Ursina* (henceforth referred to as Schafberg Ursina) is a glacial cirque located NE of Pontresina, in the Upper Engadine (Eastern Swiss Alps) at 46°29′53″ N, 9° 55′35″ E and within an elevation range between around 2,600 and 3,000 m a.s.l (Fig. 28.1). The site is characterized by steep slopes and by the presence of permanently frozen and unstable ice-bearing substrates. The combination of steep topography and potentially unstable permafrost is problematic for Pontresina: the village is located 900 m directly below the site, at 1,800 m a.s.l., and faces the challenge of dealing with rock fall and debris flow hazards and snow avalanches in winter. The characteristics of the permafrost terrain and the types of mass movement occurring at the site are described below. In addition, the technical measures used to manage different potential hazards originating at the Ursina site are shown. The aim of the chapter is to demonstrate the use of specially adapted protection measures at sites affected by different types of natural hazards. These structural hazard mitigation measures are specifically designed for complex Alpine terrain with a combination of dynamic geomorphological features and processes.

28.2 Geographical and Geological Setting

The Schafberg Ursina site is a westward oriented glacier corrie at the base of Piz Muragl and has a surface area of around 0.5 km^2 (Figs. 28.1 and 28.2). It is characterized by having a slightly continental climate (around 1500 mm annual precipitation). Air temperatures at the site range between −25 and 20 °C, with a mean annual air temperature of −3 °C (measured at 2,970 m a.s.l since 2004). The site is mainly affected by southerly air masses. Maximum annual snow depths have ranged between 50 and 300 cm in the measurement period between 2004 and 2013.

Bedrock consists of gneiss from the upper Austroalpine Languard nappe and is mainly visible in the rock walls and peaks surrounding the Ursina corrie (e.g. the peaks Las Sours and Piz Muragl and the Muot da Barba Peider ridge). The slopes are covered with scree and blocks originating from the surrounding steep rock walls, whereas the base of the cirque is occupied by three rock glaciers (Fig. 28.1).

M. Phillips (✉) · R. Kenner
WSL Institute for Snow and Avalanche Research SLF, Flüelastrasse 11, 7260 Davos Dorf, Switzerland
e-mail: phillips@slf.ch

R. Kenner
e-mail: robert.kenner@slf.ch

E. Reynard (ed.), *Landscapes and Landforms of Switzerland*, World Geomorphological Landscapes,
https://doi.org/10.1007/978-3-030-43203-4_28

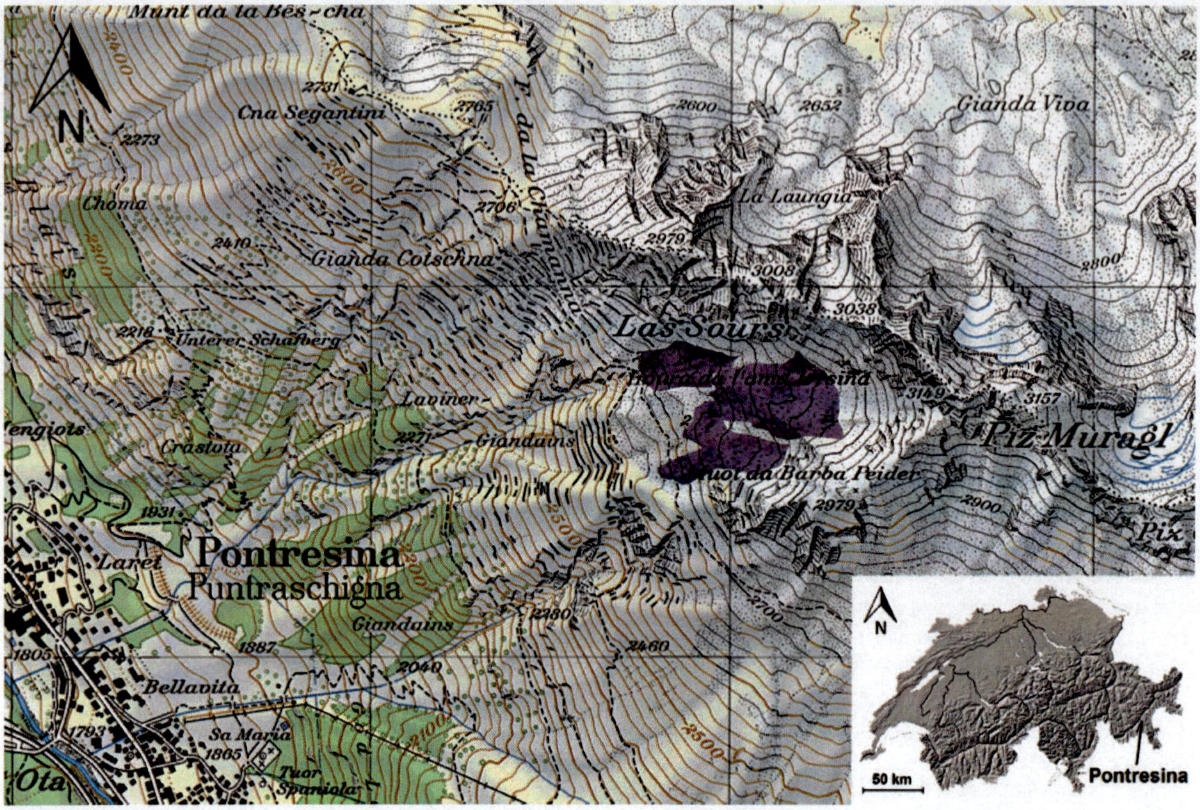

Fig. 28.1 Topographical map of the landscape above and East of Pontresina (swisstopo). The Schafberg Ursina corrie is located at the western base of Piz Muragl. The Laviner Giandains gully runs westwards from the corrie towards Pontresina. The Schafberg Ursina rock glaciers are highlighted in purple. The map grid scale is 1 km. The location of Pontresina in Switzerland is shown on the small map (bottom right inset)

There is very little vegetation on this irregular and unstable mineral surface (Fig. 28.3). Stable areas are characterized by the presence of lichen on the rocks. The rough nature of the ground surface causes a delay in the formation of a continuous snow cover in winter, which allows the ground to cool efficiently in autumn and early winter and thus helps to maintain cold ground temperatures. The presence of permafrost is confirmed by long-term temperature measurements and geophysical soundings, which have been carried out since the 1990s in several boreholes at the site (Haeberli et al. 1999; Zenklusen Mutter and Phillips 2012).

The slopes below the Ursina corrie and above Pontresina are of particular interest regarding the historical development of avalanche defence infrastructure in the Alps. They are covered with over 15 km of avalanche defence structures of varying age and type (Fig. 28.2). These are designed to retain snow in steep avalanche starting zones and their efficiency has increased with technical adaptations over time. They include dry stone walls built at the end of the nineteenth and in the early twentieth century, concrete structures erected in the 1960s, steel snow bridges built during the past few decades and recently erected steel snow nets. In addition, a large avalanche—and debris flow retention dam was built in 2003 at the base of the Laviner Giandains gully and just above the village, at 1,880 m a.s.l (Fig. 28.1). There is very little forest on these steep grassy slopes due to over-grazing by sheep in the course of the nineteenth and twentieth centuries (PH Bern and Schweizer Luftwaffe 2012) and hence there is a widespread avalanche problem in winter, particularly in the steep, ice-rich and unstable gullies where avalanche defence structures cannot be built.

Several hiking trails cross these slopes, for example from Muottas Muragl to Alp Languard via the Segantini hut. These paths are very popular due to the spectacular views

Fig. 28.2 Overview of the Ursina site with the surrounding peaks Las Sours, Piz Muragl in the cloud and the Muot da Barba Peider ridge. The Ursina rock glaciers are in the corrie above the Giandains gully. Different types of avalanche defence structures are visible: stone walls (**a**), steel snow bridges (**b**) and snow nets (**c**) (photo M. Phillips)

they allow of the Engadine valley and the Bernina massif. In summer dozens of hikers daily cross the zones through the Laviner Giandains gully and below the Ursina cirque.

28.3 Geomorphological Features and Related Hazards

28.3.1 Cirque and Rock Glaciers

According to the first edition of the swisstopo Siegfried map (www.map.geo.admin.ch), there was a cirque glacier at the western foot of Piz Muragl in 1875. Its moraines are still visible adjacent to the uppermost rock glacier, but have also been partly overrun by it. Perennial snow patches formed by avalanche deposits are now often present at the base of the steep rock walls in summer (Fig. 28.4). These are sometimes covered with rock fall debris in summer, leading to a sandwich-type formation of alternating firn and rock layers, which ultimately freeze and creep downslope.

The frozen material at the base of Las Sours, Piz Muragl and Muot da Barba Peider forms the roots of the three distinct active rock glaciers (Schafberg Ursina I-III), which can be recognized within the Ursina corrie (Kenner et al. 2014; Fig. 28.5). These ice-rich, creeping permafrost features have been investigated in detail with borehole measurements (Arenson et al. 2002) and remote sensing techniques since the 1990s in the context of various research projects on mountain permafrost (Hoelzle et al. 1998), but also from the perspective of hazard management for Pontresina (Keller et al. 2002). The lowermost rock glacier tongue (Schafberg Ursina II) creeps towards and into the head of the Laviner Giandains gully above the village. This conveyor belt-like system is slowly but steadily transporting loose rock material towards an area where the terrain characteristics are ideal for the formation of debris flows (Fig. 28.6).

Fig. 28.3 Rock walls, gullies, steep scree slopes and rock glaciers in the western sector of the Schafberg Ursina site. Several rows of nineteenth century dry stone walls can be seen on the steep slopes above Pontresina in the left part of the photograph (photo M. Phillips)

The volumetric ice content of the tongue of Schafberg Ursina II is around 80 % (Hoelzle et al. 1998). The ice-rich nature of the substrate causes over-steepening of the material in the gully but also prevents rapid temperature changes and melt due to the latent heat effect and the large amount of energy required to melt ice. Long-term borehole temperature measurements in Ursina II at the top of the Laviner Giandains gully indicate that active layer thickness is practically constant at 5 m and that the underlying permafrost is just below 0 °C (Zenklusen Mutter and Phillips 2012).

Terrestrial laser scans of the three Schafberg Ursina rock glaciers carried out yearly between 2009 and 2014 (Kenner et al. 2014) have shown that the mean annual creep rates attain several decimetres per year for certain sectors of all three rock glaciers (Fig. 28.7). When compared with the creep rates of <5 cm per year for the period 1971–1991 (Hoelzle et al. 1998), obtained on the basis of aerial photogrammetry, it is evident that these rock glaciers are currently accelerating. This is possibly a result of warming, snowmelt infiltration and increased precipitation in summer, as has recently been confirmed by first in-situ permanent GPS measurements on Schafberg Ursina I, which indicate that the rock glacier rapidly reacts by acceleration to snowmelt infiltration, rainfall and high air temperatures.

Terrestrial laser scans of the steep Schafberg Ursina II tongue at the top of the Laviner Giandains gully indicate that rock fall is the main process to occur at present in summer and that a volumetric increase of 40 m^3 has occurred here since 2009. The gradual over-steepening of the front of rock glacier Schafberg Ursina II causes sporadic rock falls across the two hiking trails below. An acceleration of the creep rates could increase the predisposition for debris flow activity in future, and creep rates to the N and NE of the upper edge of the gully are already quite high (Fig. 28.7).

Fig. 28.4 Perennial snow patches and scree slopes at the western base of Piz Muragl and in the root zone of rock glacier Schafberg Ursina I (photo M. Phillips)

28.3.2 Scree Slopes

The slopes extending between the Schafberg Ursina rock glaciers and the rock walls around the top of the cirque are up to 40° steep and covered in scree. The scree particles are sorted inversely downwards, as fines are washed through the coarse blocks at the surface—and particle size at the surface increases towards the base of the slopes, as does the thickness of the scree layer on the bedrock. Borehole temperature measurements in the NW oriented scree slope Muot da Barba Peider indicate that the thaw depth of the active layer in the scree is 2–3 m in summer (Zenklusen Mutter and Phillips 2012). Underneath, the ground is permanently frozen and has a volumetric ice content of approximately 10 %. Annual downslope creep rates range between 5 and 20 cm in the scree, as measured with a borehole inclinometer and with annual terrestrial surveys of the scree slope using a theodolite. Scree slope deformation is initially triggered by the infiltration of snowmelt water in spring and continues throughout the summer months (Rist and Phillips 2005). Intense rainfall causes debris flows in the steep gullies in the rock walls and accumulation of material on the underlying scree slopes can clearly be seen and quantified on the terrestrial laser scans. These unstable scree slopes are technically challenging for the construction and long-term maintenance of avalanche defence structures.

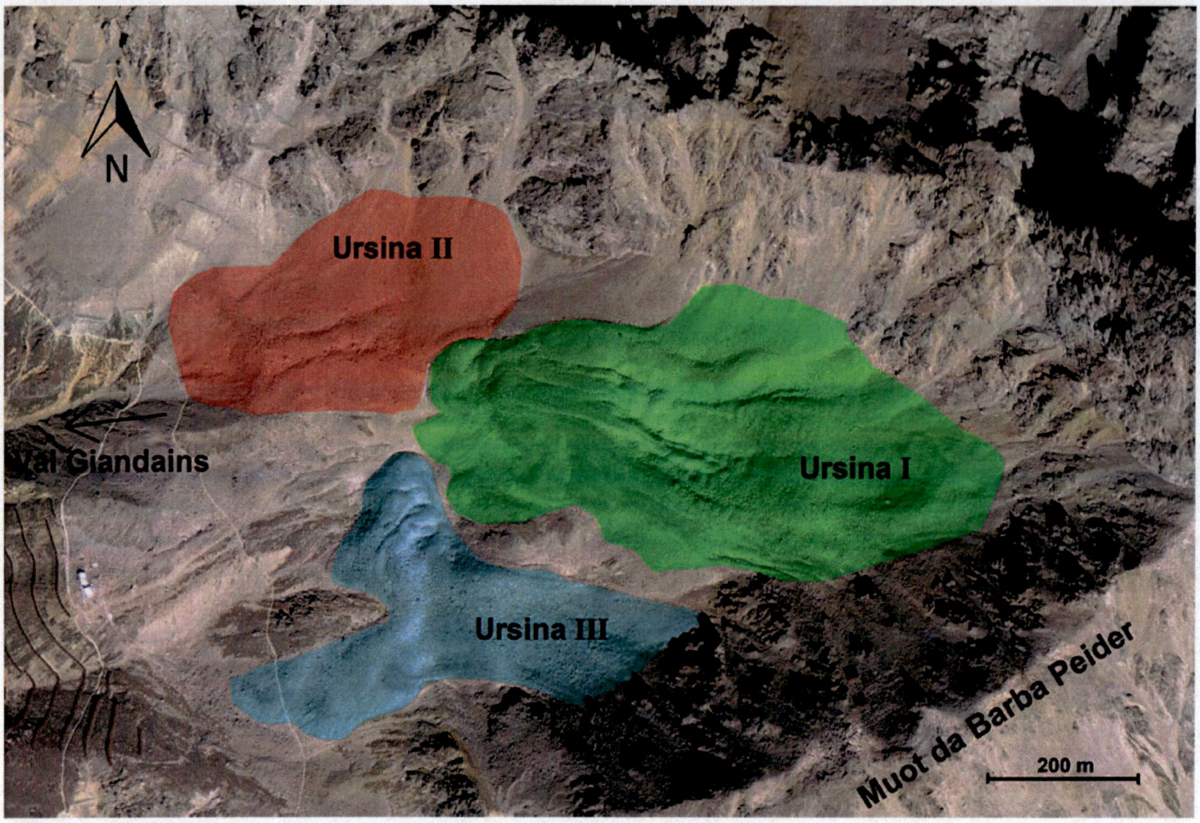

Fig. 28.5 Aerial photograph showing the position of the three Schafberg Ursina rock glaciers in the corrie, the top of the Laviner Giandains gully and the Muot da Barba Peider ridge. Several rows of steel avalanche defence structures (bottom left) and stone walls (top left) are visible (aerial photograph swisstopo)

28.4 Protection Measures and Adaptation Strategies

28.4.1 Avalanche and Debris Flow Retention Dam

Prior to 2003, large snow avalanches triggered in the top of the over-steepened Laviner Giandains gully damaged infrastructure in Pontresina and even reached the main street on a regular basis. The front of rock glacier Schafberg Ursina II, which constitutes the avalanche starting zone, was too ice-rich and unstable for the construction of snow supporting structures. The only alternative was to build a protection structure in the avalanche runout zone. A large retention dam was therefore built at the base of the slope in 2003, just above Pontresina (Fig. 28.8). The dam has the advantage of providing dual protection against snow avalanches on the one hand and debris flows on the other, both potentially originating on the front of the Schafberg Ursina II rock glacier and in the Laviner Giandains gully. It was designed to retain 240,000 m^3 of snow or 100,000 m^3 of debris flow material (Keller et al. 2002). These potential volumes were determined on the basis of previous observations of avalanche size and calculations of the potential volume of unfrozen sediments available, taking into account an active layer thickness of around 5 m in the debris flow starting zone. There have been no avalanches or debris flows in the Laviner Giandains gully since the construction of the dam in 2003, so it has not yet been put to the test.

Fig. 28.6 Rock glacier Schafberg Ursina II at the top of the steep Laviner Giandains gully. Two hiking trails cross the front of the rock glacier (centre and bottom right) (photo M. Phillips)

28.4.2 Avalanche Protection in Avalanche Starting Areas

Avalanche defence structures are steel, fence-like structures built in rows, perpendicular to the slope. They are designed to retain snow in avalanche starting zones, i.e. on slopes exceeding 30°. Their anchors are typically between 3 and 10 m long. Slow mass movements in frozen scree slopes present a restrictive technical challenge for the service-life of these structures, which are prone to damage if annual creep rates are greater than 5 cm (Phillips and Margreth 2008). Movements exceeding this magnitude lead to anchor deformation and exposure, disturbance of the row geometry and in the worst case, anchor rupture and structure failure.

Various types of experimental avalanche defence structures were built on the scree slope of the NW flank of Muot da Barba Peider in 1997 to assess their performance in creeping permafrost terrain. These included rigid steel snow bridges and flexible steel snow nets. Long-term monitoring of these test structures has led to the development of Swiss technical guidelines (Margreth 2007), which stipulate that flexible steel snow nets with "floating" foundations must be used for avalanche defence in unstable permafrost terrain. Steel plates lying on the ground surface are used under the flexible supporting pillars rather than anchors, which allows geometrical corrections of net rows in the event of deformation. This solution has proven to be ideal at several avalanche starting zones in permafrost in the Swiss Alps and

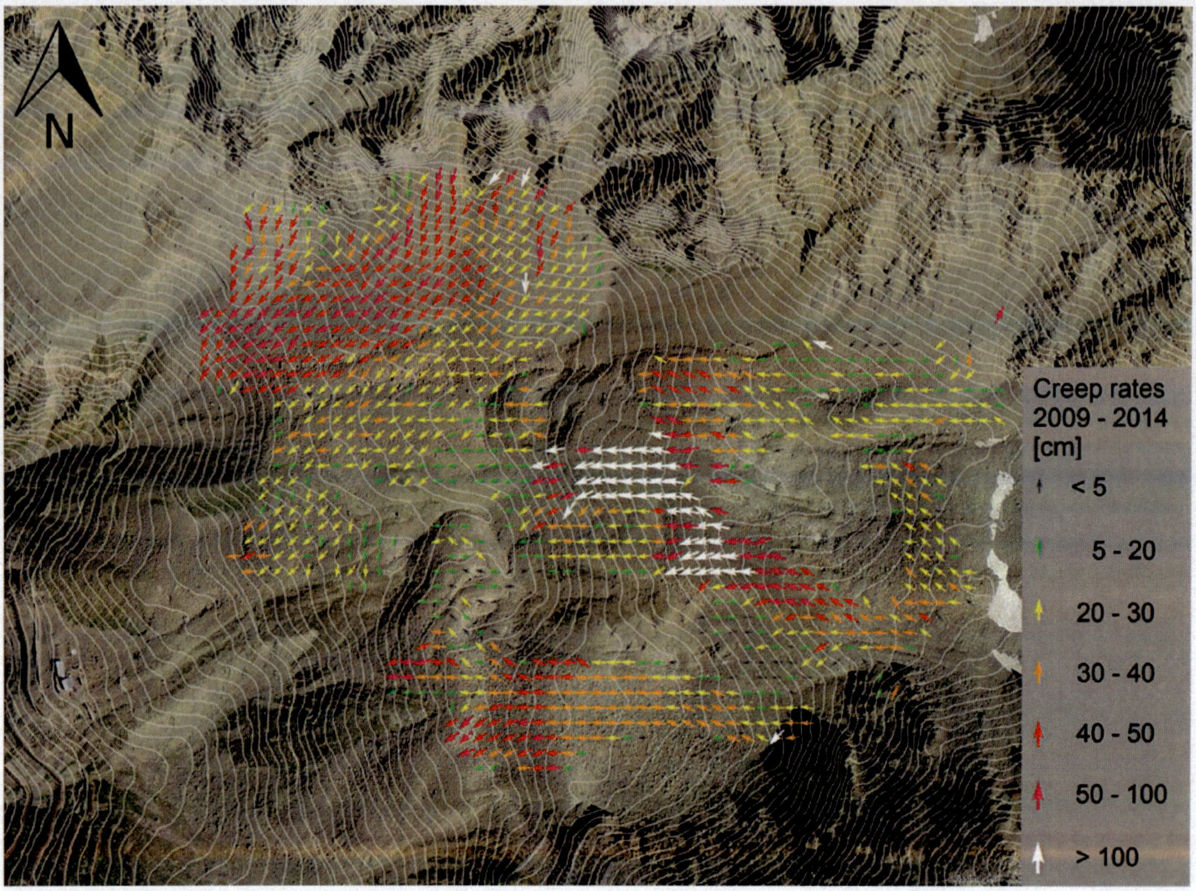

Fig. 28.7 Creep rates at the Schafberg Ursina site measured with a terrestrial laser scanner between 2009 and 2014. The highest rates were measured on the Schafberg Ursina I rock glacier (>100 cm) (aerial photograph swisstopo)

was also used for the most recent construction project on the western flank of Muot da Barba Peider, where four rows of flexible snow nets were built in ice-bearing terrain in 2011 (Fig. 28.9). Snow nets have the additional advantage of being less visible in the landscape than steel snow bridges.

28.5 Conclusions

The landscape above Pontresina is characterized by complex terrain displaying dynamic processes, which are linked to the triggering of natural hazards in the form of snow avalanches and possible debris flows. Various generations of technical solutions have been built to mitigate their impact on Pontresina and have been adapted over time. Steep, creeping and ice-rich permafrost terrain is a challenge for the construction of efficient avalanche protection measures. Specially adapted structures for creeping permafrost have therefore been tested above Pontresina, and these flexible steel nets are now widely used in other avalanche starting zones located in permafrost terrain in the Swiss Alps. Despite these developments, avalanche defence structures cannot be built in avalanche starting zones with creep rates exceeding 5 cm/year or on ice-rich substrates—as is the case in the Laviner Giandains avalanche

Fig. 28.8 Avalanche—and debris flow retention dam above Pontresina (photo M. Heggli)

starting zone at the front of rock glacier Schafberg Ursina II above Pontresina. To solve this problem, a retention dam was built in the runout zone at the base of the slope. This structure protects Pontresina against snow avalanches and debris flows. Borehole measurements and terrestrial laser scans in the source zone at the Ursina site allow to monitor changes in ground temperature and to quantify slope deformation and surface dynamics. This data has contributed towards the development of an ideal combination of practical solutions to protect Pontresina against natural hazards and towards a better understanding of the geomorphological processes taking place above the village.

Acknowledgments The authors warmly thank the WSL Institute for Snow and Avalanche Research (SLF) electronics and mechanics group for their long-lasting contribution to the research at the Ursina site and are very grateful to Canton Graubünden, the authorities in Pontresina, the Federal Office for the Environment (FOEN) and the Swiss Permafrost Monitoring Network (PERMOS) for funding the long-term permafrost monitoring programme. The HeliBernina team is thanked for safe and efficient transport in challenging terrain. We are grateful to Wilfried Haeberli, Emmanuel Reynard and Piotr Migoń for carefully reviewing the manuscript.

Fig. 28.9 Flexible steel snow nets built on the western flank of Muot da Barba Peider in 2011 with "floating" foundations (steel plates at the ground surface), specially designed for creeping permafrost conditions (photo M. Phillips)

References

Arenson L, Hoelzle M, Springman S (2002) Borehole deformation measurements and internal structure of some rock glaciers in Switzerland. Permafr Periglac Process 13:117–135. https://doi.org/10.1002/ppp.414

Haeberli W, Kääb A, Hoelzle M, Bösch H, Funk M, Vonder Mühll D, Keller F (1999) Eisschwund und Naturkatastrophen im Hochgebirge. Schlussbericht NFP 31, v/d/f Hochschulverlag ETH Zürich, 190 pp

Hoelzle M, Wagner S, Kääb A, Vonder Mühll D (1998) Surface movement and internal deformation of ice-rock mixtures within rock glaciers at Pontresina-Schafberg, Upper Engadin, Switzerland. In: Lewkowicz AG, Allard M (eds) Seventh International Conference on Permafrost. Centre d'études nordiques, Université Laval, Yellowknife, Canada, pp 465–471

Keller F, Haeberli W, Rickenmann D, Rigendinger H (2002) Dämme gegen Naturgefahren. tec21 17:13–17

Kenner R, Bühler Y, Delaloye R, Ginzler C, Phillips M (2014) Monitoring of high alpine mass movements combining laser scanning with digital airborne photogrammetry. Geomorphology 206:492–504. https://doi.org/10.1016/j.geomorph.2013.10.020

Margreth S (2007) Defense structures in avalanche starting zones. Technical guideline as an aid to enforcement. Environment in practice. Bern and Davos, FOEN and SLF

PH-Bern, Schweizer Luftwaffe (2012) Permafrost und Lawinenschutz in Pontresina. In: Luftbilder der Schweiz. Pedagogische Hochschule Bern (ed), Bern, pp 1–13

Phillips M, Margreth S (2008) Effects of ground temperature and slope deformation on the service life of snow-supporting structures in mountain permafrost: Wisse Schijen, Randa, Swiss Alps. In: Kane DL, Hinkel DL (eds) 9th International Conference on Permafrost. Institute of Northern Engineering, University of Alaska Fairbanks, Fairbanks, Alaska, pp 1417–1422

Rist A, Phillips M (2005) First results of investigations on hydrothermal processes within the active layer above alpine permafrost in steep terrain. Norsk Geogr Tidsskr 59:177–183. https://doi.org/10.1080/00291950510020574

Zenklusen Mutter E, Phillips M (2012) Active layer characteristics at ten borehole sites in Alpine permafrost terrain, Switzerland. Permafr Periglac Process 23(2):138–151. https://doi.org/10.1002/ppp.1738

Robert Kenner is a Geodetic Engineer specialized in terrestrial laser scanning and remote sensing techniques in unstable mountain permafrost terrain. His main research interests include rock glacier kinematics, natural hazards and Alpine geomorphology. He works in the Permafrost Research Group at the WSL Institute for Snow and Avalanche Research SLF in Davos.

Marcia Phillips leads the Research Group Permafrost at the WSL Institute for Snow and Avalanche Research SLF in Davos. Her main research interests include long-term permafrost monitoring, the construction and maintenance of Alpine infrastructure and rock slope failures in mountain permafrost. She is a member of the Swiss Cryosphere Expert Commission (SCNAT) and of the Swiss Society for Snow, Ice and Permafrost (SIP).

Part III
Geoheritage

29 Geoheritage, Geoconservation and Geotourism in Switzerland

Emmanuel Reynard, Thomas Buckingham, Simon Martin, and Géraldine Regolini

Abstract

Switzerland has a rich and diversified geoheritage. It marks the origin of nature protection in Switzerland, with the earliest protected erratic boulders dating back to 1838. At present, however, the significance of the geological richness and geodiversity of the country is largely unknown in public and political circles. Due to the country's federal structure, which assigns nature conservation to the cantons, geoconservation differs from one canton to the other. At the federal level, geosites are the only type of natural heritage that has not been inventoried in the framework of the Nature and Landscape Protection Act. Nevertheless, an inventory was conducted by the Swiss Academy of Sciences; it includes 322 geosites of national value and, despite its lack of a legal status, provides a basis for planning. Several cantons already have an active geoconservation policy, while others have been very passive so far. Many geosites also enjoy indirect protection through other forms of protection (biotope inventories, UNESCO World Heritage sites, etc.), but the geoparks programme is still undeveloped in Switzerland, with a national programme in place only since January 2020. No region was allowed to apply to be recognised as a UNESCO Global Geopark until 2020. Geoheritage is often at the heart of the tourist offers proposed by Swiss tourist sites (glaciers, waterfalls, mountains, valleys and a variety of landscapes of geological origin) and geotourism products do exist (guided tours, panels, discovery trails, apps, etc.), but except in the three existing natural UNESCO World Heritage sites and in parks of national importance, the offers are fragmented.

Keywords

Geoheritage • Geosites • Geoconservation • Geotourism • UNESCO • Switzerland

E. Reynard (✉)
Institute of Geography and Sustainability and Interdisciplinary Centre for Mountain Research, Géopolis, University of Lausanne, 1015 Lausanne, Switzerland
e-mail: emmanuel.reynard@unil.ch

T. Buckingham
UNESCO-World Heritage Swiss Tectonic Arena Sardona, Städtchenstrasse 45, 7320 Sargans, Switzerland
e-mail: thomas.buckingham@unesco-sardona.ch

S. Martin · G. Regolini
Bureau Relief, Ch. des Oisillons 9, 1860 Aigle, Switzerland
e-mail: simon.martin@bureau-relief.ch

G. Regolini
e-mail: geraldine.regolini@bureau-relief.ch

29.1 Introduction

Switzerland has a long history of protecting geosites and the first erratic block—the *Pierre-à-Bot*, in Neuchâtel—was protected as early as 1838 under the impulsion of glaciologist Louis Agassiz. However, like in most European countries, the current situation is far from optimal and geoheritage does not enjoy extensive legal protection, despite an inventory of geosites of national importance conducted in the 1990s and the fact that several cantons have included geosite protection in their legislation. Although Switzerland has a wide variety of geosites with high tourist potential, geotourism offers are scattered, often little known and information is rarely passed on by the traditional promoters of tourism. The purpose of this chapter is to present the organisation and development of geoconservation in Switzerland, as well as the situation regarding geoparks and geotourism.

29.2 Geoconservation and Geotourism: A Long History

Following the development of glacial theory during the first third of the nineteenth century, Switzerland was a pioneer in the protection of geoheritage. Indeed, the Alps of western

E. Reynard (ed.), *Landscapes and Landforms of Switzerland*, World Geomorphological Landscapes,
https://doi.org/10.1007/978-3-030-43203-4_29

Switzerland—the Bagnes Valley, after the glacial lake outburst flood of Giétro in 1818, Lake Geneva region and the foothills of the Jura—were at the heart of the controversy over the origin of erratic boulders (see for example Venetz 1821; Agassiz 1840; de Charpentier 1841) and more generally over the old extensions of glaciers. Thus, to preserve the evidence of ancient glaciers, the *Pierre-à-Bot*, an erratic boulder located close to the city of Neuchâtel (Fig. 29.1a), was one of the first natural sites in Switzerland to be placed under protection, at the instigation of one of the fathers of glacial theory, Louis Agassiz (Vischer 1946; Bachmann 1999; Reynard 2004a). Several blocks were inscribed with dedications to renowned naturalists, such as Jean de Charpentier (Fig. 29.1b, c), and were later given to naturalist associations (Aubert 1989; Reynard 2004a). They often feature inscriptions recalling the role of pioneers of glaciology (Jean-Pierre Perraudin, Ignaz Venetz, Jean de Charpentier, Louis Agassiz; Fig. 29.1a, c), making them geocultural sites (Reynard and Giusti 2018), which bear witness to both the history of glaciers and the history of science.

In the mid-nineteenth century, when the urbanisation of Switzerland began, with the associated need for new building stones, geologists Alphonse Favre (University of Geneva) and Bernhard Studer (University of Bern) called on the Swiss people to protect the erratic blocks (Favre and Studer 1867), many of which had already been destroyed. This was followed by a vast countrywide movement to inventory blocks carried out by schools and amateur naturalists, leading to the protection of many blocks (Reynard 2004a). In 1884, Alphonse Favre published a map of the ancient glacier extensions and inventoried erratic boulders (Fig. 29.2). Between 1872 and 1876, erratic boulders, *roches moutonnées* (rounded glacially shaped rocks) and glacial potholes were excavated in Lucerne during the construction of a cellar and were recognised to be evidence of the glaciations; under the impetus of geologists F. J. Kaufmann and Albert Heim, they were protected from destruction and formed the heart of the Glacier Garden in Lucerne (see Keller, this volume), one of the city's tourist attractions. In 1905, the right to exploit one of the largest erratic boulders in Switzerland, the *Pierre des*

Fig. 29.1 Examples of protected erratic boulders. **a** The *Pierre-à-Bot*, Neuchâtel, protected since 1838. **b** The *Pierre a Dzo*, Monthey, an erratic boulder used by Jean de Charpentier to explain the glacial origin of erratic boulders. **c** Inscription on the *Pierre a Dzo* recording the pioneers of glaciology Jean-Pierre Perraudin and Jean de Charpentier. **d** The *Pierre des Marmettes*, Monthey, property of the Swiss Academy of Sciences since 1908 (photos E. Reynard)

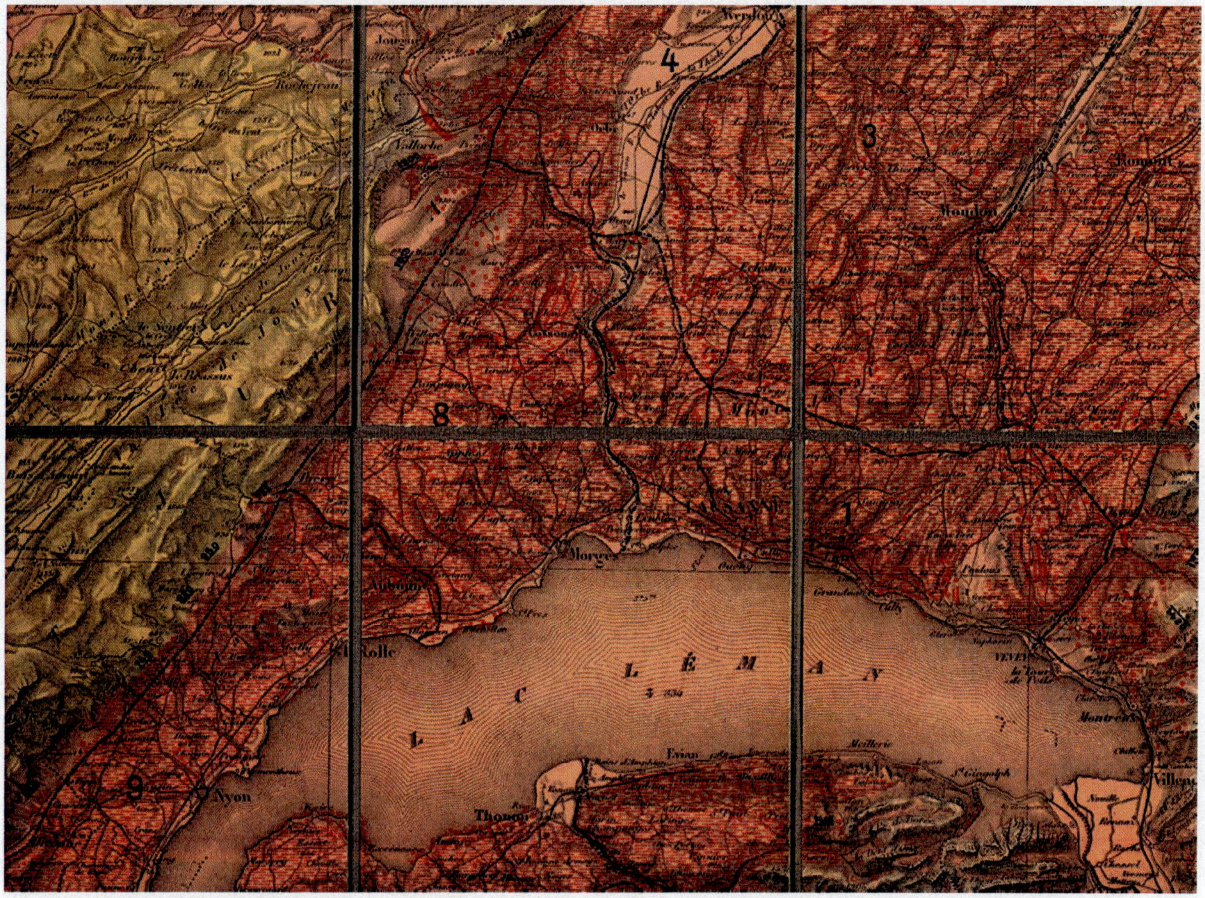

Fig. 29.2 Extract of the map of erratic boulders published by Alphonse Favre in 1884 showing the situation in the Geneva Lake area. Each red dot represents an erratic boulder (Favre 1884, swisstopo)

Marmettes (Fig. 29.1d), located on privately owned land (Reynard 2005) in the municipality of Monthey, was granted to a quarry operator for the production of dimension stones; the scientific community mobilised to save it and finally, in 1908, it was bought back with public funds in a bidding procedure and given to the Swiss Society of Natural Sciences (now the Swiss Academy of Sciences, SCNAT). Strewn throughout the country by the former glaciers (see Schlüchter et al., this volume) and owing to their importance as witnesses of climate history recognised early by some scientists, several erratic blocks were saved from destruction thanks to the civic and political involvement of scientists of the time, and thus represent the foundation of nature protection in Switzerland (Bachmann 1999).

From the point of view of tourism, like in other regions of Europe (Hose 2016), since the end of the eighteenth century, some Swiss geosites have become destinations prized by travellers for their geological interest. This is the case of glaciers (especially the Rhone glacier; Fig. 29.3a) and waterfalls, such as the Staubbach waterfall in the Lauterbrunnen valley (Fig. 29.3b), which have attracted travellers since the Romantic period (see Reichler 2002 in particular). Renowned travellers described or illustrated (see https://www.unil.ch/viaticalpes, accessed 15.08.2019) multiple geosites throughout the country including the thermal springs in the Tamina gorge near Bad Ragaz and Leukerbad, the Gemmi Pass, the falls of the Lauterbrunnen valley, the Rhone glacier, the Reuss gorge in the Gotthard and the Rigi. Among these travellers, the naturalist and poet J. W. von Goethe played a leading role; he described several geosites during his three trips to Switzerland, texts that bear witness to advances in Earth sciences and the geocultural importance of Goethe (Reynard et al. 2009) that continues to this day. These travellers and their accounts can be considered as the first wave of geotourism in Switzerland, whose history has yet to be written. From the second half of the nineteenth century on, the discovery of Switzerland as a tourist destination was driven equally by sports associations, such as the Swiss Alpine Club (created in 1863), and later the Swiss Federation of Pedestrian Tourism (created in 1934), and Touring clubs, such

a

b

Fig. 29.3 Two popular sites of Romantic tourism. **a** The Rhone glacier (Jean-François Albanis Beaumont, 1800; Viatimages/Bibliothèque cantonale et universitaire—Lausanne). **b** The Lauterbrunnen U-shaped glacial valley and the Jungfrau (William Henry Bartlett, 1836; Viatimages/Bibliothèque cantonale et universitaire—Lausanne)

as the Swiss Touring Club (created in 1896). The network of railways and postal routes is also part of the movement to discover Switzerland and, in the 1930s, Swiss Post published several geological routes along the Alpine postal routes.

After the Second World War, tourism started to take off worldwide, and alpine Switzerland was not left behind. This was most obvious in the increase in new mountain aerial railways, reaching new regions and peaks (http://mobile.hls-dhs-dss.ch/ article Bergbahen, accessed 15.08.2019)—with a strong focus on winter tourism. These growing tourism destinations and communities were initially not interested in outdoor geotouristic activities in the summer. It is only since the 2000s that the pressure on these leading Alpine tourism organisations has grown, and mostly economic considerations led some to rethink and reposition themselves (PwC 2018), particularly because climate change affected these high Alpine regions first and most directly. One consequence is a stronger focus on summer tourism, especially hiking and biking, which also created new opportunities for professional geotouristic offers, and for geological phenomena and landscapes to act as "beacons" to attract new customers.

Another reason for the lack of interest in geological features following World War II was the specialisation in particular branches of natural sciences in universities but also in departments of the federal state and cantons, in a move away from a holistic approach to nature and the integration of protection and use of it (nature conservation organisations rarely work with tourism), which has led to a decline in interest in geological topics in society, education, tourism and politics. What is more, only a handful of geologists and geographers are actively engaged in spreading the word to the Swiss general public and are actually able to create holistic offers for tourism, focusing on the links between geological, biological and cultural aspects.

29.3 Legal Framework and Actors

29.3.1 Nature Protection in Switzerland and Geoconservation

Switzerland does not have specific legislation for the protection of geosites (Jordan 1999; Reynard 2012; Stürm 2012). The protection of geological and geomorphological objects is incorporated in the Nature and Landscape Protection Act of 1966 and the Spatial Planning Act of 1979 (Jordan 1999). As Switzerland's political framework is federal, the Nature and Landscape Protection Act divides competences between the Confederation (central state) and the cantons (regional states). The cantons are responsible for the protection of nature in their territory and the Confederation is responsible for protecting sites of national value. To this end, the Confederation conducts inventories of objects of national importance that are worthy of protection. The inventories concern the protection of landscapes, different biotopes (marshes, dry grasslands, amphibian breeding sites, alluvial zones) and cultural heritage (built sites, historical paths), in particular. Despite the efforts of the scientific community to encourage the establishment of an official geosite inventory, no federal inventory of geoheritage has been catalogued by the Confederation to date.

However, indirect protection of some geosites is possible through federal biotope inventories. Thus, many biotopes are closely related to geosites; this is the case of many wetland biotopes (marshes, alluvial areas; Fig. 29.4a), where the protection of the biotope requires that the geosite also be protected. However, it should be noted that the criteria that led to the selection of sites are mainly biological, with the exception of two inventories: the inventory of wetlands of national importance and part of the inventory of alluvial areas of national importance (Reynard 2004b). In the first case, marshlands were selected through a procedure which

Fig. 29.4 Two cases of indirect protection of geoheritage. **a** The *Creux du Croue*, Jura Mountains, an anticline depression including a protected marsh biotope. **b** The *Pyramides d'Euseigne*, weathering landforms protected as a landscape of national importance (photos E. Reynard)

included geomorphological criteria (Stuber 1993). The federal inventory of alluvial areas of national importance was conducted in two stages: lowland alluvial zones were evaluated based solely on biological criteria; the selection of Alpine alluvial zones and proglacial margins was based on biological and geomorphological criteria (Gerber 1995). Many sites listed in these two inventories are thus very valuable from a geoheritage perspective.

The Federal Inventory of Landscapes, Sites and Natural Monuments of National Importance, which includes 162 sites, covers 19 % of Switzerland. It was implemented between 1977 and 1998, and it protects four categories of sites: unique landscapes, typically Swiss landscapes, vast landscapes for relaxation and natural monuments. The last category contains many geosites, particularly erratic boulders, rock outcrops or spectacular landforms (e.g. Pyramids of Euseigne; Fig. 29.4b). In 2017, the ordinance regulating the inventory was amended and the fact sheets were completed. The ordinance specifies the objectives of protection and explicitly mentions the protection of geomorphological and tectonic forms and outstanding geological formations (geotopes) (Article 5).

Other legal texts provide some protection for geoheritage (Reynard 2003). Articles 701 and 714 of the Swiss Civil Code of 1912 make it possible to limit private property rights in order to protect objects of great natural or scientific value. Indirect protection may also be achieved through Environmental Impact Assessment (EIA) procedure within the framework of the Environment Protection Act of 1982 or by applying Article 22 of the Law on the Use of Hydraulic Power of 1916, which requires the preservation of beautiful sites when hydropower plants are built. Finally, the protection of geoheritage is also mentioned in some federal directives. This is the case with the Swiss Landscape Conception (OFEFP 1998), which provides for a measure to protect geosites (measure 7.09; Jordan 1999).

In 2007, the Nature and Landscape Protection Act was amended to support the creation of parks of national importance. In addition to the Swiss National Park, which has existed since 1914 and has been governed by its own law since 1980, three new categories of parks have been created: national parks, regional nature parks and periurban nature parks (https://www.parks.swiss, accessed 15.08.2019). The "Swiss Park" label is awarded to the park belonging to the three categories for a period of 10 years and the parks are now organised in a network, which had 18 members in 2019 (Fig. 29.5): the Swiss National Park, 15 regional nature parks, one periurban nature park and one possible candidate for a periurban nature park. A large percentage of these parks is valuable in terms of geoheritage, represented in particular by the presence of geosites of national value (Darbellay 2017; Fig. 29.6). However, this heritage is sometimes ignored by park promoters and as a result, insufficiently highlighted and its potential insufficiently exploited (Fontana and Reynard 2012), although the situation is now improving (Darbellay 2017).

As nature protection is the responsibility of the cantons, the federal rules may be supplemented by specific cantonal provisions. Some cantons have developed a very proactive policy in favour of geoconservation, for example through their cantonal planning. This is the case in the cantons of Fribourg, Jura and Thurgau, for example. Other cantons have failed to advance in this respect and the result is highly differentiated situations across the country.

29.3.2 Geoheritage and International Protection

Geoconservation in Switzerland also benefits from international protection programmes. Of the 12 sites inscribed on the UNESCO World Heritage List (https://whc.unesco.org/fr/etatsparties/ch, accessed 15.05.2019), all three natural sites were classified for geological reasons (see Migoń 2018). The Swiss Alps Jungfrau-Aletsch site (see Holzhauser, this volume) was included in 2001 following criteria (vii), (viii) and (ix). Monte San Giorgio with its unique palaeontological values, located on the border between Switzerland and Italy, was included in 2003 following criterion (viii). The Swiss Tectonic Arena Sardona (see Buckingham and Pfiffner, this volume) was included in 2008 following criterion (viii) due to its unique visibility of alpine mountain formation and the geohistorical values of scientific debate thereof. Lavaux Vineyard Terraces, classified for cultural reasons (criteria (iii), (iv), (v), is of particular geomorphological interest as a geocultural site (see Reynard and Giusti 2018 and Reynard and Estoppey, this volume).

Geoheritage benefits from some indirect protection through the Biosphere Reserve and Ramsar sites programmes. Two Biosphere reserves are recognised by UNESCO in Switzerland; in both cases, the sites are of geomorphological interest. The Entlebuch Biosphere Reserve contains three national geosites, as does the Val Müstair Biosphere Reserve/Swiss National Park (Fig. 29.6). Switzerland has 11 sites designated as Wetlands of International Importance (Ramsar Sites) (https://www.ramsar.org/wetland/switzerland, accessed 15.08.2019). Several of these Ramsar sites are geomorphosites, such as braided rivers (Allondon River, see Bätz et al., this volume), deltas (Bolle di Magadino at Lake Lugano, Rhone delta at Lake Geneva in Les Grangettes), lakeshores (southern and eastern shore of Lake Neuchâtel) and glacier forelands (Rhone glacier in Gletsch, Roseg glacier in the Bernina Massif).

At the time of writing, there are no UNESCO Global Geoparks in Switzerland.

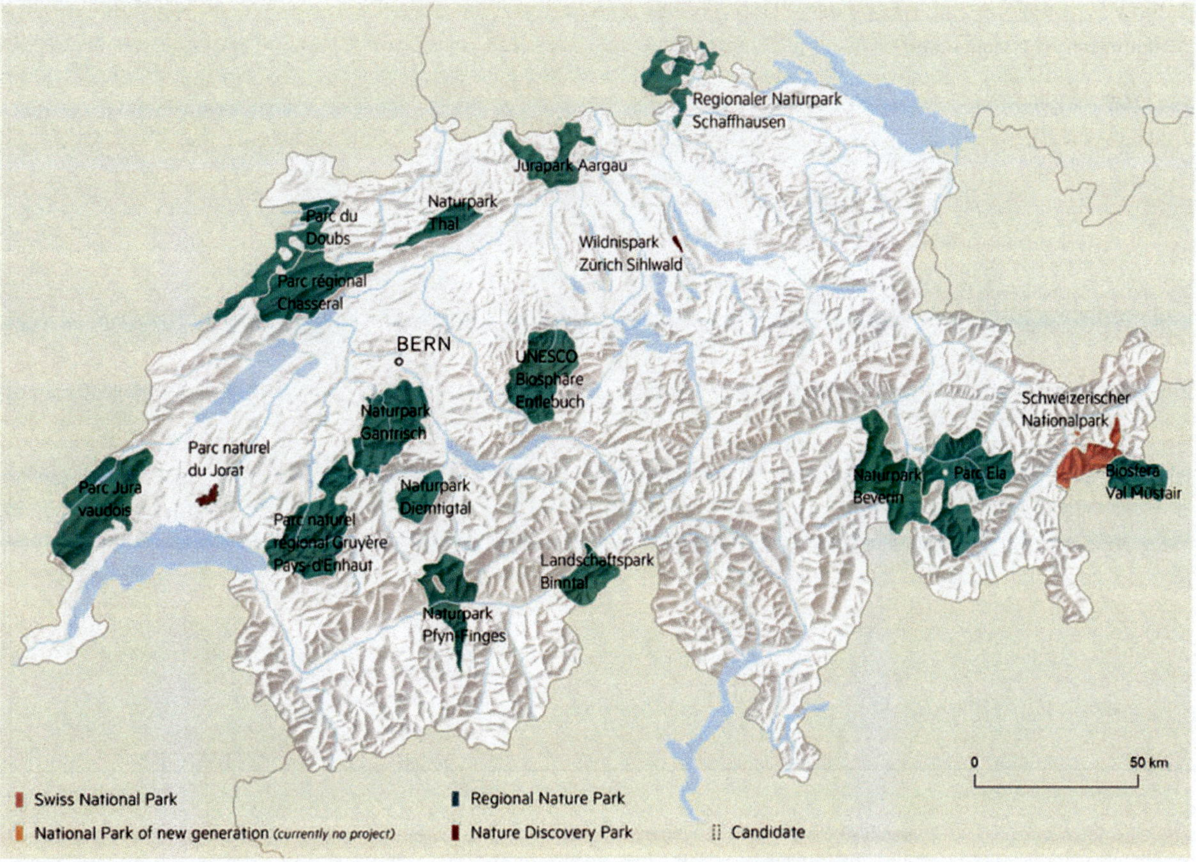

Fig. 29.5 Protected areas in the network of Swiss Parks (© Swiss Parks Network 08/2019, www.parks.swiss; data: Federal Office for the Environment FOEN, swisstopo)

29.3.3 The Main Actors

In addition to universities and private agencies specialised in geoconservation and geotourism, three main actors deal with geoconservation and geotourism issues. The Swiss Academy of Sciences is the main actor. Historically, the academy has played an important role in the protection of erratic boulders and through its affiliated regional natural science associations, it participates in the dissemination of knowledge on geology and geomorphology to the general public. The "Working group for the protection of geotopes"—currently named "Working Group Geotopes and Geoparks", part of the Geosciences Platform of the academy (https://naturalsciences.ch/organisations/geosciences/projects, accessed 15.08.2019)—was created in 1993. In 1994, the working group launched a national inquiry that revealed significant differences in geoconservation efforts among the 26 Swiss cantons (Stürm 2012), and one year later, the working group published a strategic report to improve the protection of geoheritage (Arbeitsgruppe Geotopschutz Schweiz 1995). In 1999, it published the first inventory of 401 geosites (SAS 1999). This inventory was digitised and revised between 2006 and 2012 (Reynard et al. 2012). The working group also published a strategic report on geoparks (Reynard et al. 2007) and it promotes the creation of UNESCO Global Geoparks in Switzerland. To this end, it drew up a list of significant geological values in Switzerland (Buckingham et al. 2018) that should help identify the regions with the highest potential for the creation of UNESCO geoparks in Switzerland, from a scientific point of view. The study allowed the delineation of 20 Geo-Focus regions, i.e. areas of high geological value. A total of 290 out of the 322 national geosites are located in these regions (Fig. 29.7).

The Swiss Geological Survey and the Federal Office of Topography (Swisstopo) are also active in geoconservation and geotourism as part of the "Geology for all" programme (https://www.swisstopo.admin.ch/en/knowledge-facts/geology/geology-everyday/geology-for-all.html, accessed 15.08.2019). As part of the International Year of Planet Earth 2007–2009, several geo-itineraries were published along the famous hiking itinerary ViaGeoAlpina. Four itineraries with

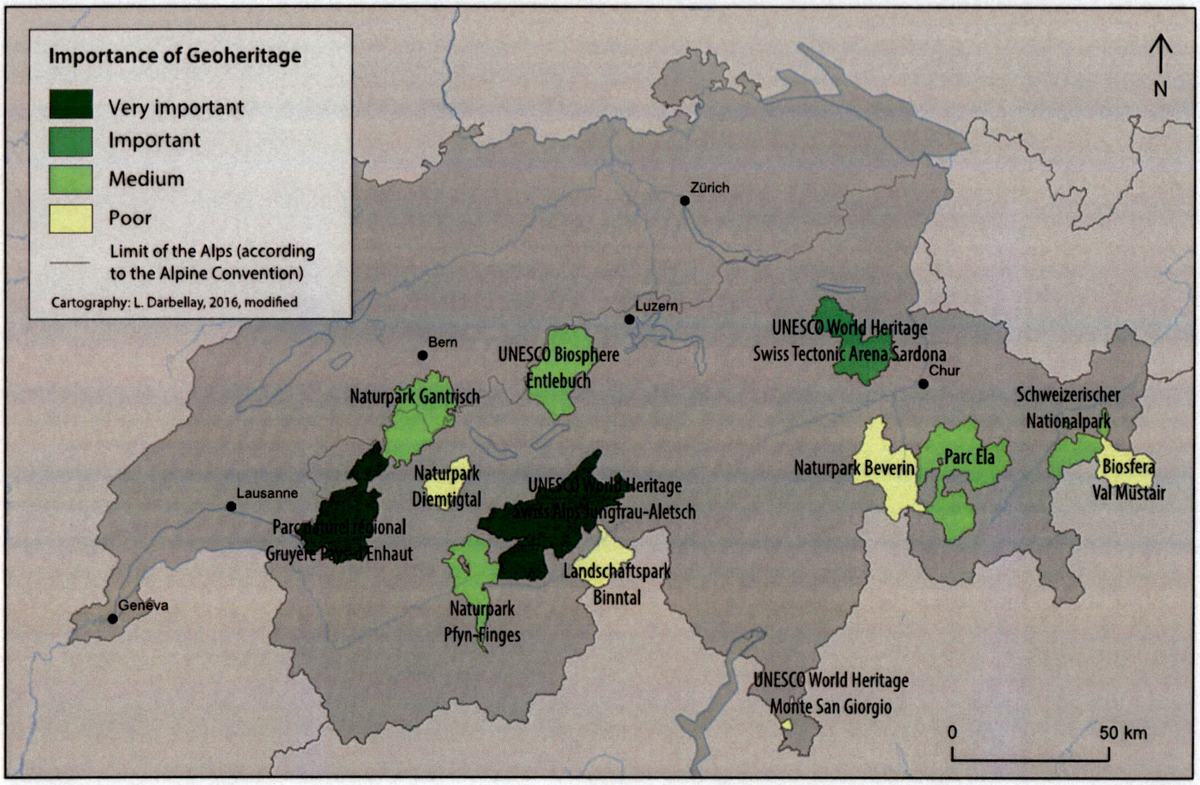

Fig. 29.6 Importance of geoheritage in nature protected areas in the Swiss Alps (from Darbellay 2017, modified). This map is based on the presence of Swiss geosites (according to Reynard et al. 2012; see Fig. 29.8) in the protected area. The importance of geoheritage is calculated as follows: 0–1 geosites = poor; 2–3 geosites = medium; 4–6 geosites = important; more than 6 = very important. The score for the paleontological site of international importance of Monte San Giorgio appears to be "poor"; this is because the protected area itself is very small (32 km^2) thus allowing the presence of only one geosite of national importance (the whole site)

extended geological and landscape descriptions are available in the Western Helvetic Alps (Champéry to Derborence), the Bernese Oberland (Grindelwald to Adelboden), the World Heritage Tectonic Arena Sardona area and the Ticino Alps (Personico to Cavergno) (http://www.viageoalpina.eu/, accessed 15.08.2019). In 2013, a brochure meant for the wider public was published. It describes various geological datasets in Switzerland and relevant publications or information on how to access them via the internet (SGS and Swisstopo 2013). Swisstopo also publishes simplified geological maps for visitors. Such maps have already been published for the World Heritage sites Tectonic Arena Sardona and Lavaux, and for the Bagnes Valley, in western Switzerland.

The association "Erlebnis Geologie" (*Experience Geology*) is another important actor, whose aim is to gather information on activities related to geology (called GeoEvents) in Switzerland and to provide an overview of all museums, visitor centres, mines, quarries and other organisations/locations with a focus of simplifying geology for the public. Various activities are organised throughout Switzerland by geologists and lay people from all backgrounds (universities, museums, private firms, hiking guides, and mountain geoguides) who are committed to sharing their passion for geology with the public (www.erlebnis-geologie.ch, accessed 15.08.2019). Yearly workshops bring the active players of this network together and every three years, a Geology Festival is held throughout Switzerland.

29.4 Geosite Inventories

29.4.1 The Inventory of Swiss Geosites

There is no inventory of geosites that falls under the Federal Nature and Landscape Protection Act. However, the Working Group Geotopes and Geoparks has been inventorying geosites since the mid-1990s (Reynard 2012). The first published list (SAS 1999) was not a real inventory, based on a standard methodology, rather a list of proposals by experts from various parts of the country, different fields (palaeontology, mineralogy, geomorphology, etc.) and institutions (cantonal administrations, nature museums, universities,

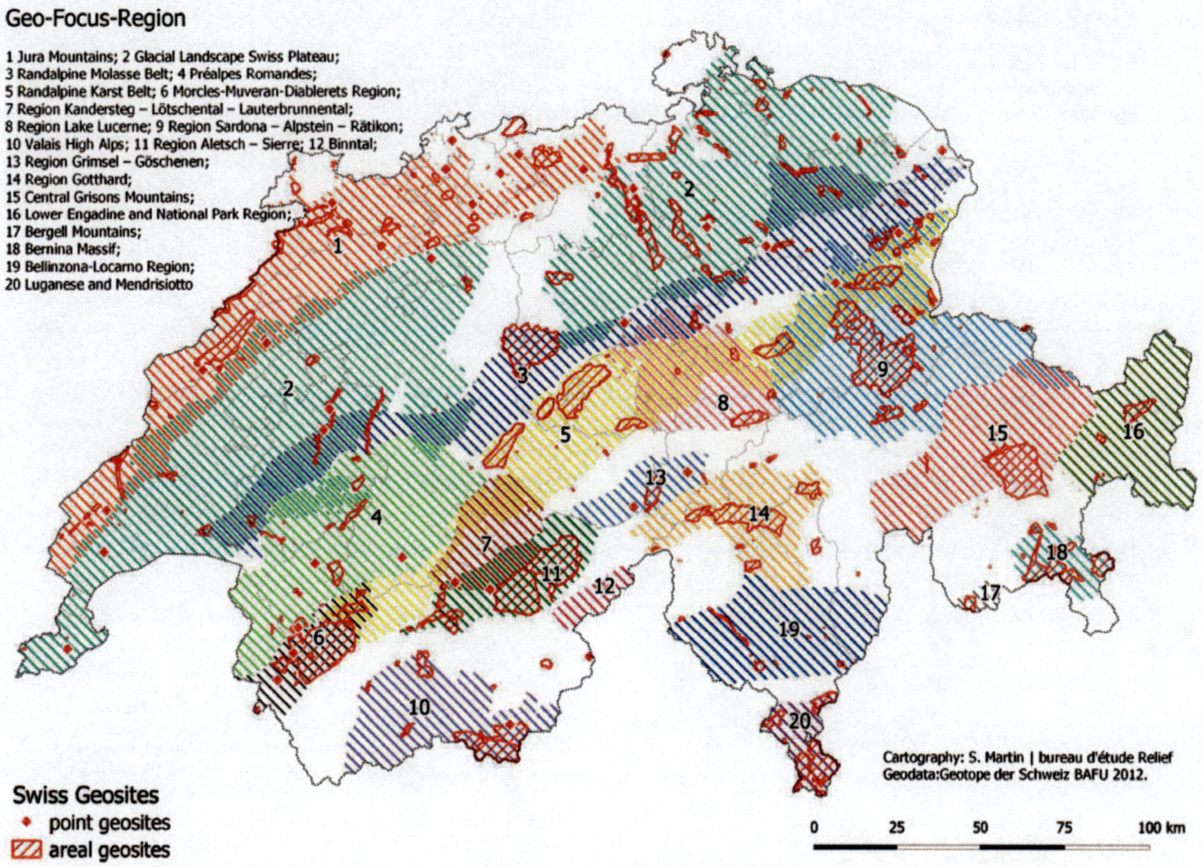

Fig. 29.7 Geo-Focus regions and Swiss geosites (after Buckingham et al. 2018, modified)

etc.). More than 800 proposals were received and after evaluation by the members of the working group, 401 geosites were selected for the list (SAS 1999).

The list contained objects considered to have high geoscientific value, but it had no legal value. Thus, in 2000, a working group composed of scientists and federal authorities studied the desirability of conducting an official inventory based on the Nature and Landscape Protection Act. However, due to financial restrictions and absence of real motivation by policymakers, the inventory was never made. Another problem with the list adopted in 1999 was the heterogeneity of both its content and form. Additionally, there was no information in digital form.

A revision—both formal and digital—of the inventory was carried out between 2006 and 2012 by a group of scientists of various disciplines. This revision was financed by the Swiss Academy of Sciences (SAS) and the Federal Office for the Environment (FOEN). Finally, a list of 322 geosites was published (Reynard et al. 2012; Fig. 29.8) and made available on the Swiss geoportal (https://map.geo.admin.ch/, accessed 15.08.2019). In terms of content, the revision of the inventory necessitated homogenisation (Reynard 2012). Large amounts of data had to be added, because most of the geosites in the 1999 inventory were described in very limited terms, and several sites were merged. Others were left out because their national importance was not obvious. On the other hand, a large survey was carried out to add new sites, especially in the regions and in the branches of Earth sciences that were poorly represented in the 1999 list. Of the 322 sites selected, 142 are of primary geomorphological interest, including Quaternary outcrops and surface karst landforms; 141 are of geological interest (stratigraphic, sedimentological, paleontological or structural); 39 are underground karst sites. Of course, many geological or speleological sites are also of geomorphological interest and vice versa.

29.4.2 Geoconservation at Cantonal Level

Because Switzerland has a federal structure and nature protection is a cantonal task, the level of geoconservation differs considerably from one part of the country to another. Some cantons have conducted a detailed and well-documented

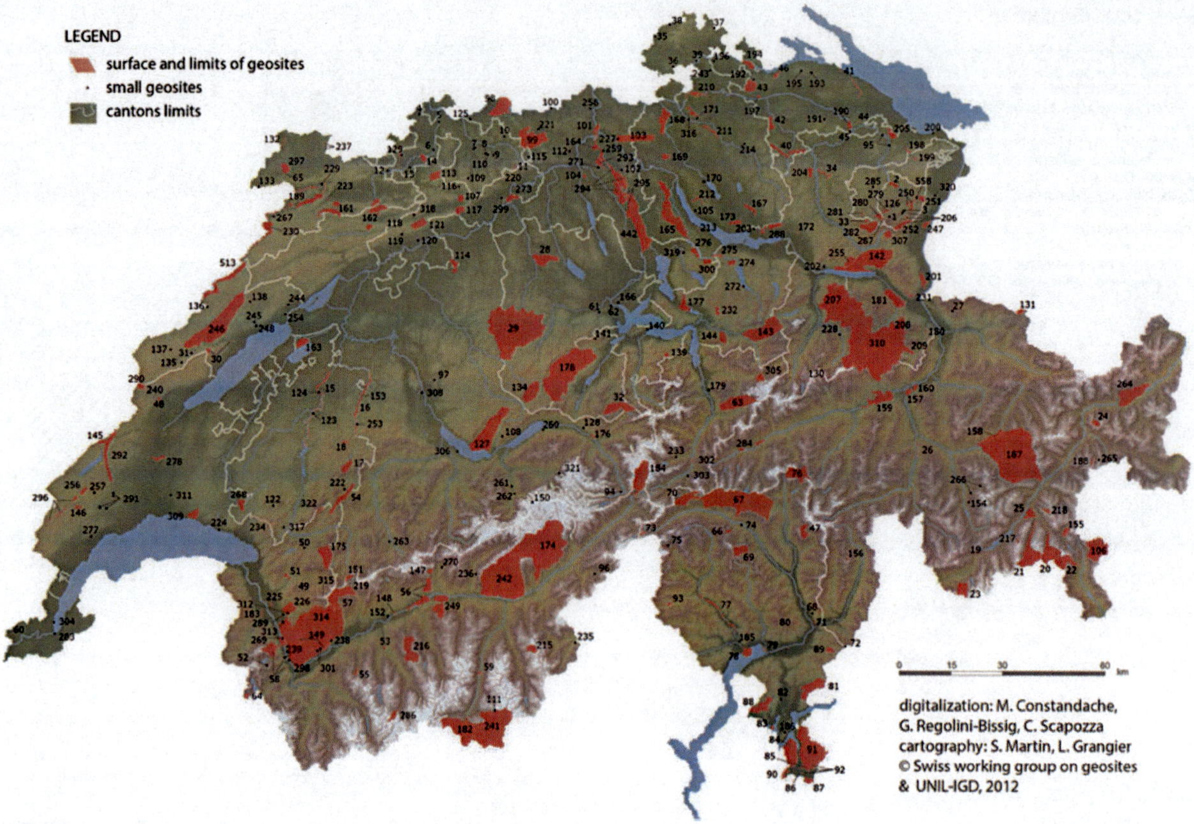

Fig. 29.8 The 322 Swiss geosites (after Reynard et al. 2012, modified). Numbers refer to the geosites; the full list is available in Reynard et al. (2012)

inventory, while in other cantons, no information on geoheritage has been collected or the work is far from complete. Currently, an inventory of geosites is available in 17 cantons, while nine cantons have no inventory at all (Fig. 29.9). An ongoing project of the Working Group Geotopes and Geoparks with the aim to gather geodata from the cantonal inventories of geosites revealed that some are under revision (AG, FR) or in preparation (TI) (Regolini and Martin 2019). The status of the geosites inventoried is very variable: in some cases, it is just a list of suggestions, with no legal value, whereas in some cantons, the inventory data have been incorporated in spatial planning documents and therefore carry legal weight.

29.5 Geoparks

Several initiatives for the creation of geoparks have been launched in different regions of Switzerland (Fig. 29.10) and a strategic report for the development of geoparks in Switzerland has been drafted (Reynard et al. 2007; Reynard 2012). Most of these initiatives failed. There are currently only two geoparks: the *Parco delle Gole della Breggia* (https://www.parcobreggia.ch, accessed 15.08.2019), which is too small to meet the criteria required for recognition as a UNESCO Global Geopark, and the Geopark Sardona, which includes the Swiss Tectonic Arena Sardona featured on the World Heritage List (Fig. 29.11). The latter began in 1999 as a regional development project and in 2004 the Geopark Sardona Association was founded, newly reorganised as a promotional club for the World Heritage and the Geopark Sardona. Its activities are closely linked to those of the World Heritage site and supplement its work.

Substantial work was recently done to identify possible regions for the creation of UNESCO Global Geoparks in Switzerland. The Federal Office for the Environment (FOEN) in collaboration with the Federal Department of Foreign Affairs (FDFA), its affiliated Swiss UNESCO Commission and the Working Group Geotopes and Geoparks of SCNAT have drafted a paper and framework for the implementation of UNESCO Global Geoparks using existing management structures such as UNESCO World Heritage sites, Swiss Parks or other protected sites with management structures in place. Whilst there is real interest

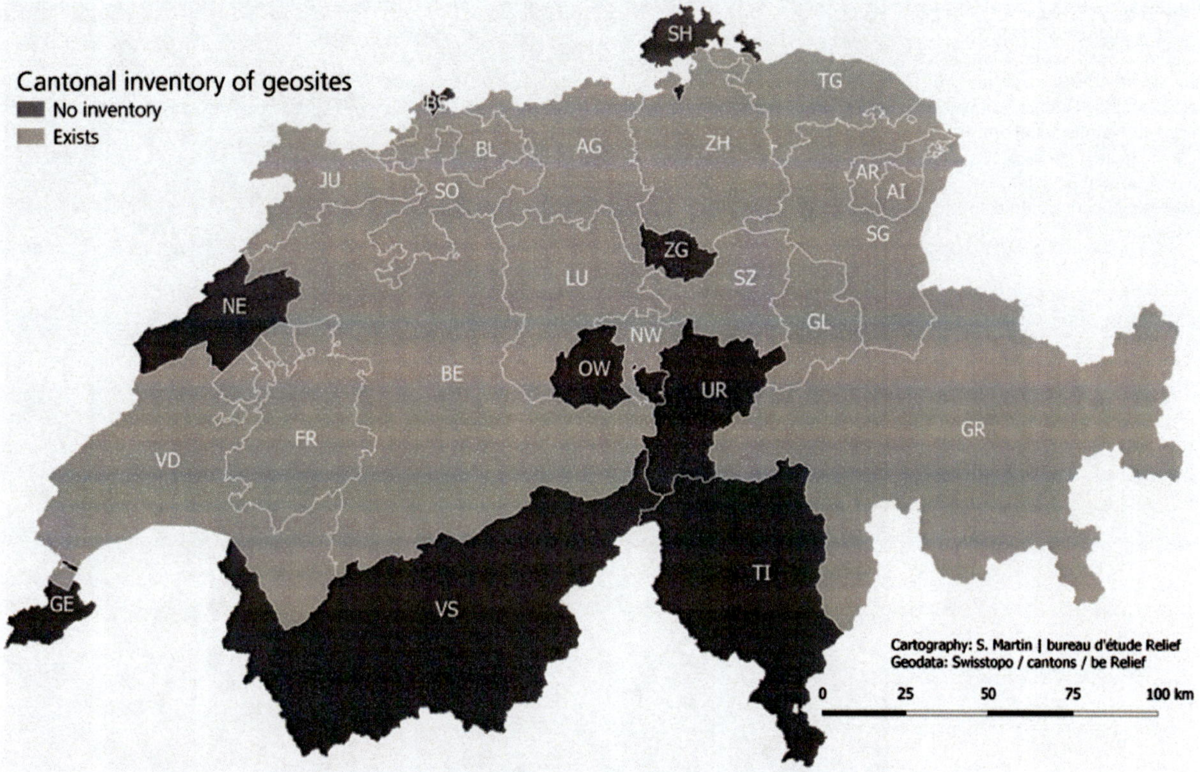

Fig. 29.9 State of geoheritage inventorying in the cantons. An inventory of geosites is available in 17 cantons, while 9 cantons have no inventory at all

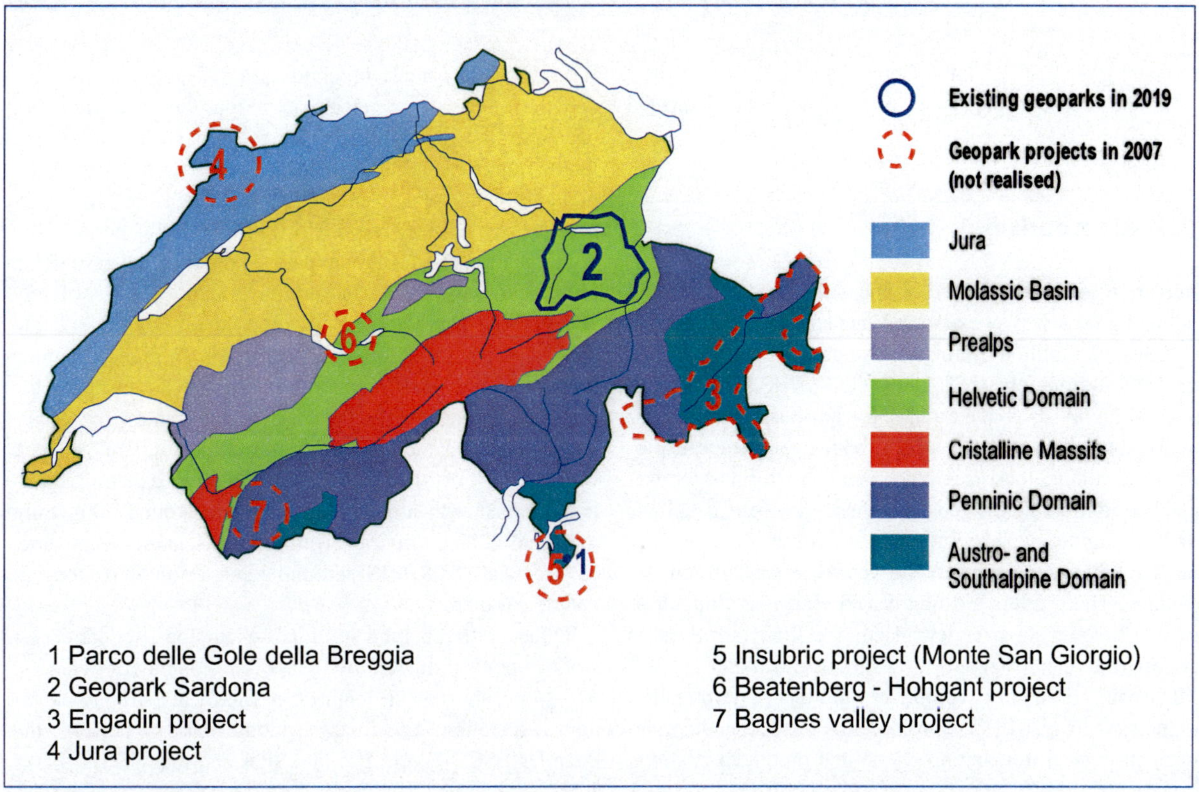

Fig. 29.10 Geoparks in Switzerland

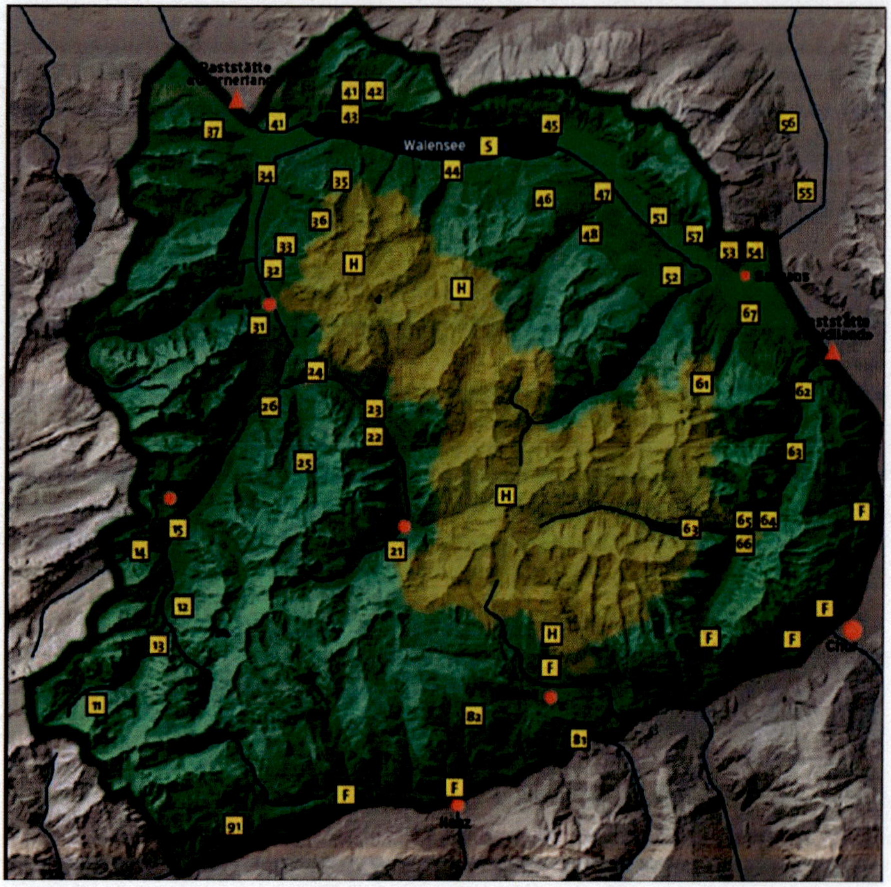

Fig. 29.11 The Sardona Geopark and the Sardona Tectonic Arena World Heritage site. The World Heritage site is in yellow and the Geopark surface is in green. Numbers and letters in yellow squares represent geological points of interest (*Source* Geopark Sardona Association)

in pursuing UNESCO Global Geoparks in at least two sites, the ball is currently in the hands of parks and other protected areas.

29.6 Geotourism

Tourism is an important economic sector in Switzerland, accounting for 4.6 % of export revenue and 2.4 % of GDP. It generates 50.4 million overnight stays per year and maintains 163,750 full-time jobs (FST 2017; OFS 2018). The tourism sector is highly diversified, ranging from business and conference tourism to winter tourism, hiking, landscape exploration and cultural tourism. Geotourism—as a form of tourism aimed at discovering the geological heritage (Hose 2012)—is relatively undeveloped, but like geoconservation, many touristic offers can indirectly be considered as geotourism activities. These offers are fragmented, and it is difficult to provide a complete overview. Here we limit ourselves to outlining the main categories of geotourism activities.

The first category comprises indirect interpretation products (Martin 2013), i.e. projects that provide geotourism information to visitors but no one to interpret them. Various formats exist (panels, brochures, digital applications, etc.) and the offers are extremely numerous and scattered. For example, the Bex Tourist Office has produced several interpretation panels (Fig. 29.12) showing the geological heritage of the region and organised a treasure hunt (geocaching) in geosites related to the last glacial period. The multiplication of educational paths and panels has resulted in projects in which scientific and/or graphic quality was not always on a par with similar offers in other domains, and content is often too complicated for non-specialised public. Some cantons have created programmes to evaluate the proposed offers; this is the case of the canton of Valais, which launched the "Nature, culture and tourism" project, with the aim of identifying and promoting existing educational paths: with a database containing around 200 nature trails and 200 culture trails (http://www.sentiers-decouverte.ch, accessed 15.08.2019), around a quarter of which concern geological themes.

Direct interpretation (guided excursions, conferences, workshops) is promoted by the Swiss Academy of Sciences, Swisstopo and the "Erlebnis Geologie" (*Experience Geology*) association. The Swiss National Park in Grisons, the three UNESCO World Heritage sites and many Swiss Parks

Fig. 29.12 Example of interpretation panels in the commune of Bex, Vaud canton (photo Bureau Relief)

also offer direct or indirect interpretation services that promote the discovery of geoheritage and landscapes. From 2013 to 2018, two New Regional Politics (NRP) projects, funded by the Federal state and the cantonal Economy departments, developed a wide range of direct interpretation products surrounding the World Heritage Sardona in eastern Switzerland and established close relationships with tourism stakeholders.

Finally, geotourism products and activities are offered at several specific geosites and are very often part of wider offers to experience nature, not specifically focusing or mentioning geology.

29.7 Conclusions

Switzerland has a highly diversified geoheritage. It was at the heart of the first actions to protect nature and heritage in the mid-nineteenth century, but then fell into oblivion. In the last two decades, many initiatives have been launched by the scientific community, particularly by the Swiss Academy of Sciences, to encourage recognition of the value of our geoheritage, protect the most vulnerable sites and promote geosites to the Swiss population and visitors. The heritage movement is well under way, but for the time being, political recognition is still limited, although it has improved in recent years. Geoconservation is also fragmented, varying from one canton to another, due to the country's federal structure. Natural sites, particularly geomorphological ones, are at the heart of tourism promotion in Switzerland; on the other hand, apart from a few initiatives, such as ViaGeoAlpina, geotourism offers remain limited. There is still work to be done to achieve broad societal recognition of the quality of Swiss geoheritage.

Acknowledgments We are grateful to Erica Baumann (Swiss Parks Network) for Fig. 29.5 and Lucie Darbellay for Fig. 29.6.

References

Agassiz L (1840) Etudes sur les glaciers. Jent et Gassmann, Soleure, 346 pp

Arbeitsgruppe Geotopschutz Schweiz (1995) Geotope und der Schutz erdwissenschaftlicher Objekte in der Schweiz: ein Strategiebericht. Arbeitsgruppe Geotopschutz Schweiz, Freiburg, 27 pp

Aubert D (1989) La protection des blocs erratiques dans le canton de vaud. Bull Soc Vaud Sc nat 79(3):185–207

Bachmann S (1999) Zwischen Patriotismus und Wissenschaft. Die schweizerischen Naturschutzpioniere (1900-1938). Chronos Verlag, Zürich, 461 pp

Bätz N, Iorgulescu J, Lane SN (this volume) The Allondon River: Decadal planform changes under changing boundary conditions

Buckingham T, Pfiffner OA (this volume) Mountain building and valley formation in the UNESCO World Heritage Tectonic Arena Sardona Region

Buckingham T, Meyer J, Regolini G (2018) International signifikante geologische Werte der Schweiz. Arbeitsgruppe Geotope und Geoparks, Bern, 89 pp

Regolini G, Martin S (2019) Bericht kantonale Geotopinventare. Bureau d'étude Relief, Aigle

Darbellay L (2017) Prise en compte du géopatrimoine dans les stratégies de conservation de la nature dans l'arc alpin. Master thesis, University of Lausanne, 153 pp

De Charpentier J (1841) Essai sur les glaciers et sur le terrain erratique du bassin du Rhône. Ducloux, Lausanne, 363 pp

Favre A (1884) Carte du phénomène erratique et des anciens glaciers du versant nord des Alpes suisses et de la chaîne du Mont-Blanc. Commission géologique, Berne

Favre A, Studer B (1867) Appel aux Suisses pour les engager à conserver les blocs erratiques. Actes de la Société helvétique des sciences naturelles 51:153–160

Fontana G, Reynard E (2012) La gestion du patrimoine géologique dans les parcs naturels en Suisse. In: Giusti C (ed) Géomorphosites 2009: imagerie, inventaire, mise en valeur et valorisation du patrimoine géomorphologique. Volume des actes. Université Paris-Sorbonne, pp 133–139

FST (2017) Le tourisme suisse en chiffres 2016. Fédération suisse du tourisme, Berne. https://www.stv-fst.ch/sites/default/files/2017-10/StiZ_fr.pdf

Gerber B (1995) Das Projekt IGLES. Inventar der Gletschervorfelder und alpinen Schwemmebenen als Ergänzung des Bundesinventars der Auengebiete von nationaler Bedeutung. Regio Basilensis 36 (2):201–207

Holzhauser H (this volume) The Aletsch region with the majestic Grosser Aletschgletscher

Hose TA (2012) 3G's for modern geotourism. Geoheritage 4:7–24

Hose TA (2016) (ed) Geoheritage and geotourism: a European perspective. The Boydell Press, Woodbridge, 336 pp

Jordan P (1999) Geotopschutz – die rechtliche Situation in der Schweiz. Geol Insubrica 4:55–58

Keller B. (this volume) Lake Lucerne and its spectacular landscape

Martin S (2013) Valoriser le géopatrimoine par la médiation indirecte et la visualisation des objets géomorphologiques. PhD thesis, University of Lausanne, 273 pp

Migoń P (2018) Geoheritage and world heritage sites. In: Reynard E, Brilha J (eds) Geoheritage. Assessment, protection, and management. Elsevier, Amsterdam, pp 237–249

OFEFP (1998) Conception Paysage suisse. Office fédéral de l'environnement, des forêts et du paysage, Berne, 133 pp

OFS (2018) Tourisme: Panorama. Office fédéral de la statistique, Neuchâtel. https://www.bfs.admin.ch/bfs/fr/home/statistiques/tourisme.html

PwC (2018) Bergbahnen. Geschäftsmodelle, Kooperationen und Digitalisierung. https://www.pwc.ch/de/publications/2018/PwC%20Studie-Bergbahnen-Hoch%20hinaus-DE.pdf

Reichler C (2002) La découverte des Alpes et la question du paysage. Georg, Genève, 256 pp

Reynard E (2003) Öffentliche Politik, Eigentumsverhältnisse und Schutz von Geomorphologischen Geotope. Schrifftenreihe der Deutschen Geologischen Gesellschaft 31:94–101

Reynard E (2004a) Protecting stones: conservation of erratic blocks in Switzerland. In: Prikryl R (ed) Dimension stone 2004. New perspectives for a traditional building material. Balkema, Leiden, pp 3–7

Reynard E (2004b) L'évaluation des géotopes géomorphologiques en Suisse. In: Reynard E, Pralong JP (eds) Paysages géomorphologiques. Institut de Géographie, Lausanne, Travaux et Recherches 27, pp 137–149

Reynard E (2005) Geomorphological sites, public policies and property rights. Conceptualization and examples from Switzerland. Il Quaternario 18(1):321–330

Reynard E (2012) Geoheritage protection and promotion in Switzerland. Eur Geol 34:44–47

Reynard E, Giusti C (2018) The landscape and the cultural value of geoheritage. In: Reynard E, Brilha J (eds) Geoheritage. Assessment, protection, and management. Elsevier, Amsterdam, pp 147–166

Reynard E, Estoppey E (this volume) The Lavaux World Heritage terraced vineyard

Reynard E, Baillifard F, Berger JP, Felber M, Heitzmann P, Hipp R, Jeannin PY, Vavrecka-Sidler D, Von Salis K (2007) Les géoparcs en Suisse: un rapport stratégique. Groupe de travail pour les géotopes en Suisse, Berne, 18 pp

Reynard E, Regolini-Bissig G, Kozlik L, Benedetti S (2009) Assessment and promotion of cultural geomorphosites in the Trient Valley (Switzerland). Mem Descr Carta Geol Italia 87:181–189

Reynard E, Berger JP, Constandache M, Felber M, Grangier L, Häuselmann P, Jeannin PY, Martin S (2012) Révision de l'inventaire des géotopes suisses: rapport final. Groupe de travail pour les géotopes en Suisse, Lausanne, 45 pp

SAS (1999) Inventory of geotopes of national significance. Geol Insubrica 4(1):31–48

Schlüchter C, Akçar N, Ivy-Ochs S (this volume) The Quaternary Period in Switzerland

SGS and swisstopo (2013) Geology of Switzerland—Knowledge from underground. Swiss Geological Survey and Federal Office of Topography swisstopo, Bern, 131 pp

Stuber A (1993) La géomorphologie dans les domaines de la protection de la nature et du paysage. Exemples d'applications cartographiques. In: Schoeneich P, Reynard E (eds) Cartographie géomorphologique, cartographie des risques. Institut de Géographie, Lausanne, Travaux et recherches 9, pp 45–51

Stürm B (2012) Switzerland. In: Wimbledon WAP, Smith-Meyer S (eds) Geoheritage in Europe and its conservation. ProGEO, Oslo, pp 356–365

Vischer W (1946) Naturschutz in der Schweiz. SBN Verlag, Basel, 380 pp

Venetz I (1821) Mémoire sur les variations de la température dans les Alpes de la Suisse. Sion, 38 pp

Emmanuel Reynard is Professor of Physical Geography and Director of the Interdisciplinary Centre on Mountain Research at the University of Lausanne, and a member of the International Association of Geomorphologists (IAG) Executive Committee. He has been chairman of the IAG Working Group on Geomorphosites (2001–2013) and president of the Working Group on Geotopes of the Swiss Academy of Sciences (2006–2017). His research concerns mountain geomorphology, geomorphological heritage, landscape geohistorical analyses and water management in mountains. He has worked in the Alps, the Romanian Carpathians and in the Maghreb (Tunisia, Morocco).

Thomas Buckingham is an M.Sc. in Earth Sciences working in education, awareness raising and public relations for the World Heritage Swiss Tectonic Arena Sardona in eastern Switzerland. He is Co-president of the Swiss Working Group on Geotopes and Geoparks and Co-president of the Association Erlebnis Geologie. His main scientific interests are the interaction between tectonics and erosion and the links to geomorphology and natural hazards.

Simon Martin is Dr. in Geosciences and Co-founder of the private office Relief. He is a member of the Swiss Working Group on Geotopes and Geoparks and of the Geotourism Group which brings together Swiss professionals interested in geotourism. He is involved in several projects dealing with the creation of geointerpretive products, the development of geotourism in Alpine areas, cartography and environmental education.

Géraldine Regolini is Dr. in Geosciences and Co-founder of the private office Relief. She is a board member of the Swiss Geomorphological Society (SGS), a member of the Swiss Working Group on Geotopes and Geoparks and the IAG Working Group on Geomorphosites (secretary from 2005 to 2011). Her main professional interests are geoheritage evaluation and interpretation. She has contributed to the development of a quality assessment method of geotourist products and she develops interpretation plans and geotourist products of all types.

Index

E. Reynard (ed.), *Landscapes and Landforms of Switzerland*, World Geomorphological Landscapes,
https://doi.org/10.1007/978-3-030-43203-4